PARPs, PAR and NAD Metabolism and Their Inhibitors in Cancer

PARPs, PAR and NAD Metabolism and Their Inhibitors in Cancer

Editors

Nicola Curtin
Péter Bay

MDPI • Basel • Beijing • Wuhan • Barcelona • Belgrade • Manchester • Tokyo • Cluj • Tianjin

Editors

Nicola Curtin
Newcastle University Centre for Cancer
Newcastle upon Tyne
UK

Péter Bay
University of Debrecen
Debrecen
Hungary

Editorial Office
MDPI
St. Alban-Anlage 66
4052 Basel, Switzerland

This is a reprint of articles from the Special Issue published online in the open access journal *Cancers* (ISSN 2072-6694) (available at: https://www.mdpi.com/journal/cancers/special_issues/parp2019).

For citation purposes, cite each article independently as indicated on the article page online and as indicated below:

LastName, A.A.; LastName, B.B.; LastName, C.C. Article Title. *Journal Name* **Year**, *Volume Number*, Page Range.

ISBN 978-3-0365-8156-9 (Hbk)
ISBN 978-3-0365-8157-6 (PDF)

Contents

cancers

MDPI

Editorial

PARPs, PAR and NAD Metabolism and Their Inhibitors in Cancer

Nicola Curtin [1,*] and Péter Bai [2,3,4,*]

1 Newcastle University Centre for Cancer, Translational and Clinical Research Institute, Faculty of Medical Sciences, Newcastle University, Newcastle upon Tyne NE2 4HH, UK
2 Department of Medical Chemistry, Faculty of Medicine, University of Debrecen, 4032 Debrecen, Hungary
3 MTA-DE Lendület Laboratory of Cellular Metabolism, 4032 Debrecen, Hungary
4 Research Center for Molecular Medicine, Faculty of Medicine, University of Debrecen, 4032 Debrecen, Hungary
* Correspondence: nicola.curtin@newcastle.ac.uk (N.C.); baip@med.unideb.hu (P.B.); Tel.: +36-52-412-345 (P.B.); Fax: +36-52-412-566 (P.B.)

Received: 15 September 2020; Accepted: 19 October 2020; Published: 24 November 2020

The role of poly(ADP-ribose) polymerase-1 (PARP1) in DNA repair and as a potential target for anticancer therapy has been under investigation for more than 50 years. The field has expanded over the decades to include not only a family of ADP-ribosylating enzymes (PARPs/ARTDs), but also interacting and polymer-degrading proteins. In this special issue of *Cancers* primary research articles and reviews describe various aspects of PARP biology along with therapeutic targeting. Some historical perspective is reviewed along with the development of the PARP inhibitor (PARPi), rucaparib/Rubraca®, including the identification of the synthetic lethality with homologous recombination repair (e.g., BRCA) defects [1]. Further consideration of the mode of action of several PARPi and their interaction with cytotoxic chemotherapy is described by Min and Im [2], and the evidence for a unique mode-of-action of another PARPi, PJ34 [3] is summarized. Several other aspects of PARPi therapy are described, including research indicating the therapeutic potential of the PARPi, olaparib, as a single agent in myelodysplastic syndromes, not only through cytotoxic and cytostatic effects but also through the induction of differentiation [4], along with the augmentation of UVB-induced DNA damage and mitochondrial alterations resulting in reduced proliferation and viability of keratinocytes [5]. On the other side of the coin, deficiency of the enzyme that degrades the PARP-generated ADP-ribose polymers, PARG, in ES cells resulted delayed tumour growth when they were implanted SC and increased antitumour activity of X-rays [6].

Moving PARPi therapy beyond BRCA mutated cancers requires the use of biomarkers to predict PARPi sensitivity as reviewed by Singh et al. [7]. The role of defects in the G1 cell cycle checkpoint signaling kinase, ATM, as a determinant of sensitivity to PARPi was reviewed, with the finding that PARPi alone are cytostatic in ATM defective cancer cells but require the addition of an inhibitor of the S/G2 cell cycle checkpoint kinase, ATR, to induce cell death [8]. An investigation of the synergy between the PARPi, olaparib, and the ATR inhibitor VE-821 in a panel of neuroblastoma cells, revealed that it was independent of MYCN or ATM status in these cells [9] and similar studies identified that the synergy between an inhibitor of the S/G2 cell cycle checkpoint kinase, CHK1, and the PARPi, rucaparib was largely through the impairment of homologous recombination repair by the CHK1 inhibitor [10]. Depletion of p60/150 CAF-1 also impaired homologous recombination repair thereby inducing sensitivity to PARPi and irradiation in head and neck cancers with therapeutic implications [11]. Interestingly, depletion of NMNAT1, which is involved in the synthesis of NAD+, PARP's substrate, induced DNA damage and sensitized cells to cisplatin but exhibited redundancy with PARPi in this respect [12]. Inhibition of EGFR or Syk (which mediates EGFR signaling) was synergistically cytotoxic in combination with olaparib in squamous cell carcinoma cells suggesting therapeutic potential [13].

Of course, resistance to PARPi is an emerging problem clinically and the role of the PI3K-AKT pathway in protection from PARPi-induced cytotoxicity and its significance in shock, inflammation, ischemia-reperfusion injury and cancer is reviewed by Gallyas et al. [14]. Interestingly, gastric cancer cell lines made resistant to olaparib were found to be cross-resistant to cisplatin, but had increased sensitivity to irinotecan due to upregulation of TOP1 and TDP1, which has therapeutic implications for patients who develop PARPi resistance [15].

PARP has been investigated as a regulator of transcription and key roles regarding the role of PARP both in nucleolar function in relation to cancer biology and the role PARylation plays in the regulation of transcription were reviewed [16,17]. With original research showing that PARylation activates the histone acetyl transferase, EP300, contributing to its regulation of transcription of DNA repair and proliferation genes [18].

An important factor in cancer biology is the tumour microenvironment and the interaction between PARP and key features of the tumour microenvironment such as autophagy, hypoxia and angiogenesis was reviewed [19]. There is significant interest in the immune microenvironment and the roles of both PARP1 and PARP2 in modulating both the innate and adaptive immune system was reviewed by Yelamos et al. [20], with the therapeutic potential of the combination of PARPi with immune checkpoint inhibitors, including translational and clinical studies were reviewed by Peyraud and Italiano [21].

Feijs and colleagues [22] reviewed the roles of ADP-ribosyl hydrolases, also involved in the degradation of ADP-ribose and poly(ADP-ribose) chains, MACROD1, MACROD2 and TARG1 in carcinogenesis. The role of ARH1, another degradatory enzyme, was reviewed by Ishiwata-Endo and colleagues [23] in cancer and non-cancer diseases.

The roles for poly(ADP-ribosyl)ation in biochemistry, cell biology, physiology and pathophysiology are rapidly expanding, the findings discussed in the "PARPs, PAR and NAD Metabolism and Their Inhibitors in Cancer" special issue of Cancers will surely provide a better understanding of these processes and widen the scope of our appreciation of poly(ADP-ribosyl)ation.

Funding: P.B. was supported by grants from NKFIH (K123975 and GINOP-2.3.2-15-2016-00006) and by the Higher Education Institutional Excellence Programme (NKFIH-1150-6/2019) of the Ministry of Innovation and Technology in Hungary, within the framework of the Biotechnology thematic programme of the University of Debrecen. NC is funded by Newcastle University.

Conflicts of Interest: N.C. has worked for many years on the development of rucaparib resulting in 2 patents WO 2005/012305 A2 andWO/2006/033006 with royalties paid to CRUK and Newcastle University through which she receives a percentage. She has received research funding foe PARP-related studies from Agouron Pharmaceuticals, Pfizer, BioMarin and consulted for Abbvie, Eisai and Tesaro.

References

1. Curtin, N.J. The Development of Rucaparib/Rubraca®: A Story of the Synergy between Science and Serendipity. *Cancers* **2020**, *12*, 564. [CrossRef] [PubMed]
2. Min, A.; Im, S.A. PARP Inhibitors as Therapeutics: Beyond Modulation of PARylation. *Cancers* **2020**, *12*, 394. [CrossRef] [PubMed]
3. Cohen-Armon, M. The Modified Phenanthridine PJ34 Unveils an Exclusive Cell-Death Mechanism in Human Cancer Cells. *Cancers* **2020**, *12*, 1628. [CrossRef] [PubMed]
4. Faraoni, I.; Consalvo, M.I.; Aloisio, F.; Fabiani, E.; Giansanti, M.; Di Cristino, F.; Falconi, G.; Tentori, L.; Di Veroli, A.; Curzi, P.; et al. Cytotoxicity and Differentiating Effect of the Poly(ADP-Ribose) Polymerase Inhibitor Olaparib in Myelodysplastic Syndromes. *Cancers* **2019**, *11*, 1373. [CrossRef] [PubMed]
5. Hegedűs, C.; Boros, G.; Fidrus, E.; Kis, G.N.; Antal, M.; Juhász, T.; Janka, E.A.; Jankó, L.; Paragh, G.; Emri, G.; et al. PARP1 Inhibition Augments UVB-Mediated Mitochondrial Changes—Implications for UV-Induced DNA Repair and Photocarcinogenesis. *Cancers* **2019**, *12*, 5. [CrossRef] [PubMed]
6. Sonoda, Y.; Sasaki, Y.; Gunji, A.; Shirai, H.; Araki, T.; Imamichi, S.; Onodera, T.; Rydén, A.M.; Watanabe, M.; Itami, J.; et al. Reduced Tumorigenicity of Mouse ES Cells and the Augmented Anti-Tumor Therapeutic Effects under Parg Deficiency. *Cancers* **2020**, *12*, 1056. [CrossRef]

7. Singh, N.; Pay, S.L.; Bhandare, S.B.; Arimpur, U.; Motea, E.A. Therapeutic Strategies and Biomarkers to Modulate PARP Activity for Targeted Cancer Therapy. *Cancers* **2020**, *12*, 972. [CrossRef]
8. Jette, N.R.; Kumar, M.; Radhamani, S.; Arthur, G.; Goutam, S.; Yip, S.; Kolinsky, M.; Williams, G.J.; Bose, P.; Lees-Miller, S.P. ATM-Deficient Cancers Provide New Opportunities for Precision Oncology. *Cancers* **2020**, *12*, 687. [CrossRef]
9. Southgate, H.E.D.; Chen, L.; Tweddle, D.A.; Curtin, N.J. ATR Inhibition Potentiates PARP Inhibitor Cytotoxicity in High Risk Neuroblastoma Cell Lines by Multiple Mechanisms. *Cancers* **2020**, *12*, 1095. [CrossRef]
10. Smith, H.L.; Prendergast, L.; Curtin, N.J. Exploring the Synergy between PARP and CHK1 Inhibition in Matched BRCA2 Mutant and Corrected Cells. *Cancers* **2020**, *12*, 878. [CrossRef]
11. Morra, F.; Merolla, F.; Picardi, I.; Russo, D.; Ilardi, G.; Varricchio, S.; Liotti, F.; Pacelli, R.; Palazzo, L.; Mascolo, M.; et al. CAF-1 Subunits Levels Suggest Combined Treatments with PARP-Inhibitors and Ionizing Radiation in Advanced HNSCC. *Cancers* **2019**, *11*, 1582. [CrossRef] [PubMed]
12. Kiss, A.; Ráduly, A.P.; Regdon, Z.; Polgár, Z.; Tarapcsák, S.; Sturniolo, I.; El-Hamoly, T.; Virág, L.; Hegedűs, C. Targeting Nuclear NAD(+) Synthesis Inhibits DNA Repair, Impairs Metabolic Adaptation and Increases Chemosensitivity of U-2OS Osteosarcoma Cells. *Cancers* **2020**, *12*, 1180. [CrossRef] [PubMed]
13. Huang, D.Y.; Chen, W.Y.; Chen, C.L.; Wu, N.L.; Lin, W.W. Synergistic Anti-Tumour Effect of Syk Inhibitor and Olaparib in Squamous Cell Carcinoma: Roles of Syk in EGFR Signalling and PARP1 Activation. *Cancers* **2020**, *12*, 489. [CrossRef] [PubMed]
14. Gallyas, F., Jr.; Sumegi, B.; Szabo, C. Role of Akt Activation in PARP Inhibitor Resistance in Cancer. *Cancers* **2020**, *12*, 532. [CrossRef]
15. Kim, J.W.; Min, A.; Im, S.A.; Jang, H.; Kim, Y.J.; Kim, H.J.; Lee, K.H.; Kim, T.Y.; Lee, K.W.; Oh, D.Y.; et al. TDP1 and TOP1 Modulation in Olaparib-Resistant Cancer Determines the Efficacy of Subsequent Chemotherapy. *Cancers* **2020**, *12*, 334. [CrossRef]
16. Engbrecht, M.; Mangerich, A. The Nucleolus and PARP1 in Cancer Biology. *Cancers* **2020**, *12*, 1813. [CrossRef]
17. Páhi, Z.G.; Borsos, B.N.; Pantazi, V.; Ujfaludi, Z.; Pankotai, T. PARylation during Transcription: Insights into the Fine-Tuning Mechanism and Regulation. *Cancers* **2020**, *12*, 183. [CrossRef]
18. Sobczak, M.; Pitt, A.R.; Spickett, C.M.; Robaszkiewicz, A. PARP1 Co-Regulates EP300-BRG1-Dependent Transcription of Genes Involved in Breast Cancer Cell Proliferation and DNA Repair. *Cancers* **2019**, *11*, 1539. [CrossRef]
19. Martí, J.M.; Fernández-Cortés, M.; Serrano-Sáenz, S.; Zamudio-Martinez, E.; Delgado-Bellido, D.; Garcia-Diaz, A.; Oliver, F.J. The Multifactorial Role of PARP-1 in Tumor Microenvironment. *Cancers* **2020**, *12*, 739. [CrossRef]
20. Yélamos, J.; Moreno-Lama, L.; Jimeno, J.; Ali, S.O. Immunomodulatory Roles of PARP-1 and PARP-2: Impact on PARP-Centered Cancer Therapies. *Cancers* **2020**, *12*, 392. [CrossRef]
21. Peyraud, F.; Italiano, A. Combined PARP Inhibition and Immune Checkpoint Therapy in Solid Tumors. *Cancers* **2020**, *12*, 1502. [CrossRef] [PubMed]
22. Feijs, K.L.H.; Cooper, C.D.O.; Žaja, R. The Controversial Roles of ADP-Ribosyl Hydrolases MACROD1, MACROD2 and TARG1 in Carcinogenesis. *Cancers* **2020**, *12*, 604. [CrossRef] [PubMed]
23. Ishiwata-Endo, H.; Kato, J.; Stevens, L.A.; Moss, J. ARH1 in Health and Disease. *Cancers* **2020**, *12*, 479. [CrossRef] [PubMed]

cancers

MDPI

Article

Targeting Nuclear NAD$^+$ Synthesis Inhibits DNA Repair, Impairs Metabolic Adaptation and Increases Chemosensitivity of U-2OS Osteosarcoma Cells

Alexandra Kiss [1,2], Arnold Péter Ráduly [1], Zsolt Regdon [1], Zsuzsanna Polgár [1], Szabolcs Tarapcsák [3], Isotta Sturniolo [1], Tarek El-Hamoly [1,4], László Virág [1,5,*] and Csaba Hegedűs [1,*]

1 Department of Medical Chemistry, Faculty of Medicine, University of Debrecen, H-4032 Debrecen, Hungary; kissalexandra@med.unideb.hu (A.K.); radulyarnold27@gmail.com (A.P.R.); regdon.zsolt@med.unideb.hu (Z.R.); polgar.zsuzsanna@med.unideb.hu (Z.P.); isotta.sturniolo@med.unideb.hu (I.S.); tahamoly@hotmail.com (T.E.-H.)

2 Doctoral School of Molecular Medicine, University of Debrecen, H-4032 Debrecen, Hungary

3 Department of Biophysics and Cell Biology, Faculty of Medicine, University of Debrecen, H-4032 Debrecen, Hungary; tarapcsakszabolcs@gmail.com

4 Drug Radiation Research Department, National Center for Radiation Research and Technology, Atomic Energy Authority, Cairo 113701, Egypt

5 MTA-DE Cell Biology and Signaling Research Group, H-4032 Debrecen, Hungary

* Correspondence: lvirag@med.unideb.hu (L.V.); hcsaba@med.unideb.hu (C.H.)

Received: 11 February 2020; Accepted: 4 May 2020; Published: 7 May 2020

Abstract: Osteosarcoma (OS) is the most common bone tumor in children and adolescents. Modern OS treatment, based on the combination of neoadjuvant chemotherapy (cisplatin + doxorubicin + methotrexate) with subsequent surgical removal of the primary tumor and metastases, has dramatically improved overall survival of OS patients. However, further research is needed to identify new therapeutic targets. Here we report that expression level of the nuclear NAD synthesis enzyme, nicotinamide mononucleotide adenylyltransferase-1 (NMNAT1), increases in U-2OS cells upon exposure to DNA damaging agents, suggesting the involvement of the enzyme in the DNA damage response. Moreover, genetic inactivation of NMNAT1 sensitizes U-2OS osteosarcoma cells to cisplatin, doxorubicin, or a combination of these two treatments. Increased cisplatin-induced cell death of NMNAT1$^{-/-}$ cells showed features of both apoptosis and necroptosis, as indicated by the protective effect of the caspase-3 inhibitor z-DEVD-FMK and the necroptosis inhibitor necrostatin-1. Activation of the DNA damage sensor enzyme poly(ADP-ribose) polymerase 1 (PARP1), a major consumer of NAD$^+$ in the nucleus, was fully blocked by NMNAT1 inactivation, leading to increased DNA damage (phospho-H2AX foci). The PARP inhibitor, olaparib, sensitized wild type but not NMNAT1$^{-/-}$ cells to cisplatin-induced anti-clonogenic effects, suggesting that impaired PARP1 activity is important for chemosensitization. Cisplatin-induced cell death of NMNAT1$^{-/-}$ cells was also characterized by a marked drop in cellular ATP levels and impaired mitochondrial respiratory reserve capacity, highlighting the central role of compromised cellular bioenergetics in chemosensitization by NMNAT1 inactivation. Moreover, NMNAT1 cells also displayed markedly higher sensitivity to cisplatin when grown as spheroids in 3D culture. In summary, our work provides the first evidence that NMNAT1 is a promising therapeutic target for osteosarcoma and possibly other tumors as well.

Keywords: NAD$^+$; NMNAT1; cisplatin; chemotherapy; apoptosis; PARP1; osteosarcoma; cancer

1. Introduction

Bone tumors constitute 4–7% of all cancers in children aged 0–14 years and 7–8% in children aged 15–19 years [1]. Osteosarcoma is the most prevalent neoplasm of the bone in children and adolescents, with approximately 850,000 cases reported each year in the US. Tumors are typically diagnosed between 10 and 30 years of age and mostly arise in the femur, tibia, and humerus. Although most osteosarcoma cases are sporadic, a higher incidence of osteosarcoma has been reported in certain hereditary cancer syndromes. For example, mutations of the tumor suppressor retinoblastoma protein (Rb) in retinoblastoma [2] or p53 in Li-Fraumeni syndrome [3], as well as the DNA repair helicase Werner protein are also associated with osteosarcoma [4]. Previous radiotherapy of other tumors may also lead, in the long term, to osteosarcoma development.

Treatment of osteosarcoma usually begins with neoadjuvant chemotherapy followed by surgical removal of the primary tumor and metastases (typically localized to the lungs). Agents used for osteosarcoma chemotherapy include cisplatin, methotrexate, and doxorubicin [5]. In the European Union, the liposome-encapsulated synthetic muramyl dipeptide analog, mifamurtide, is also approved by the European Medicines Agency for the immunotherapy of osteosarcoma [6]. The five-year survival rate depends on the stage, type, and subtype of osteosarcoma and ranges from 60% to 75% [7]. The rather poor therapeutic responsiveness indicates that new treatment modalities are clearly needed to improve the disease-free survival of osteosarcoma patients.

Neoplastic transformation is accompanied by fundamental rearrangements of metabolic pathways. A key hallmark of cancer metabolism is the reliance of tumor cells on glycolysis, even if oxygen is available (aerobic glycolysis, also known as the Warburg effect) [8]. NAD^+ is a central metabolite of energy production that serves as an electron carrier in glycolysis and the tricarboxylic acid (TCA) cycle. In addition to its role as a metabolic cofactor, NAD^+ also has signaling roles [9]. NAD^+-derived signaling messengers include the calcium-mobilizing agents cyclic ADP-ribose [10] and nicotinic acid adenine dinucleotide phosphate (NAADP) [11]. Moreover, members of the ADP-ribosyltransferase (ART) and sirtuin enzyme families also use NAD^+ as a substrate converting it to ADP-ribose and nicotinamide [12]. Sirtuins are NAD^+-dependent histone deacetylases that transfer the removed acetyl group onto NAD^+-derived ADP-ribose to generate O-acetyl-ADP-ribose, whereas ART enzymes cleave NAD^+ to nicotinamide and ADP-ribose and transfer the latter onto protein acceptors [13]. Some ARTs function as mono-ADP-ribosyltransferases while others cleave multiple NAD^+s and polymerize the resulting ADP-ribose moieties onto suitable protein acceptors, resulting in protein poly(ADP-ribosyl)ation (PARylation) [14,15]. This latter group includes poly(ADP-ribose) polymerase-1 (PARP1; also known as ARTD1), PARP2 (ARTD2), and tankyrases (ARTD5-6) [16]. PARP1, a major NAD^+ consumer, plays a central role in tumor biology [17]. On one hand, PARP1 regulates DNA replication, gene expression, DNA repair, cell adhesion, and migration [18]. On the other hand, PARP1 is the primary target for a novel cancer therapeutic modality based on the synthetic lethality paradigm [19,20]. Clinically used PARP inhibitors (PARPi) take advantage of the vulnerability of cancer cells (e.g., ovarian and breast carcinomas) that carry mutations in the DNA repair genes BRCA1/2, rendering them sensitive to DNA repair inhibitors, such as PARPi compounds. PARP inhibitors have also been evaluated as chemosensitizers (e.g., in combination with the DNA alkylating agent temozolomide or the topoisomerase inhibitor topotecan) [21–23], but myelotoxicity prevented the successful translation of promising preclinical results into the clinic.

The question arises whether and how NAD^+ metabolism can be targeted for cancer therapy. NAD^+ is synthesized through multiple routes [24]. In humans, the central pathway relies on nucleobases with special preference for nicotinamide (Nam), although nicotinic acid (NA) can also serve as starting material for NAD^+ synthesis [25]. Nicotinamide phosphoribosyltransferase (NAMPT) enzymes can use Nam and phosphoribosyl pyrophosphate (PRPP) to make nicotinamide mononucleotide (NMN) [26]. In turn, nicotinamide mononucleotide adenosyltransferases (NMNAT) catalyze the final step by forming NAD^+ from NMN and ATP. Tissue and subcellular distribution of the three NMNAT isoforms (NMNAT1-3) show distinct patterns [27]. Besides the nuclear (NMNAT1) and cytoplasmic/Golgi

(NMNAT2) isoforms, a mitochondrial NAD^+ synthase (NMNAT3) has also been described; however, the mitochondrial localization and contribution of NMNAT3 to the mitochondrial NAD^+ pool has been questioned [28]. Nuclear NAD^+ synthesis is of special interest because several major NAD^+-consuming enzymes (PARP1,2, SIRT1,6,7) that mediate important survival mechanisms operate in the nuclear compartment [29].

Since inhibitor molecules are available for NAMPT, our knowledge on the targetability of NAD^+ synthesis for cancer therapy comes from experiments with the NAMPT inhibitor compounds, FK866 and GMX1777. Although these drug candidates efficiently kill cancer cells, they also show some toxicity in non-transformed cells. In clinical trials, these drugs were relatively well-tolerated and displayed acceptable safety profiles [30]. 2,3-Dibromo-1,4-naphthoquinone (DBNQ) was recently identified as an inhibitor of purified NMNAT-1 [31]. However, this finding needs to be confirmed in cell-based assays.

The role of NMNAT1 in cancer cells is not known and its role in cancer cell chemosensitivity has not yet been investigated. In our current study, we aimed to investigate whether NMNAT1 plays a role in maintaining cell viability and whether it contributes to osteosarcoma cell survival following genotoxic stimuli. We report that U-2OS human osteosarcoma cells tolerate the genetic inactivation of the NMNAT1 gene and show increased susceptibility to cisplatin and doxorubicin. Increased chemosensitivity of the $NMNAT1^{-/-}$ phenotype is likely due in part to restricted DNA-damage-induced protein PARylation and consequently enhanced DNA damage. U-2OS cells unable to respond sufficiently to cisplatin treatment undergo caspase-mediated apoptotic and caspase-independent necroptotic death, suggesting that NMNAT1 may be worth investigating further as a potential target in cancer therapy with special regard to osteosarcoma treatment.

2. Results

2.1. DNA Damaging Drugs Induce NMNAT1 Expression in Tumor Cell Lines

Up to date NMNAT1 mRNA expression levels for osteosarcoma cells (U-2OS, SAOS-2) were not available. That is why we checked the mRNA levels in osteosarcoma cells and compared with other human tumor cell lines of different origins. Eleven human tumor cell lines were tested for NMNAT-1 mRNA expression (Figure 1A). The transcript could be detected, although at different levels, in all cell lines. Compared to the average expression level, A431 cells displayed significantly higher mRNA expression, while significantly lower expression was detected in A549, Capan2, MCF7 and HepG2 cell lines (Figure 1A). U-2OS osteosarcoma cells which showed average levels of NMNAT1 expression were chosen for further investigation. Of the three human NMNAT isoforms, NMNAT-1 is most highly expressed, in U-2OS cells, while NMNAT2 and NMNAT-3 are expressed at lower levels (Figure S1A). The DNA damaging chemotherapeutic agents doxorubicin and cisplatin significantly upregulated NMNAT1 mRNA expression (Figure 1B), suggesting that the enzyme may be a survival factor in DNA damage situations. Cisplatin, one of the drugs most commonly used for the treatment of osteosarcoma, caused a marked elevation in NMNAT1 protein expression in U-2OS cells (Figure S1B). Cisplatin treatment caused a concentration-dependent cytotoxicity in U-2OS cells (Figure 1C). Based on the viability data, we chose the 6.25 μg/mL concentration for the following experiments. At this concentration, cisplatin caused no significant change in NAD^+ levels (Figure 1D).

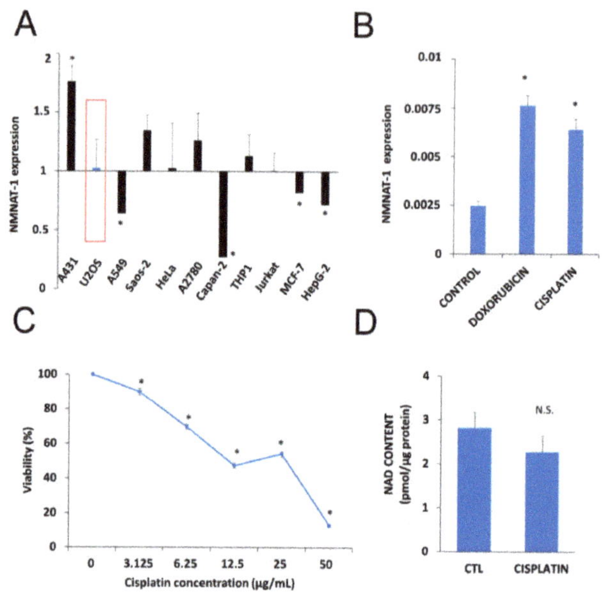

Figure 1. Nicotinamide mononucleotide adenylyltransferase-1 (NMNAT1) mRNA expression and induction by antitumor agents. The expression of NMNAT1 mRNA was determined in eleven human tumor cell lines. The y-axis shows the average expression level of the tested cell lines. Bars marked with asterisks are significantly different from the average expression (Student–Newman–Keuls method; * $p < 0.05$) (**A**). NMNAT1 expression in the U-2OS cell line was induced 24 h after cisplatin (6.25 µg/mL) or doxorubicin (2 µg/mL) treatment. Bars marked with asterisks are significantly different from the control (Dunnett test; * $p < 0.05$) (**B**). Calcein acetoxymethyl (Calcein AM) assay, indicating the concentration-dependent cytotoxic effect of cisplatin (3.125–50 µg/mL) on U-2OS cells, was measured 24 h after cisplatin treatment. Bars marked with asterisks are significantly different from the control (Bonferroni test; * $p < 0.05$) (**C**). Total NAD^+ content was measured in cell lysates 24 h after cisplatin (6.25 µg/mL) treatment and normalized to protein content. Bars marked with asterisks are significantly different from the control (Student's *t* test; * $p < 0.05$, N.S.: not significant) (**D**). Data plotted are means ± SEM ($n = 3$).

2.2. Generation and Characterization of an NMNAT1$^{-/-}$ Cell Line

To investigate the role of NMNAT1 in the survival of cisplatin-treated cells, we inactivated the gene for NMNAT1 using CRISPR-Cas9 technology. Single cell clones were obtained by cell sorting from cultures of NMNAT1$^{-/-}$ cells. We tested all the clones and all of them lacked NMNAT1 mRNA (Figure 2A). Clone 1B6 was selected for downstream experiments. Western blotting proved that NMNAT1 protein was missing from this clone (Figure 2B). Morphological properties of wild type and NMNAT1$^{-/-}$ cells (Figure S2A) revealed a significant reduction in the nuclear size and cell size (Figure S2B and C). The nuclear and cellular roundness was also slightly but significantly affected by the absence of a functional NMNAT1 protein (Figure S2D,E). The NMNAT1 deficient U-2OS cell line showed unaltered cell viability, as determined using the Calcein acetoxymethyl (Calcein AM) method (Figure 2C). However, clonogenic activity was impaired in the absence of a functional enzyme (Figure 2D). Despite elevated NMNAT-2 expression (Figure S1A), total cellular NAD^+ levels dropped to approximately one third of the control cell line (Figure 2E), indicating that NMNAT1 plays a dominant role in cellular NAD^+ synthesis. Interestingly, lower NAD^+ levels in NMNAT1$^{-/-}$ cells did not suppress ATP levels (Figure 2F) or impair cellular respiration, as indicated by the unchanged oxygen consumption rate (Figure 2G). Extracellular acidification rate (ECAR), a measure of glycolysis, showed higher values in the absence of NMNAT1 compared to the parent cell line (Figure 2H).

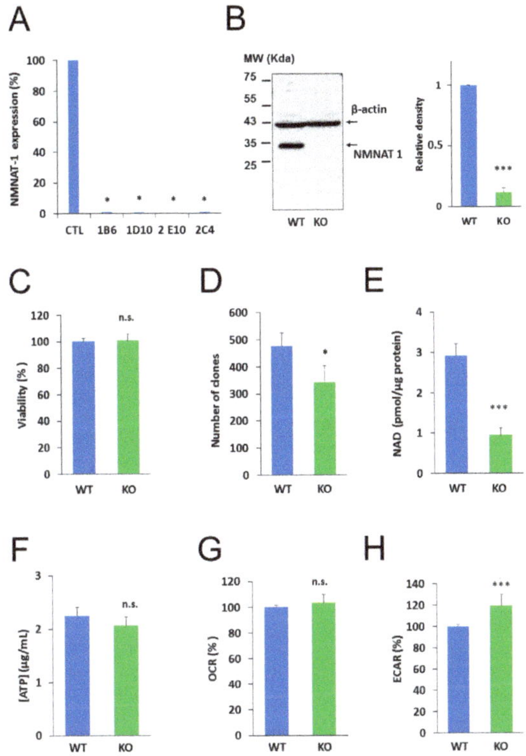

Figure 2. Characterization of NMNAT 1 KO cell line. NMNAT1 knockout cell lines were generated with CRISPR-CAS9 technology. Puromycin resistant cells were sorted and single cell colonies were grown. NMNAT1 mRNA levels were measured with RT-QPCR in each colony. Results are expressed as a percentage of NMNAT1 expression of the wild type U-2OS cell line (control). Bars marked with asterisks are significantly different from the control (Dunnett test; * $p < 0.05$) (**A**). Clone 1B6 was chosen for further investigation. NMNAT1 protein was measured in cell lysates of wild type U-2OS and the 1B6 clone with Western blot (**B**). Full WB image can be found in Supplementary Material. The following experiments compare the basic characteristics of wild type (WT) and NMNAT1 knockout (KO) cells. Viability was measured with a Calcein AM viability assay. Data points marked with asterisks are significantly different from the control (Student's t test; * $p < 0.05$, N.S.: not significant) (**C**). Clonogenic activity was assessed on day 6 by counting crystal violet stained colonies. Bars marked with asterisks are significantly different from the control (Student's *t* test; * $p < 0.05$, N.S.: not significant) (**D**). Basal total NAD^+ (**E**) and ATP (**F**) levels were assayed from cell lysates and normalized to protein content. Bars marked with asterisks are significantly different from the control (Student's *t* test; *** $p < 0.001$, N.S.: not significant) Metabolic parameters such as oxygen consumption rate (OCR)/oxidative phosphorylation (**G**) and extracellular acidification rate (ECAR)/glycolytic activity (**H**) were measured with a Seahorse metabolic analyzer. Bars marked with asterisks are significantly different from the control (Student's *t* test; *** $p < 0.001$, n.s.: not significant). Data plotted are means ± SEM ($n = 3$).

2.3. NMNAT-1 Inactivation Increases Chemosensitivity of the U-2OS Osteosarcoma Cells

Next, we investigated the role of NMNAT1 in the survival/death of cisplatin-treated cells. We found that the absence of NMNAT1 sensitized cells to the toxic effects of cisplatin (Figure 3A). For example, 5 μg/mL cisplatin did not cause significant toxicity in the parent cell line but killed 45% of NMNAT1$^{-/-}$ cells (Figure 3A). Morphological phenotyping of cisplatin-treated cells was performed to understand the mechanism of chemosensitization in NMNAT1$^{-/-}$ cells.

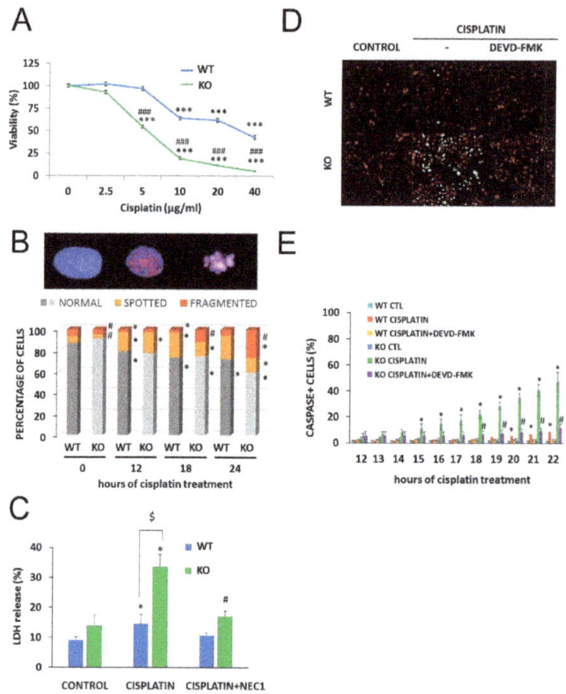

Figure 3. Chemosensitivity of NMNAT-1 KO cell line. Chemosensitivity of wild type (WT) and knockout (KO) cells was compared in a Calcein-AM viability assay 24 h after treatment with different concentrations of cisplatin (0.625–10 µg/mL). A new batch of cisplatin was used for this and all subsequent experiments. Results are expressed as percentages, compared to the vehicle (medium) treated samples. Data points marked with asterisks are significantly different from the control (Bonferroni test; *** $p < 0.001$). Data points marked with hashmarks are significantly different from the corresponding treatment of the wild type cells (Bonferroni test; ### $p < 0.001$). (**A**). Cisplatin-induced DNA damage was assessed by quantifying the P-H2AX signal after immunofluorescent staining. High-content analysis detected three different cell morphologies, "normal", "spotted" and "fragmented", representative images are shown (**B**). Chart shows the percentage of the three types of morphologies in untreated and in cisplatin-treated samples at 12, 18, and 24 h after cisplatin treatment. Bars marked with asterisks are significantly different from the control (0 hours of cisplatin treatment) (Student–Newman–Keuls method; * $p < 0.05$). Bars marked with hashmarks are significantly different from the corresponding treatment of the wild type cells (Student–Newman–Keuls method; # $p < 0.05$) (**B**). Error bars are not shown on panel B because of the 3D presentation. Lactate dehydrogenase (LDH) release was determined from the supernatants at 24 h after cisplatin treatment and expressed as a percentage of the positive control (lysed cells). Cells were treated with cisplatin or with the combination of cisplatin with the necroptosis inhibitor, concentration necrostatin-1 (NEC1). Bars marked with asterisks are significantly different from the control (not shown) (Dunnett test; * $p < 0.05$). Bars marked with hashmarks are significantly different from the cisplatin-treated samples (Dunnett test; # $p < 0.05$). Bars marked with $ are significantly different from the corresponding treatment of the wild type cells (Dunnett test; $ $p < 0.05$) (**C**). Apoptotic cell death was measured with a fluorogenic caspase-3 substrate in a kinetic assay on a Perkin Elmer Opera Phenix High Content Analyzer. Representative images (**D**) show the caspase activity signal (green) with digital phase contrast (red) at 22 h after cisplatin treatment. Images were taken every hour between 12 and 22 h after cisplatin treatment. Some samples were pretreated with the caspase-3 inhibitor z-DEVD-FMK as shown. The numbers of caspase positive cells are shown as a percentage of the actual number of cells. Data points marked with asterisks are significantly different from the control (12 h of cisplatin treatment) (Student–Newman–Keuls method; * $p < 0.05$). Data points marked with hashmarks are significantly different from the cisplatin treated samples Student–Newman–Keuls method; # $p < 0.05$). (**E**). Means ± SEM are shown ($n = 3$).

Cisplatin is a DNA crosslinking agent that does not induce DNA breaks directly. γH2AX foci, indicating DNA double strand breaks (DSB), have been reported in cisplatin-treated cells and reflect activation of DNA repair mechanisms, such as nucleotide excision repair or non-homologs end joining [32]. High-content analysis detected three different cell morphologies: "normal" cells contained no P-H2AX signal, "spotted" cell type contained P-H2AX foci, while the "fragmented" type showed a condensed morphology, with a diffuse P-H2AX signal. We measured γH2AX and analyzed staining in combination with cell fragmentation, a marker of cell death that indicates failed DNA repair. In line with previous reports [32], cisplatin caused H2AX phosphorylation (γH2AX formation) in wild type cells. Significantly more cells were positive for γH2AX formation and/or showed signs of nuclear fragmentation in cisplatin-treated NMNAT1 deficient cells compared to their wild type counterparts (Figure 3B).

Cisplatin-induced cell death occurred via multiple pathways in the absence of NMNAT1. Loss of plasma membrane integrity, a sign of necroptosis, was observed, as indicated by increased lactate dehydrogenase (LDH) release in cisplatin-treated NMNAT-1$^{-/-}$ cells. Necrostatin-1 (NEC1) abolished LDH release, suggesting that it was indeed the consequence of necroptosis (Figure 3C). The involvement of the caspase-mediated apoptotic cell death pathway was also clearly demonstrated with a fluorogenic caspase-3/7 substrate and inhibition of the signal by z-DEVD-FMK (Figure 3D,E).

Metabolic reprogramming is a hallmark of cancer. Moreover, anticancer agents, including cisplatin, have been reported to cause metabolic perturbations [33,34]. Considering that NAD$^+$ is a central energy metabolite and NMNAT1 knockout resulted in a significant drop in cellular NAD$^+$, we hypothesized that reduced NAD$^+$ availability may have an impact on cell metabolism in cisplatin-treated U-2OS cells. Basal NAD$^+$ levels were significantly lower in NMNAT1$^{-/-}$ cells and slightly but significantly decreased by cisplatin treatment (Figure 4A). A more dramatic change in ATP levels was observed. Cisplatin caused a marked drop in cellular ATP content in NMNAT1 deficient cells while no change could be seen in the wild type cells (Figure 4B). Cell death-associated impairment of energy production was partly responsible for the drop in ATP level, as both the caspase inhibitor z-DEVD-FMK and the necroptosis inhibitor NEC1 significantly prevented ATP loss (Figure 4B).

To characterize cellular energy-producing pathways, we measured oxygen consumption rate, (OCR) as a measure of cellular respiration, and extracellular acidification rate (ECAR), as a measure of glycolysis. No major difference could be observed in basal respiration between the two cell lines (Figure 4C). A mitochondrial stress test, conducted by sequential addition of different mitochondrial toxins (see Materials and Methods section for details), also showed similar responses in both cell lines. However, cisplatin treatment completely reduced mitochondrial respiratory reserve capacity in NMNAT1$^{-/-}$ cells and cells were unable to recover even basal respiratory activity after oligomycin treatment (Figure 4C). Wild type cells, on the other hand, displayed unaltered respiratory adaptation in response to cisplatin. The basal glycolytic activity proved to be significantly higher in the NMNAT1$^{-/-}$ cell line (Figure 4D). However, no significant differences between the two cell lines could be observed in glycolytic stress tests with or without the addition of cisplatin. This finding suggests that glycolytic activity remains mostly unaffected by cisplatin regardless of the NMNAT1 status. Cellular metabolism can be characterized by the OCR/ECAR ratio, based on the basal rates of the two parameters. Knockout cells have a reduced OCR/ECAR ratio indicating that they are more glycolytic, compared to the WT cells (Figure 4E). Furthermore, cisplatin treatment caused a more pronounced drop in the basal OCR/ECAR ratios in the NMNAT1$^{-/-}$ cells, demonstrating even more reliance on glycolysis rather than respiratory energy production.

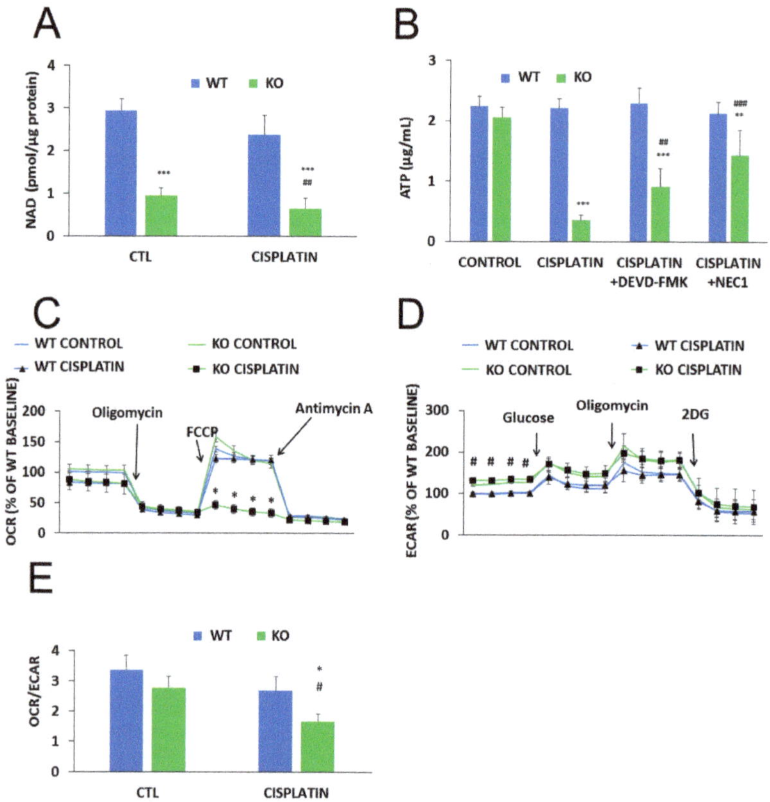

Figure 4. Metabolic alterations in NMNAT-1 KO cell line. Following cisplatin (5 µg/mL) treatment (24 h), total NAD$^+$ levels were determined from cell lysates of WT and KO cells and normalized to protein content. Bars marked with asterisks are significantly different from the control (Bonferroni test; *** $p < 0.001$). Data points marked with hashmarks are significantly different from the corresponding treatment of the wild type cells (Bonferroni test; ## $p < 0.01$) (**A**). Cellular ATP was assayed 24 h after cisplatin treatment (**B**). Cells were treated with cisplatin alone or in combination with apoptosis (z-DEVD-FMK, 100 µM) or necroptosis (NEC1, 30 µM) inhibitors. Inhibitors were added 30 minutes before cisplatin treatment. Bars marked with asterisks are significantly different from the control (not shown) (Dunnett test; ** $p < 0.01$, *** $p < 0.001$). Bars marked with hashmarks are significantly different from the cisplatin-treated samples (Dunnett test; ## $p < 0.01$, ### $p < 0.001$). (**B**). Oxygen consumption rate (OCR)/oxidative phosphorylation (**C**) and extracellular acidification rate (ECAR)/glycolytic activity (**D**) were determined using a Seahorse metabolic analyzer and expressed as a percentage of the WT baseline. Both OCR/oxidative phosphorylation and ECAR/glycolytic activity were monitored in specific stress tests 13 h after cisplatin treatment. Mitochondrial stress tests included oligomycin (2 µM), FCCP (0.5 µM), and antimycin A (1 µM) for OCR (**C**) and glucose (10 mM), oligomycin (2 µM) and 2-deoxyglucose (50 mM) for ECAR (**D**). Data points marked with asterisks are significantly different from the corresponding vehicle treated data points at the same phase of the graph (Student–Newman–Keuls method; * $p < 0.05$). Data points marked with hashmarks are significantly different from the corresponding data points of wild type cells at the same phase of the graph (Student-Newman-Keuls method; # $p < 0.05$). Ratios of the two metabolic routes were determined from the baseline OCR and ECAR values 13 hours after cisplatin treatment. Bars marked with asterisks are significantly different from the corresponding vehicle treated group (Student-Newman-Keuls method; * $p < 0.05$). Bars marked with hashmarks are significantly different from the corresponding group of wild type cells (Student-Newman-Keuls method; # $p < 0.05$) (**E**). Data plotted are means ± SEM ($n = 3$).

2.4. The Role of PARylation in the Increased Cisplatin Sensitivity of NMNAT1$^{-/-}$ K.O. Cells

Cisplatin-induced DNA damage activates PARP1 [35] and PARP1 activation contributes to the repair of cisplatin-crosslinked DNA lesions and cell survival. Thus, we hypothesized that NMNAT1 deficiency and consequent nuclear NAD$^+$ scarcity may compromise PARP1's ability to initiate the DNA damage response and prevent cell death. We found that cisplatin caused a time dependent poly(ADP-ribose) (PAR) formation in WT cells, while no PAR formation could be detected in the NMNAT1$^{-/-}$ cells (Figure 5A,B). Moreover, cisplatin treatment caused only a mild and non-significant decrease in the clonogenic activity of wild type cells, but significantly reduced the clonogenic activity of NMNAT1$^{-/-}$ cells. Combined treatment with cisplatin and the PARP inhibitor (olaparib) caused a significantly lower proliferation in wild type cells, compared to the cisplatin-treated cells, but no further decrease could be detected in the NMNAT1$^{-/-}$ cells (Figure 5C). These data suggest that cisplatin-induced DNA damage is unable to activate PARP1 in NMNAT1$^{-/-}$ cells and substrate deprivation of PARP1 limits the capacity of the cells to cope efficiently with DNA injury.

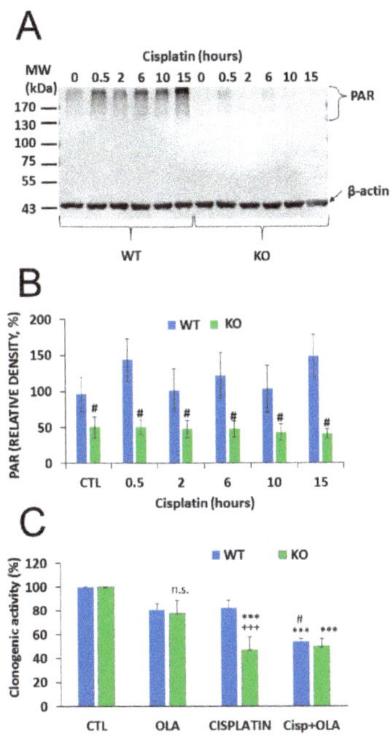

Figure 5. Impaired PARylation contributes to increased cisplatin sensitivity of NMNAT1$^{-/-}$ cells. The level of poly(ADP-ribose) (PAR) polymer formation was detected using western blot in wild type (WT) and in NMNAT1 $^{-/-}$ (KO) cells at the indicated timepoints (0.5, 2, 6, 10, and 15 h) after cisplatin (5 µg/mL) treatment (**A**). Full WB image can be found in Supplementary Material. Actin was used as a loading control. Relative densities (normalized to beta actin and compared to the untreated WT control) are shown on panel (**B**). Bars marked with hashmarks are significantly different from the corresponding samples of wild type cells (Bonferroni test; # $p < 0.05$). The role of possible of PARP-1-NMNAT1 interaction in clonogenic activity was determined by counting crystal violet-stained colonies at day 6 (**C**). Cells were treated with vehicle (CTL), cisplatin, or the combination of cisplatin (5 µg/mL) and olaparib (10 µM) (Cisp+ OLA) and individual colonies were counted in each sample. Results are expressed as a percentage of the number of colonies in the vehicle treated samples. Bars marked with asterisks are significantly different from the corresponding control (Bonferroni test; *** $p < 0.001$). Bars marked with hashmarks are significantly different from the corresponding cisplatin treated samples (Bonferroni test; # $p < 0.05$). Bars marked with + are significantly different from the corresponding samples of wild type cells (Bonferroni test; +++ $p < 0.001$). Data plotted are means ± SEM ($n = 3$).

2.5. Chemosensitization in 3D Cultures and in Combination Treatment

To demonstrate the role of NMNAT1 in cellular models that more closely resemble in vivo conditions or clinical settings, we investigated a) chemosensitization to cisplatin by NMNAT1 deficiency in a 3D model (as opposed to 2D cultures) and b) a treatment protocol based on a combination of chemotherapeutic drugs. Wild type and NMNAT1$^{-/-}$ U-2OS cells formed spheroids under favorable conditions (see "Materials and Methods" section for detail). Cell-to-cell contacts in spheroids render cells resistant to toxic stimuli [36]; therefore, we used higher concentrations of cisplatin in the spheroid experiments. While spheroids of the wild type cells only became less compact at 50 μg/mL cisplatin concentration, NMNAT1$^{-/-}$ U-2OS cells displayed a significant decrease in size, or a complete disintegration of the spheroids under the same conditions (Figure 6A,B). The inner region of cisplatin-treated spheroids show a marked elevation in Annexin V positivity, a more dramatic elevation can be detected in the case of NMNAT1$^{-/-}$ spheroids (Figure 6C).

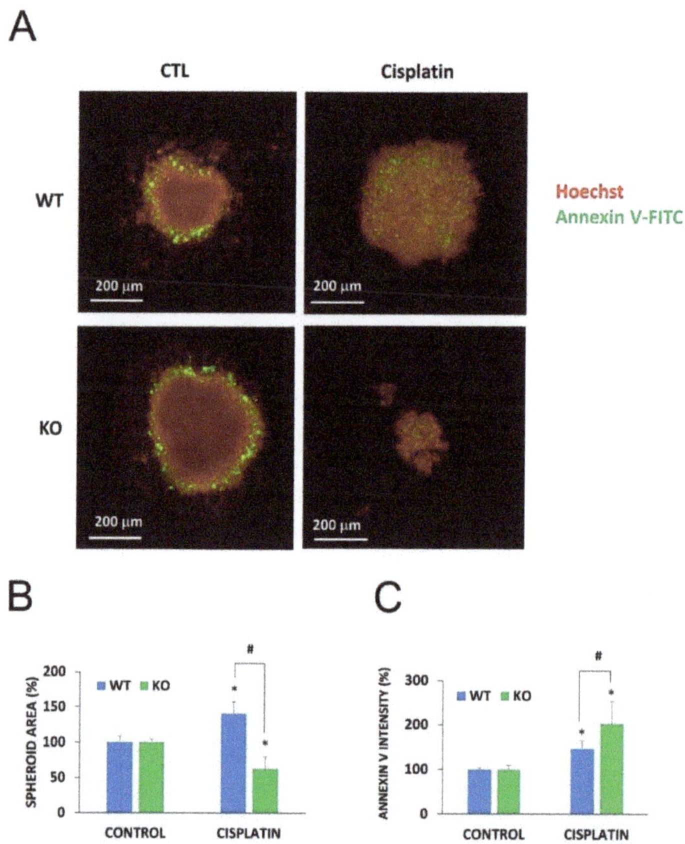

Figure 6. Chemosensitivity of spheroids to cisplatin treatment. Using WT and NMNAT-1$^{-/-}$ U-2OS cells, 3D cell cultures (spheroids) were generated. Spheroids were treated with cisplatin (50 μg/mL) for 6 days. Cells were stained without fixation with Hoechst (red pseudocolor) and Annexin V (green). Representative images are shown in panel **A**. Images of 3 spheroids/treatment were taken and analyzed for area (**B**) and the intensity of the inner region (**C**). Results are expressed as percentages of the control (vehicle treated) samples. Bars marked with asterisks are significantly different from the vehicle treated control (Student-Newman-Keuls method; * $p < 0.05$). Bars marked with hashmarks are significantly different from the cisplatin treated wild type spheroids (Student-Newman-Keuls method; # $p < 0.05$). Data plotted are means ± SEM ($n = 3$).

Moreover, a combination of cisplatin with doxorubicin resulted in more efficient killing of cancer cells (Figure 7). A concentration series of doxorubicin was used to explore differences in the sensitivity of wild type and NMNAT1$^{-/-}$ U-2OS cells. At concentrations higher than 750 ng/mL, knockout cells were significantly more sensitive to doxorubicin (Figure 7A). Cells were exposed to different concentrations of cisplatin and a fixed doxorubicin concentration of 150 ng/ml. Combination treatment was significantly more efficient in killing NMNAT1$^{-/-}$ cells than wild type cells (Figure 7B).

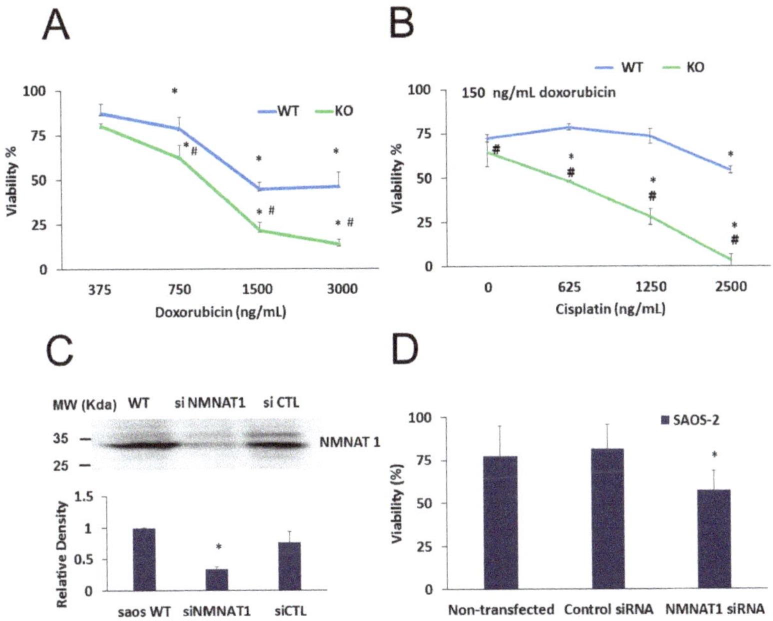

Figure 7. Combined treatments on U-2OS WT and NMNAT1$^{-/-}$ cells, chemosensitivity of SAOS-2. Wild type (WT) and NMNAT1$^{-/-}$ cells were treated with the indicated concentrations of doxorubicin (DOX) and viability was determined with the Calcein-AM assay 24 hours after DOX treatment. Data points marked with asterisks are significantly different from the corresponding vehicle treated data points (not shown) (Dunnett test; * $p < 0.05$). Data points marked with hashmarks are significantly different from the corresponding data points of wild type cells (Student–Newman–Keuls method; # $p < 0.05$) (**A**). DOX was also applied in combination with cisplatin (**B**). Panel B shows the viability of WT and KO at constant (150 ng/mL) DOX and increasing cisplatin concentrations. Data points marked with asterisks are significantly different from the corresponding vehicle treated data points (Dunnett test; * $p < 0.05$). Data points marked with hashmarks are significantly different from the corresponding data points of wild type cells (Student–Newman–Keuls method; # $p < 0.05$). SAOS-2 cells were transfected with control or NMNAT-1 specific siRNAs. Efficiency of NMNAT-1 silencing was verified by Western blot (**C**). Full WB image can be found in Supplementary Material. Cells were treated with cisplatin for 48 hours and a Calcein-AM viability assay was performed (**D**). Data points marked with asterisks are significantly different from the vehicle-treated sample (Dunnett test; * $p < 0.05$). Data plotted are means ± SEM ($n = 3$).

To confirm the sensitizing effect of the absence of NMNAT1 in cisplatin-induced cell death, cells of another osteosarcoma cell line, SAOS-2, were used. NMNAT1 could be effectively silenced in SAOS-2 cells (Figure 7C). Similarly to NMNAT-1 knockout U-2OS cells, NMNAT-1 silenced SAOS-2 cells also showed significantly higher sensitivity to cisplatin treatment (Figure 7D).

3. Discussion

Similar to most neoplastic diseases, osteosarcoma poses a great challenge to modern medicine. The therapeutic approaches are limited and are based on the combination of chemotherapy and surgical removal of the primary tumor and metastasis. Two of the most efficient chemotherapeutics used for the therapy of osteosarcoma (cisplatin and doxorubicin) cause DNA injury. Methotrexate, the third component of the most common chemococktails, is an anti-metabolite folate analog that interferes with the synthesis of purine and pyrimidine bases and, therefore, DNA and RNA synthesis. Of note, the anticancer effect of methotrexate has also been reported to be due to DNA breakage [37]. Thus, it seems that the DNA damage response (DDR) might be the Achilles heel of osteosarcomas. Targeting the DNA damage response is a novel anticancer approach in cancers displaying DNA repair defects. The best known example is PARPi treatment of BRCA-negative ovarian cancers [38]. Nuclear NAD^+ synthesis may be an unlikely target of DNA repair in cancer therapy. However, as the substrate of PARP enzymes, NAD^+ may be a limiting factor for PARylation activity. Moreover, the nuclear SIRT enzymes, another major NAD^+ consumer group, are also known to promote cell survival and are potential cancer targets (see more details below).

Previous experiments with NAMPT inhibitors suggested that inhibition of the salvage pathway of NAD^+ synthesis may have beneficial effects for cancer therapy [39]. Due to the lack of specific NMNAT1 inhibitors, however, the role of NMNAT1 in cancer has not yet been investigated. In the present study, we have investigated the role of NMNAT1 as a potential target in osteosarcoma. All cancer cell lines that were tested expressed NMNAT1 mRNA. Compared to the other assayed cell lines, U-2OS osteosarcoma cells displayed an average level of NMNAT1 mRNA. Our mRNA data also show that in U-2OS cells, the NMNAT-1 has the highest expression, while NMNAT2 and NMNAT-3 have lower contributions. These cells tolerated inactivation of the NMNAT1 gene without showing decreased viability or marked morphological alterations. Notable phenotypic differences between $NMNAT1^{-/-}$ and wild type cells included reduced proliferation capacity (clonogenicity) and moderately increased glycolytic activity. The most prominent feature of the $NMNAT1^{-/-}$ cells was a markedly reduced NAD^+ level, which did not compromise cellular energy production in the resting state, as reflected by the unaltered ATP content. The two-thirds reduction in cellular NAD^+ in $NMNAT1^{-/-}$ cells highlights the dominant role of NMNAT1 in cellular NAD^+ synthesis, as elevated NMNAT-2 expression cannot substitute the role of NMNAT1 in NAD+ production of NMNAT1 KO cells. A cross-compartment exchange of NAD^+ between nuclear and cytoplasmic pools is possible, as demonstrated in a model of adipogenic signaling where NMNAT2 activation drained nuclear NAD^+ while competing for the common substrate NMN [40]. Similar nuclear and cytoplasmic NAD^+ concentrations also suggest relatively barrier-free equilibration of NAD^+ between these compartments [41]. However, deciphering mechanisms underlying the marked drop of total NAD^+ in $NMNAT1^{-/-}$ cells and characterization of the effect of NMNAT1 inactivation on the main NAD^+ compartments require further investigation.

A key question asked in our study was whether NMNAT1 contributes to the cells' ability to cope with DNA damage caused by chemotherapeutic agents used in osteosarcoma therapy. Induction of NMNAT1 by cisplatin and doxorubicin suggests that NAD^+-dependent processes may be involved in the cellular survival response to these chemotherapeutics. Moreover, a major finding of our paper is that NMNAT1 targeting is a potent chemosensitizing strategy. $NMNAT1^{-/-}$ U-2OS cells displayed increased sensitivity to cisplatin, as indicated by reduced viability and increased cell death compared to wild type U-2OS cells. Cisplatin-induced cell death of NMNAT1 knockout cells showed features of both apoptosis and necroptosis. Activation of caspase3/7-like proteases and the cytoprotective effect of caspase-3/7 inhibition suggest that caspase-mediated apoptosis is involved in cell death in our model. Furthermore, the protective effect of the necroptosis inhibitor Nec1 indicates that the necroptotic pathway is also activated and contributes to cell death. If cisplatin-induced killing of NMNAT1 deficient cells also has a necroptosis-like component in vivo, NMNAT1 inhibition may enhance the anticancer immune response or may promote metastasis [42]. A detailed characterization of cell death modalities involved in the killing of osteosarcoma cells with reduced nuclear NAD^+

production and the effects on antitumor immunity versus metastasis formation goes beyond the scope of the current paper.

An important question raised by our study is whether the anticancer potential of NMNAT1 targeting is restricted to osteosarcoma. It is quite likely that targeting NMNAT1 may also be beneficial in other types of tumors. In support of this statement, we found that NMNAT1 is likely to be important for the progression of various tumors. RNAsec data from the kmplot database [43] show that low NMNAT1 expression correlates with better survival of patients with sarcomas, liver hepatocellular carcinoma, bladder carcinoma, breast cancer, esophageal adenocarcinoma, kidney renal papillary cell carcinoma, pancreatic ductal adenocarcinoma and uterine corpus endometrial carcinoma (Figure S3). Whether NMNAT1 inhibition provides therapeutic benefits in these and other tumors requires further investigation.

Our data strongly support the role of PARP1 in the chemosensitization process. We hypothesize that nuclear NAD^+ levels may be too low to support PARP activity in $NMNAT1^{-/-}$ cells. In fact, cisplatin stimulated PARylation in wild type cells. However, PAR formation was absent in $NMNAT1^{-/-}$ cells. Moreover, the potent PARP inhibitor olaparib enhanced the anti-clonogenic effect of cisplatin in the wild type but not the NMNAT1-deficient cells. The role of PARP-1 in the anti-clonogenic effect is also supported by our unpublished data from experiments with PARP-1 and PARP-2 silenced osteosarcoma cells. These results suggest that PARP1 activation acts as a survival factor via assisting DNA repair, in line with our current understanding of the role of PARylation in the repair of cisplatin-induced DNA damage [44]. PARP1 is unable to contribute to cell survival in $NMNAT1^{-/-}$ cells (due to the shortage of NAD^+). Thus, a key factor in the chemosensitizing effect of NMNAT1 inactivation is reduced PARylation and, consequently, impaired DNA repair. It is tempting to hypothesize that NMNAT1 and PARP1 interact directly at sites of DNA damage, similar to the direct interaction between NMNAT1 and PARP1 on PARP1-dependent promoters [45]. This interaction might involve the binding of NMNAT1 to PAR polymers [46]. This hypothesis, however, needs experimental verification. The possible role of PARP1 in the chemosensitization of $NMNAT1^{-/-}$ osteosarcoma cells was also supported by previous preclinical studies. These papers reported (a) correlation between PARP1 expression level and osteosarcoma cell survival; (b) osteosarcoma cell killing or chemosensitization by PARP1 knockout or PARP inhibition [47–49]. Ongoing and finished clinical trials (e.g., ClinicalTrials.gov ID: NCT01583543, NCT01858168 NCT02044120) also set out to investigate the effects of olaparib and other PARP inhibitors in osteosarcoma. Our data suggest that, at least under our current experimental settings, much of the chemosensitization observed in NMNAT1 cells is due to indirect inhibition of PARylation. It remains to be seen, however, how potent NMNAT1 inhibitors (which are not yet available) will compare with clinically used PARPi compounds in terms of therapeutic spectrum, efficiency, tolerability and side effects in different types of tumors. We hypothesize that PARPi and NMNAT1i will necessarily differ in one or more of these aspects.

The hypothesis that impaired metabolic adaptation may also play a role in the chemosensitivity of NMNAT1-deficient cells is plausible, considering the central role of NAD^+ in energy metabolism. Under basal conditions, a slight shift from respiration towards glycolysis could be detected in $NMNAT1^{-/-}$ cells but energy production was balanced, as indicated by the normal ATP levels. DNA damage revealed the vulnerability of the respiratory system, as the mitochondrial stress test identified severely impaired respiratory reserve capacity. Multiple lines of evidence highlight the importance of metabolic regulation in tumor behavior and chemosensitivity. For example, analysis of integrated transcriptomic and metabolomics datasets identified altered glycolysis-related mRNAs and metabolites in osteosarcoma samples [50]. Moreover, various microRNAs regulate both osteosarcoma growth (proliferation) and glycolysis either positively (e.g., miRNA543) [51] or negatively (e.g., miRNA186) [52]. Furthermore, glycolysis also appears to correlate with chemosensitivity (e.g., cisplatin sensitivity [53]) or viability of osteosarcoma cells [54,55].

The extent to which SIRT inhibition contributes to chemosensitization under NAD^+ scarcity is worth investigating. The nucleus is home to several sirtuin (SIRT) enzymes (SIRT1,6,7). SIRT1 plays a role in the DNA damage response [56]. Furthermore, the central role of sirtuins in energy homeostasis [57] and metabolic regulation [58] support the possible involvement of sirtuins in shifting metabolic balance (e.g., respiration vs glycolysis) after DNA damage. Nuclear SIRTs facilitate cell survival under various stress conditions, underlining their potential role in chemosensitization. For example, SIRT6 is downregulated in DOX-treated liver cancer cells, contributing to DOX-induced cell death. The pro-survival role of SIRT6 in this model, as demonstrated by the overexpression of the enzyme, is mediated by Foxo3 degradation [59]. Inhibition of SIRT6 also sensitizes cancer cells to chemotherapy [60]. Similarly, SIRT7 has been identified as a survival factor via complex mechanisms including direct p53 diacetylation [61] and the promotion of autophagy [62]. In a transcription regulatory setting, a requirement for NMNAT1 and NAMPT for SIRT1 activity has already been demonstrated [63] and similar interplay between nuclear SIRTs and enzymes of the salvage pathway of NAD^+ synthesis is likely to take place in the DNA damage response as well. Nevertheless, clarifying the role of SIRTs in the chemosensitizing effect of NMNAT1 inactivation goes beyond the scope of this paper.

In summary, NMNAT1 has been identified as a possible target for the treatment of osteosarcoma. DNA damaging chemotherapeutic agents are likely to cause synergistic toxicity with NMNAT1 inhibitors, if NMNAT1 inhibitors are developed. Key mechanisms underlying this synergy include the impairment of PARylation-dependent DNA repair processes and metabolic adaptation. The chemosensitization by NMNAT1 inactivation could also be observed under more clinically relevant conditions, i.e., in 3D spheroids and combination chemotherapy. However, the effect of NMNAT1 targeting in non-transformed cells is a critical issue for the development of anticancer strategies targeting NMNAT1.

4. Materials and Methods

4.1. *D Cell Culture*

Human U-2 OS cells were grown in Dulbecco's modified Eagle's medium (DMEM, #12-604F, Lonza, Basel, Switzerland), containing 10% fetal bovine serum (#10500-064, GIBCO, ThermoFisher Waltham, MA, USA) L-glutamine, and penicillin-streptomycin, under standard cell culture conditions (humidified atmosphere, 5% CO_2). The cells were routinely tested for mycoplasma contamination.

4.2. *D Cell Culture (Spheroids)*

Spheroids were grown as previously described [64] with modifications as follows. Flat-bottom cell culture microplates were coated with low melting point agarose to form a U-shaped, cell-repellent bottom. Cells were seeded into the wells and grown for 2 days to form spheroids.

4.3. *Generation of NMNAT1$^{-/-}$ Cells by Crispr-cas9*

NMNAT1 knock-out U-2OS cells were made by CRISPR-Cas9 technology using reagents from Santa Cruz Biotechnology, following the manufacturer's instructions. Cells were grown to 40–80% confluence and then trypsinized in antibiotic-free standard growth medium (DMEM, #12-604F, Lonza, Basel, Switzerland). Cells were counted, cell number was adjusted to 5.33×10^5 cells/mL, and a 2 mL cell suspension was centrifuged (100× *g*, 10 min). UltraCruz® Transfection Reagent (18 µL; sc-395739, Santa Cruz Biotechnology, Dallas, TX, USA) was diluted with 82 µL of plasmid transfection medium (sc-108062, Santa Cruz Biotechnology, Dallas, TX, USA) to bring the final volume to 100 µL. The medium was carefully aspirated and the diluted transfection reagent was pipetted onto the cells. Next, 2.5 µL of NMNAT1 CRISPR Plasmid and 2.5 µL of HDR Plasmid were added to the tube. Cells were immediately transferred to transfection cuvettes and transfected using an Amaxa Nucleofector II (Lonza, Basel, Switzerland). Transfected cells were pipetted into 6-well plates, which contained pre-warmed cell culture media (DMEM, #12-604F, Lonza, Basel, Switzerland) supplemented with 10% fetal bovine serum (#10500-064, GIBCO, ThermoFisher Waltham, MA, USA), 5% L-glutamine, and 5% penicillin-streptomycin. Cells were incubated for 24–72 hours under regular conditions and transfection efficiency was visually confirmed by the detection of red fluorescent protein (RFP) via fluorescent microscopy. Transfected cells were selected for puromycin (2.5 μg/mL) resistance for 3 weeks and sorted for RFP fluorescence with a BD LSR II Cell Sorter (Franklin Lakes, NJ, USA. Single-cell colonies were grown and used for the experiments. The NMNAT1 expression of the clones was checked with qPCR for mRNA-level and Western blot for protein level (see below).

4.4. Western Blot

Western blotting was carried out as previously described [65–67], with modifications as follows. U-2OS cells were sonicated in RIPA lysis buffer supplemented with protease inhibitor cocktail (#M221, VWR International, Radnor, PA, USA) and phosphatase inhibitor (PMSF, 1:100; # PMSF-RO, Merck, Darmstadt, Germany). Lysates were centrifuged at 16,100× *g* for 10 min at 4 °C. Protein concentrations of the supernatants were measured with a Direct Detect infrared spectrometer (EMD Millipore Corporation, Burlington, MA, USA). Equal amounts of the proteins from each extract were separated on SDS-polyacrylamide gels (10%) in SDS-Tris-glycine running buffer (10× glycine, Tris, SDS). The proteins were then transferred electrophoretically to nitrocellulose membranes in transfer buffer (5× Tris, glycine, in dH_2O). Membranes were blocked for 1 hour at room temperature with 5% skim milk-powder (#70166, Merck, Darmstadt, Germany) dissolved in 0.01% PBS-Tween20. Primary antibodies were applied in blocking solution overnight at 4 °C. Horseradish peroxidase (HRP)-coupled secondary antibodies were applied in the same type of solution as primary antibodies for 2 hours at room temperature. ECL-based chemiluminescence (Super Signal, West Pico Plus, Luminol/Enhancer #1863098, Peroxidase Solution #1863099, ThermoFisher Waltham, MA, USA) was used for detection. Images were acquired with a Bio-Rad ChemiDoc Imager (Bio-RAD, Hercules, CA, USA) and ImageLab 6.0 software was used for protein quantification. The antibodies used for western blotting are shown in Table S1.

4.5. High Content Analysis (HCA)

4.5.1. HCA on Fixed Cells: Quantification of γH2AX

Cells (2×10^4, 100 μL/well) were seeded into sterile microplates (Cell Carrier-96 ultra, PerkinElmer, Waltham, MA, USA) and grown for 24 h. After applying the indicated cell treatments (see figure legends), cells were fixed in 3% formaldehyde/PBS solution for 15 min at room temperature, washed 3× with PBS, and incubated in blocking solution (5% BSA in PBS) for 15 min at room temperature. The anti-phospho-H2AX antibody (Table S2) was diluted in blocking solution and added to wells (50 μL/well, 2 h at room temperature). The secondary antibody (Alexa Fluor 488, Table S2) was diluted in blocking solution and incubated for 1 hour. After antibody incubations, cells were washed twice with PBS and incubated for 5 min with PBS containing 4′,6-diamidino-2-phenylindole dihydrochloride (DAPI, Table S2) at room temperature for nuclear staining. Cells were then washed three times with PBS and were kept in 100 μL of PBS until imaging. Images were acquired using an Opera Phenix High Content Analyzer (PerkinElmer, Waltham, MA, USA) with a 10× air objective (NA 0.3). Image analysis was performed with the built in Harmony software (version 4.8).

4.5.2. HCA on Live Cells: Caspase Activation

Cells (2×10^4 cells, 100 μL/well) were seeded into 96-well plates (Perkin Elmer, Waltham, MA, USA) and grown for 24 h. The following day, cells were pretreated with 100 μM z-DEVD-FMK caspase inhibitor (#S7312, Selleckchem, Houston, TX, USA) and/or treated with 5 μM cisplatin (Accord, Warsaw, Poland). Then, the cells were stained with 50 μL CellEvent™ Caspase-3/7 Green Detection Reagent (Table S2) in 7 μM final concentration. The plates were placed into an Opera Phenix High-Content Analyzer (Perkin Elmer, Waltham, MA, USA) with environment control (5% CO_2, 37 °C) and fluorescence was measured every hour for 11 h at 502/530 nm. The analysis was performed with the Harmony software (Perkin Elmer, Waltham, MA, USA).

4.5.3. HCA on Live Cells: Cell Death in Spheroids

Spheroids were transferred to glass bottom microplates (Cell Carrier-96 ultra, PerkinElmer, Waltham, MA, USA) and treated with cisplatin (25, 50, and 100 μg/mL) on day 2 and day 4. Afterward, the cells were stained with Hoechst (Table S2) and Annexin V-FITC (Table S2.) for 1 hour in growth medium. Images were acquired using an Opera Phenix High Content Analyzer (Perkin Elmer, Waltham, MA, USA). Fluorescence intensity was detected at 350 (Hoechst) and 488 nm (Annexin V-FITC).

4.5.4. HCA on Live Cells: Cell Morphology

Cells (2×10^4 cells, 100 μL/well) were seeded into 96-well plates (Perkin Elmer, Waltham, MA, USA) and grown for 24 h. The following day, cells were stained with 50 μL DRAQ5 (Table S2.) to obtain a final concentration of 2.5 μM. Plates were placed into an Opera Phenix High-Content Analyzer with environment control (5% CO_2, 37 °C), fluorescence was measured at 488 to 647 nm, and the analysis was performed with the Harmony software (version 4.8).

4.6. Calcein-AM Viability Assay

This assay was carried out as previously described [68], with modifications as follows. Cells (2×10^4 cells, 100 μL/well) were seeded into 96-well plates and were grown for 24 h. The next day, cells were treated with the indicated treatments (see figure legends). Two different batches of cisplatin were used for the experiments. The first batch was used for the experiments presented on Figure 1 (Figure 1B–D), whereas a second batch was used for the experiments presented in Figure 3, Figure 4, Figure 5, Figure 6, and Figure 7. Then the cells were stained by adding 50 μL of Calcein-AM (#17783, Merck, Darmstadt, Germany) solution in a final concentration of 1 μM. After incubating cells for 1 h at 37 °C, the fluorescent signal was measured with a Tecan Spark 20M (Tecan, Männedorf, Switzerland) multimode reader (Ex/Em = 485/530 nm with). Viability was expressed as a percentage of the untreated control.

4.7. Clonogenic Survival Assay

U-2OS cells were seeded into 6-well plates (#92006, TPP, Trasadingen, Switzerland) at a density of 1×10^3 cells/mL. After 24 h the cells were treated with 10 μM PJ34 (P4365, Merck, Darmstadt, Germany) or 10 μM Olaparib (S1060, Selleckchem, Houston, TX, USA) for 30 minutes, and then with 5 μg/mL cisplatin (Accord Healthcare Inc., Durham, UK). Cells were incubated for 6 days and then counted manually after staining with 0.5% crystal violet dissolved in 20% ethanol [69].

4.8. Lactate Dehydrogenase (LDH) Release Assay

Cell death was assessed by determining the activity of LDH released into the culture medium. The LDH assay kit (#786-210, GBiosiences, St. Louis, MO, USA) was used as described by the manufacturer.

4.9. Measurement of Total Cellular NAD$^+$ and ATP

NAD$^+$ content was measured with an NAD$^+$ assay kit following the manufacturer's instructions (NAD$^+$/NADH Quantitation Kit. #MAK037-1KT, Merck, Darmstadt, Germany). The NAD$^+$ content of the cells was measured with a plate reader (Tecan Spark 20M, Tecan, Männedorf, Switzerland) at 450 nm.

ATP content was determined using an ATP assay kit following the manufacturer's instructions (#110M6101, Merck, Darmstadt, Germany). Luminescence was measured with a plate reader (Tecan Spark 20M, Tecan, Männedorf, Switzerland). NAD$^+$ content and ATP content were normalized to protein content.

4.10. *Metabolic Analysis with Seahorse Metabolic Analyzer*

U-2OS cells were seeded (1 × 10^5 cells/well) in DMEM in Seahorse XF96 cell culture microplates (Agilent Technologies, Inc., Santa Clara, CA, USA) and incubated overnight at 5% CO_2 and 37 °C. The sensor cartridge was prepared by adding 200 μL of dH_2O overnight. Then, Seahorse Bioscience XF96 calibrant solution (pH 7.4) (Part No.: 100-840-000) was added to each well of a Seahorse 96-well utility plate. The sensors with the calibrant solution were incubated for 2 hours at 37 °C without CO_2. The measurement was performed using the Seahorse XF96 Analyzer (Agilent Technologies, Inc., Santa Clara, CA, USA).

XF Cell Mito Stress analyses were performed as described in [70], with modifications as follows. The mitochondrial inhibitors were applied at the following final concentrations: 2 μM oligomycin, 0.5 μM FCCP, and 1 μM antimycin-A. Glycolysis stress was based on [70].

4.11. *Sulforhodamine B (SRB) Assay*

The SRB assay was performed as described in Kovács, P. et al. [71].

4.12. *RNA Extraction and Quantitative PCR*

RNA was purified with TRIzol reagent (Tri-RNA reagent, #FATRR001, Amplicon, Odense, Denmark) as described in the manufacturer's instructions. Quantitative real-time PCR was performed with a LightCycler 480 thermocycler (Roche, Basel, Switzerland) using SYBR Green (SyberGreen, #4472908, Applied Biosystems, Foster City, CA, USA) according to the manufacturer's protocol. Reactions were carried out in triplicate and data were normalized to the geometric mean of housekeeping genes (36B4 and cyclophilin). Sequences of primers are given in Table S3. hNMNAT1, hNMNAT2, hNMNAT3, hATP5B, hHSP60, hPKM2 primers were ordered from IDT (Coralville, IA, USA); and h36B4 and hcyclophilin from Merck (St. Louis, MO, USA).

4.13. *Statistical Analysis*

Experiments were repeated at least three times and data are expressed as mean ± SEM. Based on the type of experiment and the distribution of data, different kinds of statistical analysis were used. The statistical tests used are indicated in the figure legends.

5. Conclusions

Genetic inactivation of NMNAT1 sensitizes U-2OS osteosarcoma cells to cisplatin, doxorubicin, or a combination of these two treatments. Upon cisplatin treatment, an impaired PARP1 activity, insufficient DNA repair, a marked drop in cellular ATP and a limited mitochondrial reserve capacity could be observed, which all contribute to the higher chemosensitivity of NMNAT-1$^{-/-}$ cells. Increased cell death of NMNAT1$^{-/-}$ cells shows features of both apoptosis and necroptosis. NMNAT1$^{-/-}$ cells also displayed markedly higher sensitivity to cisplatin when grown as spheroids in 3D culture. Our results suggest that NMNAT1 may be worth investigating further as a potential target in cancer therapy.

Supplementary Materials: The following are available online at http://www.mdpi.com/2072-6694/12/5/1180/s1, Figure S1: mRNA expressions of NMNAT enzymes and induction of NMNAT1 protein expression in U-2OS cells, Figure S2: Cell morphology analysis, Figure S3: Tumors with better patient survival at low NMNAT1 expression, Table S1: Antibodies used for Western blotting, Table S2: Antibodies and dyes used for High Content Analysis and Table S3: Sequences of primers, used in quantitative PCR experiments. Supplementary materials also contain the full images of Western Blots.

Author Contributions: Conceptualization, L.V. and C.H.; Data curation, C.H.; Formal analysis, A.K., Z.R. and C.H.; Funding acquisition, L.V. and C.H.; Investigation, A.K., A.P.R., Z.R., S.T., I.S., T.E.-H. and C.H.; Methodology, Z.P., S.T. and C.H.; Supervision, L.V. and C.H.; Writing—original draft, L.V. and C.H.; Writing—review and editing, L.V. All authors have read and agreed to the published version of the manuscript.

Funding: L.V. received funding from the National Research, Development and Innovation Office grants GINOP-2.3.2-15-2016-00020 TUMORDNS", GINOP-2.3.2-15-2016-00048-STAYALIVE, OTKA K132193. CH received funding from the National Research, Development and Innovation Office grant PD 116845; Bolyai postdoctoral fellowship (BO/00468/17/8); and was supported by the ÚNKP-19-4-DE-299 New National Excellence Program of the Ministry of Human Capacities.

Acknowledgments: The authors thank Karen Uray for careful English language editing; Erzsébet Herbály and Erika Gulyás for their excellent technical assistance.

Conflicts of Interest: The authors declare no conflict of interest. The funding agencies had no role in the design of the study; in the collection, analyses, or interpretation of data; in the writing of the manuscript, or in the decision to publish the results.

References

1. Steliarova-Foucher, E.; Colombet, M.; Ries, L.A.G.; Moreno, F.; Dolya, A.; Bray, F.; Hesseling, P.; Shin, H.Y.; A Stiller, C.; Bouzbid, S.; et al. International incidence of childhood cancer, 2001–10: A population-based registry study. *Lancet Oncol.* **2017**, *18*, 719–731. [CrossRef]
2. Mendoza, P.R.; Grossniklaus, H.E. The Biology of Retinoblastoma. *Prog. Mol. Biol. Transl. Sci.* **2015**, *134*, 503–516. [CrossRef]
3. Santibáñez-Koref, M.; Birch, J.; Hartley, A.; Crowther, D.; Harris, M.; Jones, P.M.; Kelsey, A.; Craft, A.; Eden, T. p53 germline mutations in Li-Fraumeni syndrome. *Lancet* **1991**, *338*, 1490–1491. [CrossRef]
4. Lu, L.; Jin, W.; Liu, H.; Wang, L.L. RECQ DNA Helicases and Osteosarcoma. *Adv. Exp. Med. Biol.* **2014**, *804*, 129–145. [CrossRef] [PubMed]
5. Buondonno, I.; Gazzano, E.; Tavanti, E.; Chegaev, K.; Kopecka, J.; Fanelli, M.; Rolando, B.; Fruttero, R.; Gasco, A.; Hattinger, C.M.; et al. Endoplasmic reticulum-targeting doxorubicin: A new tool effective against doxorubicin-resistant osteosarcoma. *Cell. Mol. Life Sci.* **2018**, *76*, 609–625. [CrossRef] [PubMed]
6. Jimmy, R.; Stern, C.; Lisy, K.; White, S. Effectiveness of mifamurtide in addition to standard chemotherapy for high-grade osteosarcoma. *JBI Database Syst. Rev. Implement. Rep.* **2017**, *15*, 2113–2152. [CrossRef]
7. Osteosarcoma—Childhood and Adolescence: Statistics. 2020. Available online: https://www.cancer.net/cancer-types/osteosarcoma-childhood-and-adolescence/statistics (accessed on 30 January 2020).
8. Pavlova, N.; Thompson, C.B. The Emerging Hallmarks of Cancer Metabolism. *Cell Metab.* **2016**, *23*, 27–47. [CrossRef]
9. Kim, M.Y.; Zhang, T.; Kraus, W.L. Poly(ADP-ribosyl)ation by PARP-1: 'PAR-laying' NAD+ into a nuclear signal. *Genes Dev.* **2005**, *19*, 1951–1967. [CrossRef]
10. Lee, H.C.; Aarhus, R. A Derivative of NADP Mobilizes Calcium Stores Insensitive to Inositol Trisphosphate and Cyclic ADP-ribose. *J. Boil. Chem.* **1995**, *270*, 2152–2157. [CrossRef]
11. Chini, E.; Dousa, T. Enzymatic-Synthesis and Degradation of Nicotinate Adenine Dinucleotide Phosphate (NAADP), a Ca2+-Releasing Agonist, in Rat Tissues. *Biochem. Biophys. Res. Commun.* **1995**, *209*, 167–174. [CrossRef]
12. Cantó, C.; Sauve, A.A.; Bai, P. Crosstalk between poly(ADP-ribose) polymerase and sirtuin enzymes. *Mol. Asp. Med.* **2013**, *34*, 1168–1201. [CrossRef] [PubMed]
13. Eisemann, T.; Pascal, J.M. Poly(ADP-ribose) polymerase enzymes and the maintenance of genome integrity. *Cell. Mol. Life Sci.* **2019**, *77*, 19–33. [CrossRef] [PubMed]
14. Virág, L. 50Years of poly(ADP-ribosyl)ation. *Mol. Asp. Med.* **2013**, *34*, 1043–1045. [CrossRef] [PubMed]
15. Hegedűs, C.; Virág, L. Inputs and outputs of poly(ADP-ribosyl)ation: Relevance to oxidative stress. *Redox Boil.* **2014**, *2*, 978–982. [CrossRef]
16. Hottiger, M.O.; O Hassa, P.; Lüscher, B.; Schüler, H.; Koch-Nolte, F. Toward a unified nomenclature for mammalian ADP-ribosyltransferases. *Trends Biochem. Sci.* **2010**, *35*, 208–219. [CrossRef]
17. Wang, L.; Liang, C.; Li, F.; Guan, D.; Wu, X.; Fu, X.; Lu, A.; Zhang, G. PARP1 in Carcinomas and PARP1 Inhibitors as Antineoplastic Drugs. *Int. J. Mol. Sci.* **2017**, *18*, 2111. [CrossRef]
18. Bürkle, A.; Virág, L. Poly(ADP-ribose): PARadigms and PARadoxes. *Mol. Asp. Med.* **2013**, *34*, 1046–1065. [CrossRef]
19. Bryant, H.E.; Schultz, N.; Thomas, H.D.; Parker, K.M.; Flower, D.; Lopez, E.; Kyle, S.; Meuth, M.; Curtin, N.J.; Helleday, T. Specific killing of BRCA2-deficient tumours with inhibitors of poly(ADP-ribose) polymerase. *Nature* **2005**, *434*, 913–917. [CrossRef]

20. Farmer, H.; McCabe, N.; Lord, C.J.; Tutt, A.N.J.; Johnson, D.A.; Richardson, T.B.; Santarosa, M.; Dillon, K.J.; Hickson, I.; Knights, C.; et al. Targeting the DNA repair defect in BRCA mutant cells as a therapeutic strategy. *Nature* **2005**, *434*, 917–921. [CrossRef]
21. Calabrese, C.R.; Almassy, R.; Barton, S.; Batey, M.A.; Calvert, A.H.; Canan-Koch, S.; Durkacz, B.W.; Hostomsky, Z.; Kumpf, R.A.; Kyle, S.; et al. Anticancer chemosensitization and radiosensitization by the novel poly(ADP-ribose) polymerase-1 inhibitor AG14361. *J. Natl. Cancer Inst.* **2004**, *96*, 56–67. [CrossRef]
22. Plummer, R.; Lorigan, P.C.; Steven, N.; Scott, L.; Middleton, M.R.; Wilson, R.H.; Mulligan, E.; Curtin, N.J.; Wang, D.; Dewji, R.; et al. A phase II study of the potent PARP inhibitor, Rucaparib (PF-01367338, AG014699), with temozolomide in patients with metastatic melanoma demonstrating evidence of chemopotentiation. *Cancer Chemother. Pharmacol.* **2013**, *71*, 1191–1199. [CrossRef] [PubMed]
23. Curtin, N.J.; Szabo, C. Therapeutic applications of PARP inhibitors: Anticancer therapy and beyond. *Mol. Asp. Med.* **2013**, *34*, 1217–1256. [CrossRef] [PubMed]
24. Gossmann, T.I.; Ziegler, M. Sequence divergence and diversity suggests ongoing functional diversification of vertebrate NAD metabolism. *DNA Repair* **2014**, *23*, 39–48. [CrossRef] [PubMed]
25. Chiarugi, A.; Dölle, C.; Felici, R.; Ziegler, M. The NAD metabolome — a key determinant of cancer cell biology. *Nat. Rev. Cancer* **2012**, *12*, 741–752. [CrossRef] [PubMed]
26. Bajrami, I.; Kigozi, A.; Van Weverwijk, A.; Brough, R.; Frankum, J.; Lord, C.J.; Ashworth, A. Synthetic lethality of PARP and NAMPT inhibition in triple-negative breast cancer cells. *EMBO Mol. Med.* **2012**, *4*, 1087–1096. [CrossRef]
27. Berger, F.; Lau, C.; Dahlmann, M.; Ziegler, M. Subcellular Compartmentation and Differential Catalytic Properties of the Three Human Nicotinamide Mononucleotide Adenylyltransferase Isoforms. *J. Boil. Chem.* **2005**, *280*, 36334–36341. [CrossRef]
28. Yamamoto, M.; Hikosaka, K.; Mahmood, A.; Tobe, K.; Shojaku, H.; Inohara, H.; Nakagawa, T. Nmnat3 Is Dispensable in Mitochondrial NAD Level Maintenance In Vivo. *PLoS ONE* **2016**, *11*, e0147037. [CrossRef]
29. Michishita, E.; Park, J.Y.; Burneskis, J.M.; Barrett, J.C.; Horikawa, I. Evolutionarily Conserved and Nonconserved Cellular Localizations and Functions of Human SIRT Proteins. *Mol. Boil. Cell* **2005**, *16*, 4623–4635. [CrossRef]
30. Barraud, M.; Garnier, J.; Loncle, C.; Gayet, O.; Lequeue, C.; Vasseur, S.; Bian, B.; Duconseil, P.; Gilabert, M.; Bigonnet, M.; et al. A pancreatic ductal adenocarcinoma subpopulation is sensitive to FK866, an inhibitor of NAMPT. *Oncotarget* **2016**, *7*, 53783–53796. [CrossRef]
31. Haubrich, B.A.; Ramesha, C.; Swinney, D.C. Development of a Bioluminescent High-Throughput Screening Assay for Nicotinamide Mononucleotide Adenylyltransferase (NMNAT). *SLAS Discov. Adv. Life Sci. R & D* **2019**, *25*, 33–42. [CrossRef]
32. Huang, X.; Okafuji, M.; Traganos, F.; Luther, E.; Holden, E.; Darzynkiewicz, Z. Assessment of histone H2AX phosphorylation induced by DNA topoisomerase I and II inhibitors topotecan and mitoxantrone and by the DNA cross-linking agent cisplatin. *Cytom. Part A J. Int. Soc. Anal. Cytol.* **2004**, *58*, 99–110. [CrossRef] [PubMed]
33. Spincemaille, P.; Alborzinia, H.; Dekervel, J.; Windmolders, P.; Van Pelt, J.; Cassiman, D.; Cheneval, O.; Craik, D.J.; Schur, J.; Ott, I.; et al. The Plant Decapeptide OSIP108 Can Alleviate Mitochondrial Dysfunction Induced by Cisplatin in Human Cells. *Molecules* **2014**, *19*, 15088–15102. [CrossRef] [PubMed]
34. Rytelewski, M.; Tong, J.G.; Buensuceso, A.; Leong, H.; Vareki, S.M.; Figueredo, R.; Di Cresce, C.; Wu, S.Y.; Herbrich, S.; Baggerly, K.A.; et al. BRCA2 inhibition enhances cisplatin-mediated alterations in tumor cell proliferation, metabolism, and metastasis. *Mol. Oncol.* **2014**, *8*, 1429–1440. [CrossRef] [PubMed]
35. Gunn, A.R.; Banos-Pinero, B.; Paschke, P.; Sanchez-Pulido, L.; Ariza, A.; Day, J.; Emrich, M.; Leys, D.; Ponting, C.P.; Ahel, I.; et al. The role of ADP-ribosylation in regulating DNA interstrand crosslink repair. *J. Cell Sci.* **2016**, *129*, 3845–3858. [CrossRef] [PubMed]
36. An, S.S.A.; Seo, O.W.; Lee, J.; Hulme, J.; Baek, N. Real-time monitoring of cisplatin cytotoxicity on three-dimensional spheroid tumor cells. *Drug Des. Dev. Ther.* **2016**, *10*, 2155–2165. [CrossRef]
37. Xie, L.; Zhao, T.; Cai, J.; Su, Y.; Wang, Z.; Dong, W. Methotrexate induces DNA damage and inhibits homologous recombination repair in choriocarcinoma cells. *Onco Targets Ther.* **2016**, *9*, 7115–7122. [CrossRef]
38. López-Camarillo, C.; Rincón, D.G.; Ruiz-García, E.; De La Vega, H.A.; Marchat, L.A. DNA Repair Proteins as Therapeutic Targets in Ovarian Cancer. *Curr. Protein Pept. Sci.* **2019**, *20*, 316–323. [CrossRef]

39. Cole, J.; Guiot, M.-C.; Gravel, M.; Bernier, C.; Shore, G.C.; Roulston, A. Novel NAPRT specific antibody identifies small cell lung cancer and neuronal cancers as promising clinical indications for a NAMPT inhibitor/niacin co-administration strategy. *Oncotarget* **2017**, *8*, 77846–77859. [CrossRef]
40. Ryu, K.W.; Nandu, T.; Kim, J.; Challa, S.; DeBerardinis, R.J.; Kraus, W.L. Metabolic regulation of transcription through compartmentalized NAD+biosynthesis. *Science* **2018**, *360*, eaan5780. [CrossRef]
41. Cambronne, X.A.; Stewart, M.L.; Kim, N.; Jones-Brunette, A.M.; Morgan, R.K.; Farrens, D.L.; Cohen, M.S.; Goodman, R.H. Biosensor reveals multiple sources for mitochondrial NAD+. *Science* **2016**, *352*, 1474–1477. [CrossRef]
42. Gong, Y.; Fan, Z.; Luo, G.; Yang, C.; Huang, Q.; Fan, K.; Cheng, H.; Jin, K.; Ni, Q.; Yu, X.-J.; et al. The role of necroptosis in cancer biology and therapy. *Mol. Cancer* **2019**, *18*, 100. [CrossRef] [PubMed]
43. Nagy, A.; Lánczky, A.; Menyhart, O.; Győrffy, B. Validation of miRNA prognostic power in hepatocellular carcinoma using expression data of independent datasets. *Sci. Rep.* **2018**, *8*, 9227. [CrossRef] [PubMed]
44. McQuade, R.; Stojanovska, V.; De Leiris, J.; Nurgali, K. PARP inhibition in platinum-based chemotherapy: Chemopotentiation and neuroprotection. *Pharmacol. Res.* **2018**, *137*, 104–113. [CrossRef] [PubMed]
45. Zhang, T.; Berrocal, J.G.; Yao, J.; Dumond, M.E.; Krishnakumar, R.; Ruhl, D.D.; Ryu, K.W.; Gamble, M.J.; Kraus, W.L. Regulation of Poly(ADP-ribose) Polymerase-1-dependent Gene Expression through Promoter-directed Recruitment of a Nuclear NAD+ Synthase. *J. Boil. Chem.* **2012**, *287*, 12405–12416. [CrossRef] [PubMed]
46. Berger, F.; Lau, C.; Ziegler, M. Regulation of poly(ADP-ribose) polymerase 1 activity by the phosphorylation state of the nuclear NAD biosynthetic enzyme NMN adenylyl transferase 1. *Proc. Natl. Acad. Sci. USA* **2007**, *104*, 3765–3770. [CrossRef] [PubMed]
47. Park, H.J.; Bae, J.S.; Kim, K.M.; Moon, Y.J.; Park, S.-H.; Ha, S.H.; Hussein, U.K.; Zhang, Z.; Park, H.S.; Park, B.-H.; et al. The PARP inhibitor olaparib potentiates the effect of the DNA damaging agent doxorubicin in osteosarcoma. *J. Exp. Clin. Cancer Res.* **2018**, *37*, 107. [CrossRef]
48. Li, S.; Cui, Z.; Meng, X. Knockdown of PARP-1 Inhibits Proliferation and ERK Signals, Increasing Drug Sensitivity in Osteosarcoma U2OS Cells. *Oncol. Res. Featur. Preclin. Clin. Cancer Ther.* **2016**, *24*, 279–286. [CrossRef]
49. Engert, F.; Kovac, M.; Baumhoer, D.; Nathrath, M.; Fulda, S. Osteosarcoma cells with genetic signatures of BRCAness are susceptible to the PARP inhibitor talazoparib alone or in combination with chemotherapeutics. *Oncotarget* **2016**, *8*, 48794–48806. [CrossRef]
50. Chen, K.; Zhu, C.; Cai, M.; Fu, N.; Cheng, B.; Cai, Z.; Li, G.; Liu, J. Integrative metabolome and transcriptome profiling reveals discordant glycolysis process between osteosarcoma and normal osteoblastic cells. *J. Cancer Res. Clin. Oncol.* **2014**, *140*, 1715–1721. [CrossRef]
51. Zhang, H.; Guo, X.; Feng, X.; Wang, T.; Hu, Z.; Que, X.; Tian, Q.; Zhu, T.; Guo, G.; Huang, W.; et al. MiRNA-543 promotes osteosarcoma cell proliferation and glycolysis by partially suppressing PRMT9 and stabilizing HIF-1α protein. *Oncotarget* **2016**, *8*, 2342–2355. [CrossRef]
52. Xiao, Q.; Wei, Z.; Li, Y.; Zhou, X.; Chen, J.; Wang, T.; Shao, G.; Zhang, M.; Zhang, Z. miR-186 functions as a tumor suppressor in osteosarcoma cells by suppressing the malignant phenotype and aerobic glycolysis. *Oncol. Rep.* **2018**, *39*, 2703–2710. [CrossRef] [PubMed]
53. Song, Y.-D.; Li, D.-D.; Guan, Y.; Wang, Y.-L.; Zheng, J. miR-214 modulates cisplatin sensitivity of osteosarcoma cells through regulation of anaerobic glycolysis. *Cell. Mol. Boil.* **2017**, *63*, 75–79. [CrossRef] [PubMed]
54. Han, X.; Yang, Y.; Sun, Y.; Qin, L.; Yang, Y. LncRNA TUG1 affects cell viability by regulating glycolysis in osteosarcoma cells. *Gene* **2018**, *674*, 87–92. [CrossRef] [PubMed]
55. Mizushima, E.; Tsukahara, T.; Emori, M.; Murata, K.; Akamatsu, A.; Shibayama, Y.; Hamada, S.; Watanabe, Y.; Kaya, M.; Hirohashi, Y.; et al. Osteosarcoma-initiating cells show high aerobic glycolysis and attenuation of oxidative phosphorylation mediated by LIN28B. *Cancer Sci.* **2019**, *111*, 36–46. [CrossRef] [PubMed]
56. Gorospe, M.; De Cabo, R. AsSIRTing the DNA damage response. *Trends Cell Boil.* **2008**, *18*, 77–83. [CrossRef] [PubMed]
57. Feige, J.N.; Auwerx, J. Transcriptional coregulators in the control of energy homeostasis. *Trends Cell Boil.* **2007**, *17*, 292–301. [CrossRef]
58. Abdellatif, M. Sirtuins and pyridine nucleotides. *Circ. Res.* **2012**, *111*, 642–656. [CrossRef]
59. Hu, J.; Deng, F.; Hu, X.; Zhang, W.; Zeng, X.; Tian, X. Histone deacetylase SIRT6 regulates chemosensitivity in liver cancer cells via modulation of FOXO3 activity. *Oncol. Rep.* **2018**, *40*, 3635–3644. [CrossRef]

60. Sociali, G.; Galeno, L.; Parenti, M.D.; Grozio, A.; Bauer, I.; Passalacqua, M.; Boero, S.; Donadini, A.; Millo, E.; Bellotti, M.; et al. Quinazolinedione SIRT6 inhibitors sensitize cancer cells to chemotherapeutics. *Eur. J. Med. Chem.* **2015**, *102*, 530–539. [CrossRef]
61. Zhao, J.; Wozniak, A.; Adams, A.; Cox, J.; Vittal, A.; Voss, J.; Bridges, B.; Weinman, S.A.; Li, Z. SIRT7 regulates hepatocellular carcinoma response to therapy by altering the p53-dependent cell death pathway. *J. Exp. Clin. Cancer Res.* **2019**, *38*, 252. [CrossRef]
62. Jiang, Y.; Han, Z.; Wang, Y.; Hao, W. Depletion of SIRT7 sensitizes human non-small cell lung cancer cells to gemcitabine therapy by inhibiting autophagy. *Biochem. Biophys. Res. Commun.* **2018**, *506*, 266–271. [CrossRef] [PubMed]
63. Zhang, T.; Berrocal, J.G.; Frizzell, K.M.; Gamble, M.J.; Dumond, M.E.; Krishnakumar, R.; Yang, T.; Sauve, A.A.; Kraus, W.L. Enzymes in the NAD+ Salvage Pathway Regulate SIRT1 Activity at Target Gene Promoters*. *J. Boil. Chem.* **2009**, *284*, 20408–20417. [CrossRef] [PubMed]
64. Lakatos, P.; Hegedűs, C.; Ayestarán, N.S.; Juarranz, Á.; Kövér, K.E.; Szabo, E.; Virág, L. The PARP inhibitor PJ-34 sensitizes cells to UVA-induced phototoxicity by a PARP independent mechanism. *Mutat. Res. Mol. Mech. Mutagen.* **2016**, *790*, 31–40. [CrossRef] [PubMed]
65. Hegedűs, C.; Lakatos, P.; Oláh, G.; Tóth, B.I.; Gergely, S.; Szabo, E.; Bíró, T.; Szabó, C.; Virág, L. Protein kinase C protects from DNA damage-induced necrotic cell death by inhibiting poly(ADP-ribose) polymerase-1. *FEBS Lett.* **2008**, *582*, 1672–1678. [CrossRef]
66. Regdon, Z.; Robaszkiewicz, A.; Kovács, K.; Rygielska, Ż.; Hegedűs, C.; Bodoor, K.; Szabó, É.; Virág, L. LPS protects macrophages from AIF-independent parthanatos by downregulation of PARP1 expression, induction of SOD2 expression, and a metabolic shift to aerobic glycolysis. *Free. Radic. Boil. Med.* **2019**, *131*, 184–196. [CrossRef] [PubMed]
67. Hegedűs, C.; Lakatos, P.; Kiss-Szikszai, A.; Patonay, T.; Gergely, S.; Gregus, A.; Bai, P.; Haskó, G.; Szabo, E.; Virág, L. Cytoprotective dibenzoylmethane derivatives protect cells from oxidative stress-induced necrotic cell death. *Pharmacol. Res.* **2013**, *72*, 25–34. [CrossRef]
68. Gergely, S.; Hegedűs, C.; Lakatos, P.; Kovács, K.; Gáspár, R.; Csont, T.; Virág, L. High Throughput Screening Identifies a Novel Compound Protecting Cardiomyocytes from Doxorubicin-Induced Damage. *Oxidative Med. Cell. Longev.* **2015**, *2015*, 1–12. [CrossRef]
69. Erdélyi, K.; Bai, P.; Kovács, I.; Szabó, É.; Mocsár, G.; Kakuk, A.; Szabó, C.; Gergely, P.; Virág, L. Dual role of poly(ADP-ribose) glycohydrolase in the regulation of cell death in oxidatively stressed A549 cells. *FASEB J.* **2009**, *23*, 3553–3563. [CrossRef]
70. Aladdin, A.; Király, R.; Boto, P.; Regdon, Z.; Tar, K. Juvenile Huntington's Disease Skin Fibroblasts Respond with Elevated Parkin Level and Increased Proteasome Activity as a Potential Mechanism to Counterbalance the Pathological Consequences of Mutant Huntingtin Protein. *Int. J. Mol. Sci.* **2019**, *20*, 5338. [CrossRef]
71. Kovács, P.; Csonka, T.; Kovács, T.; Sári, Z.; Ujlaki, G.; Adrienn, S.; Karányi, Z.; Szeőcs, D.; Hegedűs, C.; Uray, K.; et al. Lithocholic Acid, a Metabolite of the Microbiome, Increases Oxidative Stress in Breast Cancer. *Cancers* **2019**, *11*, 1255. [CrossRef]

cancers

MDPI

Article

ATR Inhibition Potentiates PARP Inhibitor Cytotoxicity in High Risk Neuroblastoma Cell Lines by Multiple Mechanisms

Harriet E. D. Southgate [1], Lindi Chen [1], Deborah A. Tweddle [1,*] and Nicola J. Curtin [2,*]

1 Wolfson Childhood Cancer Research Centre, Newcastle Centre for Cancer, Translational and Clinical Research Institute, Faculty of Medical Sciences, Newcastle University, Newcastle Upon Tyne NE1 7RU, UK; h.southgate2@newcastle.ac.uk (H.E.D.S.); lindi.chen@newcastle.ac.uk (L.C.)

2 Newcastle Centre for Cancer, Translational and Clinical Research Institute, Faculty of Medical Sciences, Newcastle University, Newcastle Upon Tyne NE2 4HH, UK

* Correspondence: deborah.tweddle@newcastle.ac.uk (D.A.T.); nicola.curtin@newcastle.ac.uk (N.J.C.)

Received: 30 March 2020; Accepted: 23 April 2020; Published: 28 April 2020

Abstract: *Background:* High risk neuroblastoma (HR-NB) is one the most difficult childhood cancers to cure. These tumours frequently present with DNA damage response (DDR) defects including loss or mutation of key DDR genes, oncogene-induced replication stress (RS) and cell cycle checkpoint dysfunction. Aim: To identify biomarkers of sensitivity to inhibition of Ataxia telangiectasia and Rad3 related (ATR), a DNA damage sensor, and poly (ADP-ribose) polymerase (PARP), which is required for single strand break repair. We also hypothesise that combining ATR and PARP inhibition is synergistic. *Methods:* Single agent sensitivity to VE-821 (ATR inhibitor) and olaparib (PARP inhibitor), and the combination, was determined using cell proliferation and clonogenic assays, in HR-NB cell lines. Basal expression of DDR proteins, including ataxia telangiectasia mutated (ATM) and ATR, was assessed using Western blotting. CHK1^{S345} and H2AX^{S129} phosphorylation was assessed using Western blotting to determine ATR activity and RS, respectively. RS and homologous recombination repair (HRR) activity was also measured by γH2AX and Rad51 foci formation using immunofluorescence. *Results:* *MYCN* amplification and/or low ATM protein expression were associated with sensitivity to VE-821 ($p < 0.05$). VE-821 was synergistic with olaparib (CI value 0.04–0.89) independent of *MYCN* or ATM status. Olaparib increased H2AX^{S129} phosphorylation which was further increased by VE-821. Olaparib-induced Rad51 foci formation was reduced by VE-821 suggesting inhibition of HRR. *Conclusion:* RS associated with *MYCN* amplification, ATR loss or PARP inhibition increases sensitivity to the ATR inhibitor VE-821. These findings suggest a potential therapeutic strategy for the treatment of HR-NB.

Keywords: poly (ADP-ribose) polymerase inhibitors (PARPi); ataxia telangiectasia and Rad3 related inhibitors (ATRi); replication stress; cell cycle checkpoints

---

1. Introduction

Inhibitors of ataxia telangiectasia and Rad3 related (ATR) kinase are currently being tested in early phase clinical trials for patients with adult cancers. ATR is a DNA damage sensor kinase which has a pivotal role in recovery from replication stress (RS). Activated ATR signals to many pathways involved in regulation of origin firing, and stabilisation and restart of stalled replication forks (reviewed in: [1,2]). ATR signals to S and G2/M cell cycle checkpoint control by inhibiting CDK2 and CDK1 via its primary target CHK1 [3,4]. ATR inhibition abrogates S and G2/M checkpoint arrest, leading to cell death by mitotic catastrophe [5,6].

Two ATR inhibitors, AZD6738 (Astra Zeneca) and M6620 (formerly VX-970, Merck), have entered phase II trials, with a third, BAY1895344 (Bayer) in phase I (listed on https://clinicaltrials.gov/).

Various determinants of ATR inhibitor sensitivity have been proposed, including RS, loss of G1 checkpoint control and loss of the ataxia telangiectasia mutated (ATM) protein [7,8].

RS is a common feature of cancer cells and leads to genetic instability. There are many causes of RS including overexpression of proliferation-driving oncogenes (*Ras, Myc, cyclin E* etc.), deregulation of replication origin firing, limited nucleotide pools or essential replication factors and replication through fragile sites or damaged DNA regions [9,10].

Loss of G1 checkpoint control also contributes to RS and is common in cancer through loss of tumour suppressors such as p53, pRB and ATM, imbalance of cyclins, cyclin-dependent kinases and their inhibitors and expression of oncogenes [11]. G1 checkpoint deficiency results in a reliance on the S and G2/M checkpoints to maintain genome integrity and prevent replication of damaged DNA/mitotic catastrophe [12–14].

Neuroblastoma (NB) is a rare embryonal tumour derived from cells of the developing sympathetic nervous system. Around 100 cases are diagnosed a year in the UK, of which 50% are classified as high risk, but accounts for ~10% of paediatric cancer deaths [15,16]. Long term survival of high-risk neuroblastoma (HR-NB) (metastatic disease over 1 year of age- or *MYCN*-amplified disease) currently remains less than 50% at 5 years despite intensive high-dose multimodal treatment [17,18]. Survival of relapsed NB is particularly poor with less than 10% 5-year survival [19]. HR-NB frequently present with DNA damage response (DDR) defects including loss or mutation of key DDR genes, oncogene-induced RS and cell cycle checkpoint dysfunction, which suggest they would be sensitive to ATR inhibition [20–23].

Fifty percent of HR-NB have amplification of the *MYCN* oncogene, leading to RS. *MYCN*-amplified (MNA) tumours also show defective G1 checkpoint arrest [23]. A common genetic abnormality observed in non-MNA NB is allelic loss of chromosome 11q. Tumours with 11q deletion display a poor prognosis similar to MNA [24]. Many genes coding for proteins involved in the DDR are located on 11q including *ATM, CHEK1, H2AFX* and *MRE11* [20]. Together, MNA and 11q deletion occur in 70–80% of HR-NB tumours.

Although rare at diagnosis, defects in p53 signalling have been observed in up to 50% relapsed NB tumours [22,25], causing further G1 checkpoint dysfunction and abrogating the p53 dependent intrinsic apoptosis pathway.

Poly ADP-ribose polymerase (PARP) inhibitors also cause RS [26]. PARP is activated in response to DNA single strand breaks and orchestrates repair [27]. Several PARP inhibitors have been approved for ovarian and breast cancer with defective homologous recombination repair. There are currently seven clinical trials testing the use of PARP inhibitors for paediatric tumours of which only three include NB (https://clinicaltrials.gov/: NCT04236414, NCT03233204, NCT02392793).

Preclinical testing of the PARP inhibitor olaparib (Astra Zeneca) in NB shows that PARP inhibition potentiates the cytotoxic effect of a variety of chemotherapy agents and ionising radiation [28–31]. In addition, NB tumours with *MYCN* amplification or *ATM* deficiency have been shown to have increased sensitivity to single agent olaparib treatment [32,33].

We aimed to test if the DDR defects frequently observed in NB would be potential predictive biomarkers of sensitivity to ATR inhibition using VE-821 (the preclinical lead from which M6620 was developed).

We hypothesise that there will be mutual synergy between ATR and PARP inhibitors by further increasing RS, irrespective of *MYCN* or *ATM* status, by the accumulation of unrepaired single strand breaks, when PARP is inhibited, and failure to arrest in S-phase when ATR is inhibited.

In this study, we identify features of NB cell lines that determine sensitivity to ATR inhibition, for use as potential predictive biomarkers, and examine the effect of ATR inhibition on the cytotoxicity of the PARP inhibitor olaparib.

2. Results

2.1. DDR Protein Expression in NB Cell Lines

To reflect the variety of DDR defects observed in NB tumours, we chose a panel of NB cell line of varying *MYCN*, 11q and *TP53* status to interrogate what features would lead to sensitivity to ATR and PARP inhibitors. The genetic features of these cell lines are listed in Table 1.

Table 1. Cell line genetic abnormalities.

Cell Line	*MYCN* Status	11q Status (Genes Deleted)	p53 Status	Reference
SHSY5Y	Non-amp	No deletion	WT	[34]
SKNAS	Non-amp	Deletion (MRE11, *ATM, CHEK1, H2AFX)*	Mutant Deletion of intron9/exon 10	[35,36]
NGP	Amp	Deletion *(ATM, CHEK1, H2AFX)*	WT	[36]
N20_R1	Amp	Deletion ** *(ATM, CHEK1, H2AFX)*	Mutant P98H P152T	[37]
NMB *	Amp	Deletion (MRE11, *ATM, CHEK1, H2AFX)*	Mutant G245S	[36,38]
IMR32	Amp	Deletion *(ATM, CHEK1, H2AFX)* ATM mutant V2716A	WT	[39,40]
IMR32/Kat100 (Kat100)	Amp	Unknown	Mutant C135F	[41]
IGRN91	Amp	No deletion	Mutant Duplication of exons 7–9	[42,43]
SJNB1 *	Non-amp	Deletion (MRE11, *ATM, CHEK1, H2AFX)*	WT	[36]
GIMEN	Non-amp	Deletion (MRE11, *ATM, CHEK1, H2AFX)*	WT	[36]

Amp: amplified, Non-amp: non-amplified, WT: wild type; * cell line is near tetraploid [44], Chr11 LOH; ** derived from NGP, assume same.

We analysed endogenous expression of key DDR proteins (ATM, ATR, CHK1, CHK2, MYCN and p53) in these cell lines as well as baseline activity of ATR, ATM (phospho-CHK1^{S345} and phospho-CHK2^{T68} and expression, respectively) by Western blot (Figure 1A). Mean protein band intensities of MYCN, ATM and p53 from two independent experiments are shown in Figure 1B. ATM and p53 function after activation with doxorubicin was also examined (Figure 1C,D, respectively).

As expected, *MNA* cell lines show high MYCN protein expression compared to non-*MNA* cell lines (Figure 1A and Figure S3A), with the exception of SJNB1, which has high expression of MYCN in the absence of a gene amplification. In contrast, some cell lines with 11q deletion have baseline ATM expression, suggesting that ATM expression from the other allele is sufficient to produce a functional protein (Figure 1A,C). Cell lines with *TP53* mutations show stabilised p53 protein (NMB and Kat100) or no p53 protein expression (SKNAS and IGRN91). The *TP53* mutation in the NMB and Kat100 cell lines are point mutations leading to accumulation and stabilisation of the dysfunctional protein (Figure 1D and previously in [38,41]), whereas SKNAS and IGRN91 have a deletion and duplication, respectively, of whole exons and do not stabilise the protein. The IGRN91 cell line expresses a high molecular weight gene product after activation with doxorubicin (consistent with duplication of exons 7–9 [35]), which retains some function (Figure 1D).

2.2. MYCN Amplification and Low ATM Expression are Determinants of Sensitivity to ATR Inhibition in NB Cell Lines

Growth inhibition by the ATR inhibitor VE-821 was determined by XTT cell proliferation assay (Figure 2A). Cell lines were grouped based on genetic features, baseline MYCN, p53 and ATM protein expression above (high) or below (low) median expression (Figure 1B) and ATM and p53 response to doxorubicin. Growth inhibition at 10 μM for each cell line was analysed across groups

(Figure 2B–D) by Mann–Whitney *U* test. *MNA* cell lines show high MYCN protein expression (Figure 1B) and these cell lines were significantly more sensitive to 10 μM VE-821 than non-*MNA* cell lines ($p < 0.05$). Although there was no significant difference in VE-821 sensitivity between cell lines with or without *ATM* aberration (*ATM* mutation and/or 11q deletion), cell lines with low baseline ATM protein expression were significantly more sensitive than cell lines with high baseline ATM expression ($p < 0.05$). In regression analysis, MYCN and ATM protein expression was negatively and positively correlated with sensitivity to VE-821, respectively (Figure S3B,C). Cell lines with dysfunctional ATM were more sensitive to VE-821 than cell lines with functional ATM, indicated by ATM^{S1981} auto-phosphorylation after treatment with 1 μM doxorubicin (Figure 1C). There was no significant difference in sensitivity to VE-821 when cell lines were grouped according to *TP53* mutation status, p53 protein expression or p53 function. These results were confirmed by clonogenic survival assay (Figure S4).

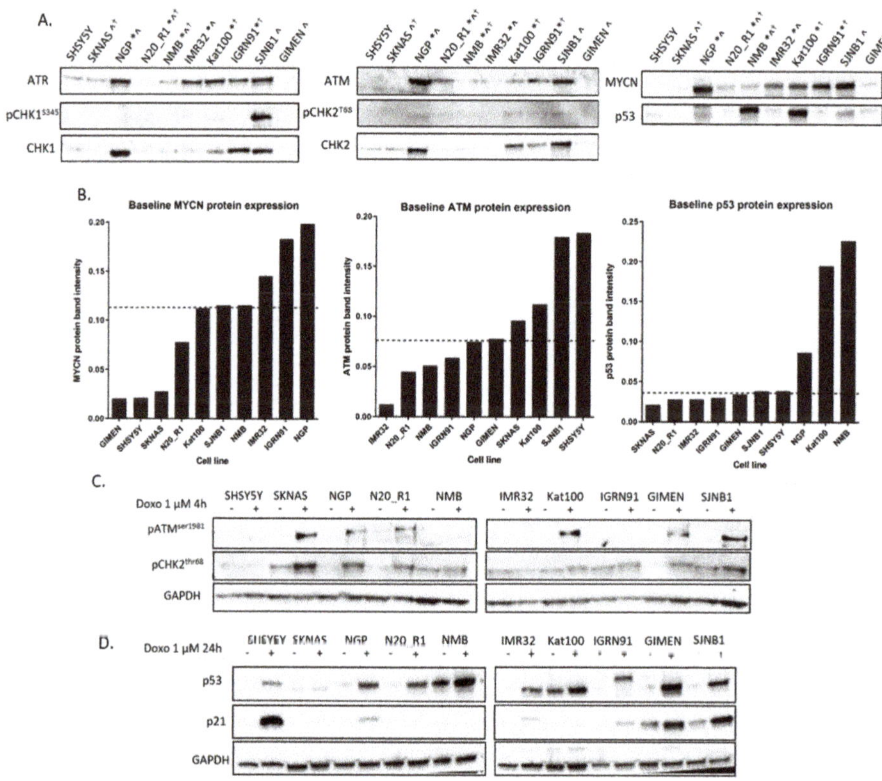

Figure 1. Expression and function of key DNA damage response (DDR) proteins in a panel of neuroblastoma (NB) cell lines. (**A**) Baseline protein expression of Rad3 related (ATR), ataxia telangiectasia mutated (ATM), CHK1, CHK2, phospho-$CHK1^{S345}$, phospho-$CHK2^{T68}$ MYCN and p53 in NB cell lines used in this study. Ponceau S stain was used as a loading control (Figure S1). *MYCN amplified, ^11q deleted, †TP53 mutant. (**B**) MYCN, ATM and p53 mean protein band intensity measured by densitometry (ImageJ) and normalised to total protein (Ponceau S) from 2 replicates. Cell lines are ordered by protein expression from low to high. The dashed line indicates median expression of each protein. (**C**) ATM function was determined by phospho-ATM^{S1981} and phospho-$CHK2^{T68}$ expression after treatment with 1 μM doxorubicin (doxo) for 4 h. (**D**) p53 function was determined by p53 and p21 expression after treatment with doxo for 24 h. Images of the uncropped Western blots for A and C, D can be found in Figures S1 and S2.

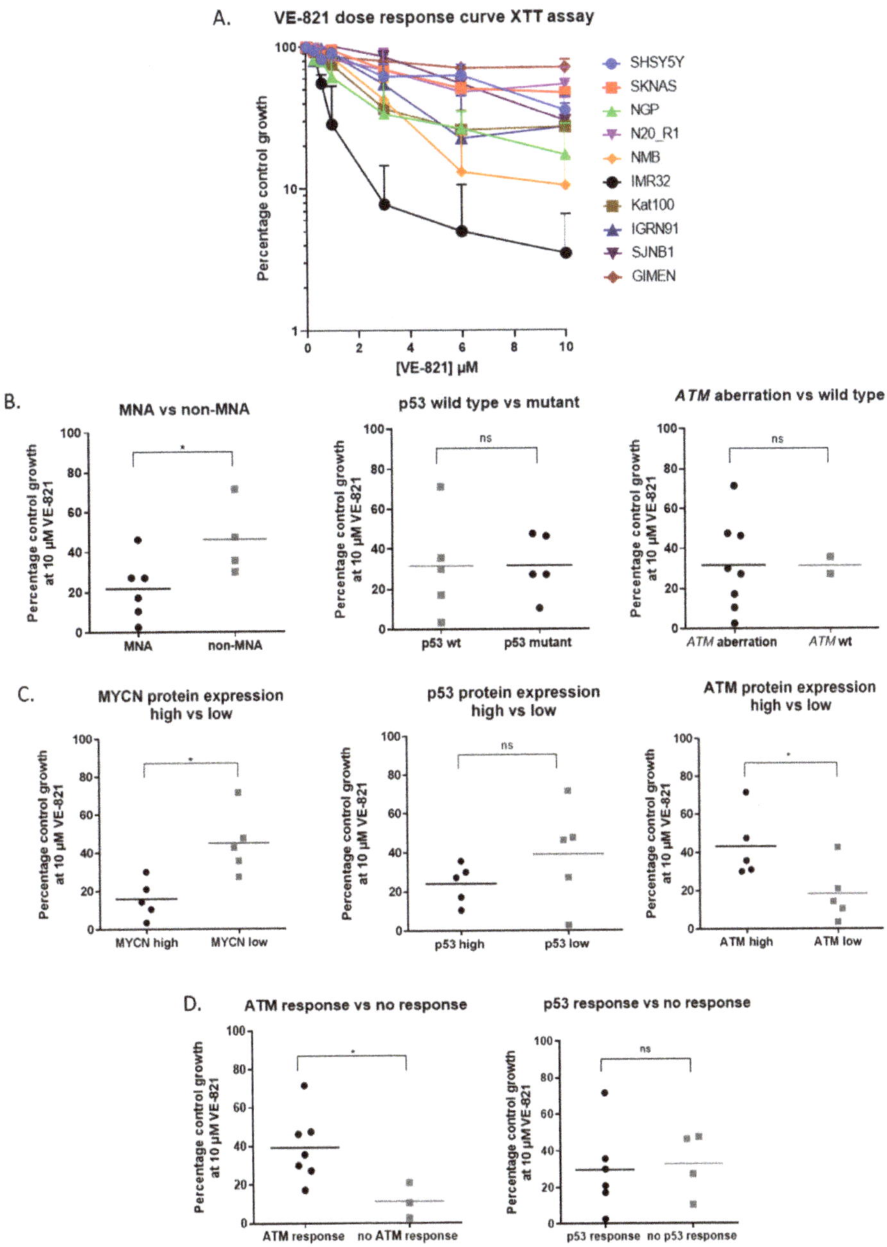

Figure 2. Determinants of ATR inhibitor sensitivity. (**A**) VE-821 (ATR inhibitor) dose response curves for NB cell lines, data are mean + SEM from 3 independent experiments. Cell lines were split into 2 groups based on (**B**) molecular features: *MYCN* amplification (MNA), p53 mutation status and ATM aberration (11q deleted or *ATM* mutated), (**C**) protein expression above (high) or below (low) median expression (Figure 1B) and (**D**) ATM and p53 responses after treatment with doxorubicin (1 μM). ATM and p53 response was determined by expression of pATM^{ser1981} and p21, respectively (Figure 1C,D). Average percentage control growth at 10 μM VE-821 was plotted for cell lines belonging to each group ($n = 3$). * $p < 0.05$ Mann Whitney U test, ns: not significant, wt: wildtype).

2.3. *MYCN Amplification and Low ATM Expression Are Not Significant Determinants of Sensitivity to PARP Inhibition in NB Cell Lines*

To see if the features that are associated with ATR inhibitor sensitivity also determine sensitivity to PARP inhibition, sensitivity to the PARP inhibitor olaparib was determined by XTT cell proliferation assay (Figure S5A). Cell lines were grouped by the molecular features above and growth inhibition at 10 μM olaparib for each cell line was analysed. Unlike for VE-821, there was no significant difference in olaparib sensitivity when cell lines were grouped according to *MYCN* amplification, MYCN protein expression, ATM protein expression or ATM function (Figure S5B–D). There was no significant difference in sensitivity to olaparib when cell lines were grouped according to *TP53* mutation status, p53 protein expression or p53 function. Single agent sensitivity of cell lines to VE-821 and olaparib by XTT cell proliferation assay is summarised in Table 2.

Table 2. Summary of VE-821 and olaparib single agent sensitivity.

Cell Line	*MYCN* Status	11q Status (Genes Deleted)	p53 Status	VE-821 GI50 (μM)	VE-821 LC50 (μM)	Olaparib GI50 (μM)	Olaparib LC50 (μM)
SHSY5Y	Non-amp	No deletion	WT	7.11	1.54	5.38	1.40
SKNAS	Non-amp	Deletion *(MRE11, ATM, CHEK1, H2AFX)*	Mut	>20	0.81	>30	1.31
NGP	Amp	Deletion *(ATM, CHEK1, H2AFX)*	WT	1.62	0.93	1.35	1.20
N20_R1	Amp	Deletion *(ATM, CHEK1, H2AFX)*	Mut	8.29	0.93	1.64	0.68
NMB	Amp	Deletion *(MRE11, ATM, CHEK1, H2AFX)*	Mut	2.36	1.91	3.88	0.92
IMR32	Amp	Deletion *(ATM, CHEK1, H2AFX) ATM mutant V2716A*	WT	0.66	0.90	1.81	0.63
Kat100	Amp	Unknown	Mut	1.88	1.50	>30	1.55
IGRN91	Amp	No deletion	Mut	3.04	1.03	>30	0.74
SJNB1	Non-amp	Deletion *(MRE11, ATM, CHEK1, H2AFX)*	WT	6.30	1.99	4.70	0.75
GIMEN	Non-amp	Deletion *(MRE11, ATM, CHEK1, H2AFX)*	WT	>20	0.87	12.25	0.37

However, when analysed by clonogenic assay (Figure S6), cell lines with high MYCN protein expression were significantly more sensitive to olaparib ($p < 0.05$). This was not the case when cell lines were analysed based on *MYCN* amplification and some cell lines, such as SJNB1, which are not *MNA* but express high levels of MYCN, were relatively sensitive to olaparib.

2.4. *PARP and ATR Inhibition Synergistically Inhibit Cell Growth*

To test if growth inhibition by olaparib and VE-821 is synergistic, an XTT cell proliferation assay was carried out testing the response of four NB cell lines to increasing concentrations of olaparib with the addition of fixed concentrations of VE-821. SHSY5Y (non-MNA, p53 wt), SKNAS (non-MNA, p53 mutant), NGP (MNA, p53 wt) and N20_R1 (MNA, p53 mut) cells were treated with 0, 0.1, 1, and 10 μM olaparib with the addition of either 0, 0.1 or 1 μM VE-821 (Figure 3A). VE-821 significantly sensitized the SKNAS, NGP and N20_R1 cell lines to olaparib, fold sensitisation 1.43-4.60 (Figure 3B,C). Although not statistically significant, the cytotoxicity of olaparib was increased by VE-821 in the SHSY5Y (non-*MNA*, *TP53* wt) cell line, fold sensitisation 1.41. Synergism between VE-821 and olaparib was determined by combination index analysis (calcusyn) where combination index (CI) <1 indicates synergy. PARP inhibition was synergistic with ATR inhibition in all cell lines, irrespective of *MYCN* or *TP53* status (Figure 3D). This was also confirmed by clonogenic survival assay (Figure S7).

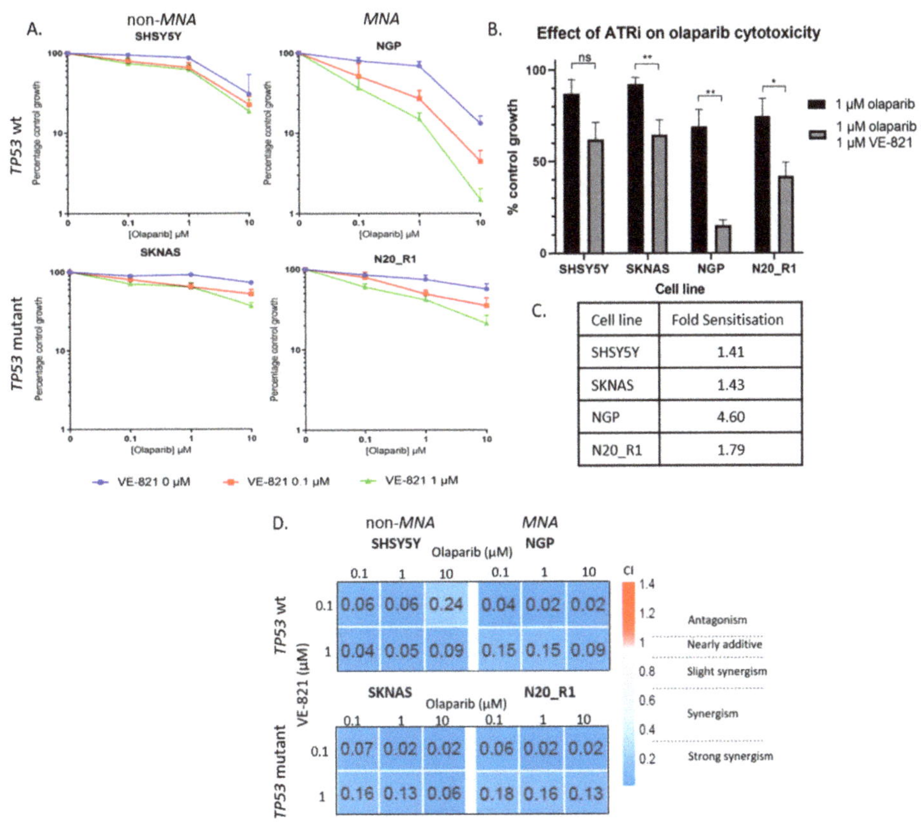

Cell line	Fold Sensitisation
SHSY5Y	1.41
SKNAS	1.43
NGP	4.60
N20_R1	1.79

Figure 3. (**A**) XTT cell proliferation of the SHSY5Y, SKNAS, NGP and N20_R1 neuroblastoma cell lines treated with 0.1, 1 and 10 μM olaparib alone and with the addition of 0 (blue), 0.1 (red) and 1 (green) μM VE-821. Percentage control growth was normalised to effect of VE-821 alone. Data shown are the mean + SEM from 4 individual experiments. (**B**) Effect of 1 μM VE-821 on cytotoxicity of 1 μM olaparib normalised to the effect of VE-821 alone. Data shown are the mean + SEM from 4 individual experiments. T-test: ns; not significant, * $p < 0.05$, ** $p < 0.01$. (**C**) Fold sensitization of 1 μM olaparib by 1 μM VE-821 for each cell line. (**D**) Combination index (CI) values were calculated using CalcuSyn and plotted in heat map.

2.5. ATR Inhibition Increases Replication Stress Caused by Olaparib by Blocking S and G2 Cell Cycle Arrest and Reducing Homologous Recombination Repair

PARP inhibition has been shown to increase RS in NB cell lines [32]. Since ATR inhibition potentiated olaparib induced growth inhibition, we investigated the effect of ATR and PARP inhibitor combination on RS markers. As most markers of RS are dependent on ATR activity, such as CHK1^{S345} and RPA2^{T21} phosphorylation, we measured phosphorylation of histone 2AX (γH2AX) as a surrogate marker of RS. γH2AX is a marker of DNA double strand breaks (DSB) and RS. In the absence of genotoxic agents, γH2AX primarily marks RS. SHSY5Y, SKNAS, NGP and N20_R1 cells were treated with olaparib 5 μM and/or VE-821 1 μM for 24 h. pCHK1^{S345} (ATR activation), pCHK2^{T68} (ATM activation) and γH2AX (RS) protein expression was measured by Western blot (Figure 4A). Protein expression was quantified by densitometry (ImageJ) and compared to DMSO control (Figure 4B). Treatment with 5 μM olaparib resulted in activation of ATR (pCHK1^{S345}) in all cell lines, which was reduced with the addition of 1 μM VE-821 as expected. pCHK2^{T68} and γH2AX expression increased after treatment with olaparib and was further increased with the addition of VE-821, suggesting increased RS and DNA damage. Another marker of RS is phosphorylation of RPA2$^{S4/S8}$ by ATM and DNA-PK [45]. Using the SHSY5Y cell line, expression of pRPA2^{S8} was measured after treatment with olaparib 5 μM +/- VE-821 1 μM over the course of 48 h (Figure 4C). The addition of VE-821 1 μM brought forward the RPA2^{S8} phosphorylation induced by olaparib 5 μM by 24 h.

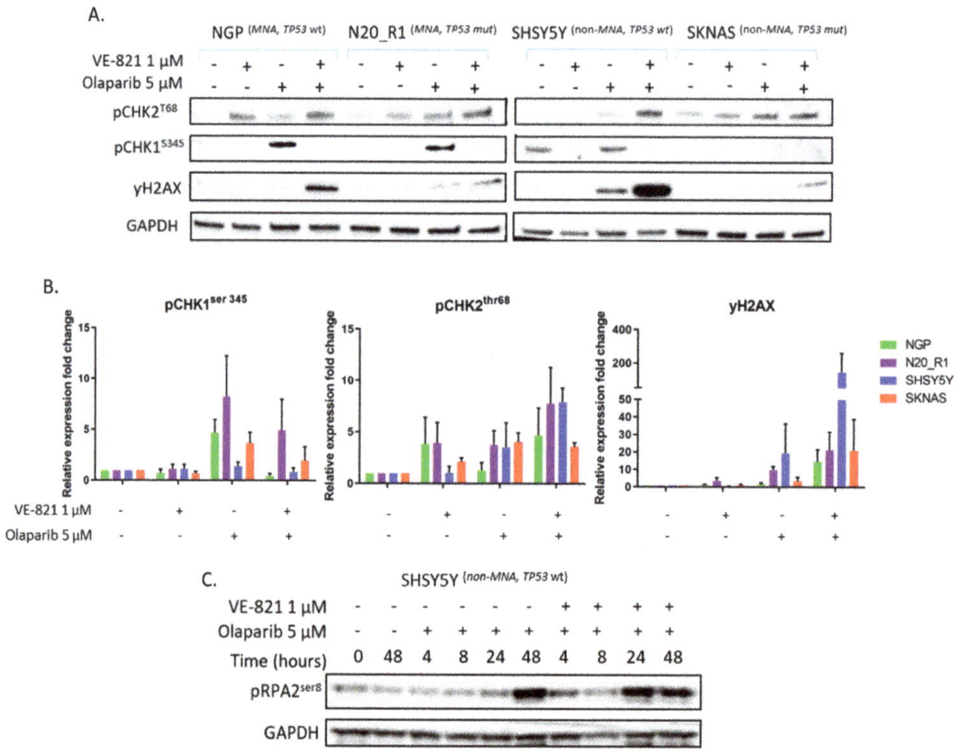

Figure 4. Effect of VE-821 and olaparib combination on replication stress. (**A**) pCHK1^{S345}, pCHK2^{T68}, γH2AX protein expression of NGP, N20_R1, SHSY5Y and SKNAS cells after incubation with 5 μM olaparib and/or 1 μM VE-821 for 24 h. (**B**) pCHK1^{S345}, pCHK2^{T68}, γH2AX mean fold change in protein band intensity measured by densitometry (ImageJ) and normalised to glyceraldehyde 3-phosphate dehydrogenase (GAPDH) loading control, data are mean + SEM from 3 independent experiments. (**C**) pRPA2^{S8} expression in SHSY5Y cell line in response to 5 μM olaparib with or without 1 μM VE-821 over 48 h. Images of the uncropped Western blots can be found in Figure S8.

ATR signals to S and G2 cell cycle arrest [4], which are abrogated when ATR is inhibited [5,6,46]. Since olaparib is known to lead to cell cycle arrest [47,48], we investigated the effect of olaparib and VE-821 combination on the cell cycle in SHSY5Y, SKNAS and NGP cell lines (Figure 5A). Olaparib treatment alone increased the proportion of cells in S and G2 phase for each cell line, consistent with response to olaparib-induced RS. This cell cycle arrest was not seen with the addition of both inhibitors, suggesting loss of S and G2 checkpoints arrest.

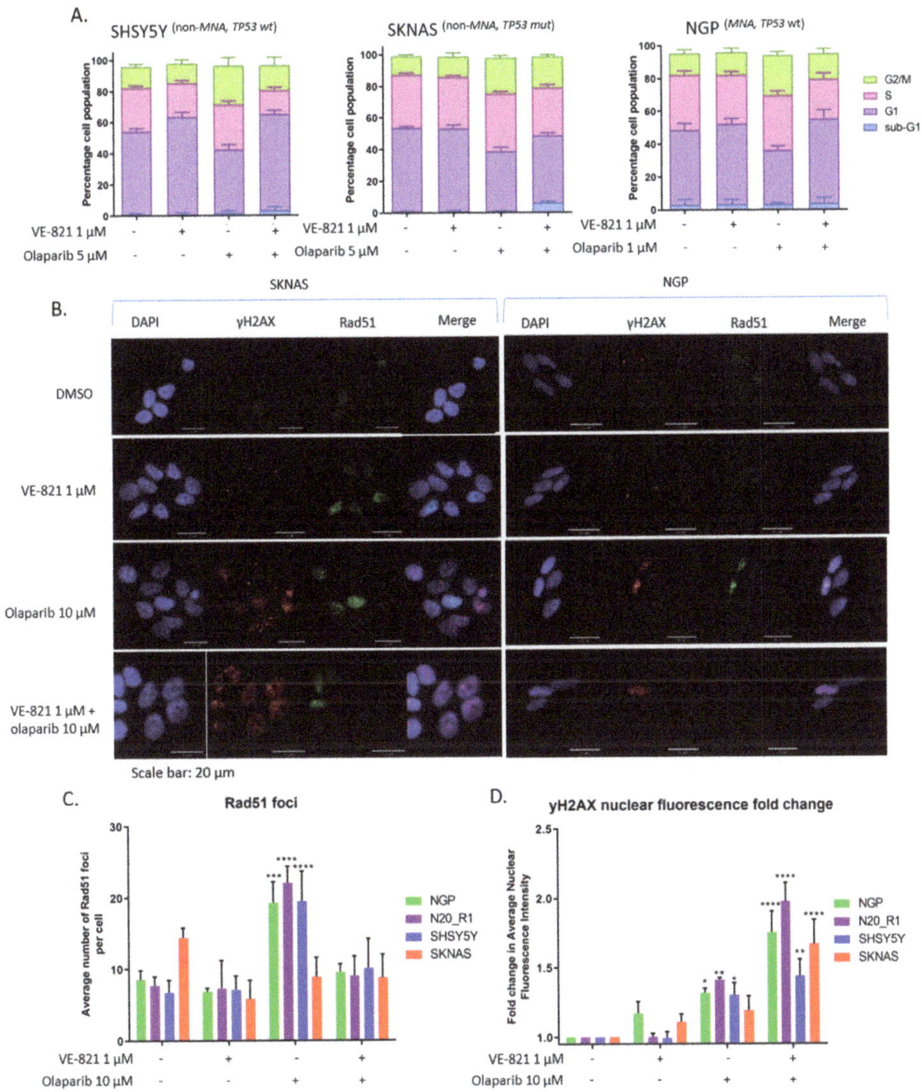

Figure 5. Effect of VE-821 and olaparib combination on cell cycle and homologous recombination repair (HRR). (**A**) Cell cycle analysis of SHSY5Y, SKNAS and NGP cell lines treated with 5 µM (SHSY5Y and SKNAS) or 1 µM (NGP) olaparib in combination with 1 µM VE-821 for 24 h. Data are mean + SEM from 3 independent experiments. (**B**) Representative γH2AX and Rad51 foci images from SKNAS and NGP cell lines treated with 1 µM VE-821, 10 µM olaparib or both for 24 h. (**C**) Average number of Rad51 foci per cell for NGP, N20_R1, SHSY5Y and SKNAS cell lines treated with 1 µM VE-821, 10 µM olaparib or both, $n = 4 +$ sd. (**D**) Average γH2AX nuclear fluorescence intensity for NGP, N20_R1, SHSY5Y and SKNAS cell lines treated with 1 µM VE-821, 10 µM olaparib or both, fold change from control (DMSO), data are mean + SEM from 4 independent experiments. * $p \leq 0.05$, ** $p \leq 0.01$, *** $p \leq 0.001$, **** $p \leq 0.0001$, 2-way ANOVA difference from control (DMSO).

We also examined the effect on homologous recombination repair (HRR). PARP inhibition is synthetically lethal with HRR deficiency [49]. ATR inhibition has been shown to reduce Rad51 foci formation (an indicator of HRR activity [50]), inducing a HRR-deficient phenotype [51,52]. We analysed the effect of 10 μM olaparib and/or 1 μM VE-821 on Rad51 and γH2AX foci in SHSY5Y, SKNAS, NGP and N20_R1 cell lines. Representative γH2AX and Rad51 foci images from SKNAS (non-*MNA*, *TP53* mutant) and NGP (*MNA*, *TP53* wt) cell lines treated with 1 μM VE-821, 10 μM olaparib or both are shown in Figure 5B. Treatment with olaparib induced an increase in Rad51 foci formation, which was suppressed with the addition of VE-821 in all cell lines except SKNAS (Figure 5C). Interestingly, SKNAS cells are relatively resistant to both olaparib and VE-821. RS was measured by γH2AX nuclear fluorescence intensity, which increased after treatment with olaparib and increased further after treatment with both inhibitors in all cell lines (Figure 5D).

Overall, the combination of olaparib and VE-821 leads to increased RS due to defective S and G2 cell cycle arrest and suppression of HRR leading to collapsed replication forks.

3. Discussion

It is widely accepted that overexpression of proliferation-inducing oncogenes leads to RS and dependency on ATR signalling [53]. In models of Ras- or MYC-driven cancer, signalling through these oncogenes has been shown to lead to sensitivity to ATR inhibition [54–56]. In NB cell lines, high MYCN expression has been shown to increase RS [32,57]. Here we have shown that amplification and overexpression of the *MYCN* oncogene in NB cell lines is associated with sensitivity to VE-821, confirming that *MYCN* driven NB cells are vulnerable to ATR inhibition.

ATM loss or dysfunction is also associated with sensitivity to VE-821. Our results confirm that reduced ATM function, which can be partially determined by protein level, confers sensitivity to ATR inhibition in NB as it does in other tumour types [5,58–60]. In addition to 11q deletion, loss of ATM can occur in *MNA* NB by upregulation of the ATM targeting micro-RNA, miR-421 [61]. ATM silencing, in addition to MYCN-driven proliferation, is likely contributing to replication stress through impaired DNA DSB signalling and repair, making these cells especially vulnerable to ATR inhibition.

MYCN and ATM protein expression levels were more powerful determinants of VE-821 sensitivity than the genetic status (amplification or deletion/mutation, respectively), and MYCN protein levels positively correlated with VE-821 sensitivity, whereas ATM protein level was negatively correlated with VE-821 sensitivity.

Mutation, expression or dysfunction of the p53 tumour suppressor protein was not associated with sensitivity to ATR inhibition. Middleton et al. also demonstrated that defective p53 signalling did not lead to sensitivity to VE-821 alone, but the addition of VE-821 showed greater potentiation of gemcitabine and ionising radiation induced cytotoxicity in $p53^{-/-}$ cell lines [46].

To summarise, MYCN and ATM protein expression levels, as well as genetic status, could provide useful predictive biomarkers to stratify NB patients who would benefit from an ATR inhibitor.

Further investigation into combining ATR inhibition with DNA-damaging chemotherapy may provide a novel therapeutic strategy for both *TP53* wt and *TP53* mutant NB.

In contrast, neither *ATM* nor *MYCN* status significantly determined sensitivity to PARP inhibition by olaparib. Colicchia et al. previously demonstrated that *MYCN*-amplification or overexpression renders NB cells especially sensitive to single agent olaparib [32]. Sensitivity of cell lines to PARP inhibitors by this group was analysed by MTS cell proliferation assay, an assay which measures metabolic activity in a similar way to XTT. Although we did not see a significant difference between MNA cell lines and non-MNA cell lines in olaparib sensitivity in either XTT or clonogenic survival assay, high MYCN expressing cell lines did show significantly increased sensitivity when analysed by clonogenic survival. The clonogenic survival assay measures the ability of single cells to divide indefinitely and relies on cell lines having good cloning efficiency. Unfortunately, some NB cell lines show poor cloning efficiency and some cell lines show stark differences in sensitivity between the two assays.

Our results highlight the importance of validating sensitivity found to novel agents in models of disease with other types of cell viability assays. In the case of NB, more evidence is required before the role of MYCN in PARP sensitivity can be established. It may be that in the context of complex genetic alterations, cells may have acquired a spectrum of features leading to resistance, as well as sensitivity to these inhibitors, making it difficult to predict response.

In addition to investigating features associated with sensitivity to these inhibitors as single agents, we provide evidence of synergy between ATR and PARP inhibition in NB cell lines. VE-821 enhanced olaparib-induced growth inhibition independently of *MYCN*, *ATM* and *TP53* status. ATR inhibition has been shown to enhance PARP inhibitor sensitivity in a variety of adult solid tumour types [51,62,63] and overcome resistance to PARP inhibitors [52,64].

Olaparib treatment alone induced markers of RS (γH2AX, pRPA2^{S8}, Figures 4 and 5) and activated ATR (pCHK1^{S345} expression, Figure 4A). When ATR was inhibited, olaparib-induced markers of RS were increased, which suggested the inhibition of ATR exacerbated RS caused by PARP inhibition. This could be because inhibition of ATR releases olaparib-induced S and G2 arrest, shown here (Figure 5A) and previously reported [63]. Loss of cell cycle checkpoints results in DNA replication through damaged areas, fork stalling, collapse and chromosome breakage, which, if left unrepaired, leads to mitotic catastrophe and cell death [44].

Loss or inhibition of ATR has been shown to restrict HRR function [51,52,65]. We confirmed this in MNA and non-MNA NB cell lines. However, one non-MNA cell line, SKNAS, did not show increased Rad51 after PARP inhibition, and would be considered HRR-defective. HRR-defective cells are usually highly sensitive to PARP inhibitors, whereas the SKNAS cell line is the most resistant to olaparib out of the cell lines analysed in this study.

Although inhibiting ATR to specifically target RS in cancer cell looks promising, other mechanisms of tolerating RS have been identified, such as overexpression of components of the fork protection complex [66] and recruitment of Y-family (translesion synthesis) DNA polymerases [67,68], which should be considered as possible resistance mechanisms. These are all factors to be considered when interpreting data from the recently initiated trials of PARPi and ATRi combinations (NCT02723864, NCT03462342, NCT04065269, NCT03787680, NCT04149145, and NCT03682289).

In conclusion, we have shown that the combination of ATR and PARP inhibition leads to increased RS and cytotoxicity in NB cell lines. This synergy is likely to be, at least in part, not only due to abrogation of cell cycle checkpoints but also inhibition of HRR, thereby inducing synthetic lethality with PARP inhibition.

4. Materials and Methods

4.1. Chemicals and Cell Lines

VE-821 and Olaparib were purchased from Stratech Scientific Ltd. (Cambridge, UK) and stored at −20 °C in stock solution of 100 mM and 20 mM, respectively, in DMSO (Sigma-Aldrich, Gillingham, Dorset, UK).

Neuroblastoma cell lines used in this study were: SHSY5Y, SKNAS, NGP, N20_R1, NMB, IMR32, Kat100, IGRN91, SJNB1, and GIMEN (Table 1). All neuroblastoma cell lines were obtained between 1996 and 2018, with the exception of N20_R1, and were validated upon receipt using cytogenetic analysis courtesy of Dr Nick Bown (Institute of Genetic Medicine, Newcastle University). N20_R1 (N_N20R1) were generated from parental NGP cell line with resistance to Nutlin-3 [37]. All cell lines were cultured in RPMI-1640 (Sigma-Aldrich) supplemented with 10% (*v*/*v*) Fetal Calf Serum (Gibco, Life Technologies Ltd., Thermo Fisher Scientific, Waltham, MA, USA) and maintained at 37 °C in a humidified incubator with 5% CO_2. Cell lines were routinely tested for Mycoplasma and were confirmed to be negative.

4.2. Growth Inhibition Assay

Cells were seeded in 96-well plates (Corning, VWR International Ltd., Lutterworth, UK), and allowed to adhere overnight before treatment with ATR or PARP inhibitors alone or in combination for 72 h. Inhibitors were added at 200× dilution to give a final DMSO concentration of 0.05%. Percentage control growth was assessed using the XTT cell proliferation assay (Roche, Burgess Hill, UK) according to the manufacturer's instructions and using the following formula: (average absorbance test/average absorbance control) × 100. Combination Index (CI) values were determined by the Chou–Talalay method using CalcuSyn v2 (Biosoft, Cambridge, UK).

4.3. Protein Extraction and Western Blotting

Cell pellets were harvested and protein extracted using PhosphoSafe™ Extraction Buffer (Novagen, Merck Millipore Ltd. Watford, UK) following the manufacturer's protocol. Protein concentration was quantified using the Pierce BCA protein assay kit (Thermo Fisher Scientific).

Proteins were separated using 3–8% Criterion™ XT Tris-Acetate Protein Gel (Bio-Rad Laboratories Ltd., Hemel Hempstead, UK) and transferred onto Hybond-C Extra membrane (GE Life Sciences, Little Chalfont, UK). Membranes were stained with Ponceau S (Sigma-Aldrich) to control for loading, destained in tris buffered saline, 0.5% tween 20 (TBST) and blocked for 1 h in 5% milk TBST.

Primary antibodies were: mouse monoclonal MYCN 1:1000 (Santa Cruz Bio-technology (SCBT), Dallas, TX, USA: A1513), rabbit polyclonal ATM 1:500 (cell signalling technology(CST), Danvers, MA, USA: 2873S), mouse phospho-ATM (ser1981) 1:1000 (CST: 4526), mouse monoclonal CHK1 1:1000 (SCBT: sc-8408), mouse monoclonal CHK2 1:1000 (SCBT: sc-17747), rabbit polyclonal phosho-CHK1 (ser345) 1:1000 (CST:2341S), rabbit monoclonal phosphor-CHK2 (thr68) 1:1000 (CST:2197S), rabbit polyclonal ATR 1:500 (CST: 2790S), p53 1:1000 (NCL-L-p53-DO7, Leica Biosystems Ltd.), rabbit monoclonal p21 Waf1/Cip1 (12D1) 1:1000 (CST: 2947S), rabbit monoclonal phosopho-H2AX (ser139) 1:2000 (CST: 2577S), rabbit monoclonal phospho-RPA2 (ser8) 1:1000 (CST: 54762S), mouse monoclonal glyceraldehyde 3-phosphate dehydrogenase (GAPDH) 1:5000 (SCBT: sc-47724). Secondary antibodies used were peroxidase-conjugated goat anti-mouse (Dako, Glostrup, Denmark) and anti-rabbit immunoglobulins (Dako) at 1:2500.

Protein detection was performed using enhanced chemiluminescence (Bio-Rad) and imaged using the ChemiDoc imaging system (Bio-Rad Laboratories Ltd., Hemel Hempstead, UK). Densitometry was performed using ImageJ image analysis software (Version 1.52p; Java 1.8.0_172 [64-bit]).

4.4. Cell Cycle Analysis

Cells were harvested post-treatment, fixed in ice-cold 70% (*v*/*v*) ethanol and stored at −20 °C. Prior to analysis, cells were washed with phosphate buffered saline (PBS), resuspended in 500 µL PBS with 50 µg/mL propidium iodide (Sigma-Aldrich) and 50 µg/mL RNAse A (Sigma-Aldrich), and incubated at 37 °C for 30 min. Samples were analysed on the Attune NxT Flow Cytometer using Invitrogen™ Attune NxT Software (Thermo Fisher Scientific). Data were analysed using FlowJo ™ (BD Biosciences, Wokingham, UK). Experiments were at least $n = 3$.

4.5. Immunofluorescence

Cells were treated for 24 h with control vehicle (DMSO) and 1 µM VE-821 with or without 10 µM olaparib. Cells were stained with mouse monoclonal anti phospho-Histone H2A.X (Ser139) antibody (SCBT) at 1:500 and rabbit monoclonal anti RAD51 antibody (CST: 8875S) at 1:250. Secondary antibodies used were Alexa 488 conjugated goat anti rabbit and Alexa 546 conjugated goat anti mouse (Invitrogen, Thermo Fisher Scientific), both at 1:1000. Cells were imaged using a Leica DM6 microscope and Leica Application Suite (LAS) X software (Leica Microsystems, Wetzlar, Germany). The number of RAD51 foci in each cell and total nuclear fluorescence intensity for γH2AX were quantified using ImageJ software (Version 1.52p; Java 1.8.0_172 (64-bit)) and data was plotted using GraphPad Prism v6 (San Diego, CA, USA).

4.6. Statistical Analysis

Mann Whitney U and 2-way ANOVA statistical tests were carried out using GraphPad Prism v6 software and $p < 0.05$ was taken to be statistically significant.

5. Conclusions

In the era of precision medicine, cancer specific DDR defects and RS have become attractive targets, with many novel agents entering clinical trials. With the introduction of PARP inhibitors into clinical trials for the treatment of paediatric cancers, including HR-NB, a disease with one of the worst long term prognoses, and ATR inhibitors being tested in adult trials, we sought to investigate which NB specific DDR defects, if any, would lead to sensitivity to these agents alone and in combination. We have shown that *MYCN*-amplification and protein overexpression and loss of functional ATM protein through deletion or mutation result in vulnerability to ATR inhibition by VE-821. The case for PARP inhibition is less clear and, although there is a trend towards MYCN overexpressing cells being more sensitive to olaparib, the role of MYCN in PARP inhibitor sensitivity needs to be investigated further. We also provide evidence of synergy between ATR and PARP inhibition in NB cell lines independent of *MYCN*, *ATM* or *TP53* status. This may be due to increased RS, compromised S and G2 checkpoint arrest and/or defective HRR in the presence of both PARP and ATR inhibitors.

Overall, our work gives exciting insights into the use of PARP and ATR inhibitors as a novel treatment strategy for HR-NB.

Supplementary Materials: The following are available online at http://www.mdpi.com/2072-6694/12/5/1095/s1, Figure S1: Full images of baseline DDR protein expression by Western blot, Figure 1A, Figure S2: Full images of Western blots in Figure 1C,D, Figure S3: Correlation of relative baseline protein expression of MYCN and ATM and VE-821 sensitivity at 10 μM, Figure S4: Determinants of ATR inhibitor sensitivity analysis by clonogenic survival assay, Figure S5: Determinants of PARP inhibitor sensitivity by XTT cell proliferation assay, Figure S6: Determinants of PARP inhibitor sensitivity analysis by clonogenic survival assay, Figure S7: Effect of VE-821 on olaparib cytotoxicity by clonogenic survival assay, Figure S8: Full images of Western blots in Figure 4.

Author Contributions: Conceptualization, H.E.D.S., L.C., D.A.T. and N.J.C.; Formal analysis, H.E.D.S.; Funding acquisition, D.A.T. and N.J.C.; Investigation, H.E.D.S.; Methodology, H.E.D.S. and L.C.; Supervision, D.A.T. and N.J.C.; Writing—original draft, H.E.D.S.; Writing—review and editing, H.E.D.S., L.C., D.A.T. and N.J.C. All authors have read and agreed to the published version of the manuscript.

Funding: This work was funded by the Children's Cancer and Leukaemia Group and Little Princess Trust (CCLGA-2017-12).

Acknowledgments: This work was funded by the Children's Cancer and Leukaemia Group and Little Princess Trust (CCLGA-2017-12)-. We thank the following for cell lines: Linda Harris (SJNB1), Penny Lovat (SHSY5Y and IMR32), Rogier Versteeg (NGP), Mirco Ponzoni (GIMEN), John Lunec (N20_R1), Jean Bénard (SKNAS and IGRN91), Garrett Brodeur (NMB) and Michelle Haber (IMR/KAT100).

Conflicts of Interest: N.J.C. has received research funding from Aguoron Pharmaceuticals and Pfizer for the development of rucaparib, BioMarin for studies on talazoparib and Tesaro for studies with niraparib and paid consultancy from Abbvie for veliparib. N.J.C. is inventor on the patent relating to the use of rucaparib in HRD cancer (WO2005012305A3) and Newcastle University receives royalty income from the sales of rucaparib that are shared with contributors to its development, including N.J.C. N.J.C. has also received research funding from Vertex Pharmaceuticals and is currently in receipt of funding from Merck KGaA for work on ATR inhibitors and holds a patent for a pharmacodynamic biomarker of ATR inhibition (WO2014055756A1). All other authors declare no conflicts of interest.

References

1. Yazinski, S.A.; Zou, L. Functions, regulation, and therapeutic implications of the atr checkpoint pathway. *Annu. Rev. Genet.* **2016**, *50*, 155–173. [CrossRef]
2. Lecona, E.; Fernandez-Capetillo, O. Targeting atr in cancer. *Nat. Rev. Cancer* **2018**, *18*, 586–595. [CrossRef] [PubMed]
3. Blackford, A.N.; Jackson, S.P. Atm, atr, and DNA-pk: The trinity at the heart of the DNA damage response. *Mol. Cell* **2017**, *66*, 801–817. [CrossRef] [PubMed]
4. Saldivar, J.; Cortez, D.; Cimprich, K. The essential kinase atr: Ensuring faithful duplication of a challenging genome. *Nat. Rev. Mol. Cell Biol.* **2017**, *18*, 622–636. [CrossRef] [PubMed]
5. Kwok, M.; Davies, N.; Agathanggelou, A.; Smith, E.; Oldreive, C.; Petermann, E.; Stewart, G.; Brown, J.; Lau, A.; Pratt, G.; et al. Atr inhibition induces synthetic lethality and overcomes chemoresistance in tp53- or atm-defective chronic lymphocytic leukemia cells. *Blood* **2016**, *127*, 582–595. [CrossRef]

6. Tu, X.; Kahila, M.M.; Zhou, Q.; Yu, J.; Kalari, K.R.; Wang, L.; Harmsen, W.S.; Yuan, J.; Boughey, J.C.; Goetz, M.P.; et al. Atr inhibition is a promising radiosensitizing strategy for triple-negative breast cancer. *Mol. Cancer Ther.* **2018**, *17*, 2462–2472. [CrossRef]
7. Reaper, P.M.; Griffiths, M.R.; Long, J.M.; Charrier, J.-D.; MacCormick, S.; Charlton, P.A.; Golec, J.M.C.; Pollard, J.R. Selective killing of atm-or p53-deficient cancer cells through inhibition of atr. *Nat. Chem. Biol.* **2011**, *7*, 428–430. [CrossRef]
8. Karnitz, L.M.; Zou, L. Molecular pathways: Targeting atr in cancer therapy. *Clin. Cancer Res. Off. J. Am. Assoc. Cancer Res.* **2015**, *21*, 4780–4785. [CrossRef]
9. Zeman, M.K.; Cimprich, K.A. Causes and consequences of replication stress. *Nat. Cell Biol.* **2014**, *16*, 2–9. [CrossRef]
10. Taylor, E.M.; Lindsay, H.D. DNA replication stress and cancer: Cause or cure? *Future Oncol.* **2016**, *12*, 221–237. [CrossRef]
11. Massague, J. G1 cell-cycle control and cancer. *Nature* **2004**, *432*, 298–306. [CrossRef] [PubMed]
12. Deckbar, D.; Birraux, J.; Krempler, A.; Tchouandong, L.; Beucher, A.; Walker, S.; Stiff, T.; Jeggo, P.; Löbrich, M. Chromosome breakage after g2 checkpoint release. *J. Cell Biol.* **2007**, *176*, 749–755. [CrossRef]
13. Stolz, A.; Bastians, H. Therapeutic s and g2 checkpoint override causes centromere fragmentation in mitosis. *Cell Cycle (Georget. Tex.)* **2013**, *12*, 1826–1827. [CrossRef]
14. Russell, K.J.; Wiens, L.W.; Demers, G.W.; Galloway, D.A.; Plon, S.E.; Groudine, M. Abrogation of the g2 checkpoint results in differential radiosensitization of g1 checkpoint-deficient and g1 checkpoint-competent cells. *Cancer Res.* **1995**, *55*, 1639–1642. [PubMed]
15. Smith, M.A.; Seibel, N.L.; Altekruse, S.F.; Ries, L.A.G.; Melbert, D.L.; O'Leary, M.; Smith, F.O.; Reaman, G.H. Outcomes for children and adolescents with cancer: Challenges for the twenty-first century. *J. Clin. Oncol. Off. J. Am. Soc. Clin. Oncol.* **2010**, *28*, 2625–2634. [CrossRef] [PubMed]
16. Stiller, C. *Childhood Cancer in Britain: Incidence, Survival, Mortality*; Oxford University Press: Oxford, UK, 2007.
17. Cohn, S.L.; Pearson, A.D.J.; London, W.B.; Monclair, T.; Ambros, P.F.; Brodeur, G.M.; Faldum, A.; Hero, B.; Iehara, T.; Machin, D. The international neuroblastoma risk group (inrg) classification system: An inrg task force report. *J. Clin. Oncol.* **2009**, *27*, 289–297. [CrossRef] [PubMed]
18. Park, J.R.; Bagatell, R.; London, W.B.; Maris, J.M.; Cohn, S.L.; Mattay, K.M.; Hogarty, M.; COG Neuroblastoma Committee. Children's oncology group's 2013 blueprint for research: Neuroblastoma. *Pediatric Blood Cancer* **2013**, *60*, 985–993. [CrossRef]
19. Basta, N.O.; Halliday, G.C.; Makin, G.; Birch, J.; Feltbower, R.; Bown, N.; Elliott, M.; Moreno, L.; Barone, G.; Pearson, A.D.J.; et al. Factors associated with recurrence and survival length following relapse in patients with neuroblastoma. *Br. J. Cancer* **2016**, *115*, 1048. [CrossRef]
20. Mlakar, V.; Mlakar, S.J.; Lopez, G.; Maris, J.M.; Ansari, M.; Gumy-Pause, F. 11q deletion in neuroblastoma: A review of biological and clinical implications. *Mol. Cancer* **2017**, *16*, 114. [CrossRef]
21. Brodeur, G.M. Neuroblastoma: Biological insights into a clinical enigma. *Nat Rev Cancer* **2003**, *3*, 203–216. [CrossRef]
22. Carr-Wilkinson, J.; O'Toole, K.; Wood, K.M.; Challen, C.C.; Baker, A.G.; Board, J.R.; Evans, L.; Cole, M.; Cheung, N.-K.V.; Boos, J.; et al. High frequency of p53/mdm2/p14arf pathway abnormalities in relapsed neuroblastoma. *Clin. Cancer Res. Off. J. Am. Assoc. Cancer Res.* **2010**, *16*, 1108–1118. [CrossRef] [PubMed]
23. Bell, E.; Premkumar, R.; Carr, J.; Lu, X.; Lovat, P.E.; Kees, U.R.; Lunec, J.; Tweddle, D. The role of mycn in the failure of mycn amplified neuroblastoma cell lines to g1 arrest after DNA damage. *Cell Cycle* **2006**, *5*, 2639–2647. [CrossRef] [PubMed]
24. Carén, H.; Kryh, H.; Nethander, M.; Sjöberg, R.-M.; Träger, C.; Nilsson, S.; Abrahamsson, J.; Kogner, P.; Martinsson, T. High-risk neuroblastoma tumors with 11q-deletion display a poor prognostic, chromosome instability phenotype with later onset. *Proc. Natl. Acad. Sci. USA* **2010**, *107*, 4323–4328. [CrossRef] [PubMed]
25. Carr, J.; Bell, E.; Pearson, A.D.J.; Kees, U.R.; Beris, H.; Lunec, J.; Tweddle, D.A. Increased frequency of aberrations in the p53/mdm2/p14arf pathway in neuroblastoma cell lines established at relapse. *Cancer Res.* **2006**, *66*, 2138–2145. [CrossRef] [PubMed]
26. Bryant, H.E.; Petermann, E.; Schultz, N.; Jemth, A.-S.; Loseva, O.; Issaeva, N.; Johansson, F.; Fernandez, S.; McGlynn, P.; Helleday, T. Parp is activated at stalled forks to mediate mre11-dependent replication restart and recombination. *EMBO J.* **2009**, *28*, 2601–2615. [CrossRef]
27. Caldecott, K.W. Single-strand break repair and genetic disease. *Nat. Rev. Genet.* **2008**, *9*, 619–631. [CrossRef]

28. Daniel, R.A.; Rozanska, A.L.; Thomas, H.D.; Mulligan, E.A.; Drew, Y.; Castelbuono, D.J.; Hostomsky, Z.; Plummer, E.R.; Boddy, A.V.; Tweddle, D.A.; et al. Inhibition of poly(adp-ribose) polymerase-1 enhances temozolomide and topotecan activity against childhood neuroblastoma. *Clin. Cancer Res.* **2009**, *15*, 1241–1249. [CrossRef]
29. Sanmartín, E.; Muñoz, L.; Piqueras, M.; Sirerol, J.A.; Berlanga, P.; Cañete, A.; Castel, V.; de Mora, J.F. Deletion of 11q in neuroblastomas drives sensitivity to parp inhibition. *Clin. Cancer Res.* **2017**, *23*, 6875–6887. [CrossRef]
30. Nile, D.L.; Rae, C.; Hyndman, I.J.; Gaze, M.N.; Mairs, R.J. An evaluation in vitro of parp-1 inhibitors, rucaparib and olaparib, as radiosensitisers for the treatment of neuroblastoma. *BMC Cancer* **2016**, *16*, 621. [CrossRef]
31. Norris, R.E.; Adamson, P.C.; Nguyen, V.T.; Fox, E. Preclinical evaluation of the parp inhibitor, olaparib, in combination with cytotoxic chemotherapy in pediatric solid tumors. *Pediatric Blood Cancer* **2014**, *61*, 145–150. [CrossRef]
32. Colicchia, V.; Petroni, M.; Guarguaglini, G.; Sardina, F.; Sahún-Roncero, M.; Carbonari, M.; Ricci, B.; Heil, C.; Capalbo, C.; Belardinilli, F. Parp inhibitors enhance replication stress and cause mitotic catastrophe in mycn-dependent neuroblastoma. *Oncogene* **2017**, *36*, 4682–4691. [CrossRef] [PubMed]
33. Takagi, M.; Yoshida, M.; Nemoto, Y.; Tamaichi, H.; Tsuchida, R.; Seki, M.; Uryu, K.; Nishii, R.; Miyamoto, S.; Saito, M. Loss of DNA damage response in neuroblastoma and utility of a parp inhibitor. *J. Natl. Cancer Inst.* **2017**, *109*, djx062. [CrossRef]
34. Ross, R.A.; Spengler, B.A.; Biedler, J.L. Coordinate morphological and biochemical interconversion of human neuroblastoma cells2. *JNCI J. Natl. Cancer Inst.* **1983**, *71*, 741–747. [PubMed]
35. Goldschneider, D.; Horvilleur, E.; Plassa, L.-F.; Guillaud-Bataille, M.; Million, K.; Wittmer-Dupret, E.; Danglot, G.; de Thé, H.; Bénard, J.; May, E.; et al. Expression of c-terminal deleted p53 isoforms in neuroblastoma. *Nucleic Acids Res.* **2006**, *34*, 5603–5612. [CrossRef]
36. Van Roy, N.; Van Limbergen, H.; Vandesompele, J.; Van Gele, M.; Poppe, B.; Salwen, H.; Laureys, G.; Manoel, N.; De Paepe, A.; Speleman, F. Combined m-fish and cgh analysis allows comprehensive description of genetic alterations in neuroblastoma cell lines. *GenesChromosomes Cancer* **2001**, *32*, 126–135. [CrossRef]
37. Drummond, C.J.; Esfandiari, A.; Liu, J.; Lu, X.; Hutton, C.; Jackson, J.; Bennaceur, K.; Xu, Q.; Makimanejavali, A.R.; Bello, F.D.; et al. Tp53 mutant mdm2-amplified cell lines selected for resistance to mdm2-p53 binding antagonists retain sensitivity to ionizing radiation. *Oncotarget* **2016**, *7*, 46203–46218. [CrossRef]
38. Goldman, S.C.; Chen, C.Y.; Lansing, T.J.; Gilmer, T.M.; Kastan, M.B. The p53 signal transduction pathway is intact in human neuroblastoma despite cytoplasmic localization. *Am. J. Pathol.* **1996**, *148*, 1381–1385.
39. Tumilowicz, J.J.; Nichols, W.W.; Cholon, J.J.; Greene, A.E. Definition of a continuous human cell line derived from neuroblastoma. *Cancer Res.* **1970**, *30*, 2110–2118.
40. Mandriota, S.J.; Valentijn, L.J.; Lesne, L.; Betts, D.R.; Marino, D.; Boudal-Khoshbeen, M.; London, W.B.; Rougemont, A.-L.; Attiyeh, E.F.; Maris, J.M. Ataxia-telangiectasia mutated (atm) silencing promotes neuroblastoma progression through a mycn independent mechanism. *Oncotarget* **2015**, *6*, 18558. [CrossRef]
41. Xue, C.; Haber, M.; Flemming, C.; Marshall, G.M.; Lock, R.B.; MacKenzie, K.L.; Gurova, K.V.; Norris, M.D.; Gudkov, A.V. P53 determines multidrug sensitivity of childhood neuroblastoma. *Cancer Res.* **2007**, *67*, 10351–10360. [CrossRef] [PubMed]
42. Blanc, E.; Goldschneider, D.; Ferrandis, E.; Barrois, M.; Le Roux, G.; Leonce, S.; Douc-Rasy, S.; Bénard, J.; Raguénez, G. Mycn enhances p-gp/mdr1 gene expression in the human metastatic neuroblastoma igr-n-91 model. *Am. J. Pathol.* **2003**, *163*, 321–331. [CrossRef]
43. Flahaut, M.; Mühlethaler-Mottet, A.; Martinet, D.; Fattet, S.; Bourloud, K.B.; Auderset, K.; Meier, R.; Schmutz, N.B.; Delattre, O.; Joseph, J.-M.; et al. Molecular cytogenetic characterization of doxorubicin-resistant neuroblastoma cell lines: Evidence that acquired multidrug resistance results from a unique large amplification of the 7q21 region. *GenesChromosomes Cancer* **2006**, *45*, 495–508. [CrossRef] [PubMed]
44. Toledo, L.; Neelsen, K.J.; Lukas, J. Replication catastrophe: When a checkpoint fails because of exhaustion. *Mol. Cell* **2017**, *66*, 735–749. [CrossRef] [PubMed]
45. Liu, S.; Opiyo, S.O.; Manthey, K.; Glanzer, J.G.; Ashley, A.K.; Amerin, C.; Troksa, K.; Shrivastav, M.; Nickoloff, J.A.; Oakley, G.G. Distinct roles for DNA-pk, atm and atr in rpa phosphorylation and checkpoint activation in response to replication stress. *Nucleic Acids Res.* **2012**, *40*, 10780–10794. [CrossRef]

46. Middleton, F.K.; Pollard, J.R.; Curtin, N.J. The impact of p53 dysfunction in atr inhibitor cytotoxicity and chemo- and radiosensitisation. *Cancers* **2018**, *10*, 275. [CrossRef]
47. Murai, J.; Huang, S.-Y.N.; Das, B.B.; Renaud, A.; Zhang, Y.; Doroshow, J.H.; Ji, J.; Takeda, S.; Pommier, Y. Trapping of parp1 and parp2 by clinical parp inhibitors. *Cancer Res.* **2012**, *72*, 5588–5599. [CrossRef]
48. Jelinic, P.; Levine, D.A. New insights into parp inhibitors' effect on cell cycle and homology-directed DNA damage repair. *Mol. Cancer Ther.* **2014**, *13*, 1645–1654. [CrossRef]
49. Lord, C.J.; Ashworth, A. Parp inhibitors: Synthetic lethality in the clinic. *Science (New York N.Y.)* **2017**, *355*, 1152–1158. [CrossRef]
50. Mukhopadhyay, A.; Elattar, A.; Cerbinskaite, A.; Wilkinson, S.J.; Drew, Y.; Kyle, S.; Los, G.; Hostomsky, Z.; Edmondson, R.J.; Curtin, N.J. Development of a functional assay for homologous recombination status in primary cultures of epithelial ovarian tumor and correlation with sensitivity to poly(adp-ribose) polymerase inhibitors. *Clin. Cancer Res.* **2010**, *16*, 2344–2351. [CrossRef]
51. Peasland, A.; Wang, L.Z.; Rowling, E.; Kyle, S.; Chen, T.; Hopkins, A.; Cliby, W.A.; Sarkaria, J.; Beale, G.; Edmondson, R.J.; et al. Identification and evaluation of a potent novel atr inhibitor, nu6027, in breast and ovarian cancer cell lines. *Br. J. Cancer* **2011**, *105*, 372–381. [CrossRef]
52. Yazinski, S.A.; Comaills, V.; Buisson, R.; Genois, M.-M.; Nguyen, H.D.; Ho, C.K.; Todorova Kwan, T.; Morris, R.; Lauffer, S.; Nussenzweig, A.; et al. Atr inhibition disrupts rewired homologous recombination and fork protection pathways in parp inhibitor-resistant brca-deficient cancer cells. *Genes Dev.* **2017**, *31*, 318–332. [CrossRef] [PubMed]
53. Kotsantis, P.; Petermann, E.; Boulton, S.J. Mechanisms of oncogene-induced replication stress: Jigsaw falling into place. *Cancer Discov.* **2018**, *8*, 537–555. [CrossRef] [PubMed]
54. Schoppy, D.W.; Ragland, R.L.; Gilad, O.; Shastri, N.; Peters, A.A.; Murga, M.; Fernandez-Capetillo, O.; Diehl, J.A.; Brown, E.J. Oncogenic stress sensitizes murine cancers to hypomorphic suppression of atr. *J. Clin. Investig.* **2012**, *122*, 241–252. [CrossRef] [PubMed]
55. Murga, M.; Campaner, S.; Lopez-Contreras, A.J.; Toledo, L.I.; Soria, R.; Montaña, M.F.; D'Artista, L.; Schleker, T.; Guerra, C.; Garcia, E. Exploiting oncogene-induced replicative stress for the selective killing of myc-driven tumors. *Nat. Struct. Mol. Biol.* **2011**, *18*, 1331–1335. [CrossRef]
56. Toledo, L.I.; Murga, M.; Zur, R.; Soria, R.; Rodriguez, A.; Martinez, S.; Oyarzabal, J.; Pastor, J.; Bischoff, J.R.; Fernandez-Capetillo, O. A cell-based screen identifies atr inhibitors with synthetic lethal properties for cancer-associated mutations. *Nat. Struct. Mol. Biol.* **2011**, *18*, 721–727. [CrossRef]
57. Petroni, M.; Sardina, F.; Infante, P.; Bartolazzi, A.; Locatelli, E.; Fabretti, F.; Di Giulio, S.; Capalbo, C.; Cardinali, B.; Coppa, A.; et al. Mre11 inhibition highlights a replication stress-dependent vulnerability of mycn-driven tumors. *Cell Death Dis.* **2018**, *9*, 895. [CrossRef]
58. Middleton, F.K.; Patterson, M.J.; Elstob, C.J.; Fordham, S.; Herriott, A.; Wade, M.A.; McCormick, A.; Edmondson, R.; May, F.E.B.; Allan, J.M.; et al. Common cancer-associated imbalances in the DNA damage response confer sensitivity to single agent atr inhibition. *Oncotarget* **2015**, *6*, 32396–32409. [CrossRef]
59. Schmitt, A.; Knittel, G.; Welcker, D.; Yang, T.-P.; George, J.; Nowak, M.; Leeser, U.; Büttner, R.; Perner, S.; Peifer, M.; et al. *Atm* deficiency is associated with sensitivity to parp1- and atr inhibitors in lung adenocarcinoma. *Cancer Res.* **2017**, *77*, 3040–3056. [CrossRef]
60. Min, A.; Im, S.-A.; Jang, H.; Kim, S.; Lee, M.; Kim, D.K.; Yang, Y.; Kim, H.-J.; Lee, K.-H.; Kim, J.W.; et al. Azd6738, a novel oral inhibitor of atr, induces synthetic lethality with atm deficiency in gastric cancer cells. *Mol. Cancer Ther.* **2017**, *16*, 566–577. [CrossRef]
61. Hu, H.; Du, L.; Nagabayashi, G.; Seeger, R.C.; Gatti, R.A. Atm is down-regulated by n-myc-regulated microrna-421. *Proc. Natl. Acad. Sci. USA* **2010**, *107*, 1506–1511. [CrossRef]
62. Mohni, K.N.; Thompson, P.S.; Luzwick, J.W.; Glick, G.G.; Pendleton, C.S.; Lehmann, B.D.; Pietenpol, J.A.; Cortez, D. A synthetic lethal screen identifies DNA repair pathways that sensitize cancer cells to combined atr inhibition and cisplatin treatments. *PLoS ONE* **2015**, *10*, e0125482. [CrossRef] [PubMed]
63. Kim, H.; George, E.; Ragland, R.; Rafail, S.; Zhang, R.; Krepler, C.; Morgan, M.; Herlyn, M.; Brown, E.; Simpkins, F. Targeting the atr/chk1 axis with parp inhibition results in tumor regression in brca-mutant ovarian cancer models. *Clin. Cancer Res. Off. J. Am. Assoc. Cancer Res.* **2017**, *23*, 3097–3108. [CrossRef] [PubMed]
64. Murai, J.; Feng, Y.; Yu, G.K.; Ru, Y.; Tang, S.-W.; Shen, Y.; Pommier, Y. Resistance to parp inhibitors by slfn11 inactivation can be overcome by atr inhibition. *Oncotarget* **2016**, *7*, 76534–76550. [CrossRef] [PubMed]

65. Buisson, R.; Niraj, J.; Rodrigue, A.; Ho, C.K.; Kreuzer, J.; Foo, T.K.; Hardy, E.J.L.; Dellaire, G.; Haas, W.; Xia, B.; et al. Coupling of homologous recombination and the checkpoint by atr. *Mol. Cell* **2017**, *65*, 336–346. [CrossRef]
66. Bianco, J.N.; Bergoglio, V.; Lin, Y.-L.; Pillaire, M.-J.; Schmitz, A.-L.; Gilhodes, J.; Lusque, A.; Mazières, J.; Lacroix-Triki, M.; Roumeliotis, T.I.; et al. Overexpression of claspin and timeless protects cancer cells from replication stress in a checkpoint-independent manner. *Nat. Commun.* **2019**, *10*, 910. [CrossRef]
67. Yang, Y.; Gao, Y.; Mutter-Rottmayer, L.; Zlatanou, A.; Durando, M.; Ding, W.; Wyatt, D.; Ramsden, D.; Tanoue, Y.; Tateishi, S.; et al. DNA repair factor rad18 and DNA polymerase polκ confer tolerance of oncogenic DNA replication stress. *J. Cell Biol.* **2017**, *216*, 3097–3115. [CrossRef]
68. Kurashima, K.; Sekimoto, T.; Oda, T.; Kawabata, T.; Hanaoka, F.; Yamashita, T. Polη, a y-family translesion synthesis polymerase, promotes cellular tolerance of myc-induced replication stress. *J. Cell Sci.* **2018**, *131*, jcs212183. [CrossRef]

cancers

MDPI

Article

Reduced Tumorigenicity of Mouse ES Cells and the Augmented Anti-Tumor Therapeutic Effects under *Parg* Deficiency

Yuki Sonoda [1,2,†], Yuka Sasaki [1,3,†], Akemi Gunji [4], Hidenori Shirai [4], Tomonori Araki [1,2], Shoji Imamichi [1,3], Takae Onodera [1,3], Anna-Margareta Rydén [4], Masatoshi Watanabe [5], Jun Itami [6], Takuya Honda [2], Kazuto Ashizawa [7], Kazuhiko Nakao [2] and Mitsuko Masutani [1,3,4,8,*]

1 Department of Molecular and Genomic Biomedicine, Nagasaki University Graduate School of Biomedical Sciences, Nagasaki 852-8523, Japan
2 Department of Gastroenterology and Hepatology, Nagasaki University Graduate School of Biomedical Sciences, Nagasaki 852-8501, Japan
3 Lab of Collaborative Research, Division of Cellular Signaling, National Cancer Center Research Institute, Tokyo 104-0045, Japan
4 Biochemistry Division, National Cancer Center Research Institute, Tokyo 104-0045, Japan
5 Department of Oncologic Pathology, Mie University Graduate School of Medicine, Mie 514-8507, Japan
6 Department of Radiation Oncology, National Cancer Center Hospital, Tokyo 104-0045, Japan
7 Department of Clinical Oncology, Nagasaki University Graduate School of Biomedical Sciences, Nagasaki 852-8523, Japan
8 Center for Bioinformatics and Molecular Medicine, Nagasaki University Graduate School of Biomedical Sciences, Nagasaki 852-8523, Japan
* Correspondence: mmasutan@nagasaki-u.ac.jp; Tel.: +81-95-819-8502
† These authors contributed equally to this work.

Received: 7 April 2020; Accepted: 21 April 2020; Published: 24 April 2020

Abstract: PolyADP-ribosylation is a post-translational modification of proteins, and poly(ADP-ribose) (PAR) polymerase (PARP) family proteins synthesize PAR using NAD as a substrate. Poly(ADP-ribose) glycohydrolase (PARG) functions as the main enzyme for the degradation of PAR. In this study, we investigated the effects of *Parg* deficiency on tumorigenesis and therapeutic efficacy of DNA damaging agents, using mouse ES cell-derived tumor models. To examine the effects of *Parg* deficiency on tumorigenesis, $Parg^{+/+}$ and $Parg^{-/-}$ ES cells were subcutaneously injected into nude mice. The results showed that *Parg* deficiency delays early onset of tumorigenesis from ES cells. All the tumors were phenotypically similar to teratocarcinoma and microscopic findings indicated that differentiation spectrum was similar between the *Parg* genotypes. The augmented anti-tumor therapeutic effects of X-irradiation were observed under *Parg* deficiency. These results suggest that *Parg* deficiency suppresses early stages of tumorigenesis and that *Parg* inhibition, in combination with DNA damaging agents, may efficiently control tumor growth in particular types of germ cell tumors.

Keywords: cancer; poly (ADP-ribose) glycohydrolase; radiosensitization; poly (ADP-ribose) polymerase; ES cells

1. Introduction

PolyADP-ribosylation is the post-translational modification of proteins, and poly(ADP-ribose) (PAR) polymerase (PARP) family proteins synthesize PAR using NAD as a substrate. Poly(ADP-ribose) glycohydrolase (PARG) is involved in the degradation of PAR as a main enzyme. PARP-1 is activated upon DNA damage and is crucial for preserving genome stability [1]. Activated PARP-1 recognizes DNA strand breaks and is involved in base-excision repair, DNA strand break repair pathways [1].

The process of polyADP-ribosylation of proteins is tightly regulated by the second enzyme in the metabolic pair: PARG and ADP-ribosyl hydrolase3 (ARH3) [2]. PARG reverses the action of PARP by hydrolyzing the glycosidic bonds of PAR to produce ADP-ribose [3,4]. The catalytic capacity of PARG keeps the polyADP-ribosylation of proteins transient; the conversion rate of PAR is measured in minutes [1].

In DNA repair process, PARG is involved in the regulation of base excision repair and DNA strand break repair through the control of XRCC1 function [5]. PARG functional inhibition leads to increase in radiosensitivity in particular types of cells [6,7]. Perturbation of the PARP-1/PARG balance by the over-expression of PARG has been also shown to alter the genome methylation pattern to that of cancer cell types [8].

The importance of PARG in normal development is underscored by the fact that disruption of PARG leads to embryonic lethality [9,10]. PARG enzyme structures have been revealed [11–13]. There are several isoforms of PARG, which are produced from a single gene by alterative splicing. The longest (110 kDa, $PARG_{110}$) and most abundant isoform is located in the nucleus, while the 60 kDa isoform is located in the cytoplasm and mitochondria [14–17]. Disruption of exon 4, which is common to all PARG isoforms, leads to cell lethality [9]. However, disruption of exons 2–3 abrogates expression of nuclear PARG, but still allows expression of the remaining isoforms [18]. Furthermore, deficiency of the full-length isoform of PARG leads to enhanced cytotoxic sensitivity by causing PAR accumulation, induced by menadione, alkylating agents, endo-toxic shock and γ-irradiation [18,19].

Upon extensive DNA damage, PARP-1 becomes excessively activated and depletes cellular NAD^+ to polyADP-ribosylate proteins. PAR accumulation and the depletion of NAD^+ severely perturb the energy balance of cells and lead to apoptosis. PAR accumulation also triggers AIF (apoptosis inducing factor) activation in mitochondria and induces parthanotos, which can be observed in neuronal and cancer cells [20,21]. In heterozygously knockout mice of *Parg*, suppression of lung cancer development has been reported [22].

As described above, the cytoprotective role of PARG through its involvement in DNA repair in normal cells has been suggested, whereas the functional inhibition of PARG is reported to cause sensitization to DNA-damaging agents in cancer cells. Therefore, PARG may be considered both a tumor suppressor and a therapeutic target of cancer. The accumulated evidence thus led us to hypothesize that PARG may impact both cancer development and cancer therapy. Meanwhile, embryonic stem (ES) cells of mice can be useful as experimental systems for tumorigenesis and teratocarcinoma. In this study, using hypomorphic *Parg* knockout ES cells, we investigated the effects of *Parg* deficiency on tumorigenesis and the therapeutic efficacy of DNA damaging agents using ES cell derived tumor models.

2. Results

2.1. *Parg*$^{-/-}$ ES Cells Show Delayed Tumor Development

We previously generated two hypomorphic *Parg*$^{-/-}$ ES cell clones, D79 and D122, which retained about 10% residual PARG activity compared to parental wild-type J1 ES cells [7]. The growth rates of these *Parg*$^{-/-}$ and J1 ES cells are similar in the absence of DNA damaging agents. To examine the effects of *Parg* deficiency on tumorigenesis, *Parg*$^{+/+}$ J1 and two D79 and D122 *Parg*$^{-/-}$ ES cells were subcutaneously injected into the flanks of nude mice. Following injection, tumor size development was observed weekly over four weeks. An initial delay of tumor growth was observed at weeks two and three in tumors derived from *Parg*$^{-/-}$ ES cells ($p < 0.01$, Figure 1). This effect was observed during only the early phase, as tumor size did not differ significantly between the genotypes at week four. These results indicate that PARG deficiency delays the early onset of tumorigenesis derived from ES cells.

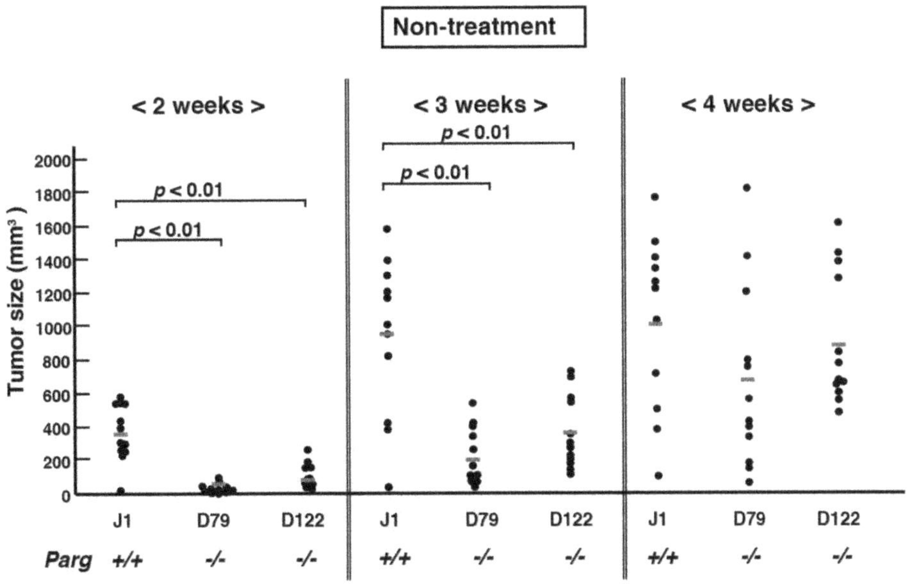

Figure 1. Effect of *Parg* deficiency on tumorigenesis from embryonic stem (ES) cells. In total, 1×10^7 ES cells were inoculated *s.c.* into nude mice and size of tumors was measured weekly. Wild-type, J1. *Parg*$^{-/-}$, D79 and D122.

2.2. Characterization of Tumor Tissues

To further characterize the tumorigenesis, histological evaluation of HE-stained tumor sections was carried out. Comparison of tissues and cell type are summarized in Table 1. Tumors observed four weeks after injection into nude mice showed ectodermal, mesodermal and endodermal tissue derivatives. Both undifferentiated and differentiated germinal components were detected. Histopathological examination at four weeks identified all tumors as immature teratoma, partially accompanying embryonal carcinoma components. Microscopically, the tumors showed heterogeneous components, especially containing immature neuroectodermal tissue in the form of primitive neuroepithelial rosettes and tubules (Figure 2B,C).

Table 1. Tissue components in tumors from *Parg*$^{+/+}$ and *Parg*$^{-/-}$ ES cells.

Tissue Type	*Parg*$^{+/+}$ (J1)	*Parg*$^{-/-}$ (D79)	*Parg*$^{-/-}$ (D122)
Embryonal carcinoma	+	+	+
Hemorrhage	−	−	−
Trophoblast giant cells	−	−	−
Ectodermal derivatives			
Primitive neuroepithelium	+	+	+
Mature neural tissue	+	+	+
Keratinized epithelium	+	+	+
Mesodermal derivatives			
Cartilage	+	+	+
Bone	+	+	+
Blood vessel	+	+	+
Lymphocyte and blood cell	+	+	+
Muscle	+	+	+
Endodermal derivatives			
Ciliated epithelium	+	+	+
Gut epithelium	+	+	+

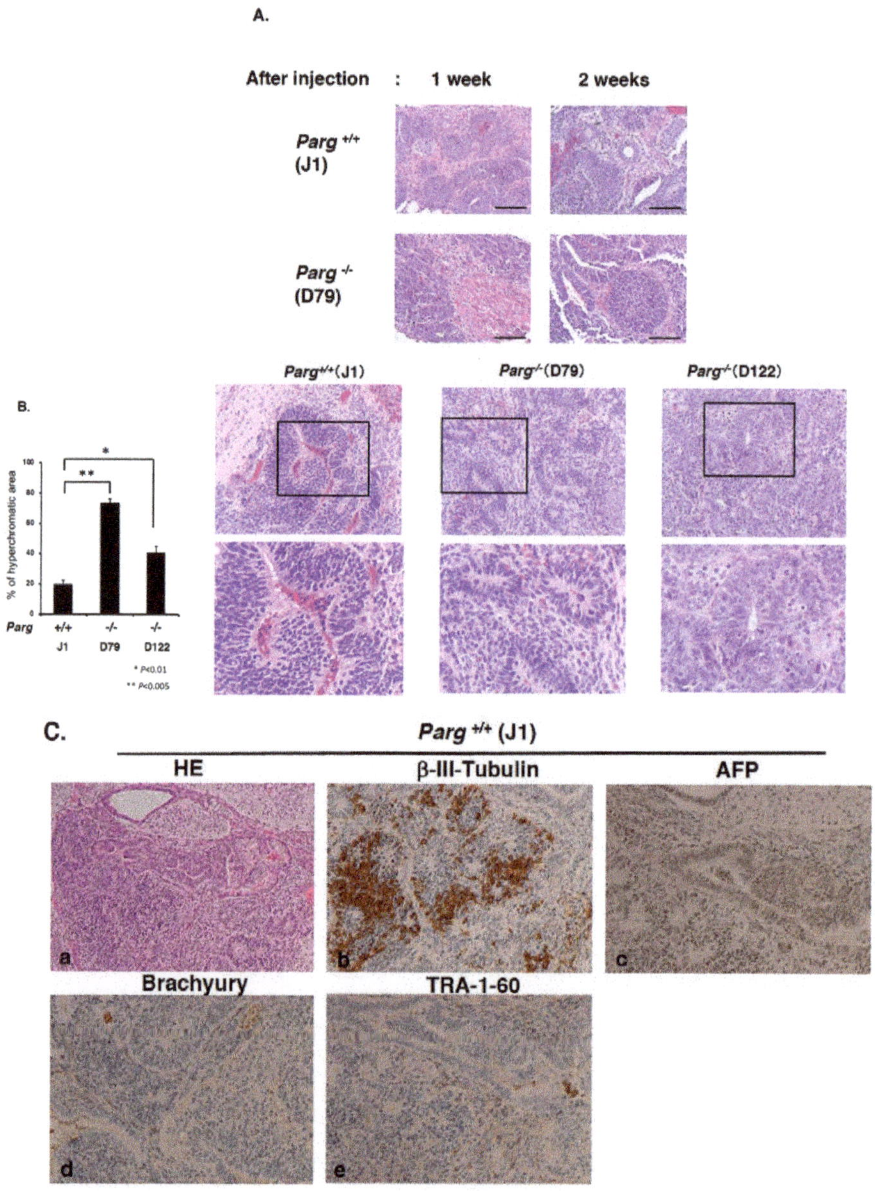

Figure 2. *Cont.*

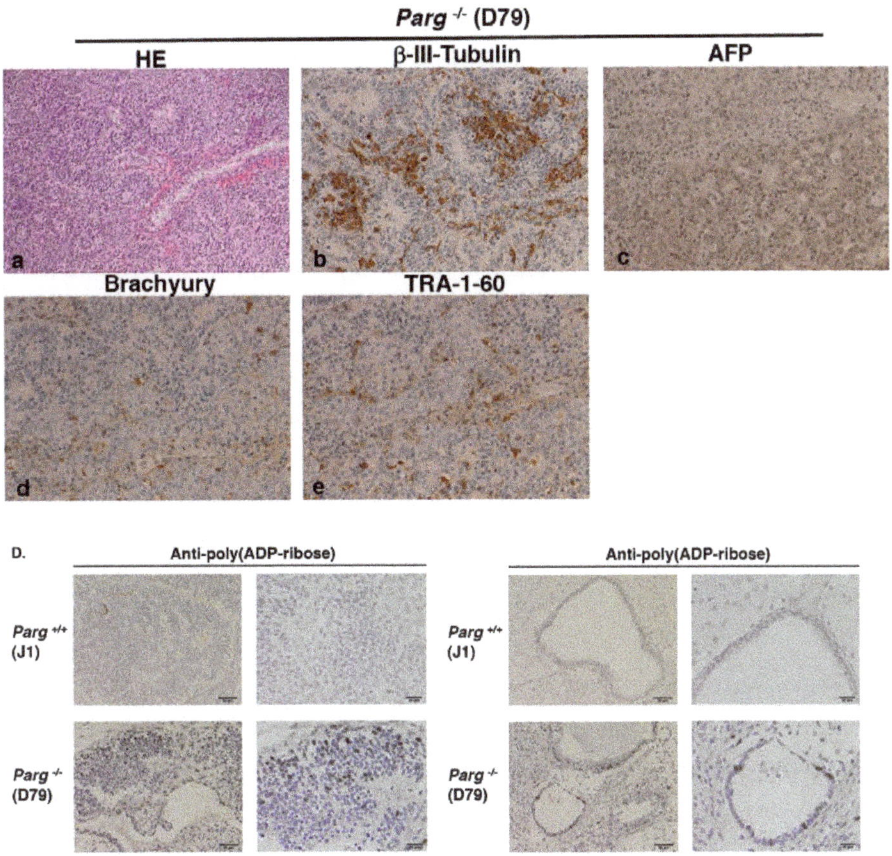

Figure 2. Hematoxylin-eosin staining of tumor tissues derived from mouse ES cells. (**A**) Hematoxylin-eosin staining of tumor tissues derived from mouse ES cells 1 and 2 weeks after injection. Upper panels, wild-type J1. Lower panels, *Parg*$^{-/-}$. (**B**) The left graph shows the percentage of hyperchromatic area of the tumors 4 weeks after injection. Percentage of hematoxylin-positive hyperchromatic area in the total area of tumor section was measured for each tumor. * $p < 0.01$, ** $p < 0.005$. Right panels show the typical hyperchromatic areas of hematoxylin-eosin staining of tumors 4 weeks after injection. Upper panels, ×20 magnification (Squares show magnified regions in the lower panels. Lower panels, ×40 magnification. The tumors showed heterogeneous cell components containing primitive neuroepithelial components and embryonal carcinoma components. (**C**) HE staining and immunostaining of the tumors at 4 weeks with antibodies against b-III-tubulin, ectoderm marker; AFP, endoderm marker (×20 magnification). Hematoxylin-eosin staining, ×10 magnification. The mixed staining pattern of ectodermal and endodermal markers was observed in hyperchromatic regions of *Parg*$^{-/-}$ tumors at 4 weeks. (**D**) Immunostaining of the tumors at 4 weeks after injection with antibody against anti-PAR. Right panels in D are magnified images, Bars, 50 mm (left panels in D), 20 mm (right panels in D). PAR staining was observed occasionally in the cell nuclei in the *Parg*$^{-/-}$ tumor but not in the *Parg*$^{+/+}$ tumor.

We previously observed that tumors derived from *Parp1*$^{-/-}$ ES cells showed differentiation into trophoblast lineages, including trophoblast giant cells [23]. Microscopic findings from the tumors derived from *Parg*$^{-/-}$ ES cells showed no such components, suggesting that in the hypomorphic *Parg* deficient state, marked differentiation alterations did not occur (Table 1).

2.3. Time Course Analysis of Tumorigenesis

To evaluate the defect in early stage tumorigenesis under *Parg* deficiency, further histological analyses were performed on sections of tumor tissues (Figure 2A,B). At one and two weeks after injection, tumors derived from *Parg*$^{-/-}$ ES cells showed a higher tendency of necrosis. The density of tumor cells and stromal cells appeared to be lower in the *Parg*$^{-/-}$ tumors. As shown in Figure 2B, comparison of percentage of hematoxylin-positive regions in the tumors at four weeks (Figure S1) showed the augmented hematoxylin-positivity, namely hyperchromatic areas (typical areas are shown as Figure 2B), in *Parg*$^{-/-}$ tumors with a statistical significance. It may suggest that the chromatin density of the cells was higher, possibly reflecting differences in the chromatin state or cell properties.

To characterize the properties of differentiated cells and hyperchromatic components further, we performed the immunostaining analysis for the tumors at four weeks with antibodies against beta-III-tubulin, ectoderm marker; AFP, endoderm marker; TRA-1-60, pluripotent marker, and Brachyury, mesoderm marker. As shown in Figure 2C, immunohistochemical analysis showed beta-III tubulin-positive staining of immature neuroepithelial tissues in both wild-type and *Parg*$^{-/-}$ tumors. It is, therefore, implied that hyperchromic regions may consist of both ectodermal and endodermal differentiated tissues. The pluripotent marker TRA-1-60 showed higher tendencies of diffused staining in the stromal components of the *Parg*$^{-/-}$ tumor compared with the wild-type tumor. For the mesoderm marker Brachyury, a higher tendency of staining was also observed in the cell components of *Parg*$^{-/-}$ tumor. On the other hand, AFP-positive staining patterns of teratocarcinoma and immature glandular components were similar between *Parg*$^{-/-}$ and wild-type tumors.

We also analyzed whether PAR accumulation occurs in *Parg*$^{-/-}$ tumors, due to hypomorphic Parg activity. As presented in Figure 2D panels, PAR staining was observed occasionally in the cell nuclei of the *Parg*$^{-/-}$ tumors but not in the *Parg*$^{+/+}$ tumor. This elevated level of PAR confirmed the defect of Parg activity.

2.4. Augmented Anti-Tumor Therapeutic Effects under Parg Deficiency

Previous reports describe that *Parg* $_{110}$ deficient animals are more sensitive to MMS treatment and γ-irradiation compared to wild-type mice [18,19]. Hypomorphic *Parg*$^{-/-}$ ES cells and particular human cancer cells with *PARG* knocked down also showed increased sensitivity to DNA-damaging agents, such as alkylating agents and γ-irradiation. Therefore, we first treated ES cells in vitro with MMS and γ-irradiation, and three hours later, the cells were subcutaneously injected to nude mice to observe the tumor growth. As shown in Figure 3A,B, reflecting the in vitro higher sensitivity of *Parg*$^{-/-}$ ES cells, the tumor growth is delayed for at least 2–3 weeks after treatment with both MMS and 6 Gy γ-irradiation.

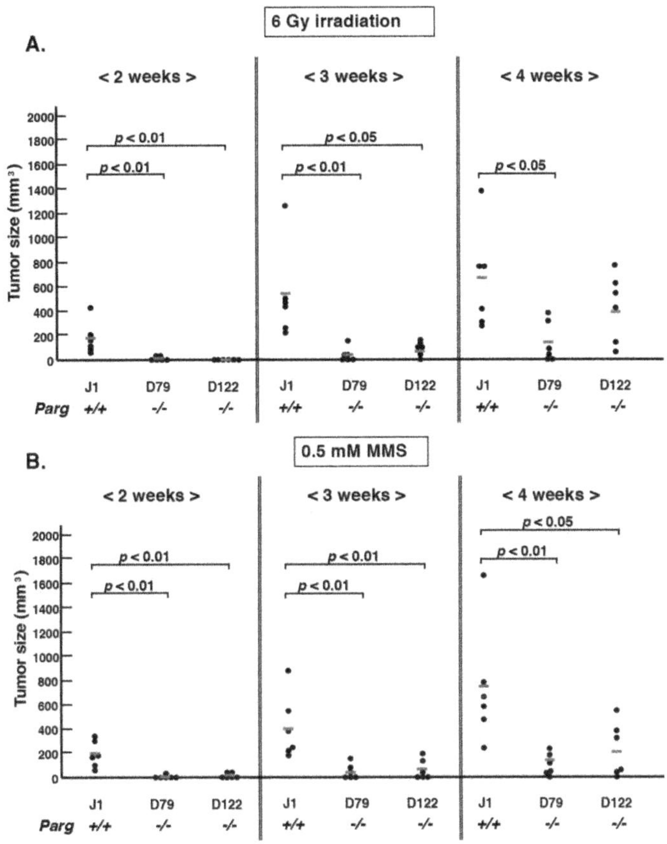

Figure 3. Effect of *Parg* deficiency on tumorigenesis from ES cells pretreated with MMS or γ–irradiation before injection into nude mice. ES cells treated with γ–irradiation at 6 Gy (**A**) or 0.5 mM MMS (**B**) and cultured for 3 h, respectively, were inoculated *s.c.* into nude mice. Tumor growth was observed for four weeks. The bars show the mean size.

Next, we carried out therapeutic models of local X-irradiation using tumors derived from wild-type and $Parg^{-/-}$ ES cells. As shown in Figure 4A–D, when the tumors derived from $Parg^{-/-}$ ES cells D79 and D122 were X-irradiated at the single dose of 7 Gy, they showed a delay in tumor growth. In contrast, when the tumors derived from wild-type J1 ES cells were X-irradiated, delay in tumor growth was not clearly observed. It is thus suggested that the therapeutic efficacy of X-irradiation could be higher in the tumors harboring *Parg* deficiency.

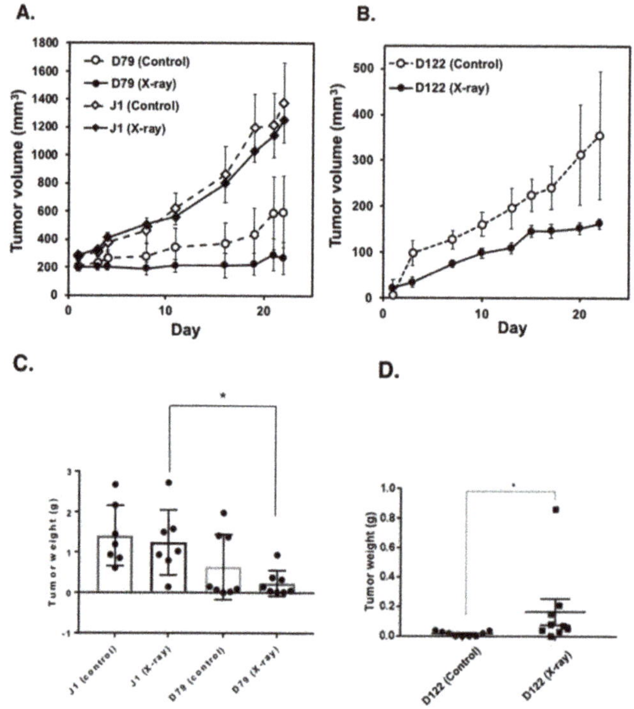

Figure 4. Effect of *Parg* deficiency on therapeutic efficacy of X–irradiation on tumors derived from ES cells. Tumors of one of the hinder leg pair were subjected to local X–irradiation with a single dose at 7 Gy ((**A**,**C**): $n = 7$ for wild-type (J1), $n = 8$ for $Parg^{-/-}$ (D79). (**B**,**D**): $n = 9$ for wild-type (J1) and $Parg^{-/-}$ (D122)). Tumor growth of non-irradiation controls (Control) was monitored for non-irradiated side of hinder leg pair. Tumor growth was observed for 22 days. (**A**,**B**) Tumor growth curve, Mean ± S. E. (**C**,**D**) Measured tumor weight 22 days after injection of ES cells, Mean ± S. E. (**C**) Statistical analysis was carried out with Turkey's test. * $p < 0.05$. (**D**) Statistical analysis was carried out with Mann–Whitney *U*-test. * $p < 0.01$.

These results suggest that Parg inhibition in combination with DNA damaging agents may efficiently control tumor growth in particular types of germ cell tumors.

3. Discussion

Our findings demonstrate that *Parg* deficiency delays the early onset of cancer in vivo in teratocarcinoma model. This suggests that PARG might be an attractive therapeutic target in cancer control of germ cell tumors. The therapeutic model used in this study showed that X-irradiation is more effective in tumors with *Parg* deficiency compared to wild-type cells.

ES cells are derived from normal cells and their tumor models are phenotypically close to teratocarcinoma and teratoma. It is reported that tumorigenicity of ES cells is driven by oncogene action of *E-Ras* gene [24], and the genes including PI(3)K [24] and *c-Jun* [25], *Cox1/2* [26] and *Oct3/4* [27] are involved in the tumorigenesis process. The present study implies that *Parg* could be a candidate target for the suppression of tumorigenesis at early stages.

In the subcutaneous injection model of ES cells used herein, ES cells should be grown in spheroids or attached to extracellular matrix in subcutaneous regions. During the stressed period of early tumorigenesis, the presence of *Parg* was suggested to enhance cell survival. In cultured $Parg^{-/-}$ ES cells, the PAR degradation activity was decreased to 10% of wild-type ES cells [7]. We observed that

PAR staining was observed occasionally in the cell nuclei of the *Parg*$^{-/-}$ tumors but not in the *Parg*$^{+/+}$ tumors. This accumulation of PAR confirmed the defect in *Parg* activity. The infrequent accumulation of PAR may explain the advance of tumor growth in *Parg*$^{-/-}$ tumors at four weeks, as efficient PAR degradation may be necessary for proliferation. The pluripotency marker staining of *Parg*$^{-/-}$ tumors at four weeks was also observed in a scattered manner, suggesting that the presence of a majority of differentiated tissues could be related to the infrequent accumulation of PAR in *Parg*$^{-/-}$ tumors, as differentiated cells are reported to have low levels of PAR compared with proliferating cells [1]. The occasional increase in PAR staining in *Parg*$^{-}$/$^{-}$ tumors could be possibly due to the presence of S-phase cells with stalled replication forks [28], leading to a lower proliferation activity. Another possibility could be an increased expression of other PAR degradation enzymes, such as ARH3 in the differentiated tumors. The clarification of the relationship between PAR accumulation and cell proliferation needs to be addressed in further studies.

When we added further stress with alkylating agents and γ-irradiation to ES cells (Figure 3A,B), tumor development was further suppressed under *Parg* deficiency. This was consistent with our previous observation that when *Parg* was inhibited in ES cells, caspase-dependent apoptosis was enhanced with S-phase arrest and PAR accumulation after MMS treatment [29].

In diseases, PARP-1 has been shown to be involved in stroke, ischemia diabetes and other inflammatory diseases [30]. PARP-1 is frequently overexpressed in various types of cancers [31–33]. As PARG has an opposing enzymatic activity to PARP-1, it is reasonable to hypothesize a similar involvement in disease categories. The involvement of PARG in cancer development has not been extensively studied. However, recent genome sequencing approach accumulated the information of *PARG* gene mutations in various types of cancer. As shown in Figure 5, the data from TCGA database (A) and CanSAR database (B) showed that *PARG* gene mutations are present in particular types of cancers and of note, deep deletions could be observed in non-seminomatous germ cell tumors, melanoma, and well-differentiated thyroid cancers, possibly indicating PARG-deficient state. On the other hand, esophageal squamous cell carcinoma, cholangiocarcinoma, bladder urothelial carcinoma, thymic epithelial tumors and diffuse glioma show amplification of *PARG* gene, possibly suggesting the overexpression of PARG. In fact, PARG activity is reported to be high in C6 glioma tumor cells [34]. With further clinical and basic studies, PARG-deficiency and PARG-overexpression may become useful biomarkers for therapeutic selection and monitoring.

Functional inhibition of PARG in pancreatic cancer cells, which has p53 pathway inactivation, enhanced the necrotic cell death after MMS treatment [29], suggesting that PARG could be a therapeutic target in certain types of cancer cells. It is reported that ovarian cancer cells show differential sensitivity to PARG and PARP inhibitors, and cells with replication vulnerability show persistent replication fork stalling and replication catastrophe, with treatment of the PARG inhibitor and sensitization to CHK1 inhibitor [35].

Whilst the hypomorphic ES cells used in this study show PARG activity of residual 10%, the cell growth was not reduced compared to wild-type ES cells [29], suggesting that ES cells may not be replication vulnerable cells. *Parg* deficient ES cells also showed a higher sensitivity to cisplatin, but not to topoisomerase I inhibitor, campthothecan and hydrogen peroxide [36].

PARG is involved not only in the DNA repair pathway but also in various cell physiological processes, including epigenetic regulation [8], microRNA regulation and RNA splicing. Pancreatic ductal adenocarcinoma cells expressing pro-oncogenic mRNA stability factor HuR (ELAVL1) show sensitization to oxaliplatin and 5-fluorouracil through persistent PARylation [37]. Silencing of PARG by shRNA decreased the proliferation rate twofold over wild type LoVo colon cancer cells [38]. PARG function was also suggested to be involved in MAP kinase cascade regulation, as implicated by the synthetic lethal effect under *PARG* and *DUSP22* double dysfunction [39]. The mechanism for delayed tumor formation process under *Parg* deficiency in ES cells should be further analyzed from multiple physiological aspects.

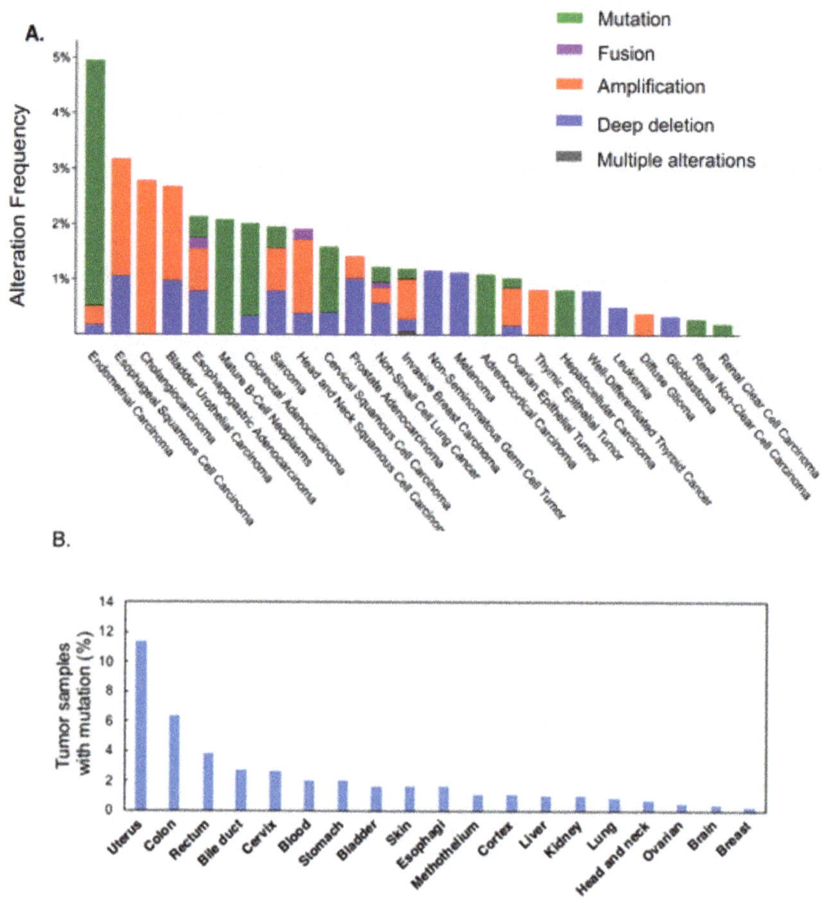

Figure 5. *PARG* mutation frequencies in human cancers. (**A**) The percentage of nonsynonymous mutation of *PARG* in each cancer in the TCGA database [40]. (**B**) The percentage of nonsynonymous substitution mutation of *PARG* in each cancer (total 165 mutations were listed in tumor samples from 8018 patients.) in the CanSAR database [41].

4. Materials and Methods

4.1. Cell Culture

The cell lines used were cultured at 37 °C in a humidified incubator, where CO_2 levels were kept at 5%. ES cells were cultured in Dulbecco's modified Eagle's medium (Life Technologies Corp., Carlsbad, CA, USA) supplemented with 20% fetal bovine serum, non-essential amino acids (Life Technologies Corp.), 55 μM β-mercaptoethanol, 0.3 mM each of adenosine, guanosine, and thymidine, 0.1 mM uridine, and 1×10^3 U/mL mouse LIF (Chemicon International Inc., Temecula, CA, USA) on gelatin-coated dishes (Asahi Glass Co. Ltd., Tokyo, Japan). Mouse ES cell J1 established from 129Sv/J mouse was provided by Dr. Ochiya of National Cancer Center Research Institute [42]. Hypomorphic two *Parg*$^{-/-}$ ES cell clones, D122 and D79, which we previously established [7], were used in this study. MMS (Sigma-Aldrich Co., St. Louis, MO, USA) was prepared in saline before use and sterilized by filtration.

4.2. Tumorigenesis Analysis

ES cells (J1, D122 and D79) cultured in the absence of a STO feeder layer [42] were trypsinized and 1×10^7 cells (Figures 1 and 3) or 5×10^6 cells (Figure 4) were subcutaneously injected into both legs of male BALB/c-nu/nu mice (CLEA Japan, Tokyo). The growth of tumors was measured continuously every 3–4 days. Four weeks after injection of ES cells, mice were euthanized and each tumor was histologically analyzed. The tumor volumes were calculated with the following formula: (smallest diameter)^2 × (largest diameter)/2. For the time course experiment, mice were euthanized at 7 and 14 days after subcutaneously injection of ES cell. Local X-ray irradiation of the tumors on hind legs were carried out with an X-ray machine (CP-160, 160-kVp, Faxitron X-Ray Corp., Tucson, AZ, USA) using radiation shield of lead. Tumors of one of the hind leg pair were subjected to local X–irradiation with a single dose at 7 Gy (Figure 4, A and C: $n = 7$ for wild-type (J1), $n = 8$ for *Parg*$^{-/-}$ (D79). B and D: $n = 9$ for wild-type (J1) and *Parg*$^{-/-}$ (D122)). Tumor growth of non-irradiation controls (Control) was monitored for non-irradiated side of hind leg pair. All animal experiments were approved by the Institutional Animal Experiment Committee of National Cancer Center Research Institute (T05-053-C01, A377-16, T17-053). All animal works were conducted according to relevant national and international guidelines for animal welfare.

4.3. Histological Analysis

After resection of the tumors, they were fixed in neutralized 10% formalin solution and embedded in paraffin blocks using standard procedures. Paraffin sections were stained with hematoxylin-eosin (HE), and histopathological analysis was performed under a light microscopic observation. Tissue sections mounted on slides were also subjected to immunostaining after deparaffinization and rehydration, and antigen retrieval, according to the manufacturer's protocol. Antibodies used in this study were anti-beta-III-tubulin (Abcam, ab264113, Cambridge, UK), anti-AFP (α-fetoprotein, Pierce, PA5-21004, Shreveport, LA, USA), anti-TRA-1-60 (Santa Cruz, sc-21705, Dallas, TX, USA), Brachyury (Santa Cruz, sc-374321, Dallas, TX, USA) and anti-PAR (BD Pharmingen, Franklin Lakes, NJ, USA). After several washes with PBS, bound antibodies were visualized using 3, 3′-diaminobenzidine according to the manufacturer's protocol. The sections were counterstained with hematoxylin and mounted.

4.4. Statistical Analysis

Statistical analysis was carried out by the One-way ANOVA test, Mann–Whitney *U* tests, Turkey's test and Student *t*-tests using either SPSS Statistics of Macintosh version (IBM corporation, Armonk, NY, USA), JMP (SAS Institute Inc. Cary, NC, USA) or GraphPad Prizm7 (GraphPad Software Inc., San Diego, CA, USA).

5. Conclusions

In conclusion, our results demonstrate that *Parg* deficiency delayed the onset of tumor formation of ES cells and augmented anti-tumor therapeutic effects of DNA-damaging agents, including alkylating agents and x- ray irradiation. These results suggest that the inhibition of PARG is likely to be an option for the therapeutic and prophylactic target of cancer control. Some specific inhibitors for PARG have been reported [43–45], although clinical studies for therapeutic agents are yet to take place. PARG inhibitors, in combination with DNA-damaging agents, may efficiently suppress tumor growth in particular types of germ cell tumors.

Supplementary Materials: The following are available online at http://www.mdpi.com/2072-6694/12/4/1056/s1, Figure S1: Hematoxylin-eosin staining of tumor tissues derived from mouse ES cells 4 weeks after injection.

Author Contributions: Conceptualization, M.M.; Methodology, A.G., H.S., Y.S. (Yuka Sasaki), T.O., S.I., M.M.; Validation, A.G., H.S., Yuka Sasaki, A.-M.R., T.A., T.O., S.I., M.W., M.M.; Investigation, Y.S. (Yuki Sonoda), Y.S. (Yuka Sasaki), A.G., H.S., T.A., T.O., S.I., M.W., M.M.; Resources, J.I., K.A., K.N., M.M.; Data curation, Y.S. (Yuki Sonoda), Y.S. (Yuka Sasaki), A.G., H.S., A.-M.R., T.A., S.I., M.M.; Writing—Original draft preparation, Y.S. (Yuki Sonoda), Y.S. (Yuka Sasaki), T.A., T.O., S.I., M.M.; Writing—Review and editing, T.H., J.I., K.A., K.N.;

Visualization, Y.S. (Yuki Sonoda), Y.S. (Yuka Sasaki), M.M.; Supervision, M.M., T.H., K.N.; Project administration, M.M.; Funding acquisition, K.A., Y.S. (Yuka Sasaki), M.M. All authors have read and agreed to the published version of the manuscript.

Funding: This research was funded in part by a Grant-in-Aid for the Second Term Comprehensive 10-Year Strategy for Cancer Control, a Grant-in-Aid for Cancer Research from the Ministry of Health, Labor and Welfare of Japan, a Grant-in-Aid for Scientific Research on Priority Areas from the Ministry of Education, Science, Sports, and Culture of Japan (15025274) and Grant-in-Aids from Foundation for Promotion of Cancer Research and Taiju Life Social Welfare Foundation.

Acknowledgments: We are grateful to Toshio Imai, Tadashige Nozaki, Shozo Takayama, M. Yanagihara, Hitoshi Nakagama, and Takashi Sugimura for valuable suggestions. We thank for technical assistance by Hideki Ogino and Masako Ichishi.

Conflicts of Interest: The authors declare no conflict of interest.

References

1. Schreiber, V.; Dantzer, F.; Ame, J.C.; de Murcia, G. Poly(ADP-ribose): Novel functions for an old molecule. *Nat. Rev. Mol. Cell Biol.* **2006**, *7*, 517–528. [CrossRef] [PubMed]
2. Oka, S.; Kato, J.; Moss, J. Identification and characterization of a mammalian 39-kDa poly(ADP-ribose) glycohydrolase. *J. Biol. Chem.* **2006**, *281*, 705–713. [CrossRef] [PubMed]
3. Ame, J.C.; Jacobson, E.L.; Jacobson, M.K. Molecular heterogeneity and regulation of poly(ADP-ribose) glycohydrolase. *Mol. Cell Biochem.* **1999**, *193*, 75–81. [CrossRef] [PubMed]
4. Miwa, M.; Sugimura, T. Splitting of the ribose-ribose linkage of poly(adenosine diphosphate-robose) by a calf thymus extract. *J. Biol. Chem.* **1971**, *246*, 6362–6364.
5. Wei, L.; Nakajima, S.; Hsieh, C.L.; Kanno, S.; Masutani, M.; Levine, A.S.; Yasui, A.; Lan, L. Damage response of XRCC1 at sites of DNA single strand breaks is regulated by phosphorylation and ubiquitylation after degradation of poly(ADP-ribose). *J. Cell Sci.* **2013**, *126*, 4414–4423. [CrossRef]
6. Nakadate, Y.; Kodera, Y.; Kitamura, Y.; Tachibana, T.; Tamura, T.; Koizumi, F. Silencing of poly(ADP-ribose) glycohydrolase sensitizes lung cancer cells to radiation through the abrogation of DNA damage checkpoint. *Biochem. Biophys. Res. Commun.* **2013**, *441*, 793–798. [CrossRef]
7. Shirai, H.; Fujimori, H.; Gunji, A.; Maeda, D.; Hirai, T.; Poetsch, A.R.; Harada, H.; Yoshida, T.; Sasai, K.; Okayasu, R.; et al. Parg deficiency confers radio-sensitization through enhanced cell death in mouse ES cells exposed to various forms of ionizing radiation. *Biochem. Biophys. Res. Commun.* **2013**, *435*, 100–106. [CrossRef]
8. Zampieri, M.; Passananti, C.; Calabrese, R.; Perilli, M.; Corbi, N.; De Cave, F.; Guastafierro, T.; Bacalini, M.G.; Reale, A.; Amicosante, G.; et al. Parp1 Localizes within the Dnmt1 Promoter and Protects Its Unmethylated State by Its Enzymatic Activity. *PLoS ONE* **2009**, *4*, e4717. [CrossRef]
9. Koh, D.W.; Dawson, V.L.; Dawson, T.M. The road to survival goes through PARG. *Cell Cycle.* **2005**, *4*, 397–399. [CrossRef]
10. Chen, L.; Gunji, A.; Uemura, A.; Fujihara, H.; Nakamoto, K.; Onodera, T.; Sasaki, Y.; Imamichi, S.; Isumi, M.; Nozaki, T.; et al. Development of renal failure in PargParp-1 null and Timm23 hypomorphic mice. *Biochem. Pharmacol.* **2019**, *167*, 116–124. [CrossRef]
11. Slade, D.; Dunstan, M.S.; Barkauskaite, E.; Weston, R.; Lafite, P.; Dixon, N.; Ahel, M.; Leys, D.; Ahel, I. The structure and catalytic mechanism of a poly(ADP-ribose) glycohydrolase. *Nature* **2011**, *477*, 616–620. [CrossRef] [PubMed]
12. Tucker, J.A.; Bennett, N.; Brassington, C.; Durant, S.T.; Hassall, G.; Holdgate, G.; McAlister, M.; Nissink, J.W.; Truman, C.; Watson, M. Structures of the human poly (ADP-ribose) glycohydrolase catalytic domain confirm catalytic mechanism and explain inhibition by ADP-HPD derivatives. *PLoS ONE* **2012**, *7*, e50889. [CrossRef] [PubMed]
13. Wang, Z.; Gagne, J.P.; Poirier, G.G.; Xu, W. Crystallographic and biochemical analysis of the mouse poly(ADP-ribose) glycohydrolase. *PLoS ONE* **2014**, *9*, e86010. [CrossRef] [PubMed]
14. Poitras, M.F.; Koh, D.W.; Yu, S.W.; Andrabi, S.A.; Mandir, A.S.; Poirier, G.G.; Dawson, V.L.; Dawson, T.M. Spatial and functional relationship between poly(ADP-ribose) polymerase-1 and poly(ADP-ribose) glycohydrolase in the brain. *Neuroscience* **2007**, *148*, 198–211. [CrossRef]

15. Meyer-Ficca, M.L.; Meyer, R.G.; Coyle, D.L.; Jacobson, E.L.; Jacobson, M.K. Human poly(ADP-ribose) glycohydrolase is expressed in alternative splice variants yielding isoforms that localize to different cell compartments. *Exp. Cell Res.* **2004**, *297*, 521–532. [CrossRef]
16. Burns, D.M.; Ying, W.; Kauppinen, T.M.; Zhu, K.; Swanson, R.A. Selective Down-Regulation of Nuclear Poly(ADP-Ribose) Glycohydrolase. *PLoS ONE* **2009**, *4*, e4896. [CrossRef]
17. Whatcott, C.J.; Meyer-Ficca, M.L.; Meyer, R.G.; Jacobson, M.K. A specific isoform of poly(ADP-ribose) glycohydrolase is targeted to the mitochondrial matrix by a N-terminal mitochondrial targeting sequence. *Exp. Cell Res.* **2009**, *315*, 3477–3485. [CrossRef]
18. Cortes, U.; Tong, W.M.; Coyle, D.L.; Meyer-Ficca, M.L.; Meyer, R.G.; Petrilli, V.; Herceg, Z.; Jacobson, E.L.; Jacobson, M.K.; Wang, Z.Q. Depletion of the 110-kilodalton isoform of poly(ADP-ribose) glycohydrolase increases sensitivity to genotoxic and endotoxic stress in mice. *Mol. Cell Biol.* **2004**, *24*, 7163–7178. [CrossRef]
19. Koh, D.W.; Lawler, A.M.; Poitras, M.F.; Sasaki, M.; Wattler, S.; Nehls, M.C.; Stoger, T.; Poirier, G.G.; Dawson, V.L.; Dawson, T.M. Failure to degrade poly(ADP-ribose) causes increased sensitivity to cytotoxicity and early embryonic lethality. *Proc. Natl. Acad. Sci. USA* **2004**, *101*, 17699–17704. [CrossRef]
20. Yu, S.W.; Wang, H.; Poitras, M.F.; Coombs, C.; Bowers, W.J.; Federoff, H.J.; Poirier, G.G.; Dawson, T.M.; Dawson, V.L. Mediation of poly(ADP-ribose) polymerase-1-dependent cell death by apoptosis-inducing factor. *Science* **2002**, *297*, 259–263. [CrossRef]
21. Andrabi, S.A.; Kim, N.S.; Yu, S.W.; Wang, H.; Koh, D.W.; Sasaki, M.; Klaus, J.A.; Otsuka, T.; Zhang, Z.; Koehler, R.C.; et al. Poly(ADP-ribose) (PAR) polymer is a death signal. *Proc. Natl. Acad. Sci. USA* **2006**, *103*, 18308–18313. [CrossRef] [PubMed]
22. Dai, W.; Fu, Y.; Deng, Y.; Zeng, Z.; Gu, P.; Liu, H.; Liu, J.; Xu, X.; Wu, D.; Luo, X.; et al. Regulation of Wnt Singaling Pathway by Poly (ADP-Ribose) Glycohydrolase (PARG) Silencing Suppresses Lung Cancer in Mice Induced by Benzo(a)pyrene Inhalation Exposure. *Front. Pharmacol.* **2019**, *10*, 338. [CrossRef] [PubMed]
23. Nozaki, T.; Masutani, M.; Watanabe, M.; Ochiya, T.; Hasegawa, F.; Nakagama, H.; Suzuki, H.; Sugimura, T. Syncytiotrophoblastic giant cells in teratocarcinoma-like tumors derived from Parp-disrupted mouse embryonic stem cells. *Proc. Natl. Acad. Sci. USA* **1999**, *96*, 13345–13350. [CrossRef] [PubMed]
24. Takahashi, K.; Mitsui, K.; Yamanaka, S. Role of ERas in promoting tumour-like properties in mouse embryonic stem cells. *Nature* **2003**, *423*, 541–545. [CrossRef]
25. Hilberg, F.; Wagner, E.F. Embryonic stem (ES) cells lacking functional c-jun: Consequences for growth and differentiation, AP-1 activity and tumorigenicity. *Oncogene* **1992**, *7*, 2371–2380.
26. Zhang, X.; Morham, S.G.; Langenbach, R.; Baggs, R.B.; Young, D.A. Lack of cyclooxygenase-2 inhibits growth of teratocarcinomas in mice. *Exp. Cell Res.* **2000**, *254*, 232–240. [CrossRef]
27. Gidekel, S.; Pizov, G.; Bergman, Y.; Pikarsky, E. Oct-3/4 is a dose-dependent oncogenic fate determinant. *Cancer Cell* **2003**, *4*, 361–370. [CrossRef]
28. Bryant, H.E.; Petermann, E.; Schultz, N.; Jemth, A.S.; Loseva, O.; Issaeva, N.; Johansson, F.; Fernandez, S.; McGlynn, P.; Helleday, T. PARP is activated at stalled forks to mediate Mre11-dependent replication restart and recombination. *EMBO J.* **2009**, *28*, 2601–2615. [CrossRef]
29. Shirai, H.; Poetsch, A.R.; Gunji, A.; Maeda, D.; Fujimori, H.; Fujihara, H.; Yoshida, T.; Ogino, H.; Masutani, M. PARG dysfunction enhances DNA double strand break formation in S-phase after alkylation DNA damage and augments different cell death pathways. *Cell Death Dis.* **2013**, *4*, e656. [CrossRef]
30. Virág, L.; Szabó, C. The Therapeutic Potential of Poly(ADP-Ribose) Polymerase Inhibitors. *Pharmacol. Rev.* **2002**, *54*, 375–429. [CrossRef]
31. Gonçalves, A.; Finetti, P.; Sabatier, R.; Gilabert, M.; Adelaide, J.; Borg, J.-P.; Chaffanet, M.; Viens, P.; Birnbaum, D.; Bertucci, F. Poly(ADP-ribose) polymerase-1 mRNA expression in human breast cancer: A meta-analysis. *Breast Cancer Res. Treat.* **2011**, *127*, 273–281. [CrossRef] [PubMed]
32. Nosho, K.; Yamamoto, H.; Mikami, M.; Taniguchi, H.; Takahashi, T.; Adachi, Y.; Imamura, A.; Imai, K.; Shinomura, Y. Overexpression of poly(ADP-ribose) polymerase-1 (PARP-1) in the early stage of colorectal carcinogenesis. *Eur. J. Cancer* **2006**, *42*, 2374–2381. [CrossRef] [PubMed]
33. Staibano, S.; Pepe, S.; Muzio, L.L.; Somma, P.; Mascolo, M.; Argenziano, G.; Scalvenzi, M.; Salvatore, G.; Fabbrocini, G.; Molea, G.; et al. Poly(adenosine diphosphate-ribose) polymerase 1 expression in malignant melanomas from photoexposed areas of the head and neck region. *Hum. Pathol.* **2005**, *36*, 724–731. [CrossRef] [PubMed]

34. Sevigny, M.B.; Silva, J.M.; Lan, W.C.; Alano, C.C.; Swanson, R.A. Expression and activity of poly(ADP-ribose) glycohydrolase in cultured astrocytes, neurons, and C6 glioma cells. *Brain Res. Mol. Brain Res.* **2003**, *117*, 213–220. [CrossRef]
35. Pillay, N.; Tighe, A.; Nelson, L.; Littler, S.; Coulson-Gilmer, C.; Bah, N.; Golder, A.; Bakker, B.; Spierings, D.C.J.; James, D.I.; et al. DNA Replication Vulnerabilities Render Ovarian Cancer Cells Sensitive to Poly(ADP-Ribose) Glycohydrolase Inhibitors. *Cancer Cell* **2019**, *35*, 519–533. [CrossRef]
36. Fujihara, H.; Ogino, H.; Maeda, D.; Shirai, H.; Nozaki, T.; Kamada, N.; Jishage, K.; Tanuma, S.; Takato, T.; Ochiya, T.; et al. Poly(ADP-ribose) Glycohydrolase deficiency sensitizes mouse ES cells to DNA damaging agents. *Curr. Cancer Drug Targets* **2009**, *9*, 953–962. [CrossRef]
37. Jain, A.; Agostini, L.C.; McCarthy, G.A.; Chand, S.N.; Ramirez, A.; Nevler, A.; Cozzitorto, J.; Schultz, C.W.; Lowder, C.Y.; Smith, K.M.; et al. Poly (ADP) Ribose Glycohydrolase Can Be Effectively Targeted in Pancreatic Cancer. *Cancer Res.* **2019**, *79*, 4491–4502. [CrossRef]
38. Fauzee, N.J.; Li, Q.; Wang, Y.L.; Pan, J. Silencing Poly (ADP-Ribose) Glycohydrolase (PARG) Expression Inhibits Growth of Human Colon Cancer Cells In Vitro via PI3K/Akt/NFkappa-B Pathway. *Pathol. Oncol. Res.* **2011**. [CrossRef]
39. Sasaki, Y.; Fujimori, H.; Hozumi, M.; Onodera, T.; Nozaki, T.; Murakami, Y.; Ashizawa, K.; Inoue, K.; Koizumi, F.; Masutani, M. Dysfunction of Poly (ADP-Ribose) Glycohydrolase Induces a Synthetic Lethal Effect in Dual Specificity Phosphatase 22-Deficient Lung Cancer Cells. *Cancer Res.* **2019**, *79*, 3851–3861. [CrossRef]
40. cBioPortal for Cancer Genomics. Available online: https://www.cbioportal.org (accessed on 30 December 2019).
41. CanSAR Database. Available online: https://cansarblack.icr.ac.uk/ (accessed on 29 December 2019).
42. Masutani, M.; Nozaki, T.; Nishiyama, E.; Ochiya, T.; Wakabayashi, K.; Suzuki, H.; and Sugimura, T. Establishment of poly(ADP-ribose) polymerase-deficient mouse embryonic stem cell lines. *Proc. Jpn. Acad.* **1998**, *74*, 233–236. [CrossRef]
43. Islam, R.; Koizumi, F.; Kodera, Y.; Inoue, K.; Okawara, T.; Masutani, M. Design and synthesis of phenolic hydrazide hydrazones as potent poly(ADP-ribose) glycohydrolase (PARG) inhibitors. *Bioorg. Med. Chem. Lett.* **2014**, *24*, 3802–3806. [CrossRef] [PubMed]
44. Gogola, E.; Duarte, A.A.; de Ruiter, J.R.; Wiegant, W.W.; Schmid, J.A.; de Bruijn, R.; James, D.I.; Llobet, S.G.; Vis, D.J.; Annunziato, S.; et al. Selective Loss of PARG Restores PARylation and Counteracts PARP Inhibitor-Mediated Synthetic Lethality. *Cancer Cell* **2019**, *35*, 950–952. [CrossRef] [PubMed]
45. Houl, J.H.; Ye, Z.; Brosey, C.A.; Balapiti-Modarage, L.P.F.; Namjoshi, S.; Bacolla, A.; Laverty, D.; Walker, B.L.; Pourfarjam, Y.; Warden, L.S.; et al. Selective small molecule PARG inhibitor causes replication fork stalling and cancer cell death. *Nat. Commun.* **2019**, *10*, 5654. [CrossRef] [PubMed]

cancers

MDPI

Article

Exploring the Synergy between PARP and CHK1 Inhibition in Matched *BRCA2* Mutant and Corrected Cells

Hannah L Smith [1], Lisa Prendergast [2] and Nicola J Curtin [1,*]

1 Newcastle Centre for Cancer, Translational and Clinical Research Institute, Faculty of Medical Sciences, Newcastle University, Newcastle upon Tyne NE2 4HH, UK; hannah.smith2@newcastle.ac.uk

2 Cancer Research UK Drug Discovery Unit, Newcastle Centre for Cancer, Translational and Clinical Research Institute, Faculty of Medical Sciences, Newcastle University, Newcastle upon Tyne NE2 4HH, UK; Lisa.Prendergast@newcastle.ac.uk

* Correspondence: nicola.curtin@newcastle.ac.uk

Received: 31 January 2020; Accepted: 31 March 2020; Published: 4 April 2020

Abstract: PARP inhibition results in the accumulation of DNA SSBs, causing replication stress (RS) and lesions that can only be resolved by homologous recombination repair (HRR). Defects in HRR, e.g., due to *BRCA2* mutation, confer profound sensitivity to PARP inhibitor (PARPi) cytotoxicity. In response to RS, CHK1 is activated to signal to S and G2/M cell cycle checkpoints and also to HRR. To determine the relative contribution of these two functions of CHK1 to survival following PARPi exposure, we investigated the effects of rucaparib (a PARPi) and PF-477736 (a CHK1 inhibitor) alone and in combination in cells with mutated and corrected *BRCA2*. The *BRCA2* mutated V-C8 cells were 1000× more sensitive to rucaparib cytotoxicity than their matched *BRCA2* corrected V-C8.B2 cells, but no more sensitive to PF-477736 despite having seven-fold higher levels of RS. PF-477736 caused a five-fold enhancement of rucaparib cytotoxicity in the V-C8.B2 cells, but no enhancement in the V-C8 cells. This differential sensitivity was not due to a difference in PARP1 or CHK1 expression or activity. PF-477736 increased rucaparib-induced RS (γH2AX foci) and completely inhibited RAD51 focus formation, indicating a profound suppression of HRR. Our data suggested that inhibition of HRR was the main mechanism of sensitisation to rucaparib, compounded with an inhibition of cell cycle checkpoints by PF-477736.

Keywords: PARP; ATR; CHK1; replication stress; homologous recombination DNA repair; cell cycle; cytotoxicity

1. Introduction

Replication stress (RS) is a key source of genomic instability, an enabling characteristic of cancer [1]. RS is increased in cancer due to an almost ubiquitous loss of G1 checkpoint control [2] and unrepaired DNA lesions encountering the replication fork. Poly(ADP-ribose) polymerase (PARP) is a key component of the DNA damage response by promoting DNA base excision repair/single-strand break repair (BER/SSBR) [3]. PARP is the first line of defence against the most common type of endogenous DNA damage, oxidative stress, a major contributor to RS. PARP inhibitors (PARPi) are a significant breakthrough in the treatment of cancer by exploiting cancer-specific defects in homologous recombination DNA repair (HRR), e.g., due to *BRCA* mutations. Toxicities associated with these drugs are generally mild [4]. Three PARPi are currently approved for the treatment of ovarian cancer, the success being largely due to the high frequency (>50%) of HRR defects in this cancer type [5–7].

The high levels of RS and loss of G1 control make cancer cells dependent on S and G2/M cell cycle checkpoint control [8]. Checkpoint kinase 1 (CHK1) is a pivotal checkpoint kinase signalling RS to cell cycle arrest through inactivation of cdc25A and cdc25C. Cdc25A and cdc25C are phosphatases that remove inactivating phosphates on CDK2 and CDK1, respectively. Since CDK2 is required for S-phase entry and progression and CDK1 is needed for mitosis, activation of CHK1 leads to S and G2/M arrest. CHK1 has also been shown to phosphorylate RAD51 and thus has key involvement in signalling to HRR, as well as halting the cell cycle to allow repair to occur [9,10].

CHK1 inhibitors have the potential to counteract HRR-mediated PARPi resistance [11]. Indeed, PARPi and CHK1 inhibitors have been shown to interact to cause increased cytotoxicity in breast and ovarian cancer cells, which was mediated by inhibition of HRR and increased DNA damage [12,13]. However, to date, no investigations have been carried out in paired HRR competent and HRR defective (HRD) cell lines to confirm this as the mechanism.

To better understand the mechanisms underlying the synergy between PARP and CHK1 inhibitors, we used paired *BRCA2* mutant (V-C8) and corrected (V-C8.B2) cells. We examined the effects of the clinically approved PARPi, rucaparib, and the CHK1 inhibitor, PF-477736, that has undergone clinical evaluation (NCT00437203) on target enzyme activity and inhibition, cell cycle control, DNA repair, and cytotoxicity. Our data suggest that CHK1 inhibition results primarily in an HRD phenotype, and this is synthetically lethal with PARP inhibition.

2. Results

2.1. *V-C8 Cells Are More Sensitive to Rucaparib, but not PF-477736, and PF-477736 only Sensitised V-C8.B2 Cells to Rucaparib*

Colony formation assays were used to determine the potency of rucaparib across V-C8 and V-C8.B2 cell lines. As expected, the HRD V-C8 cells were particularly sensitive to rucaparib (LC_{50} < 0.01 µM) and significantly more sensitive compared to matched HRR-competent V-C8.B2 cells (LC_{50} > 10 µM, $p < 0.001$) (Figure 1a). In contrast, no significant difference in cytotoxicity to PF-477736 (Figure 1b) was observed between the cell lines as both V-C8 and V-C8.B2 cells had similar LC_{50} (100.9 and 87.5 nM, respectively). This suggested that *BRCA2*/HRD status was not a determinant of sensitivity to CHK1 inhibitor PF-477736. Survival in V-C8 and V-C8.B2 cell was inhibited by less than 50% (44.0 ± 5.9 and 41.9 ± 3.5 %, respectively) at 50 nM PF-477736, so this concentration was used for further studies.

We next tested whether the CHK1 inhibitor could potentiate PARPi in HRR competent and defective cells. The survival of cells was evaluated when exposed to a range of rucaparib concentrations (V-C8, 0–0.3 µM, V-C8.B2, 0–30 µM, to account for increased sensitivity to rucaparib) with or without 50 nM PF-477736. In V-C8.B2 cells, co-incubation with PF-477736 decreased the LC_{50} of rucaparib 4.8-fold ± 2.7 (Figure 1c). PF-477736 did not sensitise HRD V-C8 cells to rucaparib (Figure 1d). This differential sensitisation of HRR functional and dysfunctional cells suggested PF-77736 was sensitising HRR competent cells by inhibiting HRR.

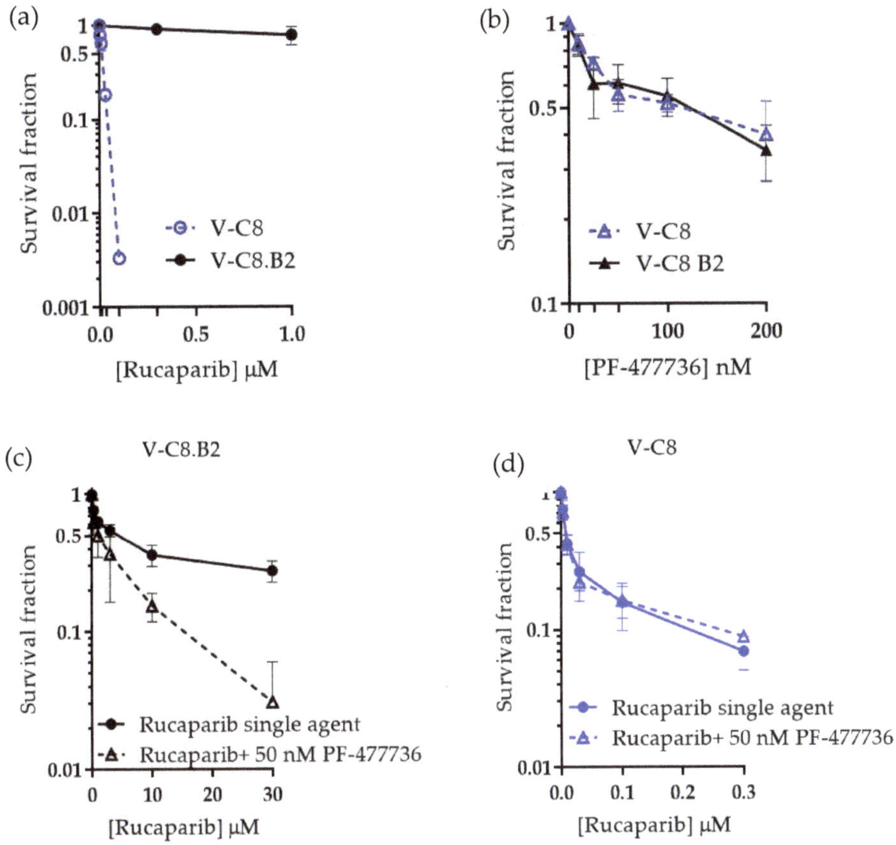

Figure 1. Differential cytotoxicity and synergy of rucaparib and PF-477736 in *BRCA2* mutant and corrected cells. V-C8 and V-C8.B2 cells were exposed to drugs at the indicated concentration for 24 h prior to replacement with drug-free medium for 7–10 days to allow colony formation. (**a**) Rucaparib, (**b**) PF-477736, (**c**) the combination of rucaparib with 50 nM PF-477736 in V-C8 B2 cells, and (**d**) the combination of rucaparib with 50 nM PF-477736 in V-C8 cells. Data are the mean and standard error of three independent experiments.

2.2. The Difference in Rucaparib Sensitivity Is not Due To Differential PARP-1 Expression between Cell Lines

To exclude differences in PARP1 expression, activity, or inhibition as factors contributing to the differential sensitisation to rucaparib by PF-477736, we measured PARP expression and activity in both cell lines. PARP1 levels were not substantially different between cell lines and were only modestly affected by rucaparib and PF-477736 (Figure 2a and Figure S1). Endogenous PAR levels were 2.3-fold higher in the HRD V-C8 cells compared to the V-C8.B2 cells (Figure 2b), possibly reflecting a higher level of endogenous DNA breakage activating PARP. There was no significant difference between the baseline PARP activity of both cell lines (Figure 2c). Rucaparib inhibited PARP activity to a similar extent in both V-C8 and V-C8.B2 cells with IC_{50} values of 59 and 53 nM, respectively (Figure 2d). In both cell lines, a concentration of 0.1 μM rucaparib inhibited PARP activity >95%. This concentration only killed 3% of V-C8.B2 cells compared with >99% of V-C8 cells, demonstrating that the differential cytotoxicity was determined by HRR status rather than the extent of PARP inhibition.

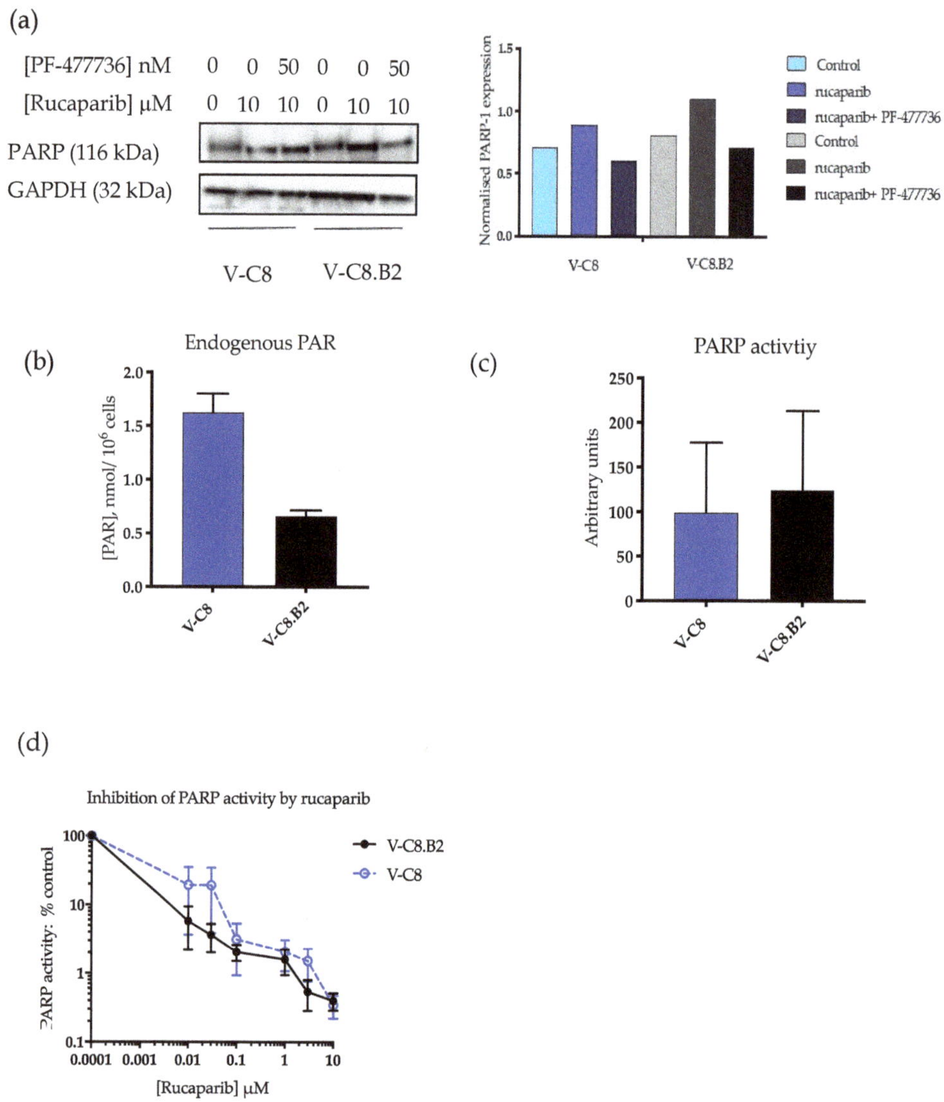

Figure 2. PARP-1 expression, activity, and inhibition by rucaparib in *BRCA2* mutant and corrected cells. (**a**) V-C8 and V-C8.B2 cells were exposed to drugs at the indicated concentration for 24 h prior to Western blot. Chemiluminescence was quantified using Syngene software. Bar charts are PARP-1 expression relative to the GAPDH loading control. Data are from a single representative experiment. (**b**) Endogenous PAR was measured in the absence of oligonucleotide and NAD+ by reference to a standard curve. Data are the mean and standard deviation of three independent experiments. (**c**) PARP activity (mean pixel values) was measured in the presence of oligonucleotide and NAD+. Data are the mean and standard deviation of three individual experiments. (**d**) PARP activity following exposure to increasing concentrations of rucaparib. Data are the mean and standard error of three individual experiments.

2.3. Rucaparib Activates CHK1 Similarly in Both Cell Lines, and This Activation Is Inhibited by PF-477736

Another possible explanation for the differential sensitisation by PF-477736 could have been a difference in the activation of CHK1 by rucaparib and a difference in the extent to which it was inhibited by PF-77736. The expression of CHK1 in the native and phosphorylated forms and downstream signalling to CDK1 were determined by Western blot (Figure 3a and Figure S2). CHK1 expression was slightly lower in V-C8 cells compared to V-C8.B2 cells, but this was not significant (Figure 3b). Rucaparib (10 μM) activated checkpoint signalling through ATR and CHK1 in both cell lines (Figure 3a). $CHK1^{S345}$ levels (indicating ATR activation) were increased 2.1-fold and 1.6-fold by rucaparib in V-C8 and V-C8.B2 cells, respectively. CHK1 activity ($CHK1^{S296}$) was increased by rucaparib to a similar extent in V-C8 and V-C8.B2 cells (1.5-fold and 1.6-fold, respectively) (Figure 3c). In both cell lines, PF-477736 (50 nM) inhibited this activation of CHK1, by 91.4% ± 12.2 in V-C8 cells and 75.6% ± 17.6 in V-C8.B2 cells. The difference in inhibition of $CHK1^{S296}$ by PF-477736 was not statistically significantly different between the two cell lines, suggesting this did not contribute towards the difference observed in sensitisation. Downstream of CHK1, PF-477736 similarly inhibited phosphorylation of CDK1 ($CDK1^{Y15}$) by 69.6% ± 24.8 in V-C8 cells and 56.88 % ± 26.1 in V-C8.B2 cells. Upstream of CHK1, PF-477736 activated ATR 2.1-fold ± 1.3 in V-C8 cells and 2.4-fold ± 1.2 in V-C8.B2 cells, suggesting an increased reliance on ATR when CHK1 was inhibited.

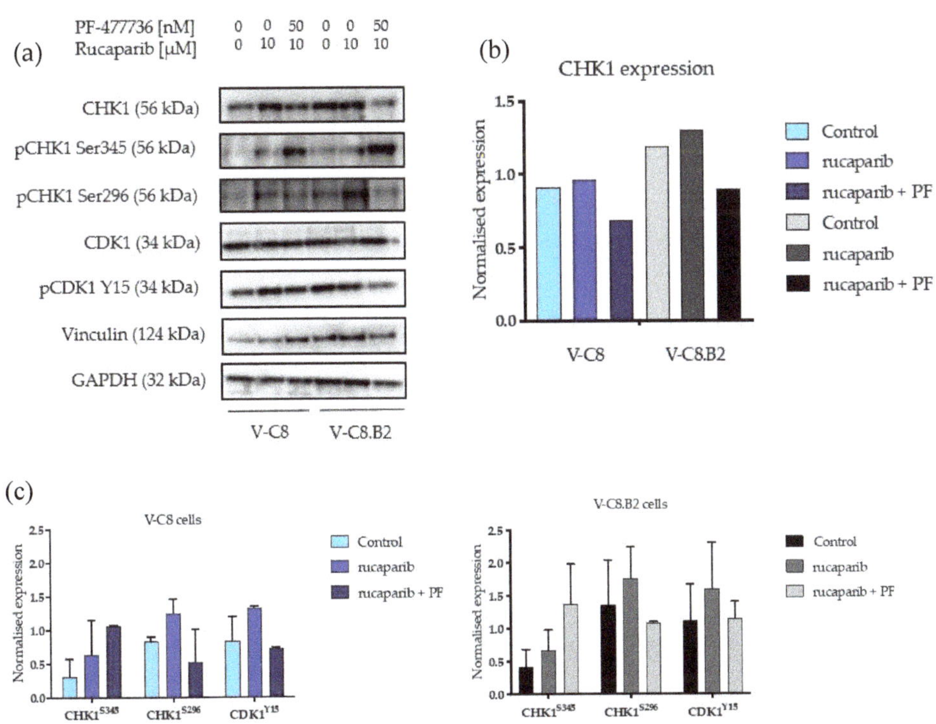

Figure 3. CHK1 expression, activation by rucaparib, and inhibition by PF-477736 in *BRCA2* mutant and corrected cells. Cells were exposed for 24 h to DMSO, rucaparib, or rucaparib and PF-477736 and analysed for $pCHK^{S345}$, CHK1, and $pCDK1^{Tyr15}$. Vinculin was used as a loading control for $pCHK^{S345}$, CHK1, $pCDK1^{Tyr15}$ and CDK1, and GAPDH was used as the loading control for $CHK1^{S296}$ (**a**) Western blot from a single representative experiment. (**b**) Densitometry was calculated using Syngene software of CHK1 expression normalised to Vinculin. (**c**) Normalised densitometry of phosphorylated CHK1 and CDK1. Data are the mean and standard deviation of three individual experiment.

2.4. *Rucaparib Causes S-phase and G2 Accumulation, Which Is Attenuated by PF-477736*

The effect of rucaparib and PF-477736 on the cell cycle was analysed to assess if the mechanism of sensitisation was by cell cycle checkpoint signalling. In V-C8.B2 cells, rucaparib caused a modest increase in S-phase (19.31%). Co-exposure to PF-477736 reduced the S-phase by 64.5% and G2/M by 33.6% compared to control cells, and this was accompanied by an increase in the sub-G1 fraction of 21.5%, suggesting that cells were progressing through S and G2/M with damaged DNA and dying, probably by mitotic catastrophe (Figure 4a).

2.5. *Rucaparib Inhibition Leads to Increased DNA Damage and HRR, and PF-477736 Inhibits HRR*

The induction of replication stress by rucaparib, resulting in collapsed replication forks following accumulation of endogenous DNA damage, is commonly measured by γH2AX phosphorylation. At baseline, untreated V-C8 cells had on average seven-fold more γH2AX foci/cell, compared to V-C8.B2 cells, reflecting their inability to resolve RS through HRR (Figure 4b). In V-C8.B2 cells, rucaparib increased γH2AX foci/cell 16-fold ($p < 0.0001$), PF-477736 caused a three-fold increase in γH2AX foci/cell ($p < 0.0001$), and the γH2AX foci/cell following the combination were higher still, but not significantly different from rucaparib alone.

In V-C8 cells, the already high levels of γH2AX foci were only increased by 20% by PF-477736 and 30% by rucaparib (Figure 4b, not significant). The combination increased the number of foci approximately three-fold ($p < 0.0001$). Due to the *BRCA2* mutation in these cells, they had very low levels of RAD51 foci, which were not increased by rucaparib or PF-477736 alone or in combination.

In marked contrast, in parallel with the induction of DNA breaks (γ-H2AX) in V-C8.B2 cells, rucaparib exposure caused an approximately 13-fold increase ($p < 0.0001$) in RAD51 foci in γ-H2AX-positive cells. Importantly, PF-477736 completely abrogated the rucaparib induction of RAD51 (significantly different from rucaparib alone, $p < 0.0001$), and the levels of foci were similar to control V-C8.B2 and the HRD V-C8 cells. These data indicated that the inhibition of CHK1 totally abolished HRR and led to a greater accumulation of collapsed replication forks.

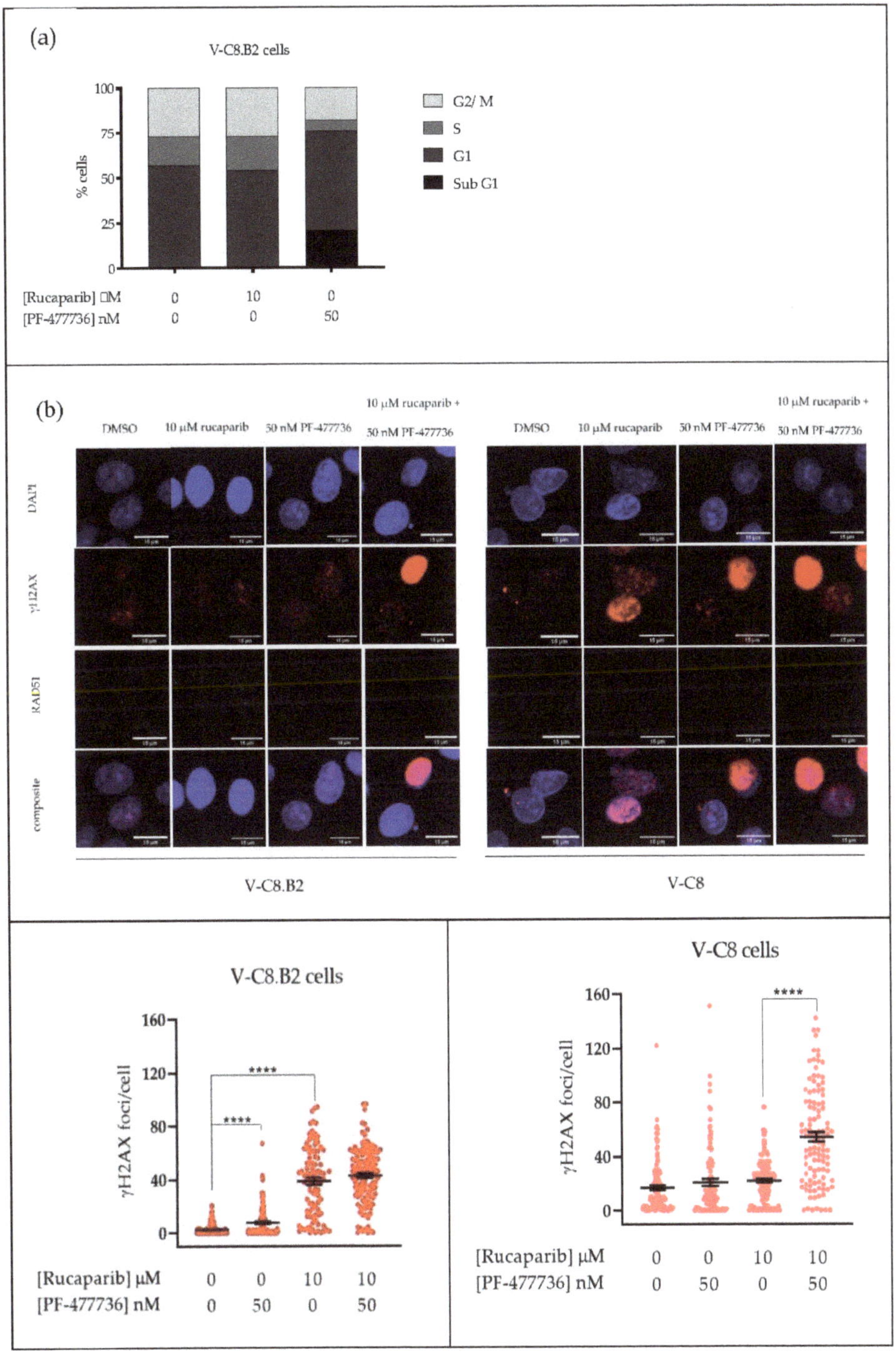

Figure 4. *Cont.*

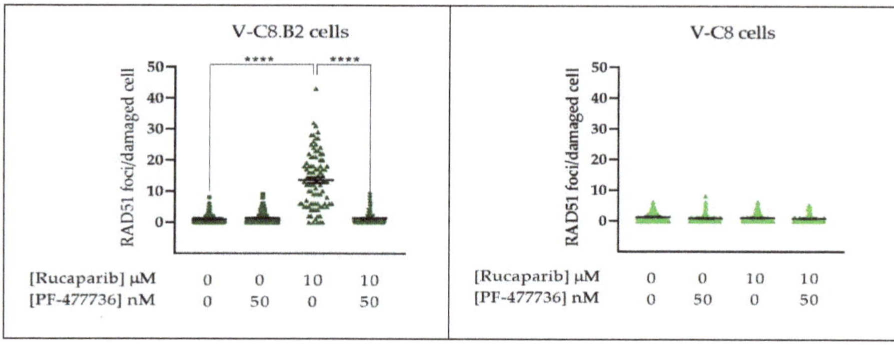

Figure 4. Impact of PF-477736 on cell cycle and homologous recombination repair (HRR) in rucaparib-treated cells. Cells were exposed for 24 h to DMSO, rucaparib, or rucaparib with PF-477736 at the indicated concentrations. (**a**) Cell cycle analysis of V-C8.B2 cells; data are from a single experiment. (**b**) Representative images are shown above quantified γH2AX and RAD51 foci, where each data point represents a single nucleus. RAD51 foci were only counted in cells with >mean γH2AX foci in untreated cells. Data are pooled from three independent experiments.

3. Discussion

Defects in either *BRCA1* or *BRCA2* have been known to result in a profound sensitivity to PARP inhibition since 2005 [14,15]. Rucaparib has previously been documented to be ≥1000× more cytotoxic to *BRCA2* mutant V-C8 cells compared to their matched *BRCA2* corrected V-C8.B2 cells (Patent WO/2005/012305), which we confirmed here. Interestingly, *BRCA2* was not a determinant of PF-477736 sensitivity. RS was thought to be a determinant of sensitivity to cell cycle checkpoint inhibitors, so this was somewhat surprising given that untreated V-C8 cell had seven-times more H2AX foci than V-C8.B2 cells. In previous studies, V-C8 cells were more sensitive to the ATR inhibitor VE-821 [16], possibly RS being a greater determinant of sensitivity to ATR than CHK1 inhibition.

Here, we confirmed that the observed difference in rucaparib sensitivity was not due to different PARP-1 expression levels or activity between the cell lines. This contrasts with previous reports of substantially higher levels of PARP activity in V-C8 compared to V-C8.B2 cells [17]. This may reflect differences in the method of detecting PARP activity. In our study, we used a GCLP-validated assay that has been used as a pharmacodynamic biomarker in PARPi clinical trials [18]. However, like this previous study, endogenous PAR levels were around two-fold higher in *BRCA2* mutant cells than wildtype, possibly indicating higher levels of endogenous DNA damage. In support of this hypothesis, there was approximately seven-times as much γH2AX foci formation in untreated *BRCA2* mutant V-C8 cells compared to *BRCA2* corrected V-C8.B2. The differential rucaparib sensitivity observed in cytotoxicity experiments was clearly due to *BRCA2* mutation and the synthetically lethal relationship between BER and HRR pathways. Therefore, it is interesting that PF-477736 only sensitised V-C8.B2 cells and not V-C8 cells to rucaparib, suggesting the CHK1 inhibitor was primarily acting on HRR.

We demonstrated that rucaparib induced RS to a higher extent in V-C8.B2 cells; however, this was most likely due to the high levels of RS in V-C8 control cells. CHK1 was activated to a similar extent by rucaparib in both V-C8 and V-C8.B2 cells, and PF-477736 caused a similar inhibition in both cell lines, so the differential sensitisation observed was not due to a greater activation or inhibition of the pathway. Rucaparib caused a similar increase in pCDK1^{Y15}, which was inhibited to a similar degree by PF-47736 in both cell lines, indicating that the downstream checkpoint signalling was intact and equally responsive to the two inhibitors in both cell lines. Therefore, this was not a factor in the differential sensitisation of VC8.B2 cells. Increased CHK^{S345} phosphorylation in both V-C8 and V-C8.B2 cells was also observed in response to CHK1 inhibition with PF-477736, suggesting upstream ATR activation as previously described [19,20].

Cell cycle analysis showed only a modest S-phase accumulation caused by rucaparib, probably associated with the increase in RS that was demonstrated by the increase in γH2AX (Figure 4). PF-477736 not only abolished this rucaparib-induced S-phase arrest, but also reduced both S- and G2-phases relative to untreated controls. This reduction in S and G2 was accompanied with a corresponding increase in the sub-G1 fraction, suggesting damaged cells were forced into mitosis before they could undertake repair, as has been reported for other PARP and CHK1 inhibitor combinations [21,22].

Although these data indicated that inhibition of cell cycle checkpoints by PF-477736 made a contribution to the cytotoxicity of rucaparib, this is unlikely to explain the differential sensitisation of the *BRCA2* corrected compared to the *BRCA2* mutant cells. We therefore tested the hypothesis that the primary mechanism of sensitisation was via PF-477736-mediated inhibition of HRR. As predicted, PF-477736 profoundly inhibited HRR, completely abolishing RAD51 foci formation. In a study of breast cancer cell lines, prexasertib synergised with the PARPi olaparib by causing S-phase arrest and inhibiting HRR/RAD51 foci [23]. In contrast, in high-grade serous *BRCA*-wildtype ovarian cancer cells, prexasertib synergised with olaparib by inhibiting G2/M arrest, as well as also inhibiting olaparib-induced RAD51 foci formation [13]. Clearly, the various molecular pathologies of these cells complicate the interpretation of the data.

4. Materials and Methods

4.1. Chemicals and Reagents

All chemicals and reagents used were obtained from Sigma-Aldrich, unless stated otherwise. Rucaparib was gifted from Pfizer Global R&D. Both rucaparib and PF-477736 (Selleckchem, Houston, TX, USA) were dissolved in dry DMSO at respective concentrations of 20 mM and 5 mM and stored at −80 °C before use.

4.2. Cell Culture

VC-8 and VC-8.B2 cells, a gift from Malgorzata Zdzienica, Leiden University [24], were maintained in exponential growth phase (<70–80% confluence) in DMEM with 10% foetal bovine serum (FBS; Thermo Fisher Scientific, Waltham, MA, USA) and incubated at 37 °C, 5% CO_2, and 95% humidity. VC8.B2 cells contained a BAC containing the *BRCA2* gene and maintained under selection with 200 µg/mL G-418 (Thermo Fisher Scientific, Waltham, MA, USA). Cells were mycoplasma free.

4.3. Cytotoxicity Assay

Exponentially growing cells were seeded at various densities in 6 well plates, allowing 3 different densities for each drug concentration estimated to give 20–200 colonies following drug treatment. After attachment, cells were exposed to various concentrations of rucaparib or PF-477736 alone or the combination of rucaparib ± 50 nM PF-477736 in DMSO or DMSO alone at a final concentration of 0.5% for 24 h. The medium was replaced with fresh medium and the cells left to incubate for 7–10 days to form colonies. The colonies were fixed in methanol:acetic acid (3:1) stained with 0.4% crystal violet, and colonies of >30 cells were counted by eye. Graphs were plotted using Graphpad Prism 6 software (San Diego, CA, USA).

4.4. Measurement of PARP Activity and Inhibition

Following incubation with rucaparib (0–10 µM) for 30 min, PARP activity was measured in permeabilised cells at 26 °C in the presence of 350 nM NAD+ and 10 mg/mL 12 mer palindromic oligonucleotide to activate the enzyme, as described previously [25,26]. Following transfer to a nitrocellulose membrane (GE Healthcare Life, Sciences, Amersham, Buckinghamshire, UK) the product, PAR, was measured with 10H anti-PAR monoclonal antibody (Enzo life sciences, Exeter, UK) overnight at 4 °C and HRP-conjugated goat anti-mouse secondary antibody (Dako, Santa Clara, CA, USA) diluted 1:1000 in PBS-Tween (PBS-T) 5%. Clarity Max ECL Western substrate (Bio-Rad, Hercules, CA,

USA) was added to the membrane, chemiluminescence imaged using the GBox and Genesys software (Syngene, Cambridge, UK), and quantified with reference to a standard curve of 0–25 pmol purified PAR (Enzo life sciences, Farmingdale, NY, USA) after subtraction of background reactions in the absence of NAD and oligonucleotide. Baseline PAR levels were measured using 50–100-times the cells used for inhibition experiments and with no oligonucleotide or NAD+ present. Purified PAR standard was diluted accordingly to a concentration of 0–25 pmol.

4.5. Measurement of CHK1 Activation and Inhibition by Western Blot

Exponentially growing cells in 100 mm dishes were exposed to media containing DMSO, 10 μM rucaparib, or 10 μM rucaparib and 50 nM PF-477736 for 24 h before extraction with 250 μL per dish of phosphosafe extraction reagent with protease cocktail inhibitor (Thermo Fisher Scientific, Waltham, MA, USA) at a 1:100 dilution per dish at 4 °C for 5–7 min, then scraped on ice into Eppendorf tubes. Following centrifugation at 8000× *g* for 10 min at 4 °C, the protein concentration of the supernatants was measured using a BCA protein assay kit (Thermo Fisher Scientific, Waltham, MA, USA) and diluted with deionised water to 0.5–1 mg/mL. XT sample buffer (Bio-Rad, Hercules, CA, USA) and XT reducing agent (Bio-Rad, Hercules, CA, USA) were added at constants of 25% and 0.5%, respectively. Samples were heated at 90 °C for 5 min, then 30 μg were loaded/well of 3–8% Criterion XT tris-acetate gels (Bio-Rad, Hercules, CA, USA), and the gel ran at 150V for approximately 1 h alongside HiMark pre-stained protein standard (Thermo Fisher Scientific, Waltham, MA, USA) in diluted 20× XT tricine with deionised water. The separated proteins were transferred onto a nitrocellulose membrane (Amersham, Buckinghamshire, United Kingdom) at 100 V for 1 h. The membrane was blocked for 1 h at room temperature in 5% milk in TBS-tween (TBST). Primary antibodies PARP-1 (#ab227244 Abcam, Cambridge, UK), vinculin (#4650), GAPDH (14C10 #2118), phospho-CHK1 (Ser296) (D3O9F #90178), phospho-Chk1 (Ser345) (#2341), and phospho-CDK1 (Tyr15) (10A11 #4539) were each diluted 1:1000 in TBS-Tween (TBS-T) and 5% bovine serum albumin (BSA) and left to incubate overnight at 4 °C (all obtained from Cell signalling, Massachusetts, United States except PARP-1). The membrane was cut into sections (approximately 30–45 kDa, 45–70 kDa, and 70–220 kDa), washed in TBS-T, before adding anti-rabbit goat polyclonal HRP secondary antibody (Cell Signaling Technology, Danvers, MA, USA) diluted 1:2000 in 5% skimmed milk in TBST at room temperature for 1 h. After washing with TBST, Clarity Max ECL Western substrate (Bio-Rad, Hercules, CA, USA) was applied to the membrane slices and chemiluminescence measured using Syngene software on the G-box.

4.6. Cell Cycle Analysis

Exponentially growing cells were exposed to DMSO, 10 μM rucaparib or 10 μM rucaparib with 50 nM PF-477736 for 24 h. Cells were washed twice with PBS, with each washing collected to ensure no loss of cells, before being trypsinised and harvested. Following centrifugation at 1500 rpm for 5 min, the supernatant was discarded, and the remaining cell pellet was resuspended in 1 mL ice cold PBS and centrifuged (3000 rpm, 5 min). The supernatant was removed, and 1 mL of 70% ethanol was added dropwise to the cell pellet. Samples remained at 4 °C for a minimum of 1 h. Prior to staining, cells were washed twice in PBS to remove ethanol before eventually being resuspended in 700 μL PBS. RNase was added to cells (12 μL of 1 mg/mL stock) alongside propidium iodide stain (20 μL of 1mg/mL stock). Cells were incubated in dark conditions at 37 °C for a minimum of 30 min. Analysis was carried out using Attune Nxt Cytometer. De Novo FCS Express 7 software was used to analyse the data.

4.7. Homologous Recombination Repair Assay

Cells were seeded at 0.5×10^5 cells/mL onto coverslips and incubated for 24 h before being exposed to either DMSO, 50 nM PF-477736, 10 μM rucaparib, or 10 μM rucaparib with 50 nM PF-477736 for 24 h. Cells were then washed with PBS and fixed with ice cold methanol for a minimum of 1 h at −20 °C. Coverslips were then washed with PBS 0.2% Triton-X-100 and blocked (2% BSA, 10% skimmed milk powder (Marvel, UK), 10% goat serum) for 1 h at room temperature before being incubated with anti-RAD51 (Abcam, Cambridge, UK) diluted 1:500 in blocking buffer at 4 °C overnight. After washing in PBS, 0.2% Triton X-100 anti-phospho-histone H2AX (Merck, Branchburg, NJ, USA) was added to coverslips at 1:1000 and incubated for 1 h. After washing, secondary antibodies, Alexa Fluor 488 and Alexa Fluor 546 (ThermoFisher Scientific, Waltham, MA, USA), were added and incubated for 1 h, in dark conditions. Coverslips were washed then exposed to 0.5 μg/mL DAPI solution. Coverslips were mounted onto microscope slides using H-1400 hard set mounting media (Vector Laboratories, Burlingame, CA, USA). Slides were stored in dark conditions and left to dry prior to imaging. Cells were imaged using the Leica DM6 LED fluorescence microscope (Leica microsystems GmbH, Wetzlar, Germany). Images were analysed using Fiji ImageJ software [27]. In DMSO-treated cells, the mean γH2AX foci were calculated. Cells (control and treated) with more than the mean number of γH2AX foci in control cells were deemed γH2AX positive. RAD51 foci were only counted in γH2AX positive cells.

5. Conclusions

On the basis of our data in matched cell lines that differed only in their BRCA2 status, we proposed the model shown in Figure 5. That is, the RS caused by the accumulation of unrepaired SSB when PARP was inhibited led to collapse of replication forks. These lesions triggered activation of the ATR-CHK1 pathway that signals to cell cycle arrest, but the major contribution that CHK1 makes was to protect cells from PARPi cytotoxicity by promoting HRR. The use of a CHK1 inhibitor may therefore be useful not only for tumours that are intrinsically PARPi resistant due to functional HRR, but may also overcome acquired resistance to PARPi in BRCA mutant tumours that have restored HRR function. Recent studies with ovarian cancer PDX models demonstrated that the combination of the CHK1 inhibitor prexasertib with olaparib caused greater tumour growth delay and survival in both olaparib sensitive and resistant tumours [28]. Our data along with this PDX study may help in the interpretation of the results of clinical trials of PARPi-CHK1 inhibitor combinations, such as NCT03057145, in which olaparib is being investigated in combination with prexasertib.

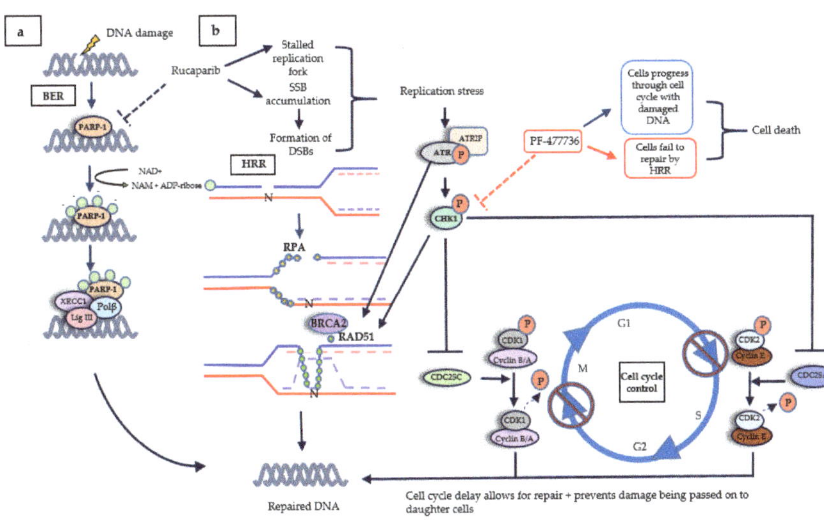

Figure 5. The DNA damage response, highlighting cross-talk between the cell cycle and DNA repair. (**a**) Base excision repair (BER) is the first line of defence when DNA damage occurs. PARP activation by SSB recruits XRCC1, Pol β, and Lig III to repair the DNA. (**b**) When PARP is inhibited, SSB accumulate, causing stalled replication forks, and single-ended DSBs, causing fork collapse. BRCA2 facilitates replacing RPA with repair protein RAD51, which subsequently forms a filament to search for the specific

homologous sequence on the sister chromatid as a template for repair and re-start. PARPi-induced replication stress activates ATR, which initiates a signalling cascade. ATR activates CHK1 by phosphorylation at serine 345, causing CHK1 to autophosphorylate at serine 296, to achieve full activation. CHK1 inactivates CDC25A/C, thereby preventing the removal of inhibitory phosphorylation on CDK2 and 1, respectively, thus preventing S-phase progression and mitosis. The pathway also promotes HRR as CHK1 activates BRCA2 and RAD51 by phosphorylation, and ATR also phosphorylates RAD51.

6. Patents

Helleday T and Curtin NJ. Therapeutic Compounds (PARP inhibitors in homologous repair/BRCA defective cancer) Patent Application Number PCT/GB2004/003183. Publication number WO 2005/012305 A2 Divisional application 16th April 2004 GB 0408524. WO2005012305A3.

Supplementary Materials: The following are available online at http://www.mdpi.com/2072-6694/12/4/878/s1: Figure S1: PARP1 expression in V-C8 and V-C8.B2 cells, Figure S2: CHK1 expression, activation by rucaaprib and inhibition by PF-477736 in V-C8 and V-C8.B2 cells.

Author Contributions: H.L.S. produced the data and wrote the manuscript; L.P. directed the research and edited the manuscript; N.J.C. conceived the project, directed the research and edited the manuscript. All authors have read and agreed to the published version of the manuscript.

Funding: This research was partly funded by the JGW Patterson foundation.

Conflicts of Interest: N.J.C. contributed to the development of rucaparib and obtained research funding from Agouron Pharmacurtical, Pfizer, and Clovis. By virtue of the two active patents and an original agreement between Cancer Research Technology, Newcastle University, and Agouron, she is in receipt of royalty payments, which she does not take personally; in the past, they have contributed to her research accounts, and a recent large sum has been used to set up a charitable fund within the local Community Foundation. https://www.communityfoundation.org.uk/group_grant/passionate-about-realising-your-potential/.

References

1. Hanahan, D.; Weinberg, R.A. Hallmarks of cancer: The next generation. *Cell* **2011**, *144*, 646–674. [CrossRef] [PubMed]
2. Massague, J. G1 cell cycle control and cancer. *Nature* **2004**, *432*, 298–306. [CrossRef] [PubMed]
3. Bai, P. Biology of Poly(ADP-Ribose) Polymerases: The Factotums of Cell Maintenance. *Mol. Cell* **2015**, *58*, 947–958. [CrossRef] [PubMed]
4. LaFargue, C.J.; Dal Molin, G.Z.; Sood, A.K.; Coleman, R.L. Exploring and comparing adverse events between PARP inhibitors. *Lancet Oncol.* **2019**, *20*, e15–e28. [CrossRef]
5. Mukhopadhyay, A.; Elattar, A.; Cerbinskaite, A.; Wilkinson, S.J.; Drew, Y.; Kyle, S.; Los, G.; Hostomsky, Z.; Edmondson, R.J.; Curtin, N.J. Development of a functional assay for homologous recombination status in primary cultures of epithelial ovarian tumor and correlation with sensitivity to poly(ADP-ribose) polymerase inhibitors. *Clin. Cancer Res.* **2010**, *16*, 2344–2351. [CrossRef]
6. Pepa, C.; Tonini, G.; Pisano, C.; Di Napoli, M.; Cecere, S.C.; Tambaro, R.; Facchini, G.; Pignata, S. Ovarian cancer standard of care: Are there real alternatives? *Chin. J. Cancer* **2015**, *34*, 17–27. [CrossRef]
7. Mirza, M.R.; Monk, B.J.; Herrstedt, J.; Oza, A.M.; Mahner, S.; Redondo, A.; Fabbro, M.; Ledermann, J.A.; Lorusso, D.; Vergote, I.; et al. Niraparib Maintenance Therapy in Platinum-Sensitive, Recurrent Ovarian Cancer. *N. Engl. J. Med.* **2016**, *375*, 2154–2164. [CrossRef]
8. Visconti, R.; Monica, R.D.; Grieco, D. Cell cycle checkpoint in cancer: A therapeutically targetable double-edged sword. *J. Exp. Clin. Cancer Res.* **2016**, *35*, 153. [CrossRef]
9. Sørensen, C.S.; Hansen, L.T.; Dziegielewski, J.; Syljuåsen, R.G.; Lundin, C.; Bartek, J.; Helleday, T. The cell-cycle checkpoint kinase Chk1 is required for mammalian homologous recombination repair. *Nat. Cell Biol.* **2005**, *7*, 195–201. [CrossRef]
10. Narayanaswamy, P.B.; Tkachuk, S.; Haller, H.; Dumler, I.; Kiyan, Y. CHK1 and RAD51 activation after DNA damage is regulated via urokinase receptor/TLR4 signaling. *Cell Death Dis.* **2016**, *7*, 2383. [CrossRef]
11. Haynes, B.; Murai, J.; Lee, J.M. Restored replication fork stabilization, a mechanism of PARP inhibitor resistance, can be overcome by cell cycle checkpoint inhibition. *Cancer Treat. Rev.* **2018**, *71*, 1–7. [CrossRef] [PubMed]

12. Booth, L.; Cruickshanks, N.; Ridder, T.; Dai, Y.; Grant, S.; Dent, P. PARP and CHK inhibitors interact to cause DNA damage and cell death in mammary carcinoma cells. *Cancer Biol. Ther.* **2013**, *14*, 458–465. [CrossRef] [PubMed]
13. Brill, E.; Yokoyama, T.; Nair, J.; Yu, M.; Ahn, Y.R.; Lee, J.M. Prexasertib, a cell cycle checkpoint kinases 1 and 2 inhibitor, increases in vitro toxicity of PARP inhibition by preventing Rad51 foci formation in BRCA wild type high-grade serous ovarian cancer. *Oncotarget* **2017**, *8*, 111026–111040. [CrossRef]
14. Bryant, H.E.; Schultz, N.; Thomas, H.D.; Parker, K.M.; Flower, D.; Lopez, E.; Kyle, S.; Meuth, M.; Curtin, N.J.; Helleday, T. Specific killing of BRCA2-deficient tumours with inhibitors of poly(ADP-ribose) polymerase. *Nature* **2005**, *434*, 913–917. [CrossRef] [PubMed]
15. Farmer, H.; McCabe, N.; Lord, C.J.; Tutt, A.N.; Johnson, D.A.; Richardson, T.B.; Santarosa, M.; Dillon, K.J.; Hickson, I.; Knights, C.; et al. Targeting the DNA repair defect in BRCA mutant cells as a therapeutic strategy. *Nature* **2005**, *434*, 917–921. [CrossRef]
16. Middleton, F.K.; Patterson, M.J.; Elstob, C.J.; Fordham, S.; Herriott, A.; Wade, M.A.; McCormick, A.; Edmondson, R.; May, F.E.; Allan, J.M.; et al. Common cancer-associated imbalances in the DNA damage response confer sensitivity to single agent ATR inhibition. *Oncotarget* **2015**, *6*, 32396–32409. [CrossRef] [PubMed]
17. Gottipati, P.; Vischioni, B.; Schultz, N.; Solomons, J.; Bryant, H.E.; Djureinovic, T.; Issaeva, N.; Sleeth, K.; Sharma, R.A.; Helleday, T. Poly(ADP-ribose) polymerase is hyperactivated in homologous recombinationdefective cells. *Cancer Res.* **2010**, *70*, 5389–5398. [CrossRef] [PubMed]
18. Plummer, R.; Jones, C.; Middleton, M.; Wilson, R.; Evans, J.; Olsen, A.; Curtin, N.; Boddy, A.; McHugh, P.; Newell, D.; et al. Phase I study of the poly(ADP-ribose) polymerase inhibitor, AG014699, in combination with temozolomide in patients with advanced solid tumors. *Clin. Cancer Res.* **2008**, *14*, 7917–7923. [CrossRef]
19. Massey, A.J. Inhibition of ATR-dependent feedback activation of Chk1 sensitises cancer cells to Chk1 inhibitor monotherapy. *Cancer Lett.* **2016**, *383*, 41–52. [CrossRef]
20. Syljuåsen, R.G.; Sørensen, C.S.; Hansen, L.T.; Fugger, K.; Lundin, C.; Johansson, F.; Helleday, T.; Sehested, M.; Lukas, J.; Bartek, J. Inhibition of human Chk1 causes increased initiation of DNA replication, phosphorylation of ATR targets, and DNA breakage. *Mol. Cell. Biol.* **2005**, *25*, 3553–3562. [CrossRef]
21. Yin, Y.; Shen, Q.; Zhang, P.; Tao, R.; Chang, W.; Li, R.; Xie, G.; Liu, W.; Zhang, L.; Kapoor, P.; et al. Chk1 inhibition potentiates the therapeutic efficacy of PARP inhibitor BMN673 in gastric cancer. *Am. J. Cancer Res.* **2017**, *7*, 473–483. [PubMed]
22. Kim, H.; George, E.; Ragland, R.; Rafail, S.; Zhang, R.; Krepler, C.; Morgan, M.; Herlyn, M.; Brown, E.; Simpkins, F. Targeting the ATR/CHK1 Axis with PARP Inhibition Results in Tumor Regression in BRCA-Mutant Ovarian Cancer Models. *Clin. Cancer Res.* **2017**, *23*, 3097–3108. [CrossRef] [PubMed]
23. Mani, C.; Jonnalagadda, S.; Lingareddy, J.; Awasthi, S.; Gmeiner, W.H.; Palle, K. Prexasertib treatment induces homologous recombination deficiency and synergizes with olaparib in triple-negative breast cancer cells. *Breast Cancer Res.* **2019**, *21*, 1–14. [CrossRef] [PubMed]
24. Kraakman-van der Zwet, M.; Overkamp, W.J.; van Lange, R.E.; Essers, J.; van Duijn-Goedhart, A.; Wiggers, I.; Swaminathan, S.; van Buul, P.P.; Errami, A.; Tan, R.T.; et al. Brca2 (XRCC11) deficiency results in radioresistant DNA synthesis and a higher frequency of spontaneous deletions. *Mol. Cell. Biol.* **2002**, *22*, 669–679. [CrossRef] [PubMed]
25. Plummer, E.R.; Middleton, M.R.; Jones, C.; Olsen, A.; Hickson, I.; McHugh, P.; Margison, G.P.; McGown, G.; Thorncroft, M.; Watson, A.J.; et al. Temozolomide pharmacodynamics in patients with metastatic melanoma: Dna damage and activity of repair enzymes O6alkylguanine alkyltransferase and poly(ADP-ribose) polymerase-1. *Clin. Cancer Res.* **2005**, *11*, 3402–3409. [CrossRef] [PubMed]
26. Canan, S.; Maegley, K.; Curtin, N.J. Strategies Employed for the Development of PARP Inhibitors. *Methods Mol. Biol.* **2017**, *1608*, 271–297.

27. Rueden, C.T.; Schindelin, J.; Hiner, M.C.; DeZonia, B.E.; Walter, A.E.; Arena, E.T.; Eliceiri, K.W. ImageJ2: ImageJ for the next generation of scientific image data. *BMC Bioinformatics* **2017**, *18*, 529. [CrossRef]
28. Parmar, K.; Kochupurakkal, B.S.; Lazaro, J.B.; Wang, Z.C.; Palakurthi, S.; Kirschmeier, P.T.; Yang, C.; Sambel, L.A.; Färkkilä, A.; Reznichenko, E.; et al. The CHK1 Inhibitor Prexasertib Exhibits Monotherapy Activity in High-Grade Serous Ovarian Cancer Models and Sensitizes to PARP Inhibition. *Clin. Cancer Res.* **2019**, *25*, 6127–6140. [CrossRef]

cancers

MDPI

Article

Synergistic Anti-Tumour Effect of Syk Inhibitor and Olaparib in Squamous Cell Carcinoma: Roles of Syk in EGFR Signalling and PARP1 Activation

Duen-Yi Huang [1], Wei-Yu Chen [2,3], Chi-Long Chen [3,4], Nan-Lin Wu [5,6,7] and Wan-Wan Lin [1,8,*]

1. Department of Pharmacology, College of Medicine, National Taiwan University, Taipei 100, Taiwan; tyh123kimo@gmail.com
2. Department of Pathology, Wan Fang Hospital, Taipei Medical University, Taipei 116, Taiwan; 1047@tmu.edu.tw
3. Department of Pathology, School of Medicine, College of Medicine, Taipei Medical University, Taipei 106, Taiwan; chencl@tmu.edu.tw
4. Department of Pathology, Taipei Medical University Hospital, Taipei 106, Taiwan
5. Department of Medicine, Mackay Medical College, New Taipei City 251, Taiwan; alvin.4200@mmh.org.tw
6. Department of Dermatology, Mackay Memorial Hospital, Taipei 104, Taiwan
7. Mackay Junior College of Medicine, Nursing, and Management, New Taipei City 252, Taiwan
8. Graduate Institute of Medical Sciences, Taipei Medical University, Taipei 106, Taiwan

* Correspondence: wwllaura1119@ntu.edu.tw; Tel.: +886-223-123-456 (ext. 88315); Fax: +886-223-513-716

Received: 20 December 2019; Accepted: 17 February 2020; Published: 19 February 2020

Abstract: Syk is a non-receptor tyrosine kinase involved in the signalling of immunoreceptors and growth factor receptors. Previously, we reported that Syk mediates epidermal growth factor receptor (EGFR) signalling and plays a negative role in the terminal differentiation of keratinocytes. To understand whether Syk is a potential therapeutic target of cancer cells, we further elucidated the role of Syk in disease progression of squamous cell carcinoma (SCC), which is highly associated with EGFR overactivation, and determined the combined effects of Syk and PARP1 inhibitors on SCC viability. We found that pharmacological inhibition of Syk could attenuate the EGF-induced phosphorylation of EGFR, JNK, p38 MAPK, STAT1, and STAT3 in A431, CAL27 and SAS cells. In addition, EGF could induce a Syk-dependent IL-8 gene and protein expression in SCC. Confocal microscopic data demonstrated the ability of the Syk inhibitor to change the subcellular distribution patterns of EGFR after EGF treatment in A431 and SAS cells. Moreover, according to Kaplan-Meier survival curve analysis, higher Syk expression is correlated with poorer patient survival rate and prognosis. Notably, both Syk and EGFR inhibitors could induce PARP activation, and synergistic cytotoxic actions were observed in SCC cells upon the combined treatment of the PARP1 inhibitor olaparib with Syk or the EGFR inhibitor. Collectively, we reported Syk as an important signalling molecule downstream of EGFR that plays crucial roles in SCC development. Combining Syk and PARP inhibition may represent an alternative therapeutic strategy for treating SCC.

Keywords: Syk; PARP; EGFR; squamous cell carcinoma

---

1. Introduction

The epidermal growth factor (EGF) receptor/ligand system plays key roles in essential cellular functions, including proliferation and migration. Ligand binding to the epidermal growth factor receptor (EGFR) induces the formation of receptor dimerisation and stimulates multiple pathways of signal transduction. To date, EGFR has been linked to several malignant phenotypes of human cancers, including proliferation, inflammatory response, DNA repair, therapeutic resistance, and poor clinical outcomes in patients with cancer [1,2]. Currently, there are several EGFR antagonists, including humanised neutralising monoclonal antibodies and tyrosine kinase inhibitors clinically used in patients with non-small-cell lung [3] and metastatic colorectal cancers [4].

Spleen tyrosine kinase (Syk) is a non-receptor tyrosine kinase that is widely expressed in hematopoietic cells. Syk contains two tandem NH2 terminal Src homology 2 (SH2) domains, multiple tyrosine phosphorylation sites, and a COOH terminal tyrosine kinase domain. The SH2 domains bind phosphorylated immunoreceptor tyrosine-based activation motifs (ITAM), and hence, they couple activated immunoreceptors to multiple downstream signalling pathways [5–7]. Anomalous regulation of this kinase could lead to different allergic disorders and antibody-mediated autoimmune diseases. Thus, Syk may serve as a therapeutically relevant target for rheumatoid arthritis, asthma, psoriasis, and allergic rhinitis [8–10]. It has been believed for years that Syk function is solely linked to hematopoietic cell signalling. However, more recent studies have indicated a ubiquitous pattern of Syk gene expression. In addition to functioning as the therapeutic target of haematological malignancies [11,12], Syk can function as either a tumour suppressor in breast, colorectal, and gastric cancers or a tumour enhancer in lung, pancreatic, ovarian, and squamous cell carcinoma (SCC) of the head and neck [13]. In cancers of non-immune cells, Syk elicits a pro-survival signal, but can also suppress tumourigenesis by restricting epithelial-mesenchymal transition, enhancing cell-cell interactions, and inhibiting migration [13]. To date, only a few studies have suggested a functional link between Syk and EGFR. In our previous study on human primary normal keratinocytes, we demonstrated that Syk is a downstream signal of EGFR and is involved in negatively regulating keratinocyte differentiation [14].

SCC is an uncontrolled growth of abnormal cells arising in the squamous cells, which compose most of the skin's upper layers (the epidermis). Similar to other cancer cell types, inhibition of EGFR activity is a potential drug regimen for SCC, which overexpresses EGFR [15–17]. A recent in vivo xenograph study suggested a potential therapy involving the use of EGFR monoclonal antibodies in SCC [18]. For better therapeutic efficacy, combination regimens are increasingly under investigation. For example, the EGFR inhibitor, gefitinib (Iressa), can enhance the apoptotic effects of cisplatin in SCC [17]. Clinical trials involving the combination regimens of anti-EGFR agents with immune checkpoint inhibitors are currently in progress for SCC of the head and neck [19,20]. In this study, we explored the role of Syk in EGFR signalling in three SCC cell lines (A431, CAL27, and SAS), analysed Syk expression in oral SCC cancer tissues, and determined the relationship between Syk activity and disease prognosis. Finally, because inhibition of poly(ADP-ribose) polymerase (PARP), a key DNA repair enzyme, could improve the efficacy of radiotherapy in human head and neck cancer [21], we determined the combined effects of the Syk and PARP inhibitors on cell viability in SCC.

2. Materials and Methods

2.1. Reagents and Antibodies

Dulbecco's Modified Eagle Medium (DMEM) and trypsin-EDTA were obtained from Gibco (Carlsbad, CA, USA). Foetal bovine serum (FBS) was procured from HyClone (Logan, UT, USA). Penicillin-streptomycin-amphotericin B solution was sourced from Biological Industries (Kibbutz Beit Haemek, Israel). Dulbecco's Phosphate Buffered Saline (PBS) was obtained from Sigma-Aldrich (St. Louis, MO, USA). Gefitinib was procured from Selleckchem (Houston, TX, USA). R406, PP2, and the p-Syk antibody were purchased from Calbiochem (San Diego, CA, USA). Olaparib was procured from Selleckchem. Recombinant human EGF was obtained from PeproTech (Rocky Hill, NJ, USA). PARP, γH2AX, p-EGFR, p-JNK, JNK, p-p38, p-ERK, p-Lyn, p-Src, p-STAT1, p-STAT3, and p-STAT5 antibodies were obtained from Cell Signaling (Beverly, MA, USA). EGFR, p38, ERK, Lyn, Fyn, Fgr, Src, Syk, STAT1, and STAT3 antibodies were purchased from Santa Cruz (Santa Cruz, CA, USA). Antibody against PAR was obtained from BD Pharmingen (San Diego, CA, USA). The β-actin antibody was procured from NOVUS (Littleton, CO, USA).

2.2. Cell Culture

Human skin SCC (A431), oral SCC (CAL27), and head and neck SCC (SAS) cell lines were cultured in DMEM supplemented with 10% FBS and 1% penicillin-streptomycin-amphotericin B. All cell lines were incubated at 37 °C under a humidified atmosphere of 5% CO_2 in the air. Prior to stimulation of cells with EGFR ligands or otherwise specified conditions, cells were maintained in the medium without FBS for at least 1 d.

2.3. Immunoblotting

Protein expression was determined in cell lysates by electrophoresis and immunoblotting as previously described [14].

2.4. Real-tTime PCR

The primer sequence pairs used for quantitative real-time PCR were human β-actin: 5′-CGGGGACCTGACTGACT ACC-3′ and 5′-AGGAAGGCTGGAAGAGTGC-3′; and human IL-8: 5′-ACTGAGAGTGATTGAGAGTGGAC-3′ and 5′-AACCCTCTGCACCCAGTTTTC-3′. The experiment and data analysis were conducted as previously described [22].

2.5. Enzyme-Linked Immunosorbent Assay (ELISA)

For the detection of secreted proteins in the cell culture supernatant, a commercial ELISA kit for IL-8 (CXCL8) (R&D Systems, Minneapolis, MN, USA) was used to determine the level of released proteins in accordance with the manufacturer's protocol.

2.6. Confocal Microscopic Analysis

All procedures were conducted at room temperature unless otherwise noted. Cells were fixed with 4% paraformaldehyde in PBS for 20 min. After being rinsed three times with PBS, the fixed cells were permeabilised with 0.2% Triton X-100 in PBS for 20 min. After fixation, the samples were blocked with 4% BSA for 1 h and then incubated with the primary antibody for 2 h at room temperature or overnight at 4 °C after aspiration of the blocking solution. The primary antibody was then removed and specimens were washed three times with PBS. Subsequently, the specimens were incubated with the fluorochrome-conjugated secondary antibody for 1 h in the dark. Following immunostaining, the coverslip was counterstained with 4′,6-diamidino-2-phenylindole (DAPI) and mounted on microscope slides in the dark. Slides were analysed using an LSM 780 confocal microscope (Carl Zeiss Micro Imaging GMbH, Jena, Germany).

2.7. Crystal Violet Assay

To evaluate the relative cell numbers, cells were seeded in 96-well plates in medium followed by the indicated treatment. After washing with PBS, cells were fixed with methanol for 10 min at room temperature and stained with 0.1% crystal violet. The relative cell number was determined by measuring the absorbance of the dissolved dye at 540 nm after elution with 33% acetic acid.

2.8. Intracellular ATP Assay

Intracellular ATP was determined by a CellTiter-Glo® Luminescent Kit (G7571, Promega Heidelberg, Germany) according to the manufacturer's protocol. After the indicated drug treatment, the reagent in the kit was added to lyse the cells, and then 150 μL solution was transferred to a 96-well plate for chemiluminescence detection using a luminometer. With background subtraction, the values were normalised to the individual control group being 100%.

2.9. MTT Assay

The viability of cells was measured by the turnover of yellow 3-(4,5-dimethyl-2-thiazolyl)-2,5-diphenyl-2*H*-tetrazolium bromide (MTT, Calbiochem) to dark blue formazan by mitochondrial reductase in living cells. A sample of 1×10^4 cells was seeded in 96-well plates for at least 16 h followed by the indicated treatment. At the end of the assay, MTT solution (5 mg/mL in PBS) was added to the medium with a 5-fold dilution for 1 h at 37 °C until a purple precipitate was visible. Cells in the medium were then collected and centrifuged at 800 × rpm for 5 min. After discarding the supernatant, the pellet was dissolved in DMSO. The absorbance (OD_{550}–OD_{630}) was detected using a spectrophotometer. The absorbance of the control group represented 100% cell viability.

2.10. Cell Death Assays by Annexin V/PI Staining and Flow Cytometry

After the indicated treatment, cells were collected and washed with ice-cold PBS. Cells were stained with Annexin V-FITC/PI according to the manufacturer's instructions (BioLegend, San Diego, CA, USA). After incubation, the cells were measured by flow cytometry (FACSCalibur, BD, Franklin Lakes, NJ, USA).

2.11. Tissue Samples and Immunohistochemistry

Pathology files of the Wan Fang Hospital (Taipei, Taiwan) from 2000 to 2010 were searched for SCC of the oral cavity. The pathological diagnoses were microscopically reconfirmed by pathologists (WYC and CLC). Tissue microarrays of oral SCC were manually constructed. Triplicate tissue cores of SCC and one tissue core from the adjacent non-neoplastic squamous epithelium were obtained for each case. In total, 105 patients were included in this study. The clinical records were reviewed for tumour recurrence and survival. For immunohistochemistry, the sections were deparaffinised, rehydrated, and blocked with 3% hydrogen peroxide. Heat-induced antigen retrieval was performed in a citric acid buffer (pH 6.0) at 121 °C for 10 min using a decloaking chamber (Biocare Medical, Concord, CA, USA). The sections were incubated with rabbit polyclonal phospho-Syk (Y525/526) antibody (Catalogue No. AP3271a, 1:300; Abgent Inc., San Diego, CA, USA) at 4 °C overnight. Next, the sections were incubated with a prediluted biotin-conjugated secondary antibody (Starr Trek Universal HRP Detection System; Biocare Medical) at room temperature for 30 min, followed by prediluted streptavidin-horseradish peroxidase complex at room temperature for 10 min. The antigen was detected by adding 3,3′-diaminobenzidine. The sections were then counterstained by haematoxylin. The appropriate positive and negative controls were included in these assays. The intensity of p-Syk expression in carcinoma cells was evaluated by immunostaining. The scoring of expression in tumour tissue was performed according to a four-tiered approach: negative (0), weak (1+), moderate (2+), and strong (3+) staining. Based on these criteria, p-Syk expression was defined as either low (scores 0 and 1+) or high expression (scores 2+ and 3+). Tissue samples were obtained and used according to protocols approved by the Taipei Medical University-Wan Fang Hospital Joint Institutional Review Board (approval no. 99049). The study was conducted according to the Declaration of Helsinki principles.

The clinicopathological characteristics and p-Syk expression were compared using the Chi-square test for categorical data and the two-tailed Student's t test for continuous data. The overall survival and disease-free curves were calculated using the Kaplan-Meier method, and the difference between the low and high p-Syk expression groups was evaluated using the log-rank test. The Cox proportional-hazard model was utilised to identify statistically significant clinicopathological factors affecting the prognosis of patients. A p value < 0.05 was considered statistically significant. All statistical calculations were performed using SPSS statistics 17.0 software (SPSS Inc. Chicago, IL, USA).

2.12. Statistical Analysis

Except immunohistochemistry data as mentioned above, all other data were presented as the mean ± standard error of the mean (S.E.M) from at least three independent experiments. Comparisons between two groups were performed using an unpaired two-tailed Student's *t* test.

3. Results

3.1. SFK-Dependent Syk Activation Mediates Downstream Signal Pathways of EGFR in SCC

After observing the crucial role of Syk in skin keratinocytes [14], we further addressed the roles of Syk in skin tumour development. To this end, we first detected the expression levels of Syk and Src family kinases (SFKs) (c-Src, Lyn, Hck, and Fgr) in A431 SCC, melanoma A375, and SK-MEL-28 cells. We found that Syk expression was highly expressed in A431 cells but not in the others. Lyn and Hck expression was also higher in A431 cells than that in melanoma cells. In contrast, c-Src expression was comparable in these cell lines, whereas Fgr was non-detectable (Figure 1A). To confirm the absence of Fgr in these cells, we used U937 monocytes as a comparison. We found that Syk and Lyn were expressed in U937 cells to the same extent as in the A431 cells; however, U937 cells expressed Fgr, but not c-Src or Hck (Figure 1A).

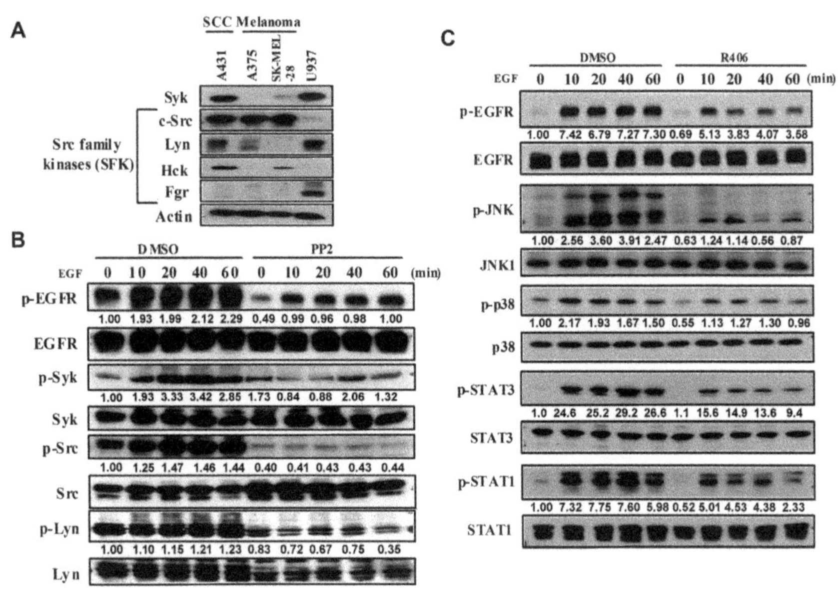

Figure 1. EGF can induce Syk activation through SFK signalling in A431 cells. (**A**) Western blot analysis of Syk and SFK expression in various cancer cell lines. (**B**) Phosphorylation of EGFR, Syk, Src, and Lyn after EGF (100 ng/mL) treatment with or without PP2 (10 μM) for 0, 10, 20, 40, and 60 min in A431 cells. (**C**) Phosphorylation of EGFR downstream signalling molecules after EGF (100 ng/mL) treatment with or without R406 (1 μM) for 0, 10, 20, 40, and 60 min in A431 cells. Data are presented from three independent experiments.

Next, we found that EGF (100 ng/mL) can increase Syk and c-Src phosphorylation in a time-dependent manner. Moreover, constitutive Lyn phosphorylation was detected but was not altered upon EGF stimulation (Figure 1B). Because of the non-availability of a specific antibody for the phosphorylated Hck, we could not determine its activation status in EGF-stimulated A431 cells. Notably, non-selective SFK inhibitor PP2 (10 μM) decreased the phosphorylation of c-Src, Lyn, Syk, and EGFR (Figure 1B). Because PP2 also reduced EGFR phosphorylation, we suggest there is a feedback loop to control EGFR activation through SFKs as previously reported in fibroblasts [23], breast [23,24], and colon cancer cells [25] in a transactivation manner. All these findings suggested that Syk is a signal molecule for EGFR, and its activation is downstream of SFKs.

Next, we used the Syk inhibitor, R406, to test its effects on EGF-induced signalling pathways. As shown in Figure 1C, R406 treatment significantly inhibited the phosphorylation extent of JNK, p38 MAPK, STAT1, and STAT3 in EGF-stimulated A431 cells. Notably, R406 also reduced EGFR phosphorylation as did PP2 (Figure 1C).

In addition to A431 cells, we also examined the role of Syk in CAL27 (oral SCC) and SAS (head and neck SCC) cells. As observed in A431 cells, EGF stimulated the phosphorylation of EGFR, c-Src, JNK, and p38 in time- and concentration-dependent manners in both the cell lines (Figure 2A). Moreover, EGF activated Syk in CAL27 (Figure 2B) and SAS cells (Figure 2C). In CAL27 cells, EGF-induced phosphorylation of EGFR, STAT3, Akt, and JNK was inhibited by R406 (Figure 2B). Similarly, EGF-induced STAT3 phosphorylation in SAS cells was reduced by R406 (Figure 2C). These findings indicated that Syk is the upstream signal molecule that transduces the downstream signalling pathways of EGFR.

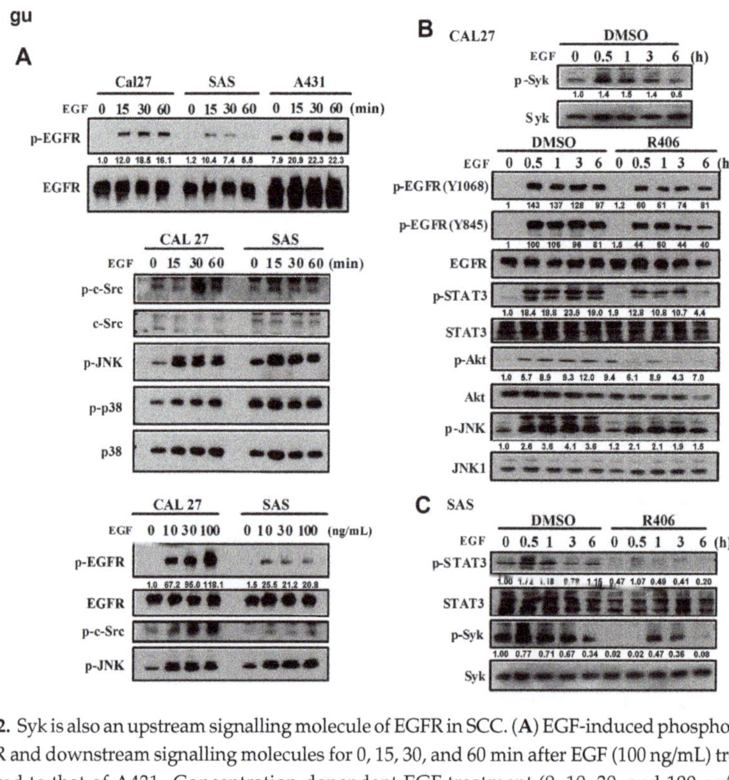

Figure 2. Syk is also an upstream signalling molecule of EGFR in SCC. (**A**) EGF-induced phosphorylation of EGFR and downstream signalling molecules for 0, 15, 30, and 60 min after EGF (100 ng/mL) treatment compared to that of A431. Concentration-dependent EGF treatment (0, 10, 30, and 100 ng/mL) for 30 min in CAL27 and SAS cells. (**B**,**C**) Phosphorylation of EGFR and downstream signalling molecules after EGF (100 ng/mL) treatment with or without R406 (1 µM) for 0, 0.5, 1, 3, and 6 h in CAL27 (**B**) and SAS cells (**C**). Data are presented from three independent experiments.

3.2. *Syk Mediates EGF-Induced IL-8 Upregulation But Is Not Involved in the Regulation of Cell Fate*

Given that IL-8 is a crucial player in cancer development [26], we examined the effects of R406 on EGF-induced IL-8 expression in A431 cells. We found that EGF can upregulate IL-8 mRNA expression and increase IL-8 protein production (Figure 3A), and these actions were inhibited by R406 and gefitinib.

Similarly, in SAS and CAL27 cells, IL-8 mRNA induced by EGF was markedly reduced by R406 and gefitinib (Figure 3B,C).

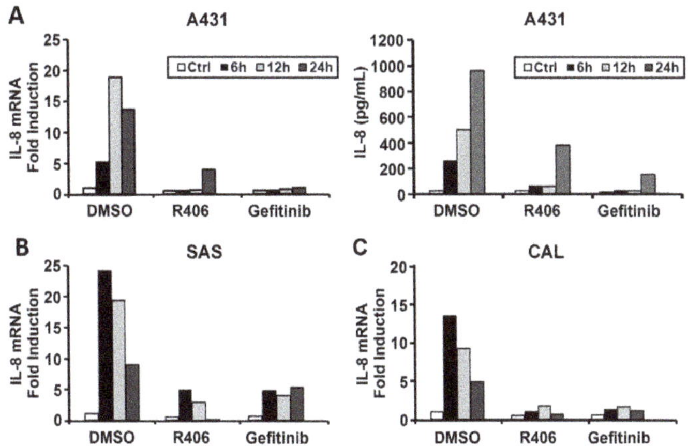

Figure 3. Syk mediates EGF-induced IL-8 mRNA and protein expression in SCC. EGF (100 ng/mL) treatment in A431 (**A**), SAS (**B**), and CAL27 (**C**) cells for 0, 6, 12, and 24 h and the addition of R406 (1 μM) or gefitinib (5 μM) as a positive control. IL-8 mRNA levels were determined by RT-PCR and protein levels by ELISA.

Using BrdU incorporation as an index of cell proliferation, our data did not show the significant effects of EGF and R406 on BrdU uptake in A431 cells (Figure 4A). We also used MTT, ATP, and crystal violet assays to assess the effects of EGF and R406 on cell viability. Our data revealed non-significant effects of both agents on the cell viability of A431 cells (Figure 4B). When Syk mRNA was silenced by the siRNA approach, EGF still could not affect cell viability in A431 cells (Figure 4C). We also tested another EGFR activator, TGFα and found that in agreement with the data for EGF, TGFα (10 and 100 ng/mL), in the absence or presence of R406, did not change cell viability as assessed by MTT (Figure 4D) and crystal violet assays (Figure 4E) in A431 cells. Similarly, MTT assays revealed that R406 did not alter the viability of SAS and CAL27 cells (Figure 4F).

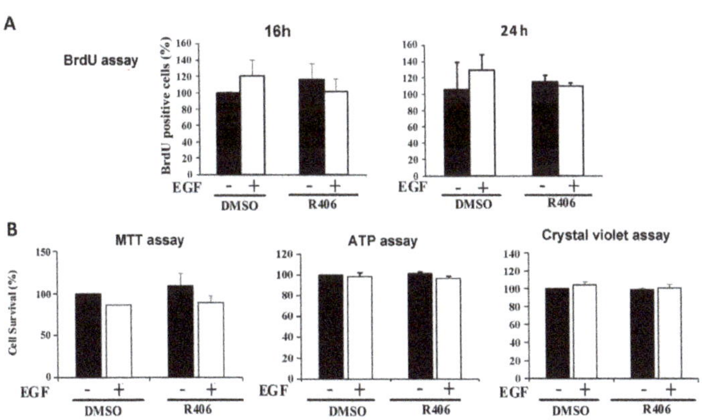

Figure 4. *Cont.*

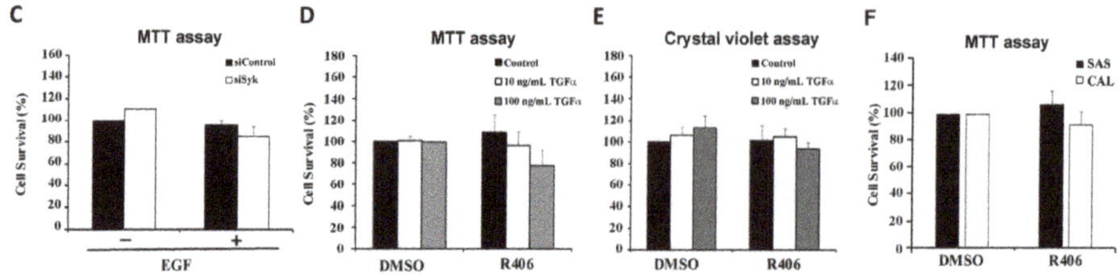

Figure 4. Syk is not involved in the proliferation and viability of SCC cells. (**A**) Effects of R406 (1 μM) with or without co-treatment with EGF (50 ng/mL) on A431 cell proliferation for 16 and 24 h were analysed by BrdU assay. (**B**) Effects of R406 (1 μM) with or without co-treatment with EGF (100 ng/mL) on A431 cell viability for 24 h analysed by MTT, ATP, and crystal violet assays. (**C**) Effects of Syk siRNA with or without EGF (50 ng/mL) for 24 h on A431 cell viability analysed by an MTT assay. (**D**,**E**) Cell viability of A431 cells after treatment with R406 (1 μM) with or without co-treatment with TGFα (10 or 100 ng/mL) for 24 h analysed by MTT and crystal violet assays. (**F**) Viability of SAS and CAL27 cells treated with R406 for 24 h determined by MTT assay. Data are expressed as the mean ± S.E.M. from three independent experiments.

3.3. Syk Regulates Intracellular EGFR Movement in a Cell Type-Specific Manner

In our previous study, in NHEK cells, we observed the co-localisation of EGFR and Syk. Moreover, Syk was moved along with EGFR trafficking intracellularly after EGF stimulation [14]. Here, we further detected the intracellular localisation of EGFR in A431 and SAS cells before and after EGF stimulation. Unlike NHEK cells, where EGFR during the resting state was apparently present in the membrane, EGFR in A431 cells was present in both the cytosol and plasma membrane during the resting state. After EGF stimulation, EGFR was translocated to the plasma membrane and peri-nuclei of a spot like. The latter site was co-stained with the early endosomal marker EEA1, suggesting the occurrence of common EGFR trafficking from the internalisation to the late endosome/lysosomal degradation and endocytotic recycling pathways. The treatment with the Syk inhibitor R406 reduced the membrane location but not the peri-nuclear translocation of EGFR induced by EGF (Figure 5A). In SAS cells, EGFR was primarily observed in the plasma membrane as NHEK and could move to the peri-nuclei early endosome upon EGF stimulation. R406 treatment appeared to induce an uneven distribution of EGFR in peri-nuclear sites after EGF stimulation in SAS cells (Figure 5B).

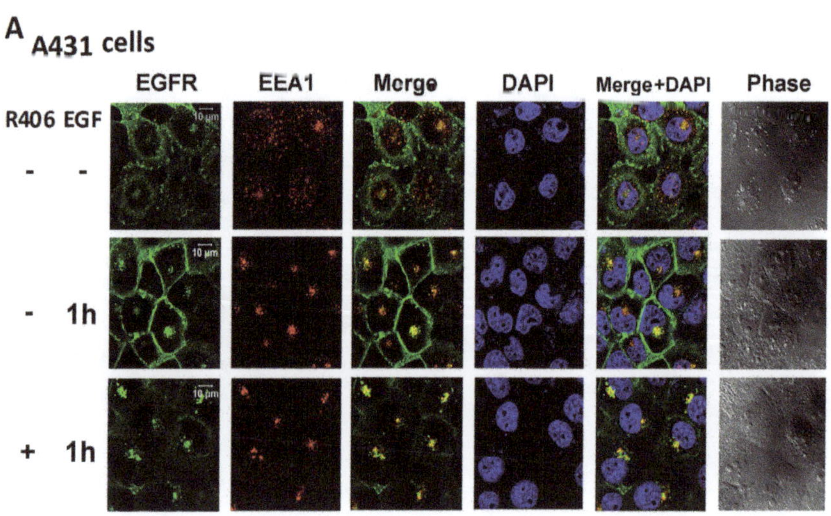

Figure 5. *Cont.*

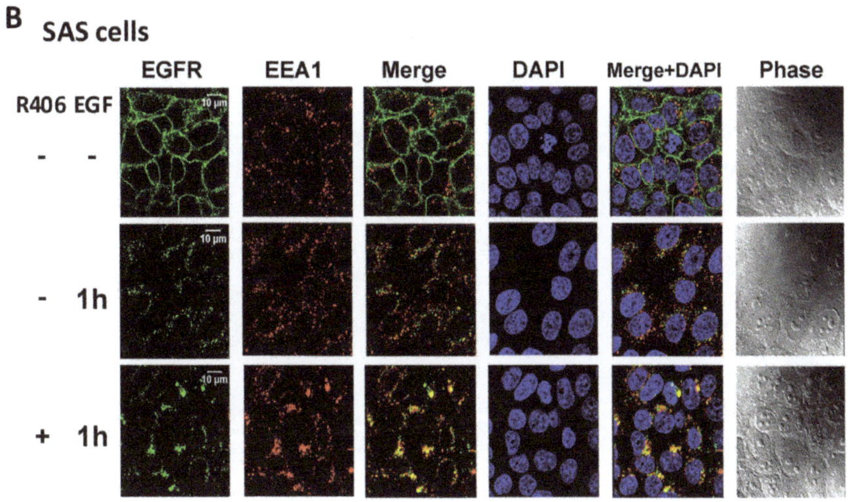

Figure 5. Syk modulates EGFR localisation under EGF treatment in A431 and SAS cells. Confocal microscopic analyses of EGFR and EEA1 (early endosome marker) localisation after treatment of EGF (50 ng/mL) for 1 h with or without the addition of R406 (1 µM) in A431 (**A**) and SAS cells (**B**). Data are from three independent experiments.

3.4. Syk Activity Is Associated with Clinicopathologic Features and Outcome of Patients with Oral SCC

To better understand the prognostic role of Syk in oral SCC, an immunohistochemical analysis for p-Syk expression was performed on the tissue microarray sections of oral SCC, and the clinicopathological phenomena and p-Syk expression in tumour tissues in 105 patients were analysed. We first compared the expression levels of p-Syk in oral SCC with that of matched non-neoplastic squamous epithelium (Figure 6A). Oral SCC had stronger p-Syk expression than normal squamous epithelium. The association between clinicopathological characteristics and p-Syk expression is shown in Table 1. Oral SCC in the high p-Syk expression group tended to be poorly differentiated in grading ($p = 0.002$). Lymphovascular invasion was more commonly detected in oral SCC with high p-Syk expression ($p = 0.018$). The associations between p-Syk and other clinicopathological factors, such as gender, age, perineural invasion, tumour size and extent, lymph node metastasis, and pathologic staging were not statistically significant. In addition, the Kaplan-Meier curve analysis revealed that the overall survival of the patients with low p-Syk expression was significantly longer than that of the patients with high p-Syk expression ($p = 0.022$) (Figure 6B). Patients with low p-Syk expression tended to have long disease-free survival; however, this was not statistically significant ($p = 0.13$) (Figure 6C). Univariate and multivariate analyses using the Cox proportional-hazard model were conducted to determine prognostic factors affecting overall survival (Tables 2 and 3). In the univariate analysis, old age (>65 years), larger tumour and/or extent (T3 + T4), presence of lymph node metastasis, high pathological staging (stage III and IV), and high p-Syk expression were associated with a poorer prognosis. The multivariate analysis showed that old age (>65 years), larger tumour and/or extent, and high p-Syk expression were independent poor prognostic factors for overall survival.

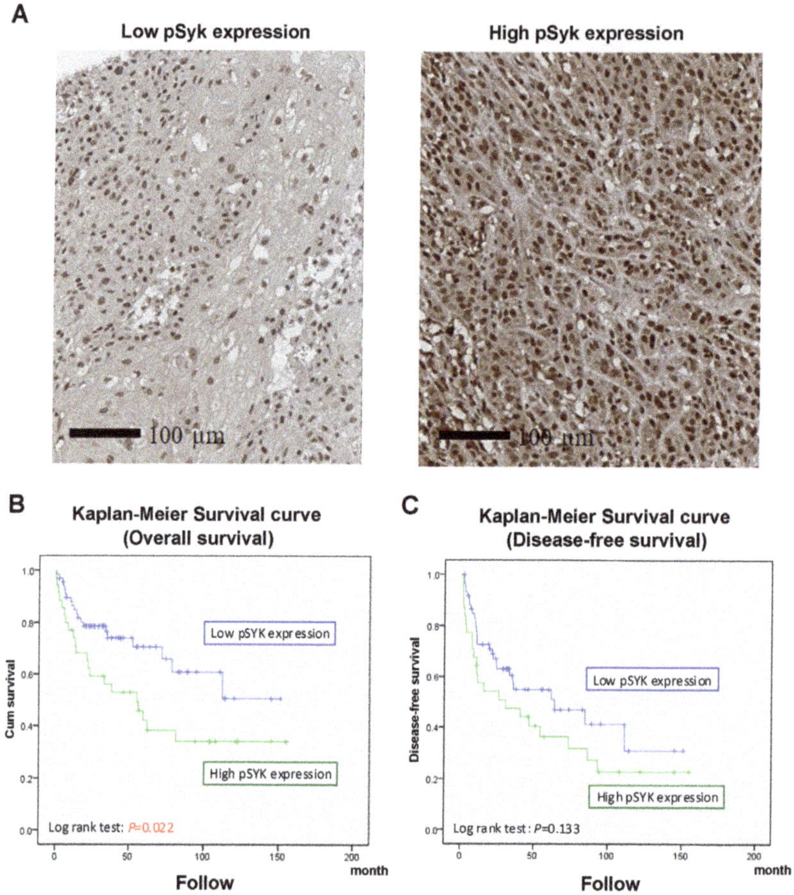

Figure 6. Syk is important for SCC development in human patients. (**A**) Immunohistochemical analysis of p-Syk expression in tissue microarray sections of oral SCC. (**B**,**C**) Correlation of Syk activity to overall survival (**B**) and disease-free survival (**C**) in patients with oral cancer.

Table 1. Association between clinicopathological characteristics and p-Syk expression.

Characteristics	**P-Syk Expression**		***p* Value**
	Low (*n* = 68)	**High (*n* = 37)**	
Gender			0.488
Male	60 (88%)	35 (95%)	
Female	8 (12%)	2 (5%)	
Mean Age (years) ± SD	53.5 ± 14.5	56.7 ± 12.1	0.255
Age			0.628
<65 years	51 (75%)	30 (81%)	
≧65 years	17 (25%)	7 (19%)	
Grading of SCC			0.002
Well to moderately differentiated	58 (85%)	21 (57%)	
Poorly differentiated	10 (15%)	16 (43%)	
Lymphovascular invasion			0.018
Negative	49 (72%)	17 (47%)	
Positive	19 (28%)	19 (53%)	

Table 1. *Cont.*

Characteristics	P-Syk Expression		p Value
	Low (n = 68)	High (n = 37)	
Perineural invasion			0.153
Negative	35 (51%)	13 (36%)	
Positive	33 (49%)	23 (64%)	
Tumor size and extent			1.000
T1+T2	46 (68%)	25 (68%)	
T3+T4	22 (32%)	12 (32%)	
Lymph node metastasis			0.669
Negative	45 (67%)	23 (62%)	
Positive	22 (33%)	14 (38%)	
Stage (AJCC 7th Ed)			1.000
I + II	33 (50%)	18 (49%)	
III + IV	33 (50%)	19 (51%)	

Abbreviation: AJCC, American Joint Committee on Cancer.

Table 2. Univariate analysis of clinicopathological factors and p-Syk expression affecting overall survival.

Variables	Overall Survival		p Value
	Hazard Ratio	95% CI	
Gender			0.388
Male	1	-	-
Female	1.509	0.593–3.840	
Age			0.013
<65 years	1	-	-
≧65 years	2.257	1.191–4.277	
Grading of SCC			0.537
Well-moderately differentiated	1	-	-
Poorly differentiated	1.216	0.653–2.266	
Lymphovascular invasion			0.608
Negative	1		
Positive	1.173	0.637–2.158	
Perineural invasion			0.423
Negative	1		
Positive	1.288	0.694–2.393	
Tumor size and extent			<0.001
T1+T2	1	-	-
T3+T4	4.636	2.562–8.389	
Lymph node metastasis			0.002
Negative	1	-	-
Positive	2.606	1.432–4.742	
Stage (AJCC 7th Ed)			<0.001
I + II	1	-	-
III + IV	4.053	2.104–7.808	
P-Syk expression			0.025
Low	1	-	-
High	2.024	1.094–3.742	

Abbreviations: CI, confidence interval; AJCC, American Joint Committee on Cancer.

Table 3. Multivariate analysis of clinicopathological factors and p-Syk expression affecting overall survival.

Variables	Overall Survival		p Value
	Hazard Ratio	95% CI	
Gender			0.279
Male	1	-	-
Female	1.887	0.597–5.960	
Age			0.001
<65 years	1	-	-
≧65 years	4.153	1.774–9.724	
Grading of SCC			0.393
Well-moderately differentiated	1	-	-
Poorly differentiated	0.710	0.324–1.558	
Lymphovascular invasion			0.809
Negative	1		
Positive	1.108	0.480–2.557	
Perineural invasion			0.586
Negative	1		
Positive	1.246	0.566–2.743	
Tumor size and extent			0.015
T1+T2	1	-	-
T3+T4	3.429	1.267–9.276	
Lymph node metastasis			0.070
Negative	1	-	-
Positive	2.572	0.927–7.139	
Stage (AJCC 7th Ed)			0.668
I + II	1	-	-
III + IV	1.356	0.337–5.447	
P-Syk expression			0.002
Low	1	-	-
High	3.393	1.581–7.283	

Abbreviation: CI, confidence interval; AJCC, American Joint Committee on Cancer.

3.5. Syk and EGFR Regulate PARP1 Activation and Syk Inhibitor Exerts a Synergistic Anti-Tumour Effect with Olaparib

Because we found that Syk is involved in EGFR signalling in SCC, we intended to determine if Syk can serve as a therapeutic target. Moreover, PARP1 is involved in DNA damage repair and its expression is negatively correlated with patient survival rate and prognosis. Given that PARP1 inhibitors (PARPi) are currently used clinically for treating ovarian and metastatic breast cancers [27,28], and PARP inhibition results in the radiosensitisation of HPV/p16-positive HNSCC cells [21,29], we also aimed to determine the combined effects of Syk and PARP inhibition. First, we found that both R406 and the EGFR inhibitor, gefitinib, could induce PARP activation and DNA damage in the three SCC cell lines as indexed by increased PAR formation and γH2AX expression, respectively. Furthermore, PARP cleavage was induced by both agents (Figure 7A). Secondly, we found that although R406 and olaparib did not significantly affect cell viability, their combination induced significant cell death (Figure 7B). Such an enhanced cytotoxic l effect was more apparent in SAS and CAL27 cells than that in A431 cells. Moreover, because targeting EGFR has been evaluated in SCC [15–17], we tested the combined effects of gefitinib and olaparib. We found that although gefitinib itself did not affect cell survival, as demonstrated in Figure 4F, co-treatment with both the agents increased cell death, whereas the efficacy was slightly weaker than that of R406/olaparib (Figure 7B). Altogether, these results suggested a promising treatment of SCC by the combined inhibition of Syk and PARP.

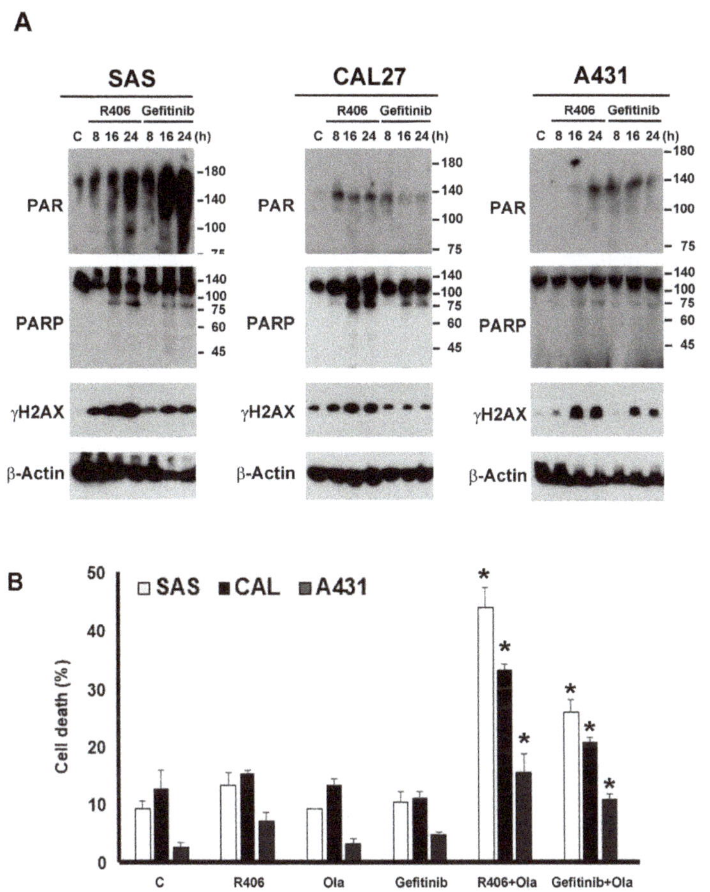

Figure 7. R406 and gefitinib induce PARP activation, and inhibition of Syk and PARP enhances cell death in SCC. (**A**) Cells were treated with R406 (1 μM) or gefitinib (5 μM) for different periods, Figure 1. γ-H2AX, and β-actin. (**B**) Cells were treated with R406 (1 μM), gefitinib (5 μM) and/or olaparib (Ola, 10 μM) for 24 h. Cell death was determined by Annexin V/PI staining and flow cytometry. Data in (**B**) are expressed as the mean ± S.E.M. from three independent experiments. * $p < 0.05$ indicates a synergistic effect of R406 + Ola and gefitinib + Ola on cytotoxicity compared to individual drug treatments.

4. Discussion

The importance of EGFR in tumourigenesis has been well established. Excessive activation of EGFR signalling by overexpression of, or mutations in, EGFR has been found in various types of human tumours, making EGFR a widely recognised target for cancer therapy [30]. Indeed, prior studies have established enhanced anti-tumour efficacy of the EGFR inhibitor (or antibody) with radiation [31] or cytotoxic drugs for A431 cells [17,32]. Nevertheless, long-term EGF treatment also leads to drug resistance, which is an urgent issue that must be overcome. This indicates the necessity to explore new targets for anti-tumour therapy. In this study, we found that Syk is a new downstream signal molecule of SFKs in activated EGFR-transduced signalling pathways, and similar to SFKs, Syk possesses a positive feedback loop to enhance EGFR phosphorylation. Moreover, the association of higher Syk activity with poor clinicopathological features of patients with oral SCC, including poor overall survival and high lymphovascular invasion, was observed. Furthermore, a combined inhibition of Syk and

PARP leading to enhanced SCC cell death may be developed into a new therapeutic regimen in the future.

The major functions of Syk in hematopoietic cells involving innate and adaptive immune systems are well characterised. Apart from the established molecular mechanisms and functional roles in autoimmunity, accumulating evidence suggests that Syk plays dual roles in cancers, i.e., either as a tumour suppressor or promoter [13]. Reportedly, the reduced expression of Syk in breast cancer and melanoma cells is associated with a higher degree of malignancy and poor prognosis. The tumour suppressor action of Syk has been associated with positive regulation of p53 activity [33] and p21 expression [34], as well as negative effects on mitotic progression [35,36]. In contrast, Syk serves as a tumour promoter in hematopoietic malignancies and could be a potential oncogenic driver in small cell lung cancer [37] and ovarian cancer [36] by inducing pro-survival signals [13]. Therefore, Syk could become a new therapeutic target in some cancer cell types. For example, a recent study indicated that a Syk inhibitor could sensitise TRAIL-induced apoptosis by downregulating Mcl-1 in breast cancer cells [38]. In this study, we strengthened our previous speculations that Syk is an upstream signalling molecule for EGFR in SCC cell lines and mediates EGF-induced IL-8 gene expression. IL-8 is a potent chemoattractant molecule that performs different pro-tumoural functions, such as proliferation, angiogenesis, and metastasis of cancer cells. Since in patients with cancer, IL-8 is mainly produced by tumour cells, its serum concentration has been shown to be correlated with the tumour burden, and IL-8 is an effective pharmacodynamic biomarker to detect an early response to immunotherapy [26]. The serum level of IL-8 is significantly increased in patients with recurrent and metastatic head and neck SCC [39]. In addition, elevated levels of IL-8 are found in the saliva of patients with oral SCC; thus, IL-8 could serve as a salivary biomarker of oral SCC [40]. In addition, using human oral SCC samples, we demonstrated the strong association of Syk activity with poor disease progression. All these data prompt us to suggest Syk inhibition as an alternative cancer treatment.

In addition to clarifying the role of Syk in EGFR signalling and the prognosis of patients with SCC, we tested the combined effects of Syk and PARP inhibition on cancer cell death. The PARP1 inhibitor has been shown to sensitise radiotherapy in head and neck SCC [21]. Moreover, because cetuximab can augment cytotoxicity with PARP inhibition in head and neck SCC [41], we also tested the anti-tumour activity of gefitinib and olaparib. Although PARP1 inhibition has been applied in clinical settings in combination with different therapeutic agents, it is still crucial to develop alternative regimens to overcome drug resistance and insufficient efficacy in cancer therapy. In addition, some recent studies have indicated the effect of EGFR activation on the regulation of PARP1 in a cell context-dependent manner. In human LNCaP prostate cancer cells, radiation-induced PARP activation is enhanced through EGFR-ERK signalling [42]. In UT-SCC5 and SAS head and neck cancer cells, PARP1 has been shown to serve as a mediator of EGFR/MEK-dependent regulation of DNA double-strand breaks [43]. In breast cancer cells, a contextual synthetic lethality may exist between combined EGFR and PARP inhibitors [44]. In glioblastoma cells, EGFRvIII overexpression causes increased ROS-dependent DNA strand break accumulation and PARP activation, and reduced DNA repair gene expression, including PARP1, results in improved patient survival [45]. In hepatocellular carcinoma, EGFR and c-MET cooperate to enhance resistance to PARP inhibitors [46]. Furthermore, a heterodimer of EGFR and MET can phosphorylate Y907 of PARP1 in the nucleus of hepatocellular carcinoma and contribute to this resistance [46]. In pancreatic cancer cells, adaptive expression of EGF following exposure to ionising radiation may induce radio-sensitisation of cells through the induction of the cyclin D1/p53/PARP pathway [47]. In contrast, EGF in combination with radiation augments the radiation effects in A431 cells by inhibiting DNA damage repair [48]. Additionally, EGFR-mutant lung cancer cells display higher sensitivity to the PARP inhibitor olaparib [49]. All these studies suggest that EGF can regulate PARP1, which might impact the efficacy of cancer therapy, particularly when DNA damage is caused by cancer therapy. However, to date, studies on the relationship between Syk and DNA breaks are limited. The only finding reported is that DNA damage and DNA double-strand breaks can suppress the expression of Syk in B [50] and NK cells [51].

Our data revealed a time-dependent effect of Syk and EGFR inhibition on DNA damage and PARP1 activation in three SCC cell lines. R406 exerts higher responses in terms of DNA damage and PARP1 activation than those exerted by gefitinib in CAL27 and A431 cells. However, such parallel events are not observed in SAS cells where R406 induces greater DNA damage but less PARylation compared to that of gefitinib. Currently, the reasons underlying this paradoxical finding are unknown. Notably, even though PARP1 cleavage was observed upon treatment of R406 or gefitinib, the cell viability assay did not reveal significant effects for either agent on cell death. We speculate that this might be due to the insufficient activation of executive caspase 3 for cell death and/or the existence of other pro-survival pathways. Another possible reason is that the PARP1 cleavage is mediated by other proteases. Previous studies showed that PARP1 can be cleaved by caspase 1 and caspase 7 to yield 85–89 kD PARP-1 fragments [52–54], and such cleavage leads to an NF-κB-dependent inflammatory gene expression [54]. Therefore, PARP1 cleavage is a hallmark of apoptosis yet not essential. In addition, we speculate the PARP1 activation evoked by R406 and gefitinib alone might participate in the cell survival, for example as previously observed for the DNA repairing process upon moderate DNA damage. However, interestingly, enhanced cell death effect was observed upon the co-inhibition of Syk and PARP. Notably, this enhanced cell death effect is greater than that induced by gefitinib and olaparib. Therefore, we suggest that Syk inhibition leads to the interruption of EGFR-dependent and -independent cellular actions that might exert synergistic effects with olaparib in terms of cell death. It is necessary to further address the molecular mechanisms underlying the coordinative cellular events mediated by Syk and PARP1 in the future. Additionally, it will be interesting to investigate and compare the efficacies between the Syk inhibitor and EGFR mAb (e.g., cetuximab) in SCC upon co-treatment with PARP1 inhibitors.

A431 cells compose a high EGFR-activating skin SCC. Notably, it is unexpected to find EGF at concentrations of 50–100 ng/mL, which is mitogenic for general cell types and cannot induce cell proliferation and viability in A431 cells within 24 h of incubation. Supporting this notion, previous studies reported a growth arrest effect of EGF in A431 cells that was observed slowly [55,56]. To date, the mechanism through which EGF induces growth arrest in A431 cells has been ascribed to the upregulation of IRF-1 through SFR-dependent STAT1 and STAT3 activation [57,58]. Similarly, cell lines that hyperexpress EGFR, such as MDA-MB-468 cells, have been documented to undergo receptor-mediated apoptosis through STAT3 [59]. Therefore, because of the overexpression of EGFR and constitutive overactivation of EGFR in A431 cells, we speculate there might be an aberrant regulation mechanism in the EGFR-dependent cellular functions. Another distinct feature in A431 is the subcellular distribution of EGFR. Unlike primary keratinocytes and SAS cells, EGFR is not only localised in the plasma membrane but is also present in the cytosol in A431 cells. Moreover, after EGF stimulation, the plasma membrane level of EGFR is increased in A431 cells, which is opposite to the general trafficking direction of activated EGFR. We found lower amounts of EGFR expression in SAS cells compared to that of A431 cells could more rapidly undergo EGFR internalisation after EGF stimulation for 1 h. All these phenomena observed in A431 cells might result from the high abundance of EGFR expression in this specific cell line. Another cell type-specific action of the Syk inhibitor is its effects on EGF-induced EGFR trafficking. We found that in A431 cells, R406 could decrease the plasma membrane level of EGFR upon EGF stimulation, whereas in SAS cells, EGF-induced EGFR trafficking to the early endosome displayed a higher level upon Syk inhibition. Because Syk can associate with EGFR in keratinocytes [14] and altered EGFR internalisation and recycling can regulate drug sensitivity in cancer cells [60,61], we plan to further determine the role of Syk in EGFR trafficking, including receptor internalisation, receptor recycling to the plasma membrane, and degradation by lysosomes in the future.

5. Conclusions

We confirmed our previous findings in keratinocytes that Syk mediates EGFR signalling in SCC and Syk is a positive contributor to disease progression in patients with oral SCC. Even though Syk and PARP inhibition cannot induce cell death individually, their combination causes enhanced cell death. Thus, our data provide the rational for co-treatment of PARP inhibition and Syk inhibition in SCC.

Author Contributions: Conceptualization, D.-Y.H. and N.-L.W.; Data curation, D.-Y.H.; Methodology, W.-Y.C. and C.-L.C.; Project administration, W.-W.L.; Resources, W.-Y.C., C.-L.C. and N.-L.W.; Writing—original draft, W.-W.L. All authors have read and agreed to the published version of the manuscript.

Acknowledgments: This work was supported by research grants from Taipei Medical University (TMUTOP103005-5), Ministry of Science and Technology (MOST 103-2320-B-002-069-MY3), Excellent Translational Medicine Research Projects of National Taiwan University College of Medicine (NSCCMOH-94-51), and National Taiwan University (NTU-CC-108L893601).

Conflicts of Interest: The authors declare no conflict of interest.

References

1. Higashiyama, S.; Iwabuki, H.; Morimoto, C.; Hieda, M.; Inoue, H.; Matsushita, N. Membrane-anchored growth factors, the epidermal growth factor family: Beyond receptor ligands. *Cancer Sci.* **2008**, *99*, 214–220. [CrossRef] [PubMed]
2. Wheeler, D.L.; Dunn, E.F.; Harari, P.M. Understanding resistance to EGFR inhibitors-impact on future treatment strategies. *Nat. Rev. Clin. Oncol.* **2010**, *7*, 493–507. [CrossRef] [PubMed]
3. Aran, V.; Omerovic, J. Current approaches in NSCLC targeting K-RAS and EGFR. *Int. J. Mol. Sci.* **2019**, *20*, 5701. [CrossRef] [PubMed]
4. Giordano, G.; Remo, A.; Porras, A.; Pancione, M. Immune resistance and EGFR antagonists in colorectal cancer. *Cancers* **2019**, *11*, 1089. [CrossRef] [PubMed]
5. Kerrigan, A.M.; Brown, G.D. Syk-coupled C-type lectins in immunity. *Trends Immunol.* **2011**, *32*, 151–156. [CrossRef]
6. Lin, Y.C.; Huang, D.Y.; Chu, C.L.; Lin, Y.L.; Lin, W.W. The tyrosine kinase Syk differentially regulates Toll-like receptor signaling downstream of the adaptor molecules TRAF6 and TRAF3. *Sci. Signal.* **2013**, *6*, ra71. [CrossRef]
7. Turner, M.; Schweighoffer, E.; Colucci, F.; Di Santo, J.P.; Tybulewicz, V.L. Tyrosine kinase SYK: Essential functions for immunoreceptor signalling. *Immunol. Today* **2000**, *21*, 148–154. [CrossRef]
8. Riccaboni, M.; Bianchi, I.; Petrillo, P. Spleen tyrosine kinases: Biology, therapeutic targets and drugs. *Drug Discov. Today* **2010**, *15*, 517–530. [CrossRef]
9. Kang, Y.; Jiang, X.; Qin, D.; Wang, L.; Yang, J.; Wu, A.; Huang, F.; Ye, Y.; Wu, J. Efficacy and safety of multiple dosages of fostamatinib in adult patients with rheumatoid arthritis: A systematic review and meta-analysis. *Front. Pharmacol.* **2019**, *10*, 897. [CrossRef]
10. Szilveszter, K.P.; Németh, T.; Mócsai, A. Tyrosine kinases in autoimmune and inflammatory skin diseases. *Front. Immunol.* **2019**, *10*, 1862. [CrossRef]
11. Liu, D.; Mamorska-Dyga, A. Syk inhibitors in clinical development for hematological malignancies. *J. Hematol. Oncol.* **2017**, *10*, 145. [CrossRef] [PubMed]
12. Bartaula-Brevik, S.; Lindstad Brattås, M.K.; Tvedt, T.H.A.; Reikvam, H.; Bruserud, Ø. Splenic tyrosine kinase (SYK) inhibitors and their possible use in acute myeloid leukemia. *Expert. Opin. Investig. Drugs* **2018**, *27*, 377–387. [CrossRef] [PubMed]
13. Krisenko, M.O.; Geahlen, R.L. Calling in SYK: SYK's dual role as a tumor promoter and tumor suppressor in cancer. *Biochim. Biophys. Acta* **2015**, *1853*, 254–263. [CrossRef] [PubMed]
14. Wu, N.L.; Huang, D.Y.; Wang, L.F.; Kannagi, R.; Fan, Y.C.; Lin, W.W. Spleen tyrosine kinase mediates EGFR signaling to regulate keratinocyte terminal differentiation. *J. Investig. Dermatol.* **2016**, *136*, 192–201. [CrossRef] [PubMed]
15. Kundu, S.K.; Nestor, M. Targeted therapy in head and neck cancer. *Tumour. Biol.* **2012**, *33*, 707–721. [CrossRef]
16. Wong, C.H.; Ma, B.B.; Hui, C.W.; Tao, Q.; Chan, A.T. Preclinical evaluation of afatinib (BIBW2992) in esophageal squamous cell carcinoma (ESCC). *Am. J. Cancer Res.* **2015**, *5*, 3588–3599.

17. Khalil, A.; Jameson, M.J. The EGFR inhibitor gefitinib enhanced the response of human oral squamous cell carcinoma to cisplatin in vitro. *Drugs R D* **2017**, *17*, 545–555. [CrossRef]
18. Lee, S.T.; Ji, H.; Greening, D.W.; Speirs, R.W.; Rigopoulos, A.; Pillay, V.; Murone, C.; Vitali, A.; Stühler, K.; Johns, T.G.; et al. Global protein profiling reveals anti-EGFR monoclonal antibody 806-modulated proteins in A431 tumor xenografts. *Growth Factors* **2013**, *31*, 154–164. [CrossRef]
19. Iberri, D.J.; Colevas, A.D. Balancing safety and efficacy of epidermal growth factor receptor inhibitors in patients with squamous cell carcinoma of the head and neck. *Oncologist* **2015**, *20*, 1393–1403. [CrossRef]
20. Stratigos, A.; Garbe, C.; Lebbe, C.; Malvehy, J.; del Marmol, V.; Pehamberger, H.; Peris, K.; Becker, J.C.; Zalaudek, I.; Saiag, P.; et al. Diagnosis and treatment of invasive squamous cell carcinoma of the skin: European consensus-based interdisciplinary guideline. *Eur. J. Cancer* **2015**, *51*, 1989–2007. [CrossRef]
21. Nowsheen, S.; Bonner, J.A.; Yang, E.S. The poly(ADP-ribose) polymerase inhibitor ABT-888 reduces radiation-induced nuclear EGFR and augments head and neck tumor response to radiotherapy. *Radiother. Oncol.* **2011**, *99*, 331–338. [CrossRef] [PubMed]
22. Wu, N.L.; Huang, D.Y.; Hsieh, S.L.; Hsiao, C.H.; Lee, T.A.; Lin, W.W. EGFR-driven up-regulation of decoy receptor 3 in keratinocytes contributes to the pathogenesis of psoriasis. *Biochim. Biophys. Acta* **2013**, *1832*, 1538–1548. [CrossRef] [PubMed]
23. Biscardi, J.S.; Maa, M.C.; Tice, D.A.; Cox, M.E.; Leu, T.H.; Parsons, S.J. c-Src-mediated phosphorylation of the epidermal growth factor receptor on Tyr845 and Tyr1101 is associated with modulation of receptor function. *J. Biol. Chem.* **1999**, *274*, 8335–8343. [CrossRef] [PubMed]
24. Nishioka, T.; Kim, H.S.; Luo, L.Y.; Huang, Y.; Guo, J.; Chen, C.Y. Sensitization of epithelial growth factor receptors by nicotine exposure to promote breast cancer cell growth. *Breast Cancer Res.* **2011**, *13*, R113. [CrossRef]
25. Asbagh, L.A.; Vazquez, I.; Vecchione, L.; Budinska, E.; De Vriendt, V.; Baietti, M.F.; Steklov, M.; Jacobs, B.; Hoe, N.; Singh, S.; et al. The tyrosine phosphatase PTPRO sensitizes colon cancer cells to anti-EGFR therapy through activation of SRC-mediated EGFR signaling. *Oncotarget* **2014**, *5*, 10070–10083. [CrossRef]
26. Alfaro, C.; Sanmamed, M.F.; Rodríguez-Ruiz, M.E.; Teijeira, Á.; Oñate, C.; González, Á.; Ponz, M.; Schalper, K.A.; Pérez-Gracia, J.L.; Melero, I. Interleukin-8 in cancer pathogenesis, treatment and follow-up. *Cancer Treat. Rev.* **2017**, *60*, 24–31. [CrossRef]
27. Boussios, S.; Karihtala, P.; Moschetta, M.; Abson, C.; Karathanasi, A.; Zakynthinakis-Kyriakou, N.; Ryan, J.E.; Sheriff, M.; Rassy, E.; Pavlidis, N. Veliparib in ovarian cancer: A new synthetically lethal therapeutic approach. *Investig. New Drugs* **2020**, *38*, 181–193. [CrossRef]
28. Sachdev, E.; Tabatabai, R.; Roy, V.; Rimel, B.J.; Mita, M.M. PARP inhibition in cancer: An update on clinical development. *Target Oncol.* **2019**, *14*, 657–679. [CrossRef]
29. Güster, J.D.; Weissleder, S.V.; Busch, C.J.; Kriegs, M.; Petersen, C.; Knecht, R.; Dikomey, E.; Rieckmann, T. The inhibition of PARP but not EGFR results in the radiosensitization of HPV/p16-positive HNSCC cell lines. *Radiother. Oncol.* **2014**, *113*, 345–351. [CrossRef]
30. Hynes, N.E.; Lane, H.A. ERBB receptors and cancer: The complexity of targeted inhibitors. *Nat. Rev. Cancer* **2005**, *5*, 341–354. [CrossRef]
31. Saleh, M.N.; Raisch, K.P.; Stackhouse, M.A.; Grizzle, W.E.; Bonner, J.A.; Mayo, M.S.; Kim, H.G.; Meredith, R.F.; Wheeler, R.H.; Buchsbaum, D.J. Combined modality therapy of A431 human epidermoid cancer using anti-EGFr antibody C225 and radiation. *Cancer Biother. Radiopharm.* **1999**, *14*, 451–463. [CrossRef] [PubMed]
32. Johns, T.G.; Luwor, R.B.; Murone, C.; Walker, F.; Weinstock, J.; Vitali, A.A.; Perera, R.M.; Jungbluth, A.A.; Stockert, E.; Old, L.J.; et al. Antitumor efficacy of cytotoxic drugs and the monoclonal antibody 806 is enhanced by the EGF receptor inhibitor AG1478. *Proc. Natl. Acad. Sci. USA* **2003**, *100*, 15871–15876. [CrossRef]
33. Althubiti, M. Spleen tyrosine kinase inhibition modulates p53 activity. *J. Cell Death* **2017**, *10*. [CrossRef] [PubMed]
34. Bailet, O.; Fenouille, N.; Abbe, P.; Robert, G.; Rocchi, S.; Gonthier, N.; Denoyelle, C.; Ticchioni, M.; Ortonne, J.P.; Ballotti, R.; et al. Spleen tyrosine kinase functions as a tumor suppressor in melanoma cells by inducing senescence-like growth arrest. *Cancer Res.* **2009**, *69*, 2748–2756. [CrossRef] [PubMed]
35. Zyss, D.; Montcourrier, P.; Vidal, B.; Anguille, C.; Merezegue, F.; Sahuquet, A.; Coopman, P.J. The Syk tyrosine kinase localizes to the centrosomes and negatively affects mitotic progression. *Cancer Res.* **2005**, *65*, 10872–10880. [CrossRef] [PubMed]

36. Prinos, P.; Garneau, D.; Lucier, J.F.; Gendron, D.; Couture, S.; Boivin, M.; Elela, S.A. Alternative splicing of SYK regulates mitosis and cell survival. *Nat. Struct. Mol. Biol.* **2011**, *18*, 673–679. [CrossRef] [PubMed]
37. Udyavar, A.R.; Hoeksema, M.D.; Clark, J.E.; Zou, Y.; Tang, Z.; Li, Z.; Quaranta, V. Co-expression network analysis identifies spleen tyrosine kinase (SYK) as a candidate oncogenic driver in a subset of small-cell lung cancer. *BMC Syst. Biol.* **2013**, *7* (Suppl. 5), S1. [CrossRef]
38. Kim, S.Y.; Park, S.E.; Shim, S.M.; Park, S.; Kim, K.K.; Jeong, S.Y.; Choi, E.K.; Hwang, J.J.; Jin, D.H.; Chung, C.D.; et al. Bay 61-3606 Sensitizes TRAIL-induced apoptosis by downregulating Mcl-1 in breast cancer cells. *PLoS ONE* **2015**, *10*, e0146073. [CrossRef]
39. Gokhale, A.S.; Haddad, R.I.; Cavacini, L.A.; Wirth, L.; Weeks, L.; Hallar, M.; Faucher, J.; Posner, M.R. Serum concentrations of interleukin-8, vascular endothelial growth factor, and epidermal growth factor receptor in patients with squamous cell cancer of the head and neck. *Oral Oncol.* **2005**, *41*, 70–76. [CrossRef]
40. John, M.A.; Li, Y.; Zhou, X.; Denny, P.; Ho, C.M.; Montemagno, C.; Shi, W.; Qi, F.; Wu, B.; Sinha, U.; et al. Interleukin 6 and interleukin 8 as potential biomarkers for oral cavity and oropharyngeal squamous cell carcinoma. *Arch. Otolaryngol. Head Neck Surg.* **2004**, *30*, 929–935. [CrossRef]
41. Nowsheen, S.; Bonner, J.A.; Lobuglio, A.F.; Trummell, H.; Whitley, A.C.; Dobelbower, M.C.; Yang, E.S. Cetuximab augments cytotoxicity with poly (ADP-ribose) polymerase inhibition in head and neck cancer. *PLoS ONE* **2011**, *6*, e24148. [CrossRef] [PubMed]
42. Hagan, M.P.; Yacoub, A.; Dent, P. Radiation-induced PARP activation is enhanced through EGFR-ERK signaling. *J. Cell Biochem.* **2007**, *101*, 1384–1393. [CrossRef] [PubMed]
43. Kwiatkowski, M.; Gleißner, L.; Riepen, B.; Hoffer, K.; Wurlitzer, M.; Petersen, C.; Dikomey, E.; Rothkamm, K.; Schlüter, H.; Kriegs, M. Quantitative proteomics unveiled: Regulation of DNA double strand break repair by EGFR involves PARP1. *Radiother. Oncol.* **2015**, *116*, 423–430.
44. Nowsheen, S.; Cooper, T.; Stanley, J.A.; Yang, E.S. Synthetic lethal interactions between EGFR and PARP inhibition in human triple negative breast cancer cells. *PLoS ONE* **2012**, *7*, e46614. [CrossRef] [PubMed]
45. Nitta, M.; Kozono, D.; Kennedy, R.; Stommel, J.; Ng, K.; Zinn, P.O.; Kushwaha, D.; Kesari, S.; Inda, M.M.; Wykosky, J.; et al. Targeting EGFR induced oxidative stress by PARP1 inhibition in glioblastoma therapy. *PLoS ONE* **2010**, *5*, e10767. [CrossRef]
46. Dong, Q.; Du, Y.; Li, H.; Liu, C.; Wei, Y.; Chen, M.K.; Zhao, X.; Chu, Y.Y.; Qiu, Y.; Qin, L.; et al. EGFR and c-MET cooperate to enhance resistance to PARP inhibitors in hepatocellular carcinoma. *Cancer Res.* **2019**, *79*, 819–829. [CrossRef]
47. Liu, X.; Chen, H.; Hou, Y.; Ma, X.; Ye, M.; Huang, R.; Hu, B.; Cao, H.; Xu, L.; Liu, M.; et al. Adaptive EGF expression sensitizes pancreatic cancer cells to ionizing radiation through activation of the cyclin D1/p53/PARP pathway. *Int. J. Oncol.* **2019**, *54*, 1466–1480. [CrossRef]
48. Kim, K.; Wu, H.G.; Jeon, S.R. Epidermal growth factor-induced cell death and radiosensitization in epidermal growth factor receptor-overexpressing cancer cell lines. *Anticancer Res.* **2015**, *35*, 245–253.
49. Pfäffle, H.N.; Wang, M.; Gheorghiu, L.; Ferraiolo, N.; Greninger, P.; Borgmann, K.; Settleman, J.; Benes, C.H.; Sequist, L.V.; Zou, L.; et al. EGFR-activating mutations correlate with a Fanconi anemia-like cellular phenotype that includes PARP inhibitor sensitivity. *Cancer Res.* **2013**, *73*, 6254–6263.
50. Soodgupta, D.; White, L.S.; Yang, W.; Johnston, R.; Andrews, J.M.; Kohyama, M.; Murphy, K.M.; Mosammaparast, N.; Payton, J.E.; Bednarski, J.J. RAG-Mediated DNA breaks attenuate PU.1 activity in early B cells through activation of a SPIC-BCLAF1 complex. *Cell Rep.* **2019**, *29*, 829–843. [CrossRef]
51. Pugh, J.L.; Nemat-Gorgani, N.; Norman, P.J.; Guethlein, L.A.; Parham, P. Human NK cells downregulate Zap70 and Syk in response to prolonged activation or DNA damage. *J. Immunol.* **2018**, *200*, 1146–1158. [CrossRef] [PubMed]
52. Lazebnik, Y.A.; Kaufmann, S.H.; Desnoyers, S.; Poirier, G.G.; Earnshaw, W.C. Cleavage of poly(ADP-ribose) polymerase by a proteinase with properties like ICE. *Nature* **1994**, *371*, 346–347. [CrossRef] [PubMed]
53. Malireddi, R.K.; Ippagunta, S.; Lamkanfi, M.; Kanneganti, T.D. Cutting edge: Proteolytic inactivation of poly(ADP-ribose) polymerase 1 by the Nlrp3 and Nlrc4 inflammasomes. *J. Immunol.* **2010**, *185*, 3127–3130. [CrossRef] [PubMed]
54. Erener, S.; Pétrilli, V.; Kassner, I.; Minotti, R.; Castillo, R.; Santoro, R.; Hassa, P.O.; Tschopp, J.; Hottiger, M.O. Inflammasome-activated caspase 7 cleaves PARP1 to enhance the expression of a subset of NF-κB target genes. *Mol. Cell* **2012**, *46*, 200–211. [CrossRef] [PubMed]

55. Barnes, D.W. Epidermal growth factor inhibits growth of A431 human epidermoid carcinoma in serum-free cell culture. *J. Cell Biol.* **1982**, *93*, 1–4. [CrossRef] [PubMed]
56. Gulli, L.F.; Palmer, K.C.; Chen, Y.Q.; Reddy, K.B. Epidermal growth factor-induced apoptosis in A431 cells can be reversed by reducing the tyrosine kinase activity. *Cell Growth Differ.* **1996**, *7*, 173–178.
57. Andersen, P.; Pedersen, M.W.; Woetmann, A.; Villingshøj, M.; Stockhausen, M.T.; Odum, N.; Poulsen, H.S. EGFR induces expression of IRF-1 via STAT1 and STAT3 activation leading to growth arrest of human cancer cells. *Int. J. Cancer* **2008**, *122*, 342–349. [CrossRef]
58. Suzuki, Y.; Demoliere, C.; Kitamura, D.; Takeshita, H.; Deuschle, U.; Watanabe, T. HAX-1, a novel intracellular protein, localized on mitochondria, directly associates with HS1, a substrate of Src family tyrosine kinases. *J. Immunol.* **1997**, *158*, 2736–2744.
59. Jackson, N.M.; Ceresa, B.P. EGFR-mediated apoptosis via STAT3. *Exp. Cell Res.* **2017**, *356*, 93–103. [CrossRef]
60. Mathew, M.P.; Tan, E.; Saeui, C.T.; Bovonratwet, P.; Sklar, S.; Bhattacharya, R.; Yarema, K.J. Metabolic flux-driven sialylation alters internalization, recycling, and drug sensitivity of the epidermal growth factor receptor (EGFR) in SW1990 pancreatic cancer cells. *Oncotarget* **2016**, *7*, 66491–66511. [CrossRef]
61. Hendriks, B.S.; Griffiths, G.J.; Benson, R.; Kenyon, D.; Lazzara, M.; Swinton, J.; Beck, S.; Hickinson, M.; Beusmans, J.M.; Lauffenburger, D.; et al. Decreased internalisation of erbB1 mutants in lung cancer is linked with a mechanism conferring sensitivity to gefitinib. *Syst. Biol.* **2006**, *153*, 457–466. [CrossRef] [PubMed]

cancers

MDPI

Article

TDP1 and TOP1 Modulation in Olaparib-Resistant Cancer Determines the Efficacy of Subsequent Chemotherapy

Jin Won Kim [1,2,3,†], Ahrum Min [4,5,†], Seock-Ah Im [1,3,4,5,6,*], Hyemin Jang [4], Yu Jin Kim [4], Hee-Jun Kim [7], Kyung-Hun Lee [1,4,5,6], Tae-Yong Kim [1,4,5,6], Keun Wook Lee [1,2], Do-Youn Oh [1,3,4,5,6], Jee-Hyun Kim [1,2,3] and Yung-Jue Bang [1,4,5,6]

1. Department of Internal Medicine, Seoul National University College of Medicine, Seoul 03080, Korea; jwkim@snubh.org (J.W.K.); kyunghunlee@snu.ac.kr (K.-H.L.); ktyongmd@gmail.com (T.-Y.K.); imdoctor@snu.ac.kr (K.W.L.); ohdoyoun@snu.ac.kr (D.-Y.O.); jhkimmd@snu.ac.kr (J.-H.K.); bangyj@snu.ac.kr (Y.-J.B.)
2. Division of Hematology and Medical Oncology, Department of Internal Medicine, Seoul National University Bundang Hospital, Seongnam 13620, Korea
3. Translational Medicine, Seoul National University College of Medicine, Seoul 03080, Korea
4. Cancer Research Institute, Seoul National University, Seoul 03080, Korea; mar6716@snu.ac.kr (A.M.); applee63@nate.com (H.J.); dbwls0331@nate.com (Y.J.K.)
5. Biomedical Research Institute, Seoul National University Hospital, Seoul 03080, Korea
6. Division of Hematology and Medical Oncology, Department of Internal Medicine, Seoul National University Hospital, Seoul 03080, Korea
7. Department of Internal Medicine, Chung-Ang University College of Medicine, Seoul 06973, Korea; heejun@cau.ac.kr

* Correspondence: moisa@snu.ac.kr; Tel.: +82-2-2072-0850; Fax: +82-2-765-7081

† Jin Won Kim and Ahrum Min contributed equally to this work as co-first authors.

Received: 31 December 2019; Accepted: 31 January 2020; Published: 3 February 2020

Abstract: The aim of this study was to elucidate the carryover effect of olaparib to subsequent chemotherapy and its underlying mechanisms. We generated olaparib-resistant SNU-484, SNU-601, SNU-668, and KATO-III gastric cancer cell lines and confirmed their resistance by cell viability and colony forming assays. Notably, olaparib-resistant cell lines displayed cross-resistance to cisplatin except for KATO-III. Inversely, olaparib-resistant SNU-484, SNU-668, and KATO-III were more sensitive to irinotecan than their parental cells. However, sensitivity to paclitaxel remained unaltered. There were compensatory changes in the ATM/ATR axis and p-Chk1/2 protein expression. ERCC1 was also induced in olaparib-resistant SNU-484, SNU-601, and SNU-668, which showed cross-resistance to cisplatin. Olaparib-resistant cells showed tyrosyl-DNA phosphodiesterase 1 (TDP1) downregulation with higher topoisomerase 1 (TOP1) activity, which is a target of irinotecan. These changes of TOP1 and TDP1 in olaparib-resistant cells was confirmed as the underlying mechanism for increased irinotecan sensitivity through manipulated gene expression of TOP1 and TDP1 by specific plasmid transfection and siRNA. The patient-derived xenograft model established from the patient who acquired resistance to olaparib with BRCA2 mutation showed increased sensitivity in irinotecan. In conclusion, the carryover effects of olaparib to improve antitumor effect of subsequent irinotecan were demonstrated. These effects should be considered when determining the subsequent therapy with olaparib.

Keywords: olaparib; carryover effect; TOP1 activity; TDP1; irinotecan

1. Introduction

Olaparib is an oral poly(ADP-ribose) polymerase (PARP) inhibitor that traps inactivated PARP onto single-strand DNA breaks (SSBs), leading to double-strand DNA breaks that induce cell death [1,2]. Olaparib showed antitumor effect alone or in combination with chemotherapy in tumors with deficiencies at the repair of double-strand breaks caused by BRCA1/2 mutations, ATM dysfunction, and RAD51C deficiency [3–6]. Recently, in germline BRCA-mutated advanced ovarian cancer patients who have been treated with three or more prior lines of chemotherapy, olaparib showed clinical efficacy and has been approved by U.S. Food and Drug Administration [4]. The Cancer Genome Atlas reported that 7 of 287 gastric patients (2.4%) showed genetic alterations in RAD51C gene, and it has been well established that loss of RAD51C is associated with increased cancer risk [7–9]. In our previous study, we found that RAD51C depletion led to hypersensitivity to olaparib in gastric cancer cells [6]. Therefore, PARP inhibition can be an attractive strategy for gastric cancer treatment.

In a randomized phase II trial of second-line metastatic gastric cancer, the addition of olaparib to paclitaxel showed much longer overall survival (OS) than paclitaxel plus placebo, although the progression-free survival (PFS) was marginally different between the two groups [10]. This finding was also identified in a randomized phase 3 trial of using olaparib in second-line gastric cancer [11]. It means that post-progression survival was longer in olaparib arm; however, the reason for this could not be fully explained. There is no preclinical evidence to explain how olaparib affects post-progression survival. In a previous study, our group showed the prolonged antitumor effect of olaparib even after stopping olaparib treatment in an in vivo xenograft model [6]. Based on this finding, carryover effect of olaparib, which has an effect on subsequent chemotherapy, was suggested as one of the potential explanations for this, particularly, in irinotecan, which is usually applied as a third line of treatment.

PARP-1 is activated by irinotecan-induced DNA breaks. PARP inhibitors share their predictive characters with platinum and enhance the cytotoxicity of topoisomerase 1 (TOP1) inhibitors [12]. Restoration of homologous recombination through a loss of 53BP1 or BRCA re-expression is an important mechanism for PARP inhibitor resistance in BRCA1-deficient mammary tumors [13]. A previous study has demonstrated that ATM regulates TOP2 expression. ATM loss results in increased TOP2 levels and enhances sensitivity to TOP2 inhibition [14]. Therefore, it could be possible to compensate for an impaired DNA repair pathway from olaparib by another repair-pathway related to SSB, including the TOP1 and tyrosyl-DNA phosphodiesterase 1 (TDP1) pathways, in resistant cell lines that have been exposed to olaparib for long periods. On the other hand, it was reported that TDP1 and TOP1 are predictive biomarkers for irinotecan and new therapeutic targets in the era of precision medicine [14,15].

On these backgrounds, we explored the carryover effect of olaparib, which influences the efficacy of subsequent chemotherapy after olaparib treatment. Additionally, we attempted to evaluate the mechanisms underlying this effect.

2. Results

2.1. Altered Sensitivity of Chemotherapeutic Agents in Olaparib-Resistant Cells

We established olaparib-resistant cell lines by long-term treatment with olaparib from SNU-484, SNU-601, SNU-668, and KATO-III. The resistance was confirmed by using cell growth inhibition and colony forming assays (Table 1 and Figure 1). Except for KATO-III, all olaparib-resistant cell lines became resistant to cisplatin. Inversely, the olaparib-resistant cell lines became more sensitive to irinotecan, except for SNU-601. Apoptosis also increased after irinotecan treatment in olaparib-resistant cell lines (Figure S1). Sensitivity toward paclitaxel was not altered after the acquisition of resistance to olaparib.

Table 1. Drug sensitivity in parental and olaparib-resistant cells, as assessed by using cell growth-inhibition assay.

Cell Lines *		Olaparib IC_{50} (μmol/L, mean ± SD)	Cisplatin IC_{50} (μmol/L, mean ± SD)	Paclitaxel IC_{50} (μmol/L, mean ± SD)	Irinotecan IC_{50} (μmol/L, mean ± SD)
SNU-484	Parental cells	4.16 ± 0.05	0.8 ± 0.01	2.7 ± 0.3	3.96 ± 0.2
	Olaparib-resistant cells	>10	2.02 ± 0.03	2.5 ± 0.08	1.47 ± 0.2
SNU-601	Parental cells	0.73 ± 0.006	0.75 ± 0.005	4.63 ± 0.08	1.053 ± 0.03
	Olaparib-resistant cells	7.3 ± 0.4	3.92 ± 0.05	4.92 ± 0.05	1.82 ± 0.007
SNU-668	Parental cells	11.07	5.88 ± 0.2	5.35 ± 0.1	>10
	Olaparib-resistant cells	>20	>10	5.66 ± 0.1	5.95 ± 0.1
KATO-III	Parental cells	3.56 ± 0.4	>10	5.67 ± 0.1	>10
	Olaparib-resistant cells	>10	4.13 ± 0.4	5.82 ± 0.1	5.5 ± 0.4

* MTT for 5 days.

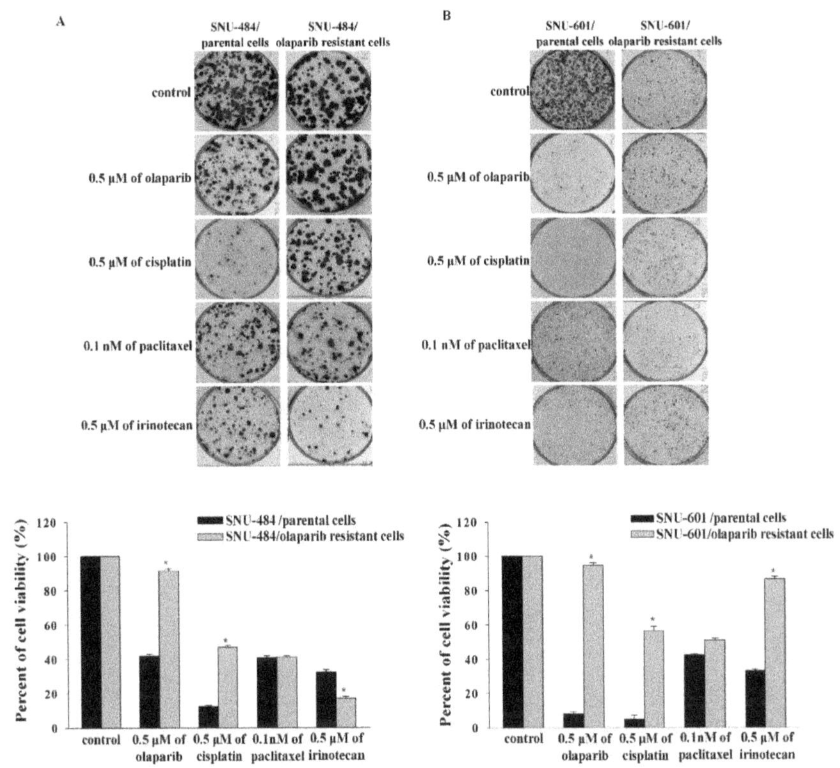

Figure 1. *Cont.*

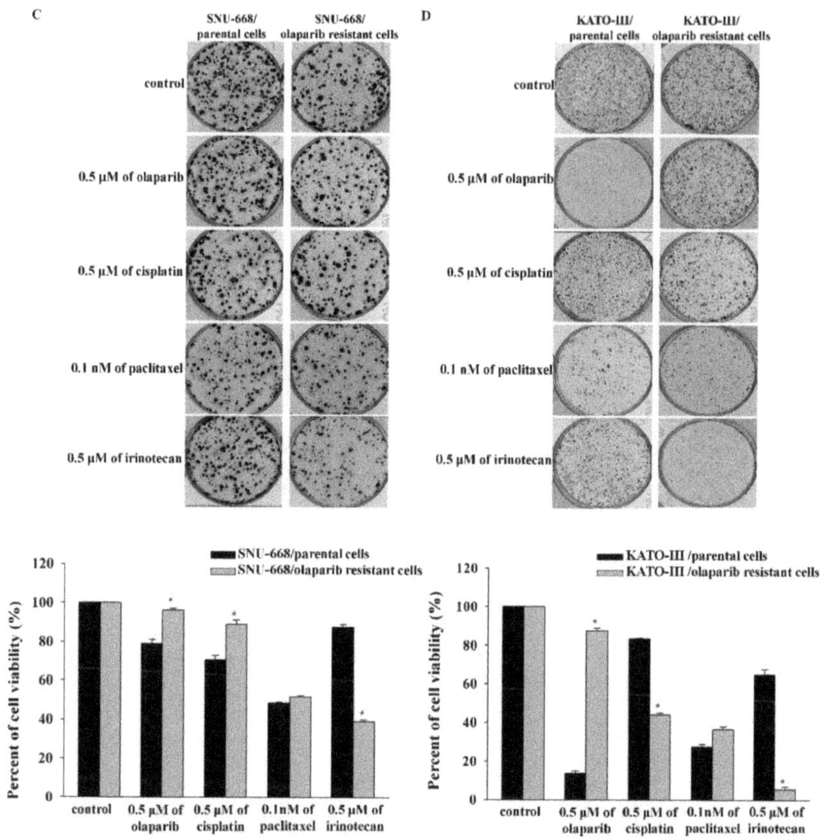

Figure 1. The sensitivity of chemotherapeutic agents was altered in olaparib-resistant cells. Cell viability was calculated through counting cell colonies by using GELCOUNT in SNU-484 (**A**), SNU-601 (**B**), SNU-668 (**C**), and KATO-III (**D**) cells. * indicates $p < 0.001$.

2.2. Changes of Proteins Related to the DNA-Damage Response in Olaparib-Resistant Cell Lines

To evaluate the mechanisms underlying the altered sensitivity toward cisplatin, changes of the DNA-damage response proteins were evaluated. p-ATR was induced in all olaparib resistant cell lines (Figure 2). p-ATM was slightly upregulated in olaparib-resistant SNU-484 and olaparib-resistant SNU-668. p-Chk1 and p-Chk2 were elevated in olaparib-resistant cell lines compared with the parental cells. Furthermore, ERCC1 was also upregulated in olaparib-resistant SNU-484, SNU-601, and SNU-668 cells, which were resistant to cisplatin. This ERCC1 induction was not detected in olaparib-resistant KATO-III cells, which were not resistant to cisplatin.

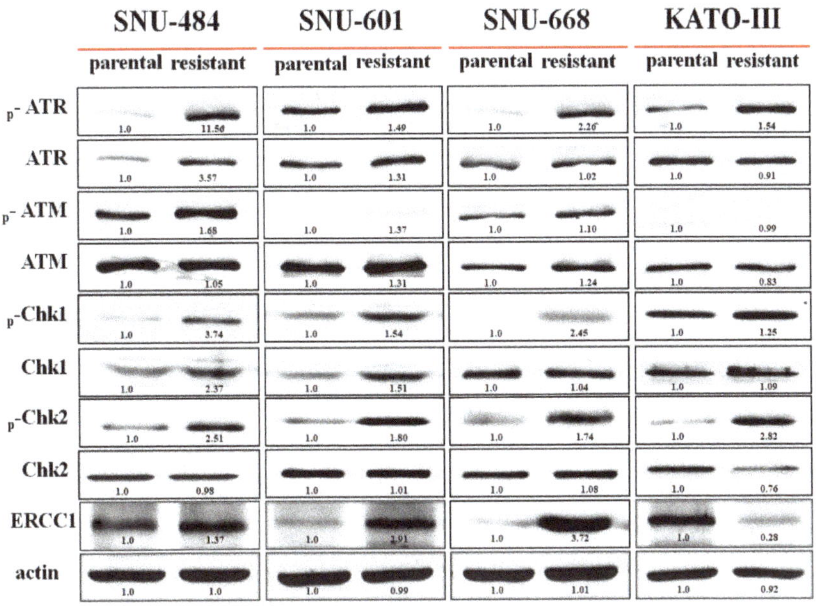

Figure 2. The expressions of DNA-damage response proteins were altered after acquisition of olaparib resistance. Western blot analysis for p-ATR, ATR, p-ATM, ATM, p-Chk1, Chk1, p-Chk2, Chk2, and ERCC1 was conducted to evaluate the altered expression after acquisition of resistance to olaparib. There were compensatory changes in the DNA-damage response proteins.

2.3. *Morphological Changes in the Olaparib-Resistant Cell Lines*

With the acquisition of resistance to olaparib, the nuclear size of the resistant cells increased, as assessed by confocal imaging (Figure 3A). Increased abnormal DNA contents indicating aneuploidy was not observed in cells with increased nuclear size (Figure 3B). In the olaparib-resistant cells, increased nuclear size indicated the relaxation of DNA condensation [16,17].

A

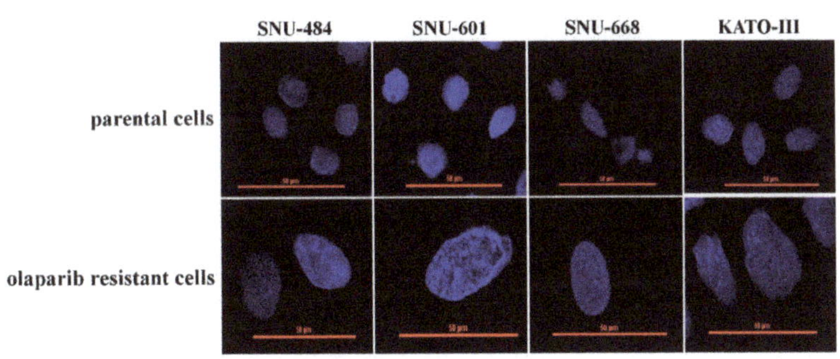

Figure 3. *Cont.*

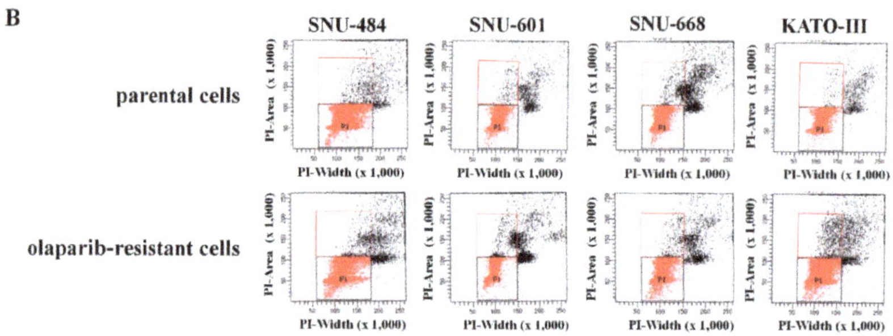

Figure 3. The nuclei were larger in olaparib-resistant cells than in parental cells. The degree of DNA condensation was examined by 4′,6-diamidino-2-phenylindole (DAPI) staining and confocal laser microscopy. Representative images are presented (80× original magnification). Scale bars represent 50 μm. In olaparib-resistant cells, increased nucleus indicates a relaxation of DNA condensation (**A**). The proportion of cells with a DNA content of more than 4n was calculated by flow cytometry with PI staining. The increased nuclear size of olaparib-resistant cells was not related to aneuploidy (**B**).

2.4. Increased TOP1 Activity and Decreased TDP1 Expression in Olaparib-Resistant Cell Lines

To evaluate the underlying mechanism of increased irinotecan sensitivity in the olaparib-resistant cell lines, protein expression, and activity of TOP1 and protein expression of TDP1, which are the target and predictive markers of irinotecan, were measured. Olaparib-resistant cells showed higher TOP1 activity than their parental cells, although protein expression of TOP1 was not altered (Figure 4A,B). In addition, TDP1 expression was decreased in olaparib-resistant cell lines except in olaparib-resistant SNU-601, which did not exhibit sensitivity to irinotecan and had a mutation in TDP1 (Table 2).

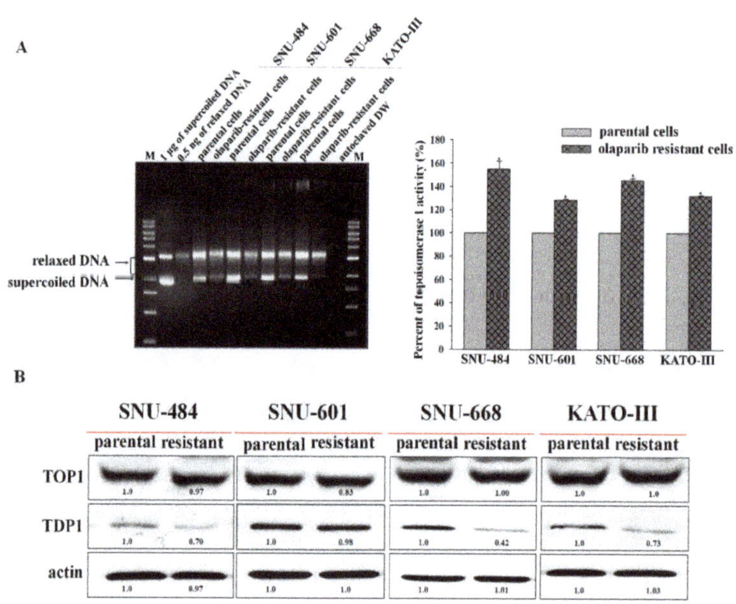

Figure 4. Topoisomerase 1 (TOP1) activity was increased and tyrosyl-DNA phosphodiesterase 1 (TDP1) expression was downregulated in olaparib-resistant cells. TOP1 activity was increased in all olaparib-resistant cells compared with parental cells (**A**), although the protein expression of TOP1 was not altered (**B**). TDP1 protein expression was downregulated in olaparib-resistant SNU-484, SNU-668, and KATO-III cells compared with their parental cells (**B**). * indicates $p < 0.001$.

Table 2. Mutation status for *TDP1* and *TOP1* in parental and olaparib-resistant cells.

Cell Line	*TDP1*	*TOP1*
SNU-484/parental cells	wild type	wild type
SNU-484/olaparib-resistant cells	wild type	wild type
SNU-601/parental cells	wild type	wild type
SNU-601/olaparib-resistant cells	A520D	wild type
SNU-668/parental cells	wild type	wild type
SNU-668/olaparib-resistant cells	wild type	wild type
KATO-III/parental cells	wild type	wild type
KATO-III/olaparib-resistant cells	wild type	wild type

TDP1, tyrosyl-DNA phosphodiesterase 1; TOP1, topoisomerase 1.

2.5. Changes of Irinotecan Sensitivity According to Manipulated TOP1 and TDP1 Expression

To determine if the manipulation of TOP1 and TDP1 could change the sensitivity of irinotecan, the sensitivity of irinotecan was evaluated according to manipulated expressions of TOP1 and TDP1 by using specific plasmid transfection and siRNA. In olaparib-resistant cells of SNU-484 and SNU-668 with increased activity of TOP1, irinotecan sensitivity was reduced when TOP1 expression was downregulated by siRNA (Figure 5A). In addition, in parental cells of SNU-484 and SNU-668, decreased TDP1 expression by siRNA induced the increased irinotecan sensitivity as same as olaparib-resistant cells. Olaparib-resistant SNU-601, in which TDP1 expression was not decreased, also showed an increased irinotecan sensitivity when TDP1 expression was knocked down (Figure 5B). Inversely, TOP1, or TDP1, expression was overexpressed by plasmid transfection in parental or olaparib-resistant cells of SNU-484 (Figure 5C). Although overexpressed TOP1 had slightly increased irinotecan sensitivity, simultaneous upregulation of TOP1 with downregulation of TDP1 in parental SNU-484 showed the significantly increased sensitivity to irinotecan compared with parental cells (Figure 5C). Furthermore, transiently overexpressed TDP1 had attenuated irinotecan sensitivity in olaparib-resistant SNU-484 cells, but olaparib-resistant SNU-484 with TOP1 downregulation and TDP1 upregulation was confirmed to attenuate the sensitivity to irinotecan dramatically (Figure 5C). The successful expression modulation by transfection was validated by western blot analysis (Figure 5C). Therefore, olaparib-resistant cell lines could be highly sensitive to subsequent irinotecan through TDP1 downregulation at the same time as the increased TOP1 activity. Furthermore, downregulation of TDP1 with TOP1 upregulation increased the nuclear size in parental cells, similar to olaparib-resistant cells (Figure 6). Thus, morphological changes of olaparib-resistant cells could be attributed to the relaxation of DNA condensation modulated by TOP1 hyperactivity and depletion of TDP1.

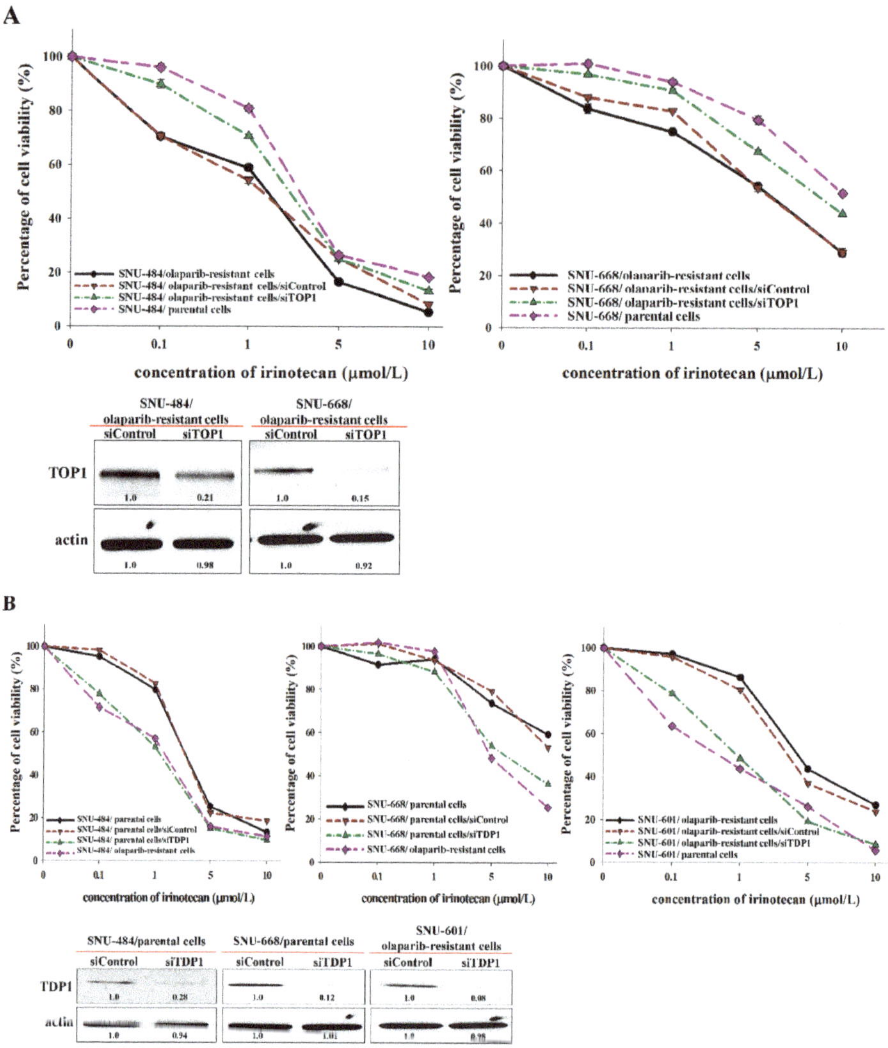

Figure 5. *Cont.*

C

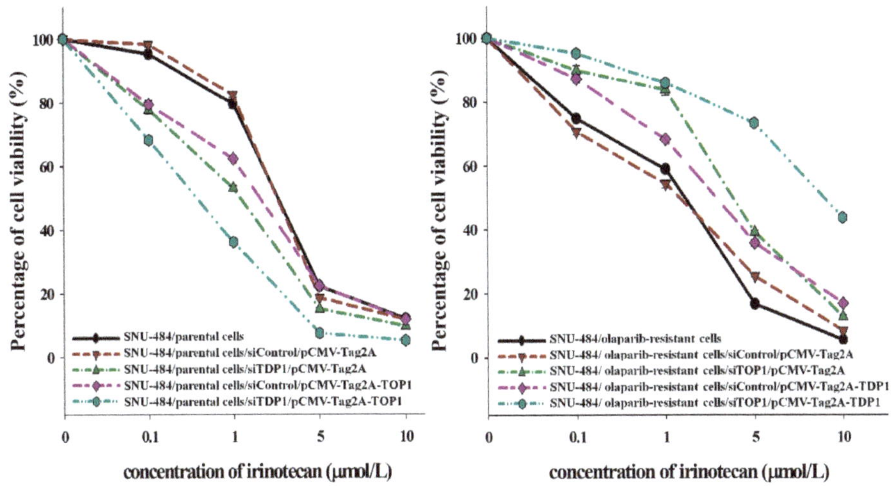

D

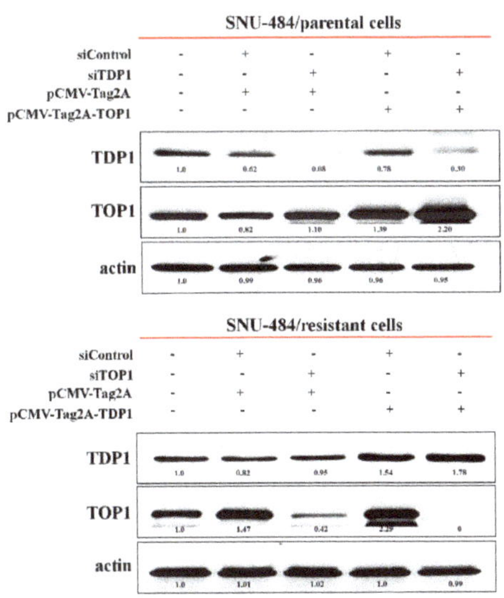

Figure 5. The sensitivities to irinotecan with modulating TOP1 and TDP1 expressions were evaluated by MTT assay. TOP1-siRNA mix, TDP1-siRNA mix, and non-targeting siRNA (160nM each) were transfected into olaparib-resistant cells or parental cells, and then treated with irinotecan for 5 d. Cell viability percentages were then measured by MTT assay, and TOP1 or TDP1 silencing was determined by western blotting (**A**,**B**). pCMV-tag2A-TOP1 or pCMV-tag2A-TDP1 and/or siTOP1 or siTDP1 were transfected into parental or olaparib-resistant cells. The sensitivity to irinotecan was calculated by MTT assay and the efficacy of transfection was confirmed by western blotting (**C**,**D**).

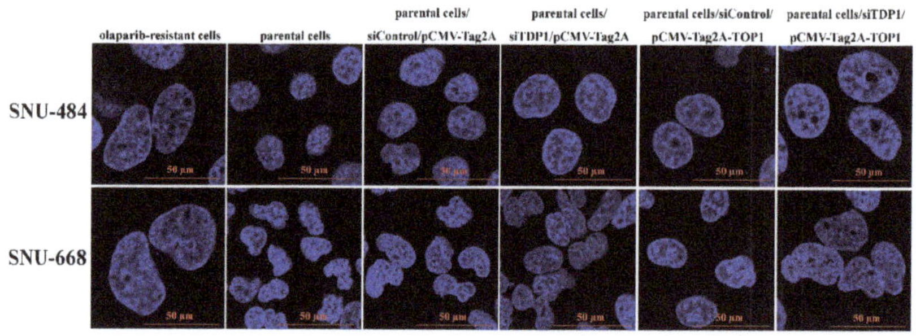

Figure 6. Parental cells were transfected with siRNA targeting TDP1 and/or pCMV-tag2A-TOP1 for 5 d. The size of nuclei was measured by DAPI staining and confocal laser microscopy. Representative images are presented (80× original magnification). Scale bars represent 50 μm.

2.6. Potent Anti-Tumor Activity of Irinotecan in Olaparib-Resistant Patient-Derived Breast Cancer Xenograft Model

To validate the increased antitumor effect of irinotecan in olaparib resistant model, we tested the antitumor activity of irinotecan in patient-derived xenograft (PDX) models established from a breast cancer patient with BRCA2 580del4 mutation who acquired resistance to olaparib. In concordance with clinical response, IMX 181 model showed no response to olaparib, but statistically significant delay of tumor growth on irinotecan treatment (Figure 7A). Both treatments were well tolerated without any sign of toxicity (Figure 7B).

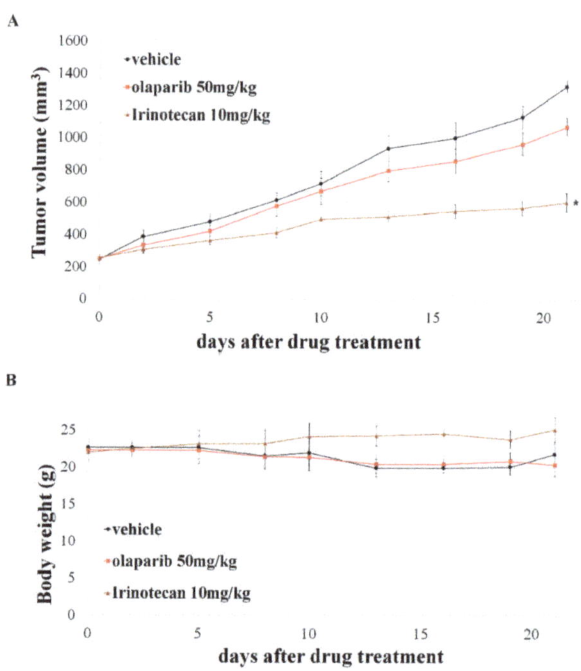

Figure 7. Irinotecan impeded tumor growth in a BRCA2 mutant PDX model with acquisition of olaparib resistance. (**A**) PDX model were treated with 50 mg/kg olaparib or 10 mg/kg irinotecan for 21 days (n = 4/group). The mean of tumor volumes was presented in a graph with the standard deviation (SD). * indicates $p < 0.001$ versus vehicle and olaparib group. (**B**) Changes in body weight were measured every three days for 21 days.

3. Discussion

This study confirmed the carryover effect of olaparib-treatment to subsequent chemotherapy, particularly in irinotecan. Through manipulating gene expressions, increased sensitivity to irinotecan in olaparib-resistant cells was confirmed to be attributed to TOP1 upregulation and TDP1 downregulation, which was shown in olaparib-resistant cells. These results could explain the higher OS improvement compared with PFS prolongation in the randomized clinical study of second-line gastric cancer, evaluating the efficacy of olaparib combined with paclitaxel.

After irinotecan treatment, SSBs are repaired by a complex consisting of TDP1, which functions in the base excision repair pathway [18]. PARP inhibitors, which inhibit base excision repair, sensitize cells to TOP1 inhibitors [19]. Therefore, irinotecan and olaparib represent a potent combination. However, concurrent treatment with both PARP inhibitors and irinotecan is too toxic for clinical development, although a preclinical study demonstrated synergistic effects [20–23]. Therefore, sequential treatment might represent a promising alternative approach. Our results suggest that the application of irinotecan after olaparib treatment may be a feasible treatment option owing to the carryover effect of olaparib.

DNA-damage response proteins function in various complex and overlapping pathways. In the present study, long-term olaparib treatment induced a compensatory alteration of the ATR/ATM axis, Chk, and ERCC1 expression. The development of olaparib resistance can be ascribed to this compensatory alteration, which also results in cisplatin resistance. Furthermore, olaparib and platinum share a common mechanism of action in the DNA repair pathway and similar predictive characters, such as the presence of BRCA mutation and RAD51 deficiency [12,24–26]. For a specific example, SNU-601 was highly sensitive to olaparib due to RAD51C-deficiency, which was also identified in several cancer types [6,25,26]. Parental SNU-601 was also sensitive to cisplatin, and olaparib-resistant SUN-601 showed resistance to cisplatin likewise. Therefore, this suggests that treatment with platinum should be avoided after olaparib failure, although clinical studies have so far not shown a decreased response to platinum after the resistance of PARP inhibitor in ovarian cancer [27]. In addition, this carryover effect might be specific to tumor cells based on synthetic lethality as a cytotoxic mechanism of PARP inhibitor. Clinically, there are no studies to report that this carryover effect could exacerbate the toxicities of subsequent chemotherapy.

TOP1 is an important cellular enzyme that allows for DNA relaxation. TOP1 cleaves DNA to create a DNA single-strand break to which it remains covalently bound to, thus allowing for rotation and relaxation of DNA. Once rotated, bound TOP1 ligates the nicked DNA and is released. In olaparib-resistant cell lines, the size of the nucleus was larger than that in the parental cell lines. This finding gave an indirect clue that TOP1 activity was increased in olaparib-resistant cells [16,17]. TOP1 activity is a predictive marker of irinotecan [28,29]. In the present study, increased TOP1 activity in olaparib-resistant cells was confirmed by using TOP1 activity assay. According to changes of TOP1 expression by transfection of siRNA, the sensitivity of irinotecan was changed, and the size of the nucleus was altered corresponding to olaparib-resistant cells. These results, therefore, provide direct evidence for the alteration of irinotecan sensitivity in olaparib-resistant cells.

TOP1 activity was increased in all olaparib-resistant cell lines, although TOP1 protein expression was not altered. TDP1 expression was downregulated in olaparib-resistant SNU-484, SNU-668, and KATO-III cells, which were more sensitive to irinotecan after the acquisition of olaparib resistance. Exceptionally, olaparib-resistant SNU-601 did not show altered expression of TDP1 and sensitivity to irinotecan. SNU-601 had TDP1 mutation (A520D). It has been reported that inactive mutant TDP1 (H263A) did not reduce DNA-damage by camptothecin, although the function of this TDP1 mutation (A520D) was unknown [30]. In colon cancer, TDP1 depletion increased the sensitivity to irinotecan in a TOP1-dependent manner [31]. In our study, in parental cells, TOP1 upregulation and TDP1 downregulation resulted in the highest sensitivity to irinotecan compared with either change alone. The mechanism underlying the carryover effect of olaparib for irinotecan could be confirmed, as these changes in olaparib-resistant cells, which included an increased TDP1 downregulation along with increased TOP1 activity.

The PDX model established from the patient who acquired resistance even harboring BRCA2 mutation exhibited sensitivity to irinotecan. These results can help to explain how olaparib showed an increase in overall survival, without increasing PFS in phase 2 trial conducted in gastric cancer, by its carryover effect on irinotecan, which is used as a third-line treatment [10].

In summary, this study demonstrated that there were carryover effects after acquisition of olaparib resistance. In olaparib-resistant cells, cisplatin resistance might occur because of compensatory alterations in the ATR/ATM axis, and Chk and ERCC1 expression. Importantly, irinotecan sensitivity was enhanced through TDP1 downregulation, concomitant with increased TOP1 activity in olaparib-resistant cells. Sensitivity to paclitaxel remained unaltered after acquisition of resistance to olaparib. Based on these results, carryover effect of olaparib to subsequent therapy should be significantly considered during the clinical use of olaparib.

4. Materials and Methods

4.1. Reagents

The PARP inhibitor, olaparib, was kindly provided by AstraZeneca, Ltd. (Macclesfield, Cheshire, UK). Cisplatin and paclitaxel were obtained from Choongwoe Co., Ltd., and Samyang Genex Co., Ltd. (Seoul, Korea). Irinotecan (cas no. 100286-90-6) was purchased from Sigma Aldrich (St. Louis, MO, USA).

4.2. Cell Lines and Cell Culture

Human gastric cancer cells (SNU-484, SNU-601, SNU-668, and KATO-III) were purchased from the Korean Cell Line Bank (Seoul, Korea). The identities of cell lines were validated by DNA fingerprinting analysis [32]. The cells were cultured at 37 °C in an atmosphere containing 5% CO_2 in Roswell Park Memorial Institute (RPMI) 1640 supplemented with 10% fetal bovine serum (FBS) and 10 μg/mL gentamicin.

4.3. Generation of Olaparib-Resistant Cells

Olaparib-resistant SNU-484, SNU-601, SNU-668, and KATO-III cells were established by continuous exposure to olaparib, starting with 1 μmol/L and incrementally increasing the concentration to 5 μmol/L over 7 months. Resistant cells were expanded in RPMI-1640 medium containing 10% FBS and 1 μmol/L olaparib.

4.4. Cell Growth Inhibition Assay

Cells (2–3 × 10^3 in 100 μL/well) were seeded in 96-well plates and incubated overnight at 37 °C in 5% CO_2, and then exposed to each drug at the indicated concentration for 5 days. After drug treatment, 50 μL of 3-(4,5-dimethylthiazol-2-yl)-2,5-diphenyltetrazolim bromide solution (Sigma Aldrich) was added to each well, and the plates were incubated for 4 h at 37 °C before the media was removed. After dissolving the formazan crystals with 150 μL of dimethyl sulfoxide, the absorbance of each well was measured at 540 nm with a VersaMax™ microplate reader (Molecular Devices, Sunnyvale, CA, USA). Half-maximal inhibitory concentration (IC_{50}) values were analyzed using SigmaPlot software (Statistical Package for the Social Sciences, Inc., Chicago, IL, USA).

4.5. Colony-Formation Assay

To compare the drug response between parental cells and resistant cells, the cells were seeded into six-well plates and incubated with the indicated concentration of each drug for 14 days. The cell colonies were washed in phosphate-buffered saline and stained with 0.1% Coomassie Blue solution (Sigma Aldrich) for 1 h at room temperature. The excess staining solution was then removed, and the plates were washed in PBS and air-dried. The cell colonies were counted using GELCOUNT (Oxford Optronix Ltd., Abingdon, UK), and cell viability was calculated by using SigmaPlot.

4.6. Western Blot Analysis

Whole cell proteins were extracted by using RIPA buffer, and equal amounts of protein were separated on 5%–15% SDS-PAGE gels, as described previously [33]. Primary antibodies against phosphorylated (p)-ATM, ATM, p-ATR, ATR, p-Chk1, Chk1, p-Chk2, Chk2, and ERCC1 were acquired from Cell Signaling Technology (Beverley, MA, USA). Antibodies against TOP1 and TDP1 were purchased from Abcam (Cambridge, UK). Actin antibody (Sigma Aldrich) was used as a control. The band intensity was calculated by Image J software.

4.7. Immunofluorescence Assay

The degree of DNA condensation was examined by 4′,6-diamidino-2-phenylindole (DAPI) staining and confocal laser microscopy. Cells were plated on 0.01% poly-L-lysine (Sigma Aldrich)-coated coverslips and incubated overnight. Then, the cells were stained with DAPI (300 nM; Invitrogen, Carlsbad, CA, USA) for 1 min. The coverslips were rinsed three times for 10 min in PBS and mounted on slides using Faramount aqueous mounting medium (DAKO, Denmark). Immunofluorescence was visualized using a Zeiss LSM 510 laser-scanning microscope.

4.8. Nuclear Extraction

Cells were incubated in hypotonic buffer A (10 mM HEPES (pH 7.5), 1.5 mM $MgCl_2$, 10 mM KCl, 0.2 mM EDTA, 0.5 mM DTT, 2 mM phenylmethylsulfonyl fluoride (PMSF), 1 mg/mL pepstatin A, 0.2 mM leupeptin, 10 μg/mL aprotinin, 1 mM sodium vanadate, 1 mM nitrophenylphosphate, and 5 mM benzamidine) for 10 min at 4 °C. The cells were incubated with 0.1% NP-40 for 10 min at 4 °C, and then centrifuged at 2000 rpm for 5 min at 4 °C. The supernatant was removed, and the pellets were resuspended in 1 mL of cold TEM solution (10 mM Tris-HCl (pH 7.5), 1 mM EDTA, and 4 mM $MgCl_2$) with 0.5 mM PMSF and centrifuged at 2000 rpm at 4 °C. The nuclear pellets were resuspended in a small volume of TEM, and mixed and incubated with an equal volume of 1 M NaCl for 30 min at 4 °C. Then, the supernatant was kept as the nuclear extract, following centrifugation at 13000 rpm for 15 min at 4 °C.

4.9. TOP1 Activity Assay

The TOP1 activities of the nuclear extracts from each cell line were assessed by measuring the relaxation of the supercoiled pHOT1 plasmid (TopoGEN, Inc., Buena Vista, CO, USA). Supercoiled pHOT1 plasmid (250 ng/μL) was incubated with 1 μL of nuclear extract in 10 mM Tris-HCl (pH 7.9), 150 mM NaCl, and 1 mM EDTA for 60 min at 37 °C in a final volume of 20 μL [34,35]. The reaction was terminated by the addition of 4 μL of stop buffer (1% Sarkosyl, 0.025% bromophenol blue, and 5% glycerol), and the samples were loaded onto 1% agarose gels. TOP1 activities were calculated by measuring the degree of disappearance of the supercoiled DNA and are presented in a bar graph.

4.10. Plasmid and siRNA Transfection

pME18S-FL3-TOP1 and pOTB-TDP1 plasmids were purchased from the Korean human gene bank (Daejoen, Korea) and both TOP1 and TDP1 were subcloned into pCMV-Tag2A vector. SNU-484 and SNU-668 parental cells were transfected with 4 μg of pCMV-Tag2A-TOP1 plasmid and/or 160 nM of TDP1-specific siRNA, while SNU-484 and SNU-668 olaparib resistant cells were transfected with 4μg of pCMV-Tag2A-TDP1 plasmid and/or 160 nM of TOP1-specific siRNA using Lipofectamin 2000 (Invitrogen, Carlsbad, CA, USA). TOP1- and TDP1-specific siRNAs were synthesized from Genolution (Seoul, Korea), and the sequences of the TOP1-specific siRNAs were 5′-GGGAAGGACTCCATCAGATACTATA-3′, 5′-AAGTGGAAATGGTGGGAAGAA-3′, and 5′-CGAAGAAGGTAGTAGAGTC-3′, and the sequences of the TDP-specific siRNAs were 5′-GACCATA TCTAGTAGTGAT-3′, 5′-GGAGTTAAGCCAAAGTATA-3′, and 5′-CTAGACAGTTTCAAAGTG-3′. The sequence of the non-specific siRNA was 5′-AATTCTCCGAACGTGTCACG-3′.

4.11. Library Preparation and Sequencing

Library preparations, clustering, and whole genome sequencing were conducted by Macrogen (Seoul, Korea). Libraries were prepared according to the TruSeq nano DNA library prep manual (Illumina; San Diego, CA, USA). Genomic DNA (100 ng) was sheared using an LE220 Focused-ultrasonicator (Covaris; Woburn, MA, USA) with a duty factor of 15%, peak incident power of 450 W, 200 cycles per burst for 50 s. Sheared DNA fragments of around 350 bp were obtained according to the manufacturer's instructions. DNA libraries were enriched after ligating the indexing adapters to the ends of the DNA fragments. The quality of the libraries was evaluated using TapeStation 2200 (Agilent; Santa Clara, CA, USA), quantified by using PicoGreen® dsDNA quantitation assay (Thermo Fisher Scientific; Waltham, MA, USA), and measured on a Victor3 plate reader (PerkinElmer; Waltham, MA, USA). A unique "bridged" amplification reaction was utilized for sequencing. A flow cell containing the libraries was prepared and then loaded on to the HiSeq X-10 sequencer (Illumina) for automated cycles of extension and imaging.

4.12. Patient-Derived Xenograft (PDX) Study

Animal experimentation was performed in accordance with the guidelines approved by the Institutional Animal Care and Use Committee (IACUC) and IRB (20180416-1402-054-555). Fresh human breast cancer tissue was obtained through a gun biopsy from hormone-receptor-positive, HER2-negative IDC breast cancer patient with *BRCA2* 580del4 mutation, who acquired resistance to olaparib. The obtained tissue was cut ~2-mm pieces within 1 h while covered in gauze dipped in saline and transplanted into six-week-old, severe combined immunodeficient (NOG) female mice (F0). Fresh human breast cancer tissue was cut into ~2-mm pieces and transplanted into 6-week-old, severe combined immunodeficient (NOG) female mice (F0). When the tumor volumes reached to 1.5 cm in diameter, it was dissected and re-implanted into another set of mice (F1). When the tumor volume of F1 reached 150 mm^3, the mice were randomly divided into three groups (4 mice per each group) and olaparib (50 mg/kg, p.o, once daily), irinotecan (10 mg/kg, i.p, twice a week), or vehicle (0.5% hydroxypropyl methylcellulose, p.o, once daily) were administered. Tumor volume was measured three times a week by caliper and calculated using $((\text{width})^2 \times (\text{height}))/2$. All mice were sacrificed with CO_2 on 21 days after the treatment.

4.13. Statistical Analysis

All experiments were conducted independently at least three times, and statistical analyses were performed using SigmaPlot version 9.0 (Statistical Package for the Social Sciences, Chicago, IL, USA). Two-sided Student's t-test was used when appropriate. The results are expressed as the mean ± standard error (SE). A p-value < 0.01 was considered to represent statistical significance.

4.14. Ethics Declarations

The PDX study was approved by the Institutional Animal Care and Use Committee (IACUC) and IRB (20180416-1402-054-555).

5. Conclusions

A carryover effect of olaparib-treatment, sensitizing to subsequent treatment, was suggested in a clinical trial. To explore the possibility and the mechanism of this effect, olaparib-resistant gastric cancer cells were tested with several chemotherapeutic agents, and underlying mechanisms were explored. Olaparib-resistant gastric cancer cells show the compensatory alterations in DNA-damage response pathways, and exhibit cross-resistance to cisplatin; however, these cells are highly sensitive to subsequent treatment with irinotecan through tyrosyl-DNA phosphodiesterase 1 (TDP1) downregulation with increased topoisomerase 1 (TOP1) activity. These findings have substantial implications on subsequent therapies for patients with olaparib-resistant cancers.

Supplementary Materials: The following are available online at http://www.mdpi.com/2072-6694/12/2/334/s1, Figure S1. The populations of apoptotic cells following irinotecan or olaparib treatment; Figure S2. Protein expression levels of parental and olaparib-resistant cells; Table S1. Sequences of the siRNAs.

Author Contributions: J.W.K. and A.M. designed the study, performed the experiments, interpreted the data, and wrote the manuscript. H.J. and Y.J.K. assisted the PDX experiment. H.-J.K., K.-H.L., T.-Y.K., K.W.L., D.-Y.O., J.-H.K., Y.-J.B., and S.-A.I. supplied the clinical sample, conceived of the study, and participated in coordination. All authors have read and agreed to the published version of the manuscript.

Funding: This study was supported by Korean Cancer Foundation (K20121110), and funded by a grant of the Korea Health Technology R&D Project through the Korea Health Industry Development Institute (KHIDI) and by the Ministry of Health & Welfare, Korea (grant number: HI14C1277).

Conflicts of Interest: Seock-Ah Im is a recipient of research funds from AstraZeneca Inc. and Pfizer and has consultant and advisory role for AstraZeneca, Hanmi Corp, Novartis, Pfizer, Roche, and Spectrum. The other authors declared that they have no conflict of interest.

References

1. Helleday, T. The underlying mechanism for the PARP and BRCA synthetic lethality: clearing up the misunderstandings. *Mol. Oncol.* **2011**, *5*, 387–393. [CrossRef]
2. Murai, J.; Huang, S.Y.; Das, B.B.; Renaud, A.; Zhang, Y.; Doroshow, J.H.; Ji, J.; Takeda, S.; Pommier, Y. Trapping of PARP1 and PARP2 by Clinical PARP Inhibitors. *Cancer Res.* **2012**, *72*, 5588–5599. [CrossRef]
3. Fong, P.C.; Boss, D.S.; Yap, T.A.; Tutt, A.; Wu, P.; Mergui-Roelvink, M.; Mortimer, P.; Swaisland, H.; Lau, A.; O'Connor, M.J.; et al. Inhibition of poly(ADP-ribose) polymerase in tumors from BRCA mutation carriers. *N. Engl. J. Med.* **2009**, *361*, 123–134. [CrossRef]
4. Pujade-Lauraine, E.; Ledermann, J.A.; Selle, F.; Gebski, V.; Penson, R.T.; Oza, A.M.; Korach, J.; Huzarski, T.; Poveda, A.; Pignata, S.; et al. Olaparib tablets as maintenance therapy in patients with platinum-sensitive, relapsed ovarian cancer and a BRCA1/2 mutation (SOLO2/ENGOT-Ov21): a double-blind, randomised, placebo-controlled, phase 3 trial. *Lancet. Oncol.* **2017**, *18*, 1274–1284. [CrossRef]
5. Tutt, A.; Robson, M.; Garber, J.E.; Domchek, S.M.; Audeh, M.W.; Weitzel, J.N.; Friedlander, M.; Arun, B.; Loman, N.; Schmutzler, R.K.; et al. Oral poly(ADP-ribose) polymerase inhibitor olaparib in patients with BRCA1 or BRCA2 mutations and advanced breast cancer: a proof-of-concept trial. *Lancet* **2010**, *376*, 235–244. [CrossRef]
6. Min, A.; Im, S.A.; Yoon, Y.K.; Song, S.H.; Nam, H.J.; Hur, H.S.; Kim, H.P.; Lee, K.H.; Han, S.W.; Oh, D.Y.; et al. RAD51C-deficient cancer cells are highly sensitive to the PARP inhibitor olaparib. *Mol. Cancer Ther.* **2013**, *12*, 865–877. [CrossRef]
7. Comprehensive molecular characterization of gastric adenocarcinoma. *Nature* **2014**, *513*, 202–209. [CrossRef]
8. Jonson, L.; Ahlborn, L.B.; Steffensen, A.Y.; Djursby, M.; Ejlertsen, B.; Timshel, S.; Nielsen, F.C.; Gerdes, A.M.; Hansen, T.V. Identification of six pathogenic RAD51C mutations via mutational screening of 1228 Danish individuals with increased risk of hereditary breast and/or ovarian cancer. *Breast Cancer Res. Treat.* **2016**, *155*, 215–222. [CrossRef]
9. Coulet, F.; Fajac, A.; Colas, C.; Eyries, M.; Dion-Miniere, A.; Rouzier, R.; Uzan, S.; Lefranc, J.P.; Carbonnel, M.; Cornelis, F.; et al. Germline RAD51C mutations in ovarian cancer susceptibility. *Clin. Genet.* **2013**, *83*, 332–336. [CrossRef]
10. Bang, Y.J.; Im, S.A.; Lee, K.W.; Cho, J.Y.; Song, E.K.; Lee, K.H.; Kim, Y.H.; Park, J.O.; Chun, H.G.; Zang, D.Y.; et al. Randomized, Double-Blind Phase II Trial With Prospective Classification by ATM Protein Level to Evaluate the Efficacy and Tolerability of Olaparib Plus Paclitaxel in Patients With Recurrent or Metastatic Gastric Cancer. *J. Clin. Oncol.* **2015**, *33*, 3858–3865. [CrossRef]
11. Bang, Y.J.; Xu, R.H.; Chin, K.; Lee, K.W.; Park, S.H.; Rha, S.Y.; Shen, L.; Qin, S.; Xu, N.; Im, S.A.; et al. Olaparib in combination with paclitaxel in patients with advanced gastric cancer who have progressed following first-line therapy (GOLD): a double-blind, randomised, placebo-controlled, phase 3 trial. *Lancet. Oncol.* **2017**, *18*, 1637–1651. [CrossRef]
12. Ledermann, J.; Harter, P.; Gourley, C.; Friedlander, M.; Vergote, I.; Rustin, G.; Scott, C.; Meier, W.; Shapira-Frommer, R.; Safra, T.; et al. Olaparib maintenance therapy in platinum-sensitive relapsed ovarian cancer. *N. Engl. J. Med.* **2012**, *366*, 1382–1392. [CrossRef]

13. Rottenberg, S.; Jaspers, J.E.; Kersbergen, A.; van der Burg, E.; Nygren, A.O.; Zander, S.A.; Derksen, P.W.; de Bruin, M.; Zevenhoven, J.; Lau, A.; et al. High sensitivity of BRCA1-deficient mammary tumors to the PARP inhibitor AZD2281 alone and in combination with platinum drugs. *Proc. Natl. Acad. Sci. USA* **2008**, *105*, 17079–17084. [CrossRef]
14. Tamaichi, H.; Sato, M.; Porter, A.C.; Shimizu, T.; Mizutani, S.; Takagi, M. Ataxia telangiectasia mutated-dependent regulation of topoisomerase II alpha expression and sensitivity to topoisomerase II inhibitor. *Cancer Sci.* **2013**, *104*, 178–184. [CrossRef]
15. Thomas, A.; Pommier, Y. Targeting Topoisomerase I in the Era of Precision Medicine. *Clin. Cancer Res.* **2019**, *25*, 6581–6589. [CrossRef]
16. Pommier, Y.; Leo, E.; Zhang, H.; Marchand, C. DNA topoisomerases and their poisoning by anticancer and antibacterial drugs. *Chem. Biol.* **2010**, *17*, 421–433. [CrossRef]
17. Koster, D.A.; Crut, A.; Shuman, S.; Bjornsti, M.A.; Dekker, N.H. Cellular strategies for regulating DNA supercoiling: a single-molecule perspective. *Cell* **2010**, *142*, 519–530. [CrossRef]
18. Plo, I.; Liao, Z.Y.; Barcelo, J.M.; Kohlhagen, G.; Caldecott, K.W.; Weinfeld, M.; Pommier, Y. Association of XRCC1 and tyrosyl DNA phosphodiesterase (Tdp1) for the repair of topoisomerase I-mediated DNA lesions. *DNA Repair (Amst.)* **2003**, *2*, 1087–1100. [CrossRef]
19. Smith, L.M.; Willmore, E.; Austin, C.A.; Curtin, N.J. The novel poly (ADP-Ribose) polymerase inhibitor, AG14361, sensitizes cells to topoisomerase I poisons by increasing the persistence of DNA strand breaks. *Clin Cancer Res.* **2005**, *11*, 8449–8457. [CrossRef]
20. Jaspers, J.E.; Kersbergen, A.; Boon, U.; Sol, W.; van Deemter, L.; Zander, S.A.; Drost, R.; Wientjens, E.; Ji, J.; Aly, A.; et al. Loss of 53BP1 causes PARP inhibitor resistance in Brca1-mutated mouse mammary tumors. *Cancer Discov.* **2013**, *3*, 68–81. [CrossRef]
21. Pommier, Y.; Barcelo, J.M.; Rao, V.A.; Sordet, O.; Jobson, A.G.; Thibaut, L.; Miao, Z.H.; Seiler, J.A.; Zhang, H.; Marchand, C.; et al. Repair of topoisomerase I-mediated DNA damage. *Prog. Nucleic Acid Res. Mol. Biol.* **2006**, *81*, 179–229. [PubMed]
22. Zander, S.A.; Kersbergen, A.; van der Burg, E.; de Water, N.; van Tellingen, O.; Gunnarsdottir, S.; Jaspers, J.E.; Pajic, M.; Nygren, A.O.; Jonkers, J.; et al. Sensitivity and acquired resistance of BRCA1;p53-deficient mouse mammary tumors to the topoisomerase I inhibitor topotecan. *Cancer Res.* **2010**, *70*, 1700–1710. [CrossRef] [PubMed]
23. Tahara, M.; Inoue, T.; Sato, F.; Miyakura, Y.; Horie, H.; Yasuda, Y.; Fujii, H.; Kotake, K.; Sugano, K. The use of Olaparib (AZD2281) potentiates SN-38 cytotoxicity in colon cancer cells by indirect inhibition of Rad51-mediated repair of DNA double-strand breaks. *Mol. Cancer Ther.* **2014**, *13*, 1170–1180. [CrossRef] [PubMed]
24. Kim, J.W.; Cho, H.J.; Kim, M.; Lee, K.H.; Kim, M.A.; Han, S.W.; Oh, D.Y.; Lee, H.J.; Im, S.A.; Kim, T.Y.; et al. Differing effects of adjuvant chemotherapy according to BRCA1 nuclear expression in gastric cancer. *Cancer Chemother. Pharm.* **2013**, *71*, 1435–1443. [CrossRef] [PubMed]
25. Wang, A.T.; Kim, T.; Wagner, J.E.; Conti, B.A.; Lach, F.P.; Huang, A.L.; Molina, H.; Sanborn, E.M.; Zierhut, H.; Cornes, B.K.; et al. A Dominant Mutation in Human RAD51 Reveals Its Function in DNA Interstrand Crosslink Repair Independent of Homologous Recombination. *Mol. Cell* **2015**, *59*, 478–490. [CrossRef] [PubMed]
26. Michl, J.; Zimmer, J.; Tarsounas, M. Interplay between Fanconi anemia and homologous recombination pathways in genome integrity. *EMBO J.* **2016**, *35*, 909–923. [CrossRef]
27. Ang, J.E.; Gourley, C.; Powell, C.B.; High, H.; Shapira-Frommer, R.; Castonguay, V.; De Greve, J.; Atkinson, T.; Yap, T.A.; Sandhu, S.; et al. Efficacy of chemotherapy in BRCA1/2 mutation carrier ovarian cancer in the setting of PARP inhibitor resistance: a multi-institutional study. *Clin. Cancer Res.* **2013**, *19*, 5485–5493. [CrossRef]
28. Jandu, H.; Aluzaite, K.; Fogh, L.; Thrane, S.W.; Noer, J.B.; Proszek, J.; Do, K.N.; Hansen, S.N.; Damsgaard, B.; Nielsen, S.L.; et al. Molecular characterization of irinotecan (SN-38) resistant human breast cancer cell lines. *BMC Cancer* **2016**, *16*, 34. [CrossRef]
29. Jansen, W.J.; Zwart, B.; Hulscher, S.T.; Giaccone, G.; Pinedo, H.M.; Boven, E. CPT-11 in human colon-cancer cell lines and xenografts: characterization of cellular sensitivity determinants. *Int. J. Cancer* **1997**, *70*, 335–340. [CrossRef]

30. Barthelmes, H.U.; Habermeyer, M.; Christensen, M.O.; Mielke, C.; Interthal, H.; Pouliot, J.J.; Boege, F.; Marko, D. TDP1 overexpression in human cells counteracts DNA damage mediated by topoisomerases I and II. *J. Biol. Chem.* **2004**, *279*, 55618–55625. [CrossRef]
31. Meisenberg, C.; Gilbert, D.C.; Chalmers, A.; Haley, V.; Gollins, S.; Ward, S.E.; El-Khamisy, S.F. Clinical and cellular roles for TDP1 and TOP1 in modulating colorectal cancer response to irinotecan. *Mol. Cancer Ther.* **2015**, *14*, 575–585. [CrossRef] [PubMed]
32. Ku, J.L.; Park, J.G. Biology of SNU cell lines. *Cancer Res. Treat.* **2005**, *37*, 1–19. [CrossRef] [PubMed]
33. Lee, M.; Lee, K.H.; Min, A.; Kim, J.; Kim, S.; Jang, H.; Lim, J.M.; Kim, S.H.; Ha, D.H.; Jeong, W.J.; et al. Pan-Pim Kinase Inhibitor AZD1208 Suppresses Tumor Growth and Synergistically Interacts with Akt Inhibition in Gastric Cancer Cells. *Cancer Res. Treat.* **2018**. [CrossRef] [PubMed]
34. Guo, L.; Liu, X.; Jiang, Y.; Nishikawa, K.; Plunkett, W. DNA-dependent protein kinase and ataxia telangiectasia mutated (ATM) promote cell survival in response to NK314, a topoisomerase IIalpha inhibitor. *Mol. Pharm.* **2011**, *80*, 321–327. [CrossRef]
35. Park, H.J.; Lee, H.J.; Lee, E.J.; Hwang, H.J.; Shin, S.H.; Suh, M.E.; Kim, C.; Kim, H.J.; Seo, E.K.; Lee, S.K. Cytotoxicity and DNA topoisomerase inhibitory activity of benz[f]indole-4,9-dione analogs. *Biosci. Biotechnol. Biochem.* **2003**, *67*, 1944–1949. [CrossRef]

cancers

MDPI

Article

PARP1 Inhibition Augments UVB-Mediated Mitochondrial Changes—Implications for UV-Induced DNA Repair and Photocarcinogenesis

Csaba Hegedűs [1], Gábor Boros [2], Eszter Fidrus [1], Gréta Nikoletta Kis [3], Miklós Antal [3], Tamás Juhász [3], Eszter Anna Janka [1], Laura Jankó [4,5], György Paragh [6], Gabriella Emri [1], Péter Bai [4,5,7,†] and Éva Remenyik [1,*,†]

1 Department of Dermatology, Faculty of Medicine, University of Debrecen, 4032 Debrecen, Hungary; hegeduscsaba88@gmail.com (C.H.); fepont92@gmail.com (E.F.); janka.eszter.a@gmail.com (E.A.J.); gemri@med.unideb.hu (G.E.)

2 BioNTech RNA pharmaceuticals GmbH, BioNTech AG, 55131 Mainz, Germany; borosgabor27@gmail.com

3 Department of Anatomy, Histology and Embryology, Faculty of Medicine, University of Debrecen, 4032 Debrecen, Hungary; greta@anat.med.unideb.hu (G.N.K.); antal@anat.med.unideb.hu (M.A.); juhaszt@anat.med.unideb.hu (T.J.)

4 Department of Medical Chemistry, Faculty of Medicine, University of Debrecen, 4032 Debrecen, Hungary; janko.laura90@gmail.com (L.J.); baip@med.unideb.hu (P.B.)

5 MTA-DE Lendület Laboratory of Cellular Metabolism Research Group, University of Debrecen, H-4032 Debrecen, Hungary

6 Department of Dermatology and Department of Cell Stress Biology, Roswell Park Comprehensive Cancer Center, 665 Elm St, Buffalo, NY 14203 USA; gyorgy.paragh@roswellpark.org

7 Research Center for Molecular Medicine, Faculty of Medicine, University of Debrecen, 4032 Debrecen, Hungary

* Correspondence: remenyik@med.unideb.hu; Tel.: +36-52-412-345

† Both authors are equally responsible for the paper.

Received: 26 November 2019; Accepted: 10 December 2019; Published: 18 December 2019

Abstract: Keratinocytes provide the first line of defense of the human body against carcinogenic ultraviolet (UV) radiation. Acute and chronic UVB-mediated cellular responses were widely studied. However, little is known about the role of mitochondrial regulation in UVB-induced DNA damage. Here, we show that poly (ADP-ribose) polymerase 1 (PARP1) and ataxia-telangiectasia-mutated (ATM) kinase, two tumor suppressors, are important regulators in mitochondrial alterations induced by UVB. Our study demonstrates that PARP inhibition by ABT-888 upon UVB treatment exacerbated cyclobutane pyrimidine dimers (CPD) accumulation, cell cycle block and cell death and reduced cell proliferation in premalignant skin keratinocytes. Furthermore, in human keratinocytes UVB enhanced oxidative phosphorylation (OXPHOS) and autophagy which were further induced upon PARP inhibition. Immunoblot analysis showed that these cellular responses to PARP inhibition upon UVB irradiation strongly alter the phosphorylation level of ATM, adenosine monophosphate-activated kinase (AMPK), p53, protein kinase B (AKT), and mammalian target of rapamycin (mTOR) proteins. Furthermore, chemical inhibition of ATM led to significant reduction in AMPK, p53, AKT, and mTOR activation suggesting the central role of ATM in the UVB-mediated mitochondrial changes. Our results suggest a possible link between UVB-induced DNA damage and metabolic adaptations of mitochondria and reveal the OXPHOS-regulating role of autophagy which is dependent on key metabolic and DNA damage regulators downstream of PARP1 and ATM.

Keywords: UVB; PARP; mitochondria; metabolism; biogenesis; autophagy; carcinogenesis; DNA repair

1. Introduction

Mitochondria regulate their shape, number, distribution, mass, content of mitochondrial DNA (mtDNA), and metabolic capacity in a process called mitochondrial biogenesis, which requires the orchestration of complex transcriptional control of both nuclear and mitochondrial genes [1,2]. The function of mitochondrial biogenesis is to provide quality control of mitochondria by regulating mitochondrial fission, fusion, and mitophagy [3,4] to maximize the energy utilization of mitochondria [5] to meet cellular and environmental demands. Imbalances or perturbations in these processes can lead to mitochondrial dysfunction [3,6].

Accumulating evidence suggests that mitochondria also play a central role in skin physiology. Although, involvement of other organ systems predominates in classical mitochondrial disorders, several cutaneous diseases can be linked to mitochondrial dysfunctions [7]. Interestingly, mitochondria lack functional nucleotide excision repair (NER) pathway [8,9] which is responsible for the removal of ultraviolet (UV)-induced DNA lesions including cyclobutane pyrimidine dimers (CPD). Accumulation of these DNA photoproducts in mtDNA leads to mutations and deletions resulting in mitochondrial alterations which have been associated with photoaging [10,11] and are present in melanoma [12], as well as in non-melanoma skin cancers [13,14]. The other types of mitochondrial alterations such as upregulated oxidative phosphorylation (OXPHOS), mitochondrial membrane hyperpolarization, and decreased mitophagy are frequently observed in patients with DNA repair deficiencies [15–18]. Even though, these DNA repair proteins are encoded by the nuclear genome, their absence lead to mitochondrial functional changes emphasizing the importance of nucleus-to-mitochondria (NM) signaling [19].

Growing evidence suggests that the key component of NM signaling is poly (ADP-ribose) polymerase 1 (PARP1) activation [19,20]. PARP1 is a multifunctional zinc-finger protein involved in the regulation of DNA repair, chromatin structure, cell cycle, calcium homeostasis, transcription regulation, cell death, immune response, and metabolism [21,22]. Through either direct interaction or via poly (ADP-ribose) (PAR) polymer formation PARP1 can modulate the activity of ataxia-telangiectasia mutated kinase (ATM) [23] and tumor suppressor protein 53 (p53) [24,25] involved in DNA damage response. Enhanced PARP1 activity also induces ATP depletion [26], which implies the activation of AMPK-activated protein kinase (AMPK) [27,28] that is responsible for the regulation of various cellular pathways including protein kinase B (AKT) and mammalian target of rapamycin (mTOR) [29–31]. This complex interplay between PARP1, ATM, AMKP, p53, AKT, and mTOR indicates that these DNA damage responders can fine-tune and modulate the interaction between DNA repair pathways with metabolism [32]. PARP1 and PARP2 also modulate mitochondrial activity through NAD^+ depletion [33–35]. Pharmacological inhibition or deletion of PARP1 and PARP2 improves mitochondrial function and protects against mitochondrial [36,37], metabolic [34], and neurological diseases [38]. PARP inhibitors became valuable tools in treating cancer cells harboring DNA repair defect with the combination of either radio- or chemotherapy [39–41]. In addition to several orally-administered PARP inhibitors that are under active clinical development, recently ABT-888 (veliparib) has emerged as an effective drug in treating various solid tumors [42] partially via modulating mitochondrial activity [43,44].

Although UV radiation has been shown to trigger morphological [45,46] and functional changes of mitochondria [47–49], published data led to contradictory results most likely due to the diversity of applied UV spectrum, UV dose, and cell type. Studies using UVC irradiation revealed that UVC induced mitochondrial hyperfusion and resulted in enhanced ATP synthesis via oxidative phosphorylation (OXPHOS) [50]. Other authors demonstrated that UVC irradiation caused a significant increase in mitochondrial content, oxygen consumption, and fatty acid oxidation [28]. Nevertheless, the functional consequences of mitochondrial alterations after UVB-induced DNA damage and the molecular pathways leading to mitochondrial changes remain to be elucidated.

In this study, we defined the central role of PARP1 and the linked molecular pathways in mediating UVB-induced DNA damage response and mitochondrial changes in a clinically relevant human keratinocyte cell line.

2. Results

2.1. PARP Inhibition Impairs CPD Repair, Augments UVB-Induced Cell Cycle Block, Apoptosis and Reduces Keratinocyte Proliferation

To explore the effect of UVB on PARP activation, firstly, we investigated poly (ADP-ribose) polymer (PAR) formation in a time-dependent manner in human immortalized keratinocytes exposed to mid (20 mJ/cm^2) or high (40 mJ/cm^2) dose of UVB (Figure 1A). In HaCaT cells, poly (ADP-ribosyl) ation (PARylation) signal was initially observed at and over 95 kDa at 5 min after UVB exposure and the extent of PARP activation was dose-dependent with higher PARP activation after 40 mJ/cm^2 UVB exposure. The signal was lost when cells were treated with ABT-888, a pan-PARP inhibitor (for uncut PAR western see Supplementary Figure S2). PARP1 is considered to contribute 85–90% to total PARP activity, the rest is largely the activity of PARP2 [51–53]. Moreover, the size of the PARylated band suggests PARP1 (auto) PARylation indicating the involvement of PARP1 in UVB-induced damage. Since PARP1 is involved in regulation of various DNA repair pathways, we wanted to assess how PARP inhibition (PARPi) regulates the removal of cyclobutane pyrimidine dimers (CPDs) introduced by UVB (Figure 1B). A slow decline of CPDs after UVB exposure was observed, reflective of nucleotide excision repair (NER) activity. In contrast, the relative amount of CPDs in PARP-inhibited cells remained elevated even after 24 h compared to the UVB-irradiated cells suggesting impaired efficiency of NER, similarly as described by King et al. [54]. Since unrepaired photolesions can initiate cell cycle block to prevent cells with DNA damage from entering mitosis, we performed cell cycle analysis (Figure 1C,D). A higher proportion of PARP-inhibited cells after 20 mJ/cm^2 UVB accumulated in G_2/M phase of cell cycle characteristic of enhanced DNA damage [21]. ABT-888 treatment sensitized cells to apoptosis after UVB as reflected by decreased cell viability 24 h post-UVB compared to controls (Figure 1E). Gene silencing of PARP1 showed similar changes regarding cell viability as PARP inhibition. Furthermore, long-term keratinocyte survival using May-Grünwald-Giemsa staining in clonogenic assay showed that ABT-888 treatment led to a marked decrease in the number of keratinocyte clones after UVB-irradiation (Figure 2A,B). To find out whether PARPi-induced retention of CPDs induce higher mutation rate, we performed hypoxanthine-guanine phosphoribosyltransferase (HPRT) mutation assay (Figure 2C,D). We could not adjust the method for HaCaT keratinocytes as these cells were extremely tolerant to the 6-thioguanine selection medium. Therefore, we used Chinese hamster ovary (CHO) cells which is a widely used cell line for HPRT mutation assay [55–57] and also show high level of p53 protein level due to p53 mutation [58,59] as HaCaT cells do [60]. In this case, we used lower UVB doses to allow the accumulation of mutations compared to higher UVB which potentially triggers apoptosis. Although, both 10 and 20 mJ UVB resulted in increased number of HPRT-mutated cells, interestingly PARP inhibition caused a marked reduction in the number of mutated cell colonies suggesting that PARPi initiate apoptosis of cells with high CPD content instead of allowing the accumulation of gene mutations. Interestingly, some DNA damage markers, including cell viability, CPD and colony formation, cell cycle progression showed no significant difference between the vehicle and ABT-888 treated groups after 40 mJ/cm^2. This phenomenon can be due to the fact that 40 mJ/cm^2 UVB dose in our experiments represents such high DNA damage that cannot be augmented by PARP inhibition. However, PARP1 knockdown cells displayed significantly lower cell viability compared to control siRNA-transfected cells even after 40 mJ/cm^2 UVB irradiation.

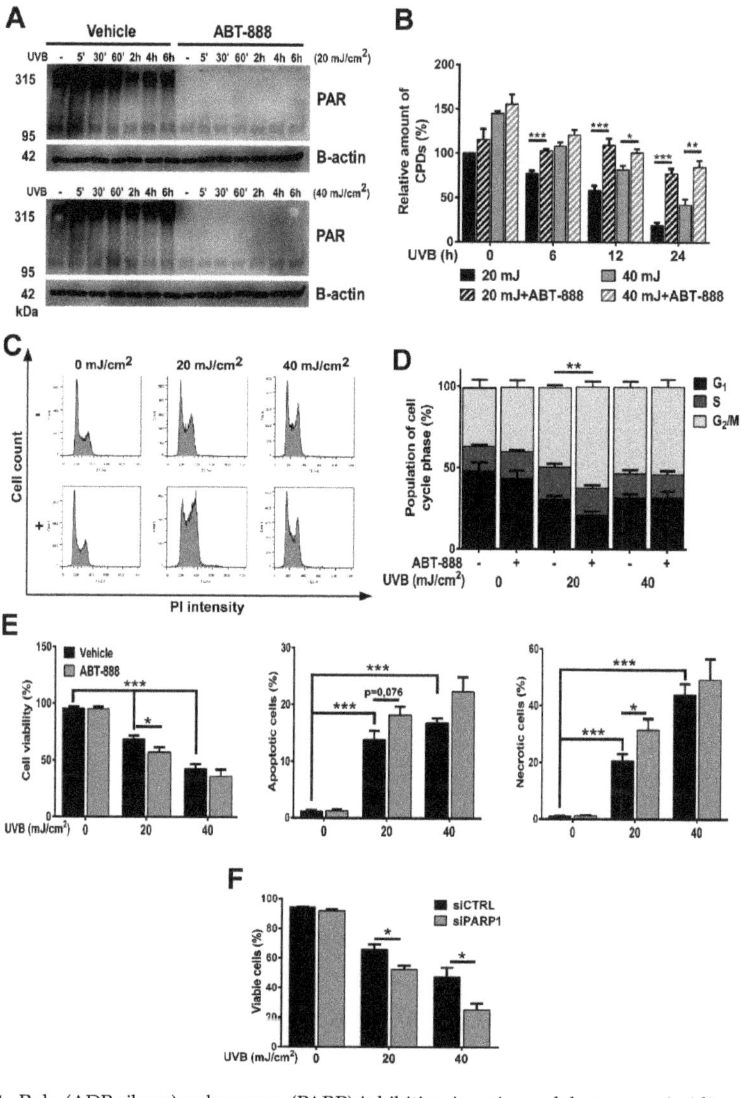

Figure 1. Poly (ADP-ribose) polymerase (PARP) inhibition impairs cyclobutane pyrimidine dimer (CPD) repair, augments ultraviolet B (UVB)-induced cell cycle block and apoptosis. (**A**) Time-course of PARP activity (PAR) after 20 and 40 mJ/cm^2 UVB exposure and 25 μM ABT-888 were analyzed by Western blot (n = 3). (**B**) Cells were exposed to a single dose of 20 or 40 mJ/cm^2 UVB and collected at various time points for DNA extraction. CPD formation was determined by ELISA (n = 4). (**C**,**D**) Cell cycle progression was evaluated by propidium iodide staining after 24 h. DNA content was analyzed in FL2-A (n = 4). (**E**) Cell viability, apoptosis, and necrosis was assessed by dual labelling with Annexin V-Alexa 488 and propidium iodide 24 h post-UVB. Double negative cells are considered as viable (n = 5). (**F**) Cell viability was measured similarly as in Figure 1E after PARP1 knockdown (n = 3). −/+ represent vehicle (−) or ABT-888 (+) treatment *; ** and *** indicate statistically significant difference at $p < 0.05$ and $p < 0.01$, $p < 0.001$, respectively. Error bars represent SEM.

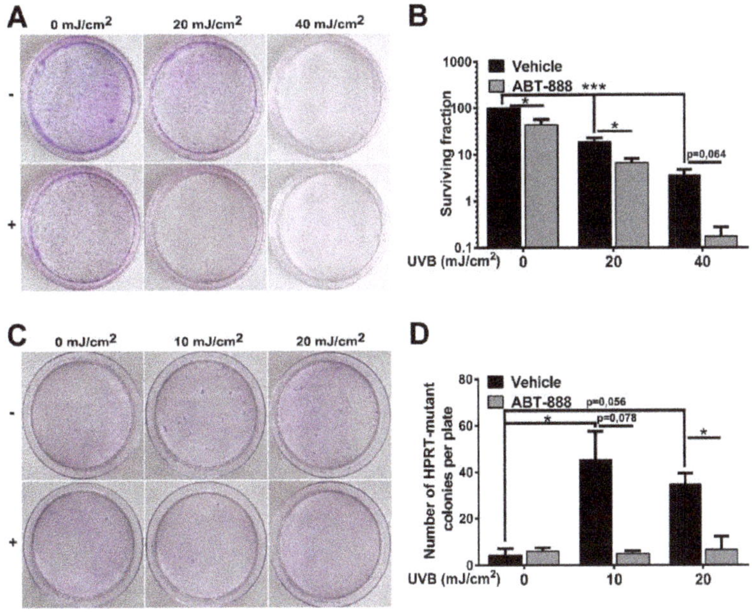

Figure 2. Poly (ADP-ribose) polymerase (PARP) inhibition decreases clone formation and ultraviolet B (UVB)-induced mutation rate. (**A**,**B**) Colony formation assay of HaCaT cells after 10 days post-UVB exposure was assessed by clonogenic assay ($n = 4$). (**C**,**D**) HPRT mutation assay was carried out on CHO cells. Preselected hypoxanthine-guanine phosphoribosyltransferase (HPRT) mutant cells were cultured for 10 days post-UVB ($n = 3$). −/+ represent vehicle (−) or ABT-888 (+) treatment * and *** indicate statistically significant difference at $p < 0.05$ and $p < 0.001$, respectively. Error bars represent SEM.

2.2. PARP Inhibition Enhances UVB-Mediated Mitochondrial Biogenesis

Mitochondrial biogenesis, by promoting the growth, formation, and assembly of newly synthesized mitochondria, has recently been linked to cancer development [61], apoptosis [62–64], and DNA damage [18,28,65]. Accumulating evidence suggest that DNA damage can initiate mitochondrial biogenesis which is accompanied by elevation in mitochondrial number, area, and mass [18,28,65,66]. Transmission electron microscopic images revealed that UVB-treated cells contain more and longer cristae than non-irradiated cells (Figure 3A). This morphological alteration became more pronounced after PARP inhibition. Similarly, UVB treatment increased both mitochondria number and total mitochondrial area (Figure 3B,C). ABT-888 treatment resulted in further increase in these parameters suggesting that PARP inhibition may boost UVB-mediated mitochondrial response. Since mitochondrial content changes with the balance between mitochondrial biogenesis and turnover, we quantified mitochondrially encoded cytochrome C oxidase I (MTCO1)/succinate dehydrogenase complex, subunit A (SDHA) ratio that is a marker of mitochondrial biogenesis. This experiment demonstrated that the mitochondrially-encoded MTCO1 show strong induction after UVB irradiation and become even more expressed after PARPi, while the expression of the nuclearly-encoded SDHA protein is unaltered (Figure 3D,E). The higher mitochondrial mass after both UVB and PARPi (Figure 3F) and the enhanced expression of the master regulators of mitochondrial biogenesis including mitochondrial transcription factor A (TFAM), nuclear respiratory factor 2 (NRF2), and estrogen-related receptor alpha (ERRA) (Figure 3G) also suggest that PARPi augments the UVB-triggered mitochondrial biogenesis.

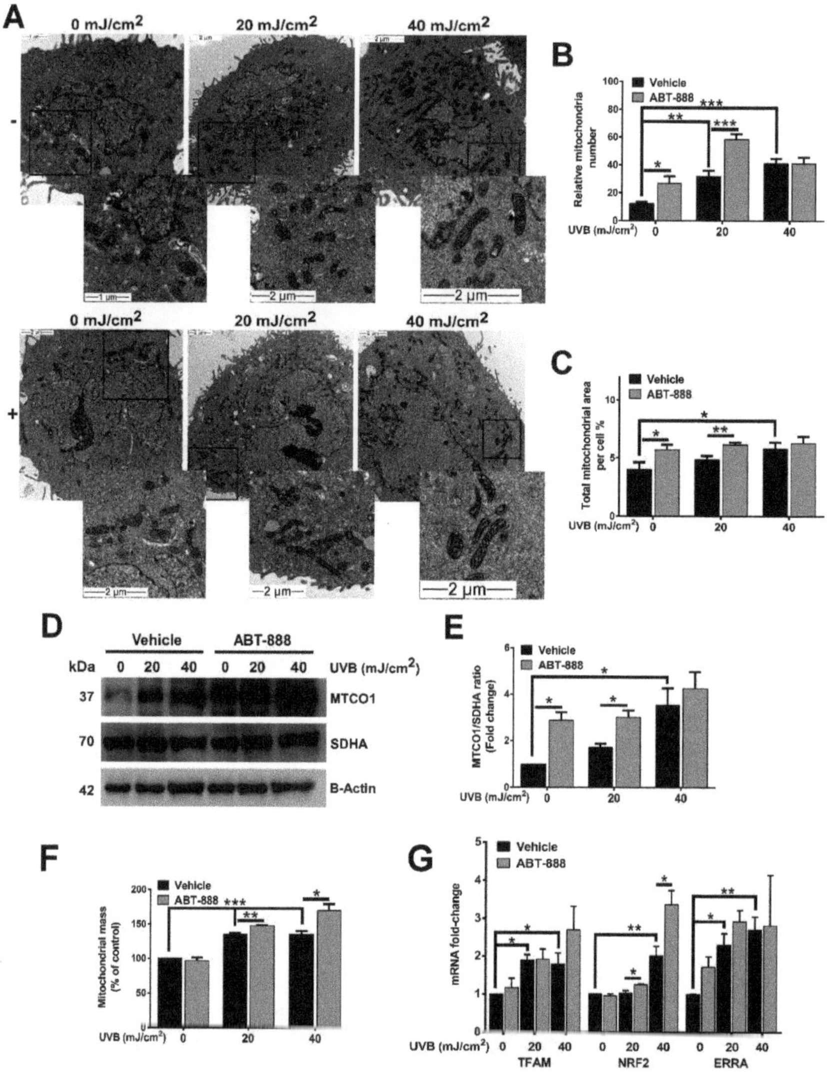

Figure 3. Poly (ADP-ribose) polymerase (PARP) inhibition enhances ultraviolet B (UVB)-mediated mitochondrial biogenesis. (**A**) Effect of UVB irradiation and PARP inhibition on mitochondrial ultrastructure visualized by transmission electron microscopy (TEM) 24 h after UVB exposure. Enlarged pictures are displayed at the right bottom corner. Scale bar is presented on the lower panels. (**B**) Mitochondrial number and (**C**) area were calculated from TEM images (minimum 7 cells were analyzed). (**D**,**E**) Mitochondrial biogenesis was quantified by the ratio of the mitochondrially encoded cytochrome C oxidase I (MTCO1) and succinate dehydrogenase complex, subunit A (SDHA) expression 24 h post UVB ($n = 3$). (**F**) Mitochondrial mass was determined by Mitotracker Green labeling 24 h after UVB irradiation ($n = 3$). (**G**) mRNA levels of master regulators of mitochondrial biogenesis were quantified by real-time PCR 24 h post-UVB (n = min. 3). −/+ represent vehicle (−) or ABT-888 (+) treatment. *; ** and *** indicate statistically significant difference at $p < 0.05$ and $p < 0.01$, $p < 0.001$, respectively. Error bars represent SEM.

2.3. PARP Inhibition Augments UVB-Mediated Mitochondrial Fusion

To identify if UVB and PARPi also alters mitochondrial dynamics, we evaluated mitochondrial morphology using confocal microscopy. Non-irradiated cells mainly contained fragmented mitochondria which normally represents low metabolic activity. After 40 mJ/cm^2 UVB, a statistically significant reduction in the frequency of fragmented mitochondria and an elevation in tubular mitochondria was detected compared to the non-irradiated control. The frequency of intermediate mitochondria was increased after PARPi at 0 mJ UVB compared to the vehicle control, and we also observed decreased fragmented mitochondrial frequency and higher percentage of tubular mitochondria after 20 mJ/cm^2 + ABT-888 treatment compared to the 20 mJ/cm^2 UVB exposed cells (Figure 4A). The dose-dependent effect of UVB in the branching aspect of mitochondria (Figure 4C) as defined by form factor ((Perimeter^2/(4π × area)) was also observed suggesting enhanced mitochondrial fusion after UVB and PARP inhibition. Finally, to confirm that the observed mitochondrial fusion are regulated by the dynamin-related proteins, we checked the expression of mitofusin-1 (Mfn1), mitofusin-2 (Mfn2), and optic atrophy 1 (OPA1) (Figure 4D,E). Similarly to mitochondrial morphological alterations, we detected enhanced protein expression of Mfn1, Mfn2, and OPA1 which support mitochondrial fusion at the outer and inner mitochondrial membrane. These results clearly indicate that besides enhancing mitochondrial biogenesis, UVB also triggers mitochondrial fusion and increases the complexity of mitochondrial network which is more prominent after PARP inhibition.

2.4. PARP Inhibition and UVB Induces Bulk Autophagy but Not Mitophagy

Several DNA-damaging agents were shown to initiate autophagy [67,68] and mitophagy [69] to remove damaged macromolecules or organelles including mitochondria, which prompted us to explore if UVB and PARPi-induced mitochondrial biogenesis and morphological changes of mitochondria affects autophagy or mitophagy. We used dual labelling of Mitotracker CMxROS to stain mitochondria and microtubule-associated proteins 1A/1B light chain 3B (LC3A/B) antibody to detect autophagosomes. Our results show the accumulation of LC3-positive cells after UVB exposure. PARP inhibition augmented autophagy not only in UVB-exposed cells, but also under non-irradiated conditions (Figure 5A,B). LC3B western blotting (Figure 5C) also confirmed these results indicating the autophagy inducer role of ABT-888. PARP1 knockdown also showed similar results but with much more pronounced autophagy induction after UVB (Figure 5E). Interestingly, the role of PARP1 in the regulation of autophagy is controversial. Both autophagy inductor [70–72] and inhibitor [73–75] role of PARP1 has been described suggesting that the autophagy-modulatory effect of PARylation might show DNA damage and cell type specificity. Even though, we detected an increased number of LC3 puncta, this type of macroautophagy cannot be considered as mitophagy since autophagic puncta show very mild colocalization with mitochondria (Figure 5D). Furthermore, a dose-dependent mitochondrial elongation was observed after UVB and PARP inhibition in Figure 4A,B indicating mitochondrial fusion. Since fused mitochondria are protected from mitophagy [76], and mitophagy-coupled elimination of mitochondria is usually preceded by fission [69] and decline in mitochondrial function [77], the here experienced induction in autophagy is considered as general autophagy and suggest that mitochondrial morphological rearrangements induced by UVB and PARPi may interfere with the initiation of mitophagy.

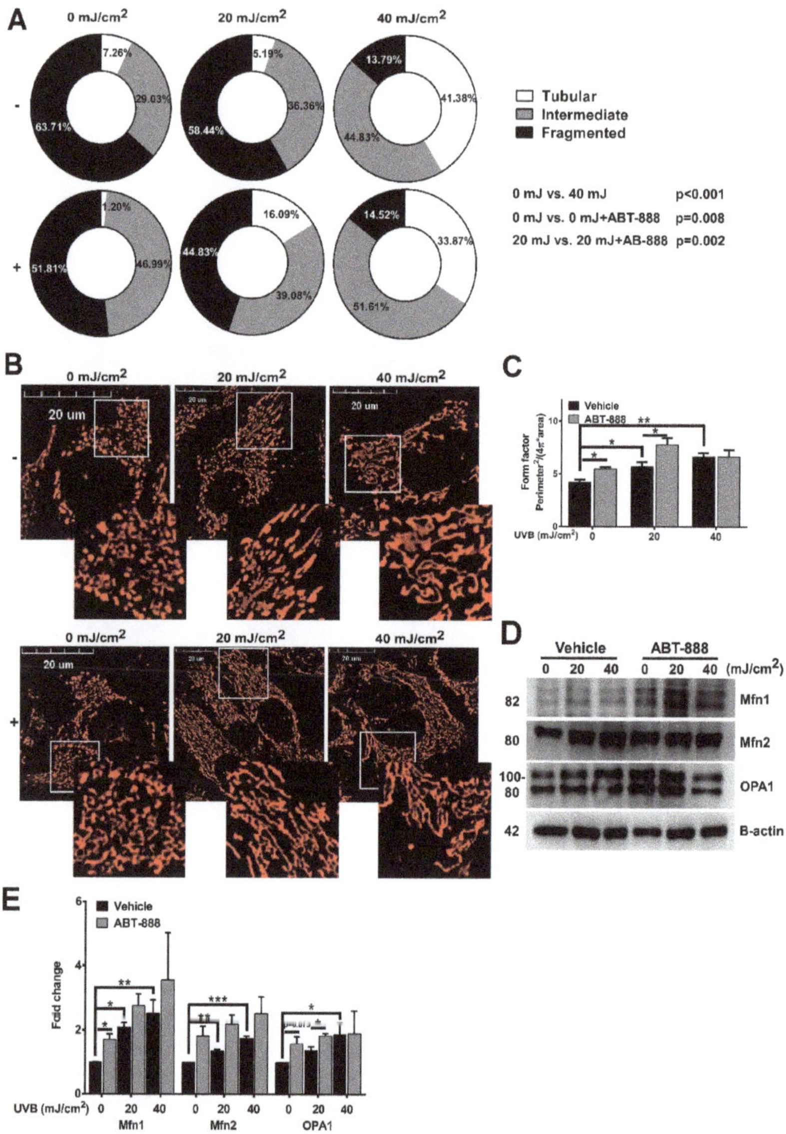

Figure 4. Poly (ADP-ribose) polymerase (PARP) inhibition augments ultraviolet B (UVB)-mediated mitochondrial fusion. (**A**) Effect of UVB irradiation and PARP inhibition on mitochondrial morphological subtypes (tubular, intermediate, and fragmented) are quantified based on confocal microscopic images 24 h post-UVB (minimum 29 cells). (**B**) Mitochondrial morphology visualized by confocal microscopy with Mitotracker Red CMXRos dye (n = 3). Enlarged pictures are displayed at the right bottom corner. Scale bar is presented on the images. (**C**) The branching aspect of mitochondria was derived from confocal microscopic images and was represented as a form factor (n = 3). (**D**) Protein expression of mitofusin-1 (Mfn1), mitofusin-2 (Mfn2), and optic atrophy 1 (OPA1) was visualized by Western blot. (**E**) Expression of mitochondrial fusion proteins were analyzed by Western blot 24 h post-UVB (n = min. 3). −/+ represent vehicle (−) or ABT-888 (+) treatment. *; ** and *** indicate statistically significant difference at $p < 0.05$ and $p < 0.01$, $p < 0.001$, respectively. Error bars represent SEM.

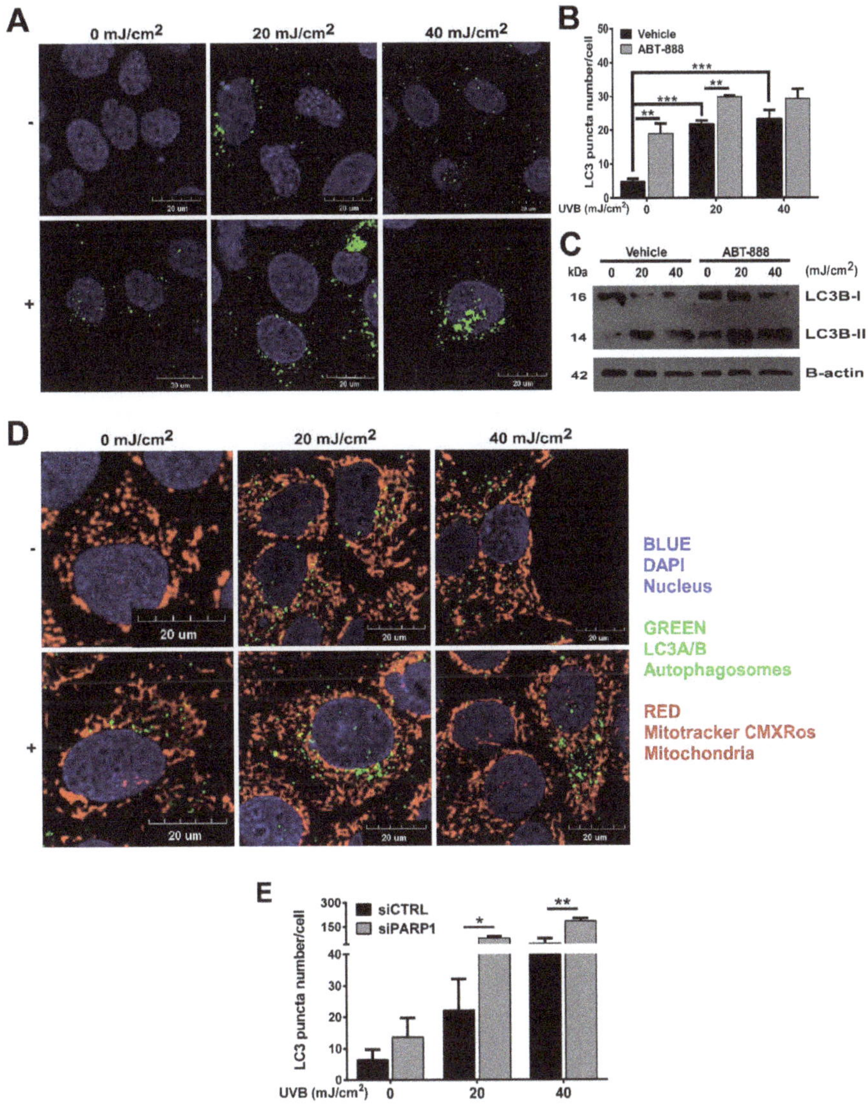

Figure 5. Poly (ADP-ribose) polymerase (PARP) inhibition and ultraviolet B (UVB) induces bulk autophagy but not mitophagy. (**A**) Expression of microtubule-associated proteins 1A/1B light chain 3B (LC3A/B) protein marker of autophagy by confocal microcopy ($n = 3$). Scale bar is presented on the figure. (**B**) Quantification of LC3A/B puncta in cells derived from confocal images in Figure 5A ($n = 3$). (**C**) LC3A/B proteins were quantified by Western blotting. Brightness and contrast were adjusted. Protein of interest were normalized to the loading control B-actin. (**D**) Dual staining of LC3-puncta and mitochondria by confocal microscopy ($n = 3$). (**E**) Quantification of LC3A/B puncta in PARP1 siRNA and control siRNA-transfected cells derived from confocal images ($n = 4$) −/+ represent vehicle (−) or ABT-888 (+) treatment *; ** and *** indicate statistically significant difference at $p < 0.05$, $p < 0.01$ and $p < 0.001$, respectively. Error bars represent SEM.

2.5. PARP Inhibition Boosts UVB-Mediated Mitochondrial Bioenergetic Changes

Alterations in the mitochondrial network is strictly controlled by intra- or extracellular signals connecting mitochondrial biogenesis and fusion with energy perturbations. To test whether increased mitochondrial biogenesis and fusion after UVB and PARP inhibition supports mitochondrial activity, we evaluated mitochondrial parameters. Our results show that UVB exposure increased mitochondrial membrane potential (Figure 6A) and total ATP level (Figure 6B). PARPi boosted the UVB-induced mitochondrial membrane hyperpolarization and it also raised ATP level compared to UVB-irradiated samples. Mitochondrial membrane potential and ATP levels change independently, and hyperpolarization of mitochondrial membrane potential does not necessarily correlate with ATP production but it may stem from decreased F_0F_1 ATP synthase activity and concomitant lower ATP production. Furthermore, increased ATP level may result either from enhanced glycolysis, increased respiration via electron flow from complex I-V, or decreased energy expenditure. In order to clarify the reasons of the elevation in ATP levels, we complemented our data with quantitative analysis of metabolic flux by XF96 oximeter and monitored extracellular acidification rate (ECAR) indicating glycolysis and oxygen consumption (OCR) representing oxidative phosphorylation with sequential addition of oligomycin and antimycin. Although basal ECAR (Figure 6C) is statistically unchanged after UVB exposure, PARP inhibition enhanced the rate of glycolysis after 20 mJ/cm^2 UVB dose and at non-irradiated conditions. In contrast, UVB dose-dependently increased oxygen consumption and ABT-888 treatment significantly augmented OXPHOS compared to 0 or 20 mJ/cm^2 UVB (Figure 6C). Citrate synthase (CS) activity, the initial enzyme in the tricarboxylic acid (TCA) cycle catalyzing the formation of citrate from oxalacetate and acetyl-CoA, showed significant elevation after UVB radiation, which was augmented by PARP inhibition (Figure 6C). This suggest that not only the distal part of respiratory chain is altered by UVB and PARPi, but also increased terminal oxidation is preceded by elevation in TCA activity as well. To exclude the possibility of the off-target effects of ABT-888, we confirmed the mitochondrial changes by PARP1 silencing. We detected increased mitochondrial mass reflecting mitochondrial biogenesis after UVB in control siRNA-transfected keratinocytes that was elevated by PARP1 knockdown (Figure 6D). Mitotracker Red CMXRos incorporation into mitochondria reflecting mitochondrial membrane potential also showed similar changes similarly after vehicle and ABT-888 treatment (Figure 6E). Although, we could not detect difference in basal ECAR between the control and PARP siRNA-transfected cells, CS activity and OXPHOS showed significant changes (Figure 6F) confirming the pivotal role of PARP1 in the UVB-induced mitochondrial changes. In summary, we can conclude that elevation in cellular ATP level after UVB is due to enhanced energy production and cells increase their energy reserves through both glycolysis, TCA cycle, and OXPHOS after PARP inhibition.

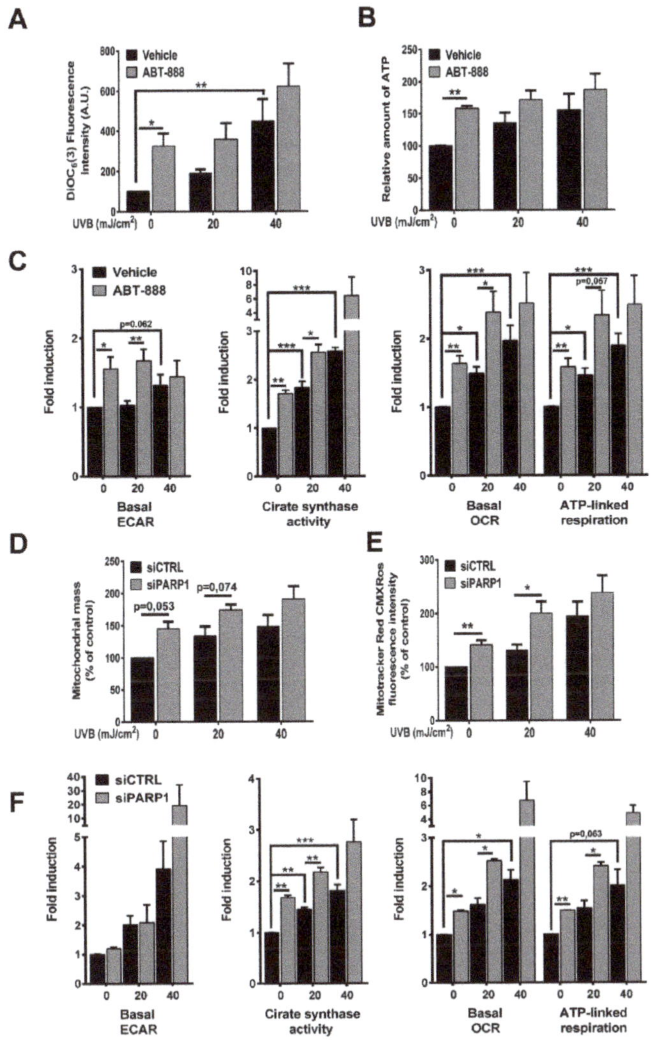

Figure 6. PARP inhibition boosts UVB-mediated mitochondrial bioenergetic changes. (**A**) For determination of mitochondrial membrane potential cells were stained with 3,3′-dihexyloxacarbocyanine iodide (DIOC$_6$(3)) 24 h post-UVB and measured by flow cytometry in FL1 channel (n = 4). (**B**) Total cellular ATP level was quantified by colorimetric assay (n = 3). (**C**) Basal extracellular acidification rate (ECAR) represents glycolysis after 24 h post-UVB. XF medium was supplemented with 10 mM glucose (n = 6). After four oxygen consumption rate (OCR) measurement oligomycin and antimycin was used to determine mitochondria-linked ATP production and basal OCR, respectively (n = 6). Citrate synthase activity, the initial enzyme of the tricarboxylic acid (TCA) cycle was measured by citrate synthase kit (n = 3). (**D**) Mitochondrial mass determined similarly as in Figure 3F (n = 4). (**E**) Mitotracker Red CMXRos mean fluorescence intensity was measured by flow cytometry (n = 4). (**F**) Metabolic parameters including glycolysis, citrate synthase activity, and oxidative phosphorylation (OPXHOS) was detected similarly as in Figure 6C (n = 2 for ECAR and OCR, n = 3 for citrate synthase (CS) activity). * $p < 0.05$, ** $p < 0.01$, *** $p < 0.001$.

2.6. *PARP Inhibition Restores NAD*$^+$ *Level and SIRTUIN Expression*

NAD$^+$ and NADH plays a central role in redox homeostasis and cellular metabolism including glycolysis via glyceraldehyde-3-phosphate dehydrogenase (GAPDH) activity [78] and OXPHOS by regulating the transfer of electrons to complex I [79]. UVB irradiation caused a slight but significant decrease in NAD$^+$ level after both 20 and 40 mJ/cm^2 UVB (Figure 7A). ABT-888 treatment efficiently restored intracellular NAD$^+$ content suggesting the role of PARP1 in UVB-mediated NAD$^+$ depletion. It has been known that a decrease in NAD$^+$ by enhanced PARP activation inhibits another NAD$^+$-consuming enzyme family, the class III histone deacetylase Sirtuins [33,34,36–38,80] which regulate diverse cellular processes [81] including mitochondrial metabolism and have intricate relationships with PARPs [82,83]. To test whether UVB modulate Sirtuin expression and PARPi can restore their expression after UVB similarly as NAD$^+$ level changes, we chose the 40 mJ/cm^2 UVB dose which caused a more significant increase in NAD$^+$ level after ABT-888 treatment compared to UVB-irradiated cells. We detected slight mRNA downregulation in SIRT1, SIRT2, SIRT3, SIRT4, SIRT5, and SIRT7 after 40 mJ/cm^2 UVB dose (Figure 7B) and PARP inhibition increased the gene expression of all Sirtuins suggesting that intracellular NAD$^+$ availability after UVB and PARPi may regulate SIRTs expression. Although, the role of Sirtuins in the UVB-mediated cellular response and skin physiology is poorly characterized [84–88], here we revealed the Sirtuin expression-modulatory effect of NAD$^+$ after UVB and PARP inhibition. To explore the potential role of NAD$^+$ and Sirtuins in supporting glycolysis or oxidative phosphorylation further experiments are needed.

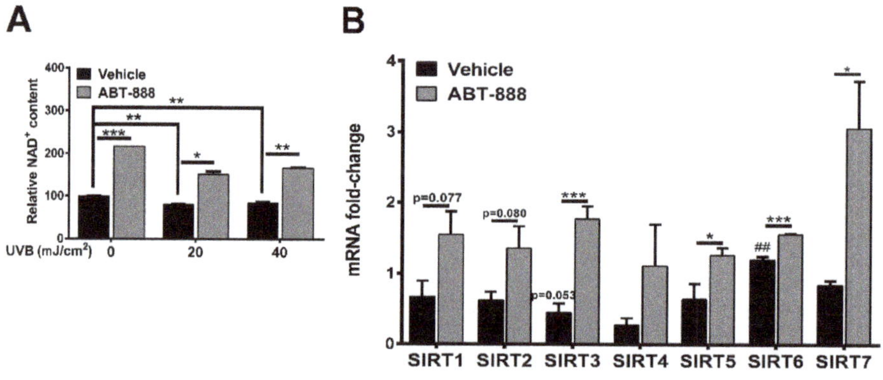

Figure 7. PARP inhibition restores NAD$^+$ level and Sirtuin expression. (**A**) Total cellular NAD$^+$ were quantified by colorimetric assay 24 h after UVB irradiation. Absorbance was measured at 490 nm (n = 2). (**B**) mRNA expression of the Sirtuin enzyme family 24 h post-UVB (40 mJ, n = 3). *; ** and *** indicate statistically significant difference at $p < 0.05$ and $p < 0.01$, $p < 0.001$, respectively. ## indicate significant difference at $p < 0.01$ compared to non-irradiated control. Error bars represent SEM.

2.7. *PARP Inhibition Enhances UVB-Mediated Upregulation of Metabolic Proteins*

To explore the potential mechanisms responsible for elevated oxidative phosphorylation after UVB and PARP inhibition, we checked the expression of metabolic proteins involved in oxidative phosphorylation by Western blot (Figure 8A,B). Consistent with their role in mitochondrial activity, we detected significant increase in ATM, p-ATM, SIRT1, peroxisome proliferator-activated receptor gamma coactivator 1-alpha (PGC1A), AMPK, p-AMPK, p53, p-p53, p-AKT, and p-p70S6K1 (mTOR activity) expression after UVB which was further enhanced by PARP inhibition. The most prominent and statistically significant upregulation in these proteins can be detected after 40 mJ/cm^2 UVB dose similarly as OXPHOS was increased to two-fold at $p < 0.001$. PARPi induced statistically significant difference at 0 and 20 mJ/cm^2 compared to vehicle control similarly as ABT-888 increased OXPHOS as well. PARP1 knockdown also induced upregulation in these proteins compared to siRNA control.

These results suggest that these proteins switch on oxidative metabolism and their elevated expression may be responsible for higher oxygen consumption after UVB and ABT-888 treatment.

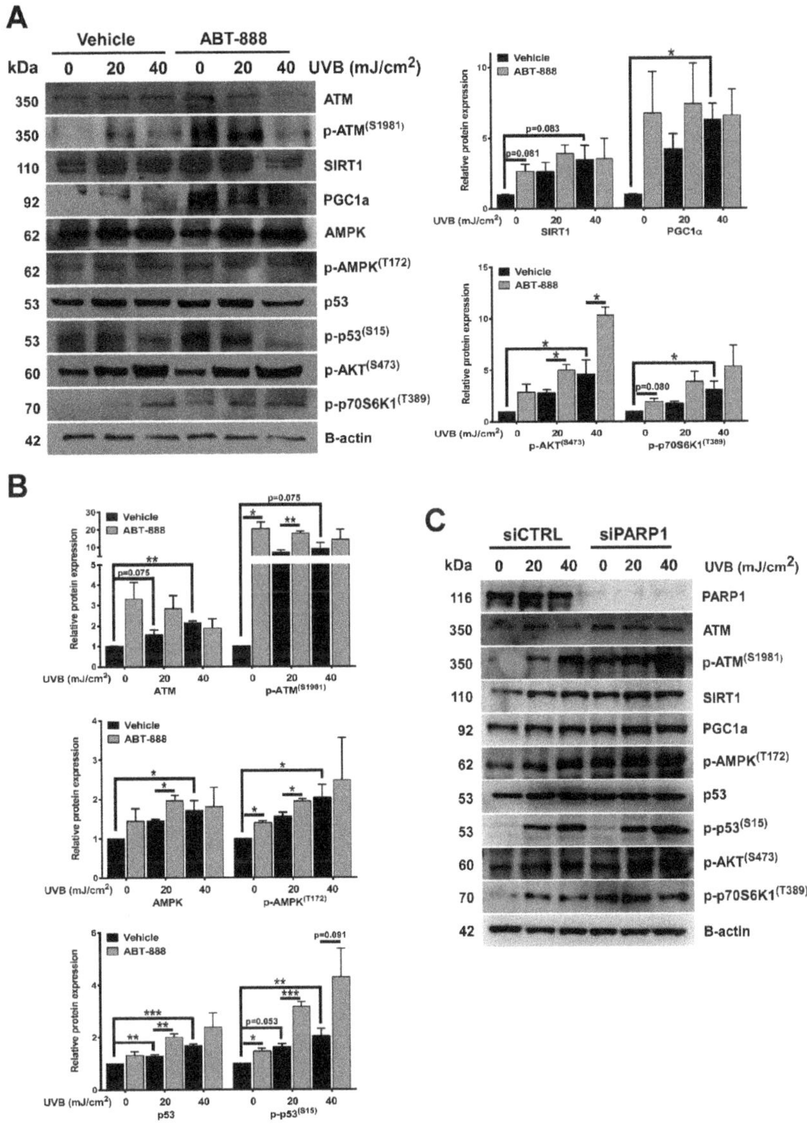

Figure 8. PARP inhibition enhances UVB-mediated upregulation of metabolic proteins. (**A**) Metabolic proteins involved in oxidative phosphorylation were analyzed in total protein lysates. Protein of interest were normalized to the loading control B-actin. Cells were harvested 24 h after UVB. Brightness and contrast were adjusted. (**B**) Densitometric representation of Figure 8A proteins (n = min. 3). (**C**) Protein expression involved in oxidative phosphorylation was confirmed by PARP1 knockdown. *; ** and *** indicate statistically significant difference at $p < 0.05$ and $p < 0.01$, $p < 0.001$, respectively. Error bars represent SEM.

2.8. PARP Inhibition and UVB-Induced Oxidative Phosphorylation and Autophagy Are Dependent on ATM, AMPK, p53, AKT, and mTOR Activation

To test their requirement for increased oxidative phosphorylation, we applied KU-60019 as an ATM inhibitor (Figure 9A) and ATM siRNA for gene silencing (Figure 10A), compound C as an AMPK inhibitor, Pifithrin-alpha-HBr as a p53 inhibitor, Wortmannin as a phosphoinositide 3-kinase (PI3Ki)/AKT inhibitor and Rapamycin as a mTOR inhibitor (Figure 9A). Chemical inhibition of these proteins led to significant reduction in OXPHOS highlighting their role in mediating metabolic alterations after UVB and PARP inhibition. ATM inhibition (Figure 9B) and ATM knockdown (Figure 10B) also decreased the phosphorylation of AMPK, p53, AKT, and p70S6K1 which suggest that elevation in OXPHOS after UVB is downstream of ATM, which is known as one of the most important players in DNA damage response besides PARP1. Since CPD removal is mediated by the energetically demanding nucleotide excision repair, keratinocytes may try to compensate their increased DNA damage with a more intense mitochondrial activity to facilitate CPD elimination. In this respect, OXPHOS seems to be a beneficial response after UVB. To confirm this hypothesis, we applied inhibitors that decreased (ATMi, AMPKi, p53i, PI3Ki, mTORi) or limit the activity of mitochondrial electron transport chain (oligomycin and rotenone) or inhibits mitochondrial protein synthesis (chloramphenicol) (Figure 9C). Our data show that all these inhibitions led to a more pronounced cell death after UVB which confirms that increased mitochondrial activity bears an adaptive and antiapoptotic response after UVB. We also wanted to explore if UVB and PARPi-triggered general autophagy (Figure 5A,B) by recycling damaged organelles and macromolecules serves to provide metabolites for OXPHOS. To confirm this hypothesis, we tested the ATM, AMPK, p53, PI3K, and mTOR inhibitors after 40 mJ/cm^2 UVB that caused the most prominent decrease in oxygen consumption (Figure 9A). Chemical inhibition of these proteins induced significant reduction in LC3B (Figure 9D) and Parkin (Figure 9E) expression similarly as UVB and PARPi-induced OXPHOS decreased after ATM, AMPK, p53, PI3K, and mTOR inhibitors indicating a parallel reduction in oxidative phosphorylation, autophagy, and PARKIN expression (a marker for mitophagy) as well. Reduction in PARKIN expression suggests impaired mitochondrial quality control, allowing the accumulation of damaged mitochondria, corrupting mitochondrial function, eventually leading to decreased OXPHOS. These results together suggest that after UVB and ABT-888 treatment, AMPK, p53, AKT, and mTOR are the main mediators of oxidative phosphorylation downstream of ATM kinase, and autophagy by recycling damaged cellular parts after stress response may provide metabolites to mitochondria and supports OXPHOS after UVB exposure and PARP inhibition.

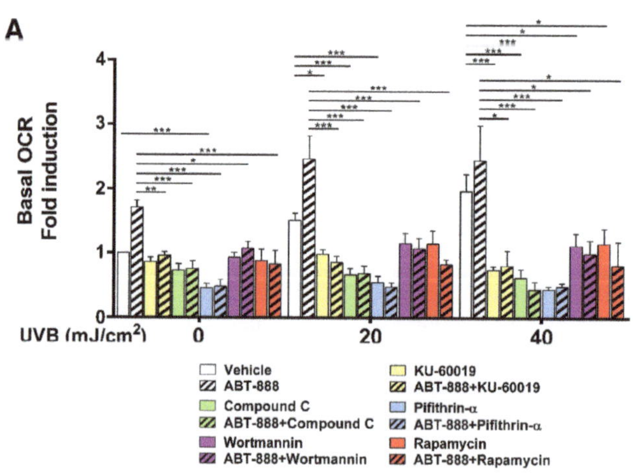

Figure 9. *Cont.*

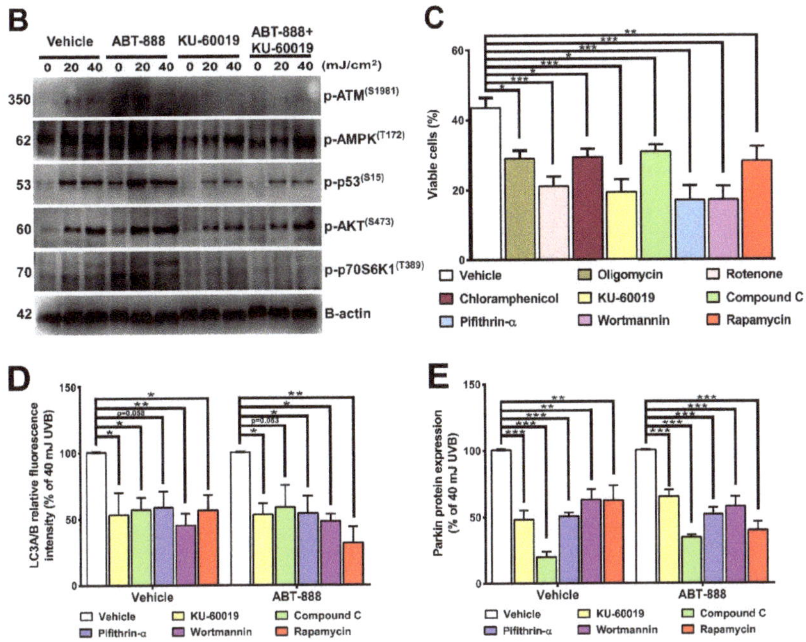

Figure 9. PARP inhibition and UVB-induced oxidative phosphorylation and autophagy are dependent on ATM, AMPK, p53, AKT, and mTOR activation. (**A**) To determine key proteins involved in mediating mitochondrial changes ATMi (KU-60019), AMPKi (Compound C), p53i (Pifithrin-alpha-HBr), PI3Ki (Wortmannin), and mTORi (Rapamycin) were added to the medium and OCR was measured as in Figure 6C (n = min.3). (**B**) ATM downstream signaling pathway was investigated by Western blot with the addition of its pharmacological inhibitors KU-60019 (n = 3). Brightness and contrast were adjusted. (**C**) Cell viability was measured by flow cytometry as in Figure 1E after 40 mJ/cm^2 UVB with oligomycin, rotenone, chloramphenicol, ATMi, AMPKi, p53i, PI3Ki, and mTORi (n = min. 4). To determine the involvement of ATM, AMPK, p53, AKT, and mTOR in the regulation of (**D**) autophagy and (**E**) PARKIN expression, we applied their respective pharmacological inhibitors after 40 mJ/cm^2 UVB. (n = 4). *; ** and *** indicate statistically significant difference at $p < 0.05$ and $p < 0.01$, $p < 0.001$, respectively. Error bars represent SEM. ATM: ataxia-telangiectasia-mutated kinase, AMPK: adenosine monophosphate-activated kinase, AKT: protein kinase B, mTOR: mammalian target of rapamycin, OCR: oxygen consumption rate.

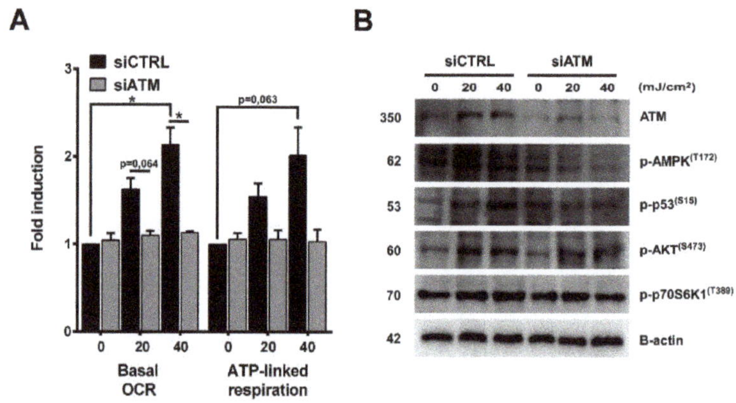

Figure 10. ATM silencing after UVB showed similar results as ATM inhibition by KU-60019. (**A**) OCR was measured as in Figure 6C ($n = 2$) after ATM knockdown. (**B**) ATM downstream signaling pathway was investigated by Western blot with ATM silencing ($n = 2$). * indicates statistically significant difference at $p < 0.05$. Error bars represent SEM.

3. Discussion

In this study, we identified morphological and functional changes of mitochondria after UVB and PARP inhibition. Early after UVB exposure, we detected PAR formation, reflecting PARP activity which was completely suppressed by PARP inhibition. We showed that loss of PARylation caused defective CPD repair compared to the UVB-irradiated cells suggesting impaired NER pathway as several PARP1-intercating partners (DDB1, DDB2, ATM, RAD51, ALC1, XPC) [89–91] have been described so far. Furthermore, PARP1 promotes the establishment of locally relaxed chromatin structure [92] to enable the removal of damaged DNA parts. Therefore, it is logical to assume that PARP inhibition renders chromatin to become compact, blocking not only replication, transcription but also the repair of UVB-induced photoproducts suggesting the particular importance of PARP1 in mediating transcription coupled nucleotide excision repair. Our results show a prolongation in G_2/M phase of cell cycle after PARPi as described earlier [93–95]. Inhibition of PARP1 was synergistic with UVB with respect to G_2/M accumulation characteristic of severe DNA damage after 20 mJ/cm^2. G_2/M accumulation can also be attributed to PARP trapping, in which inactivated PARP1 remains bound to DNA stalling replication, transcription leading to replicative stress and double-strand breaks [96,97] which make PARP inhibitors cytotoxic in combination with DNA damaging agents [98,99]. If DNA damage cannot be repaired, cells try to evade the accumulation of mutations by initiating apoptosis. Since apoptotic response after UVB cannot be ameliorated by PARPi, we can exclude the possibility that UVB-triggered cell death in our model system is PARP1-dependent, it may rather suggest that the accumulated and unrepaired CPDs augments the apoptotic response after PARPi. Long-term survival of keratinocytes revealed that the combined treatment of UVB and PARP inhibition severely reduced cell proliferation even at the non-irradiated conditions possibly suggesting the role of trapped PARP1-DNA complexes preventing cells division. We also detected that loss of PARylation decreased the frequency of mutations in UVB-treated cells as evidenced by decreased number of colonies after ABT-888 treatment suggesting that PARPi initiate apoptosis of cells with a high content of CPDs. Our results are in accordance with the anti-cancer effect of diverse PARP inhibitors and provide experimental evidence for the photosensitizing effect of PARP inhibitors experienced clinically [100,101]. Nevertheless, it is important to know that therapeutic application of PARP inhibitors might be associated with photosensitivity (sunburn) but not with an increased photocarcinogenesis risk.

Emerging evidence suggest that autophagy induction can be coupled to DNA damage response [67]. Accordingly, UVB triggered the accumulation of autophagosomes which corroborates with enhanced

LC3B expression. PARP inhibition resulted in enhanced autophagosome formation and higher LC3B protein level. The role of PARP1 in the regulation of autophagy is still poorly understood, both autophagy inducer [70] and inhibitory effect of PARP1 [75] can be found in the literature. In our model system, increased autophagosome formation after UVB and PARPi can be attributed to the elevated level of DNA damage (higher CPD content), which serves as a positive signal for autophagy since the initiation of global genome NER (GG-NER) subpathway is controlled by autophagy [67,68].

As Le Brace et al. [28] demonstrated, different DNA damaging agents including UVC which also induce CPD formation as UVB, led to AMPK activation and increased fatty acid oxidation. Similarly to their results, UVB radiation triggered mitochondrial fusion and induced mitochondrial biogenesis, which culminated in enhanced mitochondrial activity and complex metabolic events which were more prominent in PARP-inhibited cells. Increased number of mitochondria and mitochondrial area after UVB irradiation was more robust after PARPi. mRNA expression of mitochondrial biogenesis regulators, increased MTCO1/SDHA ratio, higher mitochondrial mass show similar trends suggesting that elevated mitochondrial content is indeed due to enhanced mitochondrial biogenesis and not defective mitochondrial turnover. By measuring mitochondrial function, we observed that PARP inhibition enhanced the UVB-mediated mitochondrial hyperpolarization, raised cellular ATP level and OXPHOS. Interestingly, we detected neither mitochondrial membrane depolarization nor ATP loss which normally occurs after PARP activation via NAD^+ resynthesis from ATP [102]. Since intracellular ATP concentration is also a regulator in DNA repair and a key factor in the cell's decision to die via apoptosis or necrosis [103], it is likely that this elevation in ATP level is required for the energetically costly nucleotide excision repair to remove UVB-induced photoproducts, and/or necessary for the apoptotic process as seen after staurosporine treatment [104]. Boost in mitochondrial OXPHOS was dependent on ATM, AMPK, p53, AKT, and mTOR phosphorylation since inhibition of them lead to significant reduction in oxygen consumption after both UVB and PARPi. Moreover, enhanced mitochondrial activity seems to induce a beneficial response after DNA damage, since decreasing mitochondrial biogenesis or ameliorating OXPHOS resulted in elevated cell death after UVB irradiation suggesting the protective effect of OXPHOS on cell viability. Inhibition of the ATM, AMPK, p53, PI3K, and mTOR pathways caused significant reduction in LC3A/B and Parkin expression emphasizing their role in autophagy and mitophagy. Decrease in mitophagy and the concomitant impaired mitochondrial quality control may also be an underlying cause for decreased OXPHOS after chemical inhibition of these proteins. We hypothesize that the purpose of autophagy induction after UVB-induced DNA damage is to recycle damaged molecules by fueling mitochondria with metabolites for oxidative phosphorylation as suspected by Dong et al. [105] and autophagy promotes the DNA damage recognition via NER. Since autophagy is an ATP-dependent process and cells must maintain their energy status to promote autophagy, we cannot exclude reciprocal regulation that is the shift toward oxidative metabolism regulates autophagy as described by Thomas et al. [106]. Nonetheless, in our model system it seems obvious that the purpose of autophagy and OXPHOS induction is to promote cell survival after UVB via ATM, AMPK, p53, AKT, and mTOR activation.

So how, increased mitochondrial activity is connected to DNA damage? The UVB-induced photolesions trigger local conformational changes on DNA similarly as seen after etoposide treatment [107] and UVC irradiation [108] which also resulted in increased mitochondrial biogenesis and activity [18,28]. Since, PARP-inhibited cells displayed higher level of CPDs compared to UVB-irradiated samples, it seems plausible that CPDs generated on DNA may be a trigger for mitochondrial changes. This is supported by the fact that the phosphorylation status of the main DNA damage responders (ATM, AMPK, p53, AKT, and mTOR) that are activated by UVB become more prominent after PARPi. ATM is mainly activated by DNA double-strand breaks but also a sensor of CPDs [109]. Furthermore, unrepaired CPD lesions in PARPi cells due to PARP trapping may eventually be converted into DNA breaks during replication which is an ultimate trigger for ATM activation. ATM can directly phosphorylate AMPK [110], p53 [111], and AKT [112] and via AKT and AMPK signaling ATM regulates mTOR activity [113], as well to turn on oxidative metabolism. AMPK can also phosphorylate p53 at

S15 [114], thereby potentiating its activity in enhancing mitochondrial metabolism. Furthermore, it is noteworthy that although cell type-dependently, AMPK [115], p53 [116] AKT [117], and mTOR [118] activation can trigger mitochondrial fusion which may also be a possible explanation for enhanced mitochondrial activity. Since PARP inhibition alone do not induce CPD formation, we must take into account that the PARPi-mediated changes in the non-irradiated cells are mediated by CPD-independent mechanisms either through autophagy induction and/or NAD^+ prevention. However, we can exclude the involvement of NAD^+ in the regulation of mitochondrial function after UVB, since morphological alteration of mitochondria and oxidative phosphorylation do not show similar changes as intracellular NAD^+ availability after UVB irradiation.

4. Materials and Methods

4.1. Chemicals

All chemicals were obtained from Sigma-Aldrich (St. Louis, MO, USA) unless stated otherwise.

4.2. Cell Culture

Human immortalized keratinocyte-derived (HaCaT) cell line were cultured in a T75 flask, as previously described [119] using 4500 mg/L Dulbecco's modified eagle media (DMEM) Glutamax supplement (Thermo Fisher Scientific, Waltham, MA, USA) containing 10% heat-inactivated fetal bovine serum (FBS) and 0.5% antibiotic/antimycotic solution. For HPRT mutation assay, Chinese hamster ovary (CHO) cells were used as described in Section 4.7. Cells were maintained in a humidified incubator at 37 °C with 5% CO_2 atmosphere.

4.3. Cell Treatment

Cells were harvested with trypsin-EDTA (Biosera, Budapest, Hungary) then seeded in 12-well plate in 200,000 cell/well density (unless stated otherwise), and allowed to adhere for at least 12 h. At 80% confluence, cells were pretreated with the PARP inhibitor veliparib (ABT-888) (Selleckchem, Houston, TX, USA) at a final concentration of 25 µM. For detection of cell death, autophagy, mitophagy, and OXPHOS, the following inhibitors were used: ATMi: 6 µM KU-60019 (Adooq, Irvine, CA, USA), AMPKi: 2.5 µM compound C (Selleckchem), p53i: 30 µM pifithrin-alpha-HBr (AdooQ), mTORi: 300 nM rapamycin (AdooQ), PI3Ki: 300 nM wortmannin (AdooQ), mitochondrial activity inhibitors: 5 µM oligomycin, and 500 nM rotenone. To inhibit mitochondrial protein synthesis and hamper mitochondrial function chloramphenicol (AdooQ) (25 µM) was used. For UVB irradiation, cells were covered with 400 µl pre-warmed Dulbecco's phosphate-buffered saline (DPBS) (Lonza, Walkersville, MD, USA) and were subjected to 20 or 40 mJ/cm^2 UVB using two UVB broadband tubes (TL-20W/12 RS; Philips, Eindhoven, The Netherlands). Immediately after irradiation, the old medium was placed back on the cells to evade metabolic perturbations and cells were further cultured for 24 h. Controls were covered with tin foil during irradiation. The proper UV dose was measured with a UVX digital radiometer (UVP Inc., San Gabriel, CA, USA). Detached, dead cells were excluded from the experiments except for cell viability.

4.4. Gene Silencing

On-target plus SMARTpool PARP1 and ATM siRNA sequence was purchased from Dharmacon Research, Inc. (Lafayette, CO). Non-targeting siRNA (Dharmacon) was used as a control. HaCaT cells were seeded into 12-well plates in complete high-glucose DMEM containing 10% FBS without antibiotics. DharmaFECT transfection reagent (Dharmacon) in tube 1 and siRNA in 1 × siRNA buffer (Dharmacon) in tube 2 was diluted in serum and antibiotics-free DMEM and incubated at room temperature for 5 min. The two mixtures were combined and incubated further for 20 min and added to the cells at a final siRNA concentration of 50 nM. After 48 h, cells were washed with DPBS, irradiated with UVB, and the medium was replaced with a complete growth medium. Silencing efficiency was determined by Western blotting 24 h post-UVB.

4.5. Cell Viability and Proliferation

Over the 24 h period following UVB exposure cell viability was determined by dead cell apoptosis kit containing propidium iodide/Alexa Fluor 488-conjugated Annexin V (Invitrogen, Carlsbad, CA, USA) according to the manufacturer's instruction. Labelled cells were analyzed by flow cytometry with a FACSCalibur (Becton Dickinson, San Jose, CA, USA) measuring the fluorescence emission in FL1 (530 nm) and FL3 (>575 nm). Double negative cells represent viable cells. For data collection and evaluation CellQuest software 5.2 (Becton Dickinson) and Flowjo single cell analysis software were used.

Cell proliferation was determined by clonogenic assay. Cells were seeded in 100 mm Petri dish at 5000 cells/dish and were allowed to grow for 10 days. The medium was replaced each day to monitor the long-term effect of cell treatments. Ten days later, cells were washed with PBS, fixed with 100% methanol, and stained with May-Grünwald-Giemsa solution (Histolab Products, Västra Frölunda, Sweden).

4.6. Cell Cycle Analysis

Cell cycle progression was quantified using propidium iodide (PI) staining. Briefly, cells were trypsinized, fixed with ice-cold 96% ethanol for 10 min, washed twice with PBS, and permeabilized with 0.1% Triton X-100. After extensive washing, cells were incubated in PBS solution containing 0.5 mg/mL RNase at 37 °C for 1 h and counted for cell number normalization. For cell staining, PI was used a final concentration of 20 μg/mL. Unbound PI was eliminated by washing with PBS and doublet discrimination was performed. To determine DNA content samples were analyzed on the x-axis in FL2-A channel using a FACSCalibur flow cytometer.

4.7. HPRT Mutation Assay

CHO cells were cultured in a HAT (hypoxanthine–aminopterin–thymidine) medium in T75 flask for one week to remove cells with pre-existing HPRT mutation. After a week, HAT medium was replaced with complete DMEM and cells were allowed to recover for three days. Thereafter, CHO cells were seeded into a 6-well plate and allowed to adhere for 24 h. Next day, cells were cultured with or without 25 μM ABT-888 and UVB irradiation was carried out with 10 or 20 mJ/cm^2 dose. CHO cells were cultured further for one week in DMEM with sub-culturing three times a week. After a week, cells were trypsinized, counted, distributed to 100 mm Petri dish at concentration of 5×10^4 per Petri dish in selection medium supplemented with 5 μg/mL 6-thioguanine (6-TG), and incubated for 10 days. After 10 days, cells were washed with PBS, fixed with 100% methanol for 10 min, and stained with May-Grünwald-Giemsa solution. Mutant colonies were counted.

4.8. CPD-Specific Enzyme-Linked Immunosorbent Assay (ELISA)

Genomic DNA was extracted from HaCaT cells using a Purelink Genomic DNA mini kit (Thermo Fisher Scientific) according to the manufacturer's instruction. For quantitative detection of CPDs, direct ELISA was applied as previously described [120].

4.9. Real-Time Quantitative RT-PCR

Total RNA was isolated using guanidinium isothiocyanate-phenol-chloroform extraction (TRI reagent) (MRC, Cincinnati, OH, USA) according to the protocol by Chomczynski et al. [121]. RNA concentration and purity were determined spectrophotometrically. RNA was deprived of DNA contamination by DNase I treatment (Fermentas, St. Leon-Rot, Germany). Reverse transcription (RT) was carried out using the high-capacity cDNA reverse transcription kit (Applied Biosystems, Foster

City, CA, USA) according to the manufacturer's instruction. Real-time RT-PCR analysis was performed by the SYBR green method using a Lightcyler 480 II (Roche Diagnostics). Reactions were carried out in 384-well optical plates. RNA expression values were determined by $1.8^{-\Delta CT}$ method and normalized for the housekeeping gene SDHA and phosphoglycerate kinase 1 (PGK1) [122]. Primer pairs and corresponding sequences are listed in Supplementary Table S1 "Primer pairs used in the study".

4.10. Western Blot

HaCaT cells were pelleted by mild centrifugation at 1500 rpm for 5 min at 4 °C. The pellets were lysed on ice with RIPA buffer supplemented with a protease inhibitor cocktail in 1:1000. After 5 min incubation on ice, supernatants were obtained by centrifugation of the lysates at 15,000 rpm for 10 min at 4 °C. The concentration of proteins was determined using Pierce BCA assay kit (Thermo Fisher Scientific). The lysates were mixed with 5X loading buffer (Bromophenol blue (0.25%), β-Mercaptoethanol (5%), Glycerol (50%), SDS (sodium dodecyl sulfate; 10%), Tris-HCl (0.25 M, pH 6.8)), boiled for 10 min at 100 °C and subjected to 7.5%, 10%, or 12.5% SDS polyacrylamide gel electrophoresis. Proteins were transferred onto nitrocellulose membrane (Bio-Rad, Hercules, CA, USA). The membranes were blocked in TBST (0.05% Tween 20 in TBS buffer) containing 5% bovine serum albumin (BSA). The incubation with primary antibodies was carried out overnight at 4 °C followed by washing. Horseradish peroxidase (HRP)-conjugated goat anti-mouse or anti-rabbit IgG was used as a secondary antibody (Bio-Rad) at room temperature for 1 h. Proteins were visualized by ECL Prime Western blotting detection reagent (Thermo Fisher Scientific). All the antibodies used for Western blotting were listed in Supplementary Table S2 "Primer antibodies used in the study". Bands were quantified using the ImageJ open source software [123] (version 1.51k, National Institutes of Health, Bethesda, MD). Proteins of interest were normalized for β-actin. Sample uncut blots are provided on Supplementary Figure S2.

4.11. Mitochondrial Mass

24 h post-UVB, Mitotracker Green (Thermo Fisher Scientific) were added to the culture medium at a final concentration of 100 nM and cells were incubated at 37 °C for 30 min. Following incubation, cells were washed with PBS and harvested by trypsinization and then placed on ice. The fluorescence intensity of cells stained with Mitotracker Green was analyzed in a FL1 channel using flow cytometry.

4.12. Determination of Mitochondrial Membrane Potential

HaCaT cells were seeded in a culture plate and next day subjected to UVB. 24 h post-UVB, 3,3′-dihexyloxacarbocyanine iodide ($DiOC_6(3)$) (Invitrogen) were added to the culture medium at a final concentration of 40 nM and cells were further incubated at 37 °C for 30 min. Following incubation, cells were washed with PBS and harvested by trypsinization and then placed on ice. The fluorescence intensity of cells stained with $DiOC_6(3)$ was analyzed in FL1 channel using flow cytometry.

4.13. Determination of Mitochondrial Ultrastructure, Mitochondrial Number, and Area by Transmission Electron Microscopy (TEM)

After cell treatment, cells were washed with DPBS, harvested by trypsinization, pelleted, and fixed by 3% glutaraldehyde (EMS, Hatfield, PA, USA) in 0.1 M cacodylic acid (EMS) buffer complemented by 5% sacharose for 2 h. After washing steps in 0.1 M cacodylic acid buffer, cells were osmificated in 1% OsO_4, dehydrated in ascending alcohol row, namely 50% for 2 × 10′, 70% for 2 × 10′, 96% for 2 × 15′, absolute ethanol for 3 × 20′, and propylene oxide for 2 × 10′. Then, overnight a durcupan araldite treatment was used for embedding the samples. Encapsulation occurred in an incubator for 48 h and ultrathin section was made by Leica EM UC7 ultramicrotome (Leica Microsystems, GmbH, Wetzlar, Germany). Standard contrasting was performed by uranyl acetate (EMS) and Reynolds' lead citrate solution. High resolution TEM images was made by the Jeol JEM 1010 electron microscope and software (JEOL Inc. Peabody, MA, USA).

4.14. Assessment of Mitochondrial Morphology and Autophagy by Confocal Microscopy

Cells were plated on glass slides in a 24-well plate and 24 h post-UVB cells were stained with 100 nM Mitotracker Red CMXRos (Thermo Fisher Scientific) dye at 37 °C for 30 min. Cells were washed with PBS, fixed with 3.7% paraformaldehyde solution at room temperature for 10 min and permeabilized with 0.2% Triton X-100 for 10 min. After washing with PBS, blocking was performed by 1% BSA-containing PBS at 37 °C for 1 h. Cells were incubated with an Alexa Fluor 488-conjugated LC3A/B antibody diluted in 1% BSA (1:50) at 4 °C overnight in a humid chamber. Next day, slides were washed three times with PBS. Prior to imaging, cells were stained with a mounting medium with DAPI and analyzed by confocal microscopy using 60x oil immersion objective. Background subtraction, noise reduction, local contrast enhancement, unsharp mask, and bandpass filtering were applied on raw images for better image quality and proper image evaluation [124]. Processed images were analyzed by ImageJ software. Mitochondrial complexity was calculated from confocal microscopic images and was defined as a form factor ($(\text{Perimeter}^2/(4\pi*\text{area}))$). For flow cytometric analysis of autophagy, cells were washed with PBS and harvested by trypsinization. Fixation, permeabilization, and blocking were carried out as described above. Cells were incubated with Alexa Fluor 488-conjugated LC3A/B antibody using 1:500 dilution at 4 °C overnight. Next day, after extensive washing, cells were analyzed for Alexa-488 fluorescence intensity in FL1 channel by flow cytometry.

4.15. Measurement of Citrate Synthase Activity

For determination of citrate synthase activity, a citrate synthase assay kit was used according to the manufacturer's instruction. Optical density changes by kinetic program was determined at 412 nm using microplate reader. Values were normalized to total protein concentration.

4.16. Analysis of Oxygen Consumption and Extracellular Acidification

Oxygen consumption (OCR) and extracellular acidification rate (ECAR) was measured using an XF96 extracellular flux analyzer (Seahorse Bioscience, North Billerica, MA, USA). The rate of oxygen consumption indicates oxidative phosphorylation, whereas ECAR represent lactic acid formed during glycolysis. Cells were seeded in a XF96 cell culture plate at 10,000 cell/well density. One hour prior to the assay, culture medium was replaced with unbuffered DMEM (Seahorse Bioscience) supplemented with 10 mM glucose and then cells were equilibrated in a CO_2-free incubator for 1 h. After four measurements of oxygen consumption, oligomycin and antimycin A were subsequently injected to determine mitochondria-linked ATP production and basal OCR, respectively. All OCR and ECAR values were normalized to the total protein obtained from cells lysed by 1 M NaOH.

4.17. Determination of NAD^+ Level

The colorimetric assay for NAD^+/NADH ratio determination was purchased from Biovision (Mountain View, CA, USA). Cells were washed with PBS two times and harvested by a trypsin-EDTA solution and pelleted by mild centrifugation. The pellet was extracted with NAD^+ or NADH extraction buffer and exposed to freeze/thaw cycles. Samples were spun and the supernatant was transferred to a 96-well plate for NAD^+/NADH assays. For NAD^+ determination, half of the samples were transferred into another tube and NAD^+ was decomposed by heating samples at 60 °C. Finally, working reagent was added to the samples, and optical density was determined at 450 nm. Optical density of the standard curve and samples was used to calculate NAD^+ content. Values were normalized to total protein concentration.

4.18. Measurement of ATP Content

For determination of ATP level, an ATP colorimetric/fluorometric assay purchased from Biovision was used according to the manufacturer. Optical density was determined at 570 nm using a microplate reader. Values were normalized to total protein concentration.

4.19. Statistical Analysis

Statistical analysis was carried out by the Kolmogorov–Smirnov test to assess the normality of the population. The frequency of mitochondrial morphological subtypes (fragmented, intermediate, and tubular) was calculated by chi^2 and Ficher's exact test. To assess the statistical significance between untreated and differently treated groups, ANOVA complemented by Dunnett's post-hoc test was used. The comparison of two groups was applied by an independent t-test. All data are reported as mean ± SEM. A $p < 0.05$ was considered statistically significant. Statistical significance was determined by the GraphPad Prism 7 (GraphPad Software Inc., San Diego, CA, USA) and SPSS 25 software. (SPSS package for Windows, Release 25.; SPSS, Chicago, IL, USA).

5. Conclusions

Our results provide new information about the role of ATM and PARP1 proteins in the regulation of mitochondrial morphology and function after UVB irradiation. Moreover, it suggests that UVB and PARPi-caused elevation of oxidative phosphorylation is mediated by the complex interplay of metabolic sensors including ATM, AMPK, p53, AKT, and mTOR which are the main mediators that connect DNA damage with oxidative metabolism and autophagy.

Data Availability: All primary data is uploaded to https://figshare.com/s/49fbcd8cfd5802ea15f9 (DOI: 10.6084/m9.figshare.8107727).

Supplementary Materials: The following are available online at http://www.mdpi.com/2072-6694/12/1/5/s1, Figure S1: Uncut PAR (10H) Western blot, Figure S2: Sample uncut blots, Table S1: Primer pairs used in the study, Table S2: Primary antibodies used in the study.

Author Contributions: Original draft preparation, investigation, methodology, C.H.; Writing—review and editing, G.B.; Methodology, E.F.; Investigation, methodology, G.N.K.; Resources, M.A.; Investigation, methodology, T.J.; Formal analysis, E.A.J.; Methodology, L.J.; Writing—review and editing, G.P.; Writing—review and editing, G.E.; Conceptualization, writing—review and editing, resources, P.B.; Project administration, funding acquisition, É.R. All authors have read and agreed to the published version of the manuscript.

Funding: This work was supported by the European Union and the European Regional Development Fund GINOP-2.3.2-15-2016-00005 and was supported by grants from NKFIH K123975, C129074, NKFIH K120206, Momentum program and NM-26/2019 of the Hungarian Academy of Sciences and the University of Debrecen. The project is co-financed by the University of Debrecen (OTKA Bridging Fund). Bolyai Janos Research Scholarship (J.T.). Szodoray Lajos and Magyary Zoltán Funds by Hungarian Academy of Science and the European Union and the State of Hungary, co-financed by the European Social Funding. Fund in the framework of TÁMOP 4.2.4.A/2-11-1-2012-0001 "National Excellence Program" (J.T.). The research was financed by the Higher Education Institutional Excellence Programme (NKFIH-1150-6/2019) of the Ministry of Innovation and Technology in Hungary, within the framework of the Biotechnology thematic programme of the University of Debrecen.

Conflicts of Interest: The authors declare no conflict of interest. G.P. is a consultant for ADC Therapeutics and Buffalo Biolabs. The funders had no role in the design of the study; in the collection, analyses, or interpretation of data; in the writing of the manuscript, or in the decision to publish the results.

Abbreviations

CPD	cyclobutane pyrimidine dimer
PAR	poly (ADP-ribose) polymer
PARP1	poly (ADP-ribose) polymerase 1
ATM kinase	ataxia-telangiectasia-mutated kinase
AMPK	adenosine monophosphate-activated kinase
mTOR	mammalian target of rapamycin
p70S6K1	ribosomal protein S6 kinase beta-1
HPRT	hypoxanthine-guanine phosphoribosyltransferase
PGC1A	peroxisome proliferator activated receptor gamma coactivator-1 alpha

MTCO1	mitochondrially encoded cytochrome C oxidase I
CS	citrate synthase
TCA	tricarboxylic acid
SDHA	succinate dehydrogenase complex, subunit A
PGK1	phosphoglycerate kinase 1
NRF2	nuclear respiratory factor 2
ERRA	estrogen-related receptor alpha
Tfam	mitochondrial transcription factor A
OCR	oxygen consumption rate
ECAR	extracellular acidification rate
OXPHOS	oxidative phosphorylation
CHO	Chinese hamster ovary
Mfn1	mitofusin 1
Mfn2	mitofusin 2
OPA1	optic atrophy 1
LC3A/B	microtubule-associated proteins 1A/1B light chain 3B
GAPDH	glyceraldehyde-3-phosphate dehydrogenase
PI3K	phosphoinositide 3-kinase
$DiOC_6(3)$	3,3′-dihexyloxacarbocyanine iodide

References

1. Jornayvaz, F.R.; Shulman, G.I. Regulation of mitochondrial biogenesis. *Essays Biochem.* **2010**, *47*, 69–84. [PubMed]
2. Dominy, J.E.; Puigserver, P. Mitochondrial biogenesis through activation of nuclear signaling proteins. *Cold Spring Harb. Perspect. Biol.* **2013**, *5*, a015008. [CrossRef] [PubMed]
3. Suliman, H.B.; Piantadosi, C.A. Mitochondrial quality control as a therapeutic target. *Pharmacol. Rev.* **2016**, *68*, 20–48. [CrossRef] [PubMed]
4. Mears, J.A. Mitochondrial biogenesis and quality control. In *The Structural Basis of Biological Energy Generation*; Hohmann-Marriott, M.F., Ed.; Springer: Dordrecht, The Netherlands, 2014; pp. 451–476.
5. Westermann, B. Bioenergetic role of mitochondrial fusion and fission. *Biochim. Biophys. Acta (BBA) Bioenerg.* **2012**, *1817*, 1833–1838. [CrossRef]
6. Suárez-Rivero, J.M.; Villanueva-Paz, M.; de la Cruz-Ojeda, P.; de la Mata, M.; Cotán, D.; Oropesa-Ávila, M.; de Lavera, I.; Álvarez-Córdoba, M.; Luzón-Hidalgo, R.; Sánchez-Alcázar, J.A. Mitochondrial dynamics in mitochondrial diseases. *Diseases* **2016**, *5*, 1. [CrossRef]
7. Feichtinger, R.G.; Sperl, W.; Bauer, J.W.; Kofler, B. Mitochondrial dysfunction: A neglected component of skin diseases. *Exp. Dermatol.* **2014**, *23*, 607–614. [CrossRef]
8. Clayton, D.A.; Doda, J.N.; Friedberg, E.C. The absence of a pyrimidine dimer repair mechanism in mammalian mitochondria. *Proc. Natl. Acad. Sci. USA* **1974**, *71*, 2777–2781. [CrossRef]
9. LeDoux, S.P.; Wilson, G.L.; Beecham, E.J.; Stevnsner, T.; Wassermann, K.; Bohr, V.A. Repair of mitochondrial DNA after various types of DNA damage in chinese hamster ovary cells. *Carcinogenesis* **1992**, *13*, 1967–1973. [CrossRef]
10. Krutmann, J.; Schroeder, P. Role of mitochondria in photoaging of human skin: The defective powerhouse model. *J. Investig. Dermatol. Symp. Proc.* **2009**, *14*, 44–49. [CrossRef]
11. Naidoo, K.; Hanna, R.; Birch-Machin, M.A. What is the role of mitochondrial dysfunction in skin photoaging? *Exp. Dermatol.* **2018**, *27*, 124–128. [CrossRef]
12. Takeuchi, H.; Fujimoto, A.; Hoon, D.S. Detection of mitochondrial DNA alterations in plasma of malignant melanoma patients. *Ann. N. Y. Acad. Sci.* **2004**, *1022*, 50–54. [CrossRef] [PubMed]
13. Yang, J.H.; Lee, H.C.; Chung, J.G.; Wei, Y.H. Mitochondrial DNA mutations in light-associated skin tumors. *Anticancer Res.* **2004**, *24*, 1753–1758.
14. Durham, S.E.; Krishnan, K.J.; Betts, J.; Birch-Machin, M.A. Mitochondrial DNA damage in non-melanoma skin cancer. *Br. J. Cancer* **2003**, *88*, 90. [CrossRef] [PubMed]
15. Scheibye-Knudsen, M.; Croteau, D.L.; Bohr, V.A. Mitochondrial deficiency in cockayne syndrome. *Mech. Ageing Dev.* **2013**, *134*, 275–283. [CrossRef] [PubMed]

16. Fang, E.F.; Scheibye-Knudsen, M.; Brace, L.E.; Kassahun, H.; SenGupta, T.; Nilsen, H.; Mitchell, J.R.; Croteau, D.L.; Bohr, V.A. Defective mitophagy in xpa via parp-1 hyperactivation and nad(+)/sirt1 reduction. *Cell* **2014**, *157*, 882–896. [CrossRef]
17. Croteau, D.L.; Rossi, M.L.; Canugovi, C.; Tian, J.; Sykora, P.; Ramamoorthy, M.; Wang, Z.M.; Singh, D.K.; Akbari, M.; Kasiviswanathan, R.; et al. Recql4 localizes to mitochondria and preserves mitochondrial DNA integrity. *Aging Cell* **2012**, *11*, 456–466. [CrossRef]
18. Fu, X.; Wan, S.; Lyu, Y.L.; Liu, L.F.; Qi, H. Etoposide induces atm-dependent mitochondrial biogenesis through ampk activation. *PLoS ONE* **2008**, *3*, e2009. [CrossRef]
19. Fang, E.F.; Scheibye-Knudsen, M.; Chua, K.F.; Mattson, M.P.; Croteau, D.L.; Bohr, V.A. Nuclear DNA damage signalling to mitochondria in ageing. *Nat. Rev. Mol. Cell Biol.* **2016**, *17*, 308–321. [CrossRef]
20. Hong, S.J.; Dawson, T.M.; Dawson, V.L. Nuclear and mitochondrial conversations in cell death: Parp-1 and aif signaling. *Trends Pharmacol. Sci.* **2004**, *25*, 259–264. [CrossRef]
21. Bai, P. Biology of poly(adp-ribose) polymerases: The factotums of cell maintenance. *Mol. Cell* **2015**, *58*, 947–958. [CrossRef] [PubMed]
22. Luo, X.; Kraus, W.L. On par with parp: Cellular stress signaling through poly(adp-ribose) and parp-1. *Genes Dev.* **2012**, *26*, 417–432. [CrossRef] [PubMed]
23. Aguilar-Quesada, R.; Munoz-Gamez, J.A.; Martin-Oliva, D.; Peralta, A.; Valenzuela, M.T.; Matinez-Romero, R.; Quiles-Perez, R.; Menissier-de Murcia, J.; de Murcia, G.; Ruiz de Almodovar, M.; et al. Interaction between atm and parp-1 in response to DNA damage and sensitization of atm deficient cells through parp inhibition. *BMC Mol. Biol.* **2007**, *8*, 29. [CrossRef] [PubMed]
24. Valenzuela, M.T.; Guerrero, R.; Núñez, M.I.; Ruiz de Almodóvar, J.M.; Sarker, M.; de Murcia, G.; Oliver, F.J. Parp-1 modifies the effectiveness of p53-mediated DNA damage response. *Oncogene* **2002**, *21*, 1108. [CrossRef] [PubMed]
25. Okuda, A.; Kurokawa, S.; Takehashi, M.; Maeda, A.; Fukuda, K.; Kubo, Y.; Nogusa, H.; Takatani-Nakase, T.; Okuda, S.; Ueda, K.; et al. Poly(adp-ribose) polymerase inhibitors activate the p53 signaling pathway in neural stem/progenitor cells. *BMC Neurosci.* **2017**, *18*, 14. [CrossRef] [PubMed]
26. Berger, N.A. Poly(adp-ribose) in the cellular response to DNA damage. *Radiat. Res.* **1985**, *101*, 4–15. [CrossRef] [PubMed]
27. Zhou, J.; Ng, S.; Huang, Q.; Wu, Y.T.; Li, Z.; Yao, S.Q.; Shen, H.M. Ampk mediates a pro-survival autophagy downstream of parp-1 activation in response to DNA alkylating agents. *FEBS Lett.* **2013**, *587*, 170–177. [CrossRef] [PubMed]
28. Brace, L.E.; Vose, S.C.; Stanya, K.; Gathungu, R.M.; Marur, V.R.; Longchamp, A.; Treviño-Villarreal, H.; Mejia, P.; Vargas, D.; Inouye, K.; et al. Increased oxidative phosphorylation in response to acute and chronic DNA damage. *NPJ Aging Mech. Dis.* **2016**, *2*, 16022. [CrossRef]
29. Pantovic, A.; Krstic, A.; Janjetovic, K.; Kocic, J.; Harhaji-Trajkovic, L.; Bugarski, D.; Trajkovic, V. Coordinated time-dependent modulation of ampk/akt/mtor signaling and autophagy controls osteogenic differentiation of human mesenchymal stem cells. *Bone* **2013**, *52*, 524–531. [CrossRef]
30. Tao, R.; Gong, J.; Luo, X.; Zang, M.; Guo, W.; Wen, R.; Luo, Z. Ampk exerts dual regulatory effects on the pi3k pathway. *J. Mol. Signal.* **2010**, *5*, 1. [CrossRef]
31. Bai, P.; Nagy, L.; Fodor, T.; Liaudet, L.; Pacher, P. Poly(adp-ribose) polymerases as modulators of mitochondrial activity. *Trends Endocrinol. Metab. TEM* **2015**, *26*, 75–83. [CrossRef]
32. Shimizu, I.; Yoshida, Y.; Suda, M.; Minamino, T. DNA damage response and metabolic disease. *Cell Metab.* **2014**, *20*, 967–977. [CrossRef]
33. Bai, P.; Canto, C.; Oudart, H.; Brunyanszki, A.; Cen, Y.; Thomas, C.; Yamamoto, H.; Huber, A.; Kiss, B.; Houtkooper, R.H.; et al. Parp-1 inhibition increases mitochondrial metabolism through sirt1 activation. *Cell Metab* **2011**, *13*, 461–468. [CrossRef] [PubMed]
34. Bai, P.; Canto, C.; Brunyanszki, A.; Huber, A.; Szanto, M.; Cen, Y.; Yamamoto, H.; Houten, S.M.; Kiss, B.; Oudart, H.; et al. Parp-2 regulates sirt1 expression and whole-body energy expenditure. *Cell Metab.* **2011**, *13*, 450–460. [CrossRef] [PubMed]
35. Mohamed, J.S.; Hajira, A.; Pardo, P.S.; Boriek, A.M. Microrna-149 inhibits parp-2 and promotes mitochondrial biogenesis via sirt-1/pgc-1alpha network in skeletal muscle. *Diabetes* **2014**, *63*, 1546–1559. [CrossRef] [PubMed]

36. Cerutti, R.; Pirinen, E.; Lamperti, C.; Marchet, S.; Sauve, A.A.; Li, W.; Leoni, V.; Schon, E.A.; Dantzer, F.; Auwerx, J.; et al. Nad(+)-dependent activation of sirt1 corrects the phenotype in a mouse model of mitochondrial disease. *Cell Metab.* **2014**, *19*, 1042–1049. [CrossRef]
37. Pirinen, E.; Canto, C.; Jo, Y.S.; Morato, L.; Zhang, H.; Menzies, K.J.; Williams, E.G.; Mouchiroud, L.; Moullan, N.; Hagberg, C.; et al. Pharmacological inhibition of poly(adp-ribose) polymerases improves fitness and mitochondrial function in skeletal muscle. *Cell Metab.* **2014**, *19*, 1034–1041. [CrossRef]
38. Kam, T.-I.; Mao, X.; Park, H.; Chou, S.-C.; Karuppagounder, S.S.; Umanah, G.E.; Yun, S.P.; Brahmachari, S.; Panicker, N.; Chen, R.; et al. Poly(adp-ribose) drives pathologic α-synuclein neurodegeneration in parkinson's disease. *Science* **2018**, *362*, eaat8407. [CrossRef]
39. Litton, J.K.; Rugo, H.S.; Ettl, J.; Hurvitz, S.A.; Gonçalves, A.; Lee, K.-H.; Fehrenbacher, L.; Yerushalmi, R.; Mina, L.A.; Martin, M.; et al. Talazoparib in patients with advanced breast cancer and a germline brca mutation. *New Engl. J. Med.* **2018**, *379*, 753–763. [CrossRef]
40. Mateo, J.; Carreira, S.; Sandhu, S.; Miranda, S.; Mossop, H.; Perez-Lopez, R.; Nava Rodrigues, D.; Robinson, D.; Omlin, A.; Tunariu, N.; et al. DNA-repair defects and olaparib in metastatic prostate cancer. *N. Engl. J. Med.* **2015**, *373*, 1697–1708. [CrossRef]
41. O'Shaughnessy, J.; Osborne, C.; Pippen, J.E.; Yoffe, M.; Patt, D.; Rocha, C.; Koo, I.C.; Sherman, B.M.; Bradley, C. Iniparib plus chemotherapy in metastatic triple-negative breast cancer. *New Engl. J. Med.* **2011**, *364*, 205–214. [CrossRef]
42. Wagner, L.M. Profile of veliparib and its potential in the treatment of solid tumors. *Oncotargets Ther.* **2015**, *8*, 1931–1939. [CrossRef] [PubMed]
43. Bhute, V.J.; Ma, Y.; Bao, X.; Palecek, S.P. The poly (adp-ribose) polymerase inhibitor veliparib and radiation cause significant cell line dependent metabolic changes in breast cancer cells. *Sci. Rep.* **2016**, *6*, 36061. [CrossRef] [PubMed]
44. Engert, F.; Schneider, C.; Weibeta, L.M.; Probst, M.; Fulda, S. Parp inhibitors sensitize ewing sarcoma cells to temozolomide-induced apoptosis via the mitochondrial pathway. *Mol. Cancer Ther.* **2015**, *14*, 2818–2830. [CrossRef] [PubMed]
45. Jugé, R.; Breugnot, J.; Da Silva, C.; Bordes, S.; Closs, B.; Aouacheria, A. Quantification and characterization of uvb-induced mitochondrial fragmentation in normal primary human keratinocytes. *Sci. Rep.* **2016**, *6*, 35065. [CrossRef] [PubMed]
46. Zanchetta, L.M.; Garcia, A.; Lyng, F.; Walsh, J.; Murphy, J.E. Mitophagy and mitochondrial morphology in human melanoma-derived cells post exposure to simulated sunlight. *Int. J. Radiat. Biol.* **2011**, *87*, 506–517. [CrossRef]
47. Paz, M.L.; Gonzalez Maglio, D.H.; Weill, F.S.; Bustamante, J.; Leoni, J. Mitochondrial dysfunction and cellular stress progression after ultraviolet b irradiation in human keratinocytes. *Photodermatol. Photoimmunol. Photomed.* **2008**, *24*, 115–122. [CrossRef]
48. Denning, M.F.; Wang, Y.; Tibudan, S.; Alkan, S.; Nickoloff, B.J.; Qin, J.Z. Caspase activation and disruption of mitochondrial membrane potential during uv radiation-induced apoptosis of human keratinocytes requires activation of protein kinase c. *Cell Death Differ.* **2002**, *9*, 40–52. [CrossRef]
49. Jing, L.; He, M.-T.; Chang, Y.; Mehta, S.L.; He, Q.-P.; Zhang, J.-Z.; Li, P.A. Coenzyme q10 protects astrocytes from ros-induced damage through inhibition of mitochondria-mediated cell death pathway. *Int. J. Biol. Sci.* **2015**, *11*, 59–66. [CrossRef]
50. Tondera, D.; Grandemange, S.; Jourdain, A.; Karbowski, M.; Mattenberger, Y.; Herzig, S.; Da Cruz, S.; Clerc, P.; Raschke, I.; Merkwirth, C.; et al. Slp-2 is required for stress-induced mitochondrial hyperfusion. *EMBO J.* **2009**, *28*, 1589–1600. [CrossRef]
51. Szanto, M.; Rutkai, I.; Hegedus, C.; Czikora, A.; Rozsahegyi, M.; Kiss, B.; Virag, L.; Gergely, P.; Toth, A.; Bai, P. Poly(adp-ribose) polymerase-2 depletion reduces doxorubicin-induced damage through sirt1 induction. *Cardiovasc. Res.* **2011**, *92*, 430–438. [CrossRef]
52. Shieh, W.M.; Ame, J.C.; Wilson, M.V.; Wang, Z.Q.; Koh, D.W.; Jacobson, M.K.; Jacobson, E.L. Poly(adp-ribose) polymerase null mouse cells synthesize adp-ribose polymers. *J. Biol. Chem.* **1998**, *273*, 30069–30072. [CrossRef]
53. Zarkovic, G.; Belousova, E.A.; Talhaoui, I.; Saint-Pierre, C.; Kutuzov, M.M.; Matkarimov, B.T.; Biard, D.; Gasparutto, D.; Lavrik, O.I.; Ishchenko, A.A. Characterization of DNA adp-ribosyltransferase activities of parp2 and parp3: New insights into DNA adp-ribosylation. *Nucleic Acids Res.* **2018**, *46*, 2417–2431. [CrossRef]

54. King, B.S.; Cooper, K.L.; Liu, K.J.; Hudson, L.G. Poly(adp-ribose) contributes to an association between poly(adp-ribose) polymerase-1 and xeroderma pigmentosum complementation group a in nucleotide excision repair. *J. Biol. Chem.* **2012**, *287*, 39824–39833. [CrossRef] [PubMed]
55. Dixit, M.; Kumar, A. Chapter 4—In vitro gene genotoxicity test methods. In *In Vitro Toxicology*; Dhawan, A., Kwon, S., Eds.; Academic Press: Cambridge, MA, USA, 2018; pp. 67–89.
56. Johnson, G. Mammalian cell hprt gene mutation assay: Test methods. In *Genetic Toxicology*; Springer: New York, NY, USA, 2012; Volume 817, pp. 55–67.
57. Hsie, A.W.; Couch, D.B.; O'Neill, J.P. Utilization of a Quantitative Mammalian Cell Mutation System, Cho/Hgprt, in Experimental Mutagenesis and Genetic Toxicology. Available online: https://www.osti.gov/biblio/7278027 (accessed on 13 December 2019).
58. Hu, T.; Miller, C.M.; Ridder, G.M.; Aardema, M.J. Characterization of p53 in chinese hamster cell lines cho-k1, cho-wbl, and chl: Implications for genotoxicity testing. *Mutat. Res.* **1999**, *426*, 51–62. [CrossRef]
59. Tzang, B.S.; Lai, Y.C.; Hsu, M.; Chang, H.W.; Chang, C.C.; Huang, P.C.; Liu, Y.C. Function and sequence analyses of tumor suppressor gene p53 of cho.K1 cells. *DNA Cell Biol.* **1999**, *18*, 315–321. [CrossRef] [PubMed]
60. Lehman, T.A.; Modali, R.; Boukamp, P.; Stanek, J.; Bennett, W.P.; Welsh, J.A.; Metcalf, R.A.; Stampfer, M.R.; Fusenig, N.; Rogan, E.M.; et al. P53 mutations in human immortalized epithelial cell lines. *Carcinogenesis* **1993**, *14*, 833–839. [CrossRef] [PubMed]
61. Vyas, S.; Zaganjor, E.; Haigis, M.C. Mitochondria and cancer. *Cell* **2016**, *166*, 555–566. [CrossRef] [PubMed]
62. Onishi, Y.; Ueha, T.; Kawamoto, T.; Hara, H.; Toda, M.; Harada, R.; Minoda, M.; Kurosaka, M.; Akisue, T. Regulation of mitochondrial proliferation by pgc-1α induces cellular apoptosis in musculoskeletal malignancies. *Sci. Rep.* **2014**, *4*, 3916. [CrossRef] [PubMed]
63. Saleem, A.; Adhihetty, P.J.; Hood, D.A. Role of p53 in mitochondrial biogenesis and apoptosis in skeletal muscle. *Physiol. Genom.* **2009**, *37*, 58–66. [CrossRef] [PubMed]
64. Dam, A. The Effect of Mitochondrial Biogenesis on Apoptotic Susceptibility in l6 Myoblasts. Master's Thesis, University of Waterloo, Waterloo, ON, Canada, 2010.
65. Gong, B.; Chen, Q.; Almasan, A. Ionizing radiation stimulates mitochondrial gene expression and activity. *Radiat. Res.* **1998**, *150*, 505–512. [CrossRef]
66. Yu, J.; Wang, Q.; Chen, N.; Sun, Y.; Wang, X.; Wu, L.; Chen, S.; Yuan, H.; Xu, A.; Wang, J. Mitochondrial transcription factor a regulated ionizing radiation-induced mitochondrial biogenesis in human lung adenocarcinoma a549 cells. *J. Radiat. Res.* **2013**, *54*, 998–1004. [CrossRef] [PubMed]
67. Eliopoulos, A.G.; Havaki, S.; Gorgoulis, V.G. DNA damage response and autophagy: A meaningful partnership. *Front. Genet.* **2016**, *7*, 204. [CrossRef] [PubMed]
68. Qiang, L.; Zhao, B.; Shah, P.; Sample, A.; Yang, S.; He, Y.Y. Autophagy positively regulates DNA damage recognition by nucleotide excision repair. *Autophagy* **2016**, *12*, 357–368. [CrossRef] [PubMed]
69. Wang, H.-T.; Lin, J.-H.; Yang, C.-H.; Haung, C.-H.; Weng, C.-W.; Maan-Yuh Lin, A.; Lo, Y.-L.; Chen, W.-S.; Tang, M.-S. Acrolein induces mtdna damages, mitochondrial fission and mitophagy in human lung cells. *Oncotarget* **2017**, *8*, 70406–70421. [CrossRef] [PubMed]
70. Rodríguez-Vargas, J.M.; Rodríguez, M.I.; Majuelos-Melguizo, J.; García-Diaz, Á.; González-Flores, A.; López-Rivas, A.; Virág, L.; Illuzzi, G.; Schreiber, V.; Dantzer, F.; et al. Autophagy requires poly(adp-ribosyl)ation-dependent ampk nuclear export. *Cell Death Differ.* **2016**, *23*, 2007. [CrossRef] [PubMed]
71. Rodríguez-Vargas, J.M.; Ruiz-Magaña, M.J.; Ruiz-Ruiz, C.; Majuelos-Melguizo, J.; Peralta-Leal, A.; Rodríguez, M.I.; Muñoz-Gámez, J.A.; de Almodóvar, M.R.; Siles, E.; Rivas, A.L.; et al. Ros-induced DNA damage and parp-1 are required for optimal induction of starvation-induced autophagy. *Cell Res.* **2012**, *22*, 1181–1198. [CrossRef] [PubMed]
72. Muñoz-Gámez, J.A.; Rodríguez-Vargas, J.M.; Quiles-Pérez, R.; Aguilar-Quesada, R.; Martín-Oliva, D.; de Murcia, G.; de Murcia, J.M.; Almendros, A.; de Almodóvar, M.R.; Oliver, F.J. Parp-1 is involved in autophagy induced by DNA damage. *Autophagy* **2009**, *5*, 61–74. [CrossRef] [PubMed]
73. Wyrsch, P.; Blenn, C.; Bader, J.; Althaus, F.R. Cell death and autophagy under oxidative stress: Roles of poly(adp-ribose) polymerases and Ca2+. *Mol. Cell. Biol.* **2012**, *32*, 3541–3553. [CrossRef] [PubMed]
74. Rajawat, J.; Shukla, N.; Mishra, D.P. Therapeutic targeting of poly(adp-ribose) polymerase-1 (parp1) in cancer: Current developments, therapeutic strategies, and future opportunities. *Med. Res. Rev.* **2017**, *37*, 1461–1491. [CrossRef]

75. Arun, B.; Akar, U.; Gutierrez-Barrera, A.M.; Hortobagyi, G.N.; Ozpolat, B. The parp inhibitor azd2281 (olaparib) induces autophagy/mitophagy in brca1 and brca2 mutant breast cancer cells. *Int. J. Oncol.* **2015**, *47*, 262–268. [CrossRef]
76. Rambold, A.S.; Kostelecky, B.; Elia, N.; Lippincott-Schwartz, J. Tubular network formation protects mitochondria from autophagosomal degradation during nutrient starvation. *Proc. Natl. Acad. Sci. USA* **2011**, *108*, 10190–10195. [CrossRef] [PubMed]
77. Biel, T.G.; Rao, V.A. Mitochondrial dysfunction activates lysosomal-dependent mitophagy selectively in cancer cells. *Oncotarget* **2018**, *9*, 995–1011. [CrossRef] [PubMed]
78. Eyschen, J.; Vitoux, B.; Marraud, M.; Cung, M.T.; Branlant, G. Engineered glycolytic glyceraldehyde-3-phosphate dehydrogenase binds the anti conformation of nad+ nicotinamide but does not experience a-specific hydride transfer. *Arch. Biochem. Biophys.* **1999**, *364*, 219–227. [CrossRef] [PubMed]
79. Santidrian, A.F.; Matsuno-Yagi, A.; Ritland, M.; Seo, B.B.; LeBoeuf, S.E.; Gay, L.J.; Yagi, T.; Felding-Habermann, B. Mitochondrial complex i activity and nad+/nadh balance regulate breast cancer progression. *J. Clin. Investig.* **2013**, *123*, 1068–1081. [CrossRef] [PubMed]
80. Imai, S.-I.; Guarente, L. It takes two to tango: Nad+ and sirtuins in aging/longevity control. *NPJ Aging Mech. Dis.* **2016**, *2*, 16017. [CrossRef]
81. Morris, B.J. Seven sirtuins for seven deadly diseases of aging. *Free Radic. Biol. Med.* **2013**, *56*, 133–171. [CrossRef]
82. Canto, C.; Sauve, A.A.; Bai, P. Crosstalk between poly(adp-ribose) polymerase and sirtuin enzymes. *Mol. Asp. Med.* **2013**, *34*, 1168–1201. [CrossRef]
83. Luna, A.; Aladjem, M.I.; Kohn, K.W. Sirt1/parp1 crosstalk: Connecting DNA damage and metabolism. *Genome Integr.* **2013**, *4*, 6. [CrossRef]
84. Ming, M.; Soltani, K.; Shea, C.R.; Li, X.; He, Y.Y. Dual role of sirt1 in uvb-induced skin tumorigenesis. *Oncogene* **2015**, *34*, 357–363. [CrossRef]
85. Cao, C.; Lu, S.; Kivlin, R.; Wallin, B.; Card, E.; Bagdasarian, A.; Tamakloe, T.; Wang, W.J.; Song, X.; Chu, W.M.; et al. Sirt1 confers protection against uvb- and h2o2-induced cell death via modulation of p53 and jnk in cultured skin keratinocytes. *J. Cell. Mol. Med.* **2009**, *13*, 3632–3643. [CrossRef]
86. Benavente, C.A.; Schnell, S.A.; Jacobson, E.L. Effects of niacin restriction on sirtuin and parp responses to photodamage in human skin. *PLoS ONE* **2012**, *7*, e42276. [CrossRef] [PubMed]
87. Scher, M.B.; Vaquero, A.; Reinberg, D. Sirt3 is a nuclear nad+-dependent histone deacetylase that translocates to the mitochondria upon cellular stress. *Genes Dev.* **2007**, *21*, 920–928. [CrossRef] [PubMed]
88. Garcia-Peterson, L.M.; Wilking-Busch, M.J.; Ndiaye, M.A.; Philippe, C.G.A.; Setaluri, V.; Ahmad, N. Sirtuins in skin and skin cancers. *Ski. Pharmacol. Physiol.* **2017**, *30*, 216–224. [CrossRef] [PubMed]
89. Robu, M.; Shah, R.G.; Purohit, N.K.; Zhou, P.; Naegeli, H.; Shah, G.M. Poly(adp-ribose) polymerase 1 escorts xpc to uv-induced DNA lesions during nucleotide excision repair. *Proc. Natl. Acad. Sci. USA* **2017**, *114*, E6847–e6856. [CrossRef] [PubMed]
90. Robu, M.; Shah, R.G.; Petitclerc, N.; Brind'Amour, J.; Kandan-Kulangara, F.; Shah, G.M. Role of poly(adp-ribose) polymerase-1 in the removal of uv-induced DNA lesions by nucleotide excision repair. *Proc. Natl. Acad. Sci. USA* **2013**, *110*, 1658–1663. [CrossRef] [PubMed]
91. Pines, A.; Vrouwe, M.G.; Marteijn, J.A.; Typas, D.; Luijsterburg, M.S.; Cansoy, M.; Hensbergen, P.; Deelder, A.; de Groot, A.; Matsumoto, S.; et al. Parp1 promotes nucleotide excision repair through ddb2 stabilization and recruitment of alc1. *J. Cell Biol.* **2012**, *199*, 235–249. [CrossRef] [PubMed]
92. Ciccarone, F.; Zampieri, M.; Caiafa, P. Parp1 orchestrates epigenetic events setting up chromatin domains. *Semin. Cell Dev. Biol.* **2017**, *63*, 123–134. [CrossRef] [PubMed]
93. Murai, J.; Zhang, Y.; Morris, J.; Ji, J.; Takeda, S.; Doroshow, J.H.; Pommier, Y. Rationale for poly(adp-ribose) polymerase (parp) inhibitors in combination therapy with camptothecins or temozolomide based on parp trapping versus catalytic inhibition. *J. Pharmacol. Exp. Ther.* **2014**, *349*, 408–416. [CrossRef] [PubMed]
94. Prasad, C.B.; Prasad, S.B.; Yadav, S.S.; Pandey, L.K.; Singh, S.; Pradhan, S.; Narayan, G. Olaparib modulates DNA repair efficiency, sensitizes cervical cancer cells to cisplatin and exhibits anti-metastatic property. *Sci. Rep.* **2017**, *7*, 12876. [CrossRef]
95. Jelinic, P.; Levine, D.A. New insights into parp inhibitors' effect on cell cycle and homology-directed DNA damage repair. *Mol. Cancer Ther.* **2014**, *13*, 1645–1654. [CrossRef] [PubMed]

96. Murai, J.; Huang, S.Y.; Das, B.B.; Renaud, A.; Zhang, Y.; Doroshow, J.H.; Ji, J.; Takeda, S.; Pommier, Y. Trapping of parp1 and parp2 by clinical parp inhibitors. *Cancer Res.* **2012**, *72*, 5588–5599. [CrossRef] [PubMed]
97. Helleday, T. The underlying mechanism for the parp and brca synthetic lethality: Clearing up the misunderstandings. *Mol. Oncol.* **2011**, *5*, 387–393. [CrossRef]
98. Horton, J.K.; Wilson, S.H. Strategic combination of DNA-damaging agent and parp inhibitor results in enhanced cytotoxicity. *Front. Oncol.* **2013**, *3*, 257. [CrossRef]
99. Muvarak, N.E.; Chowdhury, K.; Xia, L.; Robert, C.; Choi, E.Y.; Cai, Y.; Bellani, M.; Zou, Y.; Singh, Z.N.; Duong, V.H.; et al. Enhancing the cytotoxic effects of parp inhibitors with DNA demethylating agents—A potential therapy for cancer. *Cancer Cell* **2016**, *30*, 637–650. [CrossRef] [PubMed]
100. Plummer, R.; Dua, D.; Cresti, N.; Suder, A.; Drew, Y.; Prathapan, V.; Stephens, P.; Thornton, J.K.; de las Heras, B.; Ink, B.; et al. Phase 1 study of the parp inhibitor e7449 as a single agent in patients with advanced solid tumors or b-cell lymphoma. *J. Clin. Oncol.* **2014**, *32*, e19531. [CrossRef]
101. O'Cearbhaill, R.E. Using parp inhibitors in advanced ovarian cancer. *Oncology* **2018**, *32*, 339–343.
102. Alano, C.C.; Garnier, P.; Ying, W.; Higashi, Y.; Kauppinen, T.M.; Swanson, R.A. Nad+ depletion is necessary and sufficient for poly(adp-ribose) polymerase-1-mediated neuronal death. *J. Neurosci. Off. J. Soc. Neurosci.* **2010**, *30*, 2967–2978. [CrossRef]
103. Eguchi, Y.; Shimizu, S.; Tsujimoto, Y. Intracellular atp levels determine cell death fate by apoptosis or necrosis. *Cancer Res.* **1997**, *57*, 1835–1840.
104. Zamaraeva, M.V.; Sabirov, R.Z.; Maeno, E.; Ando-Akatsuka, Y.; Bessonova, S.V.; Okada, Y. Cells die with increased cytosolic atp during apoptosis: A bioluminescence study with intracellular luciferase. *Cell Death Differ.* **2005**, *12*, 1390–1397. [CrossRef]
105. Fu, D.; Lippincott-Schwartz, J.; Arias, I.M. Increased mitochondrial fusion and autophagy help isolated hepatocytes repolarize in collagen sandwich cultures. *Autophagy* **2013**, *9*, 2154–2155. [CrossRef]
106. Thomas, H.E.; Zhang, Y.; Stefely, J.A.; Veiga, S.R.; Thomas, G.; Kozma, S.C.; Mercer, C.A. Mitochondrial complex i activity is required for maximal autophagy. *Cell Rep.* **2018**, *24*, 2404–2417.e2408. [CrossRef] [PubMed]
107. Montecucco, A.; Zanetta, F.; Biamonti, G. Molecular mechanisms of etoposide. *EXCLI J.* **2015**, *14*, 95–108.
108. Chatterjee, N.; Walker, G.C. Mechanisms of DNA damage, repair, and mutagenesis. *Environ. Mol. Mutagenesis* **2017**, *58*, 235–263. [CrossRef] [PubMed]
109. Ray, A.; Milum, K.; Battu, A.; Wani, G.; Wani, A.A. Ner initiation factors, ddb2 and xpc, regulate uv radiation response by recruiting atr and atm kinases to DNA damage sites. *DNA Repair* **2013**, *12*, 273–283. [CrossRef]
110. Sun, Y.; Connors, K.E.; Yang, D.Q. Aicar induces phosphorylation of ampk in an atm-dependent, lkb1-independent manner. *Mol. Cell. Biochem.* **2007**, *306*, 239–245. [CrossRef] [PubMed]
111. Lavin, M.F.; Kozlov, S. Atm activation and DNA damage response. *Cell Cycle* **2007**, *6*, 931–942. [CrossRef]
112. Viniegra, J.G.; Martinez, N.; Modirassari, P.; Hernandez Losa, J.; Parada Cobo, C.; Sanchez-Arevalo Lobo, V.J.; Aceves Luquero, C.I.; Alvarez Vallina, L.; Ramon y Cajal, S.; Rojas, J.M.; et al. Full activation of pkb/akt in response to insulin or ionizing radiation is mediated through atm. *J. Biol. Chem.* **2005**, *280*, 4029–4036. [CrossRef] [PubMed]
113. Khalil, H.; Tummala, H.; Zhelev, N. ATM in focus: A damage sensor and cancer target. *Biodiscovery* **2012**, *5*. [CrossRef]
114. Jones, R.G.; Plas, D.R.; Kubek, S.; Buzzai, M.; Mu, J.; Xu, Y.; Birnbaum, M.J.; Thompson, C.B. Amp-activated protein kinase induces a p53-dependent metabolic checkpoint. *Mol. Cell* **2005**, *18*, 283–293. [CrossRef] [PubMed]
115. Zhang, J.; Wang, Y.; Liu, X.; Dagda, R.K.; Zhang, Y. How ampk and pka interplay to regulate mitochondrial function and survival in models of ischemia and diabetes. *Oxidative Med. Cell. Longev.* **2017**, *2017*, 12. [CrossRef]
116. Wang, D.B.; Kinoshita, C.; Kinoshita, Y.; Morrison, R.S. P53 and mitochondrial function in neurons. *Biochim. Biophys. Acta* **2014**, *1842*, 1186–1197. [CrossRef] [PubMed]
117. Parra, V.; Verdejo, H.E.; Iglewski, M.; Del Campo, A.; Troncoso, R.; Jones, D.; Zhu, Y.; Kuzmicic, J.; Pennanen, C.; Lopez-Crisosto, C.; et al. Insulin stimulates mitochondrial fusion and function in cardiomyocytes via the akt-mtor-nfkappab-opa-1 signaling pathway. *Diabetes* **2014**, *63*, 75–88. [CrossRef] [PubMed]

118. Morita, M.; Prudent, J.; Basu, K.; Goyon, V.; Katsumura, S.; Hulea, L.; Pearl, D.; Siddiqui, N.; Strack, S.; McGuirk, S.; et al. Mtor controls mitochondrial dynamics and cell survival via mtfp1. *Mol. Cell* **2017**, *67*, 922–935.e925. [CrossRef] [PubMed]
119. Boukamp, P.; Petrussevska, R.T.; Breitkreutz, D.; Hornung, J.; Markham, A.; Fusenig, N.E. Normal keratinization in a spontaneously immortalized aneuploid human keratinocyte cell line. *J. Cell Biol.* **1988**, *106*, 761–771. [CrossRef]
120. Boros, G.; Miko, E.; Muramatsu, H.; Weissman, D.; Emri, E.; Rozsa, D.; Nagy, G.; Juhasz, A.; Juhasz, I.; van der Horst, G.; et al. Transfection of pseudouridine-modified mrna encoding cpd-photolyase leads to repair of DNA damage in human keratinocytes: A new approach with future therapeutic potential. *J. Photochem. Photobiol. B Biol.* **2013**, *129*, 93–99. [CrossRef]
121. Chomczynski, P.; Sacchi, N. Single-step method of rna isolation by acid guanidinium thiocyanate-phenol-chloroform extraction. *Anal. Biochem.* **1987**, *162*, 156–159. [CrossRef]
122. Balogh, A.; Paragh, G., Jr.; Juhasz, A.; Kobling, T.; Torocsik, D.; Miko, E.; Varga, V.; Emri, G.; Horkay, I.; Scholtz, B.; et al. Reference genes for quantitative real time pcr in uvb irradiated keratinocytes. *J. Photochem. Photobiol. B Biol.* **2008**, *93*, 133–139. [CrossRef]
123. Rueden, C.T.; Schindelin, J.; Hiner, M.C.; DeZonia, B.E.; Walter, A.E.; Arena, E.T.; Eliceiri, K.W. Imagej2: Imagej for the next generation of scientific image data. *BMC Bioinform.* **2017**, *18*, 529. [CrossRef]
124. Merrill, R.A.; Flippo, K.H.; Strack, S. Measuring mitochondrial shape with imagej. In *Techniques to Investigate Mitochondrial Function in Neurons*; Strack, S., Usachev, Y.M., Eds.; Springer: New York, NY, USA, 2017; pp. 31–48.

cancers

MDPI

Article

CAF-1 Subunits Levels Suggest Combined Treatments with PARP-Inhibitors and Ionizing Radiation in Advanced HNSCC

Francesco Morra [1,†], Francesco Merolla [2,3,†], Ida Picardi [1], Daniela Russo [2], Gennaro Ilardi [2], Silvia Varricchio [2], Federica Liotti [1], Roberto Pacelli [2], Luca Palazzo [1], Massimo Mascolo [2], Angela Celetti [1,*] and Stefania Staibano [2,*]

1 Institute for the Experimental Endocrinology and Oncology, National Research Council, 80131 Naples, Italy; fmorra85@gmail.com (F.M.); idapicardi93@icloud.com (I.P.); fedeliotti@hotmail.com (F.L.); l.palazzo@ieos.cnr.it (L.P.)

2 Department of Advanced Biomedical Sciences, University "Federico II", Surgical Pathology Section, 80131 Naples, Italy; francesco.merolla@unimol.it (F.M.); daniela.russo@unina.it (D.R.); gennaro.ilardi@gmail.com (G.I.); silvia.varricchio@gmail.com (S.V.); pacerto@yahoo.com (R.P.); mmascol@gmail.com (M.M.)

3 Department of Medicine and Health Sciences "V. Tiberio", University of Molise, 86100 Campobasso, Italy

* Correspondence: celetti@unina.it (A.C.); staibano@unina.it (S.S.)

† These authors contributed equally to the research.

Received: 6 September 2019; Accepted: 14 October 2019; Published: 17 October 2019

Abstract: Oral (OSCC) and oropharyngeal (OPSCC) squamous cell carcinomas show high morbidity and mortality rates. We aimed to investigate the role of the "Chromatin Assembly Factor-1" (CAF-1) p60 and p150 subunits, involved in DNA repair and replication, in OSCC and OPSCC progression and in response to Poly(ADP-ribose) polymerase (PARP)-inhibitors and exposure to ionizing radiation (IR). We immunostained tissue microarrays (TMAs), including 112 OSCC and 42 OPSCC, with anti-CAF-1/p60 and anti-CAF-1/p150 specific antibodies, correlating their expression with prognosis. Moreover, we assessed the sensitivity to PARP inhibitors and the double-strand breaks repair proficiency by cell viability and HR reporter assays, respectively, in HPV-positive and HPV-negative cell lines upon CAF-1/p60 and CAF-1/p150 depletion. The immunohistochemical analysis revealed a significant prognostic value of both tissue biomarkers combined expression in OSCC but not in OPSCC. In in vitro studies, the p60/150 CAF-1 subunits' depletion impaired the proficiency of Homologous Recombination DNA damage repair, inducing sensitivity to the PARP-inhibitors, able to sensitize both the cell lines to IR. These results indicate that regardless of the prognostic meaning of p60/p150 tissue expression, the pharmacological depletion of CAF-1 complex's function, combined to PARP-inhibitors and/or IR treatment, could represent a valid therapeutic strategy for squamous cell carcinomas of head and neck region.

Keywords: head and neck squamous cell carcinoma; CAL27; SCC90; p60/p150 CAF-1 subunits; DNA repair; homologous recombination; biomarkers; personalized treatment

---

1. Introduction

Head and neck cancer (HNC) is the eighth most common malignancy in the world, with *squamous cell carcinoma* (SCC) accounting for *more than 90%* [1]. HNSCC encompass tumours of larynx, hypopharynx, paranasal sinuses, oral cavity, and oropharynx. Of all HNSCC, oral cavity SCC (OSCC) and oropharynx SCC (OPSCC) represent more than a half-million new cases per year, with an estimated incidence of 7.0 per 100,000 inhabitants, worldwide. HNSCC are characterized by a high-rate of morbidity and mortality. Furthermore, in most countries, the five-year survival rate is less than 50%

and, in the United States, more than 20,000 new cases are estimated to occur in 2019, with more than 10,000 deaths [1–3]. Multiple risk factors contribute to the initiation of HNSCC. Tobacco and heavy use of alcoholic beverages, for OSCC, and high-risk Human Papilloma Virus, mainly HR-HPV16 persistent infections, for a significant percentage of OPSCC, are the most important risk-factors, causing a substantial percentage of these tumours in the Western countries. HPV16 accounts for >80% of HPV-positive oropharyngeal SCCs compared with HPV-positive oral SCCs and laryngeal SCCs. Conversely, HPV18 is relatively rare in HPV positive oropharyngeal SCCs compared with other head and neck sites.

While OSCC are commonly observed in males, aged over 50 years, OPSCC occurs according a bi-modal pattern, respectively, in those under 40 and in elderly people [4–6], and mainly affect females. OPSCC comprise of HPV-negative (HPV−) and HPV-positive (HPV+) tumours, which represent two distinctive clinicopathological and molecular entities, with a disparate range of survival rates. HPV+ OPSCC are characterized by a significantly slow progression and a high response to chemo- and radiotherapy, while HPV− OPSCC and tobacco smoking/alcohol-related OSCC are intrinsically highly aggressive, and highly chemo- and radio-resistant when in an advanced stage. Currently, there are still no reliable prognostic and predictive biomarkers for these deadly cancers that kill about 50% of patients with metastatic disease. Genotoxic exposure to tobacco carcinogens and consequent adducts formation resulting in DNA damage [7] and genomic instability induced by the unscheduled cell replication of integrated HR-HPVs, as demonstrated in in vitro experiments [8], are thought to represent important mechanisms of carcinogenesis for HPV− and HPV+ HNSCC, respectively. At least five major DNA repair pathways—base excision repair (BER), nucleotide excision repair (NER), mismatch repair (MMR), homologous recombination (HR), and non-homologous and joining (NHEJ)—are active throughout different stages of the cell cycle, allowing the cells to repair the DNA damage in a substrate-dependent manner [9–11]. The chromatin assembly factor 1 (CAF-1), a heterotrimeric protein complex formed of three subunits (p48, p60 and p150), plays a key role in the steric organization of DNA and in the assembly of nucleosomes [12]. In particular, while p48 subunit acts on acetylation/deacetylation of histones, p150 appears to be more active in interphase DNA-damage repair process, interacting with PCNA on the damaged DNA, specifically during NER [13,14] and double-strand breaks repair [15], the CAF-1/p60 subunit is more specifically connected to controlling cell replication [16–18]. Several reports have recently shown that a deregulated expression of this subunit leads cells to incorrectly replicate DNA, consequently accumulating DNA damage [16,19]. Interestingly, CAF-1/p60 expression levels are significantly correlated with the biological aggressiveness of tumours, metastasizing behaviour and worse prognosis in breast, oral, prostate, laryngeal and salivary gland carcinomas, as well as in skin melanoma [20–25], suggesting for this protein a promising role as a new sensible prognostic marker, apparently unrelated to their histogenesis. In recent decades, despite therapeutic advances in the HNSCC treatment, patient survival has not markedly improved and the mortality is still around 40–50%. CAF-1 appears as an interesting candidate to explore in HNSCC. In the present study, we evaluated the expression of CAF-1/p60 and p150 subunits in a tissue microarray (TMA) selected series of OSCC and OPSCC. We related our data to the HPV status of primary OPSCC, evaluated by immunohistochemical expression of p16^{INK4a} protein. We also verified the prognostic value of these tissue biomarkers through follow-up analysis. We used, as models of in vitro study, in vitro cultured cells of OSCC and HPV+ OPSCC, testing the efficiency of repair mechanisms and radiation sensitivity, upon silencing of CAF-1/p60 and p150.

2. Results

2.1. Immunohistochemistry Expression of CAF-1/p60 and p150 Subunits in OSCC and OPSCC Tumour Samples

To investigate the functional role of the CAF-1/p60 and p150 subunits in HNSCC, we evaluated a Tissue microarray (TMA) case study (total number of cases $N = 154$) of non-oropharyngeal (OSCC) ($N = 112$) and oropharyngeal squamous cell carcinomas (OPSCC) ($N = 42$), the latter tested for the presence of the HPV virus (HPV+ OPSCC $N = 8$). All tumour samples were staged according to the 8th AJCC staging manual [26] (Table 1). In these samples, the immunohistochemistry staining for the p60 and p150 subunits of CAF-1 was assessed and scored as "HIGH" and "LOW" as defined in the Section 4 (representative images of staining score categories are shown in Figure 1). The p60/p150 frequency distribution data were further analysed with a classification algorithm allowing stratification of samples in three clusters, homogeneous for tissue expression of p60 and p150 according to the IHC staining. The three clusters were defined as follow: p150 HIGH/p60 HIGH; p150^{LOW}/p60^{HIGH}; and p150^{LOW}/p60^{LOW}. The category p150^{HIGH}/p60^{LOW} was not revealed since no p60LOW cases were present in the p150^{HIGH} sub-group (Table 1).

Table 1. Descriptive statistics of the studied population. NOP, non-oropharyngeal tumours; OP, oropharyngeal tumours. HPV positivity (p16 IHC) is only reported for oropharyngeal squamous cell carcinomas.

Study Population		Frequency	%
Stage	I	15	9.7
	II	30	19.5
	III	18	11.7
	IVA	56	36.4
	IVB	12	7.8
	Missing	23	14.9
Sex	F	72	46.8
	M	82	53.2
Age	Mean	63.8	
	Median	64	
	Std Dev	13.1	
	Range	57	
	Min	33	
	Max	90	
p60 score	HIGH	107	69.5
	LOW	47	30.5
p150 score	HIGH	26	16.9
	LOW	128	83.1
p60/p150 combined score	p60HIGH/p150HIGH	26	16.9
	p60HIGH/p150LOW	81	52.6
	p60LOW/p150LOW	47	30.5
Tumour site	NOP	112	72.7
	OP	42	27.3
HPV (p16)	NEG	34	81
	POS	8	19
F-up (months)	Mean	32.92	
	Median	24	
	Mode	12	
	Range	156	
	Min	1	
	Max	157	
Tot		154	100

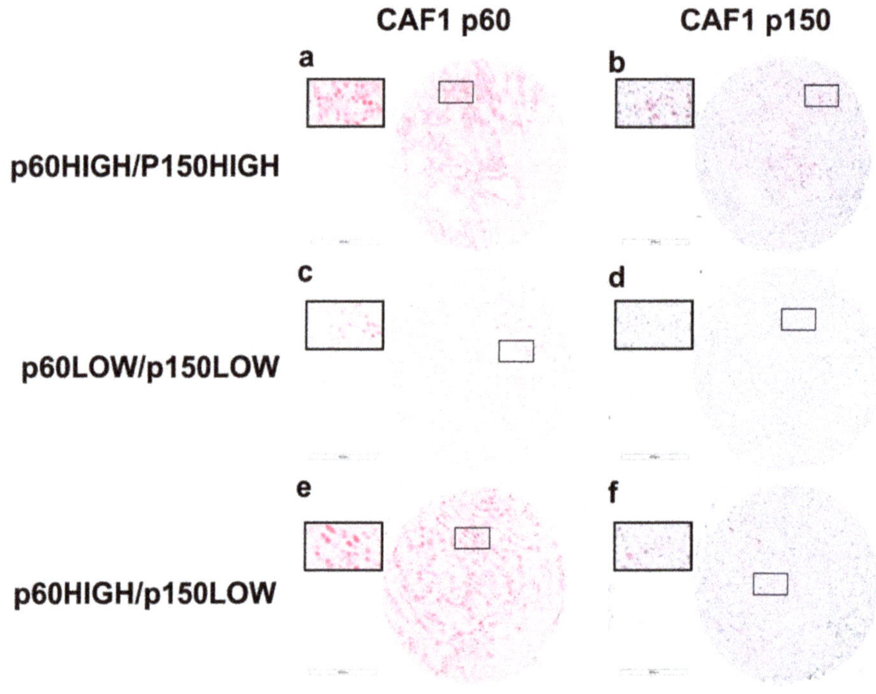

Figure 1. IHC staining of OSCC FFPE tumour samples with anti-CAF-1 p60 and anti-CAF-1 p150 antibodies. The figure shows representative images of anti-CAF-1 p60 and anti-CAF-1 p150 IHC staining intensity in OSCC FFPE tumour samples grouped according to cluster classification as resulted by cluster analysis of immunohistochemistry expression data: (**a**,**b**) p60 HIGH and p150 HIGH staining category, respectively; (**c**,**d**) p60 LOW and p150 LOW staining category, respectively; and (**e**,**f**) p60 HIGH and p150 LOW staining category, respectively. Magnification: for each panel, a 5× image of the entire core is shown and the inset shows the highlighted region.

To analyse the frequency distribution of CAF-1/p60, CAF-1/p150, and overall survival (OS) variables, we set up a contingency table.

By applying Fisher's exact test, the frequency distribution of p60 and p150 positivity, crossed by OS, proved to be statistically significant in the whole study population ($p = 0.022$), statistical significance was particularly high in OSCC group ($p = 0.013$) and no significance resulted from OPSCC samples analysis ($p = 0.485$) (Table S1).

By contingency table analysis, we could observe that, in the whole tested population, p60^{HIGH} score mostly segregates with a poor prognosis, in terms of overall survival, as expected (dead/alive ratio = 1.61 in p60^{HIGH}, 0.68 in p60^{LOW}). Moreover, the association of p60 with the worst outcome was even stronger in the p150^{LOW} score group (dead/alive ratio = 1.89). The analysis of p60 and p150 frequency distribution revealed the highest dead/alive ratio in the p60^{HIGH}/p150^{LOW} population of OSCC samples (45/18 = 2.5), while p60 expression did not correlate with outcome in OPSCC samples ($p = 0.485$). Nevertheless, out of eight HPV+ OPSCC samples, the only one presenting a poor outcome at follow-up belonged to p60^{HIGH}/p150^{LOW} subgroup.

Survival curves analysis confirmed a statistically significant difference between p60^{HIGH} and p60^{LOW} curves in the p150^{LOW} population, (log-rank test, $p = 0.0034$). A not-significant p-value was obtained evaluating the differences in OPSCC group ($p = 0.477$), irrespective of HPV status (Figure 2).

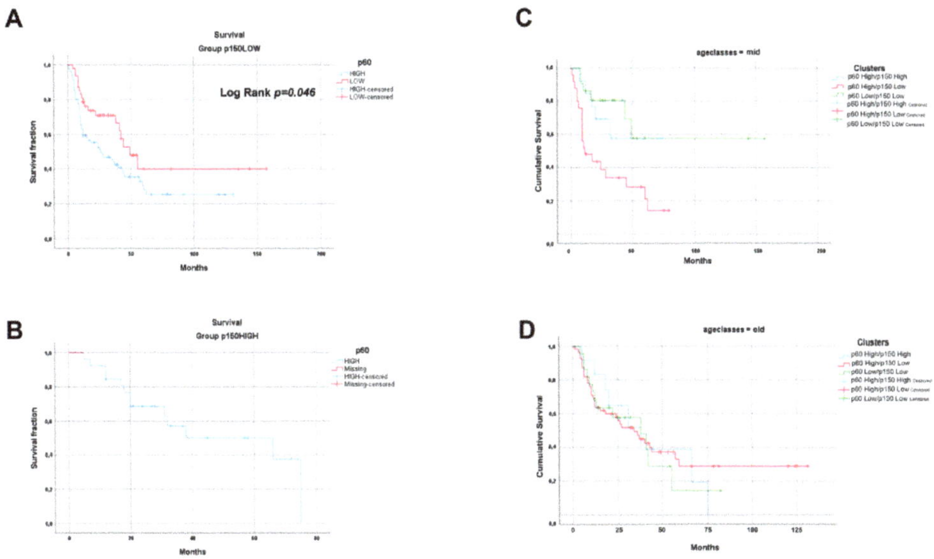

Figure 2. Survival analysis by Kaplan–Meier curves. The picture shows CAF-1 p60 HIGH and CAF-1 p60 LOW survival curves in the study population grouped by CAF-1 p150 staining score. (**A**) CAF-1 p60 HIGH and CAF-1 p60 LOW curves in p150 LOW group. (**B**) Since CAF-1 HIGH category is only associated with CAF-1 p60 HIGH staining score, only this curve is shown. Statistical differences between curves were assessed by log-rank test, where applicable. (**C**,**D**) Kaplan–Meier curves survival analysis of the three clusters stratified by age class were performed (**C**) for age class "41–60" and (**D**) for age class ">60". Difference between curves was statistically significant in mid-age group (40–60 years old) (Log-Rank test $p = 0.005$).

To further unravel the prognostic potential of p60 and p150 tissue expression in OSCC, we stratified the studied population by the age of patients at diagnosis, grouping the population in "young" (<40 years); "mid" (41–60 years); and "old" (>60 years). The survival data frequency distribution analysis, upon crossing the clusters by the overall survival results and the age groups, gave an extremely significant result in the mid-age population ($p = 0.002$). In the old age population, the distribution was not significant ($p = 0.910$), and the group of young age population was not big enough to allow statistical analysis ($N = 5$) (Table S2).

The tissue overexpression of CAF-1/p60 subunit associates with poor overall survival in OSCC, in the absence or with concomitant low expression of CAF-1/p150 subunit. We did not observe a significant association with the outcome in OPSCC tumour samples.

2.2. The Silencing of the CAF-1 Subunits Increases the Sensitivity to Ionizing Radiation in HPV-Negative and HPV-Positive Head and Neck Cancer Cell Lines

The protein "chromatin Assembly Factor-1" (CAF-1) plays a fundamental role in the steric organization of DNA and the assembly of Nucleosomes [12], thanks to its subunits, p60, p48 and p150 that make up the Heterotrimer CAF-1 [13,17,18]. CAF1 functions as "histone chaperone", and its activity is also required during the DNA damage repair [12,27]. While CAF-1/p150 subunit appears to be more active in interphase DNA damage repair processes, interacting with PCNA on the damaged DNA, during nucleotide excision repair (NER) [10,14] and double-strand breaks repair [15], CAF-1/p60 has been described to be more specifically connected to cell replication. Nevertheless, loss of p60 leads the cells to incorrectly replicate DNA, consequently accumulating DNA damage [16].

In this work, by utilizing in vitro cultured cells of HPV-negative and HPV-positive oral cavity squamous cell carcinoma, we investigated whether the stable silencing of the CAF1 subunits, p60, p150 or p60 and p150 together, could modulate the response to ionizing radiation (IR) therapy.

The CAL27 cells (OSCC, HPV-negative) and SCC90 cells (OPSCC, HPV-positive) were silenced for the large subunit p150 and for the small subunit p60 of CAF-1, alone or in combination. The efficiency of silencing was evaluated by quantitative Real-Time-PCR (qRT-PCR), using a specific set of primers for both subunits of CAF-1. The analysis of the transcripts demonstrated low mRNA levels for the two proteins upon single or combined silencing (Figure 3A). The CAF1/p60 protein levels were evaluated in the p60 and in the p60 and p150 stably silenced cells, with Western blot (Figure 3B). The detection of the silencing efficacy of the p150 CAF-1 subunit was not possible at the protein level due to the failure of anti-p150 antibody. The standard therapy for non-metastatic head and neck cancer is based on ionizing radiation and/or surgery. However, metastatic lesions are treated with chemo- and radiotherapy, with a poor prognosis. In recent years, the possibility to characterize the presence and activation of the HPV virus in head and neck tumours has allowed the stratification of such tumours by the onset site with the prediction of response to therapies. The HPV-positive tumours showed high sensitivity to radiation, whereas the HPV-negative tumours resulted in a more resistant phenotype.

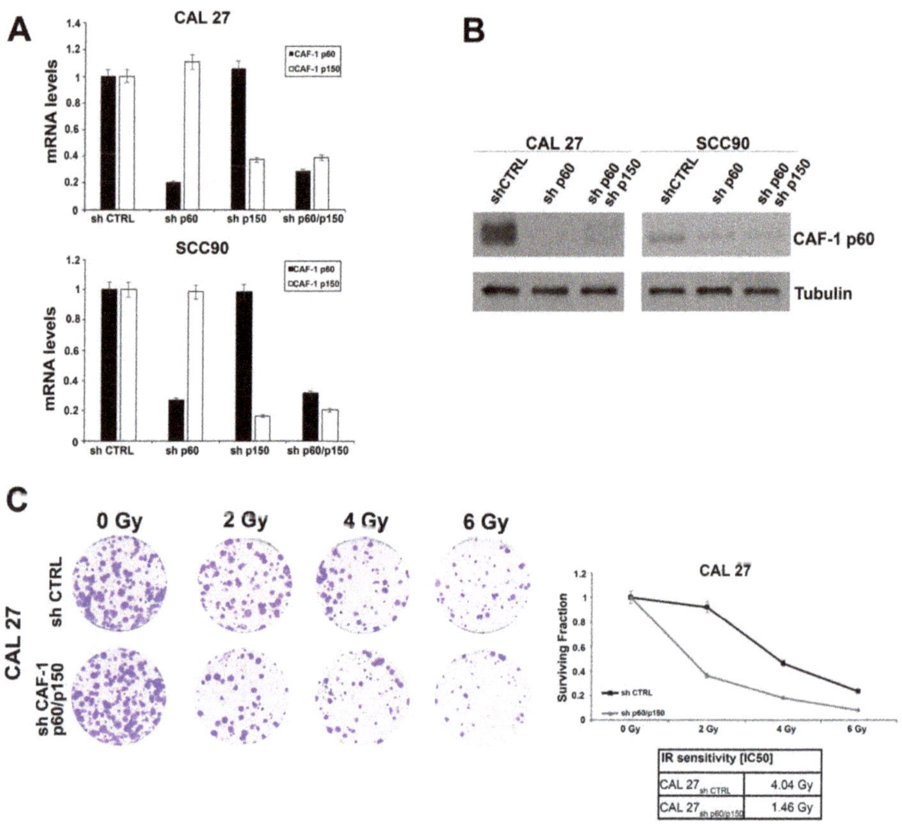

Figure 3. *Cont.*

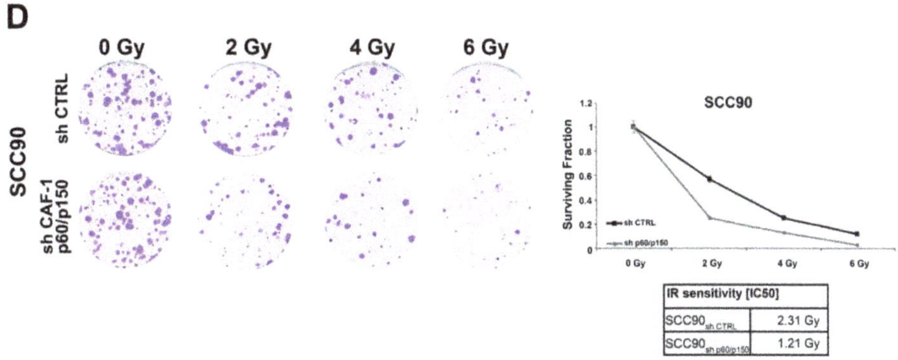

Figure 3. (**A**) Analysis of CAF-1 p60 and p150 relative mRNA expression by quantitative RT-PCR in human OSCC (CAL 27 HPV-negative) and OPSCC (SCC90 HPV-positive) cell lines, following stable trasnfection of the sh-CAF-1 p60, sh-CAF-1 p150 or both the sh-CAF1 p60/p150 vectors. (**B**) Immunoblot analysis of the CAF-1 p60 protein levels in OSCC- and OPSCC- derived cell lines. Anti-tubulin is shown as loading control. (**C**,**D**) (left) Clonogenic assay of CAL 27 and SCC90 cells after 0, 2, 4, and 6 Gy irradiation. Only colonies consisting of at least 30 cells were counted. (right) Clonogenic survival curve of CAL 27 and SCC90 cells. Error bars indicate the standard error mean. The sensitivity to ionizing radiation is expressed as IC_{50} (the value of radiation able to inhibit the cell growth of 50%).

In search of the mechanisms responsible for the different behaviours of the HPV-positive and HPV-negative tumours, we investigated the role of CAF-1 subunits in determining the sensitivity to IR in head and neck cancer. As a model, we utilized two cell lines of head and neck tumours, which were confirmed for the HPV status (CAL 27 HPV-negative and SCC90 HPV-positive).

To evaluate whether the silencing of the CAF-1 subunits p60 and p150 (as single silencing or in combination) could affect the sensitivity to IR, we treated the CAL27 and SCC90 with range doses of IR (0, 2, 4, and 6 Gy). After ten days from irradiation, we analysed the rate of survival by a Colony Forming Assay (CFA) (Figure 3C,D). A significant reduction in the doses of radiations able to decrease the survival rate in 50% of the cells was observed. These effects were more evident in the shp60/shp150-silenced cells (IC_{50} CAL 27: shp60/p150: 1.46 Gy vs. shCTRL: 4.04 Gy; IC_{50} SCC90: shp60/p150: 1.2 Gy vs. shCTRL: 2.3 Gy) (Figure 3E).

The silencing of both the subunits of CAF-1 in our HPV-negative and HPV-positive cellular models resulted in an increased sensitivity to ionizing radiations.

2.3. The Silencing of the Subunits of CAF-1 Leads to Defect in DNA Repair Mediated by Homologous Recombination and Sensitizes OSCC and OPSCC Cells to PARP-Inhibitors

CAF-1 proteins have been reported to be involved in DNA repair mechanisms [28,29]. To investigate the proficiency of Homologous Recombination DNA damage repair in CAF-1 silenced cells, CAL27 and SCC90 cells were transfected with the DR-GFP reporter plasmid alone, as a control, or together with the I-SceI plasmid, able to induce DSBs. The ability to repair the DSBs by HR was measured by flow cytometry, and the frequency of HR is reported as a percentage of GFP positive cells. The silencing of both subunits determined a significant decrease of the GFP positive cells, compared to the control, suggesting that reduction of CAF-1 levels affected the DNA repair by HR. These results were obtained in both cellular models (Figure 4A,B). The HR proficiency was also assessed in cells stably silenced with shp60 or shp150 (as single silencing), obtaining similar results. Then, as defects in DNA repair by HR are reported to increase the sensitivity to inhibitors of the Poly(ADP-ribose) polymerase 1 and 2 (PARP1 and PARP2) [30,31], we evaluated the effects of the CAF-1/p150 and p60 subunits in modulating the CAL27 and SCC90 cells' sensitivity to the PARP-inhibitor Olaparib. We treated cells with different concentrations of Olaparib and quantified the cytotoxic effect by a cell viability assay.

Silencing of both the subunits of CAF-1 significantly increased the sensitivity to Olaparib (IC_{50} CAL27 shp60/p150CAF-1: 1.42 µM vs. shCTRL: 3.01 µM; IC_{50} SCC90 shp60/p150CAF-1: 0.42 µM vs. shCTRL: 1.78 µM) (Figure 4C,D). The impact of p150/p60 silencing and Olaparib effects on DNA double strand breaks with gamma H2AX staining was also assessed in the same experimental conditions (Figure S1).

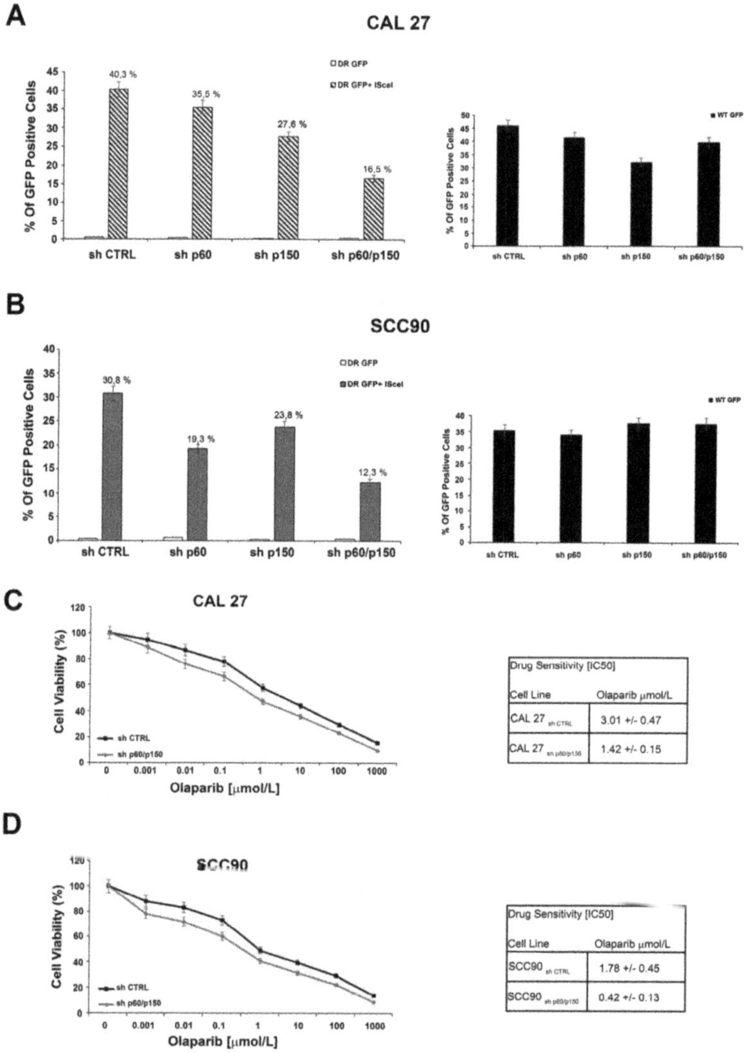

Figure 4. (**A**,**B**) Silencing of p60 and p150 suppresses homologous recombination. CAL 27 and SCC90 cells were transfected with DR-GFP alone, as control, or together with I-SceI. The percentage of GFP positive cells, compared to controls, is plotted on histograms representative of three independent experiments. Error bars indicate the standard error mean. (**C**,**D**) Silencing of p60 and p150 increases the sensitivity to Olaparib independently HPV-status. Survival fractions of CAL 27 and SCC90 cells treated with Olaparib, at the indicated doses, in presence (sh CTRL) or absence of CAF-1 (sh p60/p150) for 144 h. The sensitivity to the Olaparib was determined by the modified MTT assay (MTS), Cell Titer 96 AQueous One Solution assay, and expressed as IC_{50}, i.e., 50% of the inhibitory concentration. The values are expressed as mean ± the standard deviation.

2.4. Treatment with PARP-Inhibitors Increases the Radiosensitivity in HPV-Negative Head and Neck Cancer Cells

We finally analysed whether the silencing of CAF-1 might induce a radiosensitization effect in stably silenced cells upon PARP inhibitor treatment. The CAL 27 and SCC90 silenced cells and controls were plated and exposed to arange IR doses (0, 1, 2, 3, and 4 Gy), upon treatment with Olaparib at a fixed dose [0.1 μM]. Interestingly, by transfecting control shRNAs (shCTRL), shPARP1 and shPARP2, alone or together, in CAL27 cells (wild type or p150/p60 knockdown) and in SCC90 cells (wild type or p150/p60 knockdown), we yielded a radiosensibilizing effect similar to with PARP-inhibitors, as shown in Figure 5.

A

CAL 27 sh CTRL

	Ionizing Radiation Dose	Ionizing Radiation + Sh PARP1 Dose	DER
Fractional Effect 50%	3.30 Gy	3.49 Gy	0.94

CAL 27 sh p60/p150

	Ionizing Radiation Dose	Ionizing Radiation + Sh PARP1 Dose	DER
Fractional Effect 50%	1.26 Gy	1.08 Gy	1.17

B

CAL 27 sh CTRL

	Ionizing Radiation Dose	Ionizing Radiation + Sh PARP1+Sh PARP2 Dose	DER
Fractional Effect 50%	3.30 Gy	3.50 Gy	0.94

CAL 27 sh p60/p150

	Ionizing Radiation Dose	Ionizing Radiation + Sh PARP1+Sh PARP2 Dose	DER
Fractional Effect 50%	1.26 Gy	0.96 Gy	1.31

C

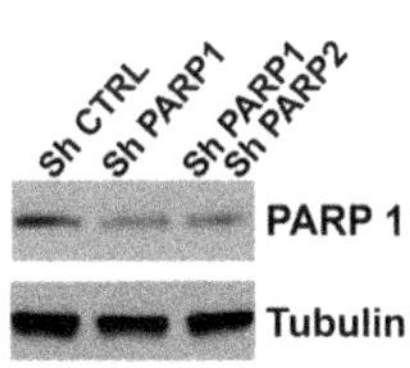

Figure 5. (**A**,**B**) The Dose Enhancement Ratio (DER) at 50% of the effect refers to the ratio between the dose with radiation alone and the dose with radiation + shPARP1 alone or in combination with shPARP2 for the same biological effect. If the DER is greater than one, then the silencing of PARP1/2 is functioning as a radiosensitizer. If the DER is less than one, the PARP silencing acts as a radioprotector. In the table are shown DER values at 50% effect for CAL 27 (sh CTRL vs. sh p60/p150). (**C**) The PARP1 depletion was assessed by Western blot. Tubulin is shown as loading control.

We calculated cell survival by viability assay at 144 h (Figure 5). The results of the combined treatment were extremely relevant, as analysed by the Dose Enhancement Ratio (DER) calculated at 50% of cells' survival [32]. The HPV-negative CAL27 cells showed a DER > 1 after cells exposure to different doses of ionizing radiation (0–4 Gy) in the presence of Olaparib [0.1 μM], which suggested a radiosensitization effect upon CAF-1 subunits depletion. However, the HPV-positive SCC90 cells, showed a moderate radiosensitization effect following a combined treatment, either in CAF-1 subunits depleted cells or in control cells (Figure 6). These data seem very promising as the SF2Gy, a standard measure of cells sensitivity to IR highly utilized in the field of radiobiology, indicates a great radiosensibilization effect upon CAF1 subunits silencing (Table S3).

A

CAL 27 sh CTRL

	Ionizing Radiation Dose	Ionizing Radiation + Olaparib [0.1μM] Dose	DER
Fractional Effect 50%	3.71 Gy	3.43 Gy	1.08

CAL 27 sh p60/p150

	Ionizing Radiation Dose	Ionizing Radiation + Olaparib [0.1μM] Dose	DER
Fractional Effect 50%	1.31 Gy	1.01 Gy	1.30

B

SCC90 sh CTRL

	Ionizing Radiation Dose	Ionizing Radiation + Olaparib [0.1μM] Dose	DER
Fractional Effect 50%	2.87 Gy	2.60 Gy	1.10

SCC90 sh p60/p150

	Ionizing Radiation Dose	Ionizing Radiation + Olaparib [0.1μM] Dose	DER
Fractional Effect 50%	1.15 Gy	1.05 Gy	1.09

Figure 6. (**A**,**B**) The Dose Enhancement Ratio (DER) at 50% of the effect refers to the ratio between the dose with radiation alone and the dose with radiation + drug for the same biological effect. If the DER is greater than one, then the addition of the drug is functioning as a radiosensitizer. If the DER is less than one, the drug acts as a radioprotector. In the table are shown DER values at 50% effect for CAL 27 (sh CTRL vs. sh p60/p150) and SCC90 (sh CTRL vs. sh p60/p150).

3. Discussion

Despite the recent advances in early diagnosis and surgical management of HNSCC, the outcome of these tumours has not substantially changed during the last decades, and the prognosis, for each HNSCC patient, remains linked to the tumour's stage at presentation. The standard of care for these patients consists of radical surgery, complemented by radiotherapy, chemotherapy and, recently, immunotherapy in the case of advanced disease. The five-year survival rate for HNSCC patients remains poor, with up to 50% of mortality rates [1]. The most relevant limits of radio- and chemotherapy are represented by systemic and/or local toxicity, and by the frequent radio/chemoresistance of these tumours with particular attention to OSCC [33].

The limited information available on the molecular determinants of the biology of HNSCC indicates the urgent need to identify new prognostic markers and/or molecular targets for personalized therapeutic strategy [34].

Oral/oropharyngeal carcinogenesis mostly depends on environmental factors, such as tobacco and alcohol abuse, particularly in the oral cavity, and persistent infection from HR-HPV in a rising percentage of oropharyngeal cancers. Thus far, only a few reliable prognostic markers have been reported, alone or in combination, for OSCC and OPSCC [35–39].

The HPV status (persistent infection by HR-HPV, mostly HPV-16), only for oropharynx SCC, has been indicated, by the new TNM classification (AJCC 8th edition) as a significant marker of a more favourable biological behaviour of tumours, as has been shown by several studies in the last decade [40–44]. In HPV-negative tumours, mutations of TP53 and amplification of the Epidermal Growth Factor Receptor (EGFR) have been reported and used for prognostic and intervention indications with dismal results [45–49].

Chromatin remodelling proteins play an important role in genome maintenance processes, including DNA repair and replication, also involved in the development and progression of several human malignant tumours. Chromatin Assembly Factor 1 (CAF-1) is a "histone chaperone", which delivers newly synthesized H3/H4 dimers to the replicative fork during the DNA synthesis phase (S) of the cell cycle [50]. CAF-1 is a heterotrimeric protein, of which p48 subunit cooperates with the Retinoblastoma protein (Rb) [51] and p150 and p60 subunits are involved in DNA repair and replication processes. Noteworthy, a leading role of p60 in sustaining the proliferative activity of cells in different malignancies has been described in the last two decades [14,52–54].

Recently, the presence of defects in genes involved in DNA damage repair by homologous recombination (HR) opened the way to the use of PARP inhibitors, alone or in association with genotoxic agents, such as ionizing radiation, in some malignant tumours [55]. The inhibition of PARP enzymes as an anticancer strategy has been established because of the biological concept of synthetic lethality, for which two genomic events, individually not lethal, become lethal when occurring together. When PARP enzymes are pharmacologically inhibited, the DNA single-strand breaks cannot be repaired and eventually progress to toxic double-strand breaks (DSBs), which result in being lethal in cells that lack HR repair capacity or have lost DNA repair genes [56–58]. In head and neck cancer cell lines characterized by a different ability to repair the DNA double-strand breaks through HR, the effectiveness of combined irradiation and PARP-inhibitor treatment in HR-deficient cells has been evaluated [59]. Very interestingly, it has been demonstrated that HPV positive head and neck tumours, carrying defects in DNA damage repair by HR, result sensitive to ionizing radiation [60].

Since HNSCC rarely carry mutations in DNA repair genes, there is considerable interest in finding alternative determinants of PARPi sensitivity. Our in vitro study, carried out by depleting CAF-1/p60 and p150 subunits (singularly or together), showed impairment of DSBs DNA repair and increased sensitivity to PARP inhibitor Olaparib in both cell lines used. This suggests that the presence of HPV, in the OPSCC cell line, did not modify the sensitivity of tumour cells with respect to the HPV-negative OSCC cell line. As a further step, we evaluated whether the treatment with Olaparib could modulate the response to ionizing radiation in the CAF-1 depleted cell lines. By combining ionizing radiation and PARP inhibitors drug, we observed an increased radio-sensitivity. Of particular interest, this effect resulted more evident in the OSCC CAL27 cells depleted of the CAF-1 p60/p150 subunits, as indicated by the DER that resulted greater than 1 (DER > 1). These data suggest that the combined treatment with ionizing radiation and PARP inhibitors can lead to the increase of the radio-sensitivity of OSCC, characteristically resistant to standard radio-treatment regimens.

All of the tumour samples included in the present study were re-examined and re-staged according to the eighth edition of the cancer staging manual of AJCC, and a "stage migration" was observed, as reported in Table S4 [61].

Classification through cluster analysis of immunohistochemical data allowed us to stratify the patients' outcome in different subgroups: (i) a p60^{low}/p150^{low} subgroup, mainly characterized by a good prognosis in both OSCC and OPSCC; (ii) a p60^{high}/p150^{low} group which showed, mainly in OSCC, the worst outcome prediction; and (iii) a p60^{high}/p150^{high} subgroup with an intermediate behaviour. These results confirm our preliminary data, obtained in the tongue tumour samples, of a prognostic role of the CAF-1/p60 subunit, which correlated with poor outcome and the importance of p60/p150 dual assessment as a prognostic determinant in OSCC tumours [20,24,25]. Interestingly, the survival data frequency distribution analysis, upon crossing the clusters by the overall survival results and the age groups, gave an extremely significant result in the mid-age population. Nevertheless, the Cox multivariate analysis, performed including age and stage variables together with p60/p150 clusters, revealed that p60/p150 staining is an independent prognostic factor (Figure S2).

Our data let us envisage novel potential therapeutic approaches for OSCC by blocking the p60 protein, hampering the CAF-1 complex function and inducing HR defects that would sensitize tumour cells to Olaparib and/or in association with ionizing radiation.

We did not obtain significant results in OPSCC cohort, probably due to the small number of HPV+ in our series of cases. Interestingly, the only tumour with poor outcome among the HPV+ OPSCC showed the more aggressive immunophenotype (p60^{high}/p150^{low}) observed in OSCC.

However, this finding is not sufficient, at present, to propose the evaluation of CAF-1/p60 and p150 protein expression for prognostic and predictive stratification also of this tumours' subset. For this reason, the evaluation of CAF-1 proteins' expression needs to be evaluated on a larger, multi-institutional case study before we can lead to a definitive conclusion. This study is currently in progress.

In conclusion, our data confirm the reliability of CAF-1/p60 subunit as prognostic marker for OSCC, indicating in addition that the combined evaluation of p60 and p150 subunit may be of particular utility in stratifying the different prognostic classes of OSCC.

In addition, we showed that CAF-1/p60 and p150 subunits are involved in HR-DDR, thus indicating the chance to induce a radiosensitizing synthetic lethality mechanism by treating tumour cells pharmacologically inhibited for CAF-1/p60 and p150 with PARP inhibitors, in OSCC patients in the worse prognostic group, in the direction of a truly personalized therapy.

4. Materials and Methods

4.1. Cell Lines and Drugs

Experiments were carried out using two human head and neck squamous cell lines, the CAL27 (Oral Squamous Cell Carcinoma, OSCC, HPV-negative) and SCC90 cells (Oropharyngeal Squamous Cell Carcinoma, OPSCC, HPV-positive). CAL27 cells were obtained by the "American Type Culture Collection" (ATCC) and SCC090 cell line was obtained by Leibniz Institute DSMZ-German Collection of Microorganisms and Cell Cultures, Germany.

Cell lines were cultured in the DMEM plus 10% of fetal bovine serum (Gibco, Paisley, UK).), L-Glutamine (2mM) and 100 U/mL of penicillin-streptomycin (Gibco, Paisley, UK) in 5% of CO_2 at 37 °C. Olaparib (AZD2281) was provided by SelleckChem (Houston, TX, USA).

4.2. Real Time PCR

PCR reactions were performed on RNA isolated from cell lines using the RNeasy Mini Kit (Qiagen, Hilden, Germany) and reverse-transcribed using MuLVRT (Invitrogen, Carlsbad, CA, USA). The qRT-PCR analysis was performed with Syber Green (Agilent, Santa Clara,

CA, USA). Primer sequences are reported in Table S5. The relative expression levels were calculated by the 2-ΔΔCT method.

4.3. Western Blotting and Antibodies

Western blotting was performed as described (Figures S3 and S4) [62,63]. Immunoblotting experiments were carried out according to standard procedures and visualized using the ECL chemiluminescence system (Amersham/Pharmacia Biotech, Little Chalfont, UK). Anti-CHAF1B (HPA021679) and anti-Tubulin were provided by SIGMA-Aldrich, Inc. Secondary antibodies were from Biorad (Hercules, CA, USA).

4.4. Plasmids and Transfection

MISSION shRNA (pLKO.1) NM_005441 (CHAF1B MISSION shRNA Plasmid DNA, cod: TRCN0000074278, TRCN0000074279, and TRCN0000074281] and NM_005483 [CHAF1A MISSION shRNA Plasmid DNA, cod: TRCN0000074273 and TRCN00000234596) and shRNA of control (sh CTRL) were utilized for stable transfection and were provided by Sigma-Aldrich, Inc. (St. Louis, Missouri, USA) For transfection, the FuGENE 6 Transfection Reagent was provided by Promega Italia S.r.l. (Milano, Italy).

4.5. Sensitivity Test and Design for Drug Combination

Antiproliferative activity was determined by a modified 3-(4,5-dimethylthiazole-2-yl)-2-5-diphenyltetrazolium bromide assay, CellTiter 96 AQueous One Solution Assay (Promega, Milano, Italy), calculated as 50% inhibitory concentration (IC_{50}) values, according to the manufacturer instructions.

Briefly, cells were plated in quintuplicate in 96-well plates at a density of 1000 cells per well, and continuously exposed to each drug for 144 h. Each assay was performed in quintuplicate and IC_{50} values were expressed as mean ± standard deviation.

The results of the combined treatment were expressed as a DER, calculated at 50% of survival. DER is a measure of how many folds each radiation dose may be reduced when administered in combination with a drug to obtain a given effect, in comparison to the same dose of radiation when given alone. A value of DER greater than one (DER > 1) indicates that a fixed dose of the drug used, in association with a range doses of IR was able to act as radiosensitizer, while a value of DER lower than one (DER < 1) indicates that a fixed dose of the drug used, in association with a range of IR doses, works as a radioprotector [32,64].

4.6. Colony Forming Assay

One-hundred-millimeter-cubed dishes of proliferating cells were exposed to 0, 2, 4, and 6 Gy of IR. After cells counting, a pre-defined number of viable cells were plated in 6-well plates, in triplicate. To receive a sufficient colony count, a different number of cells were plated for each dose of irradiation (0 Gy, 200 cells; 2 Gy, 500 cells; 4 Gy, 600 cells; and 6 Gy, 800 cells) [65]. After 10 days of incubation, prior to counting colonies, cells were stained with 0.5% Crystal Violet (10 min at room temperature). A population of at least 30 cells was scored as one survivable colony and considered for the count. The colonies' counting was performed at the optic microscope and through the open source software ImageJ-NIH [66]. The relative colony formation (surviving fraction) was expressed as the number of colonies per treatment level versus colonies that appeared in the untreated control. (mean colony counts ± standard errors are reported).

4.7. TMA and IHC

One hundred fifty-four HNSCC FFPE tumour samples (112 OSCC and 42 OPSCC, of which 8 HPV-positive), from surgical resections, were used to build tissue micro-arrays (TMAs). The study was performed according to the guidelines of the Institutional Ethic Committee, which, in agreement with the Italian law, with reference to the topics of the current research and according to the Declaration of Helsinki require, for studies based only on retrospective analyses on routine archival FFPE-tissue, a written informed consent from the living patient, following the indication of Italian DLgs No. 196/03 (Codex on Privacy), as modified by UE 2016/679 law of European parliament and Commission at the time of surgery.

The HPV positivity was confirmed through p16 immunostaining and HPV genotyping by INNO Lipa [67]. Seven TMAs were built selecting the most representative areas from each selected paraffin block, at least in duplicate. Three-millimeter tissue cores were punched from morphologically representative tissue areas of each donor block and placed into one recipient paraffin block (3 cm × 2.5 cm) using a semi-automated tissue arrayer (Galileo TMA, Milan, Italy). To ensure a sufficient representation of the tissue composition of individual cancer cases, in view of the known frequency of tumour heterogeneity in H&N SCC, we built TMAs taking 3-mm cores in at least duplicate or triplicate (in the case of bigger tumours). We already challenged our TMA protocol with whole slide assessment of tissue biomarkers expression (please see reference mentioned below) finding a high degree of agreement between the CAF-1/p60 assessment on TMAs and on routine tissue sections [54]. One section of each TMA (4 μm) was stained with hematoxylin and eosin (H&E) to check the adequacy of cores. The immunohistochemical staining with anti-CAF-1/p60 and CAF-1/p150 (ab8133 and ab126625 obtained from AbCam, Cambridge, UK) [68] were evaluated semi-quantitatively as the percentage of positive cells (with either nuclear or cytoplasmic localization) among the total number of tumour cells and classified as LOW staining (including 0 (<5% di cellule positive); and + (5% to <15%) scores) and HIGH staining (including ++ (15% to <30%); +++ (>30%) scores). All immunoassayed TMA glass slides were digitalized with an Aperio AT2 digital pathology slide scanner (Leica Biosystems Nussloch GmbH, Heidelberger, Germany) and visualized with QuPath image software analysis [69–71].

4.8. Statistical Analysis

Statistical analysis was performed using SPSS software (IBM Corp. Released 2013. IBM SPSS Statistics for Windows, Version 22.0. Armonk, NY, USA). Tissue biomarker expression scores' correlation with the overall survival variable was performed by Fisher exact test. A K-mean cluster analysis was performed to sort relatively homogeneous groups of cases based on selected characteristics (CAF-1/p60 and CAF-1/p150 IHC expression). Survival analysis was performed by test Kaplan–Meier survival curves' differences by log-rank test. p-value was considered significant at $p = 0.05$.

4.9 HR Reporter Assay

The DR-GFP reporter and pCAGGS-I-SceI plasmids were used to verify the functionality of DNA repair mechanism by HR. The DR-GFP reporter plasmid, based on a construct developed by M. Jasin [72], consists of two mutated and GFP negative expression cassettes (GFP-I, GFP-II). In Cassette I, there is a unique cutting site recognized by the restriction enzyme I-SceI. In cells transfected with the DR-GFP and the I-SceI plasmids, the expression of I-SceI induces a double-stranded break (DSB) in Cassette I. This damage can be repaired through HR, using the GFP gene present as a template in Cassette II and thus restoring the expression of GFP (Figure 1). The ability of cells to repair double-stranded damage by HR was measured by flow cytometry reporting the percentage of positive GFP cells as a measure of HR proficiency.

5. Conclusions

The immunohistochemical analysis of CAF-1/p60 and CAF-1/p150 tissue expression in OSCC revealed a significant prognostic value of the combined expression of both tissue biomarkers, while no significant prognostic value was demonstrated in OPSCC. In HPV-positive and HPV-negative cell lines, the p60/150 CAF-1 subunits depletion impaired the proficiency of Homologous Recombination (HR) DNA damage repair, inducing sensitivity to the PARP-inhibitors drugs able to sensitize both the cell lines to ionizing radiations. These results indicate that, regardless of the prognostic meaning of the expression of the two markers, the pharmacological depletion of CAF-1 complex's function, combined to PARP-inhibitors and/or radiotherapy, could represent a valid therapeutic strategy for squamous cell carcinomas of head and neck region, by the mean of a synthetic lethality mechanism.

Supplementary Materials: The following are available online at http://www.mdpi.com/2072-6694/11/10/1582/s1, Table S1: Contingency table showing frequency distribution of IHC expression data crossed with overall survival, Table S2 (A): The table shows the distribution of the three clusters frequency counts stratifying according to stage, age class and outcome, Table S2 (B): Demographic table showing the distribution of cases count frequency grouping by age class and outcome and stratifying by CAF p60/p150 clusters. The distribution was statistically significant in mid age class at Chi-squared test of significance ($p = 0.002$), Table S3: Type one (α) and type two (β) DNA damage in Cal27 Sh CTRL and CAL27 Sh p60/p150, using linear quadratic radiobiological model, Table S4: 34 out of 154 samples were upstaged in the staging score, 17/34 (50%) of the upstaged patients died at follow-up. The table shows the cunt of upstaged samples at the restaging according to the 8th edition of AJCC TNM manual. A&W = alive and well. DOD = death of disease, Table S5: Oligi sequence of the utilized primers, Figure S1: The percentage of g-H2AX positive nuclei at 1h and 4h from irradiation are in A and B. Error bars represent standard error mean. Results are representative of at least two independent experiments, Figure S2: p60/p150 tissue exprssion is an independent prognosti factor in a Cox multivariate analysis including also age and stage variables (global significance of the model $p = 0{,}008$, HR of $p60^{high}/p150^{low}$ equal to 1789.

Author Contributions: F.M. (Francesco Morra), F.M. (Francesco Merolla), I.P., S.S. and A.C. made substantial contributions to the conception and design of the experiments; F.M. (Francesco Morra), I.P., F.L. and R.P. performed the in vitro experiments; F.M. (Francesco Merolla), D.R., G.I., and S.V. contributed to methodology and software employment for data acquisition, validation and formal analysis (as in M&M); F.M. (Francesco Morra), F.M. (Francesco Merolla), I.P., G.I., D.R., F.L., S.S. and A.C. actively participated in the interpretation of data; and F.M. (Francesco Morra), F.M. (Francesco Merolla), I.P., L.P., M.M., A.C. and S.S. were involved in drafting the manuscript or revising it critically for important intellectual content. All authors gave final approval of the version to be published and agreed to be accountable for all aspects of the work in ensuring that questions related to the accuracy or integrity of any part of the work are appropriately investigated and resolved.

Funding: Our research was supported by POR Campania FESR 2014-2020 grant; "Technological Platform: eMORFORAD-Campania "grant PG/2017/0623667; POR Campania FESR 2014-2020 "SATIN" grant; POR Campania FESR 2014-2020 "RARE.PLAT.NET" grant; and "POR Campania FESR 2014-2020 "grant.

Acknowledgments: We are grateful to Mario De Felice for his countnuous support and encouragement, and to Rosaria Catalano for technical help.

Conflicts of Interest: The authors declare no conflict of interest.

References

1. Siegel, R.L.; Miller, K.D.; Jemal, A. Cancer statistics, 2019. *CA Cancer J. Clin.* **2019**, *69*, 7–34. [CrossRef]
2. Pontes, F.S.; Carneiro, J.T., Jr.; Fonseca, F.P.; da Silva, T.S.; Pontes, H.A.; Pinto Ddos, S., Jr. Squamous cell carcinoma of the tongue and floor of the mouth: Analysis of survival rate and independent prognostic factors in the Amazon region. *J. Craniofac. Surg.* **2011**, *22*, 925–930. [CrossRef]
3. Torre, L.A.; Bray, F.; Siegel, R.L.; Ferlay, J.; Lortet-tieulent, J.; Jemal, A. Global Cancer Statistics, 2012. *CA Cancer J. Clin.* **2015**, *65*, 87–108. [CrossRef]
4. Hashibe, M.; Brennan, P.; Chuang, S.C.; Boccia, S.; Castellsague, X.; Chen, C.; Curado, M.P.; Dal Maso, L.; Daudt, A.W.; Fabianova, E.; et al. Interaction between tobacco and alcohol use and the risk of head and neck cancer: Pooled analysis in the International Head and Neck Cancer Epidemiology Consortium. *Cancer Epidemiol. Biomark. Prev.* **2009**, *18*, 541–550. [CrossRef]
5. Mallen-St Clair, J.; Alani, M.; Wang, M.B.; Srivastan, E.S. Human papillomavirus in oropharyngeal cancer: The changing face of a disease. *Biochim. Biophys. Acta* **2016**, *1866*, 141–150. [CrossRef]
6. Windon, M.J.; D'Souza, G.; Rettig, E.M.; Westra, W.H.; van Zante, A.; Wang, S.J.; Ryan, W.R.; Mydlarz, W.K.; Ha, P.K.; Miles, B.A.; et al. Increasing prevalence of human papillomavirus-positive oropharyngeal cancers among older adults. *Cancer* **2018**, *124*, 2993–2999. [CrossRef]
7. Hecht, S.S. Tobacco carcinogens, their biomarkers and tobacco-induced cancer. *Nat. Rev. Cancer* **2003**, *3*, 733–744. [CrossRef]
8. Kadaja, M.; Isok-Paas, H.; Laos, T.; Ustav, E.; Ustav, M. Mechanism of genomic instability in cells infected with the high-risk human papillomaviruses. *PLoS Pathog.* **2009**, *5*, e1000397. [CrossRef]

9. Chatterjee, N.; Walker, G.C. Mechanisms of DNA damage, repair, and mutagenesis. *Environ. Mol. Mutagen.* **2017**, *58*, 235–263. [CrossRef]
10. Merolla, F.; Mascolo, M.; Ilardi, G.; Siano, M.; Russo, D.; Graziano, V.; Celetti, A.; Staibano, S. Nucleotide Excision Repair and head and neck cancers. *Front. Biosci. (Landmark Ed.)* **2016**, *21*, 55–69.
11. Wood, R.D.; Mitchell, M.; Lindahl, T. Human DNA repair genes, 2005. *Mutat. Res.* **2005**, *577*, 275–283. [CrossRef]
12. Verreault, A.; Kaufman, P.D.; Kobayashi, R.; Stillman, B. Nucleosome assembly by a complex of CAF-1 and acetylated histones H3/H4. *Cell* **1996**, *87*, 95–104. [CrossRef]
13. Ridgway, P.; Almouzni, G. CAF-1 and the inheritance of chromatin states: At the crossroads of DNA replication and repair. *J. Cell Sci.* **2000**, *113 Pt 15*, 2647–2658.
14. Taddei, A.; Roche, D.; Sibarita, J.B.; Turner, B.M.; Almouzni, G. Duplication and maintenance of heterochromatin domains. *J. Cell Biol.* **1999**, *147*, 1153–1166. [CrossRef]
15. Moggs, J.G.; Grandi, P.; Quivy, J.P.; Jonsson, Z.O.; Hubscher, U.; Becker, P.B.; Almouzni, G. A CAF-1-PCNA-mediated chromatin assembly pathway triggered by sensing DNA damage. *Mol. Cell. Biol.* **2000**, *20*, 1206–1218. [CrossRef]
16. Hoek, M.; Myers, M.P.; Stillman, B. An analysis of CAF-1-interacting proteins reveals dynamic and direct interactions with the KU complex and 14-3-3 proteins. *J. Biol. Chem.* **2011**, *286*, 10876–10887. [CrossRef]
17. Kaufman, P.D.; Kobayashi, R.; Kessler, N.; Stillman, B. The p150 and p60 subunits of chromatin assembly factor, I. A molecular link between newly synthesized histones and DNA replication. *Cell* **1995**, *81*, 1105–1114. [CrossRef]
18. Smith, S.; Stillman, B. Purification and characterization of CAF-I, a human cell factor required for chromatin assembly during DNA replication in vitro. *Cell* **1989**, *58*, 15–25. [CrossRef]
19. Nabatiyan, A.; Krude, T. Silencing of chromatin assembly factor 1 in human cells leads to cell death and loss of chromatin assembly during DNA synthesis. *Mol. Cell. Biol.* **2004**, *24*, 2853–2862. [CrossRef]
20. Staibano, S.; Mignogna, C.; Lo Muzio, L.; Mascolo, M.; Salvatore, G.; Di Benedetto, M.; Califano, L.; Rubini, C.; De Rosa, G. Chromatin assembly factor-1 (CAF-1)-mediated regulation of cell proliferation and DNA repair: A link with the biological behaviour of squamous cell carcinoma of the tongue? *Histopathology* **2007**, *50*, 911–919. [CrossRef]
21. Staibano, S.; Mascolo, M.; Mancini, F.P.; Kisslinger, A.; Salvatore, G.; Di Benedetto, M.; Chieffi, P.; Altieri, V.; Prezioso, D.; Ilardi, G.; et al. Overexpression of chromatin assembly factor-1 (CAF-1) p60 is predictive of adverse behaviour of prostatic cancer. *Histopathology* **2009**, *54*, 580–589. [CrossRef]
22. Mascolo, M.; Vecchione, M.L.; Ilardi, G.; Scalvenzi, M.; Molea, G.; Di Benedetto, M.; Nugnes, L.; Siano, M.; De Rosa, G.; Staibano, S. Overexpression of Chromatin Assembly Factor-1/p60 helps to predict the prognosis of melanoma patients. *BMC Cancer* **2010**, *10*, 63. [CrossRef]
23. Staibano, S.; Mascolo, M.; Rocco, A.; Lo Muzio, L.; Ilardi, G.; Siano, M.; Pannone, G.; Vecchione, M.L.; Nugnes, L.; Califano, L.; et al. The proliferation marker Chromatin Assembly Factor-1 is of clinical value in predicting the biological behaviour of salivary gland tumours. *Oncol. Rep.* **2011**, *25*, 13–22. [CrossRef]
24. Mascolo, M.; Ilardi, G.; Romano, M.F.; Celetti, A.; Siano, M.; Romano, S.; Luise, C.; Merolla, F.; Rocco, A.; Vecchione, M.L.; et al. Overexpression of chromatin assembly factor-1 p60, Poly(ADP-ribose) polymerase 1 and nestin predicts metastasizing behaviour of oral cancer. *Histopathology* **2012**, *61*, 1089–1105. [CrossRef]
25. Mesolella, M.; Iorio, B.; Landi, M.; Cimmino, M.; Ilardi, G.; Iengo, M.; Mascolo, M. Overexpression of chromatin assembly factor-1/p60 predicts biological behaviour of laryngeal carcinomas. *Acta Otorhinolaryngol. Ital.* **2017**, *37*, 17–24. [CrossRef]
26. Bochner, B.H.; Hansel, D.E.; Efstathiou, J.A.; Konety, B.; Lee, C.; Mckiernan, J.M.; Plimack, E.R.; Reuter, V.E.; Sridhar, S.; Vikram, R.; et al. *AJCC Cancer Staging Manual*, 8th ed.; American Joint Committee on Cancer: Chicago, IL, USA, 2017. [CrossRef]
27. Gaillard, P.; Martini, E.; Kaufman, P. Chromatin assembly coupled to DNA repair: A new role for chromatin assembly factor, I. *Cell* **1996**, *86*, 887–896. [CrossRef]
28. Polo, S.E.; Theocharis, S.E.; Grandin, L.; Gambotti, L.; Antoni, G.; Savignoni, A.; Asselain, B.; Patsouris, E.; Almouzni, G. Clinical significance and prognostic value of chromatin assembly factor-1 overexpression in human solid tumours. *Histopathology* **2010**, *57*, 716–724. [CrossRef]
29. Volk, A.; Crispino, J.D. The role of the chromatin assembly complex (CAF-1) and its p60 subunit (CHAF1b) in homeostasis and disease. *Biochim. Biophys. Acta* **2015**, *1849*, 979–986. [CrossRef]

30. Cerrato, A.; Morra, F.; Celetti, A. Use of poly ADP-ribose polymerase [PARP] inhibitors in cancer cells bearing DDR defects: The rationale for their inclusion in the clinic. *J. Exp. Clin. Cancer Res.* **2016**, *35*, 179. [CrossRef]
31. Palazzo, L.; Ahel, I. PARPs in genome stability and signal transduction: Implications for cancer therapy. *Biochem. Soc. Trans.* **2018**, *46*, 1681–1695. [CrossRef]
32. Roeske, J.C.; Nunez, L.; Hoggarth, M.; Labay, E.; Weichselbaum, R.R. Characterization of the theorectical radiation dose enhancement from nanoparticles. *Technol. Cancer Res. Treat.* **2007**, *6*, 395–401. [CrossRef]
33. Johnson, F.L. Management of advanced premalignant laryngeal lesions. *Curr. Opin. Otolaryngol. Head Neck Surg.* **2003**, *11*, 462–466. [CrossRef]
34. Schuller, D.E.; Wilson, H.E.; Smith, R.E.; Batley, F.; James, A.D. Preoperative reductive chemotherapy for locally advanced carcinoma of the oral cavity, oropharynx, and hypopharynx. *Cancer* **1983**, *51*, 15–19. [CrossRef]
35. De Tayrac, M.; Saikali, S.; Aubry, M.; Bellaud, P.; Boniface, R.; Quillien, V.; Mosser, J. Prognostic significance of EDN/RB, HJURP, p60/CAF-1 and PDLI4, four new markers in high-grade gliomas. *PLoS ONE* **2013**, *8*, e73332. [CrossRef]
36. Molofsky, A.V.; Pardal, R.; Morrison, S.J. Diverse mechanisms regulate stem cell self renewal. *Curr. Opin. Cell Biol.* **2004**, *16*, 700–707. [CrossRef]
37. Tabor, M.P.; Brakenhoff, R.H.; Ruijter-Schippers, H.J.; Van Der Wal, J.E.; Snow, G.B.; Leemans, C.R.; Braakhuis, B.J. Multiple head and neck tumours frequently originate from a single preneoplastic lesion. *Am. J. Pathol.* **2002**, *161*, 1051–1060. [CrossRef]
38. Zukerberg, L. The molecular basis of dysplasia. *Semin. Diagn. Pathol.* **2002**, *19*, 48–53.
39. Russo, D.; Merolla, F.; Varricchio, S.; Salzano, G.; Zarrilli, G.; Mascolo, M.; Strazzullo, V.; Di Crescenzo, R.M.; Celetti, A.; Ilardi, G. Epigenetics of oral and oropharyngeal cancers. *Biomed. Rep.* **2018**, *9*, 275–283. [CrossRef]
40. Rosai, J.; Carcangiu, M.L.; DeLellis, R.A. *Atlas of Tumour Pathology-Tumours of the Larynx*, 3rd ed.; Armed Force Institute of Pathology: Washington, DC, USA, 1992.
41. Gillison, M.L.; Koch, W.M.; Capone, R.B.; Spafford, M.; Westra, W.H.; Wu, L.; Zahurak, M.L.; Daniel, R.W.; Viglione, M.; Symer, D.E.; et al. Evidence for a causal association between human papillomavirus and a subset of head and neck cancers. *J. Natl. Cancer Inst.* **2000**, *92*, 709–720. [CrossRef]
42. Gillison, M.L.; D'Souza, G.; Westra, W.; Sugar, E.; Xiao, W.; Begum, S.; Viscidi, R. Distinct risk factor profiles for human papillomavirus type 16-positive and human papillomavirus type 16-negative head and neck cancers. *J. Natl. Cancer Inst.* **2008**, *100*, 407–420. [CrossRef]
43. D'Souza, G.; Kreimer, A.R.; Viscidi, R.; Pawlita, M.; Fakhry, C.; Koch, W.M.; Westra, W.H.; Gillison, M.L. Case-control study of human papillomavirus and oropharyngeal cancer. *N. Engl. J. Med.* **2007**, *356*, 1944–1956. [CrossRef]
44. Applebaum, K.M.; Furniss, C.S.; Zeka, A.; Posner, M.R.; Smith, J.F.; Bryan, J.; Eisen, E.A.; Peters, E.S.; McClean, M.D.; Kelsey, K.T. Lack of association of alcohol and tobacco with HPV16-associated head and neck cancer. *J. Natl. Cancer Inst.* **2007**, *99*, 1801–1810. [CrossRef]
45. Hunt, J.L.; Barnes, L.; Lewis, J.S.J.; Mahfouz, M.E.; Slootweg, P.J.; Thompson, L.D.R.; Cardesa, A.; Devaney, K.O.; Gnepp, D.R.; Westra, W.H.; et al. Molecular diagnostic alterations in squamous cell carcinoma of the head and neck and potential diagnostic applications. *Eur. Arch. Otorhinolaryngol.* **2014**, *271*, 211–223. [CrossRef]
46. Clauditz, T.S.; Gontarewicz, A.; Lebok, P.; Tsourlakis, M.-C.; Grob, T.J.; Munscher, A.; Sauter, G.; Bokemeyer, C.; Knecht, R.; Wilczak, W. Epidermal growth factor receptor (EGFR) in salivary gland carcinomas: Potentials as therapeutic target. *Oral Oncol.* **2012**, *48*, 991–996. [CrossRef]
47. Clauditz, T.S.; Reiff, M.; Gravert, L.; Gnoss, A.; Tsourlakis, M.-C.; Munscher, A.; Sauter, G.; Bokemeyer, C.; Knecht, R.; Wilczak, W. Human epidermal growth factor receptor 2 (HER2) in salivary gland carcinomas. *Pathology* **2011**, *43*, 459–464. [CrossRef]
48. Goffin, J.R.; Zbuk, K. Epidermal growth factor receptor: Pathway, therapies, and pipeline. *Clin. Ther.* **2013**, *35*, 1282–1303. [CrossRef]
49. Ang, K.K.; Berkey, B.A.; Tu, X.; Zhang, H.-Z.; Katz, R.; Hammond, E.H.; Fu, K.K.; Milas, L. Impact of epidermal growth factor receptor expression on survival and pattern of relapse in patients with advanced head and neck carcinoma. *Cancer Res.* **2002**, *62*, 7350–7356.
50. Mjelle, R.; Hegre, S.A.; Aas, P.A.; Slupphaug, G.; Drabløs, F.; Saetrom, P.; Krokan, H.E. Cell cycle regulation of human DNA repair and chromatin remodelling genes. *DNA Repair (Amst)* **2015**, *30*, 53–67. [CrossRef]

51. Henikoff, S. Versatile assembler. *Nature* **2003**, *423*, 814–815. [CrossRef]
52. Polo, S.E.; Theocharis, S.E.; Klijanienko, J.; Savignoni, A.; Asselain, B.; Vielh, P.; Almouzni, G. Chromatin assembly factor-1, a marker of clinical value to distinguish quiescent from proliferating cells. *Cancer Res.* **2004**, *64*, 2371–2381. [CrossRef]
53. Renan, M.J. How many mutations are required for tumourigenesis? Implications from human cancer data. *Mol. Carcinog.* **1993**, *7*, 139–146.
54. Mascolo, M.; Ilardi, G.; Merolla, F.; Russo, D.; Vecchione, M.L.; de Rosa, G.; Staibano, S. Tissue microarray-based evaluation of Chromatin Assembly Factor-1 (CAF-1)/p60 as tumour prognostic marker. *Int. J. Mol. Sci.* **2013**, *13*, 11044–11062. [CrossRef]
55. Lord, C.J.; Ashworth, A. BRCAness revisited. *Nat. Rev. Cancer* **2016**, *16*, 110–120. [CrossRef]
56. Cerrato, A.; Merolla, F.; Morra, F.; Celetti, A. CCDC6: The identity of a protein known to be partner in fusion. *Int. J. Cancer* **2018**, *142*, 1300–1308. [CrossRef]
57. Leone, V.; Langella, C.; Esposito, F.; Arra, C.; Palma, G.; Rea, D.; Paciello, O.; Merolla, F.; De Biase, D.; Papparella, S.; et al. Ccdc6 knock-in mice develop thyroid hyperplasia associated to an enhanced CREB1 activity. *Oncotarget* **2015**, *6*, 15628–15638. [CrossRef]
58. Lord, C.J.; Tutt, A.N.J.; Ashworth, A. Synthetic lethality and cancer therapy: Lessons learned from the development of PARP inhibitors. *Annu. Rev. Med.* **2015**, *66*, 455–470. [CrossRef]
59. Wurster, S.; Hennes, F.; Parplys, A.C.; Seelbach, J.I.; Mansour, W.Y.; Zielinski, A.; Petersen, C.; Clauditz, T.S.; Münscher, A.; Friedl, A.A.; et al. PARP1 inhibition radiosensitizes HNSCC cells deficient in homologous recombination by disabling the DNA replication fork elongation response. *Oncotarget* **2016**, *7*, 9732–9734. [CrossRef]
60. Nickson, C.M.; Moori, P.; Carter, R.J.; Rubbi, C.P.; Parsons, J.L. Misregulation of DNA damage repair pathways in HPV-positive head and neck squamous cell carcinoma contributes to cellular radiosensitivity. *Oncotarget* **2017**, *8*, 29963–29975. [CrossRef]
61. Mascitti, M.; Rubini, C.; De Michele, F.; Balercia, P.; Girotto, R.; Troiano, G.; Lo Muzio, L.; Santarelli, A. American Joint Committee on Cancer staging system 7th edition versus 8th edition: Any improvement for patients with squamous cell carcinoma of the tongue? *Oral Surg. Oral Med. Oral Pathol. Oral Radiol.* **2018**, *126*, 415–423. [CrossRef]
62. Laemli, U.K. Cleavage of structural proteins during the assembly of the head of bacteriophage T4. *Nature* **1970**, *227*, 680–685. [CrossRef]
63. Towbin, H.; Staehelin, T.; Gordon, J. Electrophoretic transfer of proteins from polyacrylamide gels to nitrocellulose sheets: Procedure and some applications. *Proc. Natl. Acad. Sci. USA* **1979**, *76*, 4350–4354. [CrossRef] [PubMed]
64. Chou, T.C. Theoretical basis, experimental design, and computerized simulation of synergism and antagonism in drug combination studies. *Pharmacol. Rev.* **2006**, *58*, 621–681. [CrossRef] [PubMed]
65. Kossatz, S.; Weber, W.A.; Reiner, T. Optical Imaging of PARP1 in Response to Radiation in Oral Squamous Cell Carcinoma. *PLoS ONE* **2016**, *11*, e0147752. [CrossRef] [PubMed]
66. Schneider, C.A.; Rasband, W.S.; Eliceiri, K.W. NIH Image to ImageJ: 25 years of image analysis. *Nat. Methods* **2012**, *9*, 671–675. [CrossRef]
67. Russo, D.; Merolla, F.; Mascolo, M.; Ilardi, G.; Romano, S.; Varricchio, S.; Napolitano, V.; Celetti, A.; Postiglione, L.; Di Lorenzo, P.P.; et al. FKBP51 Immunohistochemical Expression: A New Prognostic Biomarker for OSCC? *Int. J. Mol. Sci.* **2017**, *18*, 443. [CrossRef]
68. Mascolo, M.; Ayala, F.; Ilardi, G.; Balato, A.; Lembo, S. Chromatin Assembly Factor-1/p60 overexpression: A potential index of psoriasis severity. *Eur. J. Dermatol.* **2014**, *24*, 509–511. [CrossRef]
69. Bankhead, P.; Loughrey, M.B.; Fernández, J.A.; Dombrowski, Y.; McArt, D.G.; Dunne, P.D.; McQuaid, S.; Gray, R.T.; Murray, L.J.; Coleman, H.G.; et al. QuPath: Open source software for digital pathology image analysis. *Sci. Rep.* **2017**, *7*, 16878. [CrossRef]
70. Morra, F.; Merolla, F.; D'Abbiero, D.; Ilardi, G.; Campione, S.; Monaco, R.; Guggino, G.; Ambrosio, F.; Staibano, S.; Cerrato, A.; et al. Analysis of CCDC6 as a novel biomarker for the clinical use of PARP1 inhibitors in malignant pleural mesothelioma. *Lung Cancer* **2019**, *135*, 56–65. [CrossRef]
71. Morra, F.; Merolla, F.; Criscuolo, D.; Insabato, L.; Giannella, R.; Ilardi, G.; Cerrato, A.; Visconti, R.; Staibano, S.; Celetti, A. CCDC6 and USP7 expression levels suggest novel treatment options in high-grade urothelial bladder cancer. *J. Exp. Clin. Cancer Res.* **2019**, *38*, 90. [CrossRef]

72. Jasin, M. Genetic manipulation of genomes with rare-cutting endonucleases. *Trends Genet.* **1996**, *12*, 224–228. [CrossRef]

cancers

MDPI

Article

PARP1 Co-Regulates EP300–BRG1-Dependent Transcription of Genes Involved in Breast Cancer Cell Proliferation and DNA Repair

Maciej Sobczak [1], Andrew R. Pitt [2], Corinne M. Spickett [2] and Agnieszka Robaszkiewicz [1,*]

[1] Department of General Biophysics, Institute of Biophysics, Faculty of Biology and Environmental Protection, University of Lodz, Pomorska 141/143, 90-236 Lodz, Poland; maciej.sobczak@unilodz.eu

[2] School of Life & Health Sciences, Aston University, Aston Triangle, Birmingham B4 7ET, UK; a.r.pitt@aston.ac.uk (A.R.P.); c.m.spickett@aston.ac.uk (C.M.S.)

* Correspondence: agnieszka.robaszkiewicz@biol.uni.lodz.pl; Tel.: +48-42-635-4449; Fax: +48-42-635-4449 or +48-42-635-4473

Received: 17 August 2019; Accepted: 5 October 2019; Published: 11 October 2019

Abstract: BRG1, an active subunit of the SWI/SNF chromatin-remodeling complex, enables the EP300-dependent transcription of proliferation and DNA repair genes from their E2F/CpG-driven promoters in breast cancer cells. In the current study, we show that BRG1–EP300 complexes are accompanied by poly-ADP-ribose polymerase 1 (PARP1), which emerges as the functional component of the promoter-bound multiprotein units that are capable of controlling gene expression. This enzyme is co-distributed with BRG1 at highly acetylated promoters of genes such as *CDK4, LIG1,* or *NEIL3*, which are responsible for cancer cell growth and the removal of DNA damage. ADP-ribosylation is necessary to maintain active transcription, since it ensures an open chromatin structure that allows high acetylation and low histone density. PARP1-mediated modification of BRG1 and EP300 does not affect the association of enzymes with gene promoters; however, it does activate EP300, which acetylates nucleosomes, leading to their eviction by BRG1, thus allowing mRNA synthesis. Although PARP1 was found at BRG1 positive/H3K27ac negative promoters of highly expressed genes in a transformed breast cancer cell line, its transcriptional activity was limited to genes simultaneously controlled by BRG1 and EP300, indicating that the ADP-ribosylation of EP300 plays a dominant role in the regulation of BRG1–EP300-driven transcription. In conclusion, PARP1 directs the transcription of some proliferation and DNA repair genes in breast cancer cells by the ADP-ribosylation of EP300, thereby causing its activation and marking nucleosomes for displacement by BRG1. PARP1 in rapidly dividing cells facilitates the expression of genes that confer a cancer cell phenotype. Our study shows a new mechanism that links PARP1 with the removal of DNA damage in breast cancer cells via the regulation of BRG1–EP300-dependent transcription of genes involved in DNA repair pathways.

Keywords: poly-ADP-ribose polymerase 1 (PARP1); brahma-related gene 1 (BRG1); histone acetyltransferase p300 (EP300); gene transcription; cancer cell

1. Introduction

The pharmacological effects of inhibitors of poly-ADP-ribose polymerases (PARPs) in anticancer therapies are attributed to impairing DNA damage removal, since PARP1 plays a crucial role in the recruitment of repair machinery, mainly in an ADP-ribosylation-dependent manner [1,2]. Lesion recognition followed by the recruitment of poly-ADP-ribose polymerase 1 (PARP1) to sites of DNA damage and ADP-ribosylation of its automodification domain are prerequisites for the binding of XRCC1, POLB, LIG3, or ALC1, which are involved in base excision repair (BER), single-strand break repair (SSBR), and nucleotide excision repair (NER) [3,4]. However, PARP1 also facilitates alternative and conventional non-homologous end-joining (NHEJ) as well as homologous recombination (HR), therefore helping to protect genome integrity and preventing destabilization of the genome resulting from double-strand breaks [5–7]. According to the "access–repair–restore" model, nucleic acid repair is preceded by chromatin reorganization, since DNA lesions are curtained by DNA-associated proteins [8]. Thus, local chromatin rearrangements are required to allow the assembly of the multiprotein machinery that removes lesions. Recent findings identified a link between PARP1 activity, nucleosome density, and the efficiency of some repair pathways, including HR [9]. In the study referred to, auto-ADP-ribosylated PARP1 serves as an indispensable anchor that provides a platform at the damage site for the functional interaction between the nucleosome-evicting brahma-related gene 1 (BRG1; the SWI/SNF chromatin-remodeling enzyme) and the NAD-dependent deacetylase sirtuin-1 (SIRT1), which activates BRG1 by erasing lysine acetylation, thus promoting DNA end-joining. BRG1 interacts with the poly-ADP-ribose polymer through its ATPase domain rather than the N- or C-terminal tails, and is recruited to genomic regions enriched in PARP1. A previous paper reported the interaction of BRG1 with PARP1 and other histone-remodeling enzymes at the genomic level, where nucleosome-evicting ATPase cooperated with PARP1 and histone deacetylases (HDACs: HDAC2 and HDAC9) at the promoters of *α-MHC* and *β-MHC*, thereby preventing cardiac differentiation and maintaining the proliferation potential of the precursors [10]. However, the molecular mechanism that drives PARP1/BRG1-dependent up- or down-regulation of gene transcription has not yet been identified. The suggestion that PARP1 enables the binding of EP300 to the promoters of cell cycle-dependent genes in proliferating cells in an ADP-ribosylation-independent manner focused our attention on the possible role of PAR-synthesizing enzymes in the transcriptional regulation of genes controlled by BRG–EP300–HDAC1 complexes [11]. Our recent discoveries regarding the above-mentioned chromatin-remodeling units showed that these enzymes control the transcription of proliferation and DNA repair genes in two considerably different breast cancer cell lines, MCF7 and MDA-MB-231, which differ in terms of their expressed hormone and HER2 receptors [12]. The nucleosomes of E2F/CpG-driven promoters in the two studied gene groups were acetylated by histone acetyltransferase p300 (EP300), causing them to be marked for BRG1-mediated eviction, and enabling paused RNA polymerase II to become active, leading to active gene transcription. This mechanism operates only in proliferating cells, which are also characterized by a high abundance of PARP1. This is because the *PARP1* promoter is controlled by BRG1–EP300–HDAC1 complexes and is repressed with respect to the growth arrest seen in the great majority of normal primary cells [11].

Based on these premises, we aimed to discover whether PARP1 co-regulates BRG1–EP300-dependent transcription, and if PARP1 can be considered an active component of such multiprotein complexes in the studied breast cancer cell lines. We also aimed to uncover the molecular mechanism that links PARP1 with BRG1-dependent transcription and verify possible PARP1 selectivity toward functionally related genes.

2. Results

2.1. PARP1 Physically Interacts with SWI/SNF in Breast Cancer Cells

Data from three biological replicates run in duplicate for the PARP immunoprecipitates (IP) and two biological replicates in duplicate for the control IP were analyzed. A total of 76 interacting proteins were identified that fulfilled the selection criteria (confidence scores >50, fold change >2 and p-values < 0.05; the full list of PARP1-interacting proteins and associated data are shown in Table S1). Since PARP1 has been previously identified as a cofactor of the transcriptional machinery that cooperates with histone-remodeling enzymes and transcription factors [13–16], we focused on the identification of proteins that physically interacted with PARP1 within the cell nucleus, and initially assessed the interaction data for new, previously unidentified chromatin-associated proteins that could be involved in the regulation of gene expression in a PARP1-dependent fashion. The analysis of PARP1 co-immunoprecipitated proteins by mass spectrometry identified DNA-bound subunits of RNA polymerase and mediator complexes, as well as subunits of chromatin-remodeling complexes, such as Tip60, p400 (EP400), and SWI/SNF (ARID1A, SMARCC1, BRG1; Table S1), which were significantly overrepresented in PARP1 versus IgG pull-downs. Interestingly, brahma (BRM), the other ATPase subunit in SWI/SNF, was not detected among the significant number of interacting proteins identified in PARP1 immunoprecipitates in any of three biological replicates from peptide identification in Mascot. This finding may suggest that PARP1 cooperates only with BRG1-based SWI/SNF complexes. However, to conclude on PARP1–BRM physical and functional interaction, further examination is needed.

Among other histone writers, erasers, and readers, we also found HDAC1, which was recently reported by us to be a constitutive component the BRG1–EP300–HDAC1 complex and assembles at the cell cycle-driven gene promoters of, for example, DNA repair genes in human macrophages and breast cancer cells [11,12]. To verify that formaldehyde fixation of the nuclei did not lead to false positive readouts with mass spectrometry and whether PARP1 is a bona fide member of the SWI/SNF complexes, PARP1 was immunoprecipitated from intact cells, and pull-downs were tested using Western blotting for the presence of SWI/SNF components previously detected by mass spectrometry in fixed nuclei. Western blotting of PARP1 co-immunoprecipitated proteins confirmed the direct interaction of PARP1 with ARID1A and BRG1, but also with other subunits of SWI/SNF, such as SMARCC1 and SMARCC2, in the studied breast cancer cell lines (Figure 1B). Similarly, PARP1 was detected in BRG1 pull-downs (Figure 1C).

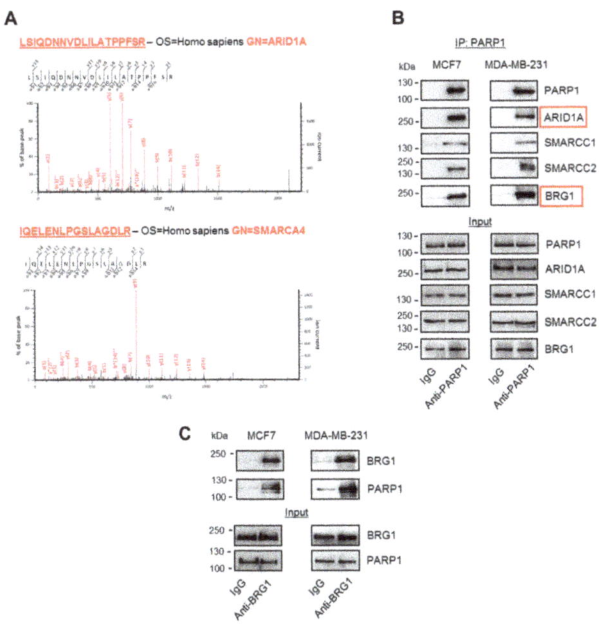

Figure 1. Poly-ADP-ribose polymerase 1 (PARP1) physically interacts with the brahma-related gene 1 (BRG1)-based SWI/SNF chromatin-remodeling complex. (**A**) Exemplar mass spectrometry (MSMS; representative peptides from three biological replicates visualized in Mascot) data for the identification of ARID1 and SMARCA4 as PARP1 interactors using PARP1 immunoenrichment. (**B**) PARP1 interaction with ARID1A, SMARCC1, SMARCC2, and SMARCA4/BRG1 was confirmed by PARP1 pull-down and protein detection by Western blotting in cell lysates of two breast cancer cell lines, MCF7 and MDA-MB-231. IgG served as an isotypic control. (**C**) PARP1 was also identified in BRG1 immunoprecipitates. Western blotting images show representative images of three biological, fully reproducible replicates.

2.2. PARP1 Is Co-Distributed with BRG1 at the Actively Transcribed Gene Promoters

To confirm the possible role of PARP1 in the regulation of SWI/SNF-dependent transcription, we first tested whether PARP1 occurrence in the genome of breast cancer cells was accompanied by BRG1. The residual signal from PARP1 and BRG1 is randomly distributed across genomes, and may reflect the antibody specificity (or lack thereof) or experimental challenges rather than (or in addition to) true protein occurrence. However, the local enrichment of these proteins can be observed, and PARP1 and BRG1-rich regions can be identified by Model-based Analysis of ChIP-Seq data (MACS). As described in the Methods section, for these proteins, we set a p-value cutoff for peak detection at 10-3 with three levels of regions around the peak region (1, 5, and 10 kbp) assessed.

BRG1 was distributed and perfectly centered on genomic regions enriched in PARP1 (Figure 2A). BRG1/PARP1 peaks appeared predominantly at the gene regulatory regions, i.e., at the promoters and enhancers (Figure 2B; definitions of terms are given in Section 4.9), where BRG1 showed a relatively strong correlation with histone modifications, which marked open chromatin and transcriptionally active regions (Figure 2C). Focusing only at the E2F/CpG-driven gene promoters (Table S2), most of the PARP1 peaks (80.2%) were localized at the BRG1/H3K27ac-featured regions adjacent (±2 kbp) to the transcription start site (Figure 2D). This agrees with our and others' reports, where BRG1 has been documented to be associated with promoters of actively transcribed genes, characterized by histone marks, which are permissive for transcription, and the presence of CpG and E2F binding motifs. Our previous findings ascribed BRG1 as a master regulator of some H3K27ac positive promoters, which is essential for the initiation of mRNA synthesis [11,12].

PARP1/BRG1/H3K27ac-positive promoters represent genes that are functionally assigned to numerous processes that are crucial for the maintenance of intracellular homeostasis (Figure 2E; Table S3). To further investigate the molecular mechanisms that underlie the possible functional cross-talk between PARP1 and the BRG1-based SWI/SNF complex, we chose genes attributed to two groups, namely, DNA repair and positive regulation of the cell cycle. These groups were chosen because we recently selected them for an in-depth examination and explained a mutual interdependence between gene expression, BRG1 activity, promoter features, and cell proliferation status [12]. Most genes in the considered groups were over-expressed in at least one of the two fast-proliferating breast cancer cell lines when compared to normal cells derived from primary breast tissue (Figure 2F and Table S4). Particular attention was paid to three of our previously studied genes, *cyclin dependent kinase 4* (*CDK4*), *DNA ligase 1* (*LIG1*), and *nei like DNA glycosylase 3* (*NEIL3*). The transcription of these genes was driven by E2F/CpG/H3K27ac-positive promoters, which were found to be enriched in PARP1 in addition to BRG1 (Figure 2G). The molecular mechanism driving the transcription in MCF7 and MDA-MB-231 cells involved cooperation between EP300, which acetylated histones at the studied gene promoters, and BRG1, which evicted marked nucleosomes in proliferating cells [11,12].

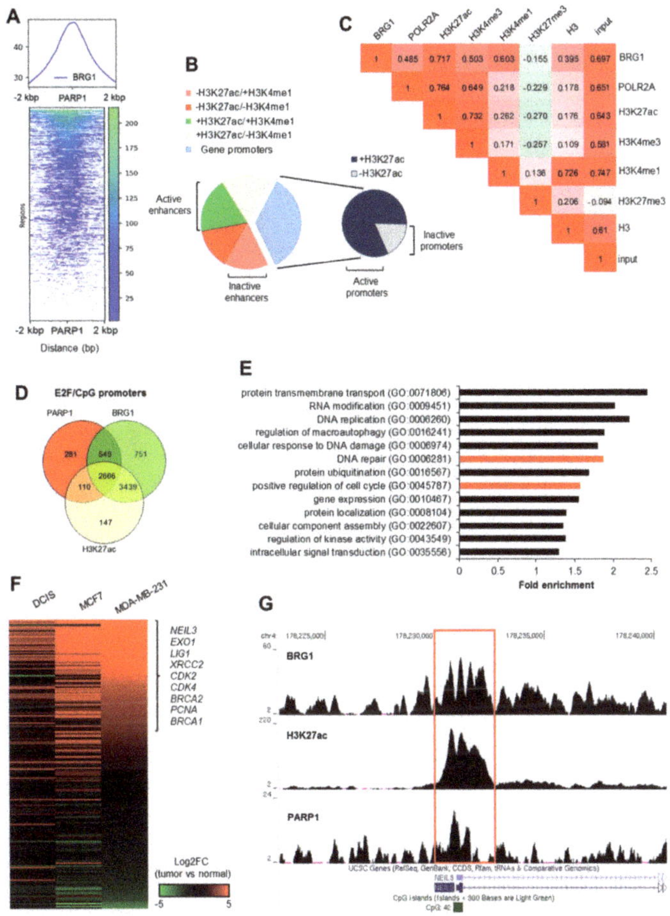

Figure 2. PARP1 co-occurs with BRG1 at the promoters of transcriptionally active genes in MDA-MB-231 cells. (**A**) An example computeMatrix/plotProfile plot of ChIP-Seq data for BRG1. BRG1 peaks are centered at PARP1-enriched regions in the genome of the MDA-MB-231 cell. (**B**) Association of PARP1/BRG1 peaks with gene regulatory regions shows a prevailing occurrence at the active gene promoters. Enhancers emerged based on H3K27ac/H3K4me1 status, whereas promoters were assumed as TSS ± 2 kbp. PARP1/BRG1-enriched regions were assigned to particular groups and quantified by bedtools Intersect intervals. (**C**) BRG1 distribution at the PARP1 peaks reveals a relatively strong correlation with histone markers and with POLR2A, which is typical for actively transcribed genes (multiBamSummary/plotCorrelation; PARP1 peaks in bed for scoring). (**D**) Venn diagram showing that a high proportion of PARP1 positive gene promoters are characterized by the presence of BRG1 and strong acetylation of H3K27. (**E**) Triple-positive gene promoters identified in (**D**) represent a functional association with numerous intracellular processes. GO-enriched terms were identified in AmiGO2. (**F**) Differential expression of genes from two selected GOs, i.e., DNA repair and positive regulation of the cell cycle (marked in red in (**E**)) in cancer cells (DCIS – ductal carcinoma in situ, pre-invasive malignancy of the breast and two breast cancer cell lines: MCF7 and MDA-MB-231) versus normal cells, as quantified by Cuffdiff. The heatmap shows Log2 of the calculated fold change (Log2FC; cancer versus normal cells). (**G**) UCSC Genome browser visualization of PARP1, BRG1, and H3K27ac enrichment using Model-based Analysis of ChIP-Seq data (MACS) (bigwig) at the NEIL3 promoter.

2.3. *PARP1 Conditions Transcriptionally Permissive Chromatin Structure at BRG1/EP300-Dependent Genes*

To verify the PARP1 contribution to the transcription of genes that are concomitantly controlled by BRG1–EP300 complexes (described by Sobczak et al. [12]), we targeted PARP1 with siRNA in both studied breast cancer cell lines and measured the mRNA levels of the selected genes representing two gene ontologies (Figure 3A). PARP1 silencing resulted in considerable suppression of most genes, with *XRCC2* being the only exception that responded to PARP1 deficiency, but not to inhibition with olaparib (iPARP, pan-PARP inhibitor) with increased transcription in MCF7 cells. Furthermore, to check whether the observed PARP1 impact on gene transcription required enzymatic activity, cells were treated with olaparib, a PARP inhibitor. Loss of this enzymatic activity phenocopied PARP1 protein deficiency for *CDK4*, *BRCA1*, *LIG1*, and *NEIL3*, indicating that poly-ADP-ribosylation plays a key role in maintaining a high transcription rate of the considered genes (Figure 3A,B, Table S5). Enzyme inhibition led to a dramatic decline in the cellular abundance of CDK4, LIG1, and NEIL3 proteins (Figure 3C). PARP1 deficiency repressed the transcription of the three studied genes: *CDK4*, *LIG1*, and *NEIL3* comparably to iSWI/SNF and iEP300 (no synergistic effect was observed according to Supplementary Figure S1 and Table S5; only *LIG1* in MDA-MB-231 cells responded with enhanced gene repression after the treatment of siPARP1 transfected cells with iEP300 and iSWI/SNF), suggesting that this enzyme operates with the same, previously studied regulatory mechanism that utilizes the activity of BRG1 and EP300 at the three gene promoters considered [11,12], and that PARP1 may positively affect at least one of the chromatin-remodeling enzymes. This set of data suggests that PARP1 may also operate independently of EP300 and BRG1 (e.g., as a repressor of *XRCC2* in MCF7 cells).

Since EP300 and BRG1 drive gene transcription by respectively acetylating and displacing histones, to allow assembly of the transcriptional machinery, we focused on nucleosome acetylation status and density as possible readouts of PARP1 activity to identify the molecular basis of the observed effect of poly-ADP-ribosylation on BRG1–EP300-dependent gene expression. PARP inhibition with olaparib led to a substantial loss of histone acetylation and was associated with an increase in histone density (Figure 3D; H3 enrichment and status of H3K27ac for each of the studied promoters can be found in Table S5: sheet: LIG1, NEIL3, CDK4 ChIP)); the *XRCC1* promoter was used as a negative control since it lacks PARP1 (Figure S2; Table S5: sheet: XRCC1 ChIP). This finding confirmed that ADP-ribosylation impacts BRG1–EP300 complexes in rapidly proliferating cells and defines the output of the considered chromatin-remodeling functional unit.

Knowing that BRG1 and EP300 co-occur at the studied gene promoters with HDAC1, the observed PARP1 effect on histone acetylation and gene transcription may result from PARP1 interaction with either of the two enzymes, since the subtle balance between acetylase and deacetylase activity determines the BRG1-dependent chromatin structure [11,12]. Thus, we tested whether poly-ADP-ribosylation inhibited HDAC1 activity at the gene promoters by comparing gene expression in the presence of HDAC and PARP inhibitors (Figure 3E). First, HDAC1 did not reduce the transcription of any of the three genes, and second, cell treatment with a mixture of both inhibitors suppressed *CDK4*, *LIG1*, and NEIL3 in a similar way to iPARP only (PARP and HDAC activities had no synergistic impact on the gene expression; Table S5). This indicates that olaparib does not inhibit HDAC1 (or any other histone deacetylase, since we used a pan-HDAC inhibitor) activity from poly-ADP-ribosylation-dependent inhibition at the studied gene promoters.

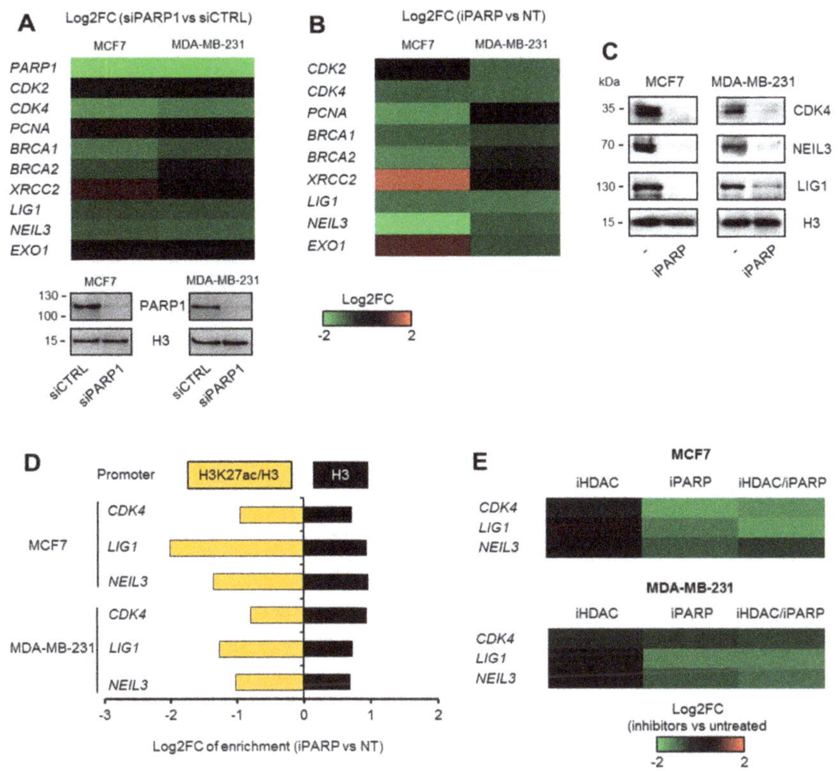

Figure 3. ADP-ribosylation confers open chromatin structure at the gene promoters. (**A**) PARP1 silencing leads to the suppression of most genes in MCF7 and MDA-MB-231 cells that feature PARP1/BRG1/H3K27ac-positive promoters. mRNA was compared 48 h after cell transfection with siCTRL and siPARP1. Log2 of the calculated fold change (Log2FC) shows gene expression in cells treated with inhibitors and normalized to untreated cells. The silencing of PARP1 was confirmed by Western blotting (below heatmap), and H3 was used as a loading control. A similar effect was observed upon PARP1 inhibition with olaparib (iPARP; 48 h) at both the mRNA (**B**) and protein level. (**C**) Representative pictures of protein detection by Western blotting. (**D**) Analysis of structure of selected PARP1-dependent gene promoters revealed a considerable loss of histone acetylation, but increased nucleosome density upon PARP1 inhibition for 24 h. Quantification was carried out by ChIP-qPCR, and data for specific antibodies were normalized first to 10% of the corresponding input and then to untreated control cells. (**E**) The iPARP effect on gene transcription with HDAC activity deficiency (cells were treated with both inhibitors for 48 h) was studied by real-time PCR. Results are presented as Log2 of the calculated fold change (inhibitor versus untreated; Log2FC).

2.4. Poly-ADP-Ribosylation of EP300 Drives BRG1–EP300-Dependent Gene Transcription

Identified PARP1 interaction with BRG1 frequently occurred at highly acetylated gene promoters (Figure 2B). The fact that at least some of them were previously confirmed to be enriched in BRG1–EP300 functional complexes [12] prompted us to check if EP300 interacted with PARP1 and if any of BRG1–EP300 components could undergo ADP-ribosylation in proliferating breast cancer cells.

Analysis of PARP1 pull-downs by Western blotting confirmed the physical interaction between PARP1 and EP300 (Figure 4A). Poly-ADP-ribose chains were detected in both immunoprecipitated enzymes, i.e., BRG1 and histone acetyltransferase, in the studied cancer lines (Figure 4B,C), thereby providing further evidence for a possible PARP1 role in the regulation of transcriptional activity of BRG1–EP300 complexes. Bearing in mind that the dependence of gene transcription on BRG1 and EP300 is conditioned by the association of enzymes with their gene promoters and then by the catalytic activity of chromatin-bound enzymes, we tested whether poly-ADP-ribosylation affected BRG1 and EP300 levels at the investigated promoters (Figure 4D). None of the studied genomic regions responded to PARP1 inhibition with substantial displacement or recruitment of chromatin remodeling enzymes, suggesting that poly-ADP-ribosylation determines the activity of enzymes rather than their occurrence at the gene promoters (Figure 4D, Table S5). The poly-ADP-ribosylation of EP300 enabled acetyltransferase activity that led to intensive nucleosome acetylation and eviction by BRG1, since cell treatment with a PARP inhibitor resulted in a dramatic loss of EP300-dependent acetylation of the studied gene promoters (Figure 3D, Table S5), without an apparent effect on the association of EP300 with the gene promoters and the HDAC1 role in gene transcription (Figures 4D and 3E).

To check if poly-ADP-ribosylation of BRG1 directly conditioned BRG1 activity and BRG1-dependent transcription, we tested the impact of the PARP1 inhibitor on the transcription of genes that are over-expressed in cancer cells and characterized by the occurrence of PARP1 and BRG1 at their promoters, but without considerable nucleosome acetylation (Figure 4D and Table S2). *IL1RL1* served as an example of repressed genes in MCF7 cells. Surprisingly, all of the genes found to be over-expressed that were considered in this experiment responded to SWI/SNF inhibition and silencing with increased transcription (Figure 4E, Table S5), suggesting that EP300 co-distribution with BRG1 might be a hallmark of gene promoters characterized by the pro-transcriptional activity of BRG1-based SWI/SNF complexes. However, this hypothesis requires further examination of a wider range of genes, especially because the considerable inhibitory role of BRG1 on gene transcription was observed in only one cell line. This finding also stresses the differences in gene transcription control in the two chosen cell lines. In any case, PARP1 was not involved in the transcriptional regulation of genes suppressed by the SWI/SNF complex; only one repressive effect was found for *RAD1* in MDA-MB-231 cells. Together, these data indicate that PARP1 co-regulates activity of promoter-bound BRG1–EP300 complexes, and that poly-ADP-ribosylation of EP300 is required to enable the BRG1-dependent eviction of acetylated nucleosome, and therefore the transcription of genes involved in key intracellular processes, such as cell division and the removal of DNA damage in breast cancer cells. This molecular mechanism of PARP1 action in BRG1–EP300 complexes is further supported by our previous findings, in which BRG1 emerged as a reader of nucleosome acetylation [11,12]. Thus, a low histone acetylation caused by PARP inhibition prevents histone eviction by BRG1.

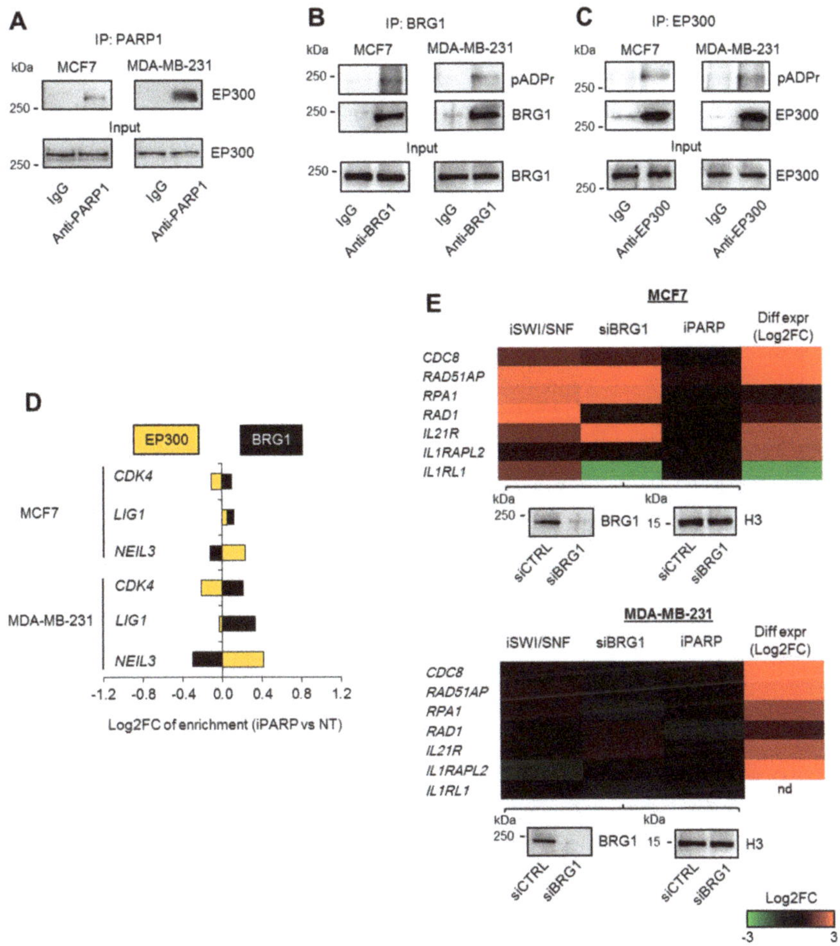

Figure 4. ADP-ribosylation of EP300 drives BRG1-dependent gene transcription. (**A**) The PARP1 interaction with EP300 was confirmed by visualization of EP300 by Western blot in PARP1 immunoprecipitates. (**B**) ADP-ribosylation of BRG1 was detected in BRG1 pull-downs by Western blot. (**C**) ADP-ribosylation of EP300 was detected as in (**B**) in EP300 pull-downs. (**D**) The effect of ADP-ribosylation on EP300 and BRG1 association with gene promoters in breast cancer cells was studied by ChIP-qPCR. Cells were supplemented with iPARP for 24 h prior to cell fixation. Data for anti-EP300 and anti-BRG1 were normalized to 10% input and then to untreated cells (Log2 enrichment vs. control cells). (**E**) The lack of contribution of ADP-ribosylation to the transcription of genes characterized by BRG1/PARP1 (no EP300) positive promoters was verified by comparing mRNA (real-time PCR) between the studied groups. For inhibitors, Log2FC was normalized to untreated cells, while data for BRG1 silencing were normalized to cells transfected with control siRNA. The efficiency of BRG1 silencing is confirmed by Western blot (below heatmaps). The column with differential expression shows gene transcription in MCF7 and MDA-MB-231 cell lines normalized to normal cells (data from RNA-Seq).

3. Discussion

Malignant transformation directs cellular changes and adaptive responses toward new requirements that cancer cells face during growth and metastasis. As long as the activation of oncogenes, loss of cell cycle checkpoint control, and impaired DNA repair capacity favor carcinogenesis, genome integrity and cell cycle re-entrance will eventually become threatened due to increased energy demand, mild but persistent oxidative stress, and the modulation of signaling cascades necessary for tumor growth and the invasion of other tissues [17,18]. The altered expression of many gene products in response to cell transformation is affected by reprogramming the epigenome, resulting in changes in the transcription and reconstruction of the cellular proteome to meet emerging needs [19]. According to our findings, the activation of *PARP1* transcription as a consequence of the transition of a cell from quiescence to proliferation may help the cancer cell gain the necessary adaptive physiology by acting at the genomic level in two ways: directly, by contributing to DNA repair machineries, and indirectly, by affecting the transcription of BRG1–EP300 targets, among others, which enable the cancer cell to rapidly divide and resist DNA damaging agents [12,20–22]. Although the first aspects of this mechanism have been relatively deeply explored and the regulatory roles of PARP1 regarding the removal of DNA lesions and the transition between consecutive cell cycle phases in various modes—for example, the modulation of SP1 activity, H1 displacement at proliferation-relevant genes, and nuclear retention of PKM2—have been documented, the detailed molecular mechanism in regard to functionally linked genes has not been identified until now [23–25]. In this study, PARP1 was shown to be an active component of the transcription machinery that drives BRG1-EP300-dependent gene expression by the poly-ADP-ribosylation of EP300 in breast cancer cells. PARP1 was highly enriched at gene promoters characterized by the occurrence of not only BRG1 and EP300, but also E2F binding motif(s) and CpG islands. Since these features apply to many functionally linked genes, the role of PARP1 in defining the breast cancer phenotype at the transcription level is likely to go far beyond cell cycle progression and the removal of DNA damage. As shown in Figure 2E, triple-positive PARP1/BRG1/H3K27ac promoters represent the genes responsible for signal transduction and autophagy. Once poly-ADP-ribosylation is proved to be a co-activator of these genes, PARP inhibitors may be important to consider for pharmacological interventions that target and suppress mediators of pro-survival cascades at the genomic level. The role of poly-ADP-ribosylation in the fine-tuning of numerous intracellular processes simultaneously allows the maximization of the effectiveness of PARP1 inhibitors in rendering cancer cells vulnerable to anticancer drugs, which challenge PARP1-dependent or concurrent intracellular routes. Bearing in mind that the BRG1/EP300 complex was shown to operate at gene promoters in proliferating breast cancer cells and human macrophages due to an association with E2F transcription factors [12,26], the same PARP1-dependent mechanism of transcription control likely applies to other tumors, since the cell cycle status conditions both PARP1 expression and the activity of BRG1–EP300 complexes. Similarly, the energy status of proliferating cells demands a high ATP concentration; thus, the NAD^+/NADH redox ratio determines, for example, that PARP1 activity is five times higher in cancer cells than in normal, non-transformed cells [27]. Therefore, it might be possible to adapt PARP1 inhibitors for the modulation of intracellular processes in a wider range of cancers.

This study revealed a new mechanism that defines cancer cell responses to DNA lesions in a poly-ADP-ribosylation-dependent manner. Although the direct contribution of PARP1 to pathway repair at sites of damage is well acknowledged [28,29], we showed for the first time an impact on the repair mechanisms that is distant, gene promoter-related, and independent of lesion location. In this context, PARP inhibitors could be used to suppress the transcription of genes characterized by promoters enriched in PARP1, BRG1, and EP300, which represent executors that are crucial for BER, SSBR, NER, MMR, HR, and alt-NHEJ, since products of only the considered genes, i.e., *BRCA1/2*, *LIG1*, and *NEIL3*, contribute to more than one repair pathway [30,31]. Another benefit of using PARP inhibitors as DNA repair modulators acting at the level of the epigenome comes from the observation that poly-ADP-ribosylation is a co-activator of cell cycle-dependent genes that are simultaneously controlled by BRG1 and EP300, and are mostly over-expressed in the studied breast cancer cells, thus providing some selectivity toward this group. PARP1 was also found at gene promoters occupied by BRG1 alone, but cell treatment with olaparib did not reveal a considerable impact on the expression of these genes. Notably, the PARP1/BRG1 promoter response to an SWI/SNF inhibitor that is capable of impeding the activity of both ATPases (Figure 4D) suggested that (a) EP300 has a co-activating role with BRG1 at the

BRG1/EP300 promoters, and/or (b) the role of particular ATPases may be determined by EP300, and the lack of acetyltransferase may switch BRG1 to BRM activity. Although the scope of iSWI/SNFs is broader than that of iPARPs, the unexpected reaction of the chromatin to the loss of BRG1 and BRM activity may result in a coinciding repression and up-regulation of genes assigned to one regulatory circuit. Discrete inhibitors of the two ATPases mentioned, as well as further study on the mutual interdependence between these two enzymes, are needed for future clinical applications. Similarly, inhibitors of EP300 modulate the transcription of a significant number of genes at different levels; at the genome level by preventing acetylation of transcription co-factors and nucleosomes, and at the signaling cascade level by modifying signal transducers [32,33]. This all results in even less specificity toward the desired gene pool, and underlines the importance of the possible applications of PARP1 inhibitors. Some PARP-1 and PARP-2 inhibitors, such as olaparib, niraparib, rucaparib, veliparib, and talazoparib, which are small-molecule NAD^+ mimetics, are currently being studied in later-stage clinical trials or are already approved for breast and ovarian cancer treatment with deleterious germline *BRCA1* and *BRCA2* mutations, which predispose women to develop triple-negative and hormone-receptor-positive, human epidermal growth factor receptor 2-negative breast cancers, respectively [34,35]. Since PARP inhibitor monotherapy strategies are effective in cancers with homologous recombination repair defects and are relatively well-tolerated by patients, they can be considered for the treatment of a wider range of cancers, both in combined therapies, due to the well-established fact that these drugs sensitize cells to DNA-damaging chemotherapy and radiation therapy, or as an alternative to taxanes and a supplement to anthracyclines [36–38]. However, numerous phase I clinical trials utilizing a combination of cytotoxic chemotherapy with PARP inhibitors failed to confirm any beneficial effects of such combinations. Therefore, the use of these drugs in adjuvant or neoadjuvant settings may need substantial revision, while also taking into consideration the myelosuppressive effects of PARP inhibitors and the careful selection of anticancer agents in combination with DNA repair inhibitor(s). Nevertheless, the long list of promoters of functionally related genes that are enriched in PARP1 presented in this manuscript suggest the likely involvement of this enzyme regarding the modulation of other intracellular processes at the transcription level. These findings open the gate for new ideas and concepts regarding anticancer approaches, which require verification first in cell and animal models.

The described contribution of PARP1 to the regulation of BRG1–EP300 activity emphasizes the role of PARP1 in chromatin remodeling. Although a number of papers have documented this enzyme as a direct or indirect regulator of chromatin structure in a context-dependent fashion, none have provided an overall mechanism for functionally linked gene sets in particular. Transcription in cells is shaped by the poly-ADP-ribosylation of nucleosomes, histone writers and erasers (KDM5B), transcription factors (e.g., C/EBPβ), or POL2 regulating co-factors (NELF), as well as physical, activity-independent interactions with gene promoters that define chromatin composition (LSD1, EP300) [13–16,39,40]. However, DNA motifs or chromatin signatures, which determine PARP1 distribution in the genome, remain unidentified. According to Gibbson, PARP1 and ADP-ribosylation correlated with histone markers (H3K4me3 and H3K27ac) featuring actively transcribed genes and with POL2 pausing machinery in embryonic fibroblasts of mice [40]. These findings agree with our own, in which PARP1 was enriched at highly acetylated CpG islands, allowing immediate POL2 pausing or release in response to received signals [11,12]. The association of PARP1 with GC-rich regions impedes the identification of the PARP1-specific motif in promoter sequences. Since only 19% of PARP1 peaks in the genome of MDA-MB-231 cells occurred at BRG1 and H3K27ac negative promoters, and less than 3% outside of BRG1 peaks, these two features of promoters, together with E2F-binding motifs and CpG islands, seem to direct the enzyme to its destination regarding chromatin, whereby the poly-ADP-ribosylation of BRG1 and EP300 enables gene expression.

4. Materials and Methods

4.1. Materials

Two epithelial, breast cancer cell lines, derived from metastatic sites, MCF7 (estrogen and progesterone receptors-positive, HER2-negative) and MDA-MB-231 (triple negative) were purchased from ATCC and Sigma Aldrich (Poznan, Poland), respectively. DMEM high glucose with L-glutamine with sodium pyruvate for MCF7, fetal bovine serum, and antibiotics (penicillin and streptomycin) were from Biowest (CytoGen, Zgierz, Poland), L15 Medium for MDA-MB-231, iSWI/SNF (PFI-3), iHDAC (sodium butyrate), anti-rabbit IgG (A0545) and anti-mouse IgG (A4416) (whole molecule)–peroxidase antibody produced in goat, BLUeye prestained protein ladder (#94964), oligonucleotides for real-time PCR, SIGMA*FAST*™ Protease Inhibitor Tablets (PIC) were from Sigma Aldrich (Poznan, Poland). iPARP (olaparib, AZD-2281) was from Cayman Chemical (Biokom, Janki/Warsaw, Poland). Lipofectamine RNAiMAX, OptiMem, Dynabeads™ Protein G, glycogen, High-Capacity cDNA Reverse Transcription Kit, SuperSignal™ West Pico Chemiluminescent Substrate, TRI Reagent™, and RNase A were from Thermofisher Scientific (Thermofisher Scientific, Warsaw, Poland). KAPA HiFi™ HotStart ReadyMix (2×) from KapaBiosystems and Takyon™No ROX SYBR Core Kit blue dTTP from Eurogentec were purchased from Polgen (Lodz, Poland). EvaGreen® Dye, 20X in water was purchased from Biotium (Corporate Place Hayward, Fremont, CA, USA). WB antibodies: anti-DNA Ligase I (sc-271678), anti-CDK4 (sc-23896), anti-NEIL3 (sc-393703), anti-pADPr (sc-56198), siPARP1 (sc-29437), and gallotannin were purchased from Santa Cruz Biotechnology (AMX, Lodz, Poland). ChIP grade antibodies: normal rabbit IgG (#2729), anti-ARID1A (#12354), anti-SMARCC2 (#12760), anti-BRG1 (#49360), anti-H3K27ac (#4353), anti-histone H3 (#4620), and anti-PARP1 (#9532) were purchased from Cell Signaling Technology (LabJOT, Warsaw, Poland). Human Cytokine and Chemokine Receptor Primer Library (HCCR-I) were from RealTime Primers (Prospecta, Warsaw, Poland). For the mass spectrometric analysis, all materials were from Thermo Fisher Scientific (Loughborough, UK) unless otherwise indicated. Porcine Trypsin (Trypsin Gold Mass Spectrometry Grade) was from Promega (Southampton, UK), and general use Protease Inhibitor Cocktail (P2174) was from Sigma-Aldrich (Poole, UK).

4.2. Cell Culture and Treatment with Inhibitors

MCF7 were cultured in DMEM supplemented with 10% FBS, penicillin/streptomycin (50 U/mL and 50 μg/mL, respectively) in 5% CO_2, whereas MDA-MB-231 was cultured in F15 medium supplemented with 15% FBS and penicillin/streptomycin (50 U/mL and 50 μg/mL, respectively) without CO_2 equilibration. After freezing and thawing, cells were cultured in DMEM as described for MCF7 cells. iSWI/SNF (10 μM; PFI-3), iPARP1 (olaparib, 1 μM), and iHDAC1 (sodium butyrate, 250 μM) were added to cells 48 h prior to analysis.

4.3. Quantification of Gene Expression

mRNA quantification was conducted as described in Pietrzak et al. [11] using Takyon™ No Rox SYBR® MasterMix dTTP blue (Eurogentec—from local distributor—Polgen, Lodz, Poland) and CFX96 C1000 Touch (BioRad, Warsaw, Poland) for real-time PCR. The median average of ACTB, GAPDH, and B2M were used for normalization. Data in figures are shown as Log2FC with respect to untreated control or to siCTRL (indicated in figures or figure legends).

For protein detection, cell lysates were processed as previously described and visualized using SuperSignal™ West Pico Chemiluminescent Substrate. Pictures were acquired with ChemiDoc-IT2 (UVP, Meranco, Poznan, Poland).

The following primer sets were used for the quantification of gene expression: *CDK2*, 5′-CAGGATGTGACCAAGCCAGT-3′ (forward) and 5′-TGAGTCCAAATAGCCCAAGG-3′ (reverse); *CDK4*, 5′-CTGGTGTTTGAGCATGTAGACC-3′ (forward) and 5′-AAACTGGCGCATCAGATCCTT-3′ (reverse), *XRCC2*, 5′-TCGCCTGGTTCTTTTTGCA-3′ (forward) and 5′-TCTGATGAGCTCGAGGCTTTC-3′ (reverse), *BRCA2*, 5′-CTTGCCCCTTTCGTCTATTTG-3′ (forward) and 5′-TACGGCCCTGAAGTACAGTCT-3′ (reverse), *LIG1*, 5′-CAGAGGGCGAGTTTGTCTTC-3′ (forward) and 5′-AGCCAGTTGTGCGATCTCTT-3′ (reverse), *EXO1*, 5′-AAACCTGAATGTGGCCGTGT-3′ (forward) and 5′CCTCATTCCCAAACAGGGACT-3′ (reverse), *NEIL3*, 5′-GGTCTCCACCCAGCTGTTAAAG-3′ (forward) and 5′-CACGTATCATTTTCATGAGGTGATG-3′ (reverse), *PCNA*, 5′-TCTGAGGGCTTCGACACCTA-3′ (forward) and 5′- TTCTCCTGGTTTGGTGCTTCA-3′ (reverse); *BRG1*, 5′-AAGAAGACTGAGCCCCGACATTC-3′ (forward) and 5′-CCGTTACTGCTAAGGCCTATGC-3′ (reverse), *BRCA1*, 5′-TGCCCACAGATCAACTGGAA-3′ (forward) and 5′-CACAGGTGCCTCACACATCT-3′ (reverse); *ACTB*, 5′- TGGCACCCAGCACAATGAA-3′ (forward) and 5′-CTAAGTCATAGTCCGCCTAGAAGCA-3′ (reverse); PARP1, 5′-AAGCCCTAAAGGCTCAGAACG-3′ and 5′-ACCATGCCATCAGCTACTCGGT-3′. *GAPDH* and *B2M* were from the Human Toll-like Receptor Signaling Primer Library (HTLR-I, RealTime Primers – from local distributor - Prospecta, Warsaw, Poland)).

4.4. PARP1 Co-Immunoprecipitation for Mass Spectrometry

2×10^7 MCF7 cells were trypsinized, washed 3× with cold PBS, and lysed on ice in hypotonic buffer (50 mM HEPES-KOH, 10 mM NaCl, 1 mM EDTA, 10% glycerol, and 0.5% NP-40, 0.25% TritonX-100). Nuclei were washed twice in PBS supplemented with protease inhibitor cocktail (10× stock added at 10% of volume, Sigma P2714 General Purpose) and fixed in 1% formaldehyde in PBS on stirrer for 10 min at room temperature. Formaldehyde was quenched by the addition of 125 mM oglycine, and after 20 min of incubation, nuclei were washed twice in 10 mM Tri-HCl (pH = 8.0), 100 mM NaCl, and 1 mM EDTA, and finally resuspended in 10 mM Tri-HCl (pH = 8.0), 150 mM NaCl, 1 mM EDTA, 0.05 Na-deoxycholate, and 0.25% N-laurosarcosine. Nuclei were sonicated until the solution was transparent, TritonX-100 was added to 1% and then centrifuged (10,000× g, 4 °C, 10 min) to remove insoluble material. The supernatant was split into two samples, to which were added either control IgG–Dynabead or anti-PARP1–Dynabead conjugates (prepared by the incubation of 5 µg of antibody and 10 µL of magnetic beads in 10 mM Tri-HCl pH 8.0, 150 mM NaCl, 1 mM EDTA, 0.05 Na-deoxycholate, and 0.25% N-laurosarcosine for at least 15 min). After overnight incubation on a roller shaker at 4 °C, the supernatant was removed, and the beads were washed 5× with 50 mM HEPES-KOH (pH = 7.5), 500 mM LiCl, 1 mM EDTA, 1% NP-40, and 0.7% Na-deoxycholate; twice with 50 mM NaCl in TE buffer (10 mM Tris pH = 8.0, 1 mM EDTA); and twice with 50 mM ammonium bicarbonate using a magnetic stand. Then, the beads were incubated with trypsin (Promega Gold, 1 µg/µL) in 50 mM ammonium bicarbonate for 6 h at 36 °C, after which the supernatant was transferred to new tubes and dried in a SpeedVac vacuum concentrator (Eppendorf, Stevenage, UK).

4.5. Mass Spectrometry Analysis of PARP1 Co-Immunoprecipitates

Tryptic digest samples were resuspended in 25 µL of 2% acetonitrile in water and 0.5% formic acid. Peptides were separated and analyzed using a U3000 nanoflow LC system (Thermo) connected to a 5600 Triple ToF mass spectrometer (Sciex, Warrington, UK). Then, 10 µL of sample was loaded onto a 0.5 × 5 mm PepMap C18 trap, washed with buffer A (2% acetonitrile 98% water containing 0.5% formic acid) for 4 min, and then separated by a 90-minute gradient from 2% to 40% buffer B (98% acetonitrile, 2% water containing 0.5% formic acid) on a 0.075 × 150 mm PepMap C18 column (Thermo Fisher Scientific, Loughborough, UK)). MSMS data were collected for precursors of 2^+ to 5^+ charge state in the range m/z 350–1250 Th using a top 10 data dependent acquisition method, collecting MS data for 200 ms and MSMS data for 100 ms, with dynamic exclusion for 15 s and a standard rolling collision energy settings. MSMS data was collected in the range of m/z 50–2000 Th. All other settings were optimized for peptides using a standard mixture. Samples were run as biological replicates. MSMS data was analyzed using Progenesis QIP (Waters, Manchester, UK) for label-free quantitative analysis and Mascot (Matrix Science, London) for protein identification. Data was loaded as .wiff files into Progenesis QIP, automatically aligned, and peak picked using the default settings; then, the alignment was manually improved where necessary. Default settings were used for peak picking, and Hi-N was used for quantification. Data was exported to Mascot using the default settings. For the Mascot analysis, the SwissProt Mammalian database (2018_09) was searched, allowing 50 ppm error for MS and 100 ppm for MSMS data, two missed cleaves, methionine oxidation as a variable modification, and an overall FDR of <1%. Data was re-imported into Progenesis QIP for further quantitative analysis. Protein identifications were deemed significant if more than two peptides were identified with an overall confidence score greater than 50, but more stringent criteria were applied for proteins to be further investigated. Quantification data was considered significant where the ANOVA *p*-value was less than 0.05, the fold change was greater than 2, and the highest mean was in the PARP immunoprecipitation.

4.6. Co-Immunoprecipitation and Western Blot

MCF7 and MDA-MB-2331 cells were lysed in 20 mM Tris-HCl (pH = 7.5), 75 mM KCl, 5 mM $MgCl_2$, 0.2 mM EDTA, 10% glycerol, 0.1% Tween20, and PIC (IP buffer); sonicated with the ultrasonic homogenizer Bandelin Sonopuls (HD 2070; 10 impulses, 60%); and centrifuged (10,000× *g*, 4 °C, 10 min). Supernatant was incubated with anti-PARP1, anti-BRG1, anty-EP300, and corresponding IgG at 4 °C for 2 h. For another 1 h, lysates were added with Dynabeads (5 µL); then, they were washed 5× with the IP buffer and once in 50 mM NaCl in TE buffer (10 mM Tris pH = 8.0, 1 mM EDTA). Beads were suspended in gel loading buffer supplemented with 5% β-mercaptoethanol, and heated at 70 °C for 10 min. Beads were collected on a magnetic stand and supernatant was separated by PAGE. BRG1, ARID1A, SMARCC1, SMARCC2, PARP1, EP300, H3, and poly-ADP-ribose were detected on nitrocellulose membranes after overnight staining with corresponding antibodies. For the detection of poly-ADP-ribosylation, cells were lysed and processed in the presence of a PARG inhibitor: 0.5 mM gallotannin.

Each immunoprecipitation followed by Western blot was repeated in three biological replicates. Each time, the striking difference in protein level being detected was observed between IgG (weak or lack of signal) versus anti-PARP1 (or anti-BRG1; strong and clear bands). Representative images were taken for figures.

4.7. Chromatin Immunoprecipitation

The immunoprecipitation of chromatin-bound proteins and histones was carried out according to the protocol previously described [11]. For the quantification of H3K27 acetylation, cells were lysed and processed in the presence of iHDAC (0.5 mM). Fragments spanning PARP1/BRG1/H3K27ac sites in selected gene promoters were amplified using KAPA HiFi™ HotStart ReadyMix supplemented with EvaGreen® Dye and 4% DMSO. Primers for *CDK4*, *LIG1*, *NEIL3*, and *XRCC1* promoters were as follows: *CDK4* prom, 5′-ATAACCAGCTCGCGAAACGA-3′ and 5′-AGAGCAATGTCAAGCGGTCA-3′, *LIG1* prom, 5′-AACACACTCAGATCCGCCAG-3′ and 5′-GCTTCCACCGATTCCTCCTC-3′, *NEIL3* prom, 5′-GTAGGGAGCGACCTCAACAG-3′ and 5′-AGTACAGCCTGGTCCTTCCA-3′, *XRCC1* prom, 5′-TGGCCAGAAGGATGAGGTAG-3′ (forward) and 5′-AGGAAACGCTCGTTGCTAAG-3′ (reverse).

4.8. Transient Gene Silencing

For PARP1 and BRG1 silencing, MCF7 and MCD-MB-23 were seeded at the density of 100,000 cells per well, transfected on the following day with RNAiMAX-siRNA complexes (3 µL of transfection reagent and 20 nmol siRNA incubated in OptiMem for 20 min). The silencing was confirmed by real-time PCR and Western blot 48 h after cell transfection.

4.9. ChIP-Seq Analysis in Galaxy Version 19.05.dev [41]

The following, publically available, generated by other groups and deposited in the *PubMed* Central Database data from MDA-MB-231 cells were taken for ChIP-Seq analysis:

PARP1—GSM1517306 (SRR1593959), BRG1—GSM1856026 (SRR2171350), GSM1856027 (SRR2171351), and GSM1856028 (SRR2171352), H3K27ac—GSM1855991 (SRR2171311) and GSM1855992 (SRR2171312); H3K4me3—GSM1700392 (SRR2044734), H3K4me1—GSM2036932 (SRR3096750 and SRR3096751), H3K27me3—GSM949581 (SRR513994), H3K9ac—GSM1619765 (SRR1820123 and SRR1820124), H3—GSM2531568 (SRR5332805), POLR2A—GSM2309434 (SRR4240635), and Input—GSM1964894 (SRR2976843). FASTQ quality formats were unified to Sanger formatted with a FASTQ Groomer [26]. Reads were aligned to Human Genome version 19 using a Map with Bowtie for Illumina, and unmapped reads were filtered out. ChIP-seq peaks were called in MACS with a *p*-value cutoff for peak detection set at 10^{-3}. BRG1 co-occurrence at the PARP1 peaks was monitored by computeMatrix/plotProfile (PARP1 peaks in bed were used as regions to plot, while mapped BRG1 reads were used for scoring) [42]. The co-distribution of BRG1, POLR2A, and histone modifications at the PARP1-enriched regions was studied by MulitBamSummary/plotCorrelation (regions of the genome were limited to PARP1 peaks in bed, mapped reads for scoring) [42]. Regions simultaneously enriched in BRG1, PARP1, H3K4me1, and H3K27ac were identified by returning intersects of the peaks in bed using bedtools Intersect intervals [43]. Regions localized outside of gene promoters and double positive for H3K4me1 and H3K27ac were assigned as active enhancers high in H3K4me1 and low in H3K27ac as inactive enhancers, while gene promoters were recognized by returning intersects of BRG1/PARP1 peaks and genomic regions ±2000 bp centered on TSS (overlapping intervals of both datasets). Promoters defined by high H3K27ac and associated with the presence of gene transcripts (RNA-Seq results for MDA-MB-231) downstream of corresponding TSS were assumed as active, while the lack of promoter acetylation marked inactive gene promoters. Genomic intervals for E2F (overlaps of E2F1 and E2F4) and CpG islands were taken from the UCSC main tables wgEncodeRegTfbsClusteredV3 and cpgIslandExt, respectively. Intersects of TSS ±2 kbp and CpG or E2F intervals were compared using bedtools Intersect intervals. The characteristics of gene promoters (occurrence of particular proteins and CpGs, histone modifications, E2F binding sites) were studied using Venn diagrams, which were created in http://www.interactivenn.net/ from gene lists. The annotation of PARP1/BRG1/H3K27ac promoters to biological processes was carried out in AmiGO2 (test type—binominal, correction—FDR) [44]. PARP1, BRG1, and H3K27ac peaks were visualized in the UCSC Genome Browser.

The following, publically available datasets for various breast cells, which have been generated by other groups, were downloaded from the PubMed Central Database data and used for ChIP-Seq analysis: data from normal breast, ductal carcinoma in situ (DCIS; pre-invasive malignancy of the breast), MCF7 and MDA-MB-231 cells were taken for RNA-Seq analysis: normal breast—GSM1695870 (SRR2040339), GSM1695872 (SRR2040341), GSM1695873 (SRR2040342), GSM1695874 (SRR2040343), GSM1695877 (SRR2040346), and GSM1695878 (SRR2040347); DCIS—GSM1695891 (SRR2040360), GSM1695898 (SRR2040367), GSM1695899 (SRR2040368), GSM1695882 (SRR2040351), GSM1695890 (SRR2040359), and GSM1695894 (SRR2040363); MCF7—GSM2422725 (SRR5094305), GSM2422726 (SRR5094306), GSM2422727 (SRR5094307), GSM2422728 (SRR5094308), GSM2422729 (SRR5094309), and GSM2422730 (SRR5094310); MDA-MB-231—GSM2422731 (SRR5094311), GSM2422732 (SRR5094312), GSM2422733 (SRR5094313), GSM2422734 (SRR5094314), GSM2422735 (SRR5094315), and GSM2422736 (SRR5094316). All samples were processed as described in Sobczak et al. [12]. Differential gene expression in cancer versus normal breast cells was calculated with Cuffdiff and shown as a heatmap for two selected GOs (positive regulation of cell cycle and DNA repair) [45].

5. Conclusions

In conclusion, our study describes a new mechanism regarding the regulation of the transcription of functionally linked genes that are significant for cancer cell physiology. Poly-ADP-ribosylation emerges as a chromatin remodeler that is capable of defining the activity of BRG1–EP300 complexes at the promoters of genes encoding cell cycle and DNA repair-promoting proteins. Although the PARP1 inhibitor olaparib emerges as a promising tool to modulate PARP1/BRG1/EP300-dependent gene expression due to its safety and well-established in vivo effects in cancer treatment, the functional impact of DNA repair gene repression in anticancer therapies requires further investigation. In any case, our study provides a basis for the search for new combinations of iPARPs with other compounds to increase the beneficial effects of anticancer approaches.

Supplementary Materials: The following are available online at http://www.mdpi.com/2072-6694/11/10/1539/s1, Table S1. Protein List identified in PARP1 immunoprecipitates by Mass Spectrometry. Table S2. PARP1, BRG1, and H3K27ac distribution at the CpG/E2F positive gene promoters. Table S3. Full list of gene enrichment analysis for PARP1/BRG1/H3K27ac/E2F/CpG positive promoters. Table S4. Differential expression of proliferation and DNA repair genes in DCIS, MCF7 and MDA-MB-231 cells versus normal breast tissue. Table S5. Statistical analysis of data. Figure S1. Effect of PARP1 silencing, SWI/SNF inhibition, and EP300 inhibition on gene expression. Figure S2. Chip-qPCR data for effect of iPARP on the *XRCC1* promoter.

Author Contributions: A.R.: conceptualization, supervision, bioinformatic analysis of DNA-bound proteins, experiments (PARP1 co-immunoprecipitation for mass spectrometry, Figures 2–4), writing; A.R.P. and C.M.S.: optimization and mass spectrometry analysis of PARP1 co-immunoprecipitates, M.S.: bioinformatic analysis of differentiation gene expression and experiments (Figures 1C and 2, Figures 3 and 4). Data shown in Figure 1A and Table S1 were collected at the Aston University in Birmingham, while all other experiments were conducted in the Department of General Biophysics, University of Lodz.

Funding: This research was funded by the Polish National Science Center, grant number DEC-2013/11/D/NZ2/00033" and A.R. was funded by Ministry of Science and Higher Education (776/STYP/11/2016).

Acknowledgments: A.R. acknowledges grants from the Polish National Science Center (DEC-2013/11/D/NZ2/00033) and Ministry of Science and Higher Education (776/STYP/11/2016); The visit of A.R. to the School of Life & Health Sciences at Aston University was funded by a FEBS Short-Term Fellowship.

Conflicts of Interest: Authors declare no conflict of interest.

References

1. Boussios, S.; Karathanasi, A.; Cooke, D.; Neille, C.; Sadauskaite, A.; Moschetta, M.; Zakynthinakis-Kyriakou, N.; Pavlidis, N. PARP Inhibitors in Ovarian Cancer: The Route to "Ithaca". *Diagnostics* **2019**, *9*, 55. [CrossRef] [PubMed]
2. Faraoni, I.; Giansanti, M.; Voso, M.T.; Lo-Coco, F.; Graziani, G. Targeting ADP-ribosylation by PARP inhibitors in acute myeloid leukaemia and related disorders. *Biochem. Pharmacol.* **2019**, *167*, 133–148. [CrossRef] [PubMed]
3. Li, M.; Yu, X. The role of poly(ADP-ribosyl)ation in DNA damage response and cancer chemotherapy. *Oncogene* **2015**, *34*, 3349–3356. [CrossRef] [PubMed]
4. Pines, A.; Vrouwe, M.G.; Marteijn, J.A.; Typas, D.; Luijsterburg, M.S.; Cansoy, M.; Hensbergen, P.; Deelder, A.; De Groot, A.; Matsumoto, S.; et al. PARP1 promotes nucleotide excision repair through DDB2 stabilization and recruitment of ALC1. *J. Cell Boil.* **2012**, *199*, 235–249. [CrossRef] [PubMed]
5. Luijsterburg, M.S.; De Krijger, I.; Wiegant, W.W.; Shah, R.G.; Smeenk, G.; De Groot, A.J.; Pines, A.; Vertegaal, A.C.; Jacobs, J.J.; Shah, G.M.; et al. PARP1 Links CHD2-Mediated Chromatin Expansion and H3.3 Deposition to DNA Repair by Non-homologous End-Joining. *Mol. Cell* **2016**, *61*, 547–562. [CrossRef] [PubMed]
6. Yang, G.; Liu, C.; Chen, S.-H.; Kassab, M.A.; Hoff, J.D.; Walter, N.G.; Yu, X. Super-resolution imaging identifies PARP1 and the Ku complex acting as DNA double-strand break sensors. *Nucleic Acids Res.* **2018**, *46*, 3446–3457. [CrossRef] [PubMed]
7. Lai, J.; Yang, H.; Zhu, Y.; Ruan, M.; Huang, Y.; Zhang, Q. MiR-7-5p-mediated downregulation of PARP1 impacts DNA homologous recombination repair and resistance to doxorubicin in small cell lung cancer. *BMC Cancer* **2019**, *19*, 602. [CrossRef] [PubMed]
8. Soria, G.; Polo, S.E.; Almouzni, G. Prime, Repair, Restore: The Active Role of Chromatin in the DNA Damage Response. *Mol. Cell* **2012**, *46*, 722–734. [CrossRef]
9. Chen, Y.; Zhang, H.; Xu, Z.; Tang, H.; Geng, A.; Cai, B.; Su, T.; Shi, J.; Jiang, C.; Tian, X.; et al. A PARP1-BRG1-SIRT1 axis promotes HR repair by reducing nucleosome density at DNA damage sites. *Nucleic Acids Res.* **2019**. [CrossRef]
10. Hang, C.T.; Yang, J.; Han, P.; Cheng, H.-L.; Shang, C.; Ashley, E.; Zhou, B.; Hang, C.T.; Chang, C.-P. Chromatin regulation by Brg1 underlies heart muscle development and disease. *Nature* **2010**, *466*, 62–67. [CrossRef]

11. Pietrzak, J.; Płoszaj, T.; Pułaski, Ł.; Robaszkiewicz, A. EP300-HDAC1-SWI/SNF functional unit defines transcription of some DNA repair enzymes during differentiation of human macrophages. *Biochim. Biophys. Acta (BBA) Bioenerg.* **2019**, *1862*, 198–208. [CrossRef] [PubMed]
12. Sobczak, M.; Pietrzak, J.; Płoszaj, T.; Robaszkiewicz, A. BRG1 emerges as the master activator of proliferation and DNA repair genes in breast cancer cells. *Cancers*. (under review).
13. Krishnakumar, R.; Kraus, W.L. PARP-1 regulates chromatin structure and transcription through a KDM5B-dependent pathway. *Mol. Cell* **2010**, *39*, 736–749. [CrossRef] [PubMed]
14. Luo, X.; Ryu, K.W.; Kim, D.S.; Nandu, T.; Medina, C.J.; Gupte, R.; Gibson, B.A.; Soccio, R.E.; Yu, Y.; Gupta, R.K.; et al. PARP-1 Controls the Adipogenic Transcriptional Program by PARylating C/EBPbeta and Modulating Its Transcriptional Activity. *Mol. Cell* **2017**, *65*, 260–271. [CrossRef] [PubMed]
15. Robaszkiewicz, A.; Wisnik, E.; Regdon, Z.; Chmielewska, K.; Virag, L. PARP1 facilitates EP300 recruitment to the promoters of the subset of RBL2-dependent genes. *Biochim. Biophys. Acta Gene Regul. Mech.* **2017**. [CrossRef] [PubMed]
16. Tokarz, P.; Płoszaj, T.; Regdon, Z.; Virág, L.; Robaszkiewicz, A. PARP1-LSD1 functional interplay controls transcription of SOD2 that protects human pro-inflammatory macrophages from death under an oxidative condition. *Free Radic. Boil. Med.* **2019**, *131*, 218–224. [CrossRef]
17. Broustas, C.G.; Lieberman, H.B. DNA Damage Response Genes and the Development of Cancer Metastasis. *Radiat. Res.* **2014**, *181*, 111–130. [CrossRef] [PubMed]
18. Selfors, L.M.; Stover, D.G.; Harris, I.S.; Brugge, J.S.; Coloff, J.L. Identification of cancer genes that are independent of dominant proliferation and lineage programs. *Proc. Natl. Acad. Sci. USA* **2017**, *114*, E11276–E11284. [CrossRef]
19. Poli, V.; Fagnocchi, L.; Zippo, A. Tumorigenic Cell Reprogramming and Cancer Plasticity: Interplay between Signaling, Microenvironment, and Epigenetics. *Stem Cells Int.* **2018**, *2018*, 4598195. [CrossRef]
20. Wiśnik, E.; Płoszaj, T.; Robaszkiewicz, A. Downregulation of PARP1 transcription by promoter-associated E2F4-RBL2-HDAC1-BRM complex contributes to repression of pluripotency stem cell factors in human monocytes. *Sci. Rep.* **2017**, *7*, 9483. [CrossRef]
21. Tempka, D.; Tokarz, P.; Chmielewska, K.; Kluska, M.; Pietrzak, J.; Rygielska, Z.; Virag, L.; Robaszkiewicz, A. Downregulation of PARP1 transcription by CDK4/6 inhibitors sensitizes human lung cancer cells to anticancer drug-induced death by impairing OGG1-dependent base excision repair. *Redox Biol.* **2018**, *15*, 316–326. [CrossRef] [PubMed]
22. Pietrzak, J.; Spickett, C.M.; Płoszaj, T.; Virag, L.; Robaszkiewicz, A. PARP1 promoter links cell cycle progression with adaptation to oxidative environment. *Redox Boil.* **2018**, *18*, 1–5. [CrossRef] [PubMed]
23. Yang, L.; Huang, K.; Li, X.; Du, M.; Kang, X.; Luo, X.; Gao, L.; Wang, C.; Zhang, Y.; Zhang, C.; et al. Identification of Poly(ADP-Ribose) Polymerase-1 as a Cell Cycle Regulator through Modulating Sp1 Mediated Transcription in Human Hepatoma Cells. *PLoS ONE* **2013**, *8*, e82872. [CrossRef] [PubMed]
24. Wright, R.H.; Castellano, G.; Bonet, J.; Le Dily, F.; Font-Mateu, J.; Ballare, C.; Nacht, A.S.; Soronellas, D.; Oliva, B.; Beato, M. CDK2-dependent activation of PARP-1 is required for hormonal gene regulation in breast cancer cells. *Genes Dev.* **2012**, *26*, 1972–1983. [CrossRef] [PubMed]
25. Li, N.; Feng, L.; Liu, H.; Wang, J.; Kasembeli, M.; Tran, M.K.; Tweardy, D.J.; Lin, S.H.; Chen, J. PARP Inhibition Suppresses Growth of EGFR-Mutant Cancers by Targeting Nuclear PKM2. *Cell Rep.* **2016**, *15*, 843–856. [CrossRef]
26. Blankenberg, D.; Gordon, A.; Von Kuster, G.; Coraor, N.; Taylor, J.; Nekrutenko, A.; Team, G. Manipulation of FASTQ data with Galaxy. *Bioinformatics* **2010**, *26*, 1783–1785. [CrossRef]
27. Moreira, J.D.V.; Hamraz, M.; Abolhassani, M.; Bigan, E.; Pérès, S.; Paulevé, L.; Nogueira, M.L.; Steyaert, J.-M.; Schwartz, L. The Redox Status of Cancer Cells Supports Mechanisms behind the Warburg Effect. *Metabolites* **2016**, *6*, 33. [CrossRef]
28. Chaudhuri, A.R.; Nussenzweig, A. The multifaceted roles of PARP1 in DNA repair and chromatin remodelling. *Nat. Rev. Mol. Cell Boil.* **2017**, *18*, 610–621. [CrossRef]
29. Pascal, J.M. The comings and goings of PARP-1 in response to DNA damage. *DNA Repair* **2018**, *71*, 177–182. [CrossRef]
30. Christmann, M.; Kaina, B. Transcriptional regulation of human DNA repair genes following genotoxic stress: Trigger mechanisms, inducible responses and genotoxic adaptation. *Nucleic Acids Res.* **2013**, *41*, 8403–8420. [CrossRef]

31. Chun, J.; Buechelmaier, E.S.; Powell, S.N. Rad51 paralog complexes BCDX2 and CX3 act at different stages in the BRCA1-BRCA2-dependent homologous recombination pathway. *Mol. Cell. Biol.* **2013**, *33*, 387–395. [CrossRef]
32. García-Carpizo, V.; Ruiz-Llorente, S.; Sarmentero, J.; Graña-Castro, O.; Pisano, D.G.; Barrero, M.J. CREBBP/EP300 bromodomains are critical to sustain the GATA1/MYC regulatory axis in proliferation. *Epigenet. Chromatin* **2018**, *11*, 30. [CrossRef] [PubMed]
33. Ebrahimi, A.; Sevinç, K.; Sevinç, G.G.; Cribbs, A.P.; Philpott, M.; Uyulur, F.; Morova, T.; Dunford, J.E.; Göklemez, S.; Arı, Ş.; et al. Bromodomain inhibition of the coactivators CBP/EP300 facilitate cellular reprogramming. *Nat. Methods* **2019**, *15*, 519–528. [CrossRef] [PubMed]
34. US National Library of Medicine. Available online: https://clinicaltrials.gov/ (accessed on 16 August 2019).
35. McCann, K.; Hurvitz, S. Advances in the use of PARP inhibitor therapy for breast cancer. *Drugs Context* **2018**, *7*, 212540. [CrossRef] [PubMed]
36. Mustacchi, G.; De Laurentiis, M. The role of taxanes in triple-negative breast cancer: Literature review. *Drug Des. Dev. Ther.* **2015**, *9*, 4303–4318. [CrossRef] [PubMed]
37. Jasra, S.; Anampa, J. Anthracycline Use for Early Stage Breast Cancer in the Modern Era: A Review. *Curr. Treat. Options Oncol.* **2018**, *19*, 30. [CrossRef] [PubMed]
38. Lorusso, V.; Manzione, L.; Silvestris, N. Role of liposomal anthracyclines in breast cancer. *Ann. Oncol.* **2007**, *18*, vi70–vi73. [CrossRef] [PubMed]
39. Martinez-Zamudio, R.; Ha, H.C. Histone ADP-Ribosylation Facilitates Gene Transcription by Directly Remodeling Nucleosomes. *Mol. Cell. Boil.* **2012**, *32*, 2490–2502. [CrossRef] [PubMed]
40. Gibson, B.A.; Zhang, Y.; Jiang, H.; Hussey, K.M.; Shrimp, J.H.; Lin, H.; Schwede, F.; Yu, Y.; Kraus, W.L. Chemical Genetic Discovery of PARP Targets Reveals a Role for PARP-1 in Transcription Elongation. *Sci.* **2016**, *353*, 45–50. [CrossRef] [PubMed]
41. Afgan, E.; Baker, D.; Batut, B.; Beek, M.V.D.; Bouvier, D.; Čech, M.; Chilton, J.; Clements, D.; Coraor, N.; A Grüning, B.; et al. The Galaxy platform for accessible, reproducible and collaborative biomedical analyses: 2018 update. *Nucleic Acids Res.* **2018**, *46*, W537–W544. [CrossRef]
42. Ramírez, F.; Ryan, D.P.; Grüning, B.; Bhardwaj, V.; Kilpert, F.; Richter, A.S.; Heyne, S.; Dündar, F.; Manke, T. deepTools2: A next generation web server for deep-sequencing data analysis. *Nucleic Acids Res.* **2016**, *44*, W160–W165. [CrossRef] [PubMed]
43. Quinlan, A.R.; Hall, I.M. BEDTools: A flexible suite of utilities for comparing genomic features. *Bioinformatics* **2010**, *26*, 841–842. [CrossRef] [PubMed]
44. Balsa-Canto, E.; Henriques, D.; Gabor, A.; Banga, J.R. AMIGO2, a toolbox for dynamic modeling, optimization and control in systems biology. *Bioinformatics* **2016**, *32*, 3357–3359. [CrossRef] [PubMed]
45. Trapnell, C.; Williams, B.A.; Pertea, G.; Mortazavi, A.; Kwan, G.; Van Baren, M.J.; Salzberg, S.L.; Wold, B.J.; Pachter, L. Transcript assembly and quantification by RNA-Seq reveals unannotated transcripts and isoform switching during cell differentiation. *Nat. Biotechnol.* **2010**, *28*, 511–515. [CrossRef] [PubMed]

cancers

MDPI

Article

Cytotoxicity and Differentiating Effect of the Poly(ADP-Ribose) Polymerase Inhibitor Olaparib in Myelodysplastic Syndromes

Isabella Faraoni [1,*], Maria Irno Consalvo [2], Francesca Aloisio [1], Emiliano Fabiani [2], Manuela Giansanti [1], Francesca Di Cristino [1], Giulia Falconi [2], Lucio Tentori [1], Ambra Di Veroli [2], Paola Curzi [3], Luca Maurillo [3], Pasquale Niscola [4], Francesco Lo-Coco [2], Grazia Graziani [1,†] and Maria Teresa Voso [2,*,†]

1 Pharmacology Section, Department of Systems Medicine, University of Rome Tor Vergata, 00133 Rome, Italy
2 Hematology Section, Department of Biomedicine and Prevention, University of Rome Tor Vergata, 00133 Rome, Italy
3 Hematology Unit, Tor Vergata Hospital, 00133 Rome, Italy
4 Hematology Unit, "S. Eugenio" Hospital, 00144 Rome, Italy
* Correspondence: faraoni@med.uniroma2.it (I.F.); voso@med.uniroma2.it (M.T.V.)
† These authors contributed equally to this work.

Received: 10 August 2019; Accepted: 10 September 2019; Published: 16 September 2019

Abstract: Myelodysplastic syndromes (MDS) are highly heterogeneous myeloid diseases, characterized by frequent genetic/chromosomal aberrations. Olaparib is a potent, orally bioavailable poly(ADP-ribose) polymerase 1 (PARP1) inhibitor with acceptable toxicity profile, designed as targeted therapy for DNA repair defective tumors. Here, we investigated olaparib activity in primary cultures of bone marrow mononuclear cells collected from patients with MDS (n = 28). A single treatment with olaparib induced cytotoxic effects in most samples, with median IC_{50} of 5.4 μM (2.0–24.8 μM), lower than plasma peak concentration reached in vivo. In addition, olaparib induced DNA damage as shown by a high proportion of γH2AX positive cells in samples with low IC_{50}s. Olaparib preferentially killed myeloid cells causing a significant reduction of blasts and promyelocytes, paralleled by an increase in metamyelocytes and mature granulocytes while sparing lymphocytes that are not part of the MDS clone. Consistently, flow cytometry analysis revealed a decrease of CD117+/CD123+ immature progenitors ($p < 0.001$) and induction of CD11b+/CD16+ ($p < 0.001$) and CD10+/CD15+ ($p < 0.01$) neutrophils. Morphological and immunophenotypic changes were associated with a dose-dependent increase of PU.1 and CEBPA transcription factors, which are drivers of granulocytic and monocytic differentiation. Moreover, the combination of olaparib with decitabine resulted in augmented cytotoxic and differentiating effects. Our data suggest that olaparib may have therapeutic potential in MDS patients.

Keywords: MDS; PARP inhibitors; olaparib; hematopoietic differentiation; PARP1; AML

1. Introduction

Myelodysplastic syndromes (MDS) are a group of highly heterogeneous diseases characterized by peripheral blood cytopenia, dysplasia in one or more hematopoietic cell lineages and a differentiation defect, with an increased risk of evolving to acute myeloid leukemia (AML) [1]. To date, only few treatment options are available in these diseases, including growth factors in lower-risk MDS and the hypomethylating agents decitabine and azacitidine in intermediate/high-risk MDS. Both agents delay progression to AML by exerting cytotoxic and differentiating effects. Moreover, azacitidine has been shown to prolong overall survival [2]. However, especially in patients with higher-risk MDS, ineligible for allogeneic stem cell transplantation, life expectancy remains dismal. Therefore, innovative treatment strategies are warranted for elderly and unfit patients.

Poly(ADP-ribose) polymerase inhibitors (PARPi) are compounds with a favorable tolerability profile, designed as targeted therapy for homologous recombination (HR)-defective tumors. The FDA and EMA have approved some PARPi as maintenance therapy for platinum-sensitive relapsed (olaparib, rucaparib, and niraparib) or newly diagnosed (olaparib) high-grade epithelial ovarian, fallopian tube or primary peritoneal cancers. Olaparib and rucaparib received market authorization also for the treatment of relapsed BRCA1 or BRCA2 mutated advanced ovarian cancer, after previous lines of chemotherapy and independently of platinum sensitivity [3]. Recently, olaparib and talazoparib were FDA approved for patients with epidermal growth factor receptor 2 (HER2)-negative metastatic breast cancer with BRCA mutations, relapsing after previous chemotherapy. These and other PARPi are also under clinical development as monotherapy or in combination with targeted agents or chemotherapy for different types of cancer. In fact, accumulating evidence suggests a potential therapeutic role for PARPi in tumors characterized by mutations in genes involved in the HR repair of DNA double strand breaks (DSBs), including RAD51, ATM, ATR, CHEK1/2, FANC family members [4,5] (reviewed by Gadducci and Guerrieri [6]), or HR upstream modulators as PTEN [7] or IDH1 and IDH2 [8] (reviewed by Nickoloff et al. [9]). Therefore, regardless of BRCA mutational status, other defects in DNA repair may induce a "BRCAness" phenotype that would increase tumor sensitivity to PARPi.

We and others recently demonstrated that olaparib and other PARPi exert anti-proliferative and proapoptotic effects in human AML blasts at clinically achievable concentrations [10–18] that do not affect the viability of normal bone marrow (BM) stem cells [10,12,18]. The low BRCA1/2 RNA and protein expression levels detected in AML blasts [10] and myeloproliferative neoplasms [19] might account for myeloid malignancies sensitivity to synthetic lethality induced by PARPi. Also, some chromosomal translocations which are recurrent in leukemia including t(8;21) (RUNX1-RUNX1T1), t(15;17) (PML-RARA) [12] and t(17;19) (TCF3-HLF) [14] can weaken the HR repair activity and sensitize AML cells to PARPi treatment (reviewed by Faraoni et al. [20]).

Besides its involvement in DNA repair, through post-translational protein modifications (PARylation), the founding member of the PARP family PARP1 has a crucial role in the regulation of chromatin structure and gene transcription [21–23] (reviewed by Hottiger [24]). Recently, we suggested that olaparib cytotoxicity on primary AML blasts is the result of drug-induced DNA damage as well as of alternative mechanisms, such as upregulation of death receptor RNAs and proteins (FAS, DR5 and DR4), which in turn requires NF-κB activation [16]. The involvement of PARP1 in modulation of gene expression suggests additional therapeutic implications of PARPi in hematological diseases. In particular, deregulated gene expression in hematopoietic stem cells is regarded as a hallmark of MDS, with involvement of different pathways depending on the disease stage [25]. Interestingly, higher-risk MDS are characterized by a decrease of apoptosis and high levels of genomic instability, likely as a consequence of the observed alterations in DNA damage response pathways [25,26].

On this basis, we investigated the effects of PARP1 inhibition in primary cultures of BM mononuclear cells (BM-MNC) collected from MDS patients. Our results indicate that olaparib is not only cytotoxic, but importantly, also stimulates differentiation of immature MDS myeloid cells, as single agent or in combination with decitabine.

2. Results

2.1. MDS Mononuclear Cells Are Sensitive to Olaparib In Vitro

To assess the sensitivity of MDS to olaparib, experiments were performed using short-term cultures of MNC freshly isolated from the BM samples of 28 MDS patients. Table 1 lists the patients' characteristics.

Table 1. Patient Characteristics.

Prognostic Indexes	Total (n = 28)
Diagnosis	
MDS-SLD	6
MDS-MLD	11
MDS-EB-1	9
MDS-EB-2	2
Age (years) median (range)	73 (58–93)
Sex (F/M)	13/15
Karyotype	
Normal	18
-Y	1
Trisomy 8	3
5q-	2
20q-	2
2 anomalies *	1
Complex **	1
BM-blasts (%) median (range)	4 (3–13)
Hb (g/dL) median (range)	11.3 (7.5–16.0)
WBC (10^9/L) median (range)	3.8 (2.0–13.7)
Neutrophils (10^9/L) median (range)	2.8 (0.4–13.1)
Platelets (10^9/L) median (range)	108 (6–608)
IPSS-R	
Very-low	0
Low	17
Intermediate	8
High	2
Very-high	1

MDS-SLD: MDS with single-lineage dysplasia; MDS-MLD: MDS with multiple-lineage dysplasia; MDS-EB: MDS with excess of blasts in BM, MDS-EB-1: 5 to 9% blasts, MDS-EB-2: 10 to 19% blasts. * inv,ins(11;9); ** 7q-,12p-,22q-. IPSS-R: Revised International Prognostic Scoring Systems.

The proliferation rate of untreated or olaparib-treated primary cells was monitored for 7 days and the results are shown by grouping samples according to the 2016 WHO classification of MDS [1]. Some of the untreated MDS cultures showed a slight increase in cell number (n = 12), while the remaining samples either did not proliferate (n = 7) or showed a reduction in cellularity (n = 9) (Figure 1, left panels). Nevertheless, viability was constantly ≥70% in control cells during the culture. No significant correlation was found between the cell growth in culture and MDS risk according to the Revised International Prognostic Scoring Systems (R-IPSS). Treatment with olaparib induced a dose-dependent decrease of cell survival in all MDS samples with a median IC_{50} of 5.5 μM (range 2.0–24.8 μM) (Figure 1, right panels). The median IC_{50} values were comparable in the four MDS subgroups analyzed (MDS-SLD, 6.1 μM; MDS-MLD, 5.4 μM; MDS-EB-1, 5.3 μM; MDS-EB-2, 3.8 μM). No statistically significant correlation was detected between the proliferation indexes of MDS cultures and olaparib IC_{50}s, suggesting that the drug sensitivity did not depend on cell ability to grow in vitro. Representative growth curves of olaparib-treated samples with comparable olaparib IC_{50} values but different proliferation rates are shown in Figure S1A. Also, we found no significant correlation between cell sensitivity, expressed as olaparib IC_{50} values at 7 days, and the MDS prognostic variables listed in Table 1. Notably, the olaparib IC_{50} values were in most cases largely below the steady-state plasma peak concentrations (C_{max} = 16–22 μM), measurable in patients with solid tumors receiving 300 mg olaparib twice daily [27,28]. Conversely, olaparib sensitivity of normal BM, CD34-enriched mobilized peripheral blood and purified CD34+ samples (IC_{50} range: 18.5–27.0 μM) was markedly higher than that of MDS cells (Figure S1B), in agreement with previous findings [5,10,12,29].

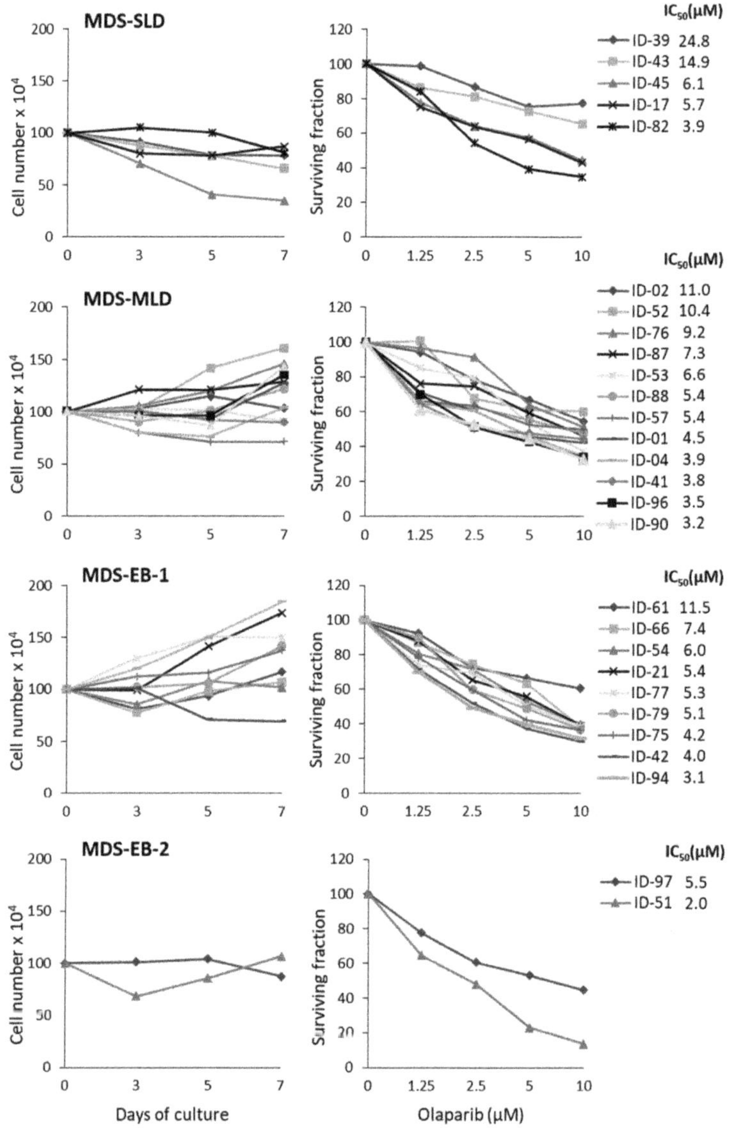

Figure 1. Olaparib exerts cytotoxic effects in primary MDS cultures. BM-MNC collected from MDS patients were cultured with IL-3, SCF and FLT3LG and treated (time 0) with increasing concentrations of olaparib. For each primary culture, cell proliferation was evaluated by counting viable cells using trypan blue exclusion at 3, 5 and 7 days. Standard deviation (SD) of four replicate counts was ≤20% and is not shown in the figure. MDS samples were grouped according to morphology, as MDS-SLD, MDS-MLD, MDS-EB-1 and MDS-EB-2. Left graphs represent the proliferation pattern of untreated primary MDS cells during 7 days of culture. Right graphs show the surviving fractions after 7 days of treatment, and the olaparib IC_{50}s for each sample calculated with respect to untreated cells cultured for the same time period.

We then investigated whether the growth inhibitory activity of olaparib in MDS cells was associated with cytotoxic effects. Apoptosis was evaluated by cell staining with annexin V/PI and FACS analysis after 7 days exposure to graded concentrations of olaparib. Cells were gated in order to separately analyze apoptosis induction within the myeloid and lymphocyte populations. Dose-dependent apoptosis was observed in the myeloid compartment of MDS samples characterized by IC_{50} values ≤ 6.1 μM (Figure 2A), without major differences among cells at different maturation stages. Pooled statistical analysis of data referring to samples with olaparib IC_{50} values ≤ 6.1 μM indicated a significant increase in the percentage of apoptotic cells at 5–10 μM olaparib concentrations (Figure 2B). On the other hand, a negligible percentage of apoptotic cells was detected in the lymphocyte population (Figure 2B). Representative plots in Figure S2A demonstrate the lack of annexin V/PI staining in CD45-positive and CD33-negative lymphocytes. Lack of apoptosis detection in the lymphocyte population was not due to a faster killing kinetics in lymphoid cells, since apoptosis induction was not observed at an earlier time point (i.e., 3 days) (Figure S2B). These data suggest that olaparib preferentially kills myeloid precursors, but spares lymphocytes that are not part of the MDS clone. To further defining the targets of olaparib cytotoxic effects, two MDS samples with cytogenetic abnormalities were FISH analyzed after 7 days exposure to olaparib. The PARPi induced a 22% and 34% decrease in the number of cells with trisomy 8 and 7q deletion, respectively. Figure 2C shows representative FISH images of MDS cells with deleted chromosome 7. These data suggest a preferential killing of malignant cells by olaparib.

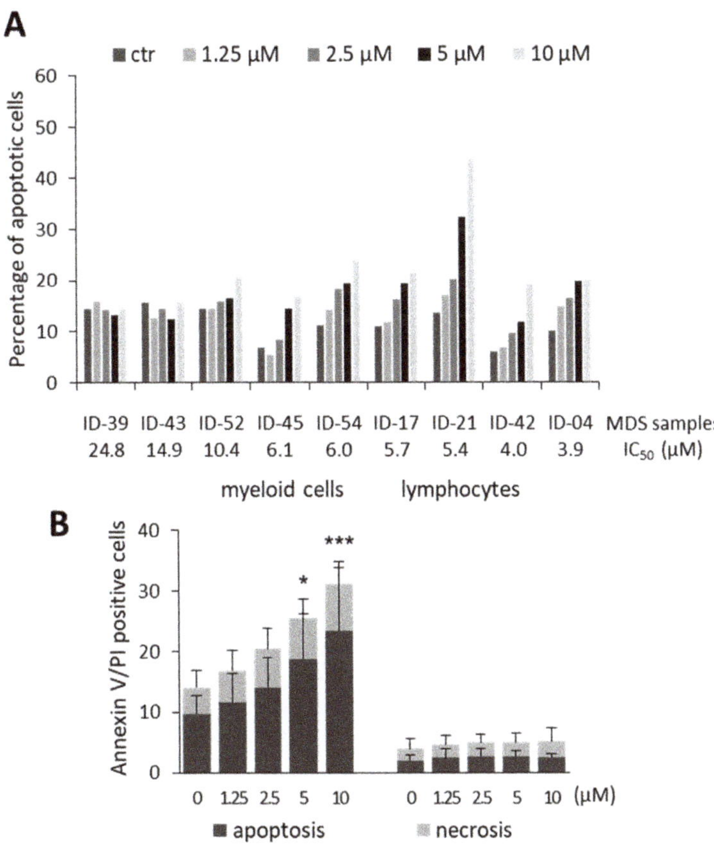

Figure 2. *Cont.*

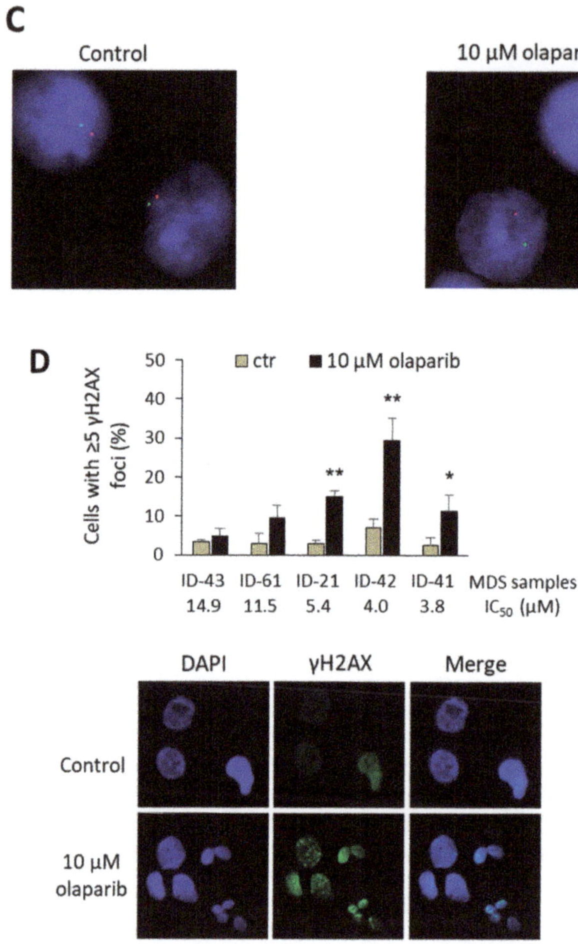

Figure 2. Apoptosis and DNA damage in primary MDS in vitro cultures treated with olaparib. (**A**) Percentages of apoptotic cells (Annexin V positive/PI negative), detected by FACS analysis, in gated myeloid cells after exposure to graded concentrations of olaparib (7 days) compared to untreated cultured cells. Samples were classified in order of decreasing olaparib IC_{50} values. (**B**) Histograms indicate the mean percentage ±SD of apoptotic and necrotic cells (Annexin V positive/PI positive), detected within the myeloid and lymphocytic populations, respectively. Data refer to six MDS samples after exposure to graded concentrations of olaparib for 7 days compared to untreated cultured cells. Olaparib induces apoptosis in the myeloid MDS cell compartment, while sparing lymphocytes. Statistical analysis was performed on apoptotic cells (Annexin V positive/PI negative cells) by Friedman followed by Dunn's test. (**C**) Interphase FISH analysis of representative MDS cells with 7q deletion (ID-75). Images were taken at 100× magnification. Nuclei with deletions involving both critical regions at 7q22 and 7q36 show one red and one green signal only, whereas cytogenetically normal cells show two red and two green signals. In the control group, 75% of cells showed the 7q deletion, whereas in the olaparib-treated group this percentage dropped to 50%. (**D**) Analysis of γH2AX foci by immunofluorescence after 72 h of exposure to 0 and 10 µM olaparib. Values are the mean percentage ± SD of the cells with ≥5 γH2AX foci in four quadrants, containing at least 50 cells each. Statistical analysis was performed by Mann-Whitney test. Representative nuclei of MDS cells untreated or exposed to 10 µM olaparib are shown. Cells were probed with γH2AX (green stain) and DAPI (blue). Images were taken at 40× magnification. * $p < 0.05$, ** $p < 0.01$, *** $p < 0.001$ vs. untreated control.

We then assessed whether the cytotoxic effects of olaparib in MDS might be attributable to DNA damage, using immunofluorescence analysis of γH2AX foci, which are markers of DSBs. The histogram in Figure 2D represents the percentage of γH2AX positive cells in five MDS samples untreated or treated with 10 μM olaparib for 72 h. The results indicate that the PARPi induced a significantly higher percentage of γH2AX positive cells in samples characterized by low IC_{50}s (ID-21, ID-42 and ID-41), as compared to the untreated control. Conversely, a negligible increase in DNA damage was observed in two other MDS samples (ID-43 and ID-61) that were more resistant to the PARPi. These data indicate that treatment with single agent olaparib induces DNA damage that accounts, at least in part, for olaparib cytotoxicity.

2.2. Differentiating Effects of Olaparib on MDS Immature Myeloid Cells

Flow cytometry analysis by side scatter (SSC) vs forward scatter (FSC) and SSC vs CD45 showed an increase in the mature cells (high-SSC) after treatment with olaparib (Figure S2A, left and middle panels), suggesting a potential PARPi effect on myeloid cell differentiation. Based on these findings, we studied morphological changes in MDS samples upon treatment with 10 μM olaparib for 7 days, after May Grunwald-Giemsa staining of cytospins. We observed a significant reduction of the proportion of blasts and promyelocytes, paralleled by an increase of metamyelocytes and mature granulocytes (Figure 3A,B). In cells from one case of MDS-EB-1, olaparib treatment induced outgrowth of 22% monolobulated megakaryocytes, as depicted in Figure 3C.

Differentiation was also analyzed by means of CD117 and CD123 antigen expression, which are markers of myeloid progenitors. The percentage of CD117+/CD123+ immature progenitors was significantly reduced in treated cells, as compared to untreated controls in 12 of 13 MDS samples (Figure 4A). Representative dot-plots of the olaparib-induced decrease in CD117+/CD123+ cell population are shown (Figure 4A, right panels). The lack of CD34 expression is known to be a common aberrancy in MDS immature myeloid compartment [30]. Indeed, the percentages of CD34+ stem cells were scored ≥1% in four of 12 samples only. Treatment with olaparib resulted in a marked reduction of CD34+ cells in three out of four samples, with no effects observed in the ID-02 sample that was the most resistant to the toxic effects of olaparib (Figure S3). Looking at neutrophil maturation patterns, samples treated with olaparib showed a significant induction of CD11b+/CD16+ cells (9 of 10) and CD10+/CD15+ (6 of 6) double positive cells, indicating differentiation towards mature neutrophils (Figure 4B). Representative dot-plots show a dose-dependent increase of CD45-positive/high SSC cells in the samples exposed to the PARPi. This effect was accompanied by a shift of the double-negative immature cells towards the double-positive mature population (Figure 4B).

In six MDS samples, we also tested whether modulation of differentiation markers was dose-dependent. Olaparib induced a reduction of CD117+/CD123+ MDS cells together with an increase of mature CD11b+/CD16+ cells that reached statistically significance at the concentration of 10 μM (Figure 4C). These data strongly suggest that inhibition of PARP1 can stimulate maturation of MDS myeloid precursors into neutrophils. A similar differentiating effect was also observed in one sample from a patient with low-blast count AML (Figure S4), which was also highly susceptible to olaparib cytotoxic effects (IC_{50} = 1.6 μM).

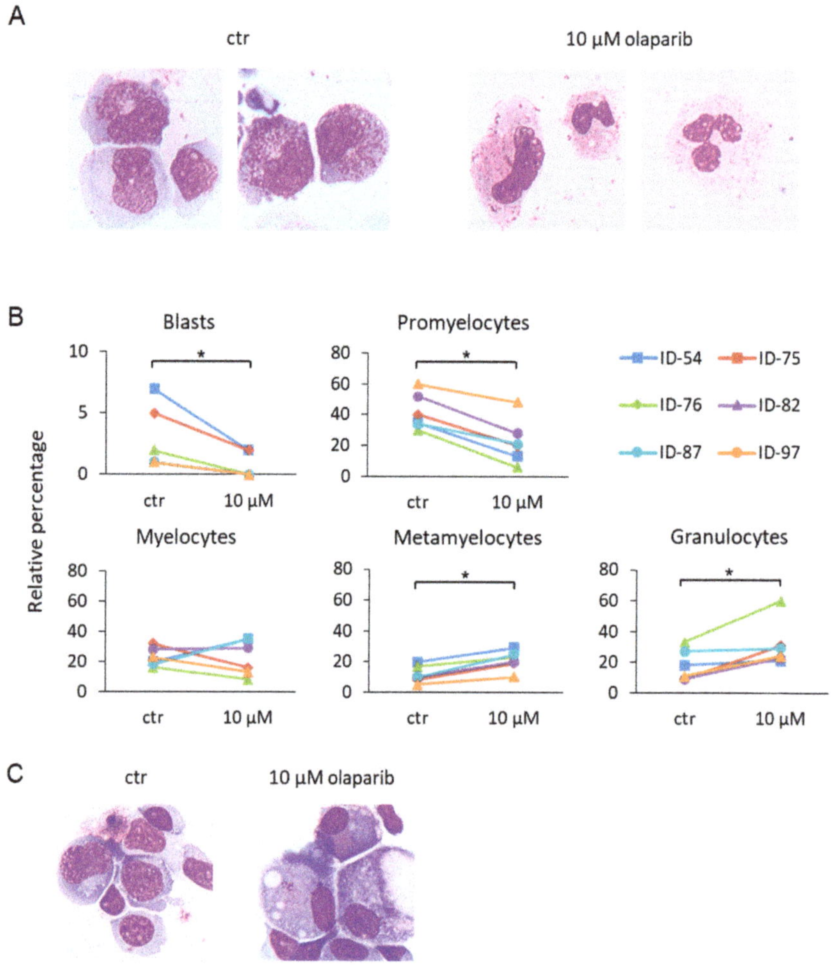

Figure 3. Morphological analysis of MDS samples treated with olaparib. May Grunwald-Giemsa staining of cytospins from six primary MDS samples after 7 days of culture. Untreated cultured cells and those treated with 10 µM olaparib are shown. (**A**) Representative pictures are shown (sample ID-54). Images were taken at 100× magnification. (**B**) Differential counts using standard microscopy were performed by two specialized hematologists (ADV and MTV). Statistical analysis by paired Wilcoxon test: * $p < 0.05$. (**C**) Sample ID-79 after 7 days of 10 µM olaparib treatment. Images were taken at 100× magnification.

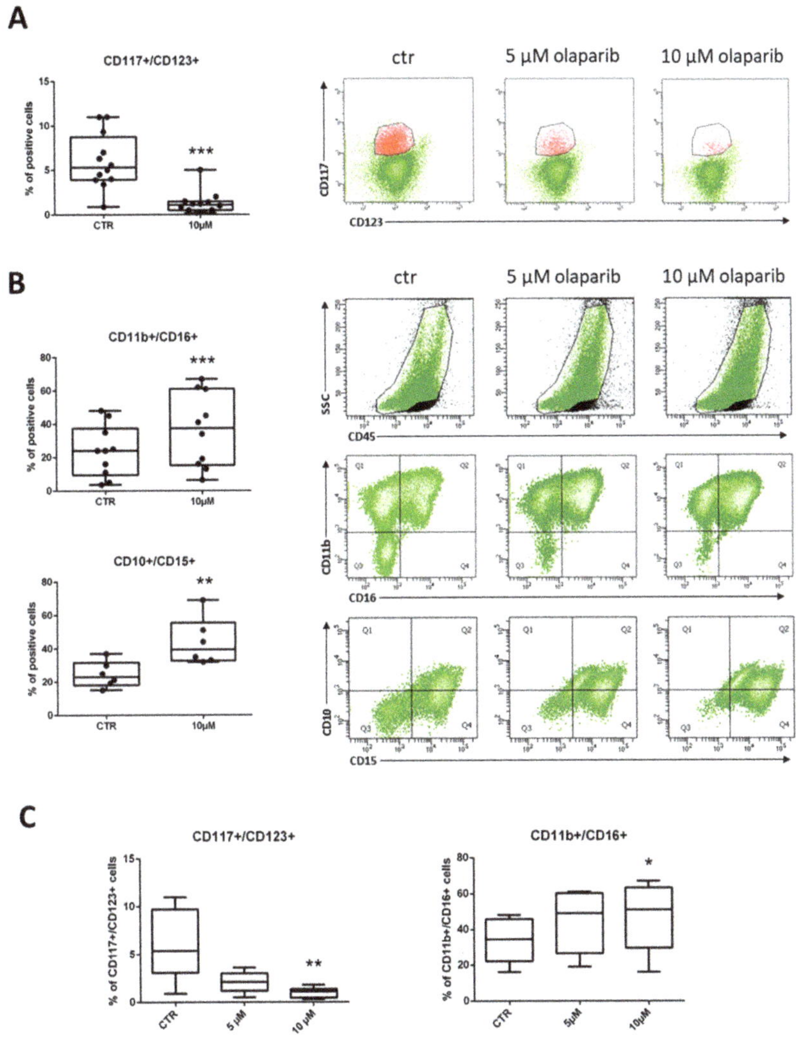

Figure 4. Immunophenotypic analysis of myeloid cells in MDS samples treated with olaparib. (**A**) Flow cytometry analysis of MDS samples treated with olaparib for 7 days vs untreated control cells cultured for the same time period. Box-whisker plots show the proportion of CD117+/CD123+ immature progenitors in MDS samples (n = 13; paired Wilcoxon test: *** p <0.001). Representative plots show the changes in the expression of CD117+/CD123+ markers (red) in sample ID-76 after treatment with 5 or 10 µM olaparib. (**B**) Box-whisker plots show the proportion of CD11b+/CD16+ (n = 10) and CD10+/CD15+ (n = 6) mature neutrophils in MDS samples (paired Wilcoxon test: ** p < 0.01, *** p < 0.001). Representative plots of flow cytometry analysis of CD45+ cells, after lymphocyte (black) exclusion, showing changes in SSC and increased expression of CD11b+/CD16+ or CD10+/CD15+ cells in ID-76 sample after treatment with 5 or 10 µM olaparib. (**C**) Box-whisker plots showing the olaparib dose-dependent effect in CD117+/CD123+ (left panel) and CD11b+/CD16+ (right panel) cells from six samples treated with 5 or 10 µM olaparib. Statistical analysis by Friedman followed by Dunn's test: * p < 0.05, ** p < 0.01.

2.3. *Olaparib Modulates the Expression of PU.1 and CEBPA Transcription Factors*

To confirm the ability of olaparib to stimulate myeloid differentiation, we studied the expression of the lineage-specific transcription factors PU.1 and CEBPA, drivers of granulocytic and monocytic differentiation. BM-MNC from six MDS patients were exposed to olaparib (5 and 10 μM), and total RNA analyzed by qRT-PCR after 7 days of treatment. We found that olaparib induced a statistically significant and dose-dependent upregulation of PU.1 RNA expression, while CEBPA was significantly upregulated at the 10 μM concentration (Figure 5A). This was associated with a dose-dependent increase of the corresponding proteins, as assessed by western blot analysis (Figure 5B).

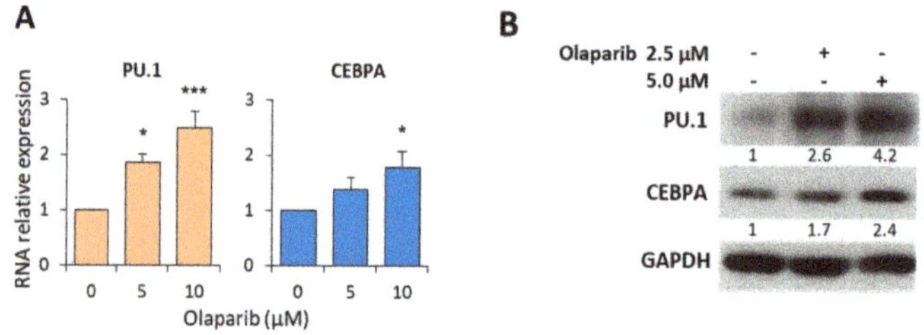

Figure 5. Olaparib modulates the expression of transcription factors associated with myeloid differentiation. (**A**) PU.1 and CEBPA mRNA expression were evaluated by qRT-PCR after 7 days of culture and treatment with 5 or 10 μM olaparib. Histograms show the mean ±SD relative expression in six MDS samples. Statistical analysis by Kruskal-Wallis followed by Dunn's test: * $p < 0.05$, *** $p < 0.001$. (**B**) Representative immunoblot showing PU.1 and CEBPA protein expression from one MDS sample exposed to the indicated concentrations of olaparib (i.e., ID-54) for 7 days. Numbers below blots refer to the densitometric analysis of the immunoreactive bands and represent the fold-increase of protein expression normalized to GAPDH.

2.4. *Cytotoxic and Differentiating Effects of Olaparib and Decitabine Combined Treatment*

To investigate the possible synergism between olaparib and decitabine, we initially used the OCI-AML2 leukemia cell line as a model. OCI-AML2 cells were treated with increasing concentrations of both drugs as single agents or in combination. After 3 days of culture, cells were analyzed by the MTS assay. Olaparib (1 to 8 μM) and decitabine (0.1 to 0.8 μM) exerted synergistic cytotoxicity on OCI-AML2 cells as assessed by the CompuSyn model [31] at all concentration ratios (Figure S5).

To assess the susceptibility to decitabine and olaparib combination (1:10), four MDS samples were exposed to equitoxic drug concentrations, including drug IC_{50} values and cytotoxicity was assayed by MTS after 7 days of treatment. In all samples, the two drugs exhibited synergistic cytotoxic effects at most of the concentrations tested (Figure 6A). In particular, for all samples the combination index (CI) values were <1 at the effective dose 50 (ED_{50}). Moreover, analysis of the dose reduction index (DRI) indicated that addition of the PARPi to decitabine allowed 3.5–20.8 fold decrease of decitabine dose at 0.5 fraction affected (Fa). The highest DRI value was observed in ID-90 cells that exhibited the lowest sensitivity to decitabine. Cytotoxicity was associated with DNA damage, as indicated by the statistically significant increase of γH2AX positive cells in co-treated samples as compared to the untreated controls (Figure 6B).

To investigate whether the synergistic activity of decitabine plus olaparib did also impact on myeloid differentiation, we evaluated by FACS analysis the percentage of CD117+/CD123+ hematopoietic progenitors and CD11b+/CD16+ neutrophils in four MDS samples, untreated or treated with the indicated concentrations of the drugs. We found that the 0.5 μM decitabine and 5 μM olaparib combination significantly decreased the proportion of immature progenitors and increased the percentage of neutrophils (Figure 6C). The expression of PU.1 and CEBPA RNA was studied in three MDS samples and at least additive effects were observed in samples treated with the drug combination (Figure 6D).

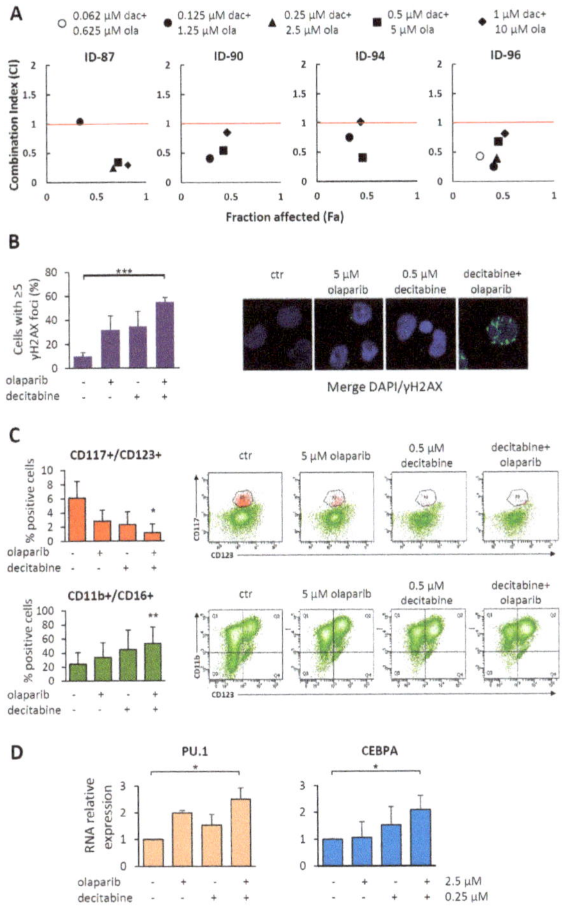

Figure 6. The olaparib and decitabine combination impairs survival and induces differentiation in primary MDS. (**A**) Primary MDS cells were treated with olaparib (ola, 0–10 µM) or decitabine (dac, 0–1 µM) or with different drug combinations at time 0. After 7 days of culture, cells were analyzed by the MTS assay to study cytotoxicity. Plots show the fraction of affected cells (Fa, x axis) and the combination index (CI, y axis) values as assessed by the CompuSyn model. Each plot indicates the CI values of a MDS samples treated at decitabine and olaparib 1:10 ratio. CI < 1, synergistic activity; CI = 1, additive effect; CI > 1, antagonism. (**B**) Analysis of γH2AX foci by immunofluorescence after 72 h of exposure of ID-90 cells to 5 µM olaparib and 0.5 µM decitabine, as single agents or in combination. Values are the mean percentage ±SD of the cells with ≥5 γH2AX foci in four quadrants, containing at least 50 cells each. Statistical analysis was performed by Kruskal-Wallis followed by Dunn's test. Representative nuclei of untreated or drug-treated MDS cells are shown. Cells were probed with γH2AX (green stain) and DAPI (blue). Images were taken at 40× magnification. (**C**) Histograms show the mean percentage ± SD of CD117+/CD123+ and CD11b+/CD16+ cells in 3 MDS samples untreated or treated with 5 µM olaparib and 0.5 µM decitabine, as single agents or in combination. Statistical analysis by Friedman followed by Dunn's test: * $p < 0.05$, ** $p < 0.01$. Representative dot plots showing data from a MDS sample (ID-88) untreated or treated with olaparib, decitabine or both drugs. A decrease of CD117+/CD123+ immature progenitors (upper plots) and increase of CD11b+/CD16+ (lower plots) are observed in treated samples as compared to the untreated control. (**D**) PU.1 and CEBPA mRNA expression evaluated by qRT-PCR after 7 days of treatment with olaparib and decitabine, as single agents or in combination. Histograms show the mean ±SD relative expression in three MDS samples. Statistical analysis by Kruskal-Wallis followed by Dunn's test: * $p < 0.05$, ** $p < 0.01$, *** $p < 0.001$.

3. Discussion

In the present study, we demonstrate for the first time that the PARPi olaparib used at clinically achievable concentrations [27,28] exerts cytotoxic and differentiating effects in short-term primary cultures of BM-MNC obtained from MDS patients.

The apoptotic effects of olaparib were observed both in actively proliferating and resting MDS cultures. These findings support the activity of PARPi also on quiescent malignant stem cells [10,15,16], that are generally more resistant to cytotoxic treatments. As recently suggested, the inefficient HR repair of DSBs in non-proliferating cells characterized by a "BRCAness" phenotype, does not account for the observed PARPi-mediated synthetic lethality, suggesting the involvement of additional repair mechanisms [15].

Interestingly, in our model, olaparib mainly targeted myeloid cells including the dysplastic cell lineages, while it spared lymphocytes, a cell subset that is not involved in the MDS clone. Consistently, normal hematopoietic cells showed lower sensitivity to olaparib as compared to MDS cells. This may indicate that PARPi-induced synthetic lethality targets MDS cells that are frequently characterized by DNA damage response defects which contribute to chromosome alterations involved in disease progression [25,32].

In our study, olaparib induced MDS cell differentiation, as indicated by the results of morphological and immunophenotypic analysis, and increased the expression of myeloid-specific transcription factors. In fact, the PARPi caused a reduction of CD117+ or CD34+ hematopoietic progenitors and an increase in maturing CD11b+/CD16+ and CD10+/CD15+ cells in almost all MDS analyzed samples. Since BM samples isolated from patients contain a mixture of normal and diseased cells with differentiation block, the observed increase in neutrophils might derive either from killing of immature MDS cells and consequent relative increase of healthy cells and/or from maturation of MDS hematopoietic progenitors. Actually, we found that olaparib-differentiated cells, as assessed by morphology, were still characterized by significant dysplastic changes, and the observed decrease of trisomy 8 and del(7)–positive cells in the two FISH-analyzed MDS samples was moderate. These data argue in favor to the hypothesis that the increase of differentiated cells in olaparib-treated cultures involves malignant cells and is not merely due to the killing of the latter and recovery of normal hematopoietic progenitors.

The regulatory effect of transcription factors determines the fate of hematopoietic stem cells from self-renewal towards generation of mature blood cells [33–35]. In our study, olaparib upregulated RNA and protein expression of the granulocytic and monocytic transcription factors PU.1 and CEBPA, both proteins reported to be frequently downregulated in myeloid malignancies [33–35]. The modulation of transcription factors involved in myeloid cell differentiation induced by olaparib might depend on the known ability of PARP1 to affect chromatin structure and RNA transcription by PARylation or direct interaction with target proteins [24,36,37]. In fact, PARP1 overexpression has been found to block differentiation in response to all-trans retinoic acid (ATRA) in leukemia cell lines [38] and a reduction of PARP1 expression was observed during differentiation of neutrophils or monocytes [39,40]. Interestingly, higher PARP1 RNA levels were observed in higher-risk MDS and correlated with inferior survival [41]. However, we cannot exclude that transcription factor modulation might be the consequence, rather than the cause, of differentiation induced by excessive DNA damage, as previously described in murine models of hematopoietic stem cells [42,43].

One clinical concern with olaparib is the risk of therapy-related MDS or AML associated to high-doses [44]. This risk has not been confirmed by a recently published phase 3 clinical trial which reported the occurrence of MDS or AML in 2% of olaparib-treated (300 mg, twice daily) vs 4% of placebo-treated patients [45]. Similar results were reported for the PARPi niraparib and rucaparib [46,47], suggesting that the burden of cytotoxic treatment prior to PARPi may be responsible for the development of therapy-related myeloid neoplasms

In our study, the cytotoxic effects of the olaparib/decitabine combination were synergistic or at least additive in the MDS cell cultures. These findings are in line with the recently reported synthetic lethality induced by the association of the two drugs in a panel of AML cell lines [48,49]. Muvarak et al.

demonstrated that low dose decitabine or azacitidine plus the PARPi talazoparib had synergistic cytotoxic effects in AML cell lines, as a consequence of increased DNA damage and delayed DNA repair [13]. In keeping with these results, the combination of the PARPi with a hypomethylating agent caused a marked antitumor response in an AML xenograft model [13]. The synergistic activity of PARPi with decitabine may also be explained by the involvement of PARP1 in the repair of decitabine-induced DNA lesions represented by randomly incorporated 5-aza-dC and trapped DNA methyltransferase 1 (DNMT1). In the presence of PARPi, DNA repair is impaired as shown by an increase of DBSs [13,48,49]. Noteworthy, the combination of olaparib and decitabine significantly decreased the proportion of immature myeloid progenitors and increased that of differentiated neutrophils. This effect was paralleled by the up-regulation of PU.1 and CEBPA transcription factors. The inhibition of DNA methylation caused by decitabine likely contributes to enhance the differentiating effect induced by olaparib. These results are of particular interest, considering that decitabine is approved by FDA for all subtypes of MDS and low-blast count AML (blasts < 30%), and by EMA for AML patients aged ≥ 65 years not candidates for standard induction chemotherapy, due to its manageable toxicity [50]. Moreover, in a clinical study with ovarian cancer patients there were no differences in the olaparib toxicity between younger and older women [51]. Thus, the favorable safety profile and good oral bioavailability of olaparib makes it a suitable agent to be combined with decitabine for MDS patients ineligible for chemotherapy or BM transplantation. This novel combination may allow dose reduction and overcome some of the limits of hypomethylating agents, by deepening the response and decreasing the burden of treatment-induced cytopenias with the related complications, ultimately leading to improved survival in MDS.

4. Materials and Methods

4.1. MDS Samples

Freshly isolated primary cells were obtained from BM aspirates of 28 adults with newly diagnosed MDS and one patient with low-blast count AML. Normal BM cells were collected from one hematopoietic stem cell donor and normal CD34+ enriched cells were obtained from the peripheral blood (PB) of three healthy donors following G-CSF-mobilization. MNC were isolated by Lynpholyte-H (Cederlane, Burlington, Canada). In the case of one CD34-enriched mobilized peripheral blood sample, CD34+ cells were purified by the MACS CD34 progenitor isolation kit using immunomagnetic beads (Miltenyi Biotech, Bergisch Gladabach, Germany). All patients and donors gave written informed consent according to institutional guidelines and the study was approved by the local institutional review board (Tor Vergata Hospital, Registry N. 12/16). Routine morphological, immunophenotypic and genetic analysis were carried out at presentation. Conventional karyotyping was performed on BM diagnostic aspirates after short-term culture and analyzed after G-banding.

4.2. Cell Culture and Drug Treatment

Freshly isolated BM-MNC were cultured at 37 °C in a humidified atmosphere of 5% CO_2 for 1 or 2 days before starting the chemosensitivity assays. Briefly, cells were seeded into culture flasks in RPMI medium (Sigma-Aldrich, St. Louis, MO, USA), supplemented with 2 mM L-glutamine (EuroClone, Pero, Milan, Italy), 1% penicillin/streptomycin (Euroclone), 20% fetal bovine serum (FBS) (Gibco, ThermoFisher Scientific, Waltham, MA, USA) and 10 ng/mL of each IL-3, SCF and FLT3LG (PeproTech, Rocky Hill, NJ, USA), which do not have differentiating effects. The OCI-AML2 leukemia cell line (DSMZ-German Collection, Braunschweig, Germany) was cultured in RPMI plus 20% FBS.

Olaparib stock (AstraZeneca, London, UK) was prepared by dissolving 10 mg in 200 µL of DMSO (Sigma-Aldrich) and then diluted with RPMI to the concentration of 2 mM. The final DMSO concentration in the cultures was ≤0.01% (*v*/*v*). The stock solution of decitabine (Cayman Chemical, Ann Arbor, MI, USA) at 2 mM concentration was prepared by dissolving the drug in PBS. Drugs were added at the beginning of the experiments and left in the medium during the entire period of incubation. Since in vitro culture of primary MDS samples could be maintained for 7–10 days only, the assays were performed at up to 7 days after drug exposure. Experiments with normal cells were performed with the same procedure.

For survival assay, cells were seeded in 48 or 24 wells culture plates (10^6 cells/mL, unless otherwise specified) in duplicate and treated with graded concentrations of olaparib (1–10 μM) or decitabine (0.1–1 μM) for the indicated times. Cells were counted by trypan blue dye exclusion in quadruplicate, and the survival fraction was calculated as proportion of the untreated control. The drug concentration capable of inhibiting 50% of cell growth (IC_{50}), compared to untreated control, was calculated with the GraphPad Prism 5 software (San Diego, CA, USA) by using linear regression. The combined effect of olaparib and decitabine was evaluated by the MTS viability test (Promega, Madison, WI, USA), according to the manufacturer's instructions. Cells were seeded in triplicate (10^4 cells/mL for OCI-AML2 and 5×10^5 cells/mL for the MDS samples), in 96 wells plates (2×10^3 cells/well and 10^5 cells/well, respectively) and treated with graded concentrations of olaparib and decitabine. Synergism was assessed by calculating the proportion of cell growth using the CompuSyn software (ComboSyn Incorporated, Paramus, NJ, USA).

4.3. Flow Cytometry and Apoptosis Assays

Apoptosis in primary MDS samples was assayed using an annexin-V apoptosis kit (GFP Certified™Apoptosis/Necrosis Detection Kit, Enzo Life Sciences, Farmingdale, NY, USA), according to the manufacturer's instructions, and analyzed by flow cytometry.

In MDS patients, BM-MNC cells are characterized by a decrease of neutrophil granularity and increased or decreased expression of differentiation antigens or their ratios, although no single specific marker is diagnostic for MDS [30]. Thus, we performed a series of analyses with different antibody combinations to evaluate the maturation pattern induced by olaparib. For immunophenotype analysis, cells were incubated with fluorochrome-tagged monoclonal antibodies anti-CD45 (# 560777), -CD33 (561157), -CD34 (#345804), -CD117 (#339217), -CD123 (#588714), -CD10 (#332776), -CD11b (#333142), -CD15 (#332778), -CD16 (#335035) (BD Biosciences). In particular, we studied the neutrophilic population by analyzing CD11b/CD16 and CD10/CD15 differentiation markers. Samples (5×10^4 cells) were acquired on a multicolor BD FACSCanto II flow cytometer and evaluated using DIVA and FlowJo softwares (BD Biosciences, San Jose, CA, USA).

4.4. Immunofluorescence, May-Grunwald Giemsa Staining and FISH Analysis

MDS cells were cytocentrifuged (10^5 cells), fixed with 4% paraformaldehyde, permeabilized in 0.3% tryton and blocked in 2% bovine serum albumin (Sigma-Aldrich). Slides were incubated with mouse anti-phospho-H2A histone family member X (γH2AX) (#05-636, Millipore, Burlington, MA, USA) followed by goat anti-mouse IgG DyLight 488 (#35502, Thermo Fisher Scientific, Rockford, IL, USA), stained with DAPI (4,6 diamidino-2-phenylindole) and mounted in Fluoromount (Sigma-Aldrich). Analysis was performed using a Leica CTR 6000 fluorescence microscope and LAS AF Lite software (Leica, Wetzlar, Germany).

After 7-day culture, cytospins containing 10^5 BM-MNC were stained using the May-Grunwald solution. Cells were then observed under a BX61 microscope (Olympus, Tokyo, Japan). and differential counts were performed by two experienced hematologists (A.D.V. and M.T.V.).

Fluorescence in situ hybridization (FISH) was performed on MDS cells cultured for 7 days in the presence of olaparib. Cytogenetic preparations were performed according to the Kreatech FISH probes manufacturer's instructions (Leica). Slide preparations were hybridized with the Kreatech FISH probes: KBI-2008G, targeting the centromere of chromosome 8, and KBI-10202 targeting chromosome 7q22/7q36 (Leica). A minimum of 200 cells were scored for each preparation by two experienced operators (P.C. and M.G.). Evaluation of the FISH signals was performed using the BX61 Olympus fluorescence microscope.

4.5. mRNA and Protein Expression

Total RNA was isolated by Trizol reagent (Invitrogen, Carlsbad, CA, USA). One μg of total RNA was reverse-transcribed (reagents from Applied Biosystem, Warrington, UK) and quantitative real-time PCR (qRT-PCR) was carried out using iQ SYBR Green Supermix, Bio-Rad (BioRad, Hercules, CA, USA). The sequence of the primers used is as follows: PU.1, FW 5′-CAGGGGATCTGACCGACTC-3′ and RV 5′-GCACCAGGTCTTCTGATGG-3′; CEBPA, FW 5′-TTGTTTGTACTGTATGCCTTC-3′ and RV 5′-GCCAGATACAAGTGTTGATAT-3′; GAPDH, FW 5′-CAGCCGAGCCACATCG-3′ and RV5′-TGAGGCTGTTGTCATACTTCTC-3′. A melting curve (62–95 °C) was generated at the end of each run to verify specificity of the reactions. Analysis was performed in triplicate on an ABI-7300 instrument (Applied Biosystems). The 2-ΔΔCt relative quantification method was used to calculate mRNA expression and RNA from untreated cells was used as calibrator. Total proteins were extracted from MDS cells as previously described [16] Immunoblot analysis was performed using anti-CEBPA (sc-166258, Santa Cruz Biotechnology, CA, USA), anti-PU.1 (PA5-17505, Thermo Fisher Scientific), and anti-GAPDH (#2118, Cell Signaling Technology, Danvers, MA, USA) antibodies. Horseradish peroxidase-conjugated IgGs were used as secondary antibodies. The autoradiograms were analyzed by densitometric analysis by ImageJ 1.45s software (National Institutes of Health, Bethesda, MD, USA).

4.6. Statistical Analysis

Statistical analysis was performed using the GraphPad Prism 5 software (GraphPad Software, San Diego, CA, USA) and data were reported as mean ± SD (unless otherwise specified). Correlations between olaparib IC_{50} values and proliferation indexes of MDS samples or the prognostic indexes listed in Table 1 were examined using the non-parametric Spearman's rank test. Differences between two groups were analyzed by the non-parametric Mann-Whitney and Wilcoxon (for matched samples) tests. For multiple comparisons, the non-parametric Kruskal-Wallis or Friedman tests followed by Dunn's post hoc test were used, as specified in the figure legends. All statistical tests were two-sided. p values below 0.05 were considered statistically significant.

5. Conclusions

Overall, our data demonstrate that single-agent olaparib, at clinically achievable concentrations, induces cell death of primary MDS cells, while sparing lymphocytes, which are not part of the MDS clone. Furthermore, in this short-term culture model, olaparib favors maturation of myeloid cells towards the neutrophilic lineage, as shown by induction of myeloid-specific transcription factors and increase of the metamyelocyte and neutrophil percentages. Our preclinical data, together with the acceptable toxicity profile of olaparib in vivo, an issue particularly relevant to the elderly MDS population, support the design of clinical trials with PARPi monotherapy or in combination with hypomethylating agents for the management of MDS patients.

Supplementary Materials: The following are available online at http://www.mdpi.com/2072-6694/11/9/1373/s1, Figure S1: Growth curves of untreated and olaparib treated primary MDS and normal progenitor cell cultures, Figure S2: Analysis of apoptosis in the myeloid and lymphocytic populations of MDS samples, Figure S3: Olaparib induces a decrease in the proportion of CD34+ cells in primary MDS samples, Figure S4: Olaparib induces myeloid differentiation in a primary low blast count AML sample, Figure S5: Olaparib and decitabine combination synergistically impairs survival in the OCI-AML2 leukemia cell line.

Author Contributions: I.F. designed the work, performed experiments, interpreted the data and wrote the paper; M.I.C., F.A. and E.F., M.G., F.D.C., G.F., L.T., P.C. performed experiments; A.D.V. read morphological slides; L.M., P.N., M.T.V., F.L.-C. followed MDS patients and provided clinical data on MDS sample; G.G., F.L.C. and M.T.V. planned the research strategy and wrote the paper. All authors reviewed and approved the manuscript.

Funding: This work was supported in part by AstraZeneca and by grants from "Associazione Italiana per la Ricerca sul Cancro" (AIRC IG N.16952 to M.T.V.; AIRC IG N.20353 to G.G. and AIRC 5 × 1000 call "Metastatic disease: the key unmet need in oncology" to MYNERVA project, #21267 http://www.progettoagimm.it). The funding sponsor had no role in the design of the study, in the collection, analyses or interpretation of data, and in the writing of the manuscript.

Acknowledgments: Dedicated to our generous mentor and friend Francesco Lo-Coco.

Conflicts of Interest: The authors declare no conflict of interest.

References

1. Arber, D.A.; Orazi, A.; Hasserjian, R.; Thiele, J.; Borowitz, M.J.; Le Beau, M.M.; Bloomfield, C.D.; Cazzola, M.; Vardiman, J.W. The 2016 revision to the World Health Organization classification of myeloid neoplasms and acute leukemia. *Blood* **2016**, *127*, 2391–2405. [CrossRef] [PubMed]
2. Gurion, R.; Vidal, L.; Gafter-Gvili, A.; Belnik, Y.; Yeshurun, M.; Raanani, P.; Shpilberg, O. 5-azacitidine prolongs overall survival in patients with myelodysplastic syndrome—A systematic review and meta-analysis. *Haematologica* **2010**, *95*, 303–310. [CrossRef] [PubMed]
3. Faraoni, I.; Graziani, G. Role of BRCA Mutations in Cancer Treatment with Poly(ADP-ribose) Polymerase (PARP) Inhibitors. *Cancers* **2018**, *10*, 487. [CrossRef] [PubMed]
4. McCabe, N.; Turner, N.C.; Lord, C.J.; Kluzek, K.; Białkowska, A.; Swift, S.; Giavara, S.; O'Connor, M.J.; Tutt, A.N.; Zdzienicka, M.Z.; et al. Deficiency in the Repair of DNA Damage by Homologous Recombination and Sensitivity to Poly(ADP-Ribose) Polymerase Inhibition. *Cancer Res.* **2006**, *66*, 8109–8115. [CrossRef]
5. Weston, V.J.; Oldreive, C.E.; Skowronska, A.; Oscier, D.G.; Pratt, G.; Dyer, M.J.S.; Smith, G.; Powell, J.E.; Rudzki, Z.; Kearns, P.; et al. The PARP inhibitor olaparib induces significant killing of ATM-deficient lymphoid tumor cells in vitro and in vivo. *Blood* **2010**, *116*, 4578–4587. [CrossRef] [PubMed]
6. Gadducci, A.; Guerrieri, M.E. PARP inhibitors alone and in combination with other biological agents in homologous recombination deficient epithelial ovarian cancer: From the basic research to the clinic. *Crit. Rev. Oncol. Hematol.* **2017**, *114*, 153–165. [CrossRef] [PubMed]
7. Majuelos-Melguizo, J.; Rodríguez, M.I.; López-Jiménez, L.; Rodríguez-Vargas, J.M.; Martín-Consuegra, J.M.M.; Serrano-Sáenz, S.; Gavard, J.; de Almodóvar, J.M.R.; Oliver, F.J. PARP targeting counteracts gliomagenesis through induction of mitotic catastrophe and aggravation of deficiency in homologous recombination in PTEN-mutant glioma. *Oncotarget* **2015**, *6*, 4790–4803. [CrossRef]
8. Sulkowski, P.L.; Corso, C.D.; Robinson, N.D.; Scanlon, S.E.; Purshouse, K.R.; Bai, H.; Liu, Y.; Sundaram, R.K.; Hegan, D.C.; Fons, N.R.; et al. 2-Hydroxyglutarate produced by neomorphic IDH mutations suppresses homologous recombination and induces PARP inhibitor sensitivity. *Sci. Transl. Med.* **2017**, *9*, eaal2463. [CrossRef]
9. Nickoloff, J.A.; Jones, D.; Lee, S.-H.; Williamson, E.A.; Hromas, R. Drugging the Cancers Addicted to DNA Repair. *J. Natl. Cancer Inst.* **2017**, *109*, 1–13. [CrossRef]
10. Faraoni, I.; Compagnone, M.; Lavorgna, S.; Angelini, D.F.; Cencioni, M.T.; Piras, E.; Panetta, P.; Ottone, T.; Dolci, S.; Venditti, A.; et al. BRCA1, PARP1 and γH2AX in acute myeloid leukemia: Role as biomarkers of response to the PARP inhibitor olaparib. *Biochim. Biophys. Acta.* **2015**, *1852*, 462–472. [CrossRef]
11. Wang, L.; Cai, W.; Zhang, W.; Chen, X.; Dong, W.; Tang, D.; Zhang, Y.; Ji, C.; Zhang, M. Inhibition of poly(ADP-ribose) polymerase 1 protects against acute myeloid leukemia by suppressing the myeloproliferative leukemia virus oncogene. *Oncotarget* **2015**, *6*, 27490–27504. [CrossRef] [PubMed]
12. Esposito, M.T.; Zhao, L.; Fung, T.K.; Rane, J.K.; Wilson, A.; Martin, N.; Gil, J.; Leung, A.Y.; Ashworth, A.; Eric So, C.W. Synthetic lethal targeting of oncogenic transcription factors in acute leukemia by PARP inhibitors. *Nat. Med.* **2015**, *21*, 1481–1490. [CrossRef] [PubMed]
13. Muvarak, N.E.; Chowdhury, K.; Xia, L.; Robert, C.; Choi, E.Y.; Cai, Y.; Bellani, M.; Zou, Y.; Singh, Z.N.; Duong, V.H.; et al. Enhancing the Cytotoxic Effects of PARP Inhibitors with DNA Demethylating Agents—A Potential Therapy for Cancer. *Cancer Cell* **2016**, *30*, 637–650. [CrossRef] [PubMed]
14. Piao, J.; Takai, S.; Kamiya, T.; Inukai, T.; Sugita, K.; Ohyashiki, K.; Delia, D.; Masutani, M.; Mizutani, S.; Takagi, M. Poly (ADP-ribose) polymerase inhibitors selectively induce cytotoxicity in TCF3-HLF–positive leukemic cells. *Cancer Lett.* **2017**, *386*, 131–140. [CrossRef] [PubMed]
15. Nieborowska-Skorska, M.; Sullivan, K.; Dasgupta, Y.; Podszywalow-Bartnicka, P.; Hoser, G.; Maifrede, S.; Martinez, E.; Di Marcantonio, D.; Bolton-Gillespie, E.; Cramer-Morales, K.; et al. Gene expression and mutation-guided synthetic lethality eradicates proliferating and quiescent leukemia cells. *J. Clin. Invest.* **2017**, *127*, 2392–2406. [CrossRef] [PubMed]

16. Faraoni, I.; Aloisio, F.; De Gabrieli, A.; Consalvo, M.I.; Lavorgna, S.; Voso, M.T.; Lo-Coco, F.; Graziani, G. The poly(ADP-ribose) polymerase inhibitor olaparib induces up-regulation of death receptors in primary acute myeloid leukemia blasts by NF-κB activation. *Cancer Lett.* **2018**, *423*, 127–138. [CrossRef] [PubMed]
17. Molenaar, R.J.; Radivoyevitch, T.; Nagata, Y.; Khurshed, M.; Przychodzen, B.; Makishima, H.; Xu, M.; Bleeker, F.E.; Wilmink, J.W.; Carraway, H.E.; et al. IDH1/2 Mutations Sensitize Acute Myeloid Leukemia to PARP Inhibition and This Is Reversed by IDH1/2-Mutant Inhibitors. *Clin. Cancer Res.* **2018**, *24*, 1705–1715. [CrossRef]
18. Maifrede, S.; Nieborowska-Skorska, M.; Sullivan-Reed, K.; Dasgupta, Y.; Podszywalow-Bartnicka, P.; Le, B.V.; Solecka, M.; Lian, Z.; Belyaeva, E.A.; Nersesyan, A.; et al. Tyrosine kinase inhibitor–induced defects in DNA repair sensitize FLT3(ITD)-positive leukemia cells to PARP1 inhibitors. *Blood* **2018**, *132*, 67–77. [CrossRef]
19. Poh, W.; Dilley, R.L.; Moliterno, A.R.; Maciejewski, J.P.; Pratz, K.W.; McDevitt, M.A.; Herman, J.G. BRCA1 Promoter Methylation Is Linked to Defective Homologous Recombination Repair and Elevated miR-155 to Disrupt Myeloid Differentiation in Myeloid Malignancies. *Clin. Cancer Res.* **2019**, *25*, 2513–2522. [CrossRef]
20. Faraoni, I.; Giansanti, M.; Voso, M.T.; Lo-Coco, F.; Graziani, G. Targeting ADP-Ribosylation by PARP inhibitors in Acute Myeloid Leukaemia and Related Disorders. *Biochem. Pharmacol.* **2019**, *167*, 133–148. [CrossRef]
21. Kraus, W.L.; Lis, J.T. PARP Goes Transcription. *Cell* **2003**, *113*, 677–683. [CrossRef]
22. Muthurajan, U.M.; Hepler, M.R.D.; Hieb, A.R.; Clark, N.J.; Kramer, M.; Yao, T.; Luger, K. Automodification switches PARP-1 function from chromatin architectural protein to histone chaperone. *Proc. Natl. Acad. Sci. USA* **2014**, *111*, 12752–12757. [CrossRef] [PubMed]
23. Martin, K.A.; Cesaroni, M.; Denny, M.F.; Lupey, L.N.; Tempera, I. Global Transcriptome Analysis Reveals That Poly(ADP-Ribose) Polymerase 1 Regulates Gene Expression through EZH2. *Mol. Cell. Biol.* **2015**, *35*, 3934–3944. [CrossRef] [PubMed]
24. Hottiger, M.O. Nuclear ADP-Ribosylation and Its Role in Chromatin Plasticity, Cell Differentiation, and Epigenetics. *Annu. Rev. Biochem.* **2015**, *84*, 227–263. [CrossRef] [PubMed]
25. Pellagatti, A.; Cazzola, M.; Giagounidis, A.; Perry, J.; Malcovati, L.; Della Porta, M.G.; Jädersten, M.; Killick, S.; Verma, A.; Norbury, C.J.; et al. Deregulated gene expression pathways in myelodysplastic syndrome hematopoietic stem cells. *Leukemia* **2010**, *24*, 756–764. [CrossRef] [PubMed]
26. Zhou, T.; Hasty, P.; Walter, C.A.; Bishop, A.J.R.; Scott, L.M.; Rebel, V.I. Myelodysplastic syndrome: An inability to appropriately respond to damaged DNA? *Exp. Hematol.* **2013**, *41*, 665–674. [CrossRef] [PubMed]
27. Fong, P.C.; Boss, D.S.; Yap, T.A.; Tutt, A.; Wu, P.; Mergui-Roelvink, M.; Mortimer, P.; Swaisland, H.; Lau, A.; O'Connor, M.J.; et al. Inhibition of Poly(ADP-Ribose) Polymerase in Tumors from BRCA Mutation Carriers. *N. Engl. J. Med.* **2009**, *361*, 123–134. [CrossRef] [PubMed]
28. Mateo, J.; Moreno, V.; Gupta, A.; Kaye, S.B.; Dean, E.; Middleton, M.R.; Friedlander, M.; Gourley, C.; Plummer, R.; Rustin, G.; et al. An Adaptive Study to Determine the Optimal Dose of the Tablet Formulation of the PARP Inhibitor Olaparib. *Target. Oncol.* **2016**, *11*, 401–415. [CrossRef]
29. Pratz, K.W.; Koh, B.; Patel, A.G.; Flatten, K.S.; Poh, W.; James, G. Poly(ADP-Ribose) Polymerase Inhibitor Hypersensitivity in Aggressive Myeloproliferative Neoplasms. *Clin. Cancer Res.* **2016**, *22*, 3894–3902. [CrossRef]
30. Westers, T.M.; Ireland, R.; Kern, W.; Alhan, C.; Balleisen, J.S.; Bettelheim, P.; Burbury, K.; Cullen, M.; Cutler, J.A.; Della Porta, M.G.; et al. Standardization of flow cytometry in myelodysplastic syndromes: A report from an international consortium and the European LeukemiaNet Working Group. *Leukemia* **2012**, *26*, 1730–1741. [CrossRef]
31. Chou, T.-C. Drug Combination Studies and Their Synergy Quantification Using the Chou-Talalay Method. *Cancer Res.* **2010**, *70*, 440–446. [CrossRef] [PubMed]
32. Puthiyaveetil, A.G.; Reilly, C.M.; Pardee, T.S.; Caudell, D.L. Non-homologous end joining mediated DNA repair is impaired in the NUP98-HOXD13 mouse model for myelodysplastic syndrome. *Leuk. Res.* **2013**, *37*, 112–116. [CrossRef] [PubMed]
33. D'Alò, F.; Di Ruscio, A.; Guidi, F.; Fabiani, E.; Greco, M.; Rumi, C.; Hohaus, S.; Voso, M.T.; Leone, G.P.U. 1 and CEBPA expression in acute myeloid leukemia. *Leuk. Res.* **2008**, *32*, 1448–1453. [CrossRef] [PubMed]
34. Burda, P.; Laslo, P.; Stopka, T. The role of PU.1 and GATA-1 transcription factors during normal and leukemogenic hematopoiesis. *Leukemia* **2010**, *24*, 1249–1257. [CrossRef] [PubMed]

35. Pundhir, S.; Bratt Lauridsen, F.K.; Schuster, M.B.; Jakobsen, J.S.; Ge, Y.; Schoof, E.M.; Rapin, N.; Waage, J.; Hasemann, M.S.; Porse, B.T. Enhancer and Transcription Factor Dynamics during Myeloid Differentiation Reveal an Early Differentiation Block in Cebpa null Progenitors. *Cell Rep.* **2018**, *23*, 2744–2757. [CrossRef]
36. Posavec Marjanović, M.; Crawford, K.; Ahel, I. PARP, transcription and chromatin modeling. *Semin. Cell Dev. Biol.* **2017**, *63*, 102–113. [CrossRef]
37. Franzese, E.; Centonze, S.; Diana, A.; Carlino, F.; Guerrera, L.P.; Di Napoli, M.; De Vita, F.; Pignata, S.; Ciardiello, F.; Orditura, M. PARP inhibitors in ovarian cancer. *Cancer Treat. Rev.* **2019**, *73*, 1–9. [CrossRef]
38. Bhatia, M.; Kirkland, J.B.; Meckling-Gill, K.A. Overexpression of poly(ADP-ribose) polymerase promotes cell cycle arrest and inhibits neutrophilic differentiation of NB4 acute promyelocytic leukemia cells. *Cell Growth Differ.* **1996**, *7*, 91–100.
39. Bhatia, M.; Kirkland, J.B.; Meckling-Gill, K.A. Modulation of poly(ADP-ribose) polymerase during neutrophilic and monocytic differentiation of promyelocytic (NB4) and myelocytic (HL-60) leukaemia cells. *Biochem. J.* **1995**, *308*, 131–137. [CrossRef]
40. Wiśnik, E.; Płoszaj, T.; Robaszkiewicz, A. Downregulation of PARP1 transcription by promoter-associated E2F4-RBL2-HDAC1-BRM complex contributes to repression of pluripotency stem cell factors in human monocytes. *Sci. Rep.* **2017**, *7*, 9483. [CrossRef]
41. Diamantopoulos, P.; Zervakis, K.; Zervakis, P.; Sofotasiou, M.; Vassilakopoulos, T.; Kotsianidis, I.; Symeonidis, A.; Pappa, V.; Galanopoulos, A.; Solomou, E.; et al. Poly (ADP-ribose) polymerase 1 mRNA levels strongly correlate with the prognosis of myelodysplastic syndromes. *Blood Cancer J.* **2017**, *7*, e533. [CrossRef] [PubMed]
42. Wang, J.; Sun, Q.; Morita, Y.; Jiang, H.; Groß, A.; Lechel, A.; Hildner, K.; Guachalla, L.M.; Gompf, A.; Hartmann, D.; et al. A Differentiation Checkpoint Limits Hematopoietic Stem Cell Self-Renewal in Response to DNA Damage. *Cell* **2012**, *148*, 1001–1014. [CrossRef] [PubMed]
43. Santos, M.A.; Faryabi, R.B.; Ergen, A.V.; Day, A.M.; Malhowski, A.; Canela, A.; Onozawa, M.; Lee, J.; Callen, E.; Gutierrez-Martinez, P.; et al. DNA-damage-induced differentiation of leukaemic cells as an anti-cancer barrier. *Nature* **2014**, *514*, 107–111. [CrossRef]
44. Mirza, M.R.; Pignata, S.; Ledermann, J.A. Latest clinical evidence and further development of PARP inhibitors in ovarian cancer. *Ann. Oncol.* **2018**, *29*, 1366–1376. [CrossRef] [PubMed]
45. Pujade-Lauraine, E.; Ledermann, J.A.; Selle, F.; Gebski, V.; Penson, R.T.; Oza, A.M.; Korach, J.; Huzarski, T.; Poveda, A.; Pignata, S.; et al. Olaparib tablets as maintenance therapy in patients with platinum-sensitive, relapsed ovarian cancer and a BRCA1/2 mutation (SOLO2/ENGOT-Ov21): A double-blind, randomised, placebo-controlled, phase 3 trial. *Lancet Oncol.* **2017**, *18*, 1274–1284. [CrossRef]
46. Mirza, M.R.; Monk, B.J.; Herrstedt, J.; Oza, A.M.; Mahner, S.; Redondo, A.; Fabbro, M.; Ledermann, J.A.; Lorusso, D.; Vergote, I.; et al. Niraparib Maintenance Therapy in Platinum-Sensitive, Recurrent Ovarian Cancer. *N. Engl. J. Med.* **2016**, *375*, 2154–2164. [CrossRef] [PubMed]
47. Coleman, R.L.; Oza, A.M.; Lorusso, D.; Aghajanian, C.; Oaknin, A.; Dean, A.; Colombo, N.; Weberpals, J.I.; Clamp, A.; Scambia, G.; et al. Rucaparib maintenance treatment for recurrent ovarian carcinoma after response to platinum therapy (ARIEL3): A randomised, double-blind, placebo-controlled, phase 3 trial. *Lancet* **2017**, *390*, 1949–1961. [CrossRef]
48. Orta, M.L.; Höglund, A.; Calderón-Montaño, J.M.; Domínguez, I.; Burgos-Morón, E.; Visnes, T.; Pastor, N.; Ström, C.; López-lázaro, M.; Helleday, T. The PARP inhibitor Olaparib disrupts base excision repair of 5-aza-2′-deoxycytidine lesions. *Nucleic Acids Res.* **2014**, *42*, 9108–9120. [CrossRef]
49. Zhao, L.; So, C.W.E. PARPi potentiates with current conventional therapy in MLL leukemia. *Cell Cycle* **2017**, *16*, 1861–1869. [CrossRef]
50. Diesch, J.; Zwick, A.; Garz, A.-K.; Palau, A.; Buschbeck, M.; Götze, K.S. A clinical-molecular update on azanucleoside-based therapy for the treatment of hematologic cancers. *Clin. Epigenetics* **2016**, *8*, 71. [CrossRef]
51. Dockery, L.E.; Tew, W.P.; Ding, K.; Moore, K.N. Tolerance and toxicity of the PARP inhibitor olaparib in older women with epithelial ovarian cancer. *Gynecol. Oncol.* **2017**, *147*, 509–513. [CrossRef] [PubMed]

cancers

MDPI

Review

The Nucleolus and PARP1 in Cancer Biology

Marina Engbrecht and Aswin Mangerich *

Molecular Toxicology Group, Department of Biology, 78457 Konstanz, Germany; marina.engbrecht@uni-konstanz.de

* Correspondence: aswin.mangerich@uni-konstanz.de; Tel.: +49-7531-88-4067

Received: 24 May 2020; Accepted: 1 July 2020; Published: 6 July 2020

Abstract: The nucleolus has been known for a long time to fulfill crucial functions in ribosome biogenesis, of which cancer cells can become addicted to in order to produce sufficient amounts of proteins for cell proliferation. Recently, the nucleolus has emerged as a central regulatory hub in many other cancer-relevant processes, including stress sensing, DNA damage response, cell cycle control, and proteostasis. This fostered the idea that nucleolar processes can be exploited in cancer therapy. Interestingly, a significant proportion of poly(ADP-ribose) polymerase 1 (PARP1) molecules are localized in the nucleolus and PARP1 also plays crucial roles in many processes that are important in cancer biology, including genome maintenance, replication, transcription, and chromatin remodeling. Furthermore, during the last years, PARP1 came into focus in oncology since it represents a promising target of pharmacological PARP inhibitors in various types of cancers. Here, we provide an overview of our current understanding on the role of PARP1 in nucleolar functions and discuss potential implications in cancer biology and therapy.

Keywords: nucleolus; poly(ADP-ribosyl)ation; PARP; ARTDs; cancer

1. Introduction into the Biology of Nucleoli

Nucleoli are self-organizing, membrane-less sub-compartments of the nucleus, which are formed around tandemly repeated clusters of 200 to 400 ribosomal DNA (rDNA) genes known as nucleolar organizing centers (NORs) on the short arms of the five human acrocentric chromosomes 13, 14, 15, 21, and 22 [1,2]. In humans, nucleoli are surrounded by peri-nucleolar heterochromatin (PNH) derived from DNA sequences located distal and proximal to NORs [3]. The main function of nucleoli is ribosome biogenesis, which is one of the most energy-demanding and highly controlled processes in a cell. It has been estimated that in proliferating cells ribosome biogenesis consumes up to 80% of the cellular energy [4]. Therefore, it is not surprising that ribosome biogenesis is tightly coupled to the availability of growth factors, nutrients and cellular energy supply [5]. Nucleolar size positively correlates with rRNA synthesis and cell proliferation, as dividing cells often possess large nucleoli, while downregulation of rRNA gene transcription is associated with a reduction in nucleolar size [2]. Interestingly, an increased number and size of nucleoli have historically been used as a biomarker for tumor development [6].

Ribosome biogenesis starts with RNA polymerase I (Pol I)-driven synthesis of 47S pre-ribosomal RNA (pre-rRNA), which is rapidly processed by more than 200 non-ribosomal proteins and small nucleolar RNAs (snoRNAs) to mature 18S, 5.8S and 28S rRNAs [7]. These RNAs are assembled with ribosomal proteins (RPs) and 5S rRNA to form the pre-40S and pre-60S ribosomal subunits that are subsequently exported to the nucleoplasm to produce mature 80S ribosomes. Since the 5S rRNA is synthesized by RNA polymerase III (Pol III) and transcription of RP genes is mediated by RNA polymerase II (Pol II), ribosome biogenesis requires the activities of all three cellular RNA polymerases.

Light and electron microscopy have revealed the tripartite structure of the nucleolus, which reflects the different stages of ribosome biogenesis and is dependent on ongoing rDNA transcription [3,8].

During interphase, nucleoli consist of one or a few fibrillar centers (FC), of which each is surrounded by a dense fibrillar component (DFC) [9]. The FC contains non-transcribed rDNA sequences and transcription factors, e.g., the upstream binding factor (UBF), which recognizes the rRNA gene promotor. Transcription by Pol I occurs at the interface between FCs and the DFCs. DFCs are enriched in pre-rRNA processing factors, e.g., small nucleolar ribonucleoproteins (snoRNPs) and fibrillarin, and are the nucleolar sites, where early rRNA processing takes place. Both the FCs and the DFCs are embedded in a large granular component (GC), which is associated with late rRNA processing. Apart from this, mature rRNAs assemble in the GC with RPs into pre-ribosomal subunits [9].

Nucleoli are membrane-less biomolecular condensates thought to be organized by liquid-liquid phase separation [10,11]. This is supported by the finding that nucleoli of *Xenopus laevis* germinal vesicles fuse and turn over rapidly, thereby displaying liquid-like properties [12,13]. Mechanistically, molecular interactions between low-complexity, disordered protein regions and RNA molecules appear to contribute to such phase separation processes [14]. For a long time, an unresolved question has been, how the three nucleolar components can coexist without fusing to a single liquid phase. Recently, this issue has been addressed by Feric et al. [15] showing that nucleolar sub-compartments represent distinct liquid phases that can co-exist due to differences in the biophysical properties of their components, in particular due to distinct droplet surface tensions. Current evidence supports a model in which nucleoli are formed by a combination of active recruitment processes of proteins, active transcription of rDNA and formation of the respective rRNA, as well as phase separation processes involving both protein and RNA components [14].

In humans, nucleoli undergo structural reorganization throughout the cell cycle resulting in a transient loss of the tripartite structure. As cells enter mitosis, transcription of rDNA genes pauses, and nucleoli disperse. Interestingly, some nucleolar proteins, e.g., UBF, remain associated with NORs throughout the cell cycle, thus functioning as a mitotic bookmark and binding platform for the remaining Pol I transcription machinery [10,16]. Components of the DFC and the GC, such as fibrillarin and nucleophosmin, dissociate from the nucleolus during mitosis [17]. Before nucleoli reassemble in late anaphase or early telophase, these processing factors coalesce in foci designated as prenucleolar bodies (PNBs) [18]. The reassembly of nucleoli is initiated when pre-rRNA processing factors associate with NORs. First, multiple small nucleoli are formed, which as the cell cycle progresses coalesce to larger mature nucleoli [19].

Intriguingly, over the past two decades the nucleolus has emerged as a regulatory hub for multiple nuclear functions [20]. Over 4500 proteins were identified to be localized in nucleoli, of which only 30% are primarily associated with ribosome biogenesis [21,22]. This led to the assumption that the nucleolus has non-canonical regulatory functions beyond ribosome biogenesis. At present, it is well accepted that the nucleolus is a multifunctional nuclear sub-compartment, with additional roles, e.g., in stress response, DNA damage signaling, telomere maintenance, cell cycle control, cell proliferation, and proteostasis [23]. Furthermore, links between nucleolar functions and complex physiological and pathophysiological cellular processes have been established. For instance, there is increasing knowledge that the nucleolus is implicated in the aging process. In this context, several studies have postulated that rDNA instability could be a causative factor for aging [24]. Recently, the nucleolar size was shown to inversely correlate with longevity, thereby identifying small nucleoli as a potential, visible hallmark for longevity and metabolic health [4,25]. Moreover, there is growing evidence, that the nucleolus is involved in the development of neurodegenerative diseases, such as Parkinson's disease [26]. Importantly, dysregulation of nucleolar functions has been linked to carcinogenesis, and consequently the nucleolus has emerged as a new target in cancer therapy [27–30]. This aspect will be discussed in more detail in Section 5 below. Before returning to this issue, in the next sections, we discuss how nucleoli respond to stress and DNA damage (Section 2), give a brief introduction into PARylation and PARP1 (Section 3), and review the role of PARP1 in nucleolar functions (Section 4).

2. How do Nucleoli Respond to Stress and DNA Damage?

Cells are constantly exposed to exogenous and endogenous sources of DNA damage [31]. To maintain genomic integrity, they possess a repertoire of repair proteins, which detect specific DNA lesions and initiate the appropriate repair pathway. Intriguingly, over 150 DNA repair proteins were identified in nucleoli under non-stress conditions, pointing to a role of nucleoli in DNA damage response [32].

As there is no structural barrier between the nucleolus and the surrounding nucleoplasm, proteins can, in principle, freely traffic from the nucleolus to the nucleoplasm and vice versa. The mechanisms by which proteins are retained in the nucleolus are still not fully understood. In general, it is assumed that nucleolar accumulation is a consequence of affinity interactions with RNA or nucleolar proteins. Some nucleolar proteins, e.g., nucleophosmin, exhibit a nucleolar localization sequence (NoLS) [33]. While there is no highly-conserved NoLS consensus sequence, about half of the amino acids (aa) inside NoLS are lysines and arginines, which render NoLS highly positively charged [23]. Since the nucleolus harbors high numbers of negatively charged RNA molecules, electrostatic interactions between NoLS and RNA molecules were postulated to be responsible for nucleolar accumulation of NoLS-bearing proteins. However, other nucleolar proteins lack a well-defined NoLS and are thought to be targeted to the nucleolus through protein-protein interactions with nucleolar anchored hub proteins, such as nucleophosmin or nucleolin [23].

The nucleolar accumulation of a variety of DNA repair proteins raised the question whether DNA repair proteins are merely sequestered in the nucleolus for storage reasons until they are required for their functional role in DNA repair, or if they also fulfill nucleolar-specific functions. Indeed, at least for some DNA repair proteins, including the base excision repair (BER) protein apurinic/apyrimidinic endonuclease 1 (APE1) and the RecQ helicase WRN, nucleolar-specific functions were postulated [32,34]. APE1 was not only shown to reside in the nucleolus and interact with nucleophosmin, there is also growing evidence that it is involved in quality control of the transcribed rRNA [35]. The gene encoding for the Werner protein (WRN) harbors mutations in Werner syndrome, an adult onset progeria characterized by premature aging and cancer development [36]. WRN is involved in a number of DNA repair pathways, including homologous recombination (HR), non-homologous end joining (NHEJ) and BER [37,38]. Under non-stress conditions, it localizes to transcriptionally active nucleoli, giving rise to the idea that WRN might also play a role in ribosome biogenesis [39]. Further studies supported this notion in demonstrating that WRN co-immunoprecipitates with the RNA Pol I subunit RPA40 and that in the absence of WRN, 18S and 28S RNA levels are reduced [40]. Although the findings require further elucidations, these data provide convincing evidence that bona fide DNA repair proteins might play dual roles in DNA repair on the one hand and ribosome biogenesis on the other hand.

Nucleoli are highly dynamic and undergo dramatic structural changes upon suffering various types of DNA damage. A plethora of cellular stressors, including UV and γ radiation, oxidative stress, genotoxic chemotherapeutic agents, hypoxia, as well as nutrient and growth factor deprivation can induce nucleolar re-organization or disruption [41]. Any stress-induced perturbation in ribosome biogenesis that ultimately leads to disruptions in cell homeostasis through activation of p53 or other stress signaling is referred to as nucleolar—or ribosomal stress [42]. Induction of nucleolar stress can occur at various steps of ribosome biogenesis, from Pol I transcription to pre-RNA processing, and eventually to assembly of ribosomal subunits and their release from the GC [42]. For instance, treatment with actinomycin D (ActD), which at low doses acts as a selective inhibitor of RNA Pol I by intercalating with G/C-rich rDNA, was shown to induce nucleolar segregation and the formation of so called nucleolar caps [14]. Nucleolar caps are bipartite structures that are excluded from the GC and reside at the nucleolar surface. They consist of FC and DFC components, with the DFC facing the nucleolar interior and the FC facing the nucleoplasm [3]. Interestingly, a similar reorganization of nucleolar structure was observed upon targeted induction of DSBs in rDNA repeats, which was shown to result in ataxia telangiectasia mutated (ATM)-dependent inhibition of RNA Pol I transcription [43–45]. Other studies reported that rDNA DSB-induced Pol I inhibition and nucleolar cap formation can also be regulated by the DNA-dependent protein kinase (DNA-PK) [46,47]. The formation of nucleolar caps was proposed to render damaged rDNA accessible for repair factors that are normally excluded from nucleoli under non-stress conditions [48]. In three recent studies, the *Physarum polycephalum* homing endonuclease I-*Ppo*I, which can cleave within the 28S rRNA coding region in human cells, was used to investigate whether rDNA DSBs are repaired by HR or NHEJ [49]. Harding et al. identified NHEJ as the predominant pathway in the repair of DSBs within rDNA, since depletion of NHEJ, but not HR factors, leads to accumulation of DSBs in the nucleolus [44]. Instead, van Sluis et al. [45] demonstrated that HR is also involved in rDNA DSB repair. Interestingly, HR factors are recruited to stress-induced nucleolar caps even in G1 cells, suggesting that in rDNA, HR occurs even in the absence of sister chromatids, possibly by using other rDNA repeats as a template. Therefore, the formation of nucleolar caps could facilitate repair of rDNA via HR by concentrating high levels of homologous sequences in close proximity [50]. Interestingly and in contrast to the general notion that HR acts as an error-free repair mechanism, it was reported that the HR-mediated repair of rDNA is error-prone, leading to a reduction of rDNA repeats, thus promoting rDNA instability [51].

Ribosomal DNA sequences are especially vulnerable to DNA damage due to their repetitive nature, the unique organization of rDNA in clusters on five different chromosomes and high transcription rates [17]. These special features render the rDNA prone to unscheduled DNA recombination events and frequent formation of DNA:RNA hybrids, which are thought to favor the generation of DSBs [52,53]. If not properly repaired, rDNA damage can give rise to rDNA instability, and therefore contribute to premature onset of disease and carcinogenesis. Indeed, 50% of lung and colorectal cancers were shown to harbor rDNA gene rearrangements [54]. The pathways involved in rDNA damage repair and their regulatory mechanisms still require further investigations to gain a deeper insight into the mechanisms, which drive rDNA instability and ultimately carcinogenesis.

Perturbations in nucleolar functions are often accompanied by the release of nucleolar and ribosomal proteins into the nucleoplasm, where they are proposed to take on secondary functions in stress response. For instance, nucleophosmin was shown to shuttle between the nucleolus, the nucleoplasm, and the cytoplasm upon induction of stress [55,56]. Nucleolar release of nucleophosmin can be induced by a variety of DNA-damaging agents, including ionizing radiation (IR), cisplatin or etoposide treatment [23]. Interestingly, it was postulated that nucleophosmin takes over moonlighting functions in the nucleoplasm by participating in several DNA repair pathways, including BER and nucleotide excision repair (NER) [57]. Additionally, nucleophosmin drives p53 signaling upon genotoxic stress. Nucleophosmin constitutively interacts with the key activator of the p53 signaling pathway ARF, thus targeting it to the nucleolus [58]. Upon genotoxic stress, modification of nucleophosmin leads to the release of ARF into the nucleoplasm, where it inhibits HDM2, the E3 ligase that negatively regulates p53 [17]. Additionally, nucleophosmin regulates the HDM2-p53 pathway in an ARF-independent manner, by direct protein-protein interactions with HDM2 [59]. It is important to note that while most proteins are released from nucleoli under stress conditions, there are some notable exceptions. For example, upon stress, the promyelocytic leukemia tumor suppressor (PML) protein sequesters Mdm2 to the nucleolus by nucleolar translocation, thereby stabilizing p53 in the nucleoplasm, and promoting apoptosis [60]. Furthermore, stress-induced ubiquitination of the NF-κB subcomponent RelA leads to its translocation to the nucleolus, by which this process elicits pro-apoptotic effects [61]. Interestingly, also Hsp70 and other chaperone proteins accumulate in the nucleolus during stress conditions, suggesting to protect nucleolar proteins from aggregation [62–65]. In this regard, a recent study by Frottin et al. [63] addressed the role of liquid-liquid demixing and phase transition processes during nucleolar stress response. The authors reported that the liquid-like GC phase of the nucleolus can act as a quality control compartment by exerting transient chaperone-like activity for nuclear proteins entering the nucleolus during stress conditions [63].

Apart from proteins that have predominantly nucleolar functions, several DNA repair factors, including APE1 and WRN, have been shown to undergo DNA damage-induced nucleolar-nucleoplasmic shuttling [23,40,66]. Despite extensive research the exact molecular mechanisms, by which proteins are released from the nucleolus remain poorly understood. In general, it is assumed that such protein dynamics are regulated by a network of PTMs and protein-protein interactions, which is likely to occur in a stress- and protein-specific manner. For instance, Karmakar et al. [67] showed that WRN is acetylated upon treatment with mitomycin C and the alkylating agent methyl methane-sulfonate (MMS), but not after UV irradiation. In addition, phosphorylation was proposed to modulate WRN's subnuclear localization [39,68]. Consistent with the idea that such DNA damage-induced protein dynamics are regulated in a stress-specific manner, we recently found that after treatment with H_2O_2 and the alkylating agent 2-chloroethyl ethyl sulfide (CEES), but not the topoisomerase inhibitor camptothecin (CPT), nucleolar-nucleoplasmic translocation of WRN was dependent on the PARP1 protein, yet independent of its enzymatic activity [69].

3. Introduction into the Biology of PARP1 and Poly(ADP-ribosyl)ation

Post-translational modifications (PTMs), such as phosphorylation, acetylation or ubiquitination, represent an extremely versatile and fast means of regulating the complexity and dynamics of cellular processes. In particular, ADP-ribosylation, which can be found in most eukaryotes and which has been shown to be essential during mammalian development [70], represents a highly dynamic and fully reversible PTM [70–73]. This modification is catalyzed by ADP-ribosyl transferases to which the family of poly(ADP-ribose) polymerases (PARPs) belongs. The PARP gene family comprises 17 members in humans and based on structural homology of their catalytic domain with the diphteria toxin, they are also referred to as ADP-ribosyltransferases diptheria toxin-like (ARTDs) [74]. By using nicotinamide adenine dinucleotide (NAD^+) as a substrate, PARPs covalently attach ADP-ribose units to a variety of aa residues of acceptor proteins, including serines, glutamates, aspartates, lysines, and tyrosines [75–82]. While most PARPs possess mono(ADP-ribosyl)transferase activity (mono(ADP-ribosyl)ation, MARylation) or are catalytically inactive, at least four of them, i.e., PARPs 1, 2, 5a, 5b, are known to synthesize negatively charged polymers of poly(ADP-ribose) (PAR), which earned them the term 'writers' of poly(ADP-ribosyl)ation (PARylation)—(PARPs 5a and 5b are also known as tankyrases 1 and 2). The resulting PAR chains are either linear or branched and may consist of more than 200 ADP-ribose moieties [83–85]. Apart from covalent modification, certain proteins—so called 'readers' of PARylation—can non-covalently interact with PAR chains via distinct PAR binding modules, which, among others, include loosely conserved PAR binding motifs (PBM), zinc finger-type structures, macrodomains, WWE domains and OB folds [86].

PARP1—the founding member of the PARP family—was extensively studied since the discovery of PARylation in 1963 [87]. It plays pivotal roles in various cellular processes, including chromatin remodeling, replication, transcription, RNA biology, energy metabolism, immunity and inflammation, as well as cell death [88–92]. In addition, one of the most critical functions of PARP1, is its role in DNA repair and genome maintenance. Thus, PARP1 participates in several DNA repair pathways, including BER, NER, as well as the two double-strand break (DSB) repair pathways, i.e., NHEJ and HR [88,93,94]. Furthermore, PARP1 was shown to recognize unligated Okazaki fragments and to stabilize stalled replication forks, owing to promote repair during replication and replicative stress [95,96]. The importance of PARP1 in DNA repair is impressively exemplified by the fact that *Parp1*-deficient mice are hypersensitive towards DNA damaging agents, which is accompanied by increased spontaneous as well as induced genomic instability and carcinogenesis [97]. On a molecular level, PARP1 acts as a sensor of DNA damage, in particular DNA single and double strand breaks, and detects those via certain zinc finger motifs of its N-terminal DNA binding domain. Binding to DNA strand breaks induces allosteric conformational changes in the protein structure, which allow binding of NAD^+ to the C-terminal catalytic domain, leading to the enzymatic activation of PARP1 [71,85,98]. Indeed, PARP1 was reported to be responsible for over 90% of DNA damage-induced PARylation, whereupon cellular PAR levels dramatically increase 10- to 500-fold compared to the PAR levels under non-stress conditions [88,99,100]. Apart from DNA damage-dependent activation, PARP1 activity is regulated by other PTMs, such as acetylation, phosphorylation and SUMOylation [101–104] as well as physical protein-protein interactions [85]. The first and main target of PARP1-dependent PARylation is PARP1 itself [105,106]. Such automodification of PARP1 provides a platform for the recruitment of downstream factors, i.e., 'readers' of PARylation, of which the DNA repair protein XRCC1 represents a prime example [107–109]. There is now considerable evidence that high-affinity, non-covalent interaction of proteins with auto-PARylated PARP1 mediates substrate specificity of PARP1 and targets these PAR-binding proteins to subsequent covalent modification by PARP1 [110–112]. In total, several hundred PARylated proteins have been identified, which are involved in diverse cellular functions, ranging from genome maintenance, DNA damage response and chromatin organization to transcription, RNA metabolism and cell cycle regulation [80,82,113].

Eventually PARP1 is released from DNA, which is assumed to occur due to steric and electrostatic repulsion of the automodified protein, yet other, more specific and so far largely unexplored, mechanisms may be conceivable as well [88,114,115]. Importantly, in many instances and in particular upon genotoxic stress, PARylation is transient, highly dynamic, and fully reversible, since after being synthesized, PAR chains are rapidly degraded in a two-step process, due to the enzymatic activities of certain 'eraser' enzymes [116,117]. Thus, the bulk of DNA damage-induced PAR can be degraded by poly(ADP-ribose) glycohydrolase (PARG), which harbors exo- as well as endo-glycosidic activities for PAR. The most proximal, protein-bound ADP-ribose moiety, however, cannot be released by PARG. This is instead carried out by several other eraser enzymes, which possess individual specificities for certain ADP-ribose acceptor sites, e.g., ARH3 can remove ADP-ribose from serine, and MacroD1/MacroD2 from glutamate or aspartate [116–121]. Moreover, the terminal ADP-ribose protein glycohydrolase 1 (TARG1/C6orf130) was shown to act on glutamate and aspartate residues, as well, by removing and releasing not only ADP-ribose, but also entire PAR chains [116,122].

In general, PARylation is thought to regulate the physicochemical properties, localization, and enzymatic activities of its target proteins in a highly controlled, spatio-temporal manner. Furthermore, PAR can serve as a signal to target proteins for degradation through the ubiquitin-proteasome system [123], which can, therefore, act as a 'kiss of death' for certain factors during DNA damage response. Another layer of complexity in the functions of PARylation is given by findings demonstrating that PAR can act as a seed for liquid-liquid demixing processes, thereby triggering the formation of biomolecular condensates [124,125]. In this regard, it has also been reported that such processes contribute to the formation of dynamic protein foci at sites of DNA damage, thereby facilitating DNA repair processes through transient and functional compartmentalization of DNA damage sites [110,124]. Moreover, there is mounting evidence that PARylation and PAR-binding regulate liquid-liquid phase separation and aggregation of several neurodegenerative disease-associated RNA-binding proteins, including α-synuclein, TDP-43 and hnRNP A1 [126]. In summary, it is assumed that PARP1 orchestrates and supports DNA damage response mechanisms and local chromatin dynamics. In addition, beyond its role in the control of protein localization and biochemistry, PARP1 is involved in the regulation of cell death and cellular energy metabolism, i.e., by using NAD^+ as a substrate, for which reason it proves itself as a global regulator of cellular physiology and pathophysiology [92,127].

Over the past decades, PARP1 came into focus as a target in clinical oncology, since PARP inhibitors were identified to act as chemosensitizers in combination with classical DNA-damaging therapies or as monotherapeutic agents to treat cancers with defects in HR repair according to the concept of synthetic lethality. In 2005, two independent groups discovered the synthetic lethal interaction between PARP1 inhibition and loss of BRCA, i.e., *BRCA1* or *BRCA2* [128,129], which spurred the development of clinical PARP inhibitors. By now, four of such small-molecule PARP inhibitors (i.e., olaparib, rucaparib, niraparib, and talazoparib) have been approved by authorities such as the EMA and the FDA to treat certain types of ovarian, breast or pancreatic cancer with germline loss-of-function mutations of *BRCA* genes [130,131]. Functional BRCA1 and BRCA2 are of critical importance for the repair of DSBs via HR [132]. Inhibition of PARP not only leads to accumulation of DNA single-strand breaks (SSBs), but also to trapping of PARP at the DNA, which may result in toxic manifestations of the damage and stalled replication forks [133]. Due to PARP inhibition and the absence of BRCA, the stalled replication forks cannot be restarted properly, leading to replication fork collapse and ultimately DSBs, which are lethal for HR-deficient cancer cells [134]. By selectively targeting certain types of cancers, PARP1 inhibitors provide a successful step towards precision medicine in oncology. In addition to their use in *BRCA*-mutated cancers, there is meanwhile good evidence for further synthetic lethal interactions of PARP inhibitors in combination with other genetic constellations. Moreover, the use of PARP inhibitors as chemosensitizers represents a promising strategy for their use in cancer treatment [130–132]. Taken together, it is expected that the area of application for the use of PARP inhibitors will further expand during the next decade and that PARP inhibitors will find their place in the range of chemotherapy regimens.

4. On the Role of PARP1 and PARylation in the Biology of the Nucleolus

Under non-stress conditions a substantial percentage of cellular PARP1 molecules (i.e., ~40%) reside within nucleoli [135]. Nucleolar accumulation of PARP1 was firstly documented in the late 1980s by using immunolabeling and was later on confirmed in proteomic studies [136,137]. PARP1, as well as PARP2, are retained in nucleoli via interaction with the multifunctional nucleolar hub protein nucleophosmin, which is implicated in multiple steps of ribosome biogenesis, including rDNA transcription and elongation, as well as rRNA processing [32,138]. Treatment with the RNA Pol I inhibitor ActD resulted in nucleolar release of PARP1 and PARP2, indicating that active nucleolar transcription is required for PARP1 and PARP2 to reside in nucleoli [139]. The presence of PARP1, particularly in transcriptionally active nucleoli, gave rise to the idea that PARP1 might be involved in canonical nucleolar functions, e.g., in regulating ribosomal biogenesis. Indeed, there is a growing body of evidence, that PARP1 and PARylation play important roles in nucleolar biology, which will be discussed in the following.

First evidence for PARP1 to play a role in nucleolar biology was proposed by Tulin et al. [140] in 2002 by using *Drosophila melanogaster* as a model system. Unlike mammals, the *Drosophila* genome contains only two PARP encoding genes, i.e., one that is highly related to mammalian PARP1, as well as one homolog of tankyrases [141,142]. Therefore, *Drosophila* is a powerful model organism to study PARP biology. In the study of Tulin et al. [140], it was shown that many of the PARylated proteins in *Drosophila* are enriched in nucleoli and in the heterochromatic chromocenter regions. Furthermore, disruption of PARP1 expression resulted in abrogated formation of nucleoli and larval lethality, suggesting that PARP1 is required for the formation of nucleoli during development [140]. As previously discussed, it was shown that liquid-liquid phase separation plays an important role in internal organization of nucleolar architecture [3]. Interestingly, PAR can function as a seed for liquid-liquid demixing in intrinsically disordered proteins, which are abundant in nucleoli [125]. Therefore, it is conceivable, that PARylation regulates the general biophysical state of nucleolar architecture by liquid-liquid demixing [124] (Figure 1). In this regard it is interesting to note that nucleoli in yeast and other lower eukaryotes exhibit a bipartite structure, i.e., lacking the FC in nucleoli [143]. The fact that PARylation is missing in yeast [144] makes this an interesting correlation, which needs to be analyzed in detail for any causative relationship in future studies.

Meanwhile, several studies have revealed an important role of PARP1 in regulating multiple steps of ribosome biogenesis. Thus, PARP1 was proposed to participate in ribosome biogenesis by controlling pre-rRNA processing, post-transcriptional modification, and assembly of pre-ribosomal subunits [145]. Disruptions of PARP1 enzymatic activity led to nucleolar disintegration and aberrant localization of nucleolar-specific proteins, such as fibrillarin, nucleophosmin and nucleolin. Therefore, the authors of this study concluded that PARP1 and PARylation are important for nucleolar integrity and the localization of nucleolar-specific proteins in proximity to pre-rRNA. Furthermore, PARP1 mutants displayed a delay in rRNA processing and increased levels of rRNA intermediates, such as 47S and 36S, resulting in decreased ribosome levels [145]. In mammalian cells, it was demonstrated that PARP1 participates in nucleolar remodeling complex (NoRC)-mediated rDNA silencing during replication [146] (Figure 1).

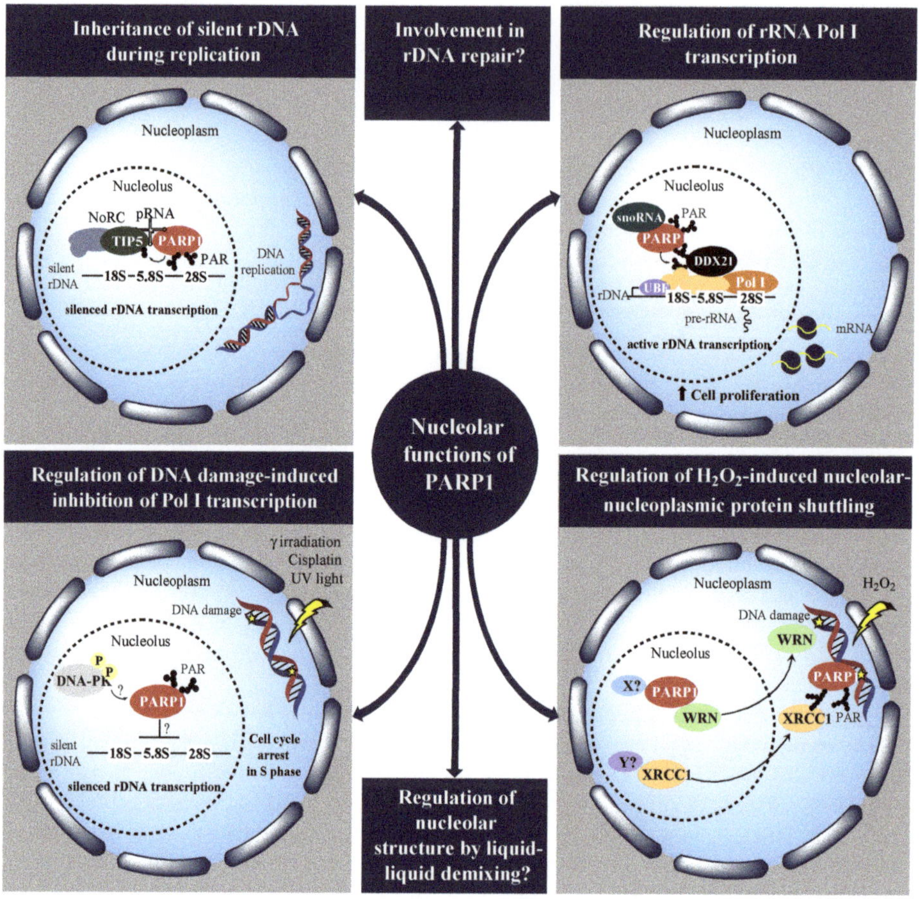

Figure 1. Functional roles of PARP1 in nucleolar biology. DDX21, DEAD-box helicase 21; DNA-PK, DNA-dependent protein kinase; NoRC, nucleolar remodeling complex; PAR, poly(ADP-ribose); PARP1, poly(ADP-ribosyl) polymerase 1; Pol I, polymerase I; pRNA, promotor-associated RNA; rDNA, ribosomal DNA; rRNA, ribosomal RNA; snoRNA, small nucleolar RNA; TIP5, TTF-1-interacting protein 5; UBF, upstream binding factor; WRN, Werner syndrome protein; XRCC1, X-ray repair cross complementing 1. For details see Section 4.

In mid-late S phase, the NoRC mediates heterochromatin formation and silencing of rDNA transcription via the recruitment of the histone acetylase HDAC1 and DNA methyltransferase (DNMT) to the rDNA promoter [147]. PARP1 associates with the rDNA repressor TTF-1-interacting protein 5 (TIP5), which is part of NoRC, at silent rRNA genes during replication [146].

Furthermore, association of PARP1 with TIP5 is mediated by the NoRC-associated noncoding RNA (pRNA, promoter-associated RNA). Interestingly, PARP1 PARylates silent chromatin and components of the NoRC complex, including TIP5. Thus, it is likely that pRNA stimulates PARP1 enzymatic activity, which is necessary to establish rDNA silencing. These findings indicate that PARP1 can modulate chromatin structure and gene expression in nucleoli, revealing a mechanism by which PARP1 ensures that silent rDNA regions are properly inherited after their disruption during DNA replication [148,149]. In a recent study, another pathway was identified by which nucleolar actions of PARP1 participate in the control of rDNA transcription and ribosome biogenesis during cell proliferation [112]. Thus, Kim et al. [112] demonstrated that PARP1's binding to snoRNAs leads to its catalytic activation in a DNA damage-independent manner. Upon activation of its catalytic activity, PARP1 ADP-ribosylates the nucleolar RNA helicase

DDX21, which results in enhanced rDNA transcription and proliferation of breast cancer cells (Figure 1). As discussed below in more detail, the pharmacological inhibition of this pathway may contribute to the effectiveness of PARP inhibitors in the treatment of certain cancers [112].

Importantly, there is emerging evidence that PARP1 and its enzymatic activity are involved in the regulation of ribosome biogenesis under certain pathological conditions. For instance, in hippocampal pyramidal neurons in Alzheimer's disease (AD) nucleolar PARP1 is significantly decreased compared to control cells [150]. It was proposed that under physiological conditions, PARP1 ADP-ribosylates DNMT1, thereby preventing rDNA methylation, which results in upregulation of rDNA transcription. Thus, in AD neurons, PARP1 mislocalization leads to hypermethylation of rDNA, reduced rDNA transcription and impaired ribosome biogenesis, which ultimately results in disruption of long-term memory formation. Interestingly, such a decrease in PARP1 staining was revealed in neurons of individuals with mild cognitive impairment (MCI), suggesting that decreased nucleolar PARP1 could act as an early biomarker of cognitive impairment [151]. For a more detailed discussion on this topic and the role of PARPs and PARylation in RNA biology in general the reader is referred to a comprehensive recent review by Kim et al. [91].

Apart from its contribution to ribosome biogenesis under non-stress conditions, PARP1 was shown to participate in the nucleolar stress response, in particular DNA damage response. A study by Calkins et al. [152] demonstrated that PARP1 regulates rDNA transcription in response to DNA damage (Figure 1). Induction of DNA damage, by using γ irradiation, UV light or the cross-linking agent cisplatin, resulted in inhibition of rDNA transcription, as well as cell cycle arrest in S phase. Inhibition of PARP1 or DNA-PK prevented silencing of rRNA synthesis, yet not the accumulation of cells in S phase. These results indicate that PARP1 and DNA-PK are involved in DNA damage-induced inhibition of rDNA transcription, however not in the accompanying cell cycle arrest. Loss of DNA-PK function prevented PARP1 from being activated and recruited to chromatin, suggesting that DNA-PK acts upstream of PARP1 to block rRNA synthesis upon DNA damage. While the exact mechanistic details by which PARP1 and DNA-PK contribute to DNA damage-induced inhibition of rDNA transcription remain to be elucidated, PARP1 may also facilitate DNA damage-induced block of rRNA synthesis through the recruitment of nucleolar proteins with roles in ribosome biogenesis to the nucleoplasm. For instance, fluorescence loss in photobleaching (FLIP) experiments showed that the macrodomain-containing protein TARG1/C6orf130 continuously undergoes nucleolar-nucleoplasmic shuttling [153]. Interestingly, the distribution of TARG1/C6orf130 between these two compartments was regulated by PARylation. Thus, in the absence of PAR, TARG1 localizes to transcriptionally active nucleoli, while in response to DNA damage-induced PAR formation, e.g., upon H_2O_2 treatment, it re-localizes to the nucleoplasm. Since TARG1 was reported to bind to RNA, ribosomal proteins, as well as proteins associated with rRNA proteins and ribosomal assembly factors, it is conceivable that in nucleoli, TARG1 plays a role in ribosome assembly or quality control, which is stalled when TARG1 is recruited to sites of DNA damage in a PAR-dependent manner [153]. With regards to a role of PARP1 and PARylation in nucleolar DNA damage response, we recently showed that upon H_2O_2 treatment of HeLa cells, WRN and XRCC1 translocate from nucleoli to the nucleoplasm, however interestingly enough, probably by different mechanisms [69]. Thus, while the release of WRN from nucleoli was purely dependent on the presence of PARP1 protein without any obvious involvement of its catalytic activity, we found that relocalization of XRCC1 upon H_2O_2-induced DNA damage was dependent on both PARP1 protein and its enzymatic activity (Figure 1). We hypothesized that in case of WRN, PARP1 and an additional unknown factor mediate the release of WRN from nucleoli, while XRCC1 requires nucleoplasmic DNA damage-bound and PARylated PARP1 as a loading platform, which leads to XRCC1 retention in the nucleoplasm until its tasks in BER are completed. This notion is supported by findings revealing that in cells without PARP1 activity, XRCC1 relocates quickly to nucleoli upon DNA damage induction [69].

Similarly to XRCC1, upon induction of DNA damage the ribosomal protein L6 (RPL6) is recruited to sites of DNA damage in a PARP-dependent manner [154]. At sites of DNA damage RPL6 appears to regulate the DNA damage response. Thus, RPL6 directly interacts with histone H2A and depletion of RPL6 impairs the recruitment of the mediator of DNA damage checkpoint 1 (MDC1), reduces ubiquitination of H2A and phosphorylated histone H2AX (γH2AX). These results exemplify that ribosomal proteins can exert PARP-dependent extraribosomal functions in DNA damage response. In general, these results support the notion that nucleolar-nucleoplasmic shuttling mechanisms are mediated by several different processes, which are highly dependent on the specific stress condition as well as the specific protein.

Another layer of complexity is added by a study of Leger et al. [155], which showed that H_2O_2 and N-methyl-N′-nitro-N-nitrosoguanidine (MNNG) induce PAR formation in nucleoli. Interestingly, the combination of H_2O_2 or MNNG treatment with low doses of ActD revealed a synergistic effect on nucleolar PAR formation. At low concentrations ActD intercalates with GC-rich sequences of rDNA downstream of rDNA transcription start sites. Thus, ActD prevents transcription at the elongation step, resulting in accumulation of short rRNA transcripts. Interestingly, this study reported that PARP2, but not PARP1, binds through its N-terminal SAP domain to these short rRNA transcripts and thereby becomes activated, which contributes to the enhanced PAR formation inside nucleoli [155].

5. Implications for Cancer Biology

The first link between nucleoli and cancer was established over a century ago, when pathologists noticed that nucleoli in cancer cells are often enlarged, irregularly shaped, as well as increased in number and therefore could serve as markers of aggressive malignancies [156]. In general, the roles of nucleolar processes in cancer biology are manifold and for a more in-depth discussion on this topic the reader is referred to comprehensive recent review articles [6,17,29,30,157]. Broadly speaking, the role of nucleoli in carcinogenesis falls into two interdependent categories: ribosome biogenesis and stress response. The current consensus is that structural abnormalities of nucleoli in cancer cells are the direct consequence of increased ribosome biogenesis, which goes along with aberrant Pol I transcription [157]. Thus, in a way, tumor cells depend on increased ribosome biogenesis to reach their demand for newly synthesized proteins during cell proliferation [6]. Interestingly, rDNA gene clusters represent hot spots of recombination in human cancer [6]. Thus, more than half of solid tumors such as lung or colorectal cancer exhibit rDNA rearrangements [54] and alterations are also frequently observed in Hodgkin's lymphoma [158].

Usually cells precisely monitor the accuracy of ribosome biogenesis, as well as nucleolar integrity, and disruption at any step of ribosome biogenesis results in activation of cellular checkpoints [27]. Thus, while the increased ribosome biogenesis was originally thought to merely reflect the increased growth and proliferation rates in cancer cells, today it is well accepted that dysregulation in ribosome biogenesis is a result of increased activity of oncogenes or inactivated tumor suppressors [28]. One of the best understood cellular surveillance mechanisms in this context is the nucleolar stress response, also referred to as the ribosomal surveillance pathway, that often results in activation of p53 [28]. The tumor suppressor p53 restrains tumor growth by inducing cell cycle arrest, senescence, or apoptosis [159]. Therefore, it is not surprising that p53 is mutated or functionally inactivated in most human cancers [160]. Targeting p53 by activating wildtype p53 or restoring the activity of loss-of-function mutants has become a promising strategy in cancer treatment. Classical chemotherapeutics, such as alkylating agents or platinum-based drugs, in general target cancer cells by directly or indirectly inducing DNA damage in rapidly proliferating cells, which leads to activation of p53 [161]. Interestingly, 21 of 36 screened chemotherapeutics, including doxorubicin and camptothecin, were shown to additionally inhibit various steps of ribosome biogenesis, leading to p53 stabilization [162]. Since classical chemotherapeutics discriminate between cancer cells and healthy cells by their different proliferative index, patients frequently suffer from side effects on other proliferating cells in the body, which includes bone marrow suppression, alopecia, mucositis, as well as toxicity to the gastrointestinal tract, skin, and heart [163]. In addition, long-term adverse effects, such as infertility or development of secondary cancers were observed in patients following chemotherapy [164]. The finding that p53 can also be activated in a non-genotoxic manner by disrupting nucleolar function and the aim to overcome adverse effects, gave rise to the development of a series of small molecule inhibitors, which selectively inhibit RNA Pol I transcription [165,166]. One of the most promising of such inhibitors is CX-5461, which impedes the selectivity factor SL1 from binding to the rDNA promoter, thereby preventing recruitment of the Pol I complex to the rDNA and initiation of transcription [167]. Apart from this, CX-5461 stabilizes G-quadruplex structures with increased toxicity in *BRCA*-deficient or PARP inhibition resistant cancer cells [168]. In vitro studies have revealed a high antiproliferative effect of CX-5461 in a wide range of human cancers, with those derived from p53 wildtype hematological malignancies being the most sensitive [167]. By using the Eμ-*Myc* mouse model of Burkitt's lymphoma, which is an aggressive type of lymphoma affecting B lymphocytes, it was shown in vivo that treatment with CX-5461 has the potential to lead to an almost complete disease remission and a significantly increased survival [165]. Importantly, the normal B cell population in those mice was maintained, suggesting that CX-5461 selectively targets cancer cells. Due to the promising results of this preclinical data, CX-5461 is currently tested in several clinical trials. A phase I clinical trial has successfully been completed in patients with hematological cancers and at present phase I/II trails are undergoing for solid tumors, including metastatic breast cancer, ovarian, and pancreas cancer [169]. Thus, selectively targeting ribosome biogenesis could provide a novel and efficient strategy in cancer therapy. Most cancer types rely on increased levels of ribosomes and therefore display higher sensitivity towards inhibition of ribosome biogenesis compared to normal cells. Thus, in contrast to existing chemotherapies, inhibition of ribosome biogenesis is likely less genotoxic to the non-tumor population of cells, which is associated with a reduced risk of adverse effects. The previously mentioned findings from studies with CX-5461 strongly support this notion. Given the high complexity of ribosome biogenesis and the variety of factors, which are involved in this process, it is likely that in future even more effective small molecule inhibitors will be identified.

As described in Section 3, PARP inhibitors have successfully entered the clinic as monotherapeutic agents, as well as in combination with cytostatic chemotherapy or radiotherapy. PARP inhibitors can act by inducing synthetic lethality in cancers that are deficient in HR-mediated DNA repair, e.g., loss-of-function mutation in *BRCA* [128,129]. More recent studies have suggested that PARP inhibitors may also induce replication stress and subsequent DNA damage [95,170,171]. Development of chemo-resistance has been proven to be a major problem in the clinical efficacy of PARP inhibitors [172]. Extensive in vitro and in vivo studies have identified several potential resistance mechanisms, including reactivation of HR, upregulation of drug efflux pumps and stabilization of replication forks [131].

Interestingly, some recent studies suggested that targeting nucleolar proteins/processes in combination with PARP inhibitor treatment may be beneficial for the treatment of some cancers, e.g., by overcoming PARP inhibitor resistance mechanisms. Thus, in a recent study CX-5461 was demonstrated to induce replication stress and activate the DNA damage response in high-grade serous ovarian cancer (HGSOC) cells [173]. CX-5461 showed significant therapeutic benefit as a single agent in HGSOC-patient-derived xenografts with reduced sensitivity to PARP inhibitors by overcoming replication fork protection, which is a well-known PARP inhibitor resistance mechanism. Importantly, the combination of CX-5461 and PARP inhibitors resulted in enhanced replication stress, DNA damage and cell death and exhibited great therapeutic efficacy, especially in HR-deficient HGSOC-patient-derived xenografts. Thus, there is evidence that combining PARP inhibitors with CX-5461 could improve treatment of HR-deficient HGSOC.

Furthermore, a novel mechanism for PARP resistance development has been reported by Sun et al. [174], which involves increased phosphorylation of the ribosomal protein S6 (RPS6). RPS6 is a component of the 40S ribosomal subunit and a well-known downstream effector of the mammalian target of rapamycin (mTOR) signaling pathway, which is the major nutrient-sensitive regulator of eukaryotic cell growth, metabolism, proliferation and survival [175]. RPS6 is phosphorylated by ribosomal protein S6 kinases (S6Ks) at five C-terminal serine sites and this modification is crucial for regulation of cell size, cell proliferation and glucose homeostasis [176]. Phosphorylation of RPS6 is greatly increased in *BRCA1*-deficient cancer cells, which are resistant to PARP inhibition [174]. Importantly, in *BRCA1*-deficient cells RPS6 phosphorylation promoted loading of the HD marker RAD51 onto DNA following IR-induced DNA damage. Thus, RPS6 phosphorylation might play a key role in PARP inhibitor resistance by regulating HR. Intriguingly, rapamycin, which is a clinically used selective inhibitor of mTOR and S6 phosphorylation, could restore sensitivity towards PARP inhibition, suggesting that combined inhibition of S6 phosphorylation and PARP could be efficient in cancers with PARP inhibitor resistance and HR defects, including *BRCA1*-deficient breast and ovarian cancers.

Several studies have reported that *BRCA* mutations or other HR-mediated DNA repair deficiencies are not mandatory for the clinical effectiveness of PARP inhibitors in cancer therapy [177–179]. In accordance with this, Kim et al. [112] revealed an alternative working mechanism of PARP inhibitors, which is independent of HR, DNA damage and replication stress. Mechanistically, it was shown that snoRNAs can stimulate PARP1 enzymatic activity in the nucleolus, resulting in ADP-ribosylation of DDX21. The DEAD-box RNA helicase DDX21 was previously reported to directly interact with rRNA and snoRNAs at the transcribed rDNA locus, thereby promoting rRNA synthesis, rRNA processing and modification [180]. Furthermore, analyses of gene expression profiles from breast cancer patients identified DDX21 as a prognostic marker in breast cancer [181]. Kim et al. [112] demonstrated that DDX21 promotes rDNA transcription and breast cancer growth upon PARP1-mediated ADP-ribosylation. PARP inhibition on the other hand resulted in reduced tumor cell growth by modulating rRNA levels, DDX21 ADP-ribosylation and DDX21 localization. These findings could be explained by the fact that some cancer types are "addicted" to ribosome biogenesis, whereas reducing ribosome biogenesis can counteract cancer cell growth. The finding that PARP inhibitors can reduce tumor growth by targeting ribosome biogenesis provides a mechanistic explanation for efficacy of PARP inhibitors in cancer cells lacking deficiencies in HR. In addition, these results further suggest that DDX21 nucleolar localization could be a predictive biomarker of clinical responses to PARP inhibition.

6. Concluding Remarks and Perspectives

Nucleolar localization of PARP1 was demonstrated almost four decades ago. For many years, scientists have been puzzled whether PARP1 accumulation in the nucleolus merely occurs for storage reasons or if PARP1 might also be implicated in nucleolar functions. As summarized in this review, there is growing evidence, supporting a role of PARP1 and PARylation in nucleolar biology. PARP1 is implicated in multiple areas of nucleolar function, including maintenance of nucleolar integrity and structure, regulation of Pol I transcription, establishment of silent rDNA chromatin, as well as regulation of DNA damage-induced nucleolar-nucleoplasmic shuttling processes of key genome maintenance factors, e.g., WRN and XRCC1 (Figure 1). Previous studies have demonstrated that PAR can function as a seed of liquid-liquid demixing processes, thereby triggering the formation of biomolecular condensates [124,125]. Therefore, it can be assumed that PARP1 and PARylation might regulate the general biophysical state of nucleolar structure. Since PARP1 plays a crucial role in several DNA repair pathways, of which at least NHEJ and HR have been reported to take place in the nucleolus, it can be anticipated that PARP1 might also be involved in the repair of rDNA [88,182]. Repetitive rDNA sequences represent one of the most unstable regions in the genome [6]. Instability of rDNA is associated with severe pathological conditions, including cancer, premature aging and neurological impairments [182]. Yet, the DNA damage repair mechanisms that govern genomic stability and maintenance in the nucleolus remain elusive. In future studies, recent advancements in CRISPR-genome engineering could provide deeper insights into the mechanisms that help cells keep their rDNA intact. Until now, in large, only PARP1 was reported to play a role in nucleolar biology. In future studies it will be interesting to investigate to what extent other members of the PARP family, e.g., PARP2, also contribute to nucleolar functions. As previously described, the tripartite organization of the nucleolus reflects the different stages of ribosome biogenesis. Therefore, identifying the exact localization of PARP1 and other PARPs, as well as PAR, in nucleolar substructures could provide further evidence for a role in certain steps of ribosome biogenesis. Importantly, PARP1 inhibitors have entered the clinic as promising chemotherapeutic agents in the treatment of various cancer types, mainly by exploiting synthetic lethal interactions between PARP inhibition and defects in genes that are responsible for HR-mediated DNA repair. As discussed above alternative working mechanisms of PARP inhibitors are conceivable, which act by preventing tumor growth through inhibition of ribosome biogenesis [112]. These data indicate that PARP inhibitors might be effective in a broader spectrum of cancer types than originally anticipated. In addition, in this study, the DEAD-box RNA helicase DDX21 was identified as a potential marker to predict sensitivity towards PARP inhibitors, which needs to be further evaluated in clinical trials. In future studies it will be important to elucidate further mechanisms by which PARP inhibitors contribute to inhibition of nucleolar functions, to pave the way for the identification of other biomarkers, which can predict response to PARP inhibitors in cancer patients.

Author Contributions: Conceptualization of the manuscript, M.E. and A.M.; writing and editing of the manuscript, M.E. and A.M. All authors have read and agreed to the published version of the manuscript.

Funding: Experimental research of our lab that is associated with the research topic reviewed in this article is funded by University of Konstanz and the German Research Foundation (DFG, grant MA4905/4-1). Funding for open access charge: University of Konstanz.

Conflicts of Interest: The authors declare no conflict of interest. The funders had no role in the design and the in the writing of the manuscript.

References

1. Henderson, A.S.; Warburton, D.; Atwood, K.C. Location of ribosomal DNA in the human chromosome complement. *Proc. Natl. Acad. Sci. USA* **1972**, *69*, 3394–3398. [CrossRef] [PubMed]
2. Bersaglieri, C.; Santoro, R. Genome Organization in and around the Nucleolus. *Cells* **2019**, *8*, 579. [CrossRef] [PubMed]
3. Mangan, H.; Gailin, M.O.; McStay, B. Integrating the genomic architecture of human nucleolar organizer regions with the biophysical properties of nucleoli. *FEBS J.* **2017**, *284*, 3977–3985. [CrossRef] [PubMed]
4. Tiku, V.; Antebi, A. Nucleolar Function in Lifespan Regulation. *Trends Cell. Biol.* **2018**, *28*, 662–672. [CrossRef]
5. Kusnadi, E.P.; Hannan, K.M.; Hicks, R.J.; Hannan, R.D.; Pearson, R.B.; Kang, J. Regulation of rDNA transcription in response to growth factors, nutrients and energy. *Gene* **2015**, *556*, 27–34. [CrossRef] [PubMed]
6. Weeks, S.E.; Metge, B.J.; Samant, R.S. The nucleolus: A central response hub for the stressors that drive cancer progression. *Cell. Mol. Life Sci.* **2019**, *76*, 4511–4524. [CrossRef]
7. Lewis, J.D.; Tollervey, D. Like attracts like: Getting RNA processing together in the nucleus. *Science* **2000**, *288*, 1385–1389. [CrossRef]
8. Farley, K.I.; Surovtseva, Y.; Merkel, J.; Baserga, S.J. Determinants of mammalian nucleolar architecture. *Chromosoma* **2015**, *124*, 323–331. [CrossRef]
9. Cerqueira, A.V.; Lemos, B. Ribosomal DNA and the Nucleolus as Keystones of Nuclear Architecture, Organization, and Function. *Trends Genet.* **2019**, *35*, 710–723. [CrossRef]
10. Nemeth, A.; Grummt, I. Dynamic regulation of nucleolar architecture. *Curr. Opin. Cell. Biol.* **2018**, *52*, 105–111. [CrossRef]
11. Banani, S.F.; Lee, H.O.; Hyman, A.A.; Rosen, M.K. Biomolecular condensates: Organizers of cellular biochemistry. *Nat. Rev. Mol. Cell. Biol.* **2017**, *18*, 285–298. [CrossRef] [PubMed]
12. Brangwynne, C.P.; Mitchison, T.J.; Hyman, A.A. Active liquid-like behavior of nucleoli determines their size and shape in Xenopus laevis oocytes. *Proc. Natl. Acad. Sci. USA* **2011**, *108*, 4334–4339. [CrossRef]
13. Hyman, A.A.; Weber, C.A.; Julicher, F. Liquid-liquid phase separation in biology. *Annu. Rev. Cell. Dev. Biol.* **2014**, *30*, 39–58. [CrossRef]
14. Latonen, L. Phase-to-Phase with Nucleoli-Stress Responses, Protein Aggregation and Novel Roles of RNA. *Front. Cell. Neurosci.* **2019**, *13*, 151. [CrossRef] [PubMed]
15. Feric, M.; Vaidya, N.; Harmon, T.S.; Mitrea, D.M.; Zhu, L.; Richardson, T.M.; Kriwacki, R.W.; Pappu, R.V.; Brangwynne, C.P. Coexisting Liquid Phases Underlie Nucleolar Subcompartments. *Cell* **2016**, *165*, 1686–1697. [CrossRef] [PubMed]

16. Grob, A.; Colleran, C.; McStay, B. Construction of synthetic nucleoli in human cells reveals how a major functional nuclear domain is formed and propagated through cell division. *Genes Dev.* **2014**, *28*, 220–230. [CrossRef]
17. Lindstrom, M.S.; Jurada, D.; Bursac, S.; Orsolic, I.; Bartek, J.; Volarevic, S. Nucleolus as an emerging hub in maintenance of genome stability and cancer pathogenesis. *Oncogene* **2018**, *37*, 2351–2366. [CrossRef]
18. Hernandez-Verdun, D. Nucleolus: From structure to dynamics. *Histochem. Cell. Biol.* **2006**, *125*, 127–137. [CrossRef]
19. Caragine, C.M.; Haley, S.C.; Zidovska, A. Nucleolar dynamics and interactions with nucleoplasm in living cells. *Elife* **2019**, *8*. [CrossRef]
20. Boisvert, F.M.; van Koningsbruggen, S.; Navascues, J.; Lamond, A.I. The multifunctional nucleolus. *Nat. Rev. Mol. Cell. Biol.* **2007**, *8*, 574–585. [CrossRef]
21. Andersen, J.S.; Lam, Y.W.; Leung, A.K.; Ong, S.E.; Lyon, C.E.; Lamond, A.I.; Mann, M. Nucleolar proteome dynamics. *Nature* **2005**, *433*, 77–83. [CrossRef] [PubMed]
22. Ahmad, Y.; Boisvert, F.M.; Gregor, P.; Cobley, A.; Lamond, A.I. NOPdb: Nucleolar Proteome Database—2008 update. *Nucleic Acids Res.* **2009**, *37*, 181–184. [CrossRef] [PubMed]
23. Iarovaia, O.V.; Minina, E.P.; Sheval, E.V.; Onichtchouk, D.; Dokudovskaya, S.; Razin, S.V.; Vassetzky, Y.S. Nucleolus: A Central Hub for Nuclear Functions. *Trends Cell. Biol.* **2019**, *29*, 647–659. [CrossRef] [PubMed]
24. Kobayashi, T. A new role of the rDNA and nucleolus in the nucleus—rDNA instability maintains genome integrity. *Bioessays* **2008**, *30*, 267–272. [CrossRef]
25. Tiku, V.; Jain, C.; Raz, Y.; Nakamura, S.; Heestand, B.; Liu, W.; Spath, M.; Suchiman, H.E.D.; Muller, R.U.; Slagboom, P.E.; et al. Small nucleoli are a cellular hallmark of longevity. *Nat. Commun.* **2017**, *8*, 16083. [CrossRef]
26. Parlato, R.; Liss, B. How Parkinson's disease meets nucleolar stress. *Biochim. Biophys. Acta* **2014**, *1842*, 791–797. [CrossRef]
27. Hein, N.; Hannan, K.M.; George, A.J.; Sanij, E.; Hannan, R.D. The nucleolus: An emerging target for cancer therapy. *Trends Mol. Med.* **2013**, *19*, 643–654. [CrossRef]
28. Woods, S.J.; Hannan, K.M.; Pearson, R.B.; Hannan, R.D. The nucleolus as a fundamental regulator of the p53 response and a new target for cancer therapy. *Biochim. Biophys. Acta* **2015**, *1849*, 821–829. [CrossRef]
29. Orsolic, I.; Jurada, D.; Pullen, N.; Oren, M.; Eliopoulos, A.G.; Volarevic, S. The relationship between the nucleolus and cancer: Current evidence and emerging paradigms. *Semin. Cancer Biol.* **2016**, *37*, 36–50. [CrossRef]
30. Stepinski, D. The nucleolus, an ally, and an enemy of cancer cells. *Histochem. Cell. Biol.* **2018**, *150*, 607–629. [CrossRef]
31. Ciccia, A.; Elledge, S.J. The DNA damage response: Making it safe to play with knives. *Mol. Cell.* **2010**, *40*, 179–204. [CrossRef] [PubMed]
32. Ogawa, L.M.; Baserga, S.J. Crosstalk between the nucleolus and the DNA damage response. *Mol. Biosyst.* **2017**, *13*, 443–455. [CrossRef] [PubMed]
33. Nishimura, Y.; Ohkubo, T.; Furuichi, Y.; Umekawa, H. Tryptophans 286 and 288 in the C-terminal region of protein B23.1 are important for its nucleolar localization. *Biosci. Biotechnol. Biochem.* **2002**, *66*, 2239–2242. [CrossRef] [PubMed]
34. Nunez Villacis, L.; Wong, M.S.; Ferguson, L.L.; Hein, N.; George, A.J.; Hannan, K.M. New Roles for the Nucleolus in Health and Disease. *Bioessays* **2018**, *40*, e1700233. [CrossRef]
35. Antoniali, G.; Lirussi, L.; Poletto, M.; Tell, G. Emerging roles of the nucleolus in regulating the DNA damage response: The noncanonical DNA repair enzyme APE1/Ref-1 as a paradigmatical example. *Antioxid. Redox. Signal* **2014**, *20*, 621–639. [CrossRef]
36. Shen, J.C.; Loeb, L.A. The Werner syndrome gene: The molecular basis of RecQ helicase-deficiency diseases. *Trends Genet.* **2000**, *16*, 213–220. [CrossRef]
37. Chen, L.; Huang, S.; Lee, L.; Davalos, A.; Schiestl, R.H.; Campisi, J.; Oshima, J. WRN, the protein deficient in Werner syndrome, plays a critical structural role in optimizing DNA repair. *Aging Cell* **2003**, *2*, 191–199. [CrossRef]
38. Harrigan, J.A.; Wilson, D.M., III; Prasad, R.; Opresko, P.L.; Beck, G.; May, A.; Wilson, S.H.; Bohr, V.A. The Werner syndrome protein operates in base excision repair and cooperates with DNA polymerase beta. *Nucleic Acids Res.* **2006**, *34*, 745–754. [CrossRef]

39. Gray, M.D.; Wang, L.; Youssoufian, H.; Martin, G.M.; Oshima, J. Werner helicase is localized to transcriptionally active nucleoli of cycling cells. *Exp. Cell Res.* **1998**, *242*, 487–494. [CrossRef]
40. Shiratori, M.; Suzuki, T.; Itoh, C.; Goto, M.; Furuichi, Y.; Matsumoto, T. WRN helicase accelerates the transcription of ribosomal RNA as a component of an RNA polymerase I-associated complex. *Oncogene* **2002**, *21*, 2447–2454. [CrossRef]
41. Rubbi, C.P.; Milner, J. Disruption of the nucleolus mediates stabilization of p53 in response to DNA damage and other stresses. *EMBO J.* **2003**, *22*, 6068–6077. [CrossRef] [PubMed]
42. Yang, K.; Yang, J.; Yi, J. Nucleolar Stress: Hallmarks, sensing mechanism and diseases. *Cell Stress* **2018**, *2*, 125–140. [CrossRef] [PubMed]
43. Kruhlak, M.; Crouch, E.E.; Orlov, M.; Montano, C.; Gorski, S.A.; Nussenzweig, A.; Misteli, T.; Phair, R.D.; Casellas, R. The ATM repair pathway inhibits RNA polymerase I transcription in response to chromosome breaks. *Nature* **2007**, *447*, 730–734. [CrossRef] [PubMed]
44. Harding, S.M.; Boiarsky, J.A.; Greenberg, R.A. ATM Dependent Silencing Links Nucleolar Chromatin Reorganization to DNA Damage Recognition. *Cell Rep.* **2015**, *13*, 251–259. [CrossRef]
45. Van Sluis, M.; McStay, B. A localized nucleolar DNA damage response facilitates recruitment of the homology-directed repair machinery independent of cell cycle stage. *Genes Dev.* **2015**, *29*, 1151–1163. [CrossRef]
46. Kuhn, A.; Gottlieb, T.M.; Jackson, S.P.; Grummt, I. DNA-dependent protein kinase: A potent inhibitor of transcription by RNA polymerase I. *Genes Dev.* **1995**, *9*, 193–203. [CrossRef]
47. Michaelidis, T.M.; Grummt, I. Mechanism of inhibition of RNA polymerase I transcription by DNA-dependent protein kinase. *Biol. Chem.* **2002**, *383*, 1683–1690. [CrossRef]
48. Van Sluis, M.; McStay, B. Nucleolar DNA Double-Strand Break Responses Underpinning rDNA Genomic Stability. *Trends Genet.* **2019**, *35*, 743–753. [CrossRef]
49. Berkovich, E.; Monnat, R.J., Jr.; Kastan, M.B. Roles of ATM and NBS1 in chromatin structure modulation and DNA double-strand break repair. *Nat. Cell. Biol.* **2007**, *9*, 683–690. [CrossRef]
50. Van Sluis, M.; McStay, B. Nucleolar reorganization in response to rDNA damage. *Curr. Opin. Cell. Biol.* **2017**, *46*, 81–86. [CrossRef]
51. Warmerdam, D.O.; van den Berg, J.; Medema, R.H. Breaks in the 45S rDNA Lead to Recombination-Mediated Loss of Repeats. *Cell Rep.* **2016**, *14*, 2519–2527. [CrossRef] [PubMed]
52. Hamperl, S.; Cimprich, K.A. The contribution of co-transcriptional RNA: DNA hybrid structures to DNA damage and genome instability. *DNA Rep.* **2014**, *19*, 84–94. [CrossRef] [PubMed]
53. Tsekrekou, M.; Stratigi, K.; Chatzinikolaou, G. The Nucleolus: In Genome Maintenance and Repair. *Int. J. Mol. Sci.* **2017**, *18*, 1411. [CrossRef] [PubMed]
54. Stults, D.M.; Killen, M.W.; Williamson, E.P.; Hourigan, J.S.; Vargas, H.D.; Arnold, S.M.; Moscow, J.A.; Pierce, A.J. Human rRNA gene clusters are recombinational hotspots in cancer. *Cancer Res.* **2009**, *69*, 9096–9104. [CrossRef] [PubMed]
55. Yung, B.Y.; Busch, H.; Chan, P.K. Translocation of nucleolar phosphoprotein B23 (37 kDa/pI 5.1) induced by selective inhibitors of ribosome synthesis. *Biochim. Biophys. Acta* **1985**, *826*, 167–173. [CrossRef]
56. Yang, K.; Wang, M.; Zhao, Y.; Sun, X.; Yang, Y.; Li, X.; Zhou, A.; Chu, H.; Zhou, H.; Xu, J.; et al. A redox mechanism underlying nucleolar stress sensing by nucleophosmin. *Nat. Commun.* **2016**, *7*, 13599. [CrossRef] [PubMed]
57. Scott, D.D.; Oeffinger, M. Nucleolin and nucleophosmin: Nucleolar proteins with multiple functions in DNA repair. *Biochem. Cell. Biol.* **2016**, *94*, 419–432. [CrossRef]
58. Korgaonkar, C.; Hagen, J.; Tompkins, V.; Frazier, A.A.; Allamargot, C.; Quelle, F.W.; Quelle, D.E. Nucleophosmin (B23) targets ARF to nucleoli and inhibits its function. *Mol. Cell. Biol.* **2005**, *25*, 1258–1271. [CrossRef]
59. Kurki, S.; Peltonen, K.; Latonen, L.; Kiviharju, T.M.; Ojala, P.M.; Meek, D.; Laiho, M. Nucleolar protein NPM interacts with HDM2 and protects tumor suppressor protein p53 from HDM2-mediated degradation. *Cancer Cell* **2004**, *5*, 465–475. [CrossRef]
60. Bernardi, R.; Scaglioni, P.P.; Bergmann, S.; Horn, H.F.; Vousden, K.H.; Pandolfi, P.P. PML regulates p53 stability by sequestering Mdm2 to the nucleolus. *Nat. Cell. Biol.* **2004**, *6*, 665–672. [CrossRef]

61. Thoms, H.C.; Loveridge, C.J.; Simpson, J.; Clipson, A.; Reinhardt, K.; Dunlop, M.G.; Stark, L.A. Nucleolar targeting of RelA (p65) is regulated by COMMD1-dependent ubiquitination. *Cancer Res.* **2010**, *70*, 139–149. [CrossRef] [PubMed]
62. Velazquez, J.M.; Lindquist, S. hsp70: Nuclear concentration during environmental stress and cytoplasmic storage during recovery. *Cell* **1984**, *36*, 655–662. [CrossRef]
63. Frottin, F.; Schueder, F.; Tiwary, S.; Gupta, R.; Korner, R.; Schlichthaerle, T.; Cox, J.; Jungmann, R.; Hartl, F.U.; Hipp, M.S. The nucleolus functions as a phase-separated protein quality control compartment. *Science* **2019**, *365*, 342–347. [CrossRef] [PubMed]
64. Welch, W.J.; Feramisco, J.R. Nuclear and nucleolar localization of the 72,000-dalton heat shock protein in heat-shocked mammalian cells. *J. Biol. Chem.* **1984**, *259*, 4501–4513.
65. Pelham, H.; Lewis, M.; Lindquist, S. Expression of a Drosophila heat shock protein in mammalian cells: Transient association with nucleoli after heat shock. *Philos. Trans. R Soc. Lond. B Biol. Sci.* **1984**, *307*, 301–307. [CrossRef] [PubMed]
66. Vascotto, C.; Fantini, D.; Romanello, M.; Cesaratto, L.; Deganuto, M.; Leonardi, A.; Radicella, J.P.; Kelley, M.R.; D'Ambrosio, C.; Scaloni, A.; et al. APE1/Ref-1 interacts with NPM1 within nucleoli and plays a role in the rRNA quality control process. *Mol. Cell. Biol.* **2009**, *29*, 1834–1854. [CrossRef]
67. Karmakar, P.; Bohr, V.A. Cellular dynamics and modulation of WRN protein is DNA damage specific. *Mech. Ageing Dev.* **2005**, *126*, 1146–1158. [CrossRef]
68. Kusumoto-Matsuo, R.; Ghosh, D.; Karmakar, P.; May, A.; Ramsden, D.; Bohr, V.A. Serines 440 and 467 in the Werner syndrome protein are phosphorylated by DNA-PK and affects its dynamics in response to DNA double strand breaks. *Aging* **2014**, *6*, 70–81. [CrossRef]
69. Veith, S.; Schink, A.; Engbrecht, M.; Mack, M.; Rank, L.; Rossatti, P.; Hakobyan, M.; Goly, D.; Hefele, T.; Frensch, M.; et al. PARP1 regulates DNA damage-induced nucleolar-nucleoplasmic shuttling of WRN and XRCC1 in a toxicant and protein-specific manner. *Sci. Rep.* **2019**, *9*, 10075. [CrossRef]
70. Menissier de Murcia, J.; Ricoul, M.; Tartier, L.; Niedergang, C.; Huber, A.; Dantzer, F.; Schreiber, V.; Ame, J.C.; Dierich, A.; LeMeur, M.; et al. Functional interaction between PARP-1 and PARP-2 in chromosome stability and embryonic development in mouse. *EMBO J.* **2003**, *22*, 2255–2263. [CrossRef]
71. Langelier, M.F.; Eisemann, T.; Riccio, A.A.; Pascal, J.M. PARP family enzymes: Regulation and catalysis of the poly (ADP-ribose) posttranslational modification. *Curr. Opin. Struct. Biol.* **2018**, *53*, 187–198. [CrossRef] [PubMed]
72. Luscher, B.; Butepage, M.; Eckei, L.; Krieg, S.; Verheugd, P.; Shilton, B.H. ADP-Ribosylation, a Multifaceted Posttranslational Modification Involved in the Control of Cell Physiology in Health and Disease. *Chem. Rev.* **2018**, *118*, 1092–1136. [CrossRef] [PubMed]
73. Cohen, M.S.; Chang, P. Insights into the biogenesis, function, and regulation of ADP-ribosylation. *Nat. Chem. Biol.* **2018**, *14*, 236–243. [CrossRef] [PubMed]
74. Hottiger, M.O.; Hassa, P.O.; Luscher, B.; Schuler, H.; Koch-Nolte, F. Toward a unified nomenclature for mammalian ADP-ribosyltransferases. *Trends Biochem. Sci.* **2010**, *35*, 208–219. [CrossRef]
75. Altmeyer, M.; Messner, S.; Hassa, P.O.; Fey, M.; Hottiger, M.O. Molecular mechanism of poly(ADP-ribosyl)ation by PARP1 and identification of lysine residues as ADP-ribose acceptor sites. *Nucleic Acids Res.* **2009**, *37*, 3723–3738. [CrossRef]
76. Daniels, C.M.; Ong, S.E.; Leung, A.K. Phosphoproteomic approach to characterize protein mono- and poly(ADP-ribosyl)ation sites from cells. *J. Proteome Res.* **2014**, *13*, 3510–3522. [CrossRef]
77. Leslie Pedrioli, D.M.; Leutert, M.; Bilan, V.; Nowak, K.; Gunasekera, K.; Ferrari, E.; Imhof, R.; Malmstrom, L.; Hottiger, M.O. Comprehensive ADP-ribosylome analysis identifies tyrosine as an ADP-ribose acceptor site. *EMBO Rep.* **2018**, *19*. [CrossRef]
78. Palazzo, L.; Leidecker, O.; Prokhorova, E.; Dauben, H.; Matic, I.; Ahel, I. Serine is the major residue for ADP-ribosylation upon DNA damage. *Elife* **2018**, *7*. [CrossRef]
79. Leidecker, O.; Bonfiglio, J.J.; Colby, T.; Zhang, Q.; Atanassov, I.; Zaja, R.; Palazzo, L.; Stockum, A.; Ahel, I.; Matic, I. Serine is a new target residue for endogenous ADP-ribosylation on histones. *Nat. Chem. Biol.* **2016**, *12*, 998–1000. [CrossRef]
80. Martello, R.; Leutert, M.; Jungmichel, S.; Bilan, V.; Larsen, S.C.; Young, C.; Hottiger, M.O.; Nielsen, M.L. Proteome-wide identification of the endogenous ADP-ribosylome of mammalian cells and tissue. *Nat. Commun.* **2016**, *7*, 12917. [CrossRef]

81. Zhang, Y.; Wang, J.; Ding, M.; Yu, Y. Site-specific characterization of the Asp- and Glu-ADP-ribosylated proteome. *Nat. Methods* **2013**, *10*, 981–984. [CrossRef] [PubMed]
82. Larsen, S.C.; Hendriks, I.A.; Lyon, D.; Jensen, L.J.; Nielsen, M.L. Systems-wide Analysis of Serine ADP-Ribosylation Reveals Widespread Occurrence and Site-Specific Overlap with Phosphorylation. *Cell Rep.* **2018**, *24*, 2493–2505.e4. [CrossRef] [PubMed]
83. Miwa, M.; Saikawa, N.; Yamaizumi, Z.; Nishimura, S.; Sugimura, T. Structure of poly(adenosine diphosphate ribose): Identification of 2′-[1′′-ribosyl-2′′-(or 3′′-) (1′′′-ribosyl)]adenosine-5′, 5′′, 5′′′-tris(phosphate) as a branch linkage. *Proc. Natl. Acad. Sci. USA* **1979**, *76*, 595–599. [CrossRef] [PubMed]
84. Alvarez-Gonzalez, R.; Jacobson, M.K. Characterization of polymers of adenosine diphosphate ribose generated in vitro and in vivo. *Biochemistry* **1987**, *26*, 3218–3224. [CrossRef] [PubMed]
85. Alemasova, E.E.; Lavrik, O.I. Poly(ADP-ribosyl)ation by PARP1: Reaction mechanism and regulatory proteins. *Nucleic Acids Res.* **2019**, *47*, 3811–3827. [CrossRef] [PubMed]
86. Kamaletdinova, T.; Fanaei-Kahrani, Z.; Wang, Z.Q. The Enigmatic Function of PARP1: From PARylation Activity to PAR Readers. *Cells* **2019**, *8*, 1625. [CrossRef] [PubMed]
87. Chambon, P.; Weill, J.D.; Mandel, P. Nicotinamide mononucleotide activation of new DNA-dependent polyadenylic acid synthesizing nuclear enzyme. *Biochem. Biophys. Res. Commun.* **1963**, *11*, 39–43. [CrossRef]
88. Ray Chaudhuri, A.; Nussenzweig, A. The multifaceted roles of PARP1 in DNA repair and chromatin remodelling. *Nat. Rev. Mol. Cell. Biol.* **2017**, *18*, 610–621. [CrossRef]
89. Kunze, F.A.; Hottiger, M.O. Regulating Immunity via ADP-Ribosylation: Therapeutic Implications and Beyond. *Trends Immunol.* **2019**, *40*, 159–173. [CrossRef]
90. Hanzlikova, H.; Caldecott, K.W. Perspectives on PARPs in S Phase. *Trends Genet.* **2019**, *35*, 412–422. [CrossRef]
91. Kim, D.S.; Challa, S.; Jones, A.; Kraus, W.L. PARPs and ADP-ribosylation in RNA biology: From RNA expression and processing to protein translation and proteostasis. *Genes Dev.* **2020**, *34*, 302–320. [CrossRef] [PubMed]
92. Hopp, A.K.; Gruter, P.; Hottiger, M.O. Regulation of Glucose Metabolism by NAD (+) and ADP-Ribosylation. *Cells* **2019**, *8*, 890. [CrossRef] [PubMed]
93. Eisemann, T.; Pascal, J.M. Poly(ADP-ribose) polymerase enzymes and the maintenance of genome integrity. *Cell. Mol. Life Sci.* **2020**, *77*, 19–33. [CrossRef] [PubMed]
94. Azarm, K.; Smith, S. Nuclear PARPs and genome integrity. *Genes Dev.* **2020**, *34*, 285–301. [CrossRef] [PubMed]
95. Hanzlikova, H.; Kalasova, I.; Demin, A.A.; Pennicott, L.E.; Cihlarova, Z.; Caldecott, K.W. The Importance of Poly(ADP-Ribose) Polymerase as a Sensor of Unligated Okazaki Fragments during DNA Replication. *Mol. Cell* **2018**, *71*, 319–331.e3. [CrossRef]
96. Berti, M.; Ray Chaudhuri, A.; Thangavel, S.; Gomathinayagam, S.; Kenig, S.; Vujanovic, M.; Odreman, F.; Glatter, T.; Graziano, S.; Mendoza-Maldonado, R.; et al. Human RECQ1 promotes restart of replication forks reversed by DNA topoisomerase I inhibition. *Nat. Struct. Mol. Biol.* **2013**, *20*, 347–354. [CrossRef]
97. Shall, S.; de Murcia, G. Poly(ADP-ribose) polymerase-1: What have we learned from the deficient mouse model? *Mutat. Res. DNA Rep.* **2000**, *460*, 1–15. [CrossRef]
98. Kruger, A.; Burkle, A.; Hauser, K.; Mangerich, A. Real-time monitoring of PARP1-dependent PARylation by ATR-FTIR spectroscopy. *Nat. Commun.* **2020**, *11*, 2174. [CrossRef]
99. Rank, L.; Veith, S.; Gwosch, E.C.; Demgenski, J.; Ganz, M.; Jongmans, M.C.; Vogel, C.; Fischbach, A.; Buerger, S.; Fischer, J.M.; et al. Analyzing structure-function relationships of artificial and cancer-associated PARP1 variants by reconstituting TALEN-generated HeLa PARP1 knock-out cells. *Nucleic Acids Res.* **2016**, *44*, 10386–10405. [CrossRef]
100. Martello, R.; Mangerich, A.; Sass, S.; Dedon, P.C.; Burkle, A. Quantification of cellular poly(ADP-ribosyl)ation by stable isotope dilution mass spectrometry reveals tissue- and drug-dependent stress response dynamics. *ACS Chem. Biol.* **2013**, *8*, 1567–1575. [CrossRef]
101. Hassa, P.O.; Haenni, S.S.; Buerki, C.; Meier, N.I.; Lane, W.S.; Owen, H.; Gersbach, M.; Imhof, R.; Hottiger, M.O. Acetylation of poly(ADP-ribose) polymerase-1 by p300/CREB-binding protein regulates coactivation of NF-kappaB-dependent transcription. *J. Biol. Chem.* **2005**, *280*, 40450–40464. [CrossRef] [PubMed]
102. Kauppinen, T.M.; Chan, W.Y.; Suh, S.W.; Wiggins, A.K.; Huang, E.J.; Swanson, R.A. Direct phosphorylation and regulation of poly(ADP-ribose) polymerase-1 by extracellular signal-regulated kinases 1/2. *Proc. Natl. Acad. Sci. USA* **2006**, *103*, 7136–7141. [CrossRef] [PubMed]

103. Messner, S.; Schuermann, D.; Altmeyer, M.; Kassner, I.; Schmidt, D.; Schar, P.; Muller, S.; Hottiger, M.O. Sumoylation of poly(ADP-ribose) polymerase 1 inhibits its acetylation and restrains transcriptional coactivator function. *FASEB J.* **2009**, *23*, 3978–3989. [CrossRef] [PubMed]
104. Piao, L.; Fujioka, K.; Nakakido, M.; Hamamoto, R. Regulation of poly(ADP-Ribose) polymerase 1 functions by post-translational modifications. *Front. Biosci.* **2018**, *23*, 13–26. [CrossRef]
105. Chapman, J.D.; Gagne, J.P.; Poirier, G.G.; Goodlett, D.R. Mapping PARP-1 auto-ADP-ribosylation sites by liquid chromatography-tandem mass spectrometry. *J. Proteome Res.* **2013**, *12*, 1868–1880. [CrossRef]
106. Gagne, J.P.; Ethier, C.; Defoy, D.; Bourassa, S.; Langelier, M.F.; Riccio, A.A.; Pascal, J.M.; Moon, K.M.; Foster, L.J.; Ning, Z.; et al. Quantitative site-specific ADP-ribosylation profiling of DNA-dependent PARPs. *DNA Repair* **2015**, *30*, 68–79. [CrossRef]
107. Mortusewicz, O.; Ame, J.C.; Schreiber, V.; Leonhardt, H. Feedback-regulated poly(ADP-ribosyl)ation by PARP-1 is required for rapid response to DNA damage in living cells. *Nucleic Acids Res.* **2007**, *35*, 7665–7675. [CrossRef]
108. Hanzlikova, H.; Gittens, W.; Krejcikova, K.; Zeng, Z.; Caldecott, K.W. Overlapping roles for PARP1 and PARP2 in the recruitment of endogenous XRCC1 and PNKP into oxidized chromatin. *Nucleic Acids Res.* **2017**, *45*, 2546–2557. [CrossRef]
109. Breslin, C.; Hornyak, P.; Ridley, A.; Rulten, S.L.; Hanzlikova, H.; Oliver, A.W.; Caldecott, K.W. The XRCC1 phosphate-binding pocket binds poly (ADP-ribose) and is required for XRCC1 function. *Nucleic Acids Res.* **2015**, *43*, 6934–6944. [CrossRef]
110. Singatulina, A.S.; Hamon, L.; Sukhanova, M.V.; Desforges, B.; Joshi, V.; Bouhss, A.; Lavrik, O.I.; Pastre, D. PARP-1 Activation Directs FUS to DNA Damage Sites to Form PARG-Reversible Compartments Enriched in Damaged DNA. *Cell Rep.* **2019**, *27*, 1809–1821.e1805. [CrossRef]
111. Fischbach, A.; Kruger, A.; Hampp, S.; Assmann, G.; Rank, L.; Hufnagel, M.; Stockl, M.T.; Fischer, J.M.F.; Veith, S.; Rossatti, P.; et al. The C-terminal domain of p53 orchestrates the interplay between non-covalent and covalent poly(ADP-ribosyl)ation of p53 by PARP1. *Nucleic Acids Res.* **2018**, *46*, 804–822. [CrossRef] [PubMed]
112. Kim, D.S.; Camacho, C.V.; Nagari, A.; Malladi, V.S.; Challa, S.; Kraus, W.L. Activation of PARP-1 by snoRNAs Controls Ribosome Biogenesis and Cell Growth via the RNA Helicase DDX21. *Mol. Cell* **2019**, *75*, 1270–1285. [CrossRef] [PubMed]
113. Jungmichel, S.; Rosenthal, F.; Altmeyer, M.; Lukas, J.; Hottiger, M.O.; Nielsen, M.L. Proteome-wide identification of poly(ADP-Ribosyl)ation targets in different genotoxic stress responses. *Mol. Cell* **2013**, *52*, 272–285. [CrossRef] [PubMed]
114. Satoh, M.S.; Lindahl, T. Role of poly(ADP-ribose) formation in DNA repair. *Nature* **1992**, *356*, 356–358. [CrossRef] [PubMed]
115. Steffen, J.D.; McCauley, M.M.; Pascal, J.M. Fluorescent sensors of PARP-1 structural dynamics and allosteric regulation in response to DNA damage. *Nucleic Acids Res.* **2016**, *44*, 9771–9783. [CrossRef] [PubMed]
116. O'Sullivan, J.; Tedim Ferreira, M.; Gagne, J.P.; Sharma, A.K.; Hendzel, M.J.; Masson, J.Y.; Poirier, G.G. Emerging roles of eraser enzymes in the dynamic control of protein ADP-ribosylation. *Nat. Commun.* **2019**, *10*, 1182. [CrossRef]
117. Rack, J.G.M.; Palazzo, L.; Ahel, I. (ADP-ribosyl) hydrolases: Structure, function, and biology. *Genes Dev.* **2020**, *34*, 263–284. [CrossRef]
118. Abplanalp, J.; Leutert, M.; Frugier, E.; Nowak, K.; Feurer, R.; Kato, J.; Kistemaker, H.V.A.; Filippov, D.V.; Moss, J.; Caflisch, A.; et al. Proteomic analyses identify ARH3 as a serine mono-ADP-ribosylhydrolase. *Nat. Commun.* **2017**, *8*, 2055. [CrossRef]
119. Jankevicius, G.; Hassler, M.; Golia, B.; Rybin, V.; Zacharias, M.; Timinszky, G.; Ladurner, A.G. A family of macrodomain proteins reverses cellular mono-ADP-ribosylation. *Nat. Struct. Mol. Biol.* **2013**, *20*, 508–514. [CrossRef]
120. Rosenthal, F.; Feijs, K.L.; Frugier, E.; Bonalli, M.; Forst, A.H.; Imhof, R.; Winkler, H.C.; Fischer, D.; Caflisch, A.; Hassa, P.O.; et al. Macrodomain-containing proteins are new mono-ADP-ribosylhydrolases. *Nat. Struct. Mol. Biol.* **2013**, *20*, 502–507. [CrossRef]
121. Fontana, P.; Bonfiglio, J.J.; Palazzo, L.; Bartlett, E.; Matic, I.; Ahel, I. Serine ADP-ribosylation reversal by the hydrolase ARH3. *Elife* **2017**, *6*, e28533. [CrossRef] [PubMed]

122. Sharifi, R.; Morra, R.; Appel, C.D.; Tallis, M.; Chioza, B.; Jankevicius, G.; Simpson, M.A.; Matic, I.; Ozkan, E.; Golia, B.; et al. Deficiency of terminal ADP-ribose protein glycohydrolase TARG1/C6orf130 in neurodegenerative disease. *EMBO J.* **2013**, *32*, 1225–1237. [CrossRef] [PubMed]
123. Vivelo, C.A.; Ayyappan, V.; Leung, A.K.L. Poly(ADP-ribose)-dependent ubiquitination and its clinical implications. *Biochem. Pharmacol.* **2019**, *167*, 3–12. [CrossRef] [PubMed]
124. Leung, A.K.L. Poly(ADP-ribose): A Dynamic Trigger for Biomolecular Condensate Formation. *Trends Cell. Biol.* **2020**, *30*, 370–383. [CrossRef]
125. Altmeyer, M.; Neelsen, K.J.; Teloni, F.; Pozdnyakova, I.; Pellegrino, S.; Grofte, M.; Rask, M.D.; Streicher, W.; Jungmichel, S.; Nielsen, M.L.; et al. Liquid demixing of intrinsically disordered proteins is seeded by poly(ADP-ribose). *Nat. Commun.* **2015**, *6*, 8088. [CrossRef]
126. Liu, C.; Fang, Y. New insights of poly(ADP-ribosylation) in neurodegenerative diseases: A focus on protein phase separation and pathologic aggregation. *Biochem. Pharmacol.* **2019**, *167*, 58–63. [CrossRef]
127. Aredia, F.; Scovassi, A.I. Poly(ADP-ribose): A signaling molecule in different paradigms of cell death. *Biochem. Pharmacol.* **2014**, *92*, 157–163. [CrossRef]
128. Bryant, H.E.; Schultz, N.; Thomas, H.D.; Parker, K.M.; Flower, D.; Lopez, E.; Kyle, S.; Meuth, M.; Curtin, N.J.; Helleday, T. Specific killing of BRCA2-deficient tumours with inhibitors of poly(ADP-ribose) polymerase. *Nature* **2005**, *434*, 913–917. [CrossRef]
129. Farmer, H.; McCabe, N.; Lord, C.J.; Tutt, A.N.; Johnson, D.A.; Richardson, T.B.; Santarosa, M.; Dillon, K.J.; Hickson, I.; Knights, C.; et al. Targeting the DNA repair defect in BRCA mutant cells as a therapeutic strategy. *Nature* **2005**, *434*, 917–921. [CrossRef]
130. Pilie, P.G.; Tang, C.; Mills, G.B.; Yap, T.A. State-of-the-art strategies for targeting the DNA damage response in cancer. *Nat. Rev. Clin. Oncol.* **2019**, *16*, 81–104. [CrossRef] [PubMed]
131. Slade, D. PARP and PARG inhibitors in cancer treatment. *Genes Dev.* **2020**, *34*, 360–394. [CrossRef] [PubMed]
132. Ashworth, A.; Lord, C.J. Synthetic lethal therapies for cancer: What's next after PARP inhibitors? *Nat. Rev. Clin. Oncol.* **2018**, *15*, 564–576. [CrossRef] [PubMed]
133. Murai, J.; Huang, S.Y.; Das, B.B.; Renaud, A.; Zhang, Y.; Doroshow, J.H.; Ji, J.; Takeda, S.; Pommier, Y. Trapping of PARP1 and PARP2 by Clinical PARP Inhibitors. *Cancer Res.* **2012**, *72*, 5588–5599. [CrossRef] [PubMed]
134. Pommier, Y.; O'Connor, M.J.; de Bono, J. Laying a trap to kill cancer cells: PARP inhibitors and their mechanisms of action. *Sci. Transl. Med.* **2016**, *8*, 362ps317. [CrossRef]
135. Rancourt, A.; Satoh, M.S. Delocalization of nucleolar poly(ADP-ribose) polymerase-1 to the nucleoplasm and its novel link to cellular sensitivity to DNA damage. *DNA Rep.* **2009**, *8*, 286–297. [CrossRef]
136. Fakan, S. Immunoelectron microscopical distribution of poly(ADP-ribose)polymerase in the mammalian cell nucleus*1. *Exp. Cell Res.* **1988**, *179*, 517–526. [CrossRef]
137. Scherl, A.; Coute, Y.; Deon, C.; Calle, A.; Kindbeiter, K.; Sanchez, J.C.; Greco, A.; Hochstrasser, D.; Diaz, J.J. Functional proteomic analysis of human nucleolus. *Mol. Biol. Cell* **2002**, *13*, 4100–4109. [CrossRef]
138. Lindstrom, M.S. NPM1/B23: A Multifunctional Chaperone in Ribosome Biogenesis and Chromatin Remodeling. *Biochem. Res. Int.* **2011**, *2011*, 195209. [CrossRef]
139. Meder, V.S.; Boeglin, M.; de Murcia, G.; Schreiber, V. PARP-1 and PARP-2 interact with nucleophosmin/B23 and accumulate in transcriptionally active nucleoli. *J. Cell Sci.* **2005**, *118*, 211–222. [CrossRef]
140. Tulin, A.; Stewart, D.; Spradling, A.C. The Drosophila heterochromatic gene encoding poly(ADP-ribose) polymerase (PARP) is required to modulate chromatin structure during development. *Genes Dev.* **2002**, *16*, 2108–2119. [CrossRef]
141. Miwa, M.; Hanai, S.; Poltronieri, P.; Uchida, M.; Uchida, K. Functional analysis of poly(ADP-ribose) polymerase in Drosophila melanogaster. *Mol. Cell. Biochem.* **1999**, *193*, 103–107. [CrossRef] [PubMed]
142. Adams, M.D.; Celniker, S.E.; Holt, R.A.; Evans, C.A.; Gocayne, J.D.; Amanatides, P.G.; Scherer, S.E.; Li, P.W.; Hoskins, R.A.; Galle, R.F.; et al. The genome sequence of Drosophila melanogaster. *Science* **2000**, *287*, 2185–2195. [CrossRef]
143. Thiry, M.; Lafontaine, D.L. Birth of a nucleolus: The evolution of nucleolar compartments. *Trends Cell. Biol.* **2005**, *15*, 194–199. [CrossRef]
144. Citarelli, M.; Teotia, S.; Lamb, R.S. Evolutionary history of the poly(ADP-ribose) polymerase gene family in eukaryotes. *BMC Evol. Biol.* **2010**, *10*, 308. [CrossRef] [PubMed]
145. Boamah, E.K.; Kotova, E.; Garabedian, M.; Jarnik, M.; Tulin, A.V. Poly(ADP-Ribose) polymerase 1 (PARP-1) regulates ribosomal biogenesis in *Drosophila nucleoli*. *PLoS Genet.* **2012**, *8*, e1002442. [CrossRef] [PubMed]

146. Guetg, C.; Scheifele, F.; Rosenthal, F.; Hottiger, M.O.; Santoro, R. Inheritance of silent rDNA chromatin is mediated by PARP1 via noncoding RNA. *Mol. Cell* **2012**, *45*, 790–800. [CrossRef]
147. Santoro, R.; Li, J.; Grummt, I. The nucleolar remodeling complex NoRC mediates heterochromatin formation and silencing of ribosomal gene transcription. *Nat. Genet.* **2002**, *32*, 393–396. [CrossRef]
148. Isabelle, M.; Gallouzi, I.E.; Poirier, G.G. PARP1 parylation promotes silent locus transmission in the nucleolus: The suspicion confirmed. *Mol. Cell* **2012**, *45*, 706–707. [CrossRef]
149. Ji, Y.; Tulin, A.V. The roles of PARP1 in gene control and cell differentiation. *Curr. Opin. Genet. Dev.* **2010**, *20*, 512–518. [CrossRef]
150. Zeng, J.; Libien, J.; Shaik, F.; Wolk, J.; Hernandez, A.I. Nucleolar PARP-1 Expression Is Decreased in Alzheimer's Disease: Consequences for Epigenetic Regulation of rDNA and Cognition. *Neural Plast.* **2016**, *2016*, 8987928. [CrossRef]
151. Regier, M.; Liang, J.; Choi, A.; Verma, K.; Libien, J.; Hernandez, A.I. Evidence for Decreased Nucleolar PARP-1 as an Early Marker of Cognitive Impairment. *Neural Plast.* **2019**, *2019*, 4383258. [CrossRef] [PubMed]
152. Calkins, A.S.; Iglehart, J.D.; Lazaro, J.B. DNA damage-induced inhibition of rRNA synthesis by DNA-PK and PARP-1. *Nucleic Acids Res.* **2013**, *41*, 7378–7386. [CrossRef] [PubMed]
153. Butepage, M.; Preisinger, C.; von Kriegsheim, A.; Scheufen, A.; Lausberg, E.; Li, J.; Kappes, F.; Feederle, R.; Ernst, S.; Eckei, L.; et al. Nucleolar-nucleoplasmic shuttling of TARG1 and its control by DNA damage-induced poly-ADP-ribosylation and by nucleolar transcription. *Sci. Rep.* **2018**, *8*, 6748. [CrossRef] [PubMed]
154. Yang, C.; Zang, W.; Ji, Y.; Li, T.; Yang, Y.; Zheng, X. Ribosomal protein L6 (RPL6) is recruited to DNA damage sites in a poly(ADP-ribose) polymerase-dependent manner and regulates the DNA damage response. *J. Biol. Chem.* **2019**, *294*, 2827–2838. [CrossRef] [PubMed]
155. Leger, K.; Bar, D.; Savic, N.; Santoro, R.; Hottiger, M.O. ARTD2 activity is stimulated by RNA. *Nucleic Acids Res.* **2014**, *42*, 5072–5082. [CrossRef] [PubMed]
156. Derenzini, M.; Montanaro, L.; Trere, D. What the nucleolus says to a tumour pathologist. *Histopathology* **2009**, *54*, 753–762. [CrossRef] [PubMed]
157. Carotenuto, P.; Pecoraro, A.; Palma, G.; Russo, G.; Russo, A. Therapeutic Approaches Targeting Nucleolus in Cancer. *Cells* **2019**, *8*, 1090. [CrossRef] [PubMed]
158. MacLeod, R.A.; Spitzer, D.; Bar-Am, I.; Sylvester, J.E.; Kaufmann, M.; Wernich, A.; Drexler, H.G. Karyotypic dissection of Hodgkin's disease cell lines reveals ectopic subtelomeres and ribosomal DNA at sites of multiple jumping translocations and genomic amplification. *Leukemia* **2000**, *14*, 1803–1814. [CrossRef]
159. Aubrey, B.J.; Kelly, G.L.; Janic, A.; Herold, M.J.; Strasser, A. How does p53 induce apoptosis and how does this relate to p53-mediated tumour suppression? *Cell Death Differ.* **2018**, *25*, 104–113. [CrossRef]
160. Wade, M.; Li, Y.C.; Wahl, G.M. MDM2, MDMX and p53 in oncogenesis and cancer therapy. *Nat. Rev. Cancer* **2013**, *13*, 83–96. [CrossRef]
161. Woods, D.; Turchi, J.J. Chemotherapy induced DNA damage response: Convergence of drugs and pathways. *Cancer Biol. Ther.* **2013**, *14*, 379–389. [CrossRef] [PubMed]
162. Burger, K.; Muhl, B.; Harasim, T.; Rohrmoser, M.; Malamoussi, A.; Orban, M.; Kellner, M.; Gruber-Eber, A.; Kremmer, E.; Holzel, M.; et al. Chemotherapeutic drugs inhibit ribosome biogenesis at various levels. *J. Biol. Chem.* **2010**, *285*, 12416–12425. [CrossRef] [PubMed]
163. Rao, B.; Lain, S.; Thompson, A.M. p53-Based cyclotherapy: Exploiting the 'guardian of the genome' to protect normal cells from cytotoxic therapy. *Br. J. Cancer* **2013**, *109*, 2954–2958. [CrossRef] [PubMed]
164. Schwartz, C.L. Long-term survivors of childhood cancer: The late effects of therapy. *Oncologist* **1999**, *4*, 45–54. [CrossRef]
165. Bywater, M.J.; Poortinga, G.; Sanij, E.; Hein, N.; Peck, A.; Cullinane, C.; Wall, M.; Cluse, L.; Drygin, D.; Anderes, K.; et al. Inhibition of RNA polymerase I as a therapeutic strategy to promote cancer-specific activation of p53. *Cancer Cell* **2012**, *22*, 51–65. [CrossRef]
166. Hannan, R.D.; Drygin, D.; Pearson, R.B. Targeting RNA polymerase I transcription and the nucleolus for cancer therapy. *Expert Opin. Ther. Target.* **2013**, *17*, 873–878. [CrossRef]
167. Drygin, D.; Lin, A.; Bliesath, J.; Ho, C.B.; O'Brien, S.E.; Proffitt, C.; Omori, M.; Haddach, M.; Schwaebe, M.K.; Siddiqui-Jain, A.; et al. Targeting RNA polymerase I with an oral small molecule CX-5461 inhibits ribosomal RNA synthesis and solid tumor growth. *Cancer Res.* **2011**, *71*, 1418–1430. [CrossRef]

168. Xu, H.; Di Antonio, M.; McKinney, S.; Mathew, V.; Ho, B.; O'Neil, N.J.; Santos, N.D.; Silvester, J.; Wei, V.; Garcia, J.; et al. CX-5461 is a DNA G-quadruplex stabilizer with selective lethality in BRCA1/2 deficient tumours. *Nat. Commun.* **2017**, *8*, 14432. [CrossRef]
169. Ferreira, R.; Schneekloth, J.S., Jr.; Panov, K.I.; Hannan, K.M.; Hannan, R.D. Targeting the RNA Polymerase I Transcription for Cancer Therapy Comes of Age. *Cells* **2020**, *9*, 266. [CrossRef]
170. Maya-Mendoza, A.; Moudry, P.; Merchut-Maya, J.M.; Lee, M.; Strauss, R.; Bartek, J. High speed of fork progression induces DNA replication stress and genomic instability. *Nature* **2018**, *559*, 279–284. [CrossRef]
171. Parsels, L.A.; Karnak, D.; Parsels, J.D.; Zhang, Q.; Velez-Padilla, J.; Reichert, Z.R.; Wahl, D.R.; Maybaum, J.; O'Connor, M.J.; Lawrence, T.S.; et al. PARP1 Trapping and DNA Replication Stress Enhance Radiosensitization with Combined WEE1 and PARP Inhibitors. *Mol. Cancer Res.* **2018**, *16*, 222–232. [CrossRef] [PubMed]
172. Noordermeer, S.M.; van Attikum, H. PARP Inhibitor Resistance: A Tug-of-War in BRCA-Mutated Cells. *Trends Cell. Biol.* **2019**, *29*, 820–834. [CrossRef] [PubMed]
173. Sanij, E.; Hannan, K.M.; Xuan, J.; Yan, S.; Ahern, J.E.; Trigos, A.S.; Brajanovski, N.; Son, J.; Chan, K.T.; Kondrashova, O.; et al. CX-5461 activates the DNA damage response and demonstrates therapeutic efficacy in high-grade serous ovarian cancer. *Nat. Commun.* **2020**, *11*, 2641. [CrossRef] [PubMed]
174. Sun, C.K.; Zhang, F.; Xiang, T.; Chen, Q.; Pandita, T.K.; Huang, Y.; Hu, M.C.; Yang, Q. Phosphorylation of ribosomal protein S6 confers PARP inhibitor resistance in BRCA1-deficient cancers. *Oncotarget* **2014**, *5*, 3375–3385. [CrossRef]
175. Saxton, R.A.; Sabatini, D.M. mTOR Signaling in Growth, Metabolism, and Disease. *Cell* **2017**, *169*, 361–371. [CrossRef]
176. Meyuhas, O. Ribosomal Protein S6 Phosphorylation: Four Decades of Research. *Int. Rev. Cell Mol. Biol.* **2015**, *320*, 41–73. [CrossRef]
177. Bitler, B.G.; Watson, Z.L.; Wheeler, L.J.; Behbakht, K. PARP inhibitors: Clinical utility and possibilities of overcoming resistance. *Gynecol. Oncol.* **2017**, *147*, 695–704. [CrossRef]
178. Evans, T.; Matulonis, U. PARP inhibitors in ovarian cancer: Evidence, experience and clinical potential. *Ther. Adv. Med. Oncol.* **2017**, *9*, 253–267. [CrossRef]
179. Scott, C.L.; Swisher, E.M.; Kaufmann, S.H. Poly (ADP-ribose) polymerase inhibitors: Recent advances and future development. *J. Clin. Oncol.* **2015**, *33*, 1397–1406. [CrossRef]
180. Calo, E.; Flynn, R.A.; Martin, L.; Spitale, R.C.; Chang, H.Y.; Wysocka, J. RNA helicase DDX21 coordinates transcription and ribosomal RNA processing. *Nature* **2015**, *518*, 249–253. [CrossRef]
181. Cimino, D.; Fuso, L.; Sfiligoi, C.; Biglia, N.; Ponzone, R.; Maggiorotto, F.; Russo, G.; Cicatiello, L.; Weisz, A.; Taverna, D.; et al. Identification of new genes associated with breast cancer progression by gene expression analysis of predefined sets of neoplastic tissues. *Int. J. Cancer* **2008**, *123*, 1327–1338. [CrossRef] [PubMed]
182. Warmerdam, D.O.; Wolthuis, R.M.F. Keeping ribosomal DNA intact: A repeating challenge. *Chromosome Res.* **2019**, *27*, 57–72. [CrossRef] [PubMed]

cancers

MDPI

Review

The Modified Phenanthridine PJ34 Unveils an Exclusive Cell-Death Mechanism in Human Cancer Cells

Malka Cohen-Armon

Sackler Faculty of Medicine, and Sagol School of Neuroscience, Tel-Aviv University, Tel-Aviv 69978, Israel; marmon@tauex.tau.ac.il

Received: 8 June 2020; Accepted: 15 June 2020; Published: 19 June 2020

Abstract: This overview summarizes recent data disclosing the efficacy of the PARP inhibitor PJ34 in exclusive eradication of a variety of human cancer cells without impairing healthy proliferating cells. Its cytotoxic activity in cancer cells is attributed to the insertion of specific un-repairable anomalies in the structure of their mitotic spindle, leading to mitotic catastrophe cell death. This mechanism paves the way to a new concept of cancer therapy.

Keywords: the PARP inhibitor PJ34; distorted mitotic spindles; exclusive eradication of human cancer cells

1. Background

In the last twenty years the modified phenanthridine PJ34 has been known for its activity as a PARP (polyADP-ribose polymerase) inhibitor [1,2] (Figure 1). Recently, PARP inhibitors attract the attention of researchers and clinicians due to their FDA approval for cancer therapy [3–6]. Several comprehensive overviews on the family of PARP proteins and their inhibitors have been published [3,7–11]. This overview is dedicated to the recently disclosed exceptional cytotoxicity of the PARP inhibitor PJ34 in human cancer cells, which does not affect human healthy cells.

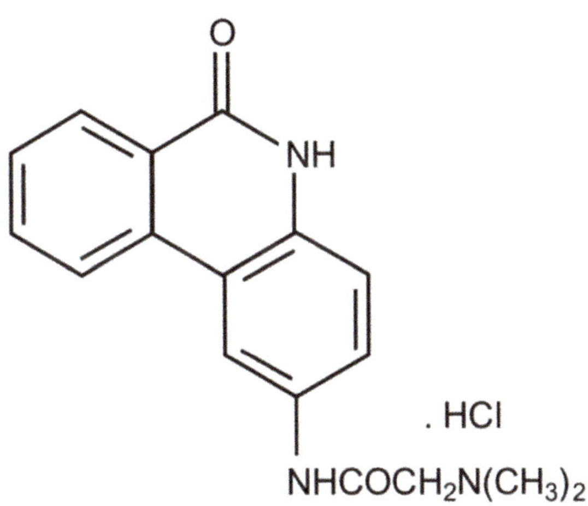

Figure 1. The chemical structure of PJ34, *N*-(6-Oxo-5,6-dihydrophenanthridin-2-yl)-(*N*,*N*-dimethylamino)acet amide hydrochloride.

The modified phenanthridine PJ34 is a stable molecule, fairly soluble in water (22 mg/mL), and permeable in the cell membrane. The potency of PJ34 to inhibit PARP proteins has been measured [8]. PJ34 is a potent inhibitor of PARP1 and PARP2 (approximate IC50, 20 nM). In addition, PJ34 inhibits tankyrase-1 and tankyrase-2 (approximate IC50, 1 μM). PJ34 hardly inhibits other PARP proteins [8]. PJ34 was originally invented to protect cells from cell death imposed by pathological stress conditions, such as ischemia [1,2,12–14].

The common concept in utilizing PARP inhibitors for cancer therapy is based on evidence associating PARP1 inhibition with the prevention of DNA break repair causing apoptotic cell death [6,15–18]. PARP1 is activated and polyADP-ribosylated in response to DNA single strand breaks frequently caused under stress conditions, including X-ray irradiation and application of DNA-damaging agents [3–6]. The binding of PARP1 to DNA single-strand breaks and its activation (polyADP-ribosylation) initiates their repair [3–6]. In addition, PARP1 is implicated in the alternative non-homologous end joining (A-NHEJ) mechanism of double strand DNA break repair [19,20]. Cancer cells have a higher percentage of DNA breaks relative to healthy cells, due to failure in arresting mitosis of cancer cells with damaged DNA [15,16]. In addition, DNA-damaging treatments are used for cancer therapy [6,16,17]. Thus, prevention of DNA repair by PARP inhibitors can be beneficial for cancer therapy, either as a monotherapy, or in combination with treatments causing DNA damage [11,16–19]. According to this concept, PARP inhibitors are offered to cancer patients carrying *BRCA* gene mutations (about 2% of the western world population). The BRCA protein is implicated in the repair of double-strand DNA breaks [17,21]. Mutations in BRCA impact the functioning of the BRCA protein, and increase DNA breaks in the cells of BRCA mutant carriers. This frequently increases the probability of mutations associated with malignancy [17,21]. Treatment with PARP inhibitors in BRCA mutant carriers is based on the interference of PARP1 inhibition with DNA repair in cells carrying a damaged DNA.

The partial efficacy of PARP inhibitors in the small population of *BRCA* mutant carriers, the resistance of a variety of cancer types to the currently approved PARP inhibitors, their side effects and the unclear impact of their chemical structure on their potency [21–26], urged a further investigation of the activity of PARP inhibitors in cancer therapy. It was observed that apart from PARP inhibition, some of these molecules target a variety of kinases implicated in signal transduction pathways in both healthy and malignant cells [27]. Unexpectedly, this research also disclosed that a group of phenanthrene derivatives acting as PARP inhibitors, exclusively kill human cancer cells without affecting benign cells [28–32]. Unlike other PARP inhibitors, these small molecules exclusively eradicated a variety of human cancer cells without affecting proliferating and non-proliferating healthy somatic cells. They did not affect human epithelial, mesenchymal and endothelial cells [28–35], nor healthy cells of mouse origin, including mouse embryonic fibroblasts (MEF), fibroblasts, neurons in the central nervous system and neuronal progenitor stem cells [28,29,31–34]. Their exclusive cytotoxic activity in human cancer cells was not shared by other potent PARP inhibitors [29–31]. Moreover, their toxicity in human cancer cells was independent of the expression of PARP1 and P53, PARP1 activity and DNA damage [29,34,35]. On the other hand, their exclusive cytotoxic activity in human cancer cells resembles the cytotoxic activity of other phenanthridines [36–38]. The modified phenanthridine PJ34, one of the molecules in this group, was the most potent in a variety of human cancer cells, including cells that are resistant to currently offered therapies [28–32]. Its specific cytotoxic activity in human cancer cells is summarized in this overview.

2. PJ34 Efficiently Eradicates a Variety of Human Cancer Cells in Tissue Cultures

After years of research based on PJ34-induced PARP inhibition in a variety of cell types under pathological conditions [1,2,8], additional activities of PJ34 have been disclosed. It was observed that PJ34 causes an irreversible cell growth arrest in cancer cells, that it interferes with angiogenesis, and, most interestingly, that PJ34 exclusively eradicates human cancer cells [29,30,39,40].

Incubation with PJ34 at higher concentrations than those inhibiting PARP1 (10–20 μM PJ34), completely eradicated within 48 h human MCF-7 breast cancer cells that are resistant to doxorubicin [28]. Furthermore, PJ34 (20–30 μM) eradicated within 72–96 h cancer cell types that are resistant to other therapies, including types of triple negative breast cancer, pancreatic cancer, ovary cancer, colon cancer and non-small lung cancer [28–32].

Gangopadhyay and colleagues found that incubation with 30 μM PJ34 for 72 h eradicates several human metastatic lung cancer cell lines: Calu-6, A549 and H460 [41]. In addition, PJ34, at higher concentrations than those inhibiting PARP1, arrested the growth of human liver cancer cell lines (HepG2 and SMMC7721) [42], and the human multiple myeloma RPMI8226 cell line [43]. PJ34 acts as a potent anti-proliferating agent in human leukemia cell lines (ATLL and transformed HTLV-I) [44], and in human ovarian cancer epithelial cells (C13 cell line) [45]. The cell death-inducing efficacy of PJ34 at higher concentrations than those inhibiting PARP activity has been also reported in a variety of breast cancer cell-lines, carrying or not BRCA mutations, and in a variety of triple-negative breast cancer cell-lines [46], as well as in melanoma cell lines and melanoma metastases [47], thyroid cancer cell lines [48], HeLa cells [49] and several glioblastoma cell lines [50]. PJ34 also efficiently prevented Helicobacter-induced gastric pre-neoplasia [51]. On the other hand, healthy proliferating cells treated with PJ34, at the same concentrations eradicating cancer cells, continued to proliferate in the presence of PJ34 as untreated cells for weeks [28–32]. Furthermore, incubation with PJ34 (20 μM) did not affect retinoic acid-induced differentiation in the human neuroblastoma cell line SHSY5Y [52], nor impaired the neuronal excitability of mouse hippocampal neurons [34].

Notably, mice treated with PJ34 for 2–3 weeks, continued to gain a similar amount of weight as untreated mice, and did not exert any visible stress or discomfort signs, as described below.

3. PJ34 Causes Eradication of Human Cancer Cells in Xenografts

The cytotoxicity of PJ34 in human cancer cells was tested in animal models as well. PJ34 (10 mg/kg) intraperitoneal (IP) injection 3 times a week for 3 weeks attenuated the growth of intracranial tumors of glioblastoma in nude mice [50]. The efficacy of daily treatment of PJ34 (30 mg/kg) using intraperitoneal injection over 14 days has been tested in xenografts of ovarian cancers. Treatment with PJ34 at this dosage most efficiently decreased the tumors' size [53]. In contrast, treatment with PJ34 in xenografts prepared from uterus and ovarian cancer of BRCA carriers showed insufficiency when PJ34 was applied per os (10 mg/kg twice a day for 16 days), in comparison to chemotherapy with carboplatin (80 mg/kg) and paclitaxel (24 mg/kg) injected IV (intravenous) once a week [54].

PJ34 prevented the development of human breast cancer MCF-7 and triple negative breast cancer MDA-MB-231 tumors in immunocompromised mice treated with a slowly released PJ34 from subcutaneous osmotic pump (Alzet) over 14 days [28]. The PJ34-treated mice that were injected subcutaneously with MCF-7, 5×10^6 cells before treatment, did not develop breast cancer tumors, and remained tumor-free during the following four months. In mice subcutaneously injected with human MDA-MB-231 triple negative breast cancer cells (5×10^6), PJ34 slow release for 14 days prevented the development of tumors in three out of five mice, and those mice remained tumor-free during the following four months. Importantly, the treatments with PJ34 did not affect the vitality, growth and weight-gain of the treated mice during the follow-up periods [28]. The therapeutic potency of PJ34 was also tested in triple negative MDA-MB-231 breast cancer xenografts after tumor development. In these experiments PJ34 (60 mg/kg) was injected IP after the tumors reached a volume >100 mm^3. PJ34 injected daily for 14 days efficiently suppressed tumor growth [31].

The efficacy of PJ34 (60 mg/kg, IV injected daily, 5 days a week, 14 intravenous injections) was tested in xenografts of human pancreas ductal adenocarcinoma PANC1 developed in immunocompromised mice. A substantial reduction of 80–90% in human cancer cells in the tumors was measured by immunohistochemistry in slices prepared from the excised tumors (16 mice) 30 days after the treatment with PJ34 has been terminated. One of the tumors disappeared after the treatment with PJ34 [32]. Benign fibroblasts infiltrated into the PANC1 tumors (stroma cells) were not impaired by the treatment with PJ34. Growth, weight-gain and behavior of the treated mice were not impaired during, and 30 days after the treatment with PJ34 has been terminated [32]. A similar cytotoxic activity of PJ34 was observed in patients-derived pancreatic cancer cells, and in patient-derived xenografts [32].

A combined treatment of PJ34 (IV) with other agents was examined, as well. A combined treatment of PJ34 with agents inducing TRAIL-mediated apoptosis in glioma xenografts reduced glioma tumor growth, and revealed minimal cytotoxicity in non-neoplastic astrocytes [55]. The combined treatment of PJ34 with TRAIL agonists was non-toxic to normal human primary glial and neuronal cells, anticipating minimal side-effects of this treatment in patients [55]. A similar effect was achieved in pancreas cancer xenografts treated with PJ34 in combination with agonists activating TRAIL-mediated apoptosis [56].

A combined treatment with PJ34 and the HDAC (histone-deacetylase) inhibitor SAHA blocked the growth of liver tumors [42]. In these experiments, HepG2 (5×10^6) cancer cells were injected subcutaneous into immunocompromised mice. These mice were treated with a combination of PJ34 (IP, 10 mg/kg) and SAHA (IP, 25 mg/kg), 3 times a week for 3 weeks. Under these conditions, tumor growth was substantially suppressed without affecting human hepatocytes [42]. Notably, SAHA at higher doses was cytotoxic to normal liver human fetal hepatocytes, while high dosage of PJ34 did not harm these cells [42].

Combining PJ34 with other anti-cancer treatments enabled reducing their cytotoxic doses and achieved efficient eradication of HL-60, Jurkat-T cells, multiple myeloma cellsPRMI8226/R and B16F10 melanoma cells [57–59].

In view of these findings, the possibility that common mechanisms are targeted by PJ34 in a variety of human cancer cell types has been examined.

4. The Mechanism of Action of PJ34 in Human Cancer Cells

In a search for the mechanism of action of PJ34 in human cancer cells, several mechanisms have been considered. The activity of PJ34 in cancer cells has been suggested to promote cell death by preventing PARP1-mediated DNA repair. This suggested mechanism is in line with PARP inhibition sensitizing cancer cells to apoptotic cell death by DNA-damaging agents [53,58–60] and by blocking DNA double-strand break repair [42]. In view of the decreased expression of PARP1 and NFkappaB in several cancer cell types eradicated by PJ34, it has been suggested that eradication by PJ34 can be attributed to the attenuation of PARP1-dependent NF-kappaB activity that promotes proliferation [45]. However, the suggested mechanisms of PARP1-dependent activity of PJ34 in cancer cells are inconsistent with its exclusive PARP1 independent cytotoxic activity in human cancer cells at a higher concentration range than that causing PARP1 inhibition [29–31,35]. Its exclusive cytotoxic activity in human cancer cells was not shared by other potent PARP inhibitors [29–31,61]. PJ34 applied at the concentration range eradicating cancer cells did not cause breaks in the DNA, nor impaired healthy cells, and its activity was not restricted to cancer cells exposed to DNA-damaging agents nor BRCA mutants [28–32,35,46].

Flow cytometry measurements revealed that PJ34 exclusively arrests mitosis in human cancer cells [28,29,32]. Incubation with PJ34 causes cell cycle arrest in a variety of healthy proliferating and cancer cells within 3–6 h [28]. However, healthy cells of mouse and human origin overcame the imposed cell-cycle arrest [28,29,31,33], and continued to proliferate in the presence of PJ34 similarly to untreated cells for weeks [28,29] (Figure 2). The tested healthy proliferating human cells included human epithelial, mesenchymal and endothelial cells [28,29,31]. Proliferating healthy cells of mouse origin included mouse embryonic fibroblasts, fibroblasts [28,29,32], and neuronal progenitor cells and astroglia [33].

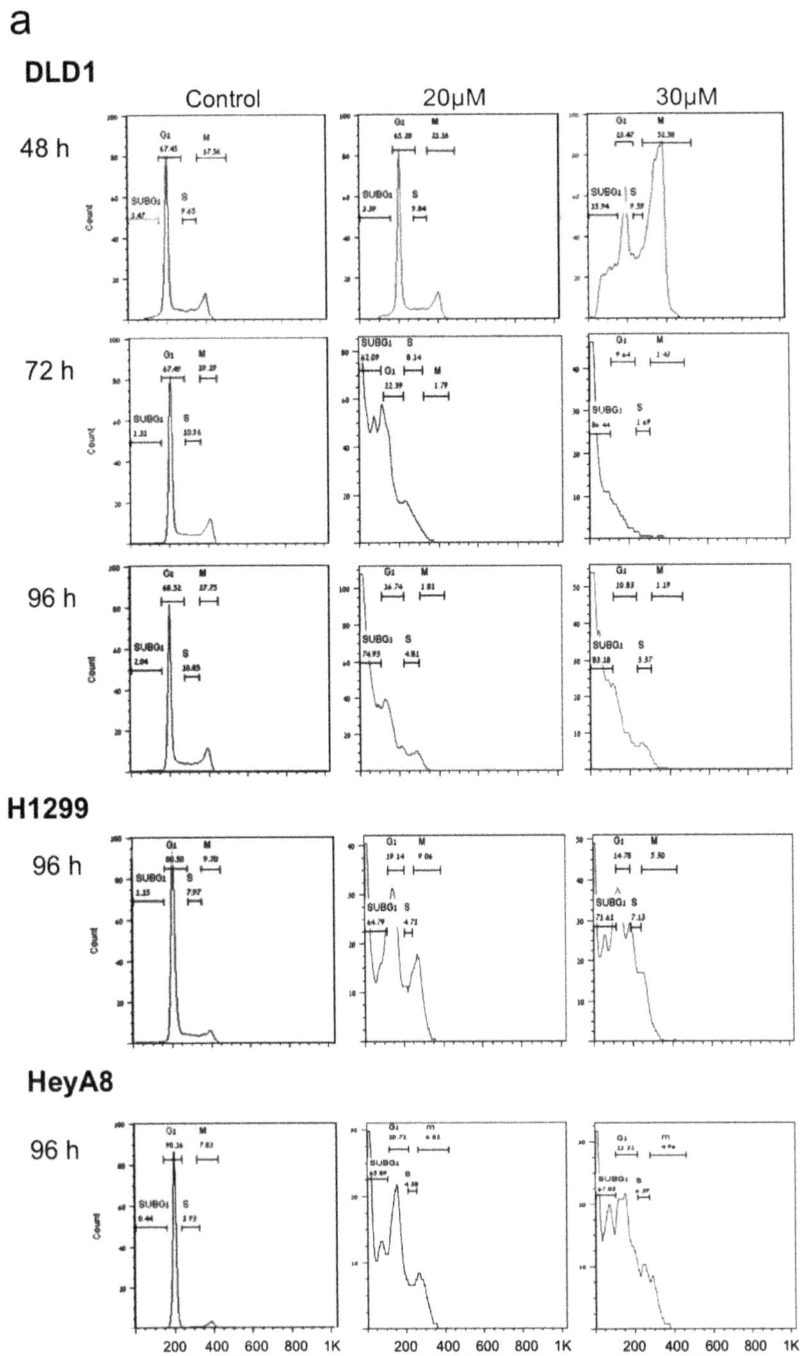

Figure 2. *Cont.*

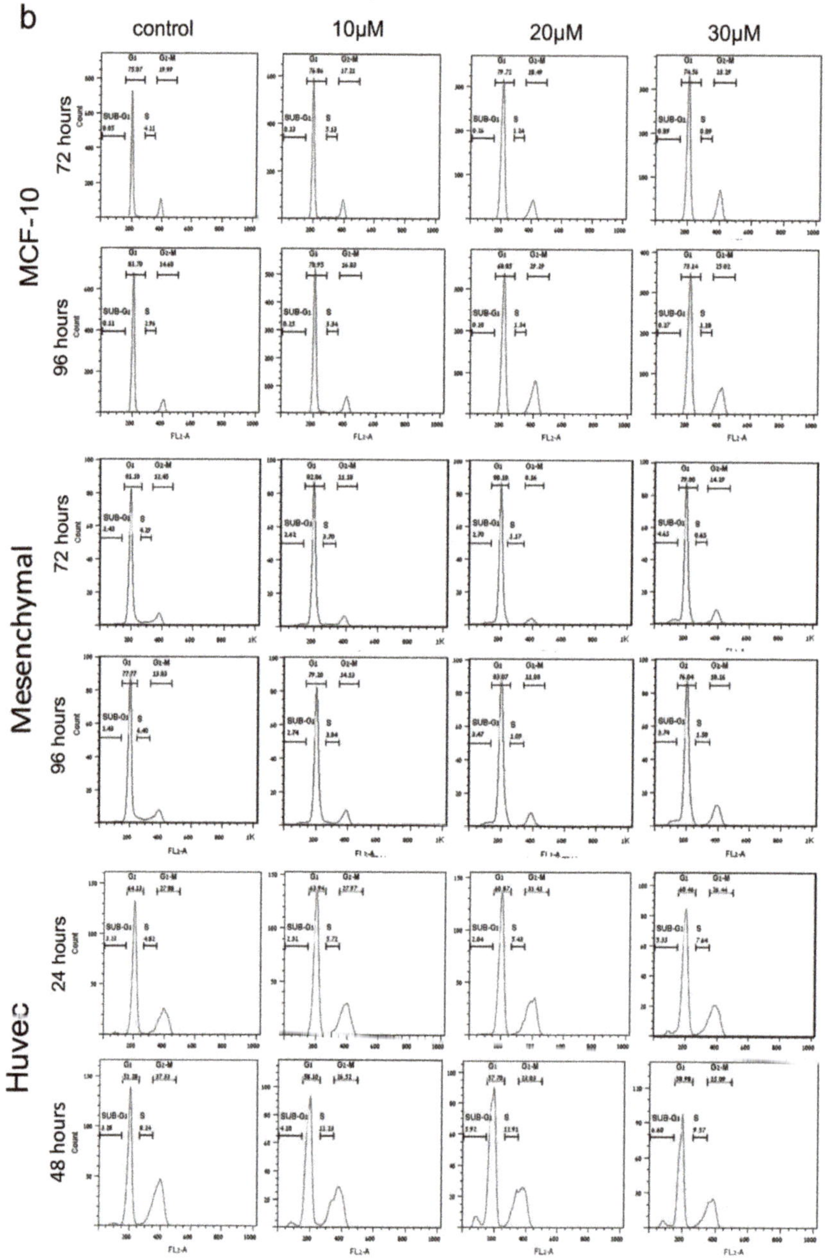

Figure 2. G2/M arrest and cell death in human cancer cells treated with PJ34. PJ34 does not impair the cell-cycle of human healthy proliferating cells. Flow cytometry of the indicated human cancer cells (**a**) and human healthy proliferating cells (**b**) is displayed. Cells were treated with PJ34 at the indicated concentrations and incubation periods. From: Castiel et al., 2011, *BMC Cancer*.

In contrast, the cell-cycle arrest in mitosis which was imposed by PJ34 in human cancer cells was irreversible in all the tested human cancer cells, and the mitosis arrest was accompanied by cell death [28–32] (Figure 2). Cell-cycle arrest preceding cell death was measured in a variety of human cancer cell types, including cancer cells prepared from human colon, ovary, lung, pancreas, T-cell leukemia and breast cancers, including triple negative breast cancer [28,29,31,32,44].

Flow cytometry measurements at the measured time intervals did not reveal G1 or S-phase arrest in the tested human healthy and cancer cells treated with PJ34 (10, 20 and 30 μM) ([28,29], Figure 2)).

Since many of the tested malignant cell types included a high percentage of extra-centrosomal cells [28–30], the cytotoxic activity of PJ34 was attributed to de-clustering of extra-centrosomes in the mitotic spindle poles of multi-centrosomal cancer cells [29,30]. According to several reports, de-clustering of centrosomes in multi-centrosomal cancer cells leads to aberrant multi-polar spindles with un-assembled chromosomes [30,62–64]. Thus, it has been suggested that multi-centrosomes de-clustering in the spindle poles activates the spindle assembly check-point (SAC) proteins, which arrest mitosis [65,66], subsequently leading to mitotic catastrophe cell death [65–68].

Since mitotic arrest was induced by PJ34 in a variety of cancer cells, including cell types with a low percentage of extra-centrosomal cells [69,70], other mechanisms inducing mitotic catastrophe cell death were examined.

In one approach, all proteins currently known to be implicated in mitosis were screened in a group of human cancer and healthy cells, in an attempt to identify different effects of PJ34 (eradicating only cancer cells) on the post-translational modifications of proteins in the cancer versus healthy cells. Changes induced by PJ34 in their post-translational modifications were measured by the shift in their isoelectric point (IP) in two-dimensional (2-D) gel electrophoresis [31]. This approach was used because PARP inhibitors may modify a variety of proteins in both healthy and malignant cells [27,61,71].

Among other proteins associated with mitosis, proteins that are implicated in the structure of the mitotic spindle were examined for a possible interference with their post-translational modification by PJ34 in human cancer cells versus the effect of PJ34 on their post-translational modification in healthy epithelial cells [31]. This analysis was conducted in four types of human cancer cell lines (glioblastoma U87, PANC1 pancreas cancer cells, lung cancer cells A549 and triple-negative breast cancer cells MDA-MB-231). All these cells are eradicated by PJ34 [28–32]. A similar analysis was used to identify possible effects of PJ34 on the post-translational modifications of the same proteins in healthy human epithelial cells, which are not impaired by PJ34 [28,29,31].

This analysis identified only three proteins in the tested cancer cells with isoelectric point significantly shifted by PJ34, while not affected in healthy cells [31]. These proteins included two motor proteins [72], human kinesins 14/HSET/kifC1 and kif18A, and the non-motor protein NuMA (nuclear mitotic apparatus protein) [73–77]. For comparison, the IP shift of these proteins in the tested cells was also measured when these cells were incubated with the PARP inhibitor ABT-888 (Veliparib), which is a potent inhibitor of PARP-1,-2,-3 proteins (with a similar affinity for PARP1 and PARP2, $Kd = 2.9$ and 5.2 nM, respectively) [8,31]. Unlike PJ34, ABT-888 did not affect the isoelectric points of human kinesins 14/HSET/kifC1 and kif18A and of NuMA. There was no difference between the IP of these proteins in ABT-888-treated versus untreated cancer cells [31]. This result may exclude possible involvement of PARP-1,-2,-3 inhibition in the IP shift induced by PJ34 in the tested cancer cells [31]. In accordance, ABT-888 neither arrested mitosis, nor eradicated the tested human cancer cells [31]. These findings suggested a possible interference of PJ34 with the construction of the mitotic spindle in the tested human cancer cells.

Numerous findings indicate the essential role of HSET/kifC1 in the spindle structure of human cancer cells [73–75,78–81]. Differences in the expression and function of HSET in cancer versus healthy cells have been reported [74,78–80]. HSET/kifC1 inhibition or silencing causes small aberrant spindles in human malignant cells [79].

The kinesin Kif18A is implicated in microtubules de-polymerization, necessary for the binding of the duplicated chromosomes to kinetochores in the spindle mid-zone. Loss of its function results in the formation of long microtubules with chromosomes un-attached to kinetochores in the mid-zone [76].

The third identified protein, NuMA, is essential for mitosis in both malignant and benign cells [77,82,83]. Recently, several reports indicated some differences between NuMA proteins and their expression in benign versus malignant cells [84,85]. In this analysis, a clear-cut difference has been disclosed by the effect of PJ34 on the post-translational modification of NuMA only in the variety of human malignant cells ([31] and Figure 3). PJ34 did not affect the isoelectric point of NuMA in the healthy epithelial cells [31]. Concomitantly, the ability of NuMA to bind proteins was lost in the PJ34-treated malignant cells [31]. Moreover, the lost ability of NuMA to bind proteins was accompanied by un-clustering of NuMA in the spindle poles of the malignant cells treated with PJ34, as disclosed by confocal imaging [31] (Figure 3). In contrast, the bi-polar clustering of NuMA in the mitotic spindles of the healthy benign cells was not affected by the treatment with PJ34 ([31] (Figure 3).

Previous findings indicated the post-translational modification of NuMA by both polyADP-ribosylation and phosphorylation, both modifications promoting the binding of NuMA to proteins [82–89]. NuMA is phosphorylated by serine-threonine kinase pim1, and NuMA phosphorylation by serine threonine kinases at a specific site is crucial for its ability to bind proteins [87,89]. In addition, polyADP-ribosylation of NuMA by tankyrase1 in cancer cells promotes the ability of NuMA to bind other proteins [31,88].

Pim kinases and tankyrase1 are both inhibited by PJ34 at the concentrations range PJ34 causes cell death in human cancer cells (measured IC50 = 3.7 μM for pim1 inhibition by PJ34, and for tankyrase1 inhibition by PJ34, IC50 = 1 μM) [90–92]. Furthermore, tankyrase1 and pim kinases are hardly expressed in healthy somatic cells, while they are highly expressed in human cancer cells [92,93].

Clustered NuMA in the spindle poles and tethering of microtubules to the clustered proteins in the spindle poles are essential for the construction of stable poles required for the binding of the chromosomes to kinetochores in the spindle mid-zone [31,77–83]. Thus, PJ34 blocking of the post-translational modification of NuMA concomitantly prevents the clustering of NuMA in the spindle poles, causing mitotic arrest and cell death by preventing the binding of chromosomes to kinetochores in the spindle mid zone [31,65–68,80–82].

The causal association of the post-translational modification of NuMA with the ability of NuMA to bind proteins [31,87–89] could explain the interference of PJ34 with NuMA clustering and binding to HSET in the spindle poles of cancer cells treated with PJ34 [31,80–83]. NuMA clustering in the poles and the tethering of microtubules to the spindle poles implicates that HSET and NuMA binding in the protein clusters causes a stable structure that is lost in the spindle poles of human cancer cells treated with PJ34, at the concentration range inhibiting pim1 and tankyrase1 [31,91,92].

An unstable structure of the spindle poles due to impaired function of HSET and NuMA, and impaired function of kif18A in the binding of chromosomes to the kinetochores may result in scattered chromosomes instead of chromosomes bound to the kinetochores in the mid-zone (Figure 3). This activates the spindle assembly control (SAC) proteins, which leads to mitosis arrest followed by mitotic catastrophe cell death when the structural anomaly is not amended [65–68]. This is exactly the phenomenon observed by confocal imaging in a variety of human cancer cells treated with PJ34 [29–31]. De-clustering of centrosomes observed in multi-centrosomal cancer cells treated with PJ34 (20 μM) could result from the un-stable aberrant spindle poles [31,78]. Thus, prevention of the post-translational modification of NuMA and kinesins HSET and Kif18A by PJ34, could evoke mitotic arrest and mitotic catastrophe cell death in human cancer cells [30–32] (Figure 2) just by inserting specific anomalies in their mitotic spindle structure ([31] and Figure 3). This effect of PJ34 was not copied by ABT-888 at concentrations inhibiting PARP-1,-2,-3 [8,31]. It was achieved by inhibiting the activity of serine-threonine kinases, and by inhibition of tankyrase polyADP-ribosylation [31]. Other tankyrase inhibitors and tankyrase1 silencing caused similar faults in the structure of spindle poles as well as G2/M arrest, which could be attributed to inhibition of NuMA polyADP-ribosylation [31,88,94,95]. These results are consistent with the consequences of the treatment with PJ34 causing aberrant spindles with dispersed chromosomes (Figure 3) arresting mitosis and killing cancer cells without impairing healthy proliferating cells (Figure 2) [28–32]. In consistence, the findings of Leber and collaborators implicated NuMA and HSET in the construction of aberrant spindle poles with un-clustered multi-centrosomes in human cancer cells [96].

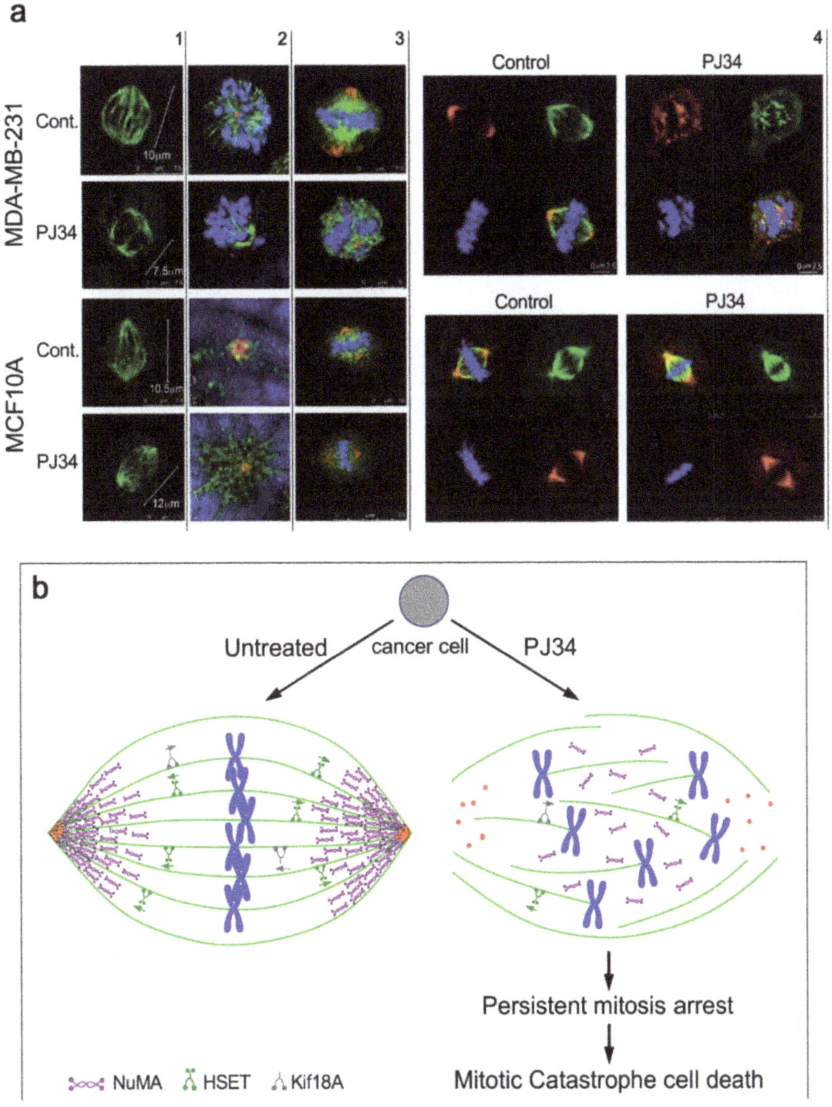

Figure 3. (**a**) Confocal images of mitotic spindles in human triple negative breast cancer cells (MDA-MB-231) and in human healthy breast epithelial cells (MCF10A), untreated or incubated with PJ34. Incubation of human cancer cells with PJ34 (20 μM, 27 h) impaired spindle poles (labeled by immunolabelling, γ-tubulin in the centrosomes—red), microtubules (labeled by immunolabelling—kinesin HSET or by immunolabelling α-tubulin—green), segregation and alignment of chromosomes (labeled by DAPI—blue), and NuMA clustering in the spindle poles (Immunolabeled NuMA—red). **Column 1**: Microtubules in spindles of healthy and cancer cells immunolabeled by the kinesin HSET in cancer and healthy cells, untreated and treated with PJ34. **Column 2**: Spindle poles labeled by γ-tubulin in healthy and cancer cells, untreated and treated with PJ34. HSET is immunolabeled in the microtubules. **Column 3**: Clustered NuMA in bipolar spindles of healthy cells either treated or not with PJ34, and in untreated cancer cells. Un-clustered NuMA in spindles of cancer cells treated with PJ34. **Column 4**: upper frame: In cancer cells—clustered NuMA in spindle poles and aligned chromosomes in the midzone of untreated cancer cells. Aberrant spindles, un-clustered NuMA

and scattered chromosomes in cancer cells treated with PJ34. Lower frame: In healthy cells—clustered NuMA in the spindle poles and segregated chromosomes aligned in the mid-zone of the mitotic spindle of healthy cells either untreated or treated with PJ34. (**b**) A schematic presentation indicating the effect of PJ34 on the spindle structure in human cancer cell. In the untreated cancer cell, normal bipolar spindles with clustered NuMA, clustered multi-centrosomes, and aligned chromosomes in the spindle mid-zone. In the PJ-34 treated cancer cell, aberrant microtubules (**green**), aberrant spindle poles, un-clustered NuMA (as indicated), dispersed chromosomes (**blue**) and un-clustered multi-centrosome (**red**), From: Visochek et al., 2017, Oncotarget.

5. The Potency of PJ34 and Other PARP Inhibitors in Preventing Metastases

PJ34 and additional PARP inhibitors inhibit the activity of matrix metalloproteinase-2 in a range of concentrations higher than that inducing PARP inhibition [97]. The measured IC50 was about 56 μM for PJ34 inhibition of MMP-2 [97], lower than the IC50 values of other tested PARP1 inhibitors [97].

In healthy tissues, matrix metalloproteinases (MMPs) are key enzymes in the development and remodeling of tissues, including in wound healing [98]. However, MMPs are also dominant in tumor angiogenesis, in tumor cells escape from the primary tumor and their dissemination to secondary sites [98,99]. Elevated expression of MMPs, including MMP-2, has been implicated in metastasis [98–100]. Metastases predict high invasive stage of the malignancy, and a poor prognosis [99,100]. MMP inhibitors prevent metastases in a variety of solid cancer types [99,100]. MMP inhibition by PARP inhibitors is their additional advantage in cancer therapy. ABT-888 hindered cancer cell migration rate measured in vitro by the scratch assay [101]. PJ34 at the concentration range inhibiting MMP-2 prevented metastasis in melanoma xenografts [102].

6. Conclusions and Future Perspectives

The potency of PJ34 to exclusively eradicate human cancer cells without impairing healthy cells can be attributed to anomalies exclusively inserted in the structure of the mitotic spindle of human cancer cells. Blocking the post-translational modification of NuMA and kinesins HSET and kif18A by PJ34 causes these anomalies, which subsequently prevent the alignment of chromosomes bound to kinetochores in the spindle mid-zone. This structural anomaly in the mitotic spindle arrests mitosis and kills the cancer cell in the pre-anaphase stage by mitotic catastrophe cell death.

Notably, the modified phenanthridine PJ34, which has been invented for PARP inhibition, exclusively eradicates a variety of human cancer cells without impairing normal healthy somatic quiescent and proliferating cells. Thus, in spite of the permeability of PJ34 in the cell membrane and its rapid distribution in the animal's tissues, treatment with PJ34 did not impair healthy tissues in the tested animals, nor their development and weight gain.

On the basis of these findings, we hope that cell death evoked by structural faults in the mitotic spindle of human cancer cells will pave the way to a new concept in cancer therapy. Inserting specific structural anomalies in the mitotic spindle of human cancer cells may specifically eradicate cancer cells while saving healthy cells and physiological functions frequently lost during the currently offered DNA-damaging cancer therapies.

Funding: This research received no external funding.

Conflicts of Interest: The author declare no conflict of interest.

References

1. Abdelkarim, G.E.; Gertz, K.; Harm, C.; Katchanov, J.; Dirnagi, U.; Szabo, C.; Enders, M. Protective effects of PJ34, a novel potent inhibitor of poly(ADP-ribose)polymerase (PARP) in in-vitro and in-vivo models of stroke. *Int. J. Mol. Med.* **2001**, *7*, 255–260. [CrossRef] [PubMed]
2. Jagtap, P.; Szabo, C. Poly(ADP-ribose)polymerase and the therapeutic effects of its inhibitors. *Nat. Rev. Drug Discov.* **2005**, *4*, 421–440. [CrossRef] [PubMed]
3. Slade, D. PARP and PARG inhibitors in cancer treatment. *Gene Dev.* **2020**, *34*, 1–35. [CrossRef] [PubMed]

4. Carden, C.P.; Yap, T.A.; Kaye, S.B. PARP inhibition: Targeting the Achilles heelof DNA repair to treat germline and sporadic ovarian cancers. *Curr. Opin. Oncol.* **2010**, *22*, 473–480. [CrossRef]
5. Plummer, R. Perspective on the pipeline of drugs being developed with modulationof DNA damage as a target. *Clin. Cancer Res.* **2010**, *16*, 4527–4531. [CrossRef]
6. Mangerich, A.; Burkle, A. How to kill tumor cells with inhibitors of poly(ADP-ribosyl)ation. *Int. J. Cancer* **2011**, *128*, 251–265. [CrossRef]
7. Hassa, P.O.; Hottiger, M.O. The diverse biological roles of mammalian PARPs, a small but powerful family of poly-ADP-ribose polymerases. *Front. Biosci.* **2008**, *13*, 3046–3082. [CrossRef]
8. Wahlberg, E.; Karlberg, T.; Kouznetsova, E.; Markova, N.; Macchiarulo, A.; Torsell, A.-G.; Pol, E.; Frostell, A.; Ekblad, T.; Öncü, D.; et al. Family-wide chemical profiling and structural analysis of PARP and tankyrase inhibitors. *Nat. Biotech.* **2012**, *30*, 283–289. [CrossRef]
9. Citarelli, M.; Teotia, S.; Lamb, R.S. Evolutionary history of the poly(ADP-ribose)polymerase gene family in eukaryotes. *BMC Evol. Biol.* **2010**, *10*, 308. [CrossRef]
10. Krishnakumar, R.; Kraus, W.L. The PARP side of the nucleus: Molecular actions, physiological outcomes, and clinical targets. *Mol. Cell* **2010**, *39*, 8–24. [CrossRef]
11. Gibson, B.A.; Kraus, W.L. New insights into the molecular and cellular functions of poly(ADP-ribose) and PARPs. *Nat. Rev. Mol. Cell Biol.* **2012**, *13*, 411–424. [CrossRef] [PubMed]
12. Kauppinen, T.M.; Swanson, R.A. The role of poly(ADP-ribose) polymerase-1 in CNS disease. *Neuroscience* **2007**, *145*, 1267–1272. [CrossRef] [PubMed]
13. Strosznajder, R.P.; Czubowicz, K.; Jesko, H.; Strosznajder, J.B. Poly(ADP-ribose) metabolism in brain and its role in ischemia pathology. *Mol. Neurobiol.* **2010**, *41*, 187–196. [CrossRef] [PubMed]
14. Ba, X.; Garg, N.J. Signaling mechanism of poly(ADP-ribose)polymerase-1 (PARP-1) in inflammatory diseases. *Am. J. Pathol.* **2011**, *178*, 946–955. [CrossRef] [PubMed]
15. Gregersen, L.H.; Jesper, J.Q. The cellular response to transcription blocking DNA damage. *Trend Biochem. Sci.* **2018**, *43*, 327–341. [CrossRef]
16. Lesueur, P.; Chevalier, F.; Austry, J.-B.; Waissi, W.; Burckel, H.; Noel, G.; Habrand, J.-L.; Saintigny, Y.; Joly, F. PolyADP-ribose) polymerase inhibitors as radiosensitizers: A systematic review of preclinical human studies. *Oncotarget* **2017**, *8*, 69105–69124. [CrossRef]
17. Bryant, H.E.; Schultz, N.; Thomas, H.D.; Parker, K.M.; Flower, D.; Lopez, E.; Kyle, S.; Meuth, M.; Curtin, N.J.; Helleday, T. Specific killing of BRCA2-deficient tumours with inhibitors of poly(ADP-ribose) polymerase. *Nature* **2005**, *434*, 913–917. [CrossRef]
18. Andrabi, S.A.; Kim, N.-S.; Yu, S.W.; Wang, H.; Koh, D.W.; Sassaki, M.; Klaus, J.A.; Otsuka, T.; Zhang, Z.; Koehler, C.K.; et al. Poly(ADP-ribose) (PAR) polymer is a death signal. *Proc. Natl. Acad. Sci. USA* **2006**, *103*, 18308–18313. [CrossRef]
19. Chang, H.H.Y.; Pannunzio, N.R.; Adachi, N.; Lieber, M.R. Non-homologous DNA end joining and alternative pathways to double-strand break repair. *Nat. Rev. Mol. Cell Biol.* **2017**, *18*, 495–506. [CrossRef]
20. Paddock, M.N.; Bauman, A.T.; Higdon, R.; Kolker, E.; Takeda, S.; Scharenberg, A.M. Competition between PARP-1 and Ku70 control the decision between high-fidelity and mutagenic DNA repair. *DNA Repair* **2011**, *10*, 338–343. [CrossRef]
21. Farmer, H.; McCabe, N.; Lord, C.J.; Tutt, A.N.; Johnson, D.A.; Richardson, T.B.; Santarosa, M.; Dillon, K.J.; Hickson, I.; Knights, C.; et al. Targeting the DNA repair defect in BRCA mutant cells as a therapeutic strategy. *Nature* **2005**, *434*, 917–921. [CrossRef] [PubMed]
22. Mak, J.P.Y.; Ma, H.T.; Poon, R.Y.C. Synergism between ATM and PARP1 inhibition involves DNA damage and abrogating the G2 DNA damage checkpoint. *Mol. Cancer Ther.* **2020**, *19*, 123–134. [CrossRef] [PubMed]
23. Zhang, R.; Hong, J.-J.; Yang, Q.; Ong, C.-T.; Li, B.-A.; Liou, Y.-C. Poly(ADP- ribosyl)ation of OVOL2 regulates aneuploidy and cell death in cancer cells. *Oncogene* **2019**, *38*, 2750–2766. [CrossRef]
24. Zhou, J.X.; Feng, L.J.; Zhang, X. Risk of severe hematologic toxicities in cancer patients treated with PARP inhibitors: A meta-analysis of randomized controlled trials. *Drug Des. Dev. Ther.* **2017**, *11*, 3009–3017. [CrossRef] [PubMed]
25. Lee, Y.R.; Yu, D.-S.; Liang, Y.-C.; Huang, K.-F.; Chou, S.-J.; Chen, T.-C.; Lee, C.-C.; Chen, C.-L.; Chiou, S.-H.; Huang, H.-S. New approaches of PARP1 inhibitors in human lung cancer cells and cancer stem-like cells by some selected anthraquinone-derived small molecules. *PLoS ONE* **2013**, *8*, e56284. [CrossRef]

26. Kishi, Y.; Fujibara, H.; Kawaguchi, K.; Yamada, H.; Nakayaima, R.; Yamamoto, N.; Fujihara, Y.; Hamada, Y.; Satomura, K.; Masutani, M. PARP inhibitor PJ34 supresses osteogenic differentiation in mouse mesenchymal stem cells by modulating BMP-2 signaling pathway. *Int. J. Mol. Sci.* **2015**, *16*, 24820–24838. [CrossRef] [PubMed]
27. Antolin, A.A.; Ameratunga, M.; Banerji, U.; Clarke, P.A.; Workman, P.; Al-Lazikani, B. The kinase polypharmacology landscape of clinical PARP inhibitors. *Sci. Rep.* **2020**, *10*, 2585. [CrossRef] [PubMed]
28. Inbar-Rozensal, D.; Visochek, L.; Castel, D.; Castiel, A.; Izraeli, S.; Dantzer, F.; Cohen-Armon, M. A selective eradication of human nonhereditary breast cancer cells by phenanthridine-derived polyADP-ribose polymerase inhibitors. *Breast Cancer Res.* **2009**, *11*, R78. [CrossRef]
29. Castiel, A.; Visochek, L.; Mittelman, L.; Dantzer, F.; Izraeli, S.; Cohen-Armon, M. A phenanthrene derived PARP inhibitor is an extra-centrosomes de-clustering agent exclusively eradicating human cancer cells. *BMC Cancer* **2011**, *11*, 412. [CrossRef]
30. Castiel, A.; Visochek, L.; Mittelman, L.; Zilberstein, Y.; Dantzer, F.; Izraeli, S.; Cohen-Armon, M. Cell death associated with abnormal mitosis observed by confocal imaging in live cancer cells. *J. Vis. Exp.* **2013**, *78*, e50568. [CrossRef]
31. Visochek, L.; Castiel, A.; Mittelman, L.; Elkin, M.; Atias, D.; Golan, T.; Izraeli, S.; Peretz, T.; Cohen-Armon, M. Exclusive destruction of mitotic spindles in human cancer cells. *Oncotarget* **2017**, *8*, 20813–20824. [CrossRef]
32. Visochek, L.; Atias, D.; Spektor, I.; Castiel, A.; Golan, T.; Cohen-Armon, M. The phenanthrene derivative PJ34 exclusively eradicates human pancreatic cancer cells in xenografts. *Oncotarget* **2019**, *10*, 6269–6282. [CrossRef] [PubMed]
33. Okuda, A.; Kurokawa, S.; Takehashi, M.; Maeda, A.; Fukuda, K.; Kubo, Y.; Nogusa, H.; Takatani-Nakase, T.; Okuda, S.; Ueda, K.; et al. Poly(ADP-ribose) polymerase inhibitors activate the p53 signaling pathway in neural stem/progenitor cells. *BMC Neurosci.* **2017**, *18*, 2–18. [CrossRef]
34. Visochek, L.; Grigoryan, G.; Kalal, A.; Milshtein-Parush, H.; Gazit, N.; Slutsky, I.; Yeheskel, A.; Shainberg, A.; Castiel, A.; Seger, R.; et al. A PARP1-ERK2 synergism is required for the induction of LTP. *Sci. Rep.* **2016**, *6*, 24950. [CrossRef] [PubMed]
35. Madison, D.L.; Stauffer, D.; Lundblad, J.R. The PARP inhibitor PJ34 causes a PARP1-independent, p21 dependent mitotic arrest. *DNA Repair* **2011**, *10*, 1003–1013. [CrossRef] [PubMed]
36. Lamoral-Theys, D.; Andolfi, A.; Van Goietsenoven, G.; Cimmino, A.; Le Calv, B.; Wauthoz, N.; Mégalizzi, V.; Gras, T.; Bruyère, C.; Dubois, J.; et al. Lycorine, the Main Phenanthridine Amaryllidaceae Alkaloid, Exhibits Significant Antitumor Activity in Cancer Cells That Display Resistance to Proapoptotic Stimuli: An Investigation of Structure-Activity Relationship and Mechanistic Insight. *J. Med. Chem.* **2009**, *52*, 6244–6256. [CrossRef]
37. Wang, Y.-Y.; Taniguchi, T.; Baba, T.; Li, Y.-Y.; Isibashi, H.; Nukaida, N. Identification of a phenanthrene derivative as a potent anticancer drug with pim kinase inhibitory activity. *Cancer Sci.* **2011**, *103*, 107–115. [CrossRef]
38. Mariappan, A.; Soni, K.; Schorpp, K.; Zhao, F.; Minakar, A.; Zheng, X.; Mandad, S.; Macheleidt, I.; Ramani, A.; Kubelka, T.; et al. Inhibition of CPAP-tubulin interaction prevents proliferation of centrosome amplified cancer cells. *EMBO J.* **2019**, *38*, e99876. [CrossRef]
39. Huang, S.H.; Xiong, M.; Chen, X.P.; Xiao, Z.Y.; Zhao, Y.F.; Huang, Z.Y. PJ34 an inhibitor of PARP1, suppresses cell growth and enhances the suppressive effects of cisplatin in liver cancer cells. *Oncol. Rep.* **2008**, *20*, 567–572.
40. Pyriochou, A.; Olah, G.; Deitch, O.; Szabo, C.; Papaetropoulos, A. Inhibition of angiogenesis by the poly(ADP-ribose) polymerase inhibitor. *J. Mol. Med.* **2008**, *22*, 113–118. [CrossRef]
41. Gangopadhyay, N.N.; Luketich, J.D.; Opest, A.; Meyer, E.M.; Landreneau, R.; Schuchert, M.J. Inhibition of Poly(ADP-Ribose) Polymerase (PARP) Induces Apoptosis in Lung Cancer Cell Lines. *Cancer Investig.* **2011**, *29*, 608–616. [CrossRef] [PubMed]
42. Liang, B.; Xiong, M.; Ji, G.; Zhang, E.; Zhang, Z.; Dong, K.; Chen, X.; Huang, Z.-Y. Synergistic suppressive effect of PARP-1 inhibitor PJ34 and HDAC inhibitor SAHA on proliferation of liver cancer cells. *J. Huazhong Univ. Sci. Technol.* **2015**, *35*, 535–540. [CrossRef] [PubMed]
43. Xiong, T.; Chen, X.; Wei, H.; Xiao, H. Influence of PJ34 on the genotoxicity induced by melphalan in human multiple myeloma cells. *Arch. Med. Sci.* **2015**, *11*, 301–306. [CrossRef]

44. Bai, X.T.; Moles, R.; Chaib-Mezrag, H.; Nicot, C. Small PARP inhibitor PJ34 induces cell-cycle arrest and apoptosis of adult T cell leukemia cells. *J. Hematol. Oncol.* **2015**, *8*, 117. [CrossRef] [PubMed]
45. Wang, Z.; Li, Y.; Lv, S.; Tian, Y. Inhibition of proliferation and invasiveness of ovarian cancer C13 cells by a poly(ADP-ribose)polymerase inhibitor and the role of nuclear factor-kappaB. *J. Int. Med. Res.* **2013**, *41*, 1577–1585. [CrossRef]
46. Keung, M.Y.; Wu, Y.; Badar, F.; Vadgama, J.V. Response of breast cancer cells to PARP inhibitors is independent of BRCA status. *J. Clin. Med.* **2020**, *9*, 940. [CrossRef]
47. Chevanne, M.; Zampieri, M.; Rizzo, A.C.R.; Ciccarone, F.; Catizone, A.; D'Angelo, C.; Guastafierro, T.; Biroccio, A.; Raele, A.; Zupi, G.; et al. Inhibition of PARP activity by PJ34 leads to growth impairment and cell death associated with aberrant mitotic pattern and nucleolar actin accumulation in M14 melanoma cell line. *J. Cell Physiol.* **2010**, *222*, 401–410. [CrossRef]
48. Lavarone, E.; Puppin, C.; Passon, N.; Filetti, S.; Russo, D.; Damante, G. The PARP inhibitor PJ34 modifies proliferation, NIS expression and epigenetic marks in thyroid cancer cell lines. *Mol. Cell Endocrinol.* **2013**, *365*, 1–10. [CrossRef]
49. Magan, N.; Isaacs, R.J.; Stowell, K.M. Treatment with the PARP-inhibitor PJ34 causes enhanced doxorubicin-mediated cell death in HeLa cells. *Anticancer Drugs* **2012**, *23*, 627–637. [CrossRef]
50. Majuelos-Melguizo, J.; Rodríguez, M.I.; López-Jiménez, L.; Jose, M.; Rodríguez-Vargas, J.; Martín-Consuegra, M.; Serrano-Sáenz, S.; Gavard, J.; Ruiz de Almodóvar, J.M.; Oliver, F.J. PARP targeting counteracts gliomagenesis through induction of mitotic catastrophe and aggravation of deficiency in homologous recombination in PTEN-mutant glioma. *Oncotarget* **2015**, *6*, 4790–4803. [CrossRef]
51. Toller, I.M.; Altmeyer, M.; Kohler, E.; Hottiger, M.O.; Muller, A. Inhibition of ADP-ribosylation prevents and cures Helicobacter-induced Gastric Preneoplasia. *Cancer Res.* **2010**, *70*, 5912–5922. [CrossRef] [PubMed]
52. Horvat, L.; Grubar, M.; Madunic, J.; Antica, M.; Matulic, M. Inhibition of PARP activity does not affect the differentiation processes caused by retinoic acid in SHSY5Y cells. *Mol. Exp. Biol. Med.* **2019**, *1*, 38–43. [CrossRef]
53. Hou, D.; Liu, Z.; Xu, X.; Liu, Q.; Zhang, X.; Kong, B.; Wei, J.-J.; Gong, Y.; Shao, C. Increased oxidative stress mediates the antitumor effect of PARP inhibition in ovarian cancer. *Redox Biol.* **2018**, *17*, 99–111. [CrossRef] [PubMed]
54. Press, J.Z.; Kenyon, J.A.; Xue, H.; Miller, M.A.; DeLuca, A.; Miller, D.M.; Huntsman, D.G.; Gilk, C.B.; McAlpine, J.N.; Wang, Y.Z. Xenografts of primary human gynecological tumors grown under the renal capsule of NOD/SCID mice show genetic stability during serial transplantation and respond to cytotoxic chemotherapy. *Gyn. Oncol.* **2008**, *109*, 256–264. [CrossRef]
55. Karpel-Massler, G.; Pareja, F.; Aime, P.; Shu, C.; Chau, L.; Westhoff, A.; Halatsch, M.-E.; Crary, J.F.; Canoll, P.; Siegelin, M. PARP inhibition restores extrinsic apoptotic sensitivity in Glioblastoma. *PLoS ONE* **2014**, *9*, e114583. [CrossRef]
56. Yuan, K.; Sun, Y.; Zhou, T.; McDonald, J.; Chen, Y. PARP1 regulates resistance of pancreatic cancer to TRAIL therapy. *Clin. Cancer Res.* **2013**, *19*, 4750–4759. [CrossRef]
57. Stepnik, M.; Spryszynska, S.; Gorzkiewicz, A.; Ferlinska, M. Cytotoxicity of anticancer drugs and PJ-34 (polyADP-ribose)polymerase-1(PARP-1) inhibitor) on HL-60 Jurkat cells. *Adv. Clin. Exp. Med.* **2017**, *26*, 379–385. [CrossRef]
58. Cseh, M.; Zsolt, F.; Quintana-Cabrera, R.; Szabo, A.; Eros, K.; Eugenia Soriano, M.; Gallyas, F.; Scorrano, L.; Sumegi, B. PARP Inhibitor PJ34 Protects mitochondria and induces DNA-damage mediated apoptosis in combination with Cisplatin or Temozolomide in B16F10 Melanoma. *Cells Front. Physiol.* **2019**, *10*, 1–15. [CrossRef]
59. Heng Wei, X.; Chen, X.; Xiao, H. PJ34, a poly(ADP-ribose) polymerase (PARP) inhibitor, reverses melphalan-resistance and inhibits repair of DNA double-strand breaks by targeting the FA/BRCA pathway in multidrug resistant multiple myeloma cell line RPMI8226/R TING. *Int. J. Oncol.* **2015**, *46*, 223.
60. Meng, W.; Koh, B.D.; Jin-San Zhang, J.-S.; Flatten, K.S.; Schneider, P.A.; Billadeau, D.D.; Hess, A.D.; Smith, B.D.; Karp, J.E.; Kaufmann, S.H. Poly(ADP-ribose) Polymerase Inhibitors Sensitize Cancer Cells to Death Receptor-mediated Apoptosis by Enhancing Death Receptor Expression. *J. Biol. Chem.* **2014**, *289*, 20543–20558. [CrossRef]

61. Passeri, D.; Camaioni, E.; Lisco, P.; Sabbatini, P.; Ferri, M.; Carroti, A.; Giacche, N.; Pellicciari, R.; Gioiello, A.; Macchiarulo, A. Concepts and molecular aspects in the polypharmacology of PARP-1 inhibitors. *ChemMedChem* **2016**, *11*, 1219–1226. [CrossRef] [PubMed]
62. Ganem, N.J.; Godinho, S.A.; Pellman, D. A mechanism linking extra centrosomes to chromosomal instability. *Nature* **2009**, *460*, 278–282. [CrossRef] [PubMed]
63. Quintyne, N.J.; Reing, J.E.; Hoffelder, D.R.; Gollin, S.M.; Saunders, W.S. Spindle multipolarity is prevented by centrosomal clustering. *Science* **2005**, *307*, 127–129. [CrossRef] [PubMed]
64. Kramer, A.; Anderhub, S.; Maier, B. Mechanisms and Consequences of centrosomes clustering in cancer cells. In *The Centrosome: Cell and Molecular Mechanisms of Functions and Disfunctions in Disease*; Schatten, E., Ed.; Humana Press Springer: Totowa, NJ, USA, 2012; pp. 255–277.
65. Rieder, C.L.; Maiato, H. Stuck in Division or Passing through: What Happens When Cells Cannot Satisfy the Spindle Assembly Checkpoint. *Dev. Cell* **2004**, *7*, 637–651. [CrossRef]
66. Etemad, B.; Kops, J.P.L.G. Kinetochore transformations and spindle checkpoint silencing. *Curr. Opin. Cell Biol.* **2016**, *39*, 101–108. [CrossRef]
67. Vakifahmetoglu, H.; Olsson, M.; Zhivotovsky, B. Death through a tragedy: Mitotic catastrophe. *Cell Death Differ.* **2008**, *15*, 1153–1162. [CrossRef]
68. Galimberti, F.; Sarah, L.; Thompson, S.L.; Ravi, S.; Compton, D.A.; Dmitrovsky, E. Anaphase Catastrophe is a Target for Cancer Therapy. *Clin. Cancer Res.* **2011**, *17*, 1218–1222. [CrossRef]
69. La Terra, S.; English, C.N.; Hergert, P.; McEwen, B.F.; Sluder, G.; Khodjakov, A. The de novo centriole assembly pathway in HeLa cells: Cell cycle progression and centriole assembly/maturation. *J. Cell Biol.* **2005**, *168*, 713–722. [CrossRef]
70. Godinho, S.A.; Pellman, D. Causes and consequences of centrosome abnormalities in cancer. *Philos. Trans. R. Soc. Lond. B Biol. Sci.* **2014**, *369*, 20130467. [CrossRef]
71. Cohen-Armon, M.; Yeheskel, A.; Pascal, J.M. Signal-induced PARP1-Erk synergism mediates IEG expression. *Sig. Transduct. Target. Ther.* **2019**, *4*, 1–8. [CrossRef]
72. Miki, H.; Okada, Y.; Hirokawa, N. Analysis of the kinesin superfamily: Insights into structure and function. *Trends Cell Biol.* **2005**, *15*, 467–476. [CrossRef]
73. Simeonov, D.R.; Kenny, K.; Seo, L.; Moyer, A.; Allen, J.; Paluh, J. Distinct kinesin-14 mitotic mechanisms in spindle polarity. *Cell Cycl.* **2009**, *8*, 3571–3583. [CrossRef] [PubMed]
74. Goshima, G.; Nedelec, F.; Vale, R.D. Mechanisms for focusing mitotic spindle poles by minus end directed motor proteins. *J. Cell Biol.* **2005**, *171*, 229–240. [CrossRef] [PubMed]
75. Verhey, K.; Hammond, J.W. Traffic control: Regulation of kinesin motors. *Nat. Rev. Mol. Cell Biol.* **2009**, *10*, 765–777. [CrossRef] [PubMed]
76. Mayr, M.I.; Hummer, S.; Bormann, J.; Gruner, T.; Adio, S.; Woehlke, G.; Mayer, T.U. The human kinesin Kif18A is a motile microtubule depolymerase essential for chromosome congression. *Curr. Biol.* **2007**, *17*, 488–498. [CrossRef] [PubMed]
77. Silk, A.D.; Holland, A.J.; Cleveland, D.W. Requirement for NuMA in maintenance and establishment of mammalian spindle poles. *J. Cell Biol.* **2009**, *184*, 677–690. [CrossRef] [PubMed]
78. Kleylein-Sohn, J.; Pollinger, B.; Ohmer, M.; Hofmann, F.; Nigg, E.A.; Hemmings, B.A.; Wartmann, M. Acentrosomal spindle organization renders cancer cells dependent on the kinesin HSET. *J. Cell Sci.* **2012**, *125*, 5391–5402. [CrossRef]
79. Cai, S.; Weaver, L.N.; Ems-McClung, S.C.; Walczak, E. Kinesin-14 Family Proteins HSET/XCTK2 Control Spindle Length by Cross-Linking and Sliding Microtubules. *Mol. Biol. Cell* **2009**, *20*, 1348–1359. [CrossRef]
80. Gordon, M.B.; Howard, L.; Compton, D.A. Chromosome Movement in Mitosis Require Microtubule Anchorage at Spindle Poles. *J. Cell Biol.* **2001**, *152*, 425–434. [CrossRef]
81. Cai, S.; Weaver, L.N.; Ems-McClung, S.C.; Walczak, C.E. Proper Organization of Microtubule Minus-Ends is needed for Midzone Stability and Cytokinesis. *Curr. Biol.* **2010**, *20*, 880–885. [CrossRef]
82. Haren, L.; Gnadt, N.; Wright, M.; Merdes, A. NuMA is required for proper spindle assembly and chromosome alignment in prometaphase. *BMC Res. Notes* **2009**, *2*, 64. [CrossRef] [PubMed]
83. Haren, L.; Andreas -Merdes, A. Direct binding of NuMA to tubulin is mediated by a novel sequence motif in the tail domain that bundles and stabilizes microtubules. *J. Cell Sci.* **2002**, *115*, 1815–1823. [PubMed]

84. Sukhai, M.A.; Wu, X.; Xuan, Y.; Zhang, T.P.P.; Dubé, K.; Rego, E.M.; Bhaumik, M.; Bailey, D.J.; Wells, R.A.; Kamel-Reid, S.; et al. Myeloid leukemia with promyelocytic features in transgenic mice expressing hCG–NuMA–RARα. *Oncogene* **2004**, *23*, 665–678. [CrossRef]
85. Zink, D.; Fischer, A.H.; Nickerson, J.A. Nuclear structure in cancer cells. *Nat. Rev. Cancer* **2004**, *4*, 677–687. [CrossRef] [PubMed]
86. Whitehurst, A.; Xie, Y.; Purinton, S.C.; Capell, K.M.; Swanik, J.T.; Larson, B.; Girard, L.; Schorge, J.O.; White, M.A. Tumor antigen Acrosin binding protein normalizes mitotic spindle function to promote cancer cell proliferation. *Cancer Res.* **2010**, *70*, 7652–7661. [CrossRef]
87. Compton, D.A.; Luo, C. Mutations in the predicted p34cdc2 phosphorylation sites in NuMA impair the assembly of the mitotic spindle and block mitosis. *J. Cell Sci.* **1995**, *108*, 621–633. [PubMed]
88. Chang, W.; Dynek, J.N.; Smith, S. NuMA is a major acceptor of polyADP-ribosylation by tankyrase1 in mitosis. *Biochem. J.* **2005**, *391*, 177–184. [CrossRef]
89. Bhattacharya, N.; Wang, Z.; Davitt, C.; McKenzie, I.F.; Xing, P.X.; Magnuson, N.S. Pim-1 associates with protein complexes necessary for mitosis. *Chromosoma* **2002**, *111*, 80–95. [CrossRef]
90. Haikarainen, T.; Narwal, M.; Joensu, P.; Lehtiö, L. Evaluation and structural basis for the inhibition of tankyrases by PARP inhibitors. *ACS Med. Chem. Lett.* **2014**, *5*, 18–22. [CrossRef]
91. Kirby, C.A.; Cheung, A.; Fazal, A.; Shultz, M.D.; Stam, T. Structure of Human tankyrase1 in complex with small-molecule inhibitors PJ34 and XAV939. *Acta Cryst.* **2012**, *68*, 115–118.
92. Antolín, A.A.; Jalencas, X.; Yelamos, J.; Mestres, J. Identification of Pim Kinases as Novel Targets for PJ34 with Confounding Effects in PARP Biology. *ACS Chem. Biol.* **2012**, *7*, 1962–1967. [CrossRef] [PubMed]
93. Haikarainen, T.; Krauss, S.; Lehtiö, L. Tankyrases: Structure, Function and Therapeutic Implications in Cancer. *Curr. Pharm. Des.* **2014**, *20*, 6472–6488. [CrossRef] [PubMed]
94. Paul Chang, P.; Margaret Coughlin, M.; Mitchison, T.J. Tankyrase-1 polymerization of poly(ADP-ribose) is required for spindle structure and function. *Nat. Cell Biol.* **2005**, *7*, 1133–1139.
95. Tian, X.H.; Hou, W.J.; Fang, Y.; Fan, J.; Tong, H.; Bai, S.-L.; Chen, Q.; Xu, H.; Li, Y. XAV939, a tankyrase 1 inhibitior, promotes cell apoptosis in neuroblastoma cell lines by inhibiting Wnt/β-catenin signaling pathway. *J. Exp. Clin. Cancer Res.* **2013**, *32*, 100. [CrossRef]
96. Leber, B.; Maler, B.; Fuchs, F.; Chi, J.; Riffel, P.; Anderhub, S.; Wagner, L.; Ho, A.D.; Salisbury, J.L.; Boutros, M.; et al. Proteins required for chromosome clustering in cancer cells. *Sci. Trans. Med.* **2010**, *2*, 1–11. [CrossRef]
97. Nicolescu, A.C.; Holt, A.; Kandasamy, A.D.; Pacher, P.; Schultz, R. Inhibition of matrix metalloproteinase-2 by PARP inhibitors. *Biochim. Biophys. Res. Commun.* **2009**, *374*, 646–650. [CrossRef]
98. Chambers, A.F.; Matrisian, L.M. Changing views on the role of metalloproteinases in metastasis. *J. Natl. Cancer Inst.* **1997**, *89*, 1260–1270. [CrossRef]
99. Deryugina, E.I.; Quigley, J.P. Matrix metalloproteinases and tumor metastasis. *Cancer Metastasis Rev.* **2006**, *25*, 9–34. [CrossRef]
100. Stuelten, C.H.; Parent, C.A.; Montell, D.J. Cell motility in cancer invasion and metastasis: Insights from simple model organisms. *Nat. Rev. Cancer* **2018**, *18*, 296–312. [CrossRef]
101. Inbar, D.; Cohen-Armon, M.; Neumann, D. Erythropoietin-driven signaling and cell migration mediated by polyADP-ribosylation. *Br. J. Cancer* **2012**, *107*, 1317–1326. [CrossRef]
102. Rodriguez, M.I.; Peralta-Leal, A.; OValle, F.; Rodriguez-Vagas, J.M.; Gonzales-Flores, A.; Majuelos-Melquizo, J.; Lopez, L.; Serrano, S.; Garcia de Herreros, A.; Rodrigues-Manzaneque, J.C.; et al. PARP1 regulates metastatic melanoma through modulation of vimentin–induced malignant transformation. *PLoS Genet.* **2013**, *9*, e1003531. [CrossRef] [PubMed]

cancers

MDPI

Review

Combined PARP Inhibition and Immune Checkpoint Therapy in Solid Tumors

Florent Peyraud [1,2] and Antoine Italiano [1,2,3,*]

1 Department of Medical Oncology, Institut Bergonié, 33000 Bordeaux, France; peyraud.florent@gmail.com
2 University of Bordeaux, 33076 Bordeaux, France
3 Early Phase Trials and Sarcoma Unit, Institut Bergonié, 33000 Bordeaux, France
* Correspondence: a.italiano@bordeaux.unicancer.fr; Tel.: +33-556-333-333

Received: 30 April 2020; Accepted: 7 June 2020; Published: 9 June 2020

Abstract: Genomic instability is a hallmark of cancer related to DNA damage response (DDR) deficiencies, offering vulnerabilities for targeted treatment. Poly (ADP-ribose) polymerase (PARP) inhibitors (PARPi) interfere with the efficient repair of DNA damage, particularly in tumors with existing defects in DNA repair, and induce synthetic lethality. PARPi are active across a range of tumor types harboring *BRCA* mutations and also *BRCA*-negative cancers, such as ovarian, breast or prostate cancers with homologous recombination deficiencies (HRD). Depending on immune contexture, immune checkpoint inhibitors (ICIs), such as anti-PD1/PD-L1 and anti-CTLA-4, elicit potent antitumor effects and have been approved in various cancers types. Although major breakthroughs have been performed with either PARPi or ICIs alone in multiple cancers, primary or acquired resistance often leads to tumor escape. PARPi-mediated unrepaired DNA damages modulate the tumor immune microenvironment by a range of molecular and cellular mechanisms, such as increasing genomic instability, immune pathway activation, and PD-L1 expression on cancer cells, which might promote responsiveness to ICIs. In this context, PARPi and ICIs represent a rational combination. In this review, we summarize the basic and translational biology supporting the combined strategy. We also detail preclinical results and early data of ongoing clinical trials indicating the synergistic effect of PARPi and ICIs. Moreover, we discuss the limitations and the future direction of the combination.

Keywords: PARP inhibitor; DNA damage response; immunotherapy; immune checkpoint inhibitor; PD-1; PD-L1; CTLA-4; combination therapy; solid tumors

1. Introduction

Over the past decade, poly(ADP-ribose) polymerase (PARP) inhibitors (PARPi) and monoclonal antibodies that block immune checkpoints, such as programmed cell-death 1 (PD-1) and cytotoxic T lymphocyte antigen 4 (CTL-4), have transformed the treatment of multiple types of cancers. Immune checkpoint inhibitors (ICIs) used as stand-alone therapeutic interventions give rise to durable objective responses in patients affected by a variety of cancers and have been approved for an ever-growing list of malignancies, including melanoma, non-small cell lung carcinoma (NSCLC), renal cell carcinoma, head and neck squamous cell carcinoma (HNSCC), and Hodgkin's lymphoma [1–7]. More recently, monotherapy with PARPi as a maintenance strategy showed significant clinical activity in several cancer types harboring germline loss-of-function BRCA mutations such as ovarian, breast and pancreatic cancer [8–11]. However, despite these substantial advancements in clinical care, the majority of patients receiving either PARPi or ICIs alone do not provide benefit and a rationale to combine these treatments has emerged [12,13].

The groundbreaking success of anticancer immunotherapy is primarily based on the features of cancer cells and their ability to potentially initiate an antitumor immune response. These notable features include gene mutations resulting in abnormal protein expression patterns, such as neoantigens or tumor-associated antigens (TAAs) [14]. TAAs represent self-antigens that are aberrantly expressed or overexpressed in tumor cells, whereas neoantigens refer to non-self-antigens arising as a result of somatic mutation [14,15]. The formation of mutation-derived TAAs and neoantigen, reflecting the mutational burden of the tumor, allow the immune system to recognize tumor cell and initiate the cancer-immunity cycle [16,17]. The subsequent antitumor immune process against neoantigens relies on several steps, including the release and presentation of cancer antigens by antigen-presenting cells (APC), priming and activation of T cells, trafficking and infiltration of T cells, and recognizing and killing cancer cells [18]. However, a subset of cancer cells can escape host immune destruction by impairing one or more steps and result in tumor progression [19]. The role of immunotherapy is to reinvigorate antitumor immune response by disrupting co-inhibitory T cell signaling, transferring additional tumor-specific T cells clones and reshaping the immunosuppressive microenvironment [20–23]. Several strategies, including ICIs, adoptive T cell transfer, and vaccination, have been put to use in multiple cancers [24–27]. Nevertheless, due to complex and constantly evolving interactions between cancer cells and the immune system, both primary and acquired resistance with ICIs monotherapy are observed [28,29]. Therefore, combination treatment with ICIs is an attractive strategy to potentiate efficacy and lower resistance.

Recent molecular profiling of DNA damage repair genes has allowed the implementation of novel therapeutic strategies. By interfering with efficient DNA damage repair, the inhibition of PARP that target the base excision repair (BER) pathway leads to insufficient DNA repair, with subsequent unsustainable DNA damage, and thus represents a synthetic lethal therapeutic approach for the treatment of cancers with compromised ability to repair double-strand DNA breaks by homologous recombination (HR), including those with defects in *BRCA1/2* [30]. The unrepaired-DNA promotes immune priming through a range of molecular mechanisms and also leads to adaptative upregulation of programmed death ligand 1 (PD-L1) expression [31]. Moreover, PARPi modulates the inflammatory immune microenvironment of tumors and reinstates a productive T_H1 immune response [31]. This multifaceted immunological effect of PARPi might be favorable to boost an antitumor immune response and enhance the efficacy of ICIs. In this review, we summarize the basic and translational biology supporting the combined strategy and provide a focus on preclinical studies and ongoing clinical trials of ICIs combined with PARPi, as well as perspectives and potential challenges of this combination strategy.

2. DNA Damage and PARP Inhibition

2.1. Role of PARP in DNA Damage Response

Cells are continuously faced with endogenous and exogenous stress that can ultimately lead to DNA damage. To preserve genomic integrity and prevent emergence of cancer, detection and repair of DNA is a critical process, managed by multiple pathways [32]. DNA single-strand break (SSB) damage is fixed by three main pathways: (1) BER, (2) nucleotide excision repair (NER), and (3) mismatch-repair (MMR). Possibly more dangerous DNA double-strand breaks (DSB) are restored by two additional pathways: (1) HR and (2) non-homologous end joining (NHEJ) [33]. Anomalies observed in DNA damage response (DDR) key genes, such as BRCA1/2 or TP53, are associated with cancer-prone phenotypes [34]. As a consequence, failure in DDR in an accurate and well-timed manner can result in the defective elimination of genome mutations and increases the risk of oncogenesis after established DNA damage events [35]. Depending on the context, cancer cells often harbor a lessened repertoire of DDR signaling competences, rendering them more reliant on a subset of DNA repair pathways and therefore more susceptible to DDR inhibition than normal cells [36].

PARP1/2 enzymes are core DNA-damage sensor and signal transducer in DDR, which bind to DNA breaks and catalyze the synthesis of poly(ADP-ribose)(PAR) chains on target proteins (PARylation) in the vicinity of the DNA break and itself (autoPARylation) [37]. These negatively charged PAR chains promote chromatin remodeling, recruit DNA repair-related protein complex and affect the replication fork progression speed [38,39]. The binding of PARP1 via zinc finger domains to sites of DNA-damage carries a conformational change in the PARP1 proteins and relieves the autoinhibitory interaction between the catalytic domain and helical domain. Then, the PARP1 co-factor nicotinamide (β-NAD$^+$) is used as a substrate at the active site of the enzyme to catalyze the transfer of ADP-ribose moieties onto target proteins. The synthesis of ADP-ribose polymeric chains on proteins in the vicinity of DNA breaks, called PARylation, likely mediates DNA repair by modifying chromatin structure and by localizing DNA repair effectors [40]. Thereafter DNA-damage restore, autoPARylation occur that rapidly dissociate PARP from damage site [41]. The role of PARP has been well identified in BER-mediated SSB repair pathways, as well as other DDR pathways [42].

2.2. The Lethal Synthetic Effect of PARP Inhibitors

2.2.1. Mechanism of Action of PARPi

Although the precise mechanism by which PARPi kill tumor cells remains to be fully clarified, the anticancer effect is attributed to catalytic inhibition of PARP that block repair of DNA SSB [43] (Figure 1a). While PARPi is well-tolerated by normal cells, this effect of PARPi is more likely observed in tumor cells with a *BRCA*-deficient background [43]. As a result of defective enzymatic function induced by PARPi, the accumulation of SSB is subsequently encountered by replication forks and generates potentially lethal DSBs that need to be fixed [43,44]. In normal cells, the accumulation of DSBs are repaired preferentially by HR rather than NHEJ [45]. HR is a high-fidelity repair pathway that utilizes the sister copy of the damaged DNA as a template, leading to the reconstitution of the original sequence [46]. In contrast, NHEJ is intrinsically error-prone, modifies the broken DNA ends, and ligates them together with little or no homology, generating deletions or insertions [47]. However, in some cancer cells lacking BRCA1 or BRCA2, two key tumor suppressor proteins involved in DSB repair by HR, loss of PARP function leads to the accumulation of DSBs that are unrepaired or unsustainably repaired by NHEJ which results in cell death [44,48]. Based on the discovery of this synthetic lethality between *BRCA* and PARP, numerous PARPi have been developed, including olaparib, rucaparib, niraparib, talazoparib, and veliparib, which are mainly applied in cancer patients with *BRCA1/2* mutations [8–11,42].

Although the greatest efficacy of PARPi has been observed in tumors with *BRCA1/2* mutations, accumulating data indicate that synthetic lethality is inadequate to explain the whole antitumor activity. First, the ability of PARPi to inhibit PARP catalytic activity is poorly correlated to its cell-killing ability in HR deficiencies (HRD) cells [43,49]. In addition, PARPi induces cytotoxicity to a greater extent than PARP depletion [50,51]. Furthermore, PARP itself is essential to the cytotoxic effects of PARPi [52]. Actually, these facts may be attributed in part to the PARP trapping potency of PARPi (Figure 1b). Although the precise mechanisms of PARP trapping remains unclear, it has been proposed that PARPi could either prevent the release of PARP1 from DNA by inhibiting autoPARylation [53]. Likewise, PARPi binding to the catalytic site could cause allosteric changes in the PARP1 structure enhancing DNA avidity [49]. The trapping DNA-PARP complex stalls the progress of replication fork and elicit cytotoxic effects primarily through the conversion of unrepaired SSBs into lethal DSBs [43,49]. Moreover, PARPi could also act via the upregulation of NHEJ pathway, which presumably leads to genomic instability and eventual lethality [54]. Finally, PARPi could suppress the role of PARP in reactivating DNA replication forks and cause cell death [43]. Additional studies further demonstrated that loss of other tumor suppressor DNA repair proteins, many of which are involved in HR, such as RAD51, ataxia telengiectasia Rad3-related (ATR), ataxia telangiectasia mutated (ATM), checkpoint kinase 1 (CHK1), checkpoint kinase 2 (CHK2), and partner and localizer of *BRCA2* (PALB2) also caused sensitization to PARPi [49,55]. These results suggested that PARPi might be a useful therapeutic strategy not only for the treatment of *BRCA*-mutated tumors but also for the treatment of a wider range of non-*BRCA*-mutated tumors that are inherently HR deficient (HRD) or "BRCAness/HRDness" [56].

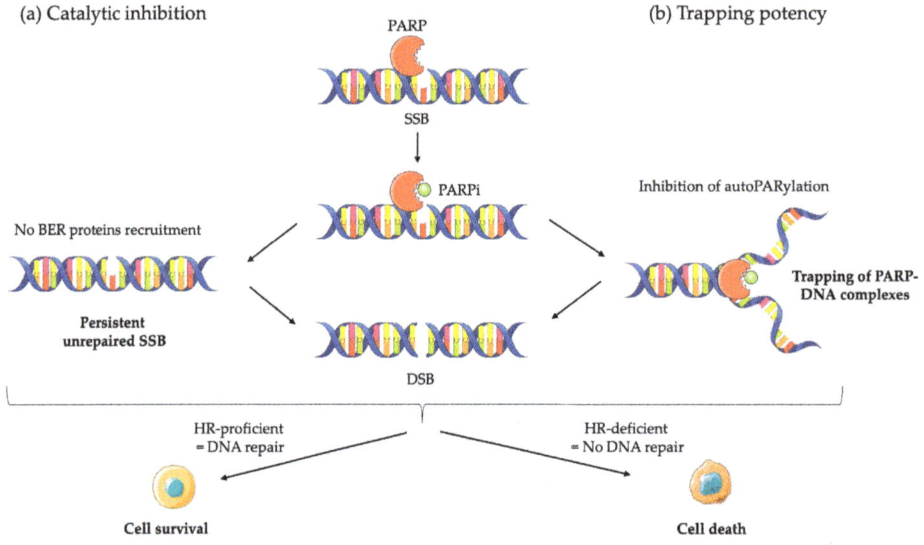

Figure 1. Mechanisms of action of PARP inhibitors (PARPi): (**a**) PARPi impedes PARP enzyme activity (or catalytic inhibition) and interferes with repair of single strand breaks (SSB) by disrupting the base excision repair (BER) pathway; (**b**) PARPi also causes trapping of PARP proteins on DNA by inhibiting autoPARylation. In homologous recombination (HR) proficient normal cells, DNA is repair and cell survive. The result in unresolved DNA double strand breaks (DSB) in HR deficient cells leads to cell death.

2.2.2. Clinical Applications of PARPi

The early development of PARPi focused initially on their use in combination with cytotoxic chemotherapy agents and radio-sensitizing drugs, but this was rapidly rejected because of excessive toxicity [57,58]. The potent antitumor effect of PARPi was originally observed in tumors harboring germline *BRCA1/2* mutations (*gBRCA1/2*m), such as familial breast and ovarian cancer [59]. This rapid translation of preclinical studies into promising clinical data triggered the development of several PARPi in different tumors types. Initially, PARPi in the clinic improved clinical benefits for germline or somatic *BRCA*-deficient ovarian cancer [60,61]. Subsequently, breast, pancreatic and prostate cancers that harbors defects in *BRCA* also demonstrated to be PARPi responsive [9–11,60–62]. More recently, it has been suggested that patients without *BRCA* mutations shared therapeutic vulnerabilities, especially tumors with deficiencies in HR. Indeed, the activity of PARPi is based on the concept of synthetic lethality, where an underlying HRD in tumor cells makes the cells highly susceptible to PARP inhibition [42]. This hypothesis has been further confirmed with multiple clinical studies showing that sensitivity to PARPi occurs in tumors beyond those with *BRCA* mutations, especially in HRD-positive tumors [63–67]. To date, five PARPi have been approved or orphan drug designed by the FDA (veliparib, rucaparib, talazoparib, niraparib, olaparib) and applied in clinical practice [42].

Despite the advances of PARPi in a particular population, acquired resistance is a common clinical phenotype. Owing to extensive preclinical studies, several resistance mechanisms have been identified that can be classified into four main categories. Firstly, numerous different mechanisms result in the reactivation of HR function. For example, secondary reversion mutations in several key HR repair (HRR) genes, such as *BRCA1/2*, *RAD51C/D*, and *PALB2*, restore the open reading frame and thus HR competency [68]. Moreover, the loss of p53-binding protein 1 (53BP1), a protein promoting NHEJ, is associated with PARPi resistance by recovery of HRR in *BRCA1*-deficient tumors [69]. By directly impacting the activity and abundance of PAR chains that decreased PARP trapping, mutations in DNA-binding domains of PARP1 and mechanisms that increase PARylation of PARP1 could also lead to PARPi resistance [70,71]. Furthermore, the cellular availability of the inhibitor is a critical step for successful therapy, as illustrated by the upregulation of ATP-binding cassette (ABC) transporters, such as the P-glycoprotein (PgP) efflux pump that have been described to reduce the efficacy of PARPi [72]. At last, restoration of replication fork protection that induces the stabilization of stalled forks may lead to PARPi resistance [73]. Indeed, fork degradation induced by PARPi is mediated by PTIP and EZH2 proteins, which upon loss lead to protection of the fork from nucleases and thereby resistance [74,75].

Intense preclinical and clinical research are ongoing in order to broadening responding patients, overcoming acquired resistance and enhancing the efficacy of PARPi [73]. The development of combination therapy encompassing PARPi is a potential approach to address these objectives. In addition to the hypothesis that patients with HRD tumors are more prone to produce neoantigen and exhibit higher mutational load, there is a preclinical rational suggesting that PARPi may promote the formation of neoantigen and generate tumor cell recognition by the immune system, making this class of drugs a potential partner for combination with ICIs [76,77].

3. The Revolution of Cancer Immunotherapy and Immune Checkpoint Inhibitors

Immunotherapy is proving to be an effective therapeutic approach in a variety of cancers [78]. In the last decade, the use of therapeutic antibodies that disrupt negative immune regulatory checkpoints and unleash pre-existing antitumor immune responses have achieved impressive clinical successes. Among the different types of cancer immunotherapy, ICIs have demonstrated the broadest impact by leveraging the cytotoxic potential of the human immune system, especially tumor-specific cytotoxic T cells. The role of T cells is critical to adaptative immunity and contribute to improved outcomes in a large range of cancers [79]. The activation of naïve T cells requires two distinct signals [80]. The generation of the first signal occurs by binding of major histocompatibility complex (MHC)-presented immunogenic peptide antigen to the heterodimeric T cell receptor (TCR). The transduction of the second signal, also referred as co-stimulation signal, arises through ligation of the T cell co-stimulatory surface receptor CD28 to its ligand CD80 (also known as B7-1) or CD86 (also known as B7-2) on the surface of professional antigen-presenting cells (APCs). Subsequent to both these signals, activated T cells begin to express co-inhibitory cell surface receptors that control T cell function, such as CTLA-4 and PD-1. The balance between co-stimulatory and co-inhibitory signals is crucial for the activation and tolerance of T cells [81]. Importantly, targeting these co-inhibitory pathways with ICIs in the context of cancer effectively shifts that balance toward activation, thereby overcoming tumor immune subversion [82].

3.1. CD80/86-CTLA-4 Signaling Pathway

The first negative regulator of T cell activation to be identified was CTLA-4, a co-inhibitory receptor that is constrictively expressed on Tregs and transiently upregulated during the course of T cell activation in peripheral lymphatic organs [83]. Bound by the same ligands (CD80/86) that provide co-stimulatory signals through CD28 but with higher affinity, CTLA-4 mainly impedes acquisition of T cell effector function by mediating transendocytosis and degradation of the ligands [84]. In addition, CTLA-4 delivers inhibitory signals that block T cell proliferation and secretion of IL-2, leading to T cell tolerance through induction of energy [85,86]. Moreover, CTLA-4 counterbalance TCR/CD3-mediated phosphorylation through immunoreceptor tyrosine-based inhibitory motif (ITIM) and impede the signal transduction of TCR [87]. Therefore, CTLA-4 engagement in numerous T lymphocyte populations operates as a cardinal immune checkpoint that ultimately hampers the acquisition of T cell effector functions and dampens the antitumor immune response.

3.2. PD-1/PD-L1 Signaling Pathway

The expression of PD-1 on activated immune cells is ubiquitous, including T cells, B cells, natural killer (NK) cells, and dendritic cells (DC), and yields inhibitory signals through binding of its two ligands, namely PD-L1 and PD-L2 [88]. Moreover, both PD-1 ligands are expressed on a wide variety of immune and non-immune cells [88]. More particularly, PD-L1 is found on a broad range of tissues and could be upregulated under inflammatory conditions such as cancers [89]. The expression of PD-L1 on the surface of tumors underlies the crucial relevance of the PD-1/PD-L1 pathway to neoplasm. Upon binding of TCR with antigen presented by MHC, PD-1 is engaged with its ligand and becomes functional. PD-1 activation leads to phosphorylation of the ITIM and immunoreceptor tyrosine-based switch motif (ITSM) in the PD-1 cytoplasmic tail and subsequently drive the recruitment of protein tyrosine phosphatase, such as Src homology region 2 domain containing phosphatase 1/2 (SHP1/2). As a consequence of dephosphorylation, these phosphatases antagonize positive signals that occur through the TCR nad CD28, affecting downstream signaling pathways. For example, TCR signaling molecules, such as Lck and ZAP-70, and co-stimulatory signaling cascades, such as PI3K-Akt-mTOR and Ras-MEK-ERK pathways, are inhibited. The impairment of these crucial signaling pathways alters the activation, proliferation, survival, cytokine production, metabolism, and epigenetic programs in T cells [90–92]. Tumors can exploit this pathway to escape T cell-mediated tumor-specific immunity.

3.3. *Clinical Application of ICI*

Since the recent success of antibodies targeting checkpoint molecules CTLA-4, PD-1, and PD-L1, the field of cancer immunotherapy has been experiencing a renaissance. The anti-CTL-4 inhibitor ipilimumab was the first ICI to obtain approval in 2011 for the treatment of metastatic melanoma [1,93]. Thereafter, ICIs have yielded broad clinical activity, leading to regulatory approval of several monoclonal antibodies in a variety of advanced and up-front disease settings, including melanoma, non-small cell lung, renal cell carcinoma, urothelial carcinoma, HNSCC, Merkel cell carcinoma, gastric carcinoma, hepatocellular carcinoma, Hodgkin's lymphoma, as well as for any MMR-deficient/microsatellite instability (MSI) positive tumors [2–6,94–105]. However, as with PARPi, only a subset of patients derives benefit and a series of biomarkers have been developed to predict efficacy of ICI and select patients before treatment beginning.

Although the current understanding of the clinical response of ICI therapy indicates that there cannot be a single predictive biomarker, several factors have been identified as the core determinants of the efficacy of ICIs, such as tumor mutation burden (TMB) and particular mutational signature, the number of tumor-infiltrating lymphocytes (TILs), PD-L1 expression, immunosuppressive microenvironment, and MMR deficiency (MMRd) [106]. For example, tumors that harbor MMRd or some specific defects in DDR pathways beyond MMR demonstrated a higher ICI response [103,104,107]. The improve outcome in these patients is believed to be a result of increased mutational load, leading to greater immunogenicity. In addition, a novel perspective has arisen with the development of ICI based combination therapy in order to improve ICI efficacy and overcome resistance. These include combinations with other checkpoint inhibitors, radiation therapy, chemotherapy, and targeted therapies, so as to foster antigen presentation, broadening T cell repertoire, impairing immunosuppressive elements, and increasing antitumor immune response [108,109].

4. Combination of PARPi and ICI Therapy

4.1. *A Rational to Combine PARPi and ICI*

4.1.1. Tumor Mutation Burden and Neoantigen

The mutational load in a tumor, termed as TMB and determined by the number of non-synonymous single nucleotide variants (nsSNVs), may impact the odds of generating immunogenic peptides and has been significantly correlated with ICI response in previous studies [110–112]. Even if the optimal TMB cut-off remains blurred across tumor types, the relationship with efficacy of ICI is robust [113,114]. TMB is considered as a surrogate of neoantigen load which predicts the therapeutic response of ICI [14,15,115]. Likewise, a growing amount of data indicate a closely association between TMB and DDR deficiency [116]. Highly mutated tumors often exhibit one or several mutations in key components of DDR or replicative pathways, including *MSH2* for MMR, *BRCA1/2* for HR and *POLE* for DNA replication, and correlate with ICI response [116]. In addition, patients with cancer harboring innate deficiencies in DDR genes, including MMR and HRR genes, achieved durable benefit from ICIs compared with patients without these deficiencies [116,117]. These results suggest that loss of normal DNA repair fidelity, such as DDR phenotype, may contribute to increased mutational load and neoantigens burden which affect the response to immunotherapy in these tumors. One relevant strategy in patients with HRD or others defects in DDR would be to combine PARPi and ICI.

Direct evidence that the targeting of DSB repair proteins with DDR inhibitors provoked and increased TMB is only beginning to emerge [118]. By affecting the HR pathway in tumor cells, impaired DNA repair induced by PARPi could subsequently generate catastrophic DNA damage that would increase the neoantigen load and TMB, thus driving a response to ICI and theoretically broadening the responding population (Figure 2a). Although tumors with non-MMR DDR genes deficiency, such as *BRCA1/2* and other HR-related genes, have increased TMB, the association is weaker than that observed in MMR deficiency. Thereby, other fundamental links in tumor immunogenicity may be involved to explain the higher response rate to ICI in these patients [119,120].

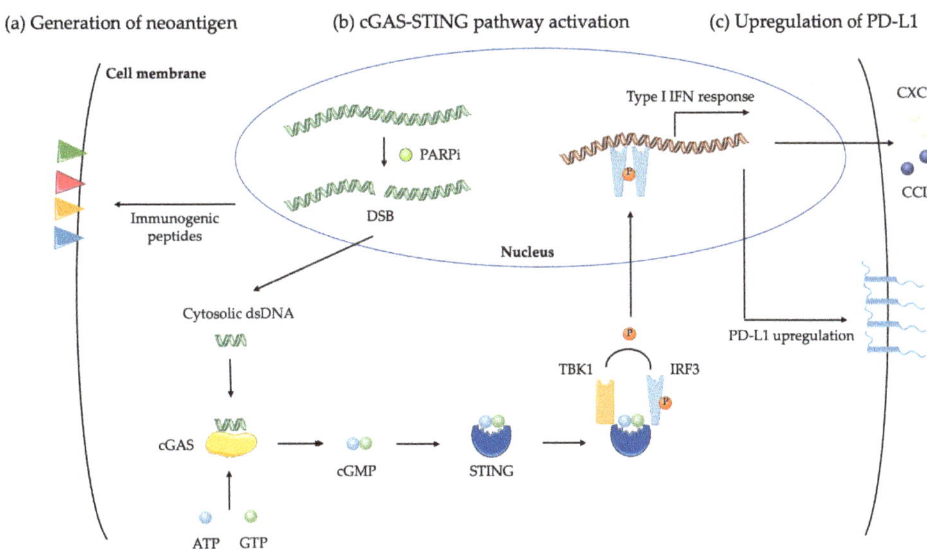

Figure 2. Biological effect of PARPi on cancer cells. (**a**) PARPi delays DNA repair that generates double-strand breaks (DSB) and catastrophic unrepaired DNA damages in tumor cells, increasing neoantigens load and tumor mutation burden; (**b**) DSB induced by PARPi in tumor cells produce double-strand DNA (dsDNA) fragments that, through cyclic GMP-AMP synthase (cGAS) binding, lead to stimulator or interferon genes (STING) activation and the generation of a type I interferon (IFN) response. This upregulates chemokines CCL5 and CXCL10 leading to T cell recruitment; (**c**) Upregulation of PD-L1 via STING pathway lead to T cell exhaustion.

4.1.2. DNA Damages and cGAS-STING Pathway

Aside from TMB, an emerging body of evidence supports a role for non-neoantigen-based mechanisms of tumor cell recognition and targeting by the host immune system. Genomic instability in tumor cells leads to the accumulation of incompletely repaired DNA damage, generating tumor-derived double-strand DNA (dsDNA) in the cytoplasm [121]. The sensing of tumor-derived dsDNA by cytosolic DNA sensor cyclic GMP-AMP synthase (cGAS) plays a major role in the activation of the stimulator of interferon (IFN) genes (STING) signaling pathway [122]. After the recognition of tumor-derived DNA within the cytosol, cGAS activates STING via the generation of 2′-5′ cyclic GMP-AMP (cGAMP). In turn,

STING prompts phosphorylation and nuclear translocation of type I IFN transcriptional regulatory factors TANK-binding kinase 1 (TBK1) and IFN regulatory factor 3 (IRF3) [123,124]. Moreover, STING activates NF-κB pathway which cooperates with IRF3. As a result, the upregulation of type I IFN promotes systemic immune response and regulates multiples components in anticancer immunity, especially T cells, NK cells and DCs [125]. According to recent studies, DNA damages and DDR deficiencies induce the activation of STING and NF-κB pathways, leading to inflammation and infiltration of tumors by immune cells across multiple types of cancers, a prerequesite of tumor-killing effect of ICI [126–130].

In clinical practice, the antitumor activity of PARPi has been observed in patients with platinum-sensitive tumors regardless of BRCA1/2 mutation or HRD status, suggesting an alternative mechanism unrelated to conventional lethal synthetic-mediated cytotoxic effects [60]. The use of PARPi treatment leads to unresolved DNA lesions and to the production of cytosolic dsDNA fragments. The accumulation of cytosolic DNA activates in turn the DNA sensing cGAS-STING pathway and boosts production of type I interferon to induce antitumor immunity independently of DNA repair deficiency [131–133] (Figure 2b). These critical changes amplify STING signaling and its associated-transcription programs, thereby promoting TILs and antitumor immunity [132,133]. Moreover, it leads to increased levels of chemokines, such as CXCL10 and CCL5, that induce the activation and function of cytotoxic $CD8^+$ T cell [132,133]. In addition, these effects of PARPi are further enhanced by ICI, providing a mechanistic rationale for the use of PARPi as immunomodulatory agents to harness the therapeutic efficacy of immunotherapy [134].

4.1.3. PD-L1 Upregulation by PARPi

A key mechanism underlying cancer immune evasion is the expression of inhibitory ligands, notably PD-L1, on the surface of cancer cells. Despite the approval of PD-L1 expression on tumor cells as a companion diagnostic for anti-PD1 therapy for patients with NSCLC, it remains an imperfect predictor of ICI response [4,135,136]. Via STING pathway, tumor-associated inflammation mainly drives the upregulation of immunosuppressive PD-L1 expression, thus reflecting the status of tumor immune microenvironment [137] (Figure 2c). In addition, defects in BRCA1/2 correlates to higher levels of PD-L1 expression [126,138]. Furthermore, the serine/threonine protein kinase glycogen synthase kinase 3β (GSK3β), a regulator of glycogen metabolism, interacts with PD-L1 and modulates its expression by inducing proteasome degradation of PD-L1 [139]. Based on the latter observation, preclinical models have unveiled that PARPi upregulates PD-L1 expression primarily through GSK3β inactivation in a dose-dependent manner, suppressing T-cell activation and increasing tumor cell killing [76]. Further explorations have shown that targeting DDR proteins PARP with PARPi significantly increased expression of PD-L1 [133]. Another report has demonstrated that PARPi-induced DSBs upregulate PD-L1 by ATM-ATR-CHK1 pathway independently of the IFN pathway [140]. Interestingly, subsequent combination therapy with ICI induced PARPi sensitization and led to a greater antitumor activity than either drug alone, putting forward a rational for combining PARPi with ICI as a useful therapeutic strategy [76].

4.1.4. Reprogramming of Immune Microenvironnement by PARPi

In addition to altering the intrinsic immunogenicity of tumor cells through modulation of surface phenotype and intracellular pathways, DNA damage and deficient DDR pathways also modify the extrinsic immunogenicity of tumors at the level of microenvironment. As aforementioned, tumors with existing defects in DNA repair promote inflammation and T_H1 immune response through a range of molecular mechanisms, leading to extrinsic tumor suppression [19]. However, despite the ability of DNA damage to contribute to tumor immune elimination, sustaining low-level DNA damage continues to foster inflammatory signaling that stimulates the infiltration by suppressive immune cells, like myeloid-derived suppressor cells (MDSCs) or tumor-associated macrophages (TAMs), which leads to further DNA damage via free radical release. This transformation boosts chronic inflammation, immunosuppression, and cancer progression [141,142]. PARPi may have the potential to shift from chronic, low level, DNA damage to more significant T_H1 immune response and create a more susceptible tumor microenvironment [143]. Nevertheless, the self-sustaining cycle of DNA damage and chronic inflammation, which is challenging to overcome with PARPi single therapeutic approach, could potentially be addressed through combination with ICIs.

In the wake of these biological findings, deficiencies in the HRR pathway and/or the use of DDR agents such as PARPi appear to activate immunosuppressive pathways, thus offering targetable immunological vulnerabilities in tumors. The interaction between DDR and immune response provides the basis of the combination therapy of ICI and PARPi. Thereby, combining ICIs and PARPi, which target HRD or induce a state of "BRCAness" in HR-proficient tumors, is a thrilling strategy, particulary as these agents have distinct and mostly non-overlapping toxicities [144,145]. Based on these assumptions, combining PARP inhibition with agents that have complementary mechanisms, such as ICIs, is currently subject to clinical testing.

4.2. Preclinical Data and Clinical Studies

4.2.1. Combination of PARPi with Anti-PD1/PD-L1 ICIs

The first preclinical study evaluating PARPi veliparib in combination with anti-PD1/PD-L1 in the BRCA1-deficient BR5 mouse ovarian cancer model observed no significant boost in T cell activity and no improvement in survival [146]. However, the disappointing lack of activity for the combination with anti-PD1/PD-L1 in this preliminary work contrasts with more recent studies. In other preclinical studies conducted on breast cancer cell lines and xenograft models, PARPi olaparib significantly upregulated PD-L1 expression independently of cGAS-STING-IFN pathway and decreased antitumor immunity by attenuating the cell-killing activity of activated human peripheral blood mononuclear cells [76]. Further investigation in EMT6 syngeneic murine models demonstrated more potent antitumor effect and higher TILs infiltration with combined PARPi olaparib with anti-PD-L1 compared to either therapy alone [76]. In another mice model bearing BRCA1-deficient ovarian tumors, anti-PD1 monotherapy exhibited a non-significant effect and PARPi monotherapy delayed tumor progression compared to control group, whereas combination therapy significantly slowed the tumor growth and prolonged survival time [147]. A recent report indicated that coadministration of PARPi niraparib and anti-PD-1 enhanced the infiltration of immune cells into tumor microenvironment and increased synergistic antitumor activities in both immunocompetent BRCA-proficient and BRCA-deficient models, including breast cancer, lung squamous cell carcinoma, colon adenocarcinoma, bladder cancer, and sarcoma [148]. Similarly, additionally to augment $CD8^+$ T cell infiltration, the association of PARPi olaparib and PD-L1 blockade induced complete tumor regression in multiple immunocompetent small cell lung carcinoma (SCLC) mice models [133]. These contrasting results may be explained by the use of different models, with disparities in immune contexture. While differences in anti-PD1/PD-L1 activities cannot be excluded, the use of PARPi with differential catalytic inhibition and various PARP trapping potencies may also explicate these discrepancies [52]. Taken together, the available translational and preclinical data clearly support the combination of PARPi and ICI.

Based on these encouraging preclinical studies, several clinical trials have been conducted and some data are available to date (Table 1). In metastatic castration-resistant prostate cancer (mCRPC), the combination therapy of olaparib and durvalumab induced PSA responses (reduction $\geq$ 50%) in eight out of 17 patients (47%)(NCT02484404) [149]. Patients with DDR mutations exhibited greater benefit than those without known alterations (12-month progression-free survival probability of DDR-deficient vs. DDR-proficient, 83.3% vs. 36.4%, p = 0.031), suggesting DDR-deficient as a predictive biomarker of response [149,150]. In heavily pretreated platinum-resistant recurrent ovarian cancer, durvalumab and olaparib had clinical activity, irrespective of BRCA mutation status (NCT02484404) [151]. Interestingly, correlative analysis of paired pre- and on-therapy fresh core biopsy and blood samples collected on the latter cohort of recurrent ovarian cancer found that treatment enhanced IFN-γ and CXCL9/CXCL10 expression, systemic IFN-γ/TNF-α production and TILs, creating a more immunostimulatory milieu [152]. While tumoral and peripheral IFN-γ increases was correlated with durable clinical benefit from combined therapy, elevated circulating VEGFR3 levels were associated with worse progression-free survival (PFS), suggesting that VEGF/VEGFR pathway may act to counterbalance immunostimulatory changes induced by PARPi and would serve as a target to further improve efficacy of the combination [152]. Despite major limitations surrounding this exploratory analysis, no significant changes in TMB, PD-L1 or STING expression were noted in ovarian cancer patients treated with olaparib and durvalumab combination, thus warranting further investigation on the underlying biological mechanism [152]. On the other hand, the results of relapsed SCLC cohort of phase 2 NCT02484404 basket study durvalumab in combination with olaparib did not meet the preset bar for efficacy [153]. The preexisting TILs level, assessed by immunohistochemistry, seemed to predict tumor responses, suggesting a contribution from an immune-mediated response. The inflamed-phenotype at baseline, defined by high TILs infiltration, may help to identify patients who are most likely to respond to ICIs [153].

The phase 2 MEDIOLA basket trial assessed the efficacy and safety of chemo-free combination of olaparib and durvalumab in patients with solid tumors, including ovarian cancer, breast cancer and gastric cancer (NCT02734004). In gBRCAm platinum-sensitive relapsed ovarian cancer, the effect of the latter combination demonstrated an overall response rate (ORR) of 63% and a 12-week disease control rate (DCR) of 81% [154]. In gBRCAm HER2 negative metastatic breast cancer, the DCR was 80% at 12 weeks and 50% at 28 weeks, with ORR of 63% [155]. Median PFS (mPFS) was 9.2 months and median overall survival (mOS) was 21.5 months [155]. Moreover, patients with no prior line of chemotherapy had higher ORR and longer OS than those with two prior lines (respectively 78% vs. 50% for ORR and 21.3 vs. 16.9 months for OS) [155]. In platinum-resistant relapsed gastric cancer, the ORR was 10% with the 12-week DCR was 26% [156].

In the phase 2 TOPACIO trial (NCT02657889), niraparib and prembrolizumab combination therapy has demonstrated clinical benefit in platinum-resistant ovarian cancers and triple-negative breast cancers, with numerically higher response rates in those with BRCA-mutated tumors only in breast cancer cohort (ORR of BRCAm vs. BRCA wild-type in breast cancer cohort, 47% vs. 11%) [157,158]. However, the ovarian cohort of the TOPACIO study did not meet its primary endpoint of ORR.

The phase 1a/b PARPi pamiparib combined with anti-PD1 tislelizumab in patients with advanced solid tumors was associated with antitumour responses and clinical benefit (ORR of 20%)(NCT02660034) [159]. Similarly, the phase 1b/2 JAVELIN PARP Medley (NCT03330405) of avelumab plus talazoparib in advanced solid tumors is ongoing but showed preliminary antitumor activity and a manageable safety profile [160].

In the recent phase 2 NEODURVARIB trial (NCT03534492), durvalumab plus olaparib administered prior surgery of resectable muscle-invasive bladder cancer (MIBC) demonstrated a pathological complete response rate of 44.5%, suggesting an active and well-tolerated neoadjuvant strategy [161].

Except for the pamiparib–tislelizumab association, where an increased rate of immune-related hepatitis was observed, all combinations were well tolerated, with toxicities in line with those detected for the relevant drugs in monotherapy settings [159]. Numerous clinical trials are ongoing in a broad range of cancers that will help to decipher the exact role of PARPi with anti-PD1/PD-L1 combination strategy (Table 2).

Table 1. Recent clinical trials of combination of PARPi and ICIs.

Tymor Type	Study Identifier	Setting	ICI Agent	PARPi Agent	Design	Patients	Primary Endpoint	Outcome
Ovarian	NCT02571725	Phase I	Tremelimumab (10 mg/kg Q4W)	Olaparib (300 mg BID)	*gBRCAm* recurrent ovarian cancer	3	Safety and RP2D	No DLT or grade 3 AE ORR 100% with 3 PRs
	NCT02484404	Phase II	Durvalumab (1500 mg Q4W)	Olaparib (300 mg BID)	Platinum-resistant recurrent ovarian cancer	35	Clinical activity (ORR)	ORR 14% with 5 PRs (irrespective of *BRCA* status), DCR 71%, mPFS 3.9 months Acceptable toxicity
	NCT02657889 (TOPACIO/KEYNOTE-162)	Phase II	Pembrolizumab (200 mg Q3W)	Niraparib (200 mg QD)	Platinum-resistant recurrent ovarian cancer	60	Clinical activity (ORR)	ORR 18% with 3 CRs and 8 PRs (irrespective of *BRCA* and HRD status), DCR 65% mPFS 3.4 months Acceptable toxicity
	NCT02734004 (MEDIOLA)	Phase II	Durvalumab (1500 mg Q4W)	Olaparib (300 mg BID)	*gBRCAm* platinum-sensitive ovarian cancer	32	Clinical activity (DCR)	12-week DCR 81%, ORR 63% with 6 CRs and 14 PRs Acceptable toxicity
Breast	NCT02657889 (TOPACIO/KEYNOTE-162)	Phase II	Pembrolizumab (200 mg Q3W)	Niraparib (200 mg QD)	Advanced/Metastatic TNBC	55	Clinical activity (ORR)	ORR 21% with 5 CRs and 5 PRs (stronger activity in *BRCA*-mutated tumors), DCR 49% Acceptable toxicity
	NCT02734004 (MEDIOLA)	Phase II	Durvalumab (1500 mg Q4W)	Olaparib (300 mg BID)	*gBRCAm* HER2 negative mBC	30 for clinical activity and 34 for safety	Clinical activity (DCR) and safety	12-week DCR 80%, 28-week DCR 80%, ORR 63%, mPFS 8.2 months, mOS 20.5 months (especially in chemotherapy-free patients) Acceptable toxicity
Prostate	NCT02484404	Phase II	Durvalumab (1500 mg Q4W)	Olaparib (300 mg BID)	Previously treated mCRPC	17	Clinical activity (rPFS) and safety	rPFS 16.1 months with 12-month rPFS 51.5% (especially in men with DDR abnomalities) Acceptable toxicity

Table 1. *Cont.*

Tymor Type	Study Identifier	Setting	ICI Agent	PARPi Agent	Design	Patients	Primary Endpoint	Outcome
SCLC	NCT02484404	Phase II	Durvalumab (1500 mg Q4W)	Olaparib (300 mg BID)	Relapsed SCLC	19	Clinical activity (ORR)	ORR 10.5% with 1 PRs and 1 CRs, clinical benefit 21.1% (preexisting TIL predictive of response) Acceptable toxicity
Bladder	NCT03534492 (NEODURVARIB)	Phase II	Durvalumab (1500 mg Q4W)	Olaparib (300 mg BID)	Resectable muscle-invasive bladder cancer	28	Clinical activity and safety	Pathological CR 44.5% Acceptable toxicity (grade 3 or higher AEs 8.3%)
Gastric	NCT02734004 (MEDIOLA)	Phase II	Durvalumab (1500 mg Q4W)	Olaparib (300 mg BID)	Platinum-resistant relapsed gastric cancer	39 for clinical activity and 40 for safety	Clinical activity (DCR) and safety	12-week DCR 26%, ORR 10% with 2 CRs and 2 PRs Unacceptable toxicity (grade 3 or higher AEs 48%)
Solid tumors	NCT02660034	Phase Ia/b	Tislelizumab (2 mk/kg Q2W)	Pamiparib (20, 40 or 60 mg BID)	Previously treated advanced solid tumors	49 patients	Safety and RP2D	DLT 8% with 23 immune-related AEs RP2D tislelizumab 200 mg and pamiparib 40 mg ORR 20% with 2CRs and 8 PRs
Solid tumors	NCT03330405 (JAVELIN PARP Medley)	Phase Ib,II	Avelumab (800 mg Q2W)	Talazoparib (1 mg QD)	Previously treated advanced solid tumors	34 patients	Safety and clinical activity (ORR)	First-cycle DLT 25% ORR 8% with 1 PR, SD 50%

Abbreviations: AE: adverse events; BID: twice a day; CR: complete response; DCR: disease control rate; DLT: dose-limiting toxicity; *gBRCAm*: germline *BRCA*-mutated; ICI: immune checkpoint inhibitor; mCRPC: metastatic castration-resistant prostate cancer; mOS: median overall survival; mPFS: median progression-free survival; ORR: overall response rate; PARPi: poly(ADP)-ribose polymerase inhibitor; PR: partial response; Q2W: every 2 weeks; Q3W: every 3 weeks; Q4W: every 4 weeks; QD: once daily; rPFS: median radiographic progression-free survival; RR: response rate according to RECIST v1 1, SD: stable disease; SCLC: small-cell lung carcinoma.

Table 2. Ongoing clinical trials of combination of PARPi and ICIs (www.clinicaltrials.gov).

PARPi	ICI Type	ICI Drug	Study Identifier	Phase	Tumor Type and Conditions	Status
	Anti-CTLA-4	Tremelimumab	NCT02571725	I/II	*gBRCAm* recurrent epithelial ovarian, fallopian tube, or primary peritoneal carcinoma	Recruiting
		Tremelimumab	NCT02485990	I/II	Recurrent or persistent epithelial ovarian, fallopian tube, or primary peritoneal carcinoma	Not recruiting
		Tremelimumab	NCT04034927	II	Platinum-sensitive advanced epithelial ovarian, fallopian tube, or primary peritoneal carcinoma	Recruiting
	Anti-PD1	Pembrolizumab	NCT04209686	II	Locally advanced or metastatic gastric carcinoma	Not yet recruiting
		Pembrolizumab	NCT04306367	II	Locally advanced or metastatic cholangiocarcinoma	Recruiting
		Pembrolizumab	NCT02861573 (KEYNOTE-365)	I	Metastatic castration-resistant prostate cancer	Recruiting
		Pembrolizumab	NCT03740165 (KEYLYNK-001	III	*BRCA*-non-mutated advanced epithelial ovarian, fallopian tube, or primary peritoneal carcinoma	Recruiting
		Pembrolizumab	NCT03976323 (KEYLINK-006)	III	Metastatic non-squamous cell lung cancer	Recruiting
		Pembrolizumab	NCT04123366 (KEYLYNK-007)	II	HRR-mutated or HRD positive advanced or metastatic solid tumors	Recruiting
		Pembrolizumab	NCT03976362 (KEYLYNK-008)	III	Metastatic squamous non-small cell lung cancer	Recruiting
		Pembrolizumab	NCT04191135 (KEYLYNK-009)	II/III	Locally advanced triple negative breast cancer	Recruiting
		Pembrolizumab	NCT03834519 (KEYLYNK-010)	III	Metastatic castration-resistant prostate cancer	Recruiting

Table 2. *Cont.*

PARPi	ICI		Study Identifier	Phase	Tumor Type and Conditions	Status
		Atezolizumab	NCT02849496	II	Locally advanced unresectable and or metastatic HER negative breast cancer	Recruiting
		Durvalumab	NCT03594396	I/II	Resectable stage II/III triple negative breast cancer	Recruiting
		Durvalumab	NCT03534492 (NEODURVARIB)	II	Resectable urothelial carcinoma	Recruiting
		Durvalumab	NCT03459846 (BAYOU)	II	Advanced or metastatic platinum-ineligible urothelial carcinoma	Not recruiting
		Durvalumab	NCT03951415 (DOMEC)	II	Recurrent, refractory or metastatic endometrial cancer or carcinosarcoma of the endometrium	Recruiting
		Durvalumab	NCT03851614 (DAPPER)	II	Locally advanced or metastatic mismatch repair proficient colorectal cancer, pancreatic adenocarcinoma or leiomyosarcoma	Recruiting
		Durvalumab	NCT03167619 (DORA)	II	Locally advanced or metastatic platinum-treated advanced triple negative breast cancer	Recruiting
		Durvalumab	NCT03544125	I	Metastatic triple negative breast cancer	Not recruiting
		Durvalumab	NCT03801369	II	Metastatic triple negative breast cancer	Recruiting
		Durvalumab	NCT04053322 (DOLAF)	II	Locally advanced or metastatic ER positive HER2 negative breast cancer	Recruiting
	Anti-PD-L1	Durvalumab	NCT03810105	II	DDR-mutated castration sensitive biochemically recurrent non-metastatic prostate cancer	Recruiting
		Durvalumab	NCT02882308 (OPHELIA)	II	Resectable squamous cell carcinoma of the head and neck	Completed
		Durvalumab	NCT03991832	II	IDH-mutated solid tumors (glioma, cholangiocarcinoma, and solid tumors)	Not yet recruiting

Table 2. *Cont.*

PARPi	ICI		Study Identifier	Phase	Tumor Type and Conditions	Status
		Durvalumab	NCT03772561 (MEDIPAC)	I	Advanced or metastatic solid tumors	Recruiting
		Durvalumab	NCT02484404	I/II	Advanced, recurrent or metastatic ovarian, triple negative breast, lung, prostate, colorectal carcinoma or solid tumors	Recruiting
		Durvalumab	NCT03579784	II	Unresectable or recurrent gastric carcinoma	Recruiting
		Durvalumab	NCT03737643 (DUO-O)	III	Newly diagnosed advanced ovarian, fallopian tube or primary peritoneal carcinoma or carcinosarcoma	Recruiting
		Durvalumab	NCT04269200 (DUO-E)	III	Newly diagnosed advanced or recurrent endometrial carcinoma	Not yet recruiting
		Durvalumab	NCT03801369	II	Metastatic triple negative breast cancer	Recruiting
		Durvalumab	NCT03775486	II	Metastatic non-squamous cell lung cancer	Recruiting
		Durvalumab	NCT02734004 (MEDIOLA)	I/II	Advanced or metastatic solid tumors	Recruiting
		Durvalumab	NCT03842228	I	DDR-mutated unresectable, advanced or metastatic solid tumors	Recruiting
	Anti-PD-L1 + Anti-CTLA-4	Durvalumab + Tremelimumab	NCT04169841 (GUIDE2REPAIR)	II	HRR-mutated advanced or metastatic solid tumors (breast, lung, head and neck, clear cell renal, endometrial, ovarian, urothelial and prostate cancer	Not yet recruiting
		Durvalumab + Tremelimumab	NCT03923270	I	Extensive small cell lung cancer	Recruiting
		Durvalumab + Tremelimumab	NCT02953457	I/II	DDR-mutated recurrent or refractory ovarian, fallopian tube or primary peritoneal carcinoma	Recruiting

Table 2. *Cont.*

PARPi	ICI		Study Identifier	Phase	Tumor Type and Conditions	Status
Niraparib	Anti-CTLA-4	Ipilimumab	NCT03404960 (Parpvax)	I/II	Advanced pancreatic cancer	Recruiting
	Anti-PD1	Cetrelimab	NCT03431350 (QUEST)	I/II	Metastatic castration-resistant prostate cancer	Recruiting
		Dostarlimab	NCT04068753 (STAR)	II	Recurrent or progressive cervix cancer	Recruiting
		Dostarlimab	NCT04313504	II	Reccurent or metastatic head and nead squamous carcinoma	Not yet recruiting
		Dostarlimab	NCT03016338	II	Recurrent or advanced endometrial cancer	Recruiting
		Dostarlimab	NCT03602859 (FIRST)	III	Stage III or IV non-mucinous epithelial ovarian, fallopian tube or primary peritoneal cancer	Recruiting
		Dostarlimab	NCT03955471 (MOONSTONE)	II	Advanced platinum-resistant ovarian, fallopian tube or primary peritoneal carcinoma	Recruiting
		Dostarlimab	NCT03806049	III	Advanced or recurrent platinum-sensitive ovarian, fallopian tube or primary peritoneal carcinoma	Not yet recruiting
		Dostarlimab	NCT03307785	I	Advanced or metastatic solid tumors	Not recruiting
		Dostarlimab	NCT03651206 (ROCSAN)	II/III	Reccurent or progressive uterine or ovarian carcinosarcoma	Not yet recruiting
		Nivolumab	NCT03404960 (Parpvax)	I/II	Advanced pancreatic cancer	Recruiting
		Pembrolizumab	NCT02657889 (TOPACIO)	I/II	Advanced or metastatic triple negative breast or ovarian cancer	Not recruiting
		PD-1 inhibitor	NCT03308942	II	Locally advanced or metastatic non-small cell lung carcinoma	Not recruiting

Table 2. *Cont.*

PARPi	ICI		Study Identifier	Phase	Tumor Type and Conditions	Status
	Anti-PD-L1	Atezolizumab	NCT03695380	I	Advanced ovarian, fallopian tube or primary peritoneal carcinoma	Recruiting
		Atezolizumab	NCT03598270 (ANITA)	III	Recurrent ovarian, fallopian tube or primary peritoneal carcinoma	Recruiting
		Atezolizumab	NCT04185831 (MEGALiT)	II	Advanced or metastatic solid tumors	Not yet recruiting
	Anti-PD-1	Nivolumab	NCT04187833	II	*BRCA*- or *BRCAness*-mutated resectable or metastatic melanoma	Recruiting
		Pembrolizumab	NCT04158336	I/II	Solid tumors	Recruiting
	Anti-PD-L1	Avelumab	NCT02912572	II	Recurrent or persistent endometrial cancer	Recruiting
		Avelumab	NCT03964532 (TALAVE)	I/II	Advanced breast cancer	Recruiting
		Avelumab	NCT03637491	II	Locally advanced or metastatic RAS-mutant solid tumors	Recruiting
		Avelumab	NCT04173507 (A LUNG-MAP)	II	*STK11*-mutated recurrent or metastatic non-squamous non-small cell lung cancer	Recruiting
		Avelumab	NCT03565991 (JAVELIN BRCA/ATM)	II	*BRCA* or *ATM*-mutated locally advanced or metastatic solid tumors	Recruiting
		Avelumab	NCT04068831	II	Locally advanced or metastatic clear-cell renal cell carcinoma	Recruiting
		Avelumab	NCT04052204	II	Locally advanced or metastatic head and neck squamous carcinoma or CRPC	Recruiting
		Avelumab	NCT03330405 (JAVELIN PARP MEDLEY)	II	Locally advanced or metastatic solid tumors	Recruiting

Table 2. *Cont.*

PARPi	ICI		Study Identifier	Phase	Tumor Type and Conditions	Status
		Avelumab	NCT03642132 (JAVELIN Ovarian PARP 100)	III	Locally advanced or metastatic ovarian cancer (NSCLC, triple negative breast cancer, HR+ breast cancer, recurrent platinum-sensitive ovarian cancer, urothelial cancer, CRPC)	Not recruiting
Veliparib	**Anti-PD-1**	Nivolumab	NCT02944396	I	Advanced or metastatic NSCLC	Completed
		Nivolumab	NCT03061188	I	Advanced, recurrent, refractory or metastatic solid tumors	Not recruiting
Rucaparib	**Anti-PD-1**	Nivolumab	NCT03572478	I/II	Metastatic CRPC or recurrent endometrial cancer	Recruiting
		Nivolumab	NCT03639935	II	Advanced or metastatic cholangiocarcinoma	Recruiting
		Nivolumab	NCT02873962	II	Relapsed ovarian, fallopian tube or peritoneal cancer	Recruiting
		Nivolumab	NCT03338790 (CheckMate 9KD)	II	Metastatic castration-resistant prostate cancer	Recruiting
		Nivolumab	NCT03522246 (ATHENA)	III	Newly diagnosed advanced ovarian, fallopian tube or primary peritoneal carcinoma or carcinosarcoma	Recruiting
		Nivolumab	NCT03824704 (ARIES)	II	Platinum-treated advanced ovarian, fallopian tube or primary peritoneal carcinoma or carcinosarcoma	Not recruiting
		Nivolumab	NCT03958045	II	Platinum-sensitive small cell lung carcinoma	Recruiting
		Nivolumab	NCT03995017 (RiME)	I/II	Unresectable or metastatic gastric or esophageal adenocarcinoma	Recruiting
		Pembrolizumab	NCT03559049	I/II	Metastatic NSCLC	Recruiting

Table 2. *Cont.*

PARPi	ICI		Study Identifier	Phase	Tumor Type and Conditions	Status
	Anti-PD-L1	Atezolizumab	NCT03101280	I	Advanced or metastatic platinum-sensitive ovarian or endometrial cancer or triple negative breast cancer	Not recruiting
		Atezolizumab	NCT04276376 (ARIANES)	II	DDR-deficient or platinum sensitive solid tumors	Recruiting
		Atezolizumab	NCT03694262 (EndoBARR)	II	Recurrent progressive endometrial carcinoma	Recruiting
Pamiparib	**Anti-PD-L1**	Tislelizumab	NCT02660034	I	Advanced or metastatic solid tumors	Recruiting

4.2.2. Combination of PARPi with Anti-CTLA-4 ICIs

Contrary to the in-depth attention paid to anti-PD1/PD-L1, the association of PARPi with anti-CTLA-4 is less studied, probably due to the misunderstood biological effect of PARPi on CTLA-4 signaling pathway and T cell effector functions. However, a previous study has unveiled that increased tumor immunogenicity modulates the response to CTLA-4 blockade [110]. Furthermore, BRCA dysfunction is associated with increased T cell recruitment to tumour site and higher expression of immune response genes [162–164]. Moreover, targeting DDR proteins through PARP inhibition may stimulate antigen presentation and immunogenicity via increased TMB and T cell cytotoxic activity [118]. Hence, one might surmise that tumors harboring BRCA dysfunction and treated with PARPi could increase tumoral immunogenicity, thus sensitizing the tumor to anti-CTLA-4 antibodies. All together, these data provide a rationale for the combination of PARPi with anti-CTLA-4 in BRCA-deficient tumours.

Initial preclinical study conducted on an immunocompetent BRCA1-deficient BR5 murine ovarian cancer model revealed that anti-CTLA-4 combined with PARPi veliparib enhanced IFN-γ production and effector/memory T cell infiltration [146]. In addition, CTLA-4 antibody synergized therapeutically with the PARPi, resulting in long-term survival in a majority of mice [146]. To date, no other preclinical study employing this combination in solid tumors has been released, and based on these data, clinical studies were developed.

Preliminary results from a phase 1 study combining olaparib and tremelimumab for the treatment of women with BRCA-deficient recurrent ovarian cancer demonstrated evidence of therapeutic effect with acceptable tolerability (NCT02571725) [165]. Ongoing clinical trials will help to figure out the promising antitumor activity of PARPi with anti-CTLA-4 monoclonal antibodies (Table 2).

4.2.3. Combination of ICI with Others DDR Inhibitors: Moving beyond PARP in Targeting the DDR

In light of the evidence that unrepaired DNA damage induced by PARPi expands the anti-tumor activity of the ICI, the therapeutic landscape of DDR-targeting agents has promptly unfolded to include inhibitors of other key mediators implied in DNA replication and repair, such as ATM, ATR, Chk1, Chk2, DNA-PK, and WEE1 [166]. The crucial roles of ATR and ATM protein kinases in DDR signaling involve the maintenance of replication fork stability and the regulation of cell cycle control checkpoints by operating together via downstream targets Chk1 and Chk2, respectively [167,168]. Additionally, the kinase activity of DNA-PK is required for NHEJ and a distinct nuclear kinase WEE1 controls mitotic entry as well as nucleotide pools in coordination with DNA damage response, making these kinase as potential targets for cancer therapy [168,169].

The role of DDR inhibitors as immunomodulatory agents that possibly potentiate ICIs has recently emerged. Recent preclinical evidence suggested that ATR or ATR inhibitors exhibit immunomodulatory functions and enhances antitumor efficacy to immune checkpoint therapy. The combination of selective ATR inhibitor, avelumab, and platinum-based chemotherapy resulted in antitumor effect in syngeneic tumor models, leading to overall survival benefit compared to any dual-combination group, and also provided protective antitumor immunity with immunological memory in cured mice [170]. Likewise, a recent study demonstrated that pharmacological ATM inhibition induced a type I IFN-mediated innate immune response in pancreatic cancer model that is further enhanced by radiation and led to increased sensitivity to anti-PD1 therapy [171]. Moreover, a preclinical model of immunocompetent SCLC in vivo observed that Chk1 inhibition, a protein kinase implicated in DSB repair, potentiated the antitumor effect of PD-L1 blockade and augmented cytotoxic T cell infiltration [133]. Moreover, a potent and selective DNA-PK inhibitor, that selectively blocks the NHEJ for repair of DSB, induced an immunomodulatory phenotype and elevated the expression of PD-L1 protein via cGAS-STING pathway activation in irradiated p53-mutant cancer cells [172]. Concordantly, combination of DNA-PK inhibitor, radiotherapy and avelumab in syngeneic mice with p53-mutant cancer cells demonstrated a superior benefit and offers a new approach to combination radio-immunotherapy of cancer [172]. All this evidence together provides a clear rationale to combine other DDR pathway inhibitor with immunotherapies.

Outside of PARPi, other DDR inhibitors are currently clinically tested in combination with ICI in several tumor types. A phase 1 modular study of ceralasertib, a potent and selective ATR inhibitor in combination with durvalumab is being evaluated in patients with advanced or metastatic cancers, including NSCLC and HNSCC (NCT02264678) [173]. Preliminary results of this study indicated acceptable tolerance with signals of activity [173]. Similarly, in the phase 1b BISCAY study, patients with metastatic MIBC with any HRD detected are being treated with durvalumab and olaparib or the WEE1 inhibitor adavosertib (NCT02546661) [174]. These ongoing trials will provide new insights into a combinatorial approach.

4.3. Future Perspectives

Although preclinical and clinical studies revealed interesting tumor responses with PARPi and ICIs in different tumors types, these combinations did not markedly improved antitumor effect compared to the individual agents alone. These disappointing outcomes suggest a lack of synergistic interaction of PARPi and ICIs. Several points could eventually explain these discrepancies.

Foremost, current animal models probably do not recapitulate the whole genomic heterogeneity or tumor microenvironment of human cancer. Indeed, data in mouse models do not predict response to the combination of PARPi and ICIs, thus limiting the transfer to Human. Better in vitro and in vivo models are needed to translate preclinical findings into clinical results. For example, the use of humanized mouse models could be a relevant strategy.

Concerning the nature and the magnitude of combination versus monotherapy benefit, most studies to date have relied on early endpoints such as ORR or DCR. The latter endpoints would be informative in case of patients with limited expected response to PARPi, such as tumors with DDR-proficient. However, in tumors where a high response to PARPi is expected, such as BRCA-mutated and other HRD phenotypes, it would be more relevant to assess the combination benefit in terms of the duration of response or survival, thus necessitating prolonged monitoring in such studies.

One another limitation in the interpretation of available data are the format of clinical trials. Indeed, current clinical results are only provided by non-randomized trials, which only allow cross-trial study comparisons. This approach of comparison is not methodologically and statistically acceptable to distinguish the specific role of each drug in terms of efficacy. To overcome this problem in the clinic, treatment strategies using DDR inhibitors with ICIs should be optimized through the use of randomized controlled multi arms phase III trials designed to enable the interpretation of the effect of each agent alone or in combination. Moreover, additional effort is required to determine the dose and schedule dependency of DNA repair modulation on the immune system.

Based on the synthetic lethality effect, the use of PARPi have been approved preferentially in tumors that harbor deficiency in the HR pathway, such as BRCA-mutated tumors and in a subsets of BRCA-negative HRD-positive cancer [175]. The assumptions that BRCA dysfunction is associated with the recruitment of T cell to tumor sites, and that PARPi may increase the immunogenicity of tumor cells have paved the way for combined strategy of ICIs and PARPi in BRCA-associated or more largely to HRD cancers. However, the combinatory effect of PARPi and ICIs in tumors without HR deficiency remains unknown. While in vitro studies in non-HRD cancer cell lines provide the rationale for a combined strategy, in vivo preclinical evidence suggesting that PARPi might increase the efficacy of ICIs has been conducted preferentially in BRCA-deficient tumor models, thus limiting the translational relevance. The question of whether PARPi and ICI should be restricted to non-HRD tumors or should be used more broadly has to be elucidated through the understanding of underlying biological mechanisms. Further work is needed to uncover the target population who are most likely to benefit from the combination strategy.

It is of note that most tumor types where the combination strategy was evaluated already demonstrated significant benefit of PARPi monotherapy, but limited activity for ICIs. Thereby, it would be more pertinent to evaluate the association in a population which cancer treatment represent a critical unmet medical need. For example, it would be interesting to focus on a subgroup which does not derive benefit or is primary/secondary resistant to either PARPi or ICI alone. Furthermore, dosing and scheduling of drugs largely differed across studies. The optimal dose and schedule of each agent needs to be determined with empirical clinical method as well as correlative analysis, including sequential tumor biopsies and serial blood collection.

In the future, a critical next step is to understand and identify the optimal patient population that will benefit the most from this combination. Biomarkers will likely play an even greater role in identifying those patients most likely to respond to PARPi. Whereas tumors with BRCA1/2 mutation or HRD are more likely to benefit from PARPi and ICI, an unmet medical need remains in the HR-proficient populations, so it is important to evaluate whether PARPi can sensitize these tumors to ICI in clinical settings. Moreover, obtaining contemporaneous tumor tissues and matched blood samples, associated with the integration of precision medicine, are key steps to better understand the mechanisms of action and the resistance pathways, and to identify novel predictive biomarkers of response. These crucial strides will allow a deeper comprehensive landscape of the interface between DNA damage and tumor immunity.

At last, although the current focus is on a combination of PARPi with anti-PD1/PD-L1 or anti-CTLA-4, other targeted agents moving beyond PARP in targeting DDR pathways as well as other promising immune-directed therapies are under development, and should be considered in the near future.

5. Conclusions

Genomic instability is a key hallmark of cancer that arises notably owing to DDR deficiencies. Major breakthroughs have been made with the successful targeting of DNA repair in clinical oncology [166]. Alike, immune-modulating therapies have also reshaped the landscape of cancer medicine [176]. However, treatment with either PARPi or ICIs alone often do not translate into benefit. While defects in DDR pathways might potentially be considered as predictive biomarkers of ICIs response, compelling evidence has provided a biological rationale, and demonstrated synergistic benefit, for combining ICIs with DDR inhibitors such as PARPi. To date, many preclinical and clinical researches focus on the identification of other tumors or molecular subtypes of cancers in which this combination will ultimately have a clinical impact, and thus turning more non-responders into responders with a strikingly boosted depth and duration of response.

While PARPi-induced a tumor HRD phenotype as well as immune modulation represent a rational approach for the association with ICIs, the combination did not markedly enhance the antitumor effect compared with individual agents in to-date clinical trials. To bring forward a critical change and improvements in patient outcomes, the development of accurate and predictive biomarker should become a priority. Moreover, unraveling the different mechanisms of resistance to PARPi and ICIs is required to pave the way for novel combination strategies. In addition, the appropriate dosing and scheduling of each agent should be determined in order to minimize adverse events while maximizing benefit and outcomes. Finally, elucidating the role of and interplay between DDR pathways, the tumor immune microenvironment and inhibitor agents, such as PARPi and ICIs, will be critical to the success and future development for this combination.

Author Contributions: F.P. and A.I. wrote the manuscript. All authors have read and agreed to the published version of the manuscript.

Funding: This research received no external funding.

Conflicts of Interest: The authors declare no conflict of interest.

References

1. Hodi, F.S.; O'Day, S.J.; McDermott, D.F.; Weber, R.W.; Sosman, J.A.; Haanen, J.B.; Gonzalez, R.; Robert, C.; Schadendorf, D.; Hassel, J.C.; et al. Improved survival with ipilimumab in patients with metastatic melanoma. *N. Engl. J. Med.* **2010**, *363*, 711–723. [CrossRef] [PubMed]

2. Robert, C.; Long, G.V.; Brady, B.; Dutriaux, C.; Maio, M.; Mortier, L.; Hassel, J.C.; Rutkowski, P.; McNeil, C.; Kalinka-Warzocha, E.; et al. Nivolumab in Previously Untreated Melanoma without BRAF Mutation. *N. Engl. J. Med.* **2015**, *372*, 320–330. [CrossRef] [PubMed]
3. Borghaei, H.; Paz-Ares, L.; Horn, L.; Spigel, D.R.; Steins, M.; Ready, N.E.; Chow, L.Q.; Vokes, E.E.; Felip, E.; Holgado, E.; et al. Nivolumab versus Docetaxel in Advanced Nonsquamous Non–Small-Cell Lung Cancer. *N. Engl. J. Med.* **2015**, *373*, 1627–1639. [CrossRef] [PubMed]
4. Reck, M.; Rodríguez-Abreu, D.; Robinson, A.G.; Hui, R.; Csőszi, T.; Fülöp, A.; Gottfried, M.; Peled, N.; Tafreshi, A.; Cuffe, S.; et al. Pembrolizumab versus Chemotherapy for PD-L1–Positive Non–Small-Cell Lung Cancer. *N. Engl. J. Med.* **2016**, *375*, 1823–1833. [CrossRef] [PubMed]
5. Motzer, R.J.; Escudier, B.; McDermott, D.F.; George, S.; Hammers, H.J.; Srinivas, S.; Tykodi, S.S.; Sosman, J.A.; Procopio, G.; Plimack, E.R.; et al. Nivolumab versus Everolimus in Advanced Renal-Cell Carcinoma. *N. Engl. J. Med.* **2015**, *373*, 1803–1813. [CrossRef]
6. Ferris, R.L.; Blumenschein, G.; Fayette, J.; Guigay, J.; Colevas, A.D.; Licitra, L.; Harrington, K.; Kasper, S.; Vokes, E.E.; Even, C.; et al. Nivolumab for Recurrent Squamous-Cell Carcinoma of the Head and Neck. *N. Engl. J. Med.* **2016**, *375*, 1856–1867. [CrossRef]
7. Ansell, S.M.; Lesokhin, A.M.; Borrello, I.; Halwani, A.; Scott, E.C.; Gutierrez, M.; Schuster, S.J.; Millenson, M.M.; Cattry, D.; Freeman, G.J.; et al. PD-1 Blockade with Nivolumab in Relapsed or Refractory Hodgkin's Lymphoma. *N. Engl. J. Med.* **2015**, *372*, 311–319. [CrossRef]
8. Moore, K.; Colombo, N.; Scambia, G.; Kim, B.-G.; Oaknin, A.; Friedlander, M.; Lisyanskaya, A.; Floquet, A.; Leary, A.; Sonke, G.S.; et al. Maintenance Olaparib in Patients with Newly Diagnosed Advanced Ovarian Cancer. *N. Engl. J. Med.* **2018**, *379*, 2495–2505. [CrossRef]
9. Robson, M.; Im, S.-A.; Senkus, E.; Xu, B.; Domchek, S.M.; Masuda, N.; Delaloge, S.; Li, W.; Tung, N.; Armstrong, A.; et al. Olaparib for Metastatic Breast Cancer in Patients with a Germline *BRCA*. Mutation. *N. Engl. J. Med.* **2017**, *377*, 523–533. [CrossRef]
10. Litton, J.K.; Rugo, H.S.; Ettl, J.; Hurvitz, S.A.; Gonçalves, A.; Lee, K.-H.; Fehrenbacher, L.; Yerushalmi, R.; Mina, L.A.; Martin, M.; et al. Talazoparib in Patients with Advanced Breast Cancer and a Germline BRCA Mutation. *N. Engl. J. Med.* **2018**, *379*, 753–763. [CrossRef]
11. Golan, T.; Hammel, P.; Reni, M.; Van Cutsem, E.; Macarulla, T.; Hall, M.J.; Park, J.-O.; Hochhauser, D.; Arnold, D.; Oh, D.-Y.; et al. Maintenance Olaparib for Germline *BRCA*. -Mutated Metastatic Pancreatic Cancer. *N. Engl. J. Med.* **2019**, *381*, 317–327. [CrossRef] [PubMed]
12. Ribas, A.; Wolchok, J.D. Cancer immunotherapy using checkpoint blockade. *Science* **2018**, *359*, 1350–1355. [CrossRef] [PubMed]
13. Patel, M.; Nowsheen, S.; Maraboyina, S.; Xia, F. The role of poly(ADP-ribose) polymerase inhibitors in the treatment of cancer and methods to overcome resistance: A review. *Cell Biosci.* **2020**, *10*, 35. [CrossRef]
14. Schumacher, T.N.; Schreiber, R.D. Neoantigens in cancer immunotherapy. *Science* **2015**, *348*, 69–74. [CrossRef]
15. Lee, C.-H.; Yelensky, R.; Jooss, K.; Chan, T.A. Update on Tumor Neoantigens and Their Utility: Why It Is Good to Be Different. *Trends Immunol.* **2018**, *39*, 536–548. [CrossRef]
16. Chen, D.S.; Mellman, I. Oncology Meets Immunology: The Cancer-Immunity Cycle. *Immunity* **2013**, *39*, 1–10. [CrossRef]
17. Goodman, A.M.; Kato, S.; Bazhenova, L.; Patel, S.P.; Frampton, G.M.; Miller, V.; Stephens, P.J.; Daniels, G.A.; Kurzrock, R. Tumor Mutational Burden as an Independent Predictor of Response to Immunotherapy in Diverse Cancers. *Mol. Cancer Ther.* **2017**, *16*, 2598–2608. [CrossRef]
18. Chen, D.S.; Mellman, I. Elements of cancer immunity and the cancer–immune set point. *Nature* **2017**, *541*, 321–330. [CrossRef]
19. Schreiber, R.D.; Old, L.J.; Smyth, M.J. Cancer immunoediting: Integrating immunity's roles in cancer suppression and promotion. *Science* **2011**, *331*, 1565–1570. [CrossRef]
20. Granier, C.; De Guillebon, E.; Blanc, C.; Roussel, H.; Badoual, C.; Colin, E.; Saldmann, A.; Gey, A.; Oudard, S.; Tartour, E. Mechanisms of action and rationale for the use of checkpoint inhibitors in cancer. *ESMO Open* **2017**, *2*, e000213. [CrossRef]
21. Yu, S.; Li, A.; Liu, Q.; Li, T.; Yuan, X.; Han, X.; Wu, K. Chimeric antigen receptor T cells: A novel therapy for solid tumors. *J. Hematol. Oncol.* **2017**, *10*, 78. [CrossRef] [PubMed]

22. Marin-Acevedo, J.A.; Dholaria, B.; Soyano, A.E.; Knutson, K.L.; Chumsri, S.; Lou, Y. Next generation of immune checkpoint therapy in cancer: New developments and challenges. *J. Hematol. Oncol.* **2018**, *11*, 39. [CrossRef] [PubMed]
23. Fucá, G.; Reppel, L.; Landoni, E.; Savoldo, B.; Dotti, G. Enhancing Chimeric Antigen Receptor T cell Efficacy in Solid Tumors. *Clin. Cancer Res.* **2020**, *26*, 2444–2451. [CrossRef]
24. Kantoff, P.W.; Higano, C.S.; Shore, N.D.; Berger, E.R.; Small, E.J.; Penson, D.F.; Redfern, C.H.; Ferrari, A.C.; Dreicer, R.; Sims, R.B.; et al. Sipuleucel-T Immunotherapy for Castration-Resistant Prostate Cancer. *N. Engl. J. Med.* **2010**, *363*, 411–422. [CrossRef]
25. June, C.H.; Sadelain, M. Chimeric Antigen Receptor Therapy. *N. Engl. J. Med.* **2018**, *379*, 64–73. [CrossRef]
26. Hargadon, K.M.; Johnson, C.E.; Williams, C.J. Immune checkpoint blockade therapy for cancer: An overview of FDA-approved immune checkpoint inhibitors. *Int. Immunopharmacol.* **2018**, *62*, 29–39. [CrossRef]
27. Lopes, A.; Vandermeulen, G.; Préat, V. Cancer DNA vaccines: Current preclinical and clinical developments and future perspectives. *J. Exp. Clin. Cancer Res.* **2019**, *38*, 146. [CrossRef] [PubMed]
28. Sharma, P.; Hu-Lieskovan, S.; Wargo, J.A.; Ribas, A. Primary, Adaptive, and Acquired Resistance to Cancer Immunotherapy. *Cell* **2017**, *168*, 707–723. [CrossRef]
29. Jenkins, R.W.; Barbie, D.A.; Flaherty, K.T. Mechanisms of resistance to immune checkpoint inhibitors. *Br. J. Cancer* **2018**, *118*, 9–16. [CrossRef]
30. Ashworth, A. A synthetic lethal therapeutic approach: Poly(ADP) ribose polymerase inhibitors for the treatment of cancers deficient in DNA double-strand break repair. *J. Clin. Oncol.* **2008**, *26*, 3785–3790. [CrossRef]
31. Stewart, R.A.; Pilié, P.G.; Yap, T.A. Development of PARP and Immune-Checkpoint Inhibitor Combinations. *Cancer Res.* **2018**, *78*, 6717–6725. [CrossRef] [PubMed]
32. Jeggo, P.A.; Pearl, L.H.; Carr, A.M. DNA repair, genome stability and cancer: A historical perspective. *Nat. Rev. Cancer* **2016**, *16*, 35–42. [CrossRef]
33. Jackson, S.P.; Bartek, J. The DNA-damage response in human biology and disease. *Nature* **2009**, *461*, 1071–1078. [CrossRef] [PubMed]
34. Ciccia, A.; Elledge, S.J. The DNA Damage Response: Making It Safe to Play with Knives. *Mol. Cell* **2010**, *40*, 179–204. [CrossRef] [PubMed]
35. Tubbs, A.; Nussenzweig, A. Endogenous DNA Damage as a Source of Genomic Instability in Cancer. *Cell* **2017**, *168*, 644–656. [CrossRef]
36. Brown, J.S.; O'Carrigan, B.; Jackson, S.P.; Yap, T.A. Targeting DNA Repair in Cancer: Beyond PARP Inhibitors. *Cancer Discov.* **2017**, *7*, 20–37. [CrossRef]
37. De Vos, M.; Schreiber, V.; Dantzer, F. The diverse roles and clinical relevance of PARPs in DNA damage repair: Current state of the art. *Biochem. Pharmacol.* **2012**, *84*, 137–146. [CrossRef]
38. Lord, C.J.; Ashworth, A. PARP inhibitors: Synthetic lethality in the clinic. *Science* **2017**, *355*, 1152–1158. [CrossRef]
39. Maya-Mendoza, A.; Moudry, P.; Merchut-Maya, J.M.; Lee, M.; Strauss, R.; Bartek, J. High speed of fork progression induces DNA replication stress and genomic instability. *Nature* **2018**, *559*, 279–284. [CrossRef]
40. Langelier, M.-F.; Eisemann, T.; Riccio, A.A.; Pascal, J.M. PARP family enzymes: Regulation and catalysis of the poly(ADP-ribose) posttranslational modification. *Curr. Opin. Struct. Biol.* **2018**, *53*, 187–198. [CrossRef]
41. Gibson, B.A.; Kraus, W.L. New insights into the molecular and cellular functions of poly(ADP-ribose) and PARPs. *Nat. Rev. Mol. Cell Biol.* **2012**, *13*, 411–424. [CrossRef] [PubMed]
42. Min, A.; Im, S.-A. PARP Inhibitors as Therapeutics: Beyond Modulation of PARylation. *Cancers* **2020**, *12*, 394. [CrossRef] [PubMed]
43. Helleday, T. The underlying mechanism for the PARP and BRCA synthetic lethality: Clearing up the misunderstandings. *Mol. Oncol.* **2011**, *5*, 387–393. [CrossRef]
44. Farmer, H.; McCabe, N.; Lord, C.J.; Tutt, A.N.J.; Johnson, D.A.; Richardson, T.B.; Santarosa, M.; Dillon, K.J.; Hickson, I.; Knights, C.; et al. Targeting the DNA repair defect in BRCA mutant cells as a therapeutic strategy. *Nature* **2005**, *434*, 917–921. [CrossRef]
45. Mao, Z.; Bozzella, M.; Seluanov, A.; Gorbunova, V. DNA repair by nonhomologous end joining and homologous recombination during cell cycle in human cells. *Cell Cycle* **2008**, *7*, 2902–2906. [CrossRef]
46. Thompson, L.H.; Schild, D. Homologous recombinational repair of DNA ensures mammalian chromosome stability. *Mutat. Res.* **2001**, *477*, 131–153. [CrossRef]

47. Lieber, M.R. The mechanism of human nonhomologous DNA end joining. *J. Biol. Chem.* **2008**, *283*, 1–5. [CrossRef] [PubMed]
48. Bryant, H.E.; Schultz, N.; Thomas, H.D.; Parker, K.M.; Flower, D.; Lopez, E.; Kyle, S.; Meuth, M.; Curtin, N.J.; Helleday, T. Specific killing of BRCA2-deficient tumours with inhibitors of poly(ADP-ribose) polymerase. *Nature* **2005**, *434*, 913–917. [CrossRef]
49. Murai, J.; Huang, S.-Y.N.; Das, B.B.; Renaud, A.; Zhang, Y.; Doroshow, J.H.; Ji, J.; Takeda, S.; Pommier, Y. Trapping of PARP1 and PARP2 by Clinical PARP Inhibitors. *Cancer Res.* **2012**, *72*, 5588–5599. [CrossRef] [PubMed]
50. Ström, C.E.; Johansson, F.; Uhlén, M.; Szigyarto, C.A.-K.; Erixon, K.; Helleday, T. Poly (ADP-ribose) polymerase (PARP) is not involved in base excision repair but PARP inhibition traps a single-strand intermediate. *Nucleic Acids Res.* **2011**, *39*, 3166–3175. [CrossRef]
51. Wang, Y.-Q.; Wang, P.-Y.; Wang, Y.-T.; Yang, G.-F.; Zhang, A.; Miao, Z.-H. An Update on Poly(ADP-ribose)polymerase-1 (PARP-1) Inhibitors: Opportunities and Challenges in Cancer Therapy. *J. Med. Chem.* **2016**, *59*, 9575–9598. [CrossRef]
52. Shen, Y.; Aoyagi-Scharber, M.; Wang, B. Trapping Poly(ADP-Ribose) Polymerase. *J. Pharmacol. Exp. Ther.* **2015**, *353*, 446–457. [CrossRef] [PubMed]
53. Hopkins, T.A.; Shi, Y.; Rodriguez, L.E.; Solomon, L.R.; Donawho, C.K.; DiGiammarino, E.L.; Panchal, S.C.; Wilsbacher, J.L.; Gao, W.; Olson, A.M.; et al. Mechanistic Dissection of PARP1 Trapping and the Impact on In Vivo Tolerability and Efficacy of PARP Inhibitors. *Mol. Cancer Res.* **2015**, *13*, 1465–1477. [CrossRef]
54. Patel, A.G.; Sarkaria, J.N.; Kaufmann, S.H. Nonhomologous end joining drives poly(ADP-ribose) polymerase (PARP) inhibitor lethality in homologous recombination-deficient cells. *Proc. Natl. Acad. Sci. USA* **2011**, *108*, 3406–3411. [CrossRef] [PubMed]
55. McCabe, N.; Turner, N.C.; Lord, C.J.; Kluzek, K.; Bialkowska, A.; Swift, S.; Giavara, S.; O'Connor, M.J.; Tutt, A.N.; Zdzienicka, M.Z.; et al. Deficiency in the repair of DNA damage by homologous recombination and sensitivity to poly(ADP-ribose) polymerase inhibition. *Cancer Res.* **2006**, *66*, 8109–8115. [CrossRef] [PubMed]
56. Lord, C.J.; Ashworth, A. BRCAness revisited. *Nat. Rev. Cancer* **2016**, *16*, 110–120. [CrossRef]
57. Oza, A.M.; Cibula, D.; Benzaquen, A.O.; Poole, C.; Mathijssen, R.H.J.; Sonke, G.S.; Colombo, N.; Špaček, J.; Vuylsteke, P.; Hirte, H.; et al. Olaparib combined with chemotherapy for recurrent platinum-sensitive ovarian cancer: A randomised phase 2 trial. *Lancet Oncol.* **2015**, *16*, 87–97. [CrossRef]
58. Dhawan, M.S.; Bartelink, I.H.; Aggarwal, R.R.; Leng, J.; Zhang, J.Z.; Pawlowska, N.; Terranova-Barberio, M.; Grabowsky, J.A.; Gewitz, A.; Chien, A.J.; et al. Differential Toxicity in Patients with and without DNA Repair Mutations: Phase I Study of Carboplatin and Talazoparib in Advanced Solid Tumors. *Clin. Cancer Res.* **2017**, *23*, 6400–6410. [CrossRef] [PubMed]
59. Fong, P.C.; Boss, D.S.; Yap, T.A.; Tutt, A.; Wu, P.; Mergui-Roelvink, M.; Mortimer, P.; Swaisland, H.; Lau, A.; O'Connor, M.J.; et al. Inhibition of Poly(ADP-Ribose) Polymerase in Tumors from BRCA Mutation Carriers. *N. Engl. J. Med.* **2009**, *361*, 123–134. [CrossRef]
60. Mirza, M.R.; Monk, B.J.; Herrstedt, J.; Oza, A.M.; Mahner, S.; Redondo, A.; Fabbro, M.; Ledermann, J.A.; Lorusso, D.; Vergote, I.; et al. Niraparib Maintenance Therapy in Platinum-Sensitive, Recurrent Ovarian Cancer. *N. Engl. J. Med.* **2016**, *375*, 2154–2164. [CrossRef]
61. Pujade-Lauraine, E.; Ledermann, J.A.; Selle, F.; Gebski, V.; Penson, R.T.; Oza, A.M.; Korach, J.; Huzarski, T.; Poveda, A.; Pignata, S.; et al. Olaparib tablets as maintenance therapy in patients with platinum-sensitive, relapsed ovarian cancer and a BRCA1/2 mutation (SOLO2/ENGOT-Ov21): A double-blind, randomised, placebo-controlled, phase 3 trial. *Lancet Oncol.* **2017**, *18*, 1274–1284. [CrossRef]
62. Diéras, V.C.; Han, H.S.; Kaufman, B.; Wildiers, H.; Friedlander, M.; Ayoub, J.-P.; Puhalla, S.L.; Bondarenko, I.; Campone, M.; Jakobsen, E.H.; et al. Phase III study of veliparib with carboplatin and paclitaxel in HER2-negative advanced/metastatic gBRCA-associated breast cancer. *Ann. Oncol.* **2019**, *30*, v857–v858. [CrossRef]
63. Mateo, J.; Carreira, S.; Sandhu, S.; Miranda, S.; Mossop, H.; Perez-Lopez, R.; Nava Rodrigues, D.; Robinson, D.; Omlin, A.; Tunariu, N.; et al. DNA-Repair Defects and Olaparib in Metastatic Prostate Cancer. *N. Engl. J. Med.* **2015**, *373*, 1697–1708. [CrossRef] [PubMed]
64. Ledermann, J.A.; Pujade-Lauraine, E. Olaparib as maintenance treatment for patients with platinum-sensitive relapsed ovarian cancer. *Ther. Adv. Med. Oncol.* **2019**, *11*. [CrossRef]

65. Ray-Coquard, I.; Pautier, P.; Pignata, S.; Pérol, D.; González-Martín, A.; Berger, R.; Fujiwara, K.; Vergote, I.; Colombo, N.; Mäenpää, J.; et al. Olaparib plus Bevacizumab as First-Line Maintenance in Ovarian Cancer. *N. Engl. J. Med.* **2019**, *381*, 2416–2428. [CrossRef]
66. Coleman, R.L.; Oza, A.M.; Lorusso, D.; Aghajanian, C.; Oaknin, A.; Dean, A.; Colombo, N.; Weberpals, J.I.; Clamp, A.; Scambia, G.; et al. Rucaparib maintenance treatment for recurrent ovarian carcinoma after response to platinum therapy (ARIEL3): A randomised, double-blind, placebo-controlled, phase 3 trial. *Lancet* **2017**, *390*, 1949–1961. [CrossRef]
67. Coleman, R.L.; Fleming, G.F.; Brady, M.F.; Swisher, E.M.; Steffensen, K.D.; Friedlander, M.; Okamoto, A.; Moore, K.N.; Efrat Ben-Baruch, N.; Werner, T.L.; et al. Veliparib with First-Line Chemotherapy and as Maintenance Therapy in Ovarian Cancer. *N. Engl. J. Med.* **2019**, *381*, 2403–2415. [CrossRef]
68. Pettitt, S.J.; Lord, C.J. Dissecting PARP inhibitor resistance with functional genomics. *Curr. Opin. Genet. Dev.* **2019**, *54*, 55–63. [CrossRef]
69. Jaspers, J.E.; Kersbergen, A.; Boon, U.; Sol, W.; van Deemter, L.; Zander, S.A.; Drost, R.; Wientjens, E.; Ji, J.; Aly, A.; et al. Loss of 53BP1 Causes PARP Inhibitor Resistance in Brca1-Mutated Mouse Mammary Tumors. *Cancer Discov.* **2013**, *3*, 68–81. [CrossRef]
70. Pettitt, S.J.; Krastev, D.B.; Brandsma, I.; Dréan, A.; Song, F.; Aleksandrov, R.; Harrell, M.I.; Menon, M.; Brough, R.; Campbell, J.; et al. Genome-wide and high-density CRISPR-Cas9 screens identify point mutations in PARP1 causing PARP inhibitor resistance. *Nat. Commun.* **2018**, *9*, 1849. [CrossRef]
71. Gogola, E.; Duarte, A.A.; de Ruiter, J.R.; Wiegant, W.W.; Schmid, J.A.; de Bruijn, R.; James, D.I.; Guerrero Llobet, S.; Vis, D.J.; Annunziato, S.; et al. Selective Loss of PARG Restores PARylation and Counteracts PARP Inhibitor-Mediated Synthetic Lethality. *Cancer Cell.* **2018**, *33*, 1078–1093.e12. [CrossRef] [PubMed]
72. Rottenberg, S.; Jaspers, J.E.; Kersbergen, A.; van der Burg, E.; Nygren, A.O.H.; Zander, S.A.L.; Derksen, P.W.B.; de Bruin, M.; Zevenhoven, J.; Lau, A.; et al. High sensitivity of BRCA1-deficient mammary tumors to the PARP inhibitor AZD2281 alone and in combination with platinum drugs. *Proc. Natl. Acad. Sci. USA* **2008**, *105*, 17079–17084. [CrossRef] [PubMed]
73. Noordermeer, S.M.; van Attikum, H. PARP Inhibitor Resistance: A Tug-of-War in BRCA-Mutated Cells. *Trends Cell Biol.* **2019**, *29*, 820–834. [CrossRef] [PubMed]
74. Ray Chaudhuri, A.; Callen, E.; Ding, X.; Gogola, E.; Duarte, A.A.; Lee, J.-E.; Wong, N.; Lafarga, V.; Calvo, J.A.; Panzarino, N.J.; et al. Replication fork stability confers chemoresistance in BRCA-deficient cells. *Nature* **2016**, *535*, 382–387. [CrossRef] [PubMed]
75. Rondinelli, B.; Gogola, E.; Yücel, H.; Duarte, A.A.; van de Ven, M.; van der Sluijs, R.; Konstantinopoulos, P.A.; Jonkers, J.; Ceccaldi, R.; Rottenberg, S.; et al. EZH2 promotes degradation of stalled replication forks by recruiting MUS81 through histone H3 trimethylation. *Nat. Cell Biol.* **2017**, *19*, 1371–1378. [CrossRef] [PubMed]
76. Jiao, S.; Xia, W.; Yamaguchi, H.; Wei, Y.; Chen, M.-K.; Hsu, J.-M.; Hsu, J.L.; Yu, W.-H.; Du, Y.; Lee, H.-H.; et al. PARP Inhibitor Upregulates PD-L1 Expression and Enhances Cancer-Associated Immunosuppression. *Clin. Cancer Res.* **2017**, *23*, 3711–3720. [CrossRef]
77. Chabanon, R.M.; Muirhead, G.; Krastev, D.B.; Adam, J.; Morel, D.; Garrido, M., Lamb, A.; Hénon, C.; Dorvault, N.; Rouanne, M.; et al. PARP inhibition enhances tumor cell-intrinsic immunity in ERCC1-deficient non-small cell lung cancer. *J. Clin. Investig.* **2019**, *129*, 1211–1228. [CrossRef]
78. Parish, C.R. Cancer immunotherapy: The past, the present and the future. *Immunol. Cell. Biol.* **2003**, *81*, 106–113. [CrossRef]
79. Becht, E.; Giraldo, N.A.; Germain, C.; de Reyniès, A.; Laurent-Puig, P.; Zucman-Rossi, J.; Dieu-Nosjean, M.-C.; Sautès-Fridman, C.; Fridman, W.H. Immune Contexture, Immunoscore, and Malignant Cell Molecular Subgroups for Prognostic and Theranostic Classifications of Cancers. *Adv. Immunol.* **2016**, *130*, 95–190. [CrossRef]
80. Bretscher, P.A. A two-step, two-signal model for the primary activation of precursor helper T cells. *Proc. Natl. Acad. Sci. USA* **1999**, *96*, 185–190. [CrossRef]
81. Chen, L.; Flies, D.B. Molecular mechanisms of T cell co-stimulation and co-inhibition. *Nat. Rev. Immunol.* **2013**, *13*, 227–242. [CrossRef]
82. Sharma, P.; Allison, J.P. Immune Checkpoint Targeting in Cancer Therapy: Toward Combination Strategies with Curative Potential. *Cell* **2015**, *161*, 205–214. [CrossRef] [PubMed]

83. Chambers, C.A.; Kuhns, M.S.; Egen, J.G.; Allison, J.P. CTLA-4-mediated inhibition in regulation of T cell responses: Mechanisms and manipulation in tumor immunotherapy. *Annu. Rev. Immunol.* **2001**, *19*, 565–594. [CrossRef] [PubMed]
84. Krummel, M.F.; Allison, J.P. CD28 and CTLA-4 have opposing effects on the response of T cells to stimulation. *J. Exp. Med.* **1995**, *182*, 459–465. [CrossRef] [PubMed]
85. Greenwald, R.J.; Boussiotis, V.A.; Lorsbach, R.B.; Abbas, A.K.; Sharpe, A.H. CTLA-4 Regulates Induction of Anergy In Vivo. *Immunity* **2001**, *14*, 145–155. [CrossRef]
86. Qureshi, O.S.; Zheng, Y.; Nakamura, K.; Attridge, K.; Manzotti, C.; Schmidt, E.M.; Baker, J.; Jeffery, L.E.; Kaur, S.; Briggs, Z.; et al. Trans-Endocytosis of CD80 and CD86: A Molecular Basis for the Cell-Extrinsic Function of CTLA-4. *Science* **2011**, *332*, 600–603. [CrossRef]
87. Rudd, C.E.; Taylor, A.; Schneider, H. CD28 and CTLA-4 coreceptor expression and signal transduction. *Immunol. Rev.* **2009**, *229*, 12–26. [CrossRef]
88. Keir, M.E.; Butte, M.J.; Freeman, G.J.; Sharpe, A.H. PD-1 and Its Ligands in Tolerance and Immunity. *Annu. Rev. Immunol.* **2008**, *26*, 677–704. [CrossRef]
89. Bardhan, K.; Anagnostou, T.; Boussiotis, V.A. The PD1:PD-L1/2 Pathway from Discovery to Clinical Implementation. *Front. Immunol.* **2016**, *7*, 550. [CrossRef]
90. Boussiotis, V.A.; Chatterjee, P.; Li, L. Biochemical signaling of PD-1 on T cells and its functional implications. *Cancer J.* **2014**, *20*, 265–271. [CrossRef]
91. Boussiotis, V.A. Molecular and Biochemical Aspects of the PD-1 Checkpoint Pathway. *N. Engl J. Med.* **2016**, *375*, 1767–1778. [CrossRef] [PubMed]
92. Sharpe, A.H.; Pauken, K.E. The diverse functions of the PD1 inhibitory pathway. *Nat. Rev. Immunol.* **2018**, *18*, 153–167. [CrossRef] [PubMed]
93. Robert, C.; Thomas, L.; Bondarenko, I.; O'Day, S.; Weber, J.; Garbe, C.; Lebbe, C.; Baurain, J.-F.; Testori, A.; Grob, J.-J.; et al. Ipilimumab plus Dacarbazine for Previously Untreated Metastatic Melanoma. *N. Engl J. Med.* **2011**, *364*, 2517–2526. [CrossRef]
94. Robert, C.; Ribas, A.; Schachter, J.; Arance, A.; Grob, J.-J.; Mortier, L.; Daud, A.; Carlino, M.S.; McNeil, C.M.; Lotem, M.; et al. Pembrolizumab versus ipilimumab in advanced melanoma (KEYNOTE-006): Post-hoc 5-year results from an open-label, multicentre, randomised, controlled, phase 3 study. *Lancet Oncol.* **2019**, *20*, 1239–1251. [CrossRef]
95. Rittmeyer, A.; Barlesi, F.; Waterkamp, D.; Park, K.; Ciardiello, F.; von Pawel, J.; Gadgeel, S.M.; Hida, T.; Kowalski, D.M.; Dols, M.C.; et al. Atezolizumab versus docetaxel in patients with previously treated non-small-cell lung cancer (OAK): A phase 3, open-label, multicentre randomised controlled trial. *Lancet* **2017**, *389*, 255–265. [CrossRef]
96. Antonia, S.J.; Villegas, A.; Daniel, D.; Vicente, D.; Murakami, S.; Hui, R.; Yokoi, T.; Chiappori, A.; Lee, K.H.; de Wit, M.; et al. Durvalumab after Chemoradiotherapy in Stage III Non-Small-Cell Lung Cancer. *N. Engl. J. Med.* **2017**, *377*, 1919–1929. [CrossRef]
97. Bellmunt, J.; de Wit, R.; Vaughn, D.J.; Fradet, Y.; Lee, J.-L.; Fong, L.; Vogelzang, N.J.; Climent, M.A.; Petrylak, D.P.; Choueiri, T.K.; et al. Pembrolizumab as Second-Line Therapy for Advanced Urothelial Carcinoma. *N. Engl. J. Med.* **2017**, *376*, 1015–1026. [CrossRef]
98. Burtness, B.; Harrington, K.J.; Greil, R.; Soulières, D.; Tahara, M.; de Castro, G.; Psyrri, A.; Basté, N.; Neupane, P.; Bratland, Å.; et al. Pembrolizumab alone or with chemotherapy versus cetuximab with chemotherapy for recurrent or metastatic squamous cell carcinoma of the head and neck (KEYNOTE-048): A randomised, open-label, phase 3 study. *Lancet* **2019**, *394*, 1915–1928. [CrossRef]
99. Kaufman, H.L.; Russell, J.; Hamid, O.; Bhatia, S.; Terheyden, P.; D'Angelo, S.P.; Shih, K.C.; Lebbé, C.; Linette, G.P.; Milella, M.; et al. Avelumab in patients with chemotherapy-refractory metastatic Merkel cell carcinoma: A multicentre, single-group, open-label, phase 2 trial. *Lancet Oncol.* **2016**, *17*, 1374–1385. [CrossRef]
100. Catenacci Daniel, V.; Wainberg, Z.; Fuchs Charles, S.; Garrido, M.; Bang, Y.-J.; Muro, K.; Savage, M.; Wang, J.; Koshiji, M.; Dalal Rita, P.; et al. KEYNOTE-059 cohort 3: Safety and efficacy of pembrolizumab monotherapy for first-line treatment of patients (pts) with PD-L1-positive advanced gastric/gastroesophageal (G/GEJ) cancer. *Ann. Oncol.* **2017**, *28* (Suppl. 3), iii153. [CrossRef]

101. El-Khoueiry, A.B.; Sangro, B.; Yau, T.; Crocenzi, T.S.; Kudo, M.; Hsu, C.; Kim, T.-Y.; Choo, S.-P.; Trojan, J.; Welling, T.H.; et al. Nivolumab in patients with advanced hepatocellular carcinoma (CheckMate 040): An open-label, non-comparative, phase 1/2 dose escalation and expansion trial. *Lancet* **2017**, *389*, 2492–2502. [CrossRef]
102. Armand, P.; Engert, A.; Younes, A.; Fanale, M.; Santoro, A.; Zinzani, P.L.; Timmerman, J.M.; Collins, G.P.; Ramchandren, R.; Cohen, J.B.; et al. Nivolumab for Relapsed/Refractory Classic Hodgkin Lymphoma After Failure of Autologous Hematopoietic Cell Transplantation: Extended Follow-Up of the Multicohort Single-Arm Phase II CheckMate 205 Trial. *JCO* **2018**, *36*, 1428–1439. [CrossRef]
103. Le, D.T.; Uram, J.N.; Wang, H.; Bartlett, B.R.; Kemberling, H.; Eyring, A.D.; Skora, A.D.; Luber, B.S.; Azad, N.S.; Laheru, D.; et al. PD-1 Blockade in Tumors with Mismatch-Repair Deficiency. *N. Engl. J. of. Med.* **2015**, *372*, 2509–2520. [CrossRef]
104. Le, D.T.; Durham, J.N.; Smith, K.N.; Wang, H.; Bartlett, B.R.; Aulakh, L.K.; Lu, S.; Kemberling, H.; Wilt, C.; Luber, B.S.; et al. Mismatch repair deficiency predicts response of solid tumors to PD-1 blockade. *Science* **2017**, *357*, 409–413. [CrossRef]
105. Overman, M.J.; McDermott, R.; Leach, J.L.; Lonardi, S.; Lenz, H.-J.; Morse, M.A.; Desai, J.; Hill, A.; Axelson, M.; Moss, R.A.; et al. Nivolumab in patients with metastatic DNA mismatch repair-deficient or microsatellite instability-high colorectal cancer (CheckMate 142): An open-label, multicentre, phase 2 study. *Lancet Oncol.* **2017**, *18*, 1182–1191. [CrossRef]
106. Havel, J.J.; Chowell, D.; Chan, T.A. The evolving landscape of biomarkers for checkpoint inhibitor immunotherapy. *Nat. Rev. Cancer* **2019**, *19*, 133–150. [CrossRef]
107. Teo, M.Y.; Seier, K.; Ostrovnaya, I.; Regazzi, A.M.; Kania, B.E.; Moran, M.M.; Cipolla, C.K.; Bluth, M.J.; Chaim, J.; Al-Ahmadie, H.; et al. Alterations in DNA Damage Response and Repair Genes as Potential Marker of Clinical Benefit from PD-1/PD-L1 Blockade in Advanced Urothelial Cancers. *J. Clin. Oncol.* **2018**, *36*, 1685–1694. [CrossRef]
108. Chowdhury, P.S.; Chamoto, K.; Honjo, T. Combination therapy strategies for improving PD-1 blockade efficacy: A new era in cancer immunotherapy. *J. Intern. Med.* **2018**, *283*, 110–120. [CrossRef]
109. Seliger, B. Combinatorial Approaches with Checkpoint Inhibitors to Enhance Anti-tumor Immunity. *Front. Immunol.* **2019**, *10*, 999. [CrossRef]
110. Snyder, A.; Makarov, V.; Merghoub, T.; Yuan, J.; Zaretsky, J.M.; Desrichard, A.; Walsh, L.A.; Postow, M.A.; Wong, P.; Ho, T.S.; et al. Genetic basis for clinical response to CTLA-4 blockade in melanoma. *N. Engl. J. Med.* **2014**, *371*, 2189–2199. [CrossRef]
111. Rizvi, N.A.; Hellmann, M.D.; Snyder, A.; Kvistborg, P.; Makarov, V.; Havel, J.J.; Lee, W.; Yuan, J.; Wong, P.; Ho, T.S.; et al. Mutational landscape determines sensitivity to PD-1 blockade in non–small cell lung cancer. *Science* **2015**, *348*, 124–128. [CrossRef] [PubMed]
112. Hellmann, M.D.; Callahan, M.K.; Awad, M.M.; Calvo, E.; Ascierto, P.A.; Atmaca, A.; Rizvi, N.A.; Hirsch, F.R.; Selvaggi, G.; Szustakowski, J.D.; et al. Tumor Mutational Burden and Efficacy of Nivolumab Monotherapy and in Combination with Ipilimumab in Small-Cell Lung Cancer. *Cancer Cell* **2018**, *33*, 853–861.e4. [CrossRef] [PubMed]
113. Yarchoan, M.; Hopkins, A.; Jaffee, E.M. Tumor Mutational Burden and Response Rate to PD-1 Inhibition. *N. Engl. J. Med.* **2017**, *377*, 2500–2501. [CrossRef]
114. Legrand, F.A.; Gandara, D.R.; Mariathasan, S.; Powles, T.; He, X.; Zhang, W.; Jhunjhunwala, S.; Nickles, D.; Bourgon, R.; Schleifman, E.; et al. Association of high tissue TMB and atezolizumab efficacy across multiple tumor types. *JCO* **2018**, *36*, 12000. [CrossRef]
115. McGranahan, N.; Furness, A.J.S.; Rosenthal, R.; Ramskov, S.; Lyngaa, R.; Saini, S.K.; Jamal-Hanjani, M.; Wilson, G.A.; Birkbak, N.J.; Hiley, C.T.; et al. Clonal neoantigens elicit T cell immunoreactivity and sensitivity to immune checkpoint blockade. *Science* **2016**, *351*, 1463–1469. [CrossRef] [PubMed]
116. Mouw, K.W.; Goldberg, M.S.; Konstantinopoulos, P.A.; D'Andrea, A.D. DNA Damage and Repair Biomarkers of Immunotherapy Response. *Cancer Discov.* **2017**, *7*, 675–693. [CrossRef]
117. Germano, G.; Lamba, S.; Rospo, G.; Barault, L.; Magrì, A.; Maione, F.; Russo, M.; Crisafulli, G.; Bartolini, A.; Lerda, G.; et al. Inactivation of DNA repair triggers neoantigen generation and impairs tumour growth. *Nature* **2017**, *552*, 116–120. [CrossRef]
118. Brown, J.S.; Sundar, R.; Lopez, J. Combining DNA damaging therapeutics with immunotherapy: More haste, less speed. *Br. J. Cancer* **2018**, *118*, 312–324. [CrossRef]

119. Chalmers, Z.R.; Connelly, C.F.; Fabrizio, D.; Gay, L.; Ali, S.M.; Ennis, R.; Schrock, A.; Campbell, B.; Shlien, A.; Chmielecki, J.; et al. Analysis of 100,000 human cancer genomes reveals the landscape of tumor mutational burden. *Genome. Med.* **2017**, *9*, 34. [CrossRef]
120. Pilié, P.G.; Gay, C.M.; Byers, L.A.; O'Connor, M.J.; Yap, T.A. PARP Inhibitors: Extending Benefit Beyond *BRCA*. -Mutant Cancers. *Clin. Cancer Res.* **2019**, *25*, 3759–3771. [CrossRef]
121. Chen, Q.; Sun, L.; Chen, Z.J. Regulation and function of the cGAS-STING pathway of cytosolic DNA sensing. *Nat. Immunol.* **2016**, *17*, 1142–1149. [CrossRef]
122. Ablasser, A.; Goldeck, M.; Cavlar, T.; Deimling, T.; Witte, G.; Röhl, I.; Hopfner, K.-P.; Ludwig, J.; Hornung, V. cGAS produces a 2′-5′-linked cyclic dinucleotide second messenger that activates STING. *Nature* **2013**, *498*, 380–384. [CrossRef] [PubMed]
123. Corrales, L.; McWhirter, S.M.; Dubensky, T.W.; Gajewski, T.F. The host STING pathway at the interface of cancer and immunity. *J. Clin. Investig.* **2016**, *126*, 2404–2411. [CrossRef] [PubMed]
124. Ma, Z.; Damania, B. The cGAS-STING Defense Pathway and Its Counteraction by Viruses. *Cell Host. Microbe* **2016**, *19*, 150–158. [CrossRef]
125. Zitvogel, L.; Galluzzi, L.; Kepp, O.; Smyth, M.J.; Kroemer, G. Type I interferons in anticancer immunity. *Nat. Rev. Immunol.* **2015**, *15*, 405–414. [CrossRef] [PubMed]
126. Strickland, K.C.; Howitt, B.E.; Shukla, S.A.; Rodig, S.; Ritterhouse, L.L.; Liu, J.F.; Garber, J.E.; Chowdhury, D.; Wu, C.J.; D'Andrea, A.D.; et al. Association and prognostic significance of BRCA1/2-mutation status with neoantigen load, number of tumor-infiltrating lymphocytes and expression of PD-1/PD-L1 in high grade serous ovarian cancer. *Oncotarget* **2016**, *7*, 13587–13598. [CrossRef]
127. Rieke, D.T.; Ochsenreither, S.; Klinghammer, K.; Seiwert, T.Y.; Klauschen, F.; Tinhofer, I.; Keilholz, U. Methylation of RAD51B, XRCC3 and other homologous recombination genes is associated with expression of immune checkpoints and an inflammatory signature in squamous cell carcinoma of the head and neck, lung and cervix. *Oncotarget* **2016**, *7*, 75379–75393. [CrossRef]
128. Green, A.R.; Aleskandarany, M.A.; Ali, R.; Hodgson, E.G.; Atabani, S.; De Souza, K.; Rakha, E.A.; Ellis, I.O.; Madhusudan, S. Clinical Impact of Tumor DNA Repair Expression and T-cell Infiltration in Breast Cancers. *Cancer Immunol. Res.* **2017**, *5*, 292–299. [CrossRef]
129. Nolan, E.; Savas, P.; Policheni, A.N.; Darcy, P.K.; Vaillant, F.; Mintoff, C.P.; Dushyanthen, S.; Mansour, M.; Pang, J.-M.B.; Fox, S.B.; et al. Combined immune checkpoint blockade as a therapeutic strategy for BRCA1-mutated breast cancer. *Sci. Transl. Med.* **2017**, *9*. [CrossRef]
130. Connor, A.A.; Denroche, R.E.; Jang, G.H.; Timms, L.; Kalimuthu, S.N.; Selander, I.; McPherson, T.; Wilson, G.W.; Chan-Seng-Yue, M.A.; Borozan, I.; et al. Association of Distinct Mutational Signatures With Correlates of Increased Immune Activity in Pancreatic Ductal Adenocarcinoma. *JAMA Oncol.* **2017**, *3*, 774–783. [CrossRef]
131. Härtlova, A.; Erttmann, S.F.; Raffi, F.A.; Schmalz, A.M.; Resch, U.; Anugula, S.; Lienenklaus, S.; Nilsson, L.M.; Kröger, A.; Nilsson, J.A.; et al. DNA damage primes the type I interferon system via the cytosolic DNA sensor STING to promote anti-microbial innate immunity. *Immunity* **2015**, *42*, 332–343. [CrossRef]
132. Pantelidou, C.; Sonzogni, O.; De Oliveria Taveira, M.; Mehta, A.K.; Kothari, A.; Wang, D.; Visal, T.; Li, M.K.; Pinto, J.; Castrillon, J.A.; et al. PARP Inhibitor Efficacy Depends on CD8+ T-cell Recruitment via Intratumoral STING Pathway Activation in BRCA-Deficient Models of Triple-Negative Breast Cancer. *Cancer Discov.* **2019**, *9*, 722–737. [CrossRef]
133. Sen, T.; Rodriguez, B.L.; Chen, L.; Corte, C.M.D.; Morikawa, N.; Fujimoto, J.; Cristea, S.; Nguyen, T.; Diao, L.; Li, L.; et al. Targeting DNA Damage Response Promotes Antitumor Immunity through STING-Mediated T-cell Activation in Small Cell Lung Cancer. *Cancer Discov.* **2019**, *9*, 646–661. [CrossRef]
134. Shen, J.; Zhao, W.; Ju, Z.; Wang, L.; Peng, Y.; Labrie, M.; Yap, T.A.; Mills, G.B.; Peng, G. PARPi Triggers the STING-Dependent Immune Response and Enhances the Therapeutic Efficacy of Immune Checkpoint Blockade Independent of BRCAness. *Cancer Res.* **2019**, *79*, 311–319. [CrossRef]
135. Topalian, S.L.; Taube, J.M.; Anders, R.A.; Pardoll, D.M. Mechanism-driven biomarkers to guide immune checkpoint blockade in cancer therapy. *Nat. Rev. Cancer* **2016**, *16*, 275–287. [CrossRef]
136. Nishino, M.; Ramaiya, N.H.; Hatabu, H.; Hodi, F.S. Monitoring immune-checkpoint blockade: Response evaluation and biomarker development. *Nat. Rev. Clin. Oncol.* **2017**, *14*, 655–668. [CrossRef]
137. Gao, Y.; Yang, J.; Cai, Y.; Fu, S.; Zhang, N.; Fu, X.; Li, L. IFN-γ-mediated inhibition of lung cancer correlates with PD-L1 expression and is regulated by PI3K-AKT signaling. *Int. J. Cancer* **2018**, *143*, 931–943. [CrossRef] [PubMed]

138. Gottlieb, C.E.; Mills, A.M.; Cross, J.V.; Ring, K.L. Tumor-associated macrophage expression of PD-L1 in implants of high grade serous ovarian carcinoma: A comparison of matched primary and metastatic tumors. *Gynecol. Oncol.* **2017**, *144*, 607–612. [CrossRef]
139. Li, C.-W.; Lim, S.-O.; Xia, W.; Lee, H.-H.; Chan, L.-C.; Kuo, C.-W.; Khoo, K.-H.; Chang, S.-S.; Cha, J.-H.; Kim, T.; et al. Glycosylation and stabilization of programmed death ligand-1 suppresses T-cell activity. *Nat. Commun.* **2016**, *7*, 1–11. [CrossRef]
140. Sato, H.; Niimi, A.; Yasuhara, T.; Permata, T.B.M.; Hagiwara, Y.; Isono, M.; Nuryadi, E.; Sekine, R.; Oike, T.; Kakoti, S.; et al. DNA double-strand break repair pathway regulates PD-L1 expression in cancer cells. *Nat. Commun.* **2017**, *8*. [CrossRef]
141. Crusz, S.M.; Balkwill, F.R. Inflammation and cancer: Advances and new agents. *Nat. Rev. Clin. Oncol.* **2015**, *12*, 584–596. [CrossRef] [PubMed]
142. Fridman, W.H.; Zitvogel, L.; Sautès–Fridman, C.; Kroemer, G. The immune contexture in cancer prognosis and treatment. *Nat. Rev. Clin. Oncol.* **2017**, *14*, 717–734. [CrossRef] [PubMed]
143. Yélamos, J.; Moreno-Lama, L.; Jimeno, J.; Ali, S.O. Immunomodulatory Roles of PARP-1 and PARP-2: Impact on PARP-Centered Cancer Therapies. *Cancers* **2020**, *12*, 392. [CrossRef] [PubMed]
144. LaFargue, C.J.; Dal Molin, G.Z.; Sood, A.K.; Coleman, R.L. Exploring and comparing adverse events between PARP inhibitors. *Lancet Oncol.* **2019**, *20*, e15–e28. [CrossRef]
145. Ramos-Casals, M.; Brahmer, J.R.; Callahan, M.K.; Flores-Chávez, A.; Keegan, N.; Khamashta, M.A.; Lambotte, O.; Mariette, X.; Prat, A.; Suárez-Almazor, M.E. Immune-related adverse events of checkpoint inhibitors. *Nat. Rev. Dis. Primers* **2020**, *6*, 38. [CrossRef]
146. Higuchi, T.; Flies, D.B.; Marjon, N.A.; Mantia-Smaldone, G.; Ronner, L.; Gimotty, P.A.; Adams, S.F. CTLA-4 Blockade Synergizes Therapeutically with PARP Inhibition in BRCA1-Deficient Ovarian Cancer. *Cancer Immunol. Res.* **2015**, *3*, 1257–1268. [CrossRef]
147. Ding, L.; Kim, H.-J.; Wang, Q.; Kearns, M.; Jiang, T.; Ohlson, C.E.; Li, B.B.; Xie, S.; Liu, J.F.; Stover, E.H.; et al. PARP Inhibition Elicits STING-Dependent Antitumor Immunity in Brca1-Deficient Ovarian Cancer. *Cell Rep.* **2018**, *25*, 2972–2980.e5. [CrossRef]
148. Wang, Z.; Sun, K.; Xiao, Y.; Feng, B.; Mikule, K.; Ma, X.; Feng, N.; Vellano, C.P.; Federico, L.; Marszalek, J.R.; et al. Niraparib activates interferon signaling and potentiates anti-PD-1 antibody efficacy in tumor models. *Sci. Rep.* **2019**, *9*, 1–12. [CrossRef]
149. Karzai, F.; Madan, R.A.; Owens, H.; Couvillon, A.; Hankin, A.; Williams, M.; Bilusic, M.; Cordes, L.M.; Trepel, J.B.; Killian, K.; et al. A phase 2 study of olaparib and durvalumab in metastatic castrate-resistant prostate cancer (mCRPC) in an unselected population. *JCO* **2018**, *36*, 163. [CrossRef]
150. Karzai, F.; VanderWeele, D.; Madan, R.A.; Owens, H.; Cordes, L.M.; Hankin, A.; Couvillon, A.; Nichols, E.; Bilusic, M.; Beshiri, M.L.; et al. Activity of durvalumab plus olaparib in metastatic castration-resistant prostate cancer in men with and without DNA damage repair mutations. *J. Immunother. Cancer* **2018**, *6*, 141. [CrossRef]
151. Lee, J.-M.; Annunziata, C.M.; Houston, N.; Kohn, E.C.; Lipkowitz, S.; Minasian, L.; Nichols, E.; Trepel, J.; Trewhitt, K.; Zia, F.; et al. A phase II study of durvalumab, a PD-L1 inhibitor and olaparib in recurrent ovarian cancer (OvCa). *Ann. Oncol.* **2018**, *29*, viii334. [CrossRef]
152. Lampert, E.J.; Zimmer, A.S.; Padget, M.R.; Cimino-Mathews, A.; Nair, J.R.; Liu, Y.; Swisher, E.M.; Hodge, J.W.; Nixon, A.B.; Nichols, E.; et al. Combination of PARP inhibitor olaparib, and PD-L1 inhibitor durvalumab, in recurrent ovarian cancer: A proof-of-concept phase 2 study. *Clin. Cancer Res.* **2020**. [CrossRef]
153. Thomas, A.; Vilimas, R.; Trindade, C.; Erwin-Cohen, R.; Roper, N.; Xi, L.; Krishnasamy, V.; Levy, E.; Mammen, A.; Nichols, S.; et al. Durvalumab in Combination with Olaparib in Patients with Relapsed SCLC: Results from a Phase II Study. *J. Thorac. Oncol.* **2019**, *14*, 1447–1457. [CrossRef]
154. Drew, Y.; de Jonge, M.; Hong, S.H.; Park, Y.H.; Wolfer, A.; Brown, J.; Ferguson, M.; Gore, M.E.; Alvarez, R.H.; Gresty, C.; et al. An open-label, phase II basket study of olaparib and durvalumab (MEDIOLA): Results in germline BRCA-mutated (gBRCAm) platinum-sensitive relapsed (PSR) ovarian cancer (OC). *Gynecol. Oncol.* **2018**, *149*, 246–247. [CrossRef]
155. Domchek, S.; Postel-Vinay, S.; Im, S.-A.; Park, Y.H.; Delord, J.-P.; Italiano, A.; Alexandre, J.; You, B.; Bastian, S.; Krebs, M.G.; et al. Phase II study of olaparib (O) and durvalumab (D) (MEDIOLA): Updated results in patients (pts) with germline BRCA-mutated (gBRCAm) metastatic breast cancer (MBC). *Ann. Oncol.* **2019**, *30*, v477. [CrossRef]

156. Bang, Y.-J.; Kaufman, B.; Geva, R.; Stemmer, S.M.; Hong, S.-H.; Lee, J.-S.; Domchek, S.M.; Lanasa, M.C.; Tang, M.; Gresty, C.; et al. An open-label, phase II basket study of olaparib and durvalumab (MEDIOLA): Results in patients with relapsed gastric cancer. *J. Clin. Oncol* **2019**. [CrossRef]
157. Konstantinopoulos, P.A.; Waggoner, S.; Vidal, G.A.; Mita, M.; Moroney, J.W.; Holloway, R.; Van Le, L.; Sachdev, J.C.; Chapman-Davis, E.; Colon-Otero, G.; et al. Single-Arm Phases 1 and 2 Trial of Niraparib in Combination with Pembrolizumab in Patients with Recurrent Platinum-Resistant Ovarian Carcinoma. *JAMA Oncol.* **2019**, *5*, 1141. [CrossRef]
158. Vinayak, S.; Tolaney, S.M.; Schwartzberg, L.; Mita, M.; McCann, G.; Tan, A.R.; Wahner-Hendrickson, A.E.; Forero, A.; Anders, C.; Wulf, G.M.; et al. Open-label Clinical Trial of Niraparib Combined with Pembrolizumab for Treatment of Advanced or Metastatic Triple-Negative Breast Cancer. *JAMA Oncol.* **2019**, *5*, 1132. [CrossRef]
159. Friedlander, M.; Meniawy, T.; Markman, B.; Mileshkin, L.; Harnett, P.; Millward, M.; Lundy, J.; Freimund, A.; Norris, C.; Mu, S.; et al. Pamiparib in combination with tislelizumab in patients with advanced solid tumours: Results from the dose-escalation stage of a multicentre, open-label, phase 1a/b trial. *Lancet Oncol.* **2019**, *20*, 1306–1315. [CrossRef]
160. Yap, T.A.; Konstantinopoulos, P.; Telli, M.L.; Saraykar, S.; Beck, J.T.; Galsky, M.D.; Abraham, J.; Wise, D.R.; Khasraw, M.; Rubovszky, G.; et al. Abstract P1-19-03: JAVELIN PARP Medley, a phase 1b/2 study of avelumab plus talazoparib: Results from advanced breast cancer cohorts. *Cancer Res.* **2020**, *80*, P1-P1-19–03. [CrossRef]
161. Rodriguez-Moreno, J.F.; de Velasco, G.; Bravo Fernandez, I.; Alvarez-Fernandez, C.; Fernandez, R.; Vazquez-Estevez, S.; Virizuela, J.A.; Gajate, P.; Font, A.; Lainez, N.; et al. Impact of the combination of durvalumab (MEDI4736) plus olaparib (AZD2281) administered prior to surgery in the molecular profile of resectable urothelial bladder cancer: NEODURVARIB Trial. *JCO* **2020**, *38*, 542. [CrossRef]
162. Clarke, B.; Tinker, A.V.; Lee, C.-H.; Subramanian, S.; van de Rijn, M.; Turbin, D.; Kalloger, S.; Han, G.; Ceballos, K.; Cadungog, M.G.; et al. Intraepithelial T cells and prognosis in ovarian carcinoma: Novel associations with stage, tumor type, and BRCA1 loss. *Mod. Pathol.* **2009**, *22*, 393–402. [CrossRef]
163. McAlpine, J.N.; Porter, H.; Köbel, M.; Nelson, B.H.; Prentice, L.M.; Kalloger, S.E.; Senz, J.; Milne, K.; Ding, J.; Shah, S.P.; et al. BRCA1 and BRCA2 mutations correlate with TP53 abnormalities and presence of immune cell infiltrates in ovarian high-grade serous carcinoma. *Mod. Pathol.* **2012**, *25*, 740–750. [CrossRef]
164. Wen, W.X.; Leong, C.-O. Association of BRCA1- and BRCA2-deficiency with mutation burden, expression of PD-L1/PD-1, immune infiltrates, and T cell-inflamed signature in breast cancer. *PLoS ONE* **2019**, *14*, e0215381. [CrossRef]
165. Adams, S.F.; Rixe, O.; Lee, J.-H.; McCance, D.J.; Westgate, S.; Eberhardt, S.C.; Rutledge, T.; Muller, C. Phase I study combining olaparib and tremelimumab for the treatment of women with BRCA-deficient recurrent ovarian cancer. *JCO* **2017**, *35*, e17052. [CrossRef]
166. Pilié, P.G.; Tang, C.; Mills, G.B.; Yap, T.A. State-of-the-art strategies for targeting the DNA damage response in cancer. *Nat. Rev. Clin. Oncol.* **2019**, *16*, 81–104. [CrossRef]
167. Smith, J.; Tho, L.M.; Xu, N.; Gillespie, D.A. The ATM–Chk2 and ATR–Chk1 Pathways in DNA Damage Signaling and Cancer. In *Advances in Cancer Research*; Elsevier: Amsterdam, The Netherland, 2010; Volume 108, pp. 73–112. ISBN 978-0-12-380888-2.
168. Blackford, A.N.; Jackson, S.P. ATM, ATR, and DNA-PK: The Trinity at the Heart of the DNA Damage Response. *Mol. Cell* **2017**, *66*, 801–817. [CrossRef]
169. Do, K.; Doroshow, J.H.; Kummar, S. Wee1 kinase as a target for cancer therapy. *Cell Cycle* **2013**, *12*, 3348–3353. [CrossRef]
170. Alimzhanov, M.; Soulard, P.; Zimmermann, A.; Schroeder, A.; Mehr, K.T.; Amendt, C.; Sim, G.C.; Blaukat, A.; Halle, J.-P.; Zenke, F.T. Abstract 2269: ATR inhibitor M6620 enhances anti-tumor efficacy of the combination of the anti-PD-L1 antibody avelumab with platinum-based chemotherapy. *Cancer Res.* **2019**, *79*, 2269. [CrossRef]
171. Zhang, Q.; Green, M.D.; Lang, X.; Lazarus, J.; Parsels, J.D.; Wei, S.; Parsels, L.A.; Shi, J.; Ramnath, N.; Wahl, D.R.; et al. Inhibition of ATM Increases Interferon Signaling and Sensitizes Pancreatic Cancer to Immune Checkpoint Blockade Therapy. *Cancer Res.* **2019**, *79*, 3940–3951. [CrossRef]

172. Carr, M.; Zimmermann, A.; Guo, Y.; Liu, X.; Steiner, P.; Hahn, S.; Zenke, F.; Blaukat, A.; Vassilev, L.T. Abstract 2923: DNA-PK inhibitor, M3814, is a potent inducer of inflammatory micronucleation in irradiated p53-deficient cancer cells: Implications for combination radio-immunotherapy. *Cancer Res.* **2019**, *79*, 2923. [CrossRef]
173. Yap, T.A.; Krebs, M.G.; Postel-Vinay, S.; Bang, Y.J.; El-Khoueiry, A.; Abida, W.; Harrington, K.; Sundar, R.; Carter, L.; Castanon-Alvarez, E.; et al. Phase I modular study of AZD6738, a novel oral, potent and selective ataxia telangiectasia Rad3-related (ATR) inhibitor in combination (combo) with carboplatin, olaparib or durvalumab in patients (pts) with advanced cancers. *Eur. J. Cancer* **2016**, *69*, S2. [CrossRef]
174. Powles, T.; Kilgour, E.; Mather, R.; Galer, A.; Arkenau, H.-T.; Farnsworth, A.; Wilde, J.; Ratnayake, J.; Landers, D. BISCAY, a phase Ib, biomarker-directed multidrug umbrella study in patients with metastatic bladder cancer. *JCO* **2016**, *34*, TPS4577. [CrossRef]
175. Yap, T.A.; Plummer, R.; Azad, N.S.; Helleday, T. The DNA Damaging Revolution: PARP Inhibitors and Beyond. In *American Society of Clinical Oncology Educational Book*; ASCO: Alexandria, VA, USA, 2019; pp. 185–195. [CrossRef]
176. Vaddepally, R.K.; Kharel, P.; Pandey, R.; Garje, R.; Chandra, A.B. Review of Indications of FDA-Approved Immune Checkpoint Inhibitors per NCCN Guidelines with the Level of Evidence. *Cancers* **2020**, *12*, 738. [CrossRef]

cancers

MDPI

Review

Therapeutic Strategies and Biomarkers to Modulate PARP Activity for Targeted Cancer Therapy

Naveen Singh [1], S. Louise Pay [1], Snehal B. Bhandare [1], Udhaya Arimpur [1] and Edward A. Motea [1,2,*]

1 School of Medicine, Department of Biochemistry and Molecular Biology, Indiana University, Indianapolis, IN 46202, USA; navsing@iu.edu (N.S.); slpay@iupui.edu (S.L.P); snebhan@iu.edu (S.B.B.); uarimpur@iu.edu (U.A.)

2 Simon Comprehensive Cancer Center, Indiana University, Indianapolis, IN 46202, USA

* Correspondence: eamotea@iu.edu

Received: 18 February 2020; Accepted: 7 April 2020; Published: 14 April 2020

Abstract: Poly-(ADP-ribose) polymerase 1 (PARP1) is commonly known for its vital role in DNA damage response and repair. However, its enzymatic activity has been linked to a plethora of physiological and pathophysiological transactions ranging from cellular proliferation, survival and death. For instance, malignancies with *BRCA1/2* mutations heavily rely on PARP activity for survival. Thus, the use of PARP inhibitors is a well-established intervention in these types of tumors. However, recent studies indicate that the therapeutic potential of attenuating PARP1 activity in recalcitrant tumors, especially where PARP1 is aberrantly overexpressed and hyperactivated, may extend its therapeutic utility in wider cancer types beyond BRCA-deficiency. Here, we discuss treatment strategies to expand the tumor-selective therapeutic application of PARP inhibitors and novel approaches with predictive biomarkers to perturb NAD^+ levels and hyperPARylation that inactivate PARP in recalcitrant tumors. We also provide an overview of genetic alterations that transform non-BRCA mutant cancers to a state of "BRCAness" as potential biomarkers for synthetic lethality with PARP inhibitors. Finally, we discuss a paradigm shift for the use of novel PARP inhibitors outside of cancer treatment, where it has the potential to rescue normal cells from severe oxidative damage during ischemia-reperfusion injury induced by surgery and radiotherapy.

Keywords: PARP Inhibitors; beta-lapachone; NQO1; PARG; NAMPT; cancer therapeutics; DNA repair; cMET

---

1. Introduction

Maintenance of genomic integrity is vital to achieve normal cellular function and to prevent the development of diseases such as cancer [1]. At the heart of this intricate biological process are DNA repair factors that work harmoniously to scan, detect, and repair potentially deleterious damage to cellular genetic information. Indeed, the disruption of one or more DNA repair pathways compromises genetic stability and is a known mechanism of cancer initiation, development, and progression. Poly-(ADP-ribose) polymerase 1 (PARP1) is the classical and founding member of at least 17 human PARP enzymes that share the ability to catalyze the transfer of ADP-ribose units to target proteins to modulate chromatin structure, transcription, replication, DNA damage response and repair [2]. PARP1 is an abundant nuclear protein that acts as a DNA damage sensor and a facilitator of DNA repair pathway choice in response to cellular stress [3–6]. Specifically, it is involved in the repair of single-stranded DNA breaks (SSB) via the base-excision repair (BER) pathway. In BER, PARP1 functions to recruit other repair factors by binding to single-stranded DNA break intermediates. Additionally, it catalyzes the synthesis of poly(ADP)-ribose (PAR) chains to the acceptor proteins (e.g., histones and XRCC1), including itself, using nicotinamide adenine dinucleotide (NAD^+) as the substrate and source of energy. The PARylated proteins then recruit and retain critical processing factors to the site of the lesion to facilitate the efficient repair of the SSB (Figure 1A) [7].

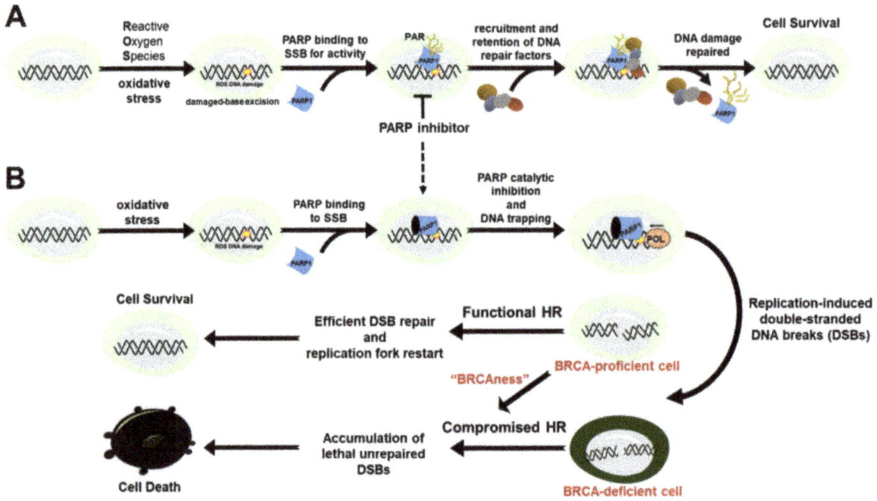

Figure 1. The role of PARP1 in DNA damage response and repair and cancer therapy. (**A**) PARP1 binds to single-strand breaks (SSB) for activity to target, recruit and retain critical DNA repair proteins at the sites of DNA lesions. (**B**) PARP inhibitors convert SSBs to lethal double-stranded breaks (DSBs) that are left unrepaired in BRCA-deficient cells due to a compromised homologous recombination (HR) repair consequently leading to cell death.

In the absence of PARP1, unrepaired SSBs are converted to double-stranded breaks (DSBs) during replication or the S-phase of the cell cycle [8]. Double-stranded DNA breaks are one of the most lethal forms of DNA damage induced by exogenous DNA damaging agents (e.g., ionizing radiation (IR) and chemotherapeutic agents) or endogenous replicative stress in fast proliferating cancer cells. Thus, accurate repair of DSBs is paramount to the growth and survival of all cells. Indeed, cells have evolved a range of DSB repair mechanisms. The two major repair pathways that have been studied extensively are homologous recombination (HR) and non-homologous end-joining (NHEJ) [9]. HR is largely error-free and more prevalently activated during and after DNA duplication when an identical chromatid is accessible as a template for repair. In contrast, NHEJ is active throughout the cell cycle and promotes direct ligation of DSB ends at the cost of small insertions, deletions, substitutions at the break, and even translocations that arise if DSBs from different parts of the genome are combined [10]. Cells utilize two mechanistically distinct end-joining pathways to process DNA DSBs [10–12]: Classical-NHEJ (c-NHEJ) leads to a minimal sequence alteration at the repair junctions, whereas alternative-NHEJ (alt-NHEJ, also known as microhomology-mediated end-joining (MMEJ) or back-up NHEJ) causes extensive genetic changes (deletions and insertions) that scar the break sites following ligation of DSB ends.

Several studies have shown that loss-of-function mutation of canonical HR factors – such as breast cancer type 1 and 2 (BRCA1/2) susceptibility proteins that are commonly associated with breast and ovarian cancer [13–15] – promotes PARP1 hyperactivation in fast replicating cancer cells [16,17]. This suggests that hyperactivation of PARP1 is essential to facilitate the repair of potentially lethal DNA breaks for the survival of HR-defective (HRD) cancers. Indeed, *BRCA1/2* mutant cancers are selectively killed by PARP inhibitors (Figure 1B), which has led to the approval of these agents to treat HR-deficient ovarian and breast cancers. However, several studies have reported that certain cancers without a BRCA deficiency have significant clinical benefits to PARP1 inhibitors positing that PARP inhibitors could be expanded to a target population beyond BRCA-deficiency (e.g., *gBRCA* mutation carriers) [18–21].

This review article highlights treatment strategies to selectively target BRCA-proficient cancers by modulating PARP activity that alters PARP binding, PAR and NAD^+ levels to induce tumor-selective cell death using predictive biomarkers for therapeutic response. We summarize how aberrant alterations of PARG, NAMPT, NQO1, cMET and "BRCAness" genes that have been shown to affect PARP activity in cancers could serve as prognostic biomarkers for targeted therapy. Finally, we briefly discuss innovative approaches for the use of novel PARP inhibitors to rescue injured normal cells from severe oxidative damage during ischemia-reperfusion injury that could be induced by surgery and radiotherapy.

2. Mechanism of Action for PARP Inhibitors

PARP inhibitors are designed to mimic the substrate-protein interactions of NAD^+ within the ADP-ribose transferase (ART) catalytic core of PARP1-3, which are key DNA damage response sensors and transducers. The inhibitors compete with NAD^+ binding site of PARP to inhibit PAR polymerization, which then hinders the recruitment and regulation of DNA repair factors and the eventual release of PARP from DNA damage. Two mechanisms are proposed to induce the lethality of PARP inhibitors: PARP catalytic inhibition and PARP trapping [22,23]. However, the relative contributions of these two pathways in mediating the lethality of PARP inhibitors remain enigmatic. For example, there is evidence suggesting that the differences in the trapping potential of PARP1 to the DNA are more efficient at killing HR-deficient cells [22,23]. While there has been no evidence that this mechanism exists with clinically-used PARP inhibitors, a recent study has demonstrated that certain non-hydrolysable NAD^+ analogs (e.g., benzamide adenine dinucleotide, BAD) binding at the catalytic site of PARP1 could greatly enhance the binding affinity of PARP1 at sites of DNA damage that prevents its release [24]. Mechanistically, this analog competes with NAD^+ at the catalytic binding site that then stabilizes a conformation that induces the DNA binding domain of PARP1 to be locked on a DNA break with a significantly better binding affinity (~10-fold) – a phenomenon known as "reverse allostery." While most PARP inhibitors (e.g., Rucaparib) bind at the catalytic site, not all of them are "created equal." In fact, the relative contribution of DNA trapping induced by PARP inhibitors contributes significantly to the toxicity induced by these agents in cancers that are deficient in HR pathways [23,25–27]. Some of them have variable inhibitory effects on the PARP isoforms other than PARP1, and some may have more off-target effects on kinases than others. Regardless, PARP inhibitors are now being explored for use in several cancers, even in cancers beyond *BRCA* mutations [25,28–30]. Interestingly, a recent study has reported a correlation between therapeutic response to PARP inhibition and the patterns of ADP-ribosylation (i.e., amount of PARylation) in a panel of ovarian cancers, suggesting that ADP-ribosylation may be a useful biomarker for HR deficiency and sensitivity to PARP inhibitors regardless of BRCA status [31]. While most of the PARP inhibitors have been suggested to specifically inhibit PARP1 and/or PARP2, our limited understanding of the overlapping and non-overlapping functions and cross-talks among all of the PARP family members still raises the problem of target specificity and the risk of unintended effects and consequences promoted by the targeting of other PARP family members.

3. Approved PARP Inhibitors for Targeted Cancer Therapy

PARP inhibitors such as Olaparib [32,33], Rucaparib [34], and Niraparib [35] have been approved by the Food and Drug Administration (FDA) and the European Medicine Agency (EMA) for the treatment of adults with deleterious or suspected deleterious germline *BRCA*-mutated ovarian cancer in different settings [36]. Moreover, these agents have also been approved for maintenance therapy in platinum-sensitive high-grade ovarian cancer (HGOC), which has a high recurrence rate and an extremely poor prognosis following relapse [37]. Currently, PARP inhibitors are being evaluated in clinical trials in other cancer settings such as nonsmall cell lung cancer, pancreatic cancer, and gastric cancer (ClinicalTrials.gov Identifiers NCT01082549, NCT02184195, NCT03427814, respectively).

Olaparib (LYNPARZA, AstraZeneca and Merck) was the first PARP inhibitor to receive FDA approval for indications in breast cancer and pancreatic cancer, and has since been approved for metastatic breast cancer in patients with germline *BRCA1* or *BRCA2* (*gBRCA1/2*) mutations. In 2009, Olaparib was deemed safer than conventional chemotherapeutics, with mild and reversible side effects during clinical trials [26]. It was subsequently determined to have anti-tumor activity in advanced ovarian cancer patients with *gBRCA1/2* mutations, particularly with primary tumors that are sensitive to platinum [18,32,38–40]. The phase III OlympiAD trial found a 60% reduction in metastatic *BRCA1/2* mutant breast tumor size with Olaparib in comparison to 29% in patients receiving conventional chemotherapy. The FDA approved Olaparib in ovarian and breast patients with *gBRCA* mutations in 2014 and 2018, respectively [27,41–43].

Rucaparib (RUBRACA, Clovis Oncology, Inc.) was granted accelerated approval by the FDA in 2016 for the treatment of adult ovarian cancer patients with germline and/or somatic *BRCA*-mutations previously treated with two or more prior lines of chemotherapy. This agent is also approved as a maintenance monotherapy for adult patients with recurrent ovarian cancer who have established complete or partial response (CR/PR) to platinum-based chemotherapy. In 2018, Rucaparib gained FDA approval for the maintenance treatment of advanced or recurrent ovarian, fallopian tube, and primary peritoneal cancer in patients who have received platinum-based chemotherapies, as well as in patients with germline or somatic *BRCA1/2* mutations following two or more treatments with conventional chemotherapy [44,45].

Niraparib (ZEJULA, Tesaro, Inc.) was approved by the FDA and EMA in 2017. Compared to Olaparib, this agent results in longer progression-free survival in patients with *BRCA1/2*-mutant tumors [35]. Niraparib also increases progression-free survival in patients with wild-type *BRCA1/2* ovarian cancers, though less effectively than in patients with *BRCA* mutations [35]. Niraparib, however, is associated with more severe side effects than other PARP inhibitors, with patients experiencing thrombocytopenia (33.8%), anemia (25.3%), and neutropenia (19.6%) [35]. In 2019, the FDA approved Niraparib for homologous recombination-deficient (HRD) patients with advanced ovarian, fallopian tube, or primary peritoneal cancer previously treated with three or more chemotherapies [46].

Talazoparib (TALZENNA, Pfizer, Inc.) is a new active PARP inhibitor approved by the FDA in 2018 for patients with deleterious or suspected deleterious *gBRCA*-mutated, HER2-negative local advanced or metastatic breast cancer [47]. Compared to other PARP inhibitors, Talazoparib is still at an early stage of clinical development in gathering evidence to support its use on the treatment of epithelial ovarian cancer.

4. Harnessing the Power of PAR and PARG Inhibition for Cancer Therapy

Poly(ADP)-ribosylation (PARylation) is a covalent and reversible posttranslational modification (PTM) of acceptor proteins catalyzed by PARPs, particularly in response to DNA damage and oxidative stress [48,49]. PARylation is also a critical signaling PTM for other biological transactions such as transcription, cell cycle regulation, and genome maintenance [50–52]. Intracellular PAR levels are very low in unstressed cells due to low enzymatic activity of PARP. When PARP1 is activated following DNA damage, PAR formation increases and consequently depletes intracellular NAD^+ and ATP levels [53]. Accumulation of PAR and loss of NAD^+ and ATP can lead to severe metabolic dysfunction and eventual cell death. Thus, the balance between ADP-ribose synthesis (i.e., PAR writers) and degradation (i.e., PAR erasers) is critical for the coordination of various cellular response pathways for survival [54].

Poly(ADP-ribose) glycohydrolase (PARG) is an endo- and exo-glycohydrolase that rapidly catalyzes the degradation of PAR polymerized by PARP1 to coordinate DNA repair [48,49]. It is comprised of an N-terminal regulatory domain required for recruitment to DNA damage sites [55], a C-terminal catalytic domain, and a central mitochondrial-targeting sequence [56]. PARG hydrolyzes the glycosidic bonds between ADP-ribose units producing free ADP-ribose and PAR oligomers (Figure 2A) [57], which play distinct biological roles in cellular processes. Free ADP-ribose is predominantly involved in energy catabolism, calcium signaling, and protein glycation [57]. The long-chain of free PAR formed by PARG cleavage has been shown to interact with the mitochondrial proteins triggering a unique intrinsic PAR-mediated cell death program known as PARthanatos (Figure 2A) [58,59]. Mechanistically, a cell commits to PARthanatos when free PAR (> 60 ADP-ribose units) migrates from the nucleus to the cytosol, which then triggers the translocation of PAR-bound apoptosis-inducing factor (AIF) from the mitochondria to the nucleus to activate the cell death process (Figure 2A) [60]. Early events in PAR-mediated cell death include loss of mitochondrial membrane potential and mitochondrial permeability transition [61]. While caspase activation has been demonstrated to act as a bystander in PAR-dependent cell death, this caspase-independent process shares cytological and morphological characteristics of both necrosis and apoptosis [58,59].

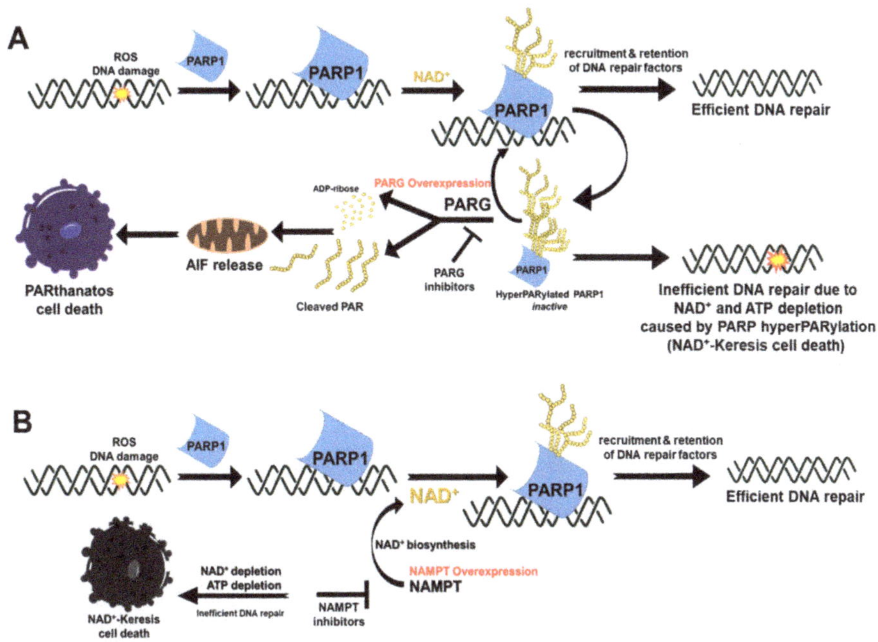

Figure 2. Pharmacological modulation of PAR ad NAD^+ in cancers for therapy. (**A**) Mechanism of PARthanatos mediated by the translocation of cleaved PAR from the nucleus to the cytosol and mitochondria to induce the release of AIF that translocates into the nucleus to initiate death in PARG-overexpressing cancers. Alternatively, breakdown and recycling of PAR can be prevented by inhibition of PARG to enhance NAD^+ depletion caused by PARP hyperactivation, ultimately starving the cell of ATP needed for various critical cellular processes (e.g., DNA repair). (**B**) In cancer cells that overexpress NAMPT, the use of NAMPT inhibitors interfere with generation of intracellular NAD^+ levels that compromise PARP activity, consequently resulting in impaired repair of DNA damage and NAD^+-Keresis.

ADP-ribose hydrolases (ARH) can catalyze PAR hydrolysis but to a lesser extent than PARG, which catalyzes the majority (~90%) of PAR catabolism [62]. PARG is aberrantly overexpressed in most human cancers, suggesting that PARG is required for tumorigenesis and cancer survival [63]. Therefore, inducing cell death via PARG inhibition and manipulation of the PAR cycle through PARP1/PARG inhibition is an attractive target for cancer therapy (Figure 2). Available PARG inhibitors consist of DNA intercalators, tannins, and substrate analogs. DNA intercalators (e.g., Tilorene, GPI116552, GPI18214, Ethacridine) indirectly inhibit PARG by forming complexes with its substrate, PAR [64–66]. Hydrolyzable tannins (e.g., Nobotanin B, Oenothenin B, gallotannin) inhibit

PARG in vitro by competing with the PAR substrate [67–69]; however, poor membrane permeability, poor bioavailability, and toxicity of these compounds hinder their usage. Adenosinediphosphate (hydroxymethyl) pyrrolidinediol (ADP-HPD), a nonhydrolyzable-analogue of ADP-ribose, has been shown to be a potent competitive PARG inhibitor [66]. Other reversible PARG inhibitors under development include modified salicylanilide pharmacophore [70] and rhodanine-based small molecule analogs (RBPIs) [70]. Recently, Pillay et al. [71] demonstrated that PARG inhibitor, PDD00017273, mimics poly(ADP-ribose) polymerase (PARP) inhibitor therapy in ovarian cancer cells by exacerbating the formation of hyperPARylated-PARP1 that severely compromises PARP activity. Mechanistically, PARG inhibitors prevent the recycling of hyperPARylated-PARP1 to its more active form for efficient DNA damage response and repair. In support of this, survival studies of certain human cancer cell lines with *PARG* knockdown synergistically enhances the lethality to DNA-damaging agents such as alkylating agents [72,73] and cisplatin [72]. Overall, pharmacological targeting of PARG is a promising therapeutic target and future investigations are required to develop effective strategies to manipulate the dynamic PAR formation and break-down process in response to DNA damage. The relative low abundance of PARG in normal cells, compared to tumors with aberrant PARG overexpression, makes it an attractive target in cancer therapeutics for tumor-selective drug response. Moreover, unlike the human PARP family with 17 related enzymes with almost similar catalytic subunit – PARG is a unique mammalian protein without any paralogs, which may offer fewer off-target effects than PARP inhibitors.

5. Targeting NAD^+ for Cancer Therapy

NAD^+ is a co-enzyme that mediates redox reactions by acting as an electron carrier and a substrate in metabolic pathways including glycolysis, the tricarboxylic acid cycle (TCA), oxidative phosphorylation, and serine biosynthesis [74]. It is a cofactor for PARP and plays a key role in energy production, cell signaling, redox homeostasis, DNA repair, gene expression, and the stress response. Many of these processes are disrupted in cancer, which alters NAD^+ production and consumption [75,76]. Nicotinamide phosphoribosyltransferase (NAMPT) and nicotinamide mononucleotide adenylyltransferase (NMNAT) enzymes regulate the salvage pathway critical for controlling intracellular NAD^+ levels [77]. Increased NAD^+ and aberrant NAMPT overexpression promotes glycolysis, which fuels the growth and survival of cancer cells [78,79]. Indeed, NAMPT inhibitors suppress cancer cell proliferation by inhibiting glycolysis [79–83].

Overexpression of NAMPT is observed in several types of malignant tumors, which supplies back-up NAD^+ to sustain cellular proliferation and promote resistance to therapeutic agents [84–89]. NAMPT as a predictive biomarker for tumor-selective targeting of NAD^+ metabolism is, therefore, a promising therapeutic strategy (Figure 2B). Mechanistically, NAMPT inhibitors induce cell death in cancer cells by depleting intracellular NAD^+ and ATP levels and inhibiting glycolysis and glucose uptake [90,91]. Examples of NAMPT inhibitors include FK866, GMX-1777/8, STF-31, STF-118804, GNE-617, GNE-618, LSN3154567, KPT-9274. Vacor adenine dinucleotide (VAD) is an NAD^+ analog that has been shown to inhibit both NAMPT and NMNAT, leading to necrotic cell death through NAD^+ depletion (NAD^+-Keresis, Figure 2B), glycolytic block, and energy failure [82,92–98]. Several NAMPT inhibitors have been investigated in the clinic as a monotherapy such as FK866 and GMX-1777/8. However, thrombocytopenia was demonstrated to be the dose-limiting toxicity in these clinical trials [99,100]. Though NAMPT inhibitors as a monotherapy are promising candidates for cancer therapy, drug resistance and toxicity to normal tissues when used at high concentration remain a major hurdle [99,101–103]. Perhaps, a combination approach with tumor-selective DNA damaging agents (e.g., NQO1-bioactivatable agents, vide infra in Section 6) may overcome this limitation to improve the tumor-selectivity of NAMPT inhibitors by severely depleting the NAD^+ needed by PARPs for efficient DNA repair and survival [82,104].

6. Leveraging NQO1 as a Biomarker for Tumor-Selective Use of PARP Inhibitors with NQO1-Bioactivatable Drugs

NAD(P)H:Quinone Oxidoreductase 1 (NQO1, DT-diaphorase) is a cytoplasmic Phase II detoxification enzyme that utilizes NAD(P)H to reduce certain quinones to stable hydroquinones via two-electron reductions that are readily conjugated by Phase III transporters for cellular efflux (Figure 3A) [105]. NQO1 is abnormally overexpressed in most malignant tumors, which contributes to drug resistance by metabolizing xenobiotics to their inactive forms or chemical structures that are excreted out of the cell for chemoprotection. Indeed, the addition of an NQO1-inhibitor (e.g., dicoumarol) in cancers could sensitize the anti-cancer effects of certain agents by inhibiting resistance due to drug efflux.

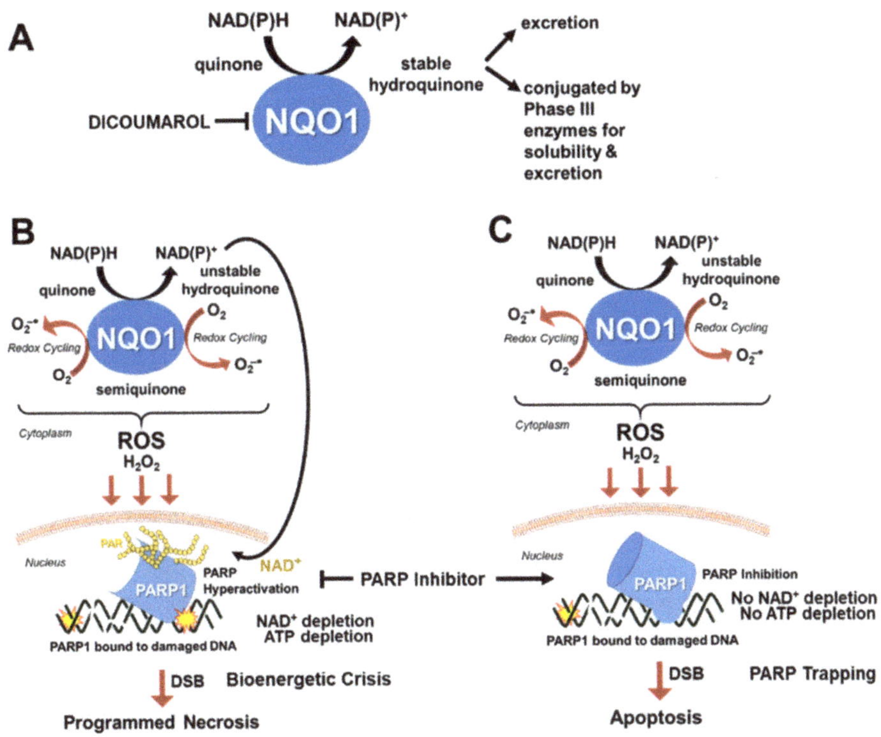

Figure 3. Strategy for tumor-selective use of PARP inhibitors in solid tumors. (**A**) The role of NQO1 in detoxifying certain quinones for cellular efflux. (**B**) The role of NQO1 in bioactivating certain quinones to induce toxicity in NQO1(+) cancer cells. Note that the stability of hydroquinones determine whether bioreduction by NQO1 leads to detoxification or toxicity. (**C**) Mechanism of tumor-selective synergistic cell death induced by combination of PARP inhibitor and NQO1-bioactivatable agents in NQO1(+) cells.

Certain quinones are bioactivated by NQO1 to create a rapid increase in reactive oxygen species (ROS), particularly hydrogen peroxide (H_2O_2), which can permeate through the nucleus to induce the formation of toxic DNA damage leading to cell death (Figure 3B) [106–110]. For example, β-lapachone (β-lap) is a soluble ortho-naphthoquinone with potent anti-tumor and radiosensitizing activity only in the presence of high NQO1 activity [108–114], which is only noted in most malignant tumors but not in normal tissues [109]. Mechanistically, the two-electron reductase capacity of NQO1 catalyzes the oxidoreduction of β-lap to an inherently unstable hydroquinone, which automatically and rapidly gets reverted back to its original quinone form as it undergoes a two-step oxidation (Figure 3B) [108]. This produces what is termed as a "futile cycle" of oxidation and produces a significant level of reactive oxygen species such as superoxide radicals (O_2^-) that are rapidly catalyzed by superoxide dismutases to hydrogen peroxide in the cytoplasm (Figure 3B) [115]. This stable form of ROS can easily penetrate through the nucleus and is converted to genotoxic hydroxyl radicals via the Fenton reaction process, which causes massive formation of oxidative DNA base damage and SSBs (Figure 3B) [109,116]. The surge of this specific types of DNA damage in the nucleus causes hyperactivation of PARP1, which is an NAD^+-dependent enzyme, to a level that consequently leads to NAD^+-Keresis death due to the depletion of critical biological energy sources, NAD^+ and ATP [82,107,109]. Another possibility is that PARP hyperactivation could lead to an NAD^+-independent glycolytic and bioenergetics crisis that is presumably caused by PAR-dependent inhibition of glycolysis that occurs through the inhibition of hexokinase leading to a specific form of cell death termed "PARthanathos" [117]. Future studies are required to firmly establish the contribution of PARP hyperactivation, NAD^+, or free PAR in specific types of cell death. Regardless, the use of NQO1-bioactivatable agents to modulate PARP activity, NAD^+ and PAR to induce cell death is a viable strategy to treat cancer. As NQO1 is typically overexpressed in tumors but not at a significant level in normal healthy cells [109], the application of NQO1-bioactivatable agents is an attractive possibility for tumor-selective generation of H_2O_2-induced DNA damages that hyperactivate PARP and deplete NAD^+ to cause eventual cell death, which is vital for developing safe and effective drugs with minimal off-target effects.

While β-lap (ARQ761 in clinical form) shows potential as a single-agent therapeutic [107,118], it has also been associated with hemolytic anemia at a dose of 450 mg/m^2 [119]. Therefore, increasing its efficacy at a lower dose is therapeutically advantageous. Due to the role of PARP hyperactivation in the mechanism of action of β-lap, PARP inhibitors have been a central focus in improving the efficacy and reducing the toxicity of β-lap for use in NQO1(+) cancers. PARP inhibition followed by β-lap treatment alters the mechanism of β-lap function by blocking the repair of damaged DNA and inhibiting PARP1 hyperactivation, slowing down the process of NAD^+ depletion as a result, and extending the cycling of β-lap by NQO1 to maximize hydrogen peroxide production and amplifying DNA damage in a tumor-selective manner (Figure 3C) [109]. Instead of dying by caspase-independent NAD^+-Keresis, cell death from exposure to β-lap following PARP1 inhibition occurs through caspase-dependent apoptosis, presumably due to the availability of NAD^+ and ATP (Figure 3C) [109]. As NQO1 will continue cycling β-lap until the drug is depleted or the required source of NAD(P)H is depleted, blocking the DNA repair process which leads to NAD^+ depletion following β-lap treatment prolongs the efficacy of the drug, allowing for lower doses to be used with the synergistic effect, minimizing the potential for hemolytic anemia.

The use of NQO1-bioactivatable drug (e.g., β-lap) with PARP inhibitors have been shown to significantly expand the clinical application of PARP inhibitors outside of *BRCA1/2* mutant cancers for which the majority of PARP inhibitors are currently approved. The prevalence of NQO1 overexpression in such a wide variety of malignant cancers (>90% in pancreatic cancers with KRAS mutation, >80% nonsmall cell lung cancer (NSCLC), >60% breast cancer, >60% prostate cancer, >45% head and neck, and >60% colon cancers [109]) suggests that this combination therapy may have significant future potential, especially in cancers in which prognosis is currently poor, such as head and neck cancer, pancreatic cancer and NSCLC [117,120–126].

7. Targeting cMET to Attenuate PARP1 Activity

Many studies have shown that most malignant tumors have increased levels of reactive oxygen species (ROS), particularly hydrogen peroxide (H_2O_2), compared to their normal counterparts. These genotoxic reactive species cause oxidative DNA damage and single-stranded DNA breaks that stimulate PARP activity in DNA repair [127]. Interestingly, Du and colleagues [128] have shown that cMET, which is a receptor tyrosine kinase, can further stimulate the activity of PARP1 in cancers to survive the lethal effects of ROS-induced DNA damage. In cancer, cMET is overexpressed and its abnormal activation can promote the development and progression of multiple cancers [129].

Mechanistically, the high level of ROS in cancer promotes the activation and translocation of cMET into the nucleus where it directly binds and phosphorylates the tyrosine residue 907 (Y907) of PARP1 to enhance its enzymatic activity (Figure 4A) [128,130,131]. Indeed, Du and colleagues have demonstrated that phosphorylated PARP1 (pY907) showed a higher level of PAR production than non-phosphorylated PARP1 in vitro, thereby highlighting the critical role of cMET in enhancing PARP1 activity in DNA repair [128,130,131]. Moreover, the phosphorylation of this specific residue (pY907) has been suggested to significantly decrease the binding of clinically-relevant PARP inhibitors due to a potential steric hindrance (Figure 4B) [128]. Indeed, c-MET-mediated phosphorylation of PARP1 at Y907 leads to PARP inhibitor resistance, which could be potentiated when combined with cMET inhibitors (e.g., crizotinib and foretinib) [128]. Thus, inhibition of cMET is an attractive strategy to indirectly block PARP1-mediated DNA repair and enhance the therapeutic effects of PARP inhibitors in cancer cells [128]. However, the mechanistic basis for requiring Y907 phosphorylation to promote H_2O_2-induced PARP1 activity remains to be firmly established. A recent report also suggests that phosphorylation of PARP1 by EGFR and cMET heterodimer contributes to PARP1 inhibitor resistance. Hence, combination treatment consisting of EGFR, cMET, and PARP1 inhibitors posits a novel therapeutic strategy in cancer treatment involving overexpression of PARP1/EGFR/cMET, which are frequent alterations in most solid tumors [132].

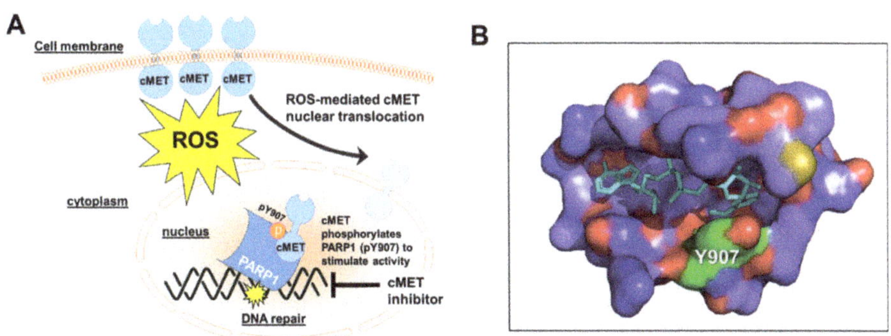

Figure 4. Attenuation of PARP Activity by cMET inhibition. (**A**) cMET enhances PARP activity during oxidative stress. Inhibition of cMET is an excellent strategy for inhibiting PARP1-mediated DNA repair and synergy with PARP inhibitors. (**B**) A snapshot of non-hydrolyzable NAD^+ analog (in cyan) bound to residues in PARP1 catalytic active site within 6Å from the ligand (PDB ID: 6 bhv, processed via PyMOL Molecular Graphics System). Phosphorylation of PARP1 residue Y907 (shown in green) inhibits binding of PARP inhibitors.

8. Targeting of PARP Activity in Non-Oncological Events

PARP inhibition or gene deletion has been shown to attenuate tissue injury associated with ischemia-reperfusion injury and inflammation that could arise during tumor surgical resection and radiation therapy [133,134]. During this non-oncological event, the catalytic activity of PARP1 becomes hyperactivated and consequently leads to NAD^+ depletion. This process forces NAD^+ replenishment through the salvage pathway, which decreases cellular ATP levels and results in bioenergetic crisis, eventually leading to necrotic cell death that is typically associated with inflammation [3,135]. Thus, inhibition of PARP activity could have protective effects by dramatically reducing NAD^+ consumption and preventing energetic failure and the consequent necrotic cell death. While the trapping mechanism induced by specific PARP inhibitors might be advantageous for the treatment of certain cancers, this mechanism of action might not be optimal to rescue normal cells from severe oxidative damage during ischemia-reperfusion injury (e.g., ischemia of the lung due to ionizing radiation [136] and cerebral ischemia during surgery [137]).

For non-oncological indications, the safety profile of most FDA-approved PARP inhibitors would be expected to be better if it had less PARP trapping activity to restore the viability of normal tissue damage from injury caused by ischemia-reperfusion. To circumvent the cytotoxic effects of PARP trapping, the Yu group [138] developed a strategy to decrease PARP activity using a novel lead compound, iRucaparib-AP6 (Figure 5), which promotes PARP degradation upon binding that mimics PARP1 genetic deletion. Mechanistically, the Rucaparib component of the small molecule specifically binds PARP at the site of damaged DNA, but it is attached to a linker with a ligand that brings the E3 ligase in close proximity to ubiquitinate PARP and subsequently gets degraded through the proteasome pathway (Figure 5C). Unlike PARP trapping, this strategy of modulating PARP activity protects cells against genotoxic stress-induced cell death [138] that could have potential clinical applications for the treatment of ischemia-reperfusion injury and neurodegeneration. However, further development is needed to firmly establish the mechanism of action and evaluate its protective effects to treat the aforementioned non-oncological indications.

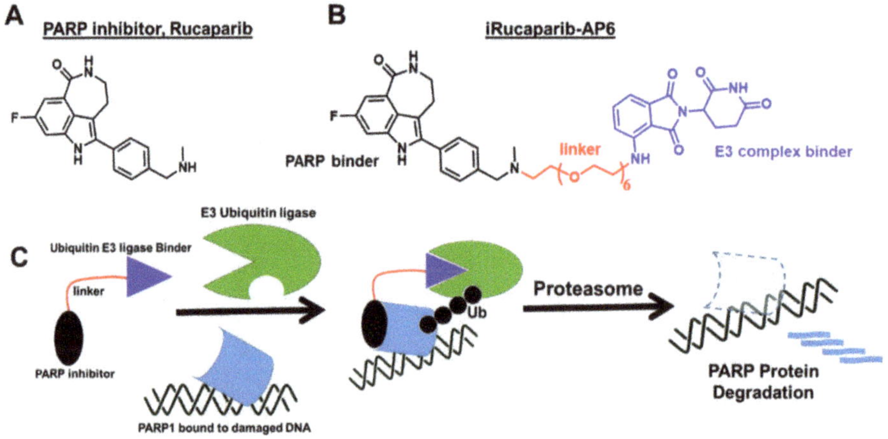

Figure 5. Development of a novel "PARP degrader" molecule as a biochemical probe and therapeutic agent. (**A**) Chemical structure of Rucaparib. (**B**) Chemical structure and elements of iRucaparib-AP6. (**C**) Simplified representation of the mechanism of action for a novel proteolysis targeting chimeric (PROTAC) molecule, iRucaparib-AP6.

To summarize, the enzymatic function of PARP1 – which is to catalyze the PARylation of a large number of PAR-acceptor proteins (including itself) – becomes hyperactivated upon detection of genotoxic stresses associated with various pathological conditions such as cancer, ischemia-reperfusion injury (e.g., myocardial infarction), inflammatory diseases (e.g., colitis, arthritis, asthma), vascular diseases (e.g., diabetic complications, atherosclerosis), and neurodegeneration. Thus, modulation of PARP activity using a novel class of PARP activity inhibitors (e.g., iRucaparib-AP6) with little to no "trapping" potential may be a promising strategy to ameliorate certain pathological conditions that are aimed to heal instead of killing the severely injured normal tissues (e.g., inflamed tissues). Indeed several studies have looked at the protective effects of PARP inhibitors in lung transplantation [139], neurodegeneration [140], renal injury [141], aging [142], and acute pancreatitis [143]. While these studies offer critical insights toward the potential therapeutic repurposing of PARP inhibitors in treating non-oncological indications, it is important to learn more about the possible side-effects of long-term treatment and whether PARP inhibition may increase the risk of mutagenesis and carcinogenesis.

9. Perspectives and Conclusions

Individualized treatments to patients with recalcitrant cancers in order to achieve cure are urgently needed. Currently, cancer patients have many therapeutic options for treatment; however, the lack of effective and tumor-selective treatment options still remains a major hurdle in the fight against cancer today. This is due to very limited predictive biomarkers that define individual risks of disease recurrence or sensitivity to treatments. For example, a significant number of cancer patients are over-treated in order to possibly attain an improved overall survival in early cancer stage. This "shotgun" approach, however, could have life-threatening consequences to cancer patients due to the emergence of drug resistance and secondary malignancies. Clearly, there is an urgent need for (1) a clear understanding of normal versus cancer cell biology to gain more informative decisions regarding the most effective therapeutic interventions, and (2) an arsenal of new treatment options based on predictive biomarkers for this disease to eliminate toxic effects to normal cells, drug resistance, and development of secondary cancers.

In less than two decades, studies on the use of PARP inhibitors for the treatment of *gBRCA1/2* mutant cancers and beyond BRCA-deficiency have increased tremendously. Of course, these were all made possible by the collaborative effort of the scientific and medical communities to understand the functional roles of breast cancer associated genes 1 and 2 (BRCA1/2) in normal physiology and breast cancer development. Accordingly, deficiencies in BRCA1 and BRCA2 genes can compromise HR-mediated repair, conferring hypersensitivity to PARP inhibitors by chemical synthetic lethality. The rarity of breast (5–10%) and ovarian (15–20%) cancers with germline *BRCA1/2* loss-of-function mutations, however, currently restricts the therapeutic utility of PARP inhibitor monotherapy. This is further complicated by several reports of potential molecular mechanisms of resistance for PARP inhibitors in BRCA-deficient breast cancers [144]. Interestingly, approximately 24% of high-grade triple negative breast cancers (TNBC) without a *BRCA* mutation showed great response to PARP inhibitor in Phase II clinical trials suggesting that the use of PARP inhibitors can be expanded beyond BRCA deficiency, and this has been reinforced by several studies in other cancers [18–21]. Indeed, genetic and pharmacological alterations of specific proteins (Table 1) or long non-coding RNAs (lncRNAs, e.g., PCAT-1 [145]) that can induce a state of "BRCAness" or HR-deficiency in BRCA-proficient cancers have the potential to broaden the application of PARP inhibitors beyond the current protocols approved by the FDA and EMA.

In conclusion, a serious concern for the use of PARP inhibitors – as with all chemotherapeutic agents – is the development of acquired drug resistance and de novo malignancies [146] due to limited information and often unclear understanding of the full extent of the agents' specificity and mechanism of action. A complete and clear knowledge of how different PARPs are activated to perform overlapping and non-overlapping functions or how PAR and NAD^+ levels are modulated to alter specific biological events toward cellular survival or death are essential to: (1) design novel PARP inhibitors that are more specific and tumor-selective; (2) develop better mechanistic-based strategies using reliable predictive biomarkers for PARP inhibitor monotherapy as well as combination treatments; and (3) identify situations that can re-sensitize recalcitrant tumor cells to PARP activity inhibition or modulation to treat cancer patients more safely and efficiently.

Table 1. Representative list of genes/proteins that have been shown to cause "BRCAness" or HR defect in BRCA-proficient tumors when expression is lost or activity is inhibited. These deficiencies could be exploited as predictive biomarkers for precision treatment with PARP inhibitors.

Protein Name	Primary Function/Activity	Association with "BRCAness"	Reference
CDK1	Cell cycle regulation	Loss of expression or activity inhibition compromises phosphorylation of BRCA1 for proper HR function	[30]
CDK12/13	Phosphorylates RNAPII CTD	Loss of expression and activity inhibition suppresses expression of specific HR proteins such as RAD51 and BRCA1	[147]
AXL	A receptor tyrosine kinase associated with metastasis, invasion and migration in many cancers	Loss of expression or activity inhibition decreases expression of specific HR genes and proteins	[148]
Kub5-Hera, RPRD1B, CREPT	Transcription termination factor	Loss of expression compromises HR by decreasing CDK1 expression	[25]
WEE1	Involved in the terminal phosphorylation and inactivation of CDK1-bound cyclin B	Activity inhibition with AZD1775 indirectly inhibits BRCA2	[149]
UCHL3	Deubiquitinase	Activity inhibition with perifosine promotes ubiquitination of RAD51 and blocks the binding of RAD51 with BRCA2	[150]
BET	Transcriptional regulators	Activity inhibition with JQ1 decreases expression of RAD51 and Ku80	[151]
PI3K	Kinase involve in cell growth, proliferation, differentiation, motility, survival and intracellular trafficking	Inhibition of activity impairs BRCA1/2 expression	[152]
Cyclin D1	Regulator of CDKs (cyclin dependent kinases), required for cell cycle G1/S transition	Loss of expression impairs recruitment of RAD51	[153,154]
AURKA	Play important role in mitosis/ regulation of cell cycle progression	Activity inhibition or loss of expression decreases expression of BRCA1 and BRCA2	[155,156]
HKMT	Regulation of histone methylation	Inhibition of activity abolishes retention of BRCA1/BARD1 complexes at sites of DSB	[157,158]
CCDC6	Tumor suppressor	Loss of expression compromises BRCC3 and DNA damage checkpoints in response to DNA damage.	[158,159]
MEK	Kinase that phosphorylates and activates MAPK	Activity inhibition or loss of expression downregulates BRCA2	[160]
HDAC	Removes acetyl groups from an amino acid on a histone	Activity inhibition with SAHA reduces BRCA1 protein levels by targeting the UHRF1/BRCA1 protein complex	[161]

Table 1. *Cont.*

Protein Name	Primary Function/Activity	Association with "BRCAness"	Reference
PAK1	Regulates cytoskeleton remodeling, phenotypic signaling and gene expression	Reduced activity and loss of expression downregulates the expression of genes involved in FA/BRCA pathway	[162]
Androgen receptor	DNA-binding transcription factor that regulates gene expression	Activity inhibition or loss of expression suppresses the expression of HR genes, thus creating HR deficiency and BRCAness	[163]
TGFβ	Involved in embryonic development, cell proliferation, motility and apoptosis, extracellular matrix production, and immunomodulation	Overexpression suppresses BRCA1, ATM, and MSH2	[164]

Author Contributions: Conceptualization: E.A.M.; resources: E.A.M.; writing–original draft preparation: E.A.M., N.S., S.L.P., S.B.B., and U.A.; writing–review, and editing: E.A.M., N.S., and S.L.P.; visualization: E.A.M.; supervision: E.A.M.; funding acquisition: E.A.M. All authors have read and agree to the final version of the manuscript.

Funding: This work was supported by Biomedical Research Grant from IUSM, Indiana CTSI Showalter Trust and NIH/NCI R01 CA210489 to E.A.M. The content is solely the responsibility of the authors and does not necessarily represent the official views of the funders.

Conflicts of Interest: The authors declare no conflict of interest.

References

1. Ciccia, A.; Elledge, S.J. The DNA damage response: Making it safe to play with knives. *Mol. Cell* **2010**, *40*, 179–204. [CrossRef] [PubMed]
2. Hottiger, M.O.; Hassa, P.O.; Luscher, B.; Schuler, H.; Koch-Nolte, F. Toward a unified nomenclature for mammalian ADP-ribosyltransferases. *Trends Biochem. Sci.* **2010**, *35*, 208–219. [CrossRef] [PubMed]
3. D'Amours, D.; Desnoyers, S.; D'Silva, I.; Poirier, G.G. Poly(ADP-ribosyl)ation reactions in the regulation of nuclear functions. *Biochem. J.* **1999**, *342 Pt 2*, 249–268. [CrossRef]
4. Gibson, B.A.; Kraus, W.L. New insights into the molecular and cellular functions of poly(ADP-ribose) and PARPs. *Nat. Rev. Mol. Cell Biol.* **2012**, *13*, 411–424. [CrossRef] [PubMed]
5. Hottiger, M.O. Nuclear ADP-Ribosylation and Its Role in Chromatin Plasticity, Cell Differentiation, and Epigenetics. *Annu. Rev. Biochem.* **2015**, *84*, 227–263. [CrossRef] [PubMed]
6. Pascal, J.M. The comings and goings of PARP-1 in response to DNA damage. *DNA Repair (Amst)* **2018**, *71*, 177–182. [CrossRef]
7. Wei, H.; Yu, X. Functions of PARylation in DNA Damage Repair Pathways. *Genomics Proteomics Bioinforma.* **2016**, *14*, 131–139. [CrossRef]
8. Hossain, M.A.; Lin, Y.; Yan, S. Single-Strand Break End Resection in Genome Integrity: Mechanism and Regulation by APE2. *Int. J. Mol. Sci.* **2018**, *19*, 2389. [CrossRef]
9. Panier, S.; Durocher, D. Push back to respond better: Regulatory inhibition of the DNA double-strand break response. *Nat. Rev. Mol. Cell Biol.* **2013**, *14*, 661–672. [CrossRef]
10. Lieber, M.R. The mechanism of double-strand DNA break repair by the nonhomologous DNA end-joining pathway. *Annu. Rev. Biochem.* **2010**, *79*, 181–211. [CrossRef]
11. Deriano, L.; Roth, D.B. Modernizing the nonhomologous end-joining repertoire: Alternative and classical NHEJ share the stage. *Annu. Rev. Genet.* **2013**, *47*, 433–455. [CrossRef] [PubMed]
12. Sfeir, A.; Symington, L.S. Microhomology-Mediated End Joining: A Back-up Survival Mechanism or Dedicated Pathway? *Trends Biochem. Sci.* **2015**, *40*, 701–714. [CrossRef] [PubMed]

13. Chae, Y.K.; Anker, J.F.; Carneiro, B.A.; Chandra, S.; Kaplan, J.; Kalyan, A.; Santa-Maria, C.A.; Platanias, L.C.; Giles, F.J. Genomic landscape of DNA repair genes in cancer. *Oncotarget* **2016**, *7*, 23312–23321. [CrossRef] [PubMed]
14. Miki, Y.; Swensen, J.; Shattuck-Eidens, D.; Futreal, P.A.; Harshman, K.; Tavtigian, S.; Liu, Q.; Cochran, C.; Bennett, L.M.; Ding, W.; et al. A strong candidate for the breast and ovarian cancer susceptibility gene BRCA1. *Science* **1994**, *266*, 66–71. [CrossRef]
15. Wooster, R.; Bignell, G.; Lancaster, J.; Swift, S.; Seal, S.; Mangion, J.; Collins, N.; Gregory, S.; Gumbs, C.; Micklem, G. Identification of the breast cancer susceptibility gene BRCA2. *Nature* **1995**, *378*, 789–792. [CrossRef]
16. Gottipati, P.; Vischioni, B.; Schultz, N.; Solomons, J.; Bryant, H.E.; Djureinovic, T.; Issaeva, N.; Sleeth, K.; Sharma, R.A.; Helleday, T. Poly(ADP-ribose) polymerase is hyperactivated in homologous recombination-defective cells. *Cancer Res.* **2010**, *70*, 5389–5398. [CrossRef]
17. Helleday, T. The underlying mechanism for the PARP and BRCA synthetic lethality: Clearing up the misunderstandings. *Mol. Oncol.* **2011**, *5*, 387–393. [CrossRef]
18. Gelmon, K.A.; Tischkowitz, M.; Mackay, H.; Swenerton, K.; Robidoux, A.; Tonkin, K.; Hirte, H.; Huntsman, D.; Clemons, M.; Gilks, B.; et al. Olaparib in patients with recurrent high-grade serous or poorly differentiated ovarian carcinoma or triple-negative breast cancer: A phase 2, multicentre, open-label, non-randomised study. *Lancet Oncol.* **2011**, *12*, 852–861. [CrossRef]
19. Evans, T.; Matulonis, U. PARP inhibitors in ovarian cancer: Evidence, experience and clinical potential. *Ther. Adv. Med. Oncol.* **2017**, *9*, 253–267. [CrossRef]
20. Ledermann, J.; Harter, P.; Gourley, C.; Friedlander, M.; Vergote, I.; Rustin, G.; Scott, C.L.; Meier, W.; Shapira-Frommer, R.; Safra, T.; et al. Olaparib maintenance therapy in patients with platinum-sensitive relapsed serous ovarian cancer: A preplanned retrospective analysis of outcomes by BRCA status in a randomised phase 2 trial. *Lancet Oncol.* **2014**, *15*, 852–861. [CrossRef]
21. Scott, C.L.; Swisher, E.M.; Kaufmann, S.H. Poly (ADP-ribose) polymerase inhibitors: Recent advances and future development. *J. Clin. Oncol.* **2015**, *33*, 1397–1406. [CrossRef] [PubMed]
22. Murai, J.; Huang, S.Y.; Das, B.B.; Renaud, A.; Zhang, Y.; Doroshow, J.H.; Ji, J.; Takeda, S.; Pommier, Y. Trapping of PARP1 and PARP2 by Clinical PARP Inhibitors. *Cancer Res.* **2012**, *72*, 5588–5599. [CrossRef] [PubMed]
23. Pommier, Y.; O'Connor, M.J.; de Bono, J. Laying a trap to kill cancer cells: PARP inhibitors and their mechanisms of action. *Sci. Transl. Med.* **2016**, *8*, 362ps17. [CrossRef] [PubMed]
24. Hopkins, T.A.; Shi, Y.; Rodriguez, L.E.; Solomon, L.R.; Donawho, C.K.; DiGiammarino, E.L.; Panchal, S.C.; Wilsbacher, J.L.; Gao, W.; Olson, A.M.; et al. Mechanistic Dissection of PARP1 Trapping and the Impact on In Vivo Tolerability and Efficacy of PARP Inhibitors. *Mol. Cancer Res.* **2015**, *13*, 1465–1477. [CrossRef]
25. Motea, E.A.; Fattah, F.J.; Xiao, L.; Girard, L.; Rommel, A.; Morales, J.C.; Patidar, P.; Zhou, Y.; Porter, A.; Xie, Y.; et al. Kub5-Hera (RPRD1B) Deficiency Promotes "BRCAness" and Vulnerability to PARP Inhibition in BRCA proficient Breast Cancers. *Clin. Cancer Res.* **2018**, *24*, 6459–6470. [CrossRef]
26. Fong, P.C.; Boss, D.S.; Yap, T.A.; Tutt, A.; Wu, P.; Mergui-Roelvink, M.; Mortimer, P.; Swaisland, H.; Lau, A.; O'Connor, M.J.; et al. Inhibition of poly(ADP-ribose) polymerase in tumors from BRCA mutation carriers. *N. Engl. J. Med.* **2009**, *361*, 123–134. [CrossRef]
27. Kaufman, B.; Shapira-Frommer, R.; Schmutzler, R.K.; Audeh, M.W.; Friedlander, M.; Balmana, J.; Mitchell, G.; Fried, G.; Stemmer, S.M.; Hubert, A.; et al. Olaparib monotherapy in patients with advanced cancer and a germline BRCA1/2 mutation. *J. Clin. Oncol.* **2015**, *33*, 244–250. [CrossRef]
28. Yang, L.; Zhang, Y.; Shan, W.; Hu, Z.; Yuan, J.; Pi, J.; Wang, Y.; Fan, L.; Tang, Z.; Li, C.; et al. Repression of BET activity sensitizes homologous recombination-proficient cancers to PARP inhibition. *Sci. Transl. Med.* **2017**, *9*. [CrossRef]
29. Mateo, J.; Carreira, S.; Sandhu, S.; Miranda, S.; Mossop, H.; Perez-Lopez, R.; Nava Rodrigues, D.; Robinson, D.; Omlin, A.; Tunariu, N.; et al. DNA-Repair Defects and Olaparib in Metastatic Prostate Cancer. *N. Engl. J. Med.* **2015**, *373*, 1697–1708. [CrossRef]
30. Johnson, N.; Li, Y.C.; Walton, Z.E.; Cheng, K.A.; Li, D.; Rodig, S.J.; Moreau, L.A.; Unitt, C.; Bronson, R.T.; Thomas, H.D.; et al. Compromised CDK1 activity sensitizes BRCA-proficient cancers to PARP inhibition. *Nat. Med.* **2011**, *17*, 875–882. [CrossRef]

31. Conrad, L.B.; Lin, K.Y.; Nandu, T.; Gibson, B.A.; Lea, J.S.; Kraus, W.L. ADP-Ribosylation Levels and Patterns Correlate with Gene Expression and Clinical Outcomes in Ovarian Cancers. *Mol. Cancer Ther.* **2020**, *19*, 282–291. [CrossRef]
32. Ledermann, J.; Harter, P.; Gourley, C.; Friedlander, M.; Vergote, I.; Rustin, G.; Scott, C.; Meier, W.; Shapira-Frommer, R.; Safra, T.; et al. Olaparib maintenance therapy in platinum-sensitive relapsed ovarian cancer. *N. Engl. J. Med.* **2012**, *366*, 1382–1392. [CrossRef] [PubMed]
33. Pujade-Lauraine, E.; Ledermann, J.A.; Selle, F.; Gebski, V.; Penson, R.T.; Oza, A.M.; Korach, J.; Huzarski, T.; Poveda, A.; Pignata, S.; et al. Olaparib tablets as maintenance therapy in patients with platinum-sensitive, relapsed ovarian cancer and a BRCA1/2 mutation (SOLO2/ENGOT-Ov21): A double-blind, randomised, placebo-controlled, phase 3 trial. *Lancet Oncol.* **2017**, *18*, 1274–1284. [CrossRef]
34. Coleman, R.L.; Oza, A.M.; Lorusso, D.; Aghajanian, C.; Oaknin, A.; Dean, A.; Colombo, N.; Weberpals, J.I.; Clamp, A.; Scambia, G.; et al. Rucaparib maintenance treatment for recurrent ovarian carcinoma after response to platinum therapy (ARIEL3): A randomised, double-blind, placebo-controlled, phase 3 trial. *Lancet* **2017**, *390*, 1949–1961. [CrossRef]
35. Mirza, M.R.; Monk, B.J.; Herrstedt, J.; Oza, A.M.; Mahner, S.; Redondo, A.; Fabbro, M.; Ledermann, J.A.; Lorusso, D.; Vergote, I.; et al. Niraparib Maintenance Therapy in Platinum-Sensitive, Recurrent Ovarian Cancer. *N. Engl. J. Med.* **2016**, *375*, 2154–2164. [CrossRef]
36. Mittica, G.; Ghisoni, E.; Giannone, G.; Genta, S.; Aglietta, M.; Sapino, A.; Valabrega, G. PARP Inhibitors in Ovarian Cancer. *Recent Pat. Anticancer Drug Discov.* **2018**, *13*, 392–410. [CrossRef]
37. Weng, C.S.; Wu, C.C.; Chen, T.C.; Chen, J.R.; Huang, C.Y.; Chang, C.L. Retrospective Analysis Of Comparative Outcomes In Recurrent Platinum-Sensitive Ovarian Cancer Treated With Pegylated Liposomal Doxorubicin (Lipo-Dox) And Carboplatin Versus Paclitaxel And Carboplatin. *Cancer Manag. Res.* **2019**, *11*, 9899–9905. [CrossRef]
38. Fong, P.C.; Yap, T.A.; Boss, D.S.; Carden, C.P.; Mergui-Roelvink, M.; Gourley, C.; De Greve, J.; Lubinski, J.; Shanley, S.; Messiou, C.; et al. Poly(ADP)-ribose polymerase inhibition: Frequent durable responses in BRCA carrier ovarian cancer correlating with platinum-free interval. *J. Clin. Oncol.* **2010**, *28*, 2512–2519. [CrossRef]
39. Tutt, A.; Robson, M.; Garber, J.E.; Domchek, S.M.; Audeh, M.W.; Weitzel, J.N.; Friedlander, M.; Arun, B.; Loman, N.; Schmutzler, R.K.; et al. Oral poly(ADP-ribose) polymerase inhibitor olaparib in patients with BRCA1 or BRCA2 mutations and advanced breast cancer: A proof-of-concept trial. *Lancet* **2010**, *376*, 235–244. [CrossRef]
40. Audeh, M.W.; Carmichael, J.; Penson, R.T.; Friedlander, M.; Powell, B.; Bell-McGuinn, K.M.; Scott, C.; Weitzel, J.N.; Oaknin, A.; Loman, N.; et al. Oral poly(ADP-ribose) polymerase inhibitor olaparib in patients with BRCA1 or BRCA2 mutations and recurrent ovarian cancer: A proof-of-concept trial. *Lancet* **2010**, *376*, 245–251. [CrossRef]
41. Olaparib Keeps Hereditary Breast Tumors in Check. *Cancer Discov.* **2017**, *7*, OF10. [CrossRef] [PubMed]
42. Robson, M.; Im, S.A.; Senkus, E.; Xu, B.; Domchek, S.M.; Masuda, N.; Delaloge, S.; Li, W.; Tung, N.; Armstrong, A.; et al. Olaparib for Metastatic Breast Cancer in Patients with a Germline BRCA Mutation. *N. Engl. J. Med.* **2017**, *377*, 523–533. [CrossRef] [PubMed]
43. Domchek, S.M.; Aghajanian, C.; Shapira-Frommer, R.; Schmutzler, R.K.; Audeh, M.W.; Friedlander, M.; Balmana, J.; Mitchell, G.; Fried, G.; Stemmer, S.M.; et al. Efficacy and safety of olaparib monotherapy in germline BRCA1/2 mutation carriers with advanced ovarian cancer and three or more lines of prior therapy. *Gynecol. Oncol.* **2016**, *140*, 199–203. [CrossRef] [PubMed]
44. Swisher, E.M.; Lin, K.K.; Oza, A.M.; Scott, C.L.; Giordano, H.; Sun, J.; Konecny, G.E.; Coleman, R.L.; Tinker, A.V.; O'Malley, D.M.; et al. Rucaparib in relapsed, platinum-sensitive high-grade ovarian carcinoma (ARIEL2 Part 1): An international, multicentre, open-label, phase 2 trial. *Lancet Oncol.* **2017**, *18*, 75–87. [CrossRef]
45. Kristeleit, R.; Shapiro, G.I.; Burris, H.A.; Oza, A.M.; LoRusso, P.; Patel, M.R.; Domchek, S.M.; Balmana, J.; Drew, Y.; Chen, L.M.; et al. A Phase I-II Study of the Oral PARP Inhibitor Rucaparib in Patients with Germline BRCA1/2-Mutated Ovarian Carcinoma or Other Solid Tumors. *Clin. Cancer Res.* **2017**, *23*, 4095–4106. [CrossRef]
46. Moore, K.N.; Secord, A.A.; Geller, M.A.; Miller, D.S.; Cloven, N.; Fleming, G.F.; Wahner Hendrickson, A.E.; Azodi, M.; DiSilvestro, P.; Oza, A.M.; et al. Niraparib monotherapy for late-line treatment of ovarian cancer (QUADRA): A multicentre, open-label, single-arm, phase 2 trial. *Lancet Oncol.* **2019**, *20*, 636–648. [CrossRef]

47. Guney Eskiler, G. Talazoparib to treat BRCA-positive breast cancer. *Drugs Today (Barc)* **2019**, *55*, 459–467. [CrossRef]
48. Okayama, H.; Edson, C.M.; Fukushima, M.; Ueda, K.; Hayaishi, O. Purification and properties of poly(adenosine diphosphate ribose) synthetase. *J. Biol. Chem.* **1977**, *252*, 7000–7005.
49. Brochu, G.; Duchaine, C.; Thibeault, L.; Lagueux, J.; Shah, G.M.; Poirier, G.G. Mode of action of poly(ADP-ribose) glycohydrolase. *Biochim. Biophys. Acta (BBA) Gene Struct. Expr.* **1994**, *1219*, 342–350. [CrossRef]
50. Hassa, P.O.; Haenni, S.S.; Elser, M.; Hottiger, M.O. Nuclear ADP-ribosylation reactions in mammalian cells: Where are we today and where are we going? *Microbiol. Mol. Biol. Rev.* **2006**, *70*, 789–829. [CrossRef]
51. Hassa, P.O.; Hottiger, M.O. The diverse biological roles of mammalian PARPS, a small but powerful family of poly-ADP-ribose polymerases. *Front. Biosci.* **2008**, *13*, 3046–3082. [CrossRef] [PubMed]
52. Kovacs, K.; Toth, A.; Deres, P.; Kalai, T.; Hideg, K.; Gallyas, F., Jr.; Sumegi, B. Critical role of PI3-kinase/Akt activation in the PARP inhibitor induced heart function recovery during ischemia-reperfusion. *Biochem. Pharmacol.* **2006**, *71*, 441–452. [CrossRef] [PubMed]
53. Juarez-Salinas, H.; Sims, J.L.; Jacobson, M.K. Poly(ADP-ribose) levels in carcinogen-treated cells. *Nature* **1979**, *282*, 740–741. [CrossRef] [PubMed]
54. O'Sullivan, J.; Tedim Ferreira, M.; Gagne, J.P.; Sharma, A.K.; Hendzel, M.J.; Masson, J.Y.; Poirier, G.G. Emerging roles of eraser enzymes in the dynamic control of protein ADP-ribosylation. *Nat. Commun.* **2019**, *10*, 1182. [CrossRef] [PubMed]
55. Botta, D.; Jacobson, M.K. Identification of a regulatory segment of poly(ADP-ribose) glycohydrolase. *Biochemistry* **2010**, *49*, 7674–7682. [CrossRef] [PubMed]
56. Meyer-Ficca, M.L.; Meyer, R.G.; Coyle, D.L.; Jacobson, E.L.; Jacobson, M.K. Human poly(ADP-ribose) glycohydrolase is expressed in alternative splice variants yielding isoforms that localize to different cell compartments. *Exp. Cell Res.* **2004**, *297*, 521–532. [CrossRef] [PubMed]
57. Braun, S.A.; Panzeter, P.L.; Collinge, M.A.; Althaus, F.R. Endoglycosidic cleavage of branched polymers by poly(ADP-ribose) glycohydrolase. *Eur. J. Biochem.* **1994**, *220*, 369–375. [CrossRef]
58. David, K.K.; Andrabi, S.A.; Dawson, T.M.; Dawson, V.L. Parthanatos, a messenger of death. *Front. Biosci. (Landmark Ed)* **2009**, *14*, 1116–1128. [CrossRef]
59. Andrabi, S.A.; Kim, N.S.; Yu, S.W.; Wang, H.; Koh, D.W.; Sasaki, M.; Klaus, J.A.; Otsuka, T.; Zhang, Z.; Koehler, R.C.; et al. Poly(ADP-ribose) (PAR) polymer is a death signal. *Proc. Natl. Acad. Sci. USA* **2006**, *103*, 18308–18313. [CrossRef]
60. Wang, Y.; Kim, N.S.; Haince, J.F.; Kang, H.C.; David, K.K.; Andrabi, S.A.; Poirier, G.G.; Dawson, V.L.; Dawson, T.M. Poly(ADP-ribose) (PAR) binding to apoptosis-inducing factor is critical for PAR polymerase-1-dependent cell death (parthanatos). *Sci. Signal.* **2011**, *4*, ra20. [CrossRef]
61. Yu, S.W.; Wang, H.; Poitras, M.F.; Coombs, C.; Bowers, W.J.; Federoff, H.J.; Poirier, G.G.; Dawson, T.M.; Dawson, V.L. Mediation of poly(ADP-ribose) polymerase-1-dependent cell death by apoptosis-inducing factor. *Science* **2002**, *297*, 259–263. [CrossRef] [PubMed]
62. Niere, M.; Mashimo, M.; Agledal, L.; Dolle, C.; Kasamatsu, A.; Kato, J.; Moss, J.; Ziegler, M. ADP-ribosylhydrolase 3 (ARH3), not poly(ADP-ribose) glycohydrolase (PARG) isoforms, is responsible for degradation of mitochondrial matrix-associated poly(ADP-ribose). *J. Biol. Chem.* **2012**, *287*, 16088–16102. [CrossRef] [PubMed]
63. Marques, M.; Jangal, M.; Wang, L.C.; Kazanets, A.; da Silva, S.D.; Zhao, T.; Lovato, A.; Yu, H.; Jie, S.; Del Rincon, S.; et al. Oncogenic activity of poly (ADP-ribose) glycohydrolase. *Oncogene* **2019**, *38*, 2177–2191. [CrossRef] [PubMed]
64. Tavassoli, M.; Tavassoli, M.H.; Shall, S. Effect of DNA intercalators on poly(ADP-ribose) glycohydrolase activity. *Biochim. Biophys. Acta* **1985**, *827*, 228–234. [CrossRef]
65. Li, Q.; Li, M.; Wang, Y.L.; Fauzee, N.J.; Yang, Y.; Pan, J.; Yang, L.; Lazar, A. RNA interference of PARG could inhibit the metastatic potency of colon carcinoma cells via PI3-kinase/Akt pathway. *Cell. Physiol. Biochem.* **2012**, *29*, 361–372. [CrossRef]
66. Putt, K.S.; Hergenrother, P.J. A nonradiometric, high-throughput assay for poly(ADP-ribose) glycohydrolase (PARG): Application to inhibitor identification and evaluation. *Anal. Biochem.* **2004**, *333*, 256–264. [CrossRef]

67. Aoki, T.; Kojima, M.; Adachi, J.; Okano, M. Effect of short-term egg exclusion diet on infantile atopic dermatitis and its relation to egg allergy: A single-blind test. *Acta Derm. Venereol. Suppl. (Stockh)* **1992**, *176*, 99–102.
68. Tsai, Y.J.; Aoki, T.; Maruta, H.; Abe, H.; Sakagami, H.; Hatano, T.; Okuda, T.; Tanuma, S. Mouse mammary tumor virus gene expression is suppressed by oligomeric ellagitannins, novel inhibitors of poly(ADP-ribose) glycohydrolase. *J. Biol. Chem.* **1992**, *267*, 14436–14442.
69. Formentini, L.; Arapistas, P.; Pittelli, M.; Jacomelli, M.; Pitozzi, V.; Menichetti, S.; Romani, A.; Giovannelli, L.; Moroni, F.; Chiarugi, A. Mono-galloyl glucose derivatives are potent poly(ADP-ribose) glycohydrolase (PARG) inhibitors and partially reduce PARP-1-dependent cell death. *Br. J. Pharmacol.* **2008**, *155*, 1235–1249. [CrossRef]
70. Finch, K.E.; Knezevic, C.E.; Nottbohm, A.C.; Partlow, K.C.; Hergenrother, P.J. Selective Small Molecule Inhibition of Poly(ADP-Ribose) Glycohydrolase (PARG). *ACS Chem. Biol.* **2012**, *7*, 563–570. [CrossRef]
71. Pillay, N.; Tighe, A.; Nelson, L.; Littler, S.; Coulson-Gilmer, C.; Bah, N.; Golder, A.; Bakker, B.; Spierings, D.C.J.; James, D.I.; et al. DNA Replication Vulnerabilities Render Ovarian Cancer Cells Sensitive to Poly(ADP-Ribose) Glycohydrolase Inhibitors. *Cancer Cell* **2019**, *35*, 519–533. [CrossRef] [PubMed]
72. Fujihara, H.; Ogino, H.; Maeda, D.; Shirai, H.; Nozaki, T.; Kamada, N.; Jishage, K.; Tanuma, S.; Takato, T.; Ochiya, T.; et al. Poly(ADP-ribose) Glycohydrolase deficiency sensitizes mouse ES cells to DNA damaging agents. *Curr. Cancer Drug Targets* **2009**, *9*, 953–962. [CrossRef] [PubMed]
73. Shirai, H.; Poetsch, A.R.; Gunji, A.; Maeda, D.; Fujimori, H.; Fujihara, H.; Yoshida, T.; Ogino, H.; Masutani, M. PARG dysfunction enhances DNA double strand break formation in S-phase after alkylation DNA damage and augments different cell death pathways. *Cell Death Dis.* **2013**, *4*, e656. [CrossRef] [PubMed]
74. Murphy, J.P.; Giacomantonio, M.A.; Paulo, J.A.; Everley, R.A.; Kennedy, B.E.; Pathak, G.P.; Clements, D.R.; Kim, Y.; Dai, C.; Sharif, T.; et al. The NAD(+) Salvage Pathway Supports PHGDH-Driven Serine Biosynthesis. *Cell Rep.* **2018**, *24*, 2381–2391. [CrossRef]
75. Verdin, E. NAD(+) in aging, metabolism, and neurodegeneration. *Science* **2015**, *350*, 1208–1213. [CrossRef]
76. Garrido, A.; Djouder, N. NAD(+) Deficits in Age-Related Diseases and Cancer. *Trends Cancer* **2017**, *3*, 593–610. [CrossRef]
77. Yaku, K.; Okabe, K.; Nakagawa, T. NAD metabolism: Implications in aging and longevity. *Ageing Res. Rev.* **2018**, *47*, 1–17. [CrossRef]
78. Yamamoto, M.; Inohara, H.; Nakagawa, T. Targeting metabolic pathways for head and neck cancers therapeutics. *Cancer Metastasis Rev.* **2017**, *36*, 503–514. [CrossRef]
79. Tan, B.; Young, D.A.; Lu, Z.H.; Wang, T.; Meier, T.I.; Shepard, R.L.; Roth, K.; Zhai, Y.; Huss, K.; Kuo, M.S.; et al. Pharmacological inhibition of nicotinamide phosphoribosyltransferase (NAMPT), an enzyme essential for NAD+ biosynthesis, in human cancer cells: Metabolic basis and potential clinical implications. *J. Biol. Chem.* **2013**, *288*, 3500–3511. [CrossRef]
80. Cerna, D.; Li, H.; Flaherty, S.; Takebe, N.; Coleman, C.N.; Yoo, S.S. Inhibition of nicotinamide phosphoribosyltransferase (NAMPT) activity by small molecule GMX1778 regulates reactive oxygen species (ROS)-mediated cytotoxicity in a p53- and nicotinic acid phosphoribosyltransferase1 (NAPRT1)-dependent manner. *J. Biol. Chem.* **2012**, *287*, 22408–22417. [CrossRef]
81. Kato, H.; Ito, E.; Shi, W.; Alajez, N.M.; Yue, S.; Lee, C.; Chan, N.; Bhogal, N.; Coackley, C.L.; Vines, D.; et al. Efficacy of combining GMX1777 with radiation therapy for human head and neck carcinoma. *Clin. Cancer Res.* **2010**, *16*, 898–911. [CrossRef] [PubMed]
82. Moore, Z.; Chakrabarti, G.; Luo, X.; Ali, A.; Hu, Z.; Fattah, F.J.; Vemireddy, R.; DeBerardinis, R.J.; Brekken, R.A.; Boothman, D.A. NAMPT inhibition sensitizes pancreatic adenocarcinoma cells to tumor-selective, PAR-independent metabolic catastrophe and cell death induced by beta-lapachone. *Cell Death Dis.* **2015**, *6*, e1599. [CrossRef] [PubMed]
83. Zerp, S.F.; Vens, C.; Floot, B.; Verheij, M.; van Triest, B. NAD(+) depletion by APO866 in combination with radiation in a prostate cancer model, results from an in vitro and in vivo study. *Radiother. Oncol.* **2014**, *110*, 348–354. [CrossRef] [PubMed]
84. Ju, H.Q.; Zhuang, Z.N.; Li, H.; Tian, T.; Lu, Y.X.; Fan, X.Q.; Zhou, H.J.; Mo, H.Y.; Sheng, H.; Chiao, P.J.; et al. Regulation of the Nampt-mediated NAD salvage pathway and its therapeutic implications in pancreatic cancer. *Cancer Lett.* **2016**, *379*, 1–11. [CrossRef] [PubMed]

85. Lucena-Cacace, A.; Otero-Albiol, D.; Jimenez-Garcia, M.P.; Munoz-Galvan, S.; Carnero, A. NAMPT Is a Potent Oncogene in Colon Cancer Progression that Modulates Cancer Stem Cell Properties and Resistance to Therapy through Sirt1 and PARP. *Clin. Cancer Res.* **2018**, *24*, 1202–1215. [CrossRef]
86. Lucena-Cacace, A.; Otero-Albiol, D.; Jiménez-García, M.P.; Peinado-Serrano, J.; Carnero, A. NAMPT overexpression induces cancer stemness and defines a novel tumor signature for glioma prognosis. *Oncotarget* **2017**, *8*, 99514–99530. [CrossRef]
87. Zhu, Y.; Guo, M.; Zhang, L.; Xu, T.; Wang, L.; Xu, G. Biomarker triplet NAMPT/VEGF/HER2 as a de novo detection panel for the diagnosis and prognosis of human breast cancer. *Oncol. Rep.* **2016**, *35*, 454–462. [CrossRef]
88. Wang, B.; Hasan, M.K.; Alvarado, E.; Yuan, H.; Wu, H.; Chen, W.Y. NAMPT overexpression in prostate cancer and its contribution to tumor cell survival and stress response. *Oncogene* **2011**, *30*, 907–921. [CrossRef]
89. Shackelford, R.E.; Bui, M.M.; Coppola, D.; Hakam, A. Over-expression of nicotinamide phosphoribosyltransferase in ovarian cancers. *Int. J. Clin. Exp. Pathol.* **2010**, *3*, 522–527.
90. Gehrke, I.; Bouchard, E.D.; Beiggi, S.; Poeppl, A.G.; Johnston, J.B.; Gibson, S.B.; Banerji, V. On-target effect of FK866, a nicotinamide phosphoribosyl transferase inhibitor, by apoptosis-mediated death in chronic lymphocytic leukemia cells. *Clin. Cancer Res.* **2014**, *20*, 4861–4872. [CrossRef]
91. Del Nagro, C.; Xiao, Y.; Rangell, L.; Reichelt, M.; O'Brien, T. Depletion of the central metabolite NAD leads to oncosis-mediated cell death. *J. Biol. Chem.* **2014**, *289*, 35182–35192. [CrossRef] [PubMed]
92. Hasmann, M.; Schemainda, I. FK866, a highly specific noncompetitive inhibitor of nicotinamide phosphoribosyltransferase, represents a novel mechanism for induction of tumor cell apoptosis. *Cancer Res.* **2003**, *63*, 7436–7442. [PubMed]
93. Olesen, U.H.; Christensen, M.K.; Björkling, F.; Jäättelä, M.; Jensen, P.B.; Sehested, M.; Nielsen, S.J. Anticancer agent CHS-828 inhibits cellular synthesis of NAD. *Biochem. Biophys. Res. Commun.* **2008**, *367*, 799–804. [CrossRef] [PubMed]
94. Watson, M.; Roulston, A.; Belec, L.; Billot, X.; Marcellus, R.; Bedard, D.; Bernier, C.; Branchaud, S.; Chan, H.; Dairi, K.; et al. The small molecule GMX1778 is a potent inhibitor of NAD+ biosynthesis: Strategy for enhanced therapy in nicotinic acid phosphoribosyltransferase 1-deficient tumors. *Mol. Cell. Biol.* **2009**, *29*, 5872–5888. [CrossRef]
95. Chan, D.A.; Sutphin, P.D.; Nguyen, P.; Turcotte, S.; Lai, E.W.; Banh, A.; Reynolds, G.E.; Chi, J.T.; Wu, J.; Solow-Cordero, D.E.; et al. Targeting GLUT1 and the Warburg effect in renal cell carcinoma by chemical synthetic lethality. *Sci. Transl. Med.* **2011**, *3*, 94ra70. [CrossRef]
96. Espindola-Netto, J.M.; Chini, C.C.S.; Tarrago, M.; Wang, E.; Dutta, S.; Pal, K.; Mukhopadhyay, D.; Sola-Penna, M.; Chini, E.N. Preclinical efficacy of the novel competitive NAMPT inhibitor STF-118804 in pancreatic cancer. *Oncotarget* **2017**, *8*, 85054–85067. [CrossRef]
97. Zhao, G.; Green, C.F.; Hui, Y.H.; Prieto, L.; Shepard, R.; Dong, S.; Wang, T.; Tan, B.; Gong, X.; Kays, L.; et al. Discovery of a Highly Selective NAMPT Inhibitor That Demonstrates Robust Efficacy and Improved Retinal Toxicity with Nicotinic Acid Coadministration. *Mol. Cancer Ther.* **2017**, *16*, 2677–2688. [CrossRef]
98. Abu Aboud, O.; Chen, C.H.; Senapedis, W.; Baloglu, E.; Argueta, C.; Weiss, R.H. Dual and Specific Inhibition of NAMPT and PAK4 By KPT-9274 Decreases Kidney Cancer Growth. *Mol. Cancer Ther.* **2016**, *15*, 2119–2129. [CrossRef]
99. Von Heideman, A.; Berglund, A.; Larsson, R.; Nygren, P. Safety and efficacy of NAD depleting cancer drugs: Results of a phase I clinical trial of CHS 828 and overview of published data. *Cancer Chemother. Pharmacol.* **2010**, *65*, 1165–1172. [CrossRef]
100. Goldinger, S.M.; Gobbi Bischof, S.; Fink-Puches, R.; Klemke, C.D.; Dreno, B.; Bagot, M.; Dummer, R. Efficacy and Safety of APO866 in Patients With Refractory or Relapsed Cutaneous T-Cell Lymphoma: A Phase 2 Clinical Trial. *JAMA Dermatol.* **2016**, *152*, 837–839. [CrossRef]
101. Holen, K.; Saltz, L.B.; Hollywood, E.; Burk, K.; Hanauske, A.R. The pharmacokinetics, toxicities, and biologic effects of FK866, a nicotinamide adenine dinucleotide biosynthesis inhibitor. *Investig. New Drugs* **2008**, *26*, 45–51. [CrossRef] [PubMed]
102. Ravaud, A.; Cerny, T.; Terret, C.; Wanders, J.; Bui, B.N.; Hess, D.; Droz, J.P.; Fumoleau, P.; Twelves, C. Phase I study and pharmacokinetic of CHS-828, a guanidino-containing compound, administered orally as a single dose every 3 weeks in solid tumours: An ECSG/EORTC study. *Eur. J. Cancer* **2005**, *41*, 702–707. [CrossRef] [PubMed]

103. Hovstadius, P.; Larsson, R.; Jonsson, E.; Skov, T.; Kissmeyer, A.M.; Krasilnikoff, K.; Bergh, J.; Karlsson, M.O.; Lonnebo, A.; Ahlgren, J. A Phase I study of CHS 828 in patients with solid tumor malignancy. *Clin. Cancer Res.* **2002**, *8*, 2843–2850. [PubMed]
104. Liu, H.Y.; Li, Q.R.; Cheng, X.F.; Wang, G.J.; Hao, H.P. NAMPT inhibition synergizes with NQO1-targeting agents in inducing apoptotic cell death in non-small cell lung cancer cells. *Chin. J. Nat. Med.* **2016**, *14*, 582–589. [CrossRef]
105. Zhang, K.; Chen, D.; Ma, K.; Wu, X.; Hao, H.; Jiang, S. NAD(P)H:Quinone Oxidoreductase 1 (NQO1) as a Therapeutic and Diagnostic Target in Cancer. *J. Med. Chem.* **2018**, *61*, 6983–7003. [CrossRef]
106. Pardee, A.B.; Li, Y.Z.; Li, C.J. Cancer therapy with beta-lapachone. *Curr. Cancer Drug Targets* **2002**, *2*, 227–242. [CrossRef]
107. Beg, M.S.; Huang, X.; Silvers, M.A.; Gerber, D.E.; Bolluyt, J.; Sarode, V.; Fattah, F.; Deberardinis, R.J.; Merritt, M.E.; Xie, X.J.; et al. Using a novel NQO1 bioactivatable drug, beta-lapachone (ARQ761), to enhance chemotherapeutic effects by metabolic modulation in pancreatic cancer. *J. Surg. Oncol.* **2017**, *116*, 83–88. [CrossRef]
108. Bey, E.A.; Bentle, M.S.; Reinicke, K.E.; Dong, Y.; Yang, C.R.; Girard, L.; Minna, J.D.; Bornmann, W.G.; Gao, J.; Boothman, D.A. An NQO1- and PARP-1-mediated cell death pathway induced in non-small-cell lung cancer cells by beta-lapachone. *Proc. Natl. Acad. Sci. USA* **2007**, *104*, 11832–11837. [CrossRef]
109. Huang, X.; Motea, E.A.; Moore, Z.R.; Yao, J.; Dong, Y.; Chakrabarti, G.; Kilgore, J.A.; Silvers, M.A.; Patidar, P.L.; Cholka, A.; et al. Leveraging an NQO1 Bioactivatable Drug for Tumor-Selective Use of Poly(ADP-ribose) Polymerase Inhibitors. *Cancer Cell* **2016**, *30*, 940–952. [CrossRef]
110. Motea, E.A.; Huang, X.; Singh, N.; Kilgore, J.A.; Williams, N.S.; Xie, X.J.; Gerber, D.E.; Beg, M.S.; Bey, E.A.; Boothman, D.A. NQO1-dependent, Tumor-selective Radiosensitization of Non-small Cell Lung Cancers. *Clin. Cancer Res.* **2019**, *25*, 2601–2609. [CrossRef]
111. Blanco, E.; Bey, E.A.; Khemtong, C.; Yang, S.G.; Setti-Guthi, J.; Chen, H.; Kessinger, C.W.; Carnevale, K.A.; Bornmann, W.G.; Boothman, D.A.; et al. Beta-lapachone micellar nanotherapeutics for non-small cell lung cancer therapy. *Cancer Res.* **2010**, *70*, 3896–3904. [CrossRef] [PubMed]
112. Dong, Y.; Bey, E.A.; Li, L.S.; Kabbani, W.; Yan, J.; Xie, X.J.; Hsieh, J.T.; Gao, J.; Boothman, D.A. Prostate cancer radiosensitization through poly(ADP-Ribose) polymerase-1 hyperactivation. *Cancer Res.* **2010**, *70*, 8088–8096. [CrossRef] [PubMed]
113. Reinicke, K.E.; Bey, E.A.; Bentle, M.S.; Pink, J.J.; Ingalls, S.T.; Hoppel, C.L.; Misico, R.I.; Arzac, G.M.; Burton, G.; Bornmann, W.G.; et al. Development of beta-lapachone prodrugs for therapy against human cancer cells with elevated NAD(P)H:quinone oxidoreductase 1 levels. *Clin. Cancer Res.* **2005**, *11*, 3055–3064. [CrossRef] [PubMed]
114. Silvers, M.A.; Deja, S.; Singh, N.; Egnatchik, R.A.; Sudderth, J.; Luo, X.; Beg, M.S.; Burgess, S.C.; DeBerardinis, R.J.; Boothman, D.A.; et al. The NQO1 bioactivatable drug, beta-lapachone, alters the redox state of NQO1+ pancreatic cancer cells, causing perturbation in central carbon metabolism. *J. Biol. Chem.* **2017**, *292*, 18203–18216. [CrossRef]
115. Pink, J.J.; Planchon, S.M.; Tagliarino, C.; Varnes, M.E.; Siegel, D.; Boothman, D.A. NAD(P)H:Quinone oxidoreductase activity is the principal determinant of beta-lapachone cytotoxicity. *J. Biol. Chem.* **2000**, *275*, 5416–5424. [CrossRef]
116. Imlay, J.A.; Chin, S.M.; Linn, S. Toxic DNA damage by hydrogen peroxide through the Fenton reaction in vivo and in vitro. *Science* **1988**, *240*, 640–642. [CrossRef]
117. Andrabi, S.A.; Umanah, G.K.; Chang, C.; Stevens, D.A.; Karuppagounder, S.S.; Gagne, J.P.; Poirier, G.G.; Dawson, V.L.; Dawson, T.M. Poly(ADP-ribose) polymerase-dependent energy depletion occurs through inhibition of glycolysis. *Proc. Natl. Acad. Sci. USA* **2014**, *111*, 10209–10214. [CrossRef]
118. Kim, S.; Lee, S.; Cho, J.Y.; Yoon, S.H.; Jang, I.J.; Yu, K.S. Pharmacokinetics and tolerability of MB12066, a beta-lapachone derivative targeting NAD(P)H: Quinone oxidoreductase 1: Two independent, double-blind, placebo-controlled, combined single and multiple ascending dose first-in-human clinical trials. *Drug Des. Dev. Ther.* **2017**, *11*, 3187–3195. [CrossRef]
119. Hartner, L.P.; Rosen, L.; Hensley, M.; Mendelson, D.; Staddon, A.P.; Chow, W.; Kovalyov, O.; Ruka, W.; Skladowski, K.; Jagiello-Gruszfeld, A.; et al. Phase 2 dose multi-center, open-label study of ARQ 501, a checkpoint activator, in adult patients with persistent, recurrent or metastatic leiomyosarcoma (LMS). *J. Clin. Oncol.* **2007**, *25* (Suppl. 18), 20521. [CrossRef]

120. Buranrat, B.; Chau-in, S.; Prawan, A.; Puapairoj, A.; Zeekpudsa, P.; Kukongviriyapan, V. NQO1 expression correlates with cholangiocarcinoma prognosis. *Asian Pac. J. Cancer Prev.* **2012**, *13*, 131–136.
121. Cui, X.; Jin, T.; Wang, X.; Jin, G.; Li, Z.; Lin, L. NAD(P)H:quinone oxidoreductase-1 overexpression predicts poor prognosis in small cell lung cancer. *Oncol. Rep.* **2014**, *32*, 2589–2595. [CrossRef] [PubMed]
122. Cui, X.; Li, L.; Yan, G.; Meng, K.; Lin, Z.; Nan, Y.; Jin, G.; Li, C. High expression of NQO1 is associated with poor prognosis in serous ovarian carcinoma. *BMC Cancer* **2015**, *15*, 244. [CrossRef] [PubMed]
123. Kolesar, J.M.; Dahlberg, S.E.; Marsh, S.; McLeod, H.L.; Johnson, D.H.; Keller, S.M.; Schiller, J.H. The NQO1*2/*2 polymorphism is associated with poor overall survival in patients following resection of stages II and IIIa non-small cell lung cancer. *Oncol. Rep.* **2011**, *25*, 1765–1772. [CrossRef] [PubMed]
124. Li, Z.; Zhang, Y.; Jin, T.; Men, J.; Lin, Z.; Qi, P.; Piao, Y.; Yan, G. NQO1 protein expression predicts poor prognosis of non-small cell lung cancers. *BMC Cancer* **2015**, *15*, 207. [CrossRef] [PubMed]
125. Lin, L.; Qin, Y.; Jin, T.; Liu, S.; Zhang, S.; Shen, X.; Lin, Z. Significance of NQO1 overexpression for prognostic evaluation of gastric adenocarcinoma. *Exp. Mol. Pathol.* **2014**, *96*, 200–205. [CrossRef] [PubMed]
126. Yang, Y.; Zhang, Y.; Wu, Q.; Cui, X.; Lin, Z.; Liu, S.; Chen, L. Clinical implications of high NQO1 expression in breast cancers. *J. Exp. Clin. Cancer Res.* **2014**, *33*, 14. [CrossRef]
127. Morales, J.; Li, L.; Fattah, F.J.; Dong, Y.; Bey, E.A.; Patel, M.; Gao, J.; Boothman, D.A. Review of poly (ADP-ribose) polymerase (PARP) mechanisms of action and rationale for targeting in cancer and other diseases. *Crit. Rev. Eukaryot. Gene Expr.* **2014**, *24*, 15–28. [CrossRef]
128. Du, Y.; Yamaguchi, H.; Wei, Y.; Hsu, J.L.; Wang, H.-L.; Hsu, Y.-H.; Lin, W.-C.; Yu, W.-H.; Leonard, P.G.; Lee IV, G.R. Blocking c-Met–mediated PARP1 phosphorylation enhances anti-tumor effects of PARP inhibitors. *Nat. Med.* **2016**, *22*, 194. [CrossRef]
129. Zhang, Y.; Xia, M.; Jin, K.; Wang, S.; Wei, H.; Fan, C.; Wu, Y.; Li, X.; Li, X.; Li, G. Function of the c-Met receptor tyrosine kinase in carcinogenesis and associated therapeutic opportunities. *Mol. Cancer* **2018**, *17*, 45. [CrossRef]
130. Chiarugi, P.; Cirri, P. Redox regulation of protein tyrosine phosphatases during receptor tyrosine kinase signal transduction. *Trends Biochem. Sci.* **2003**, *28*, 509–514. [CrossRef]
131. Chen, M.-K.; Du, Y.; Sun, L.; Hsu, J.L.; Wang, Y.-H.; Gao, Y.; Huang, J.; Hung, M.-C. H_2O_2 induces nuclear transport of the receptor tyrosine kinase c-MET in breast cancer cells via a membrane-bound retrograde trafficking mechanism. *J. Biol. Chem.* **2019**, *294*, 8516–8528. [CrossRef] [PubMed]
132. Dong, Q.; Du, Y.; Li, H.; Liu, C.; Wei, Y.; Chen, M.-K.; Zhao, X.; Chu, Y.-Y.; Qiu, Y.; Qin, L. EGFR and c-MET cooperate to enhance resistance to PARP inhibitors in hepatocellular carcinoma. *Cancer Res.* **2019**, *79*, 819–829. [CrossRef] [PubMed]
133. Crawford, R.S.; Albadawi, H.; Atkins, M.D.; Jones, J.E.; Yoo, H.J.; Conrad, M.F.; Austen, W.G., Jr.; Watkins, M.T. Postischemic poly (ADP-ribose) polymerase (PARP) inhibition reduces ischemia reperfusion injury in a hind-limb ischemia model. *Surgery* **2010**, *148*, 110–118. [CrossRef] [PubMed]
134. Zingarelli, B.; Hake, P.W.; O'Connor, M.; Denenberg, A.; Kong, S.; Aronow, B.J. Absence of poly(ADP-ribose)polymerase-1 alters nuclear factor-kappa B activation and gene expression of apoptosis regulators after reperfusion injury. *Mol. Med.* **2003**, *9*, 143–153. [CrossRef] [PubMed]
135. Blaser, H.; Dostert, C.; Mak, T.W.; Brenner, D. TNF and ROS Crosstalk in Inflammation. *Trends Cell Biol.* **2016**, *26*, 249–261. [CrossRef]
136. Johnson, P.M.; Sagerman, R.H.; Dombrowski, C.S. Ischemia of the lung due to ionizing radiation: Quantitative studies. *J. Nucl. Med.* **1970**, *11*, 491–495.
137. Zhou, Z.B.; Meng, L.; Gelb, A.W.; Lee, R.; Huang, W.Q. Cerebral ischemia during surgery: An overview. *J. Biomed. Res.* **2016**, *30*, 83–87.
138. Wang, S.; Han, L.; Han, J.; Li, P.; Ding, Q.; Zhang, Q.J.; Liu, Z.P.; Chen, C.; Yu, Y. Uncoupling of PARP1 trapping and inhibition using selective PARP1 degradation. *Nat. Chem. Biol.* **2019**, *15*, 1223–1231. [CrossRef]
139. Hatachi, G.; Tsuchiya, T.; Miyazaki, T.; Matsumoto, K.; Yamasaki, N.; Okita, N.; Nanashima, A.; Higami, Y.; Nagayasu, T. The poly(adenosine diphosphate-ribose) polymerase inhibitor PJ34 reduces pulmonary ischemia-reperfusion injury in rats. *Transplantation* **2014**, *98*, 618–624. [CrossRef]
140. Xu, J.C.; Fan, J.; Wang, X.; Eacker, S.M.; Kam, T.I.; Chen, L.; Yin, X.; Zhu, J.; Chi, Z.; Jiang, H.; et al. Cultured networks of excitatory projection neurons and inhibitory interneurons for studying human cortical neurotoxicity. *Sci. Transl. Med.* **2016**, *8*, 333ra48. [CrossRef]

141. Chatterjee, P.K.; Chatterjee, B.E.; Pedersen, H.; Sivarajah, A.; McDonald, M.C.; Mota-Filipe, H.; Brown, P.A.; Stewart, K.N.; Cuzzocrea, S.; Threadgill, M.D.; et al. 5-Aminoisoquinolinone reduces renal injury and dysfunction caused by experimental ischemia/reperfusion. *Kidney Int.* **2004**, *65*, 499–509. [CrossRef] [PubMed]
142. Mouchiroud, L.; Houtkooper, R.H.; Moullan, N.; Katsyuba, E.; Ryu, D.; Canto, C.; Mottis, A.; Jo, Y.S.; Viswanathan, M.; Schoonjans, K.; et al. The NAD(+)/Sirtuin Pathway Modulates Longevity through Activation of Mitochondrial UPR and FOXO Signaling. *Cell* **2013**, *154*, 430–441. [CrossRef] [PubMed]
143. Mota, R.; Sanchez-Bueno, F.; Berenguer-Pina, J.J.; Hernandez-Espinosa, D.; Parrilla, P.; Yelamos, J. Therapeutic treatment with poly(ADP-ribose) polymerase inhibitors attenuates the severity of acute pancreatitis and associated liver and lung injury. *Br. J. Pharmacol.* **2007**, *151*, 998–1005. [CrossRef] [PubMed]
144. Montoni, A.; Robu, M.; Pouliot, E.; Shah, G.M. Resistance to PARP-Inhibitors in Cancer Therapy. *Front. Pharmacol.* **2013**, *4*, 18. [CrossRef]
145. Prensner, J.R.; Chen, W.; Iyer, M.K.; Cao, Q.; Ma, T.; Han, S.; Sahu, A.; Malik, R.; Wilder-Romans, K.; Navone, N.; et al. PCAT-1, a long noncoding RNA, regulates BRCA2 and controls homologous recombination in cancer. *Cancer Res.* **2014**, *74*, 1651–1660. [CrossRef]
146. Livraghi, L.; Garber, J.E. PARP inhibitors in the management of breast cancer: Current data and future prospects. *BMC Med.* **2015**, *13*, 188. [CrossRef]
147. Quereda, V.; Bayle, S.; Vena, F.; Frydman, S.M.; Monastyrskyi, A.; Roush, W.R.; Duckett, D.R. Therapeutic targeting of CDK12/CDK13 in triple-negative breast cancer. *Cancer Cell* **2019**, *36*, 545–558. [CrossRef]
148. Balaji, K.; Vijayaraghavan, S.; Diao, L.; Tong, P.; Fan, Y.; Carey, J.P.; Bui, T.N.; Warner, S.; Heymach, J.V.; Hunt, K.K. AXL inhibition suppresses the DNA damage response and sensitizes cells to PARP inhibition in multiple cancers. *Mol. Cancer Res.* **2017**, *15*, 45–58. [CrossRef]
149. Garcia, T.B.; Snedeker, J.C.; Baturin, D.; Gardner, L.; Fosmire, S.P.; Zhou, C.; Jordan, C.T.; Venkataraman, S.; Vibhakar, R.; Porter, C.C. A small-molecule inhibitor of WEE1, AZD1775, synergizes with olaparib by impairing homologous recombination and enhancing DNA damage and apoptosis in acute leukemia. *Mol. Cancer Ther.* **2017**, *16*, 2058–2068. [CrossRef]
150. Song, Z.; Tu, X.; Zhou, Q.; Huang, J.; Chen, Y.; Liu, J.; Lee, S.; Kim, W.; Nowsheen, S.; Luo, K. A novel UCHL 3 inhibitor, perifosine, enhances PARP inhibitor cytotoxicity through inhibition of homologous recombination-mediated DNA double strand break repair. *Cell Death & Disease* **2019**, *10*, 398.
151. Miller, A.L.; Fehling, S.C.; Garcia, P.L.; Gamblin, T.L.; Council, L.N.; van Waardenburg, R.C.; Yang, E.S.; Bradner, J.E.; Yoon, K.J. The BET inhibitor JQ1 attenuates double-strand break repair and sensitizes models of pancreatic ductal adenocarcinoma to PARP inhibitors. *EBioMedicine* **2019**, *44*, 419–430. [CrossRef] [PubMed]
152. Ibrahim, Y.H.; García-García, C.; Serra, V.; He, L.; Torres-Lockhart, K.; Prat, A.; Anton, P.; Cozar, P.; Guzmán, M.; Grueso, J. PI3K inhibition impairs BRCA1/2 expression and sensitizes BRCA-proficient triple-negative breast cancer to PARP inhibition. *Cancer Discov.* **2012**, *2*, 1036–1047. [CrossRef] [PubMed]
153. Zhong, Q.; Hu, Z.; Li, Q.; Yi, T.; Li, J.; Yang, H. Cyclin D1 silencing impairs DNA double strand break repair, sensitizes BRCA1 wildtype ovarian cancer cells to olaparib. *Gynecol. Oncol.* **2019**, *152*, 157–165. [CrossRef] [PubMed]
154. Jirawatnotai, S.; Hu, Y.; Michowski, W.; Elias, J.E.; Becks, L.; Bienvenu, F.; Zagozdzon, A.; Goswami, T.; Wang, Y.E.; Clark, A.B.; et al. A function for cyclin D1 in DNA repair uncovered by protein interactome analyses in human cancers. *Nature* **2011**, *474*, 230–234. [CrossRef]
155. Hirst, J.; Godwin, A.K. AURKA inhibition mimics BRCAness. *Aging (Albany NY)* **2017**, *9*, 1945. [CrossRef]
156. Byrum, A.K.; Vindigni, A.; Mosammaparast, N. Defining and Modulating 'BRCAness'. *Trends Cell Biol.* **2019**. [CrossRef]
157. Wu, W.; Nishikawa, H.; Fukuda, T.; Vittal, V.; Asano, M.; Miyoshi, Y.; Klevit, R.E.; Ohta, T. Interaction of BARD1 and HP1 is required for BRCA1 retention at sites of DNA damage. *Cancer Res.* **2015**, *75*, 1311–1321. [CrossRef]
158. Criscuolo, D.; Morra, F.; Giannella, R.; Visconti, R.; Cerrato, A.; Celetti, A. New combinatorial strategies to improve the PARP inhibitors efficacy in the urothelial bladder Cancer treatment. *J. Exp. Clin. Cancer Res.* **2019**, *38*, 91. [CrossRef]
159. Cerrato, A.; Merolla, F.; Morra, F.; Celetti, A. CCDC6: The identity of a protein known to be partner in fusion. *Int. J. Cancer* **2018**, *142*, 1300–1308. [CrossRef]

160. Vena, F.; Jia, R.; Esfandiari, A.; Garcia-Gomez, J.J.; Rodriguez-Justo, M.; Ma, J.; Syed, S.; Crowley, L.; Elenbaas, B.; Goodstal, S. MEK inhibition leads to BRCA2 downregulation and sensitization to DNA damaging agents in pancreas and ovarian cancer models. *Oncotarget* **2018**, *9*, 11592. [CrossRef]
161. Yin, L.; Liu, Y.; Peng, Y.; Peng, Y.; Yu, X.; Gao, Y.; Yuan, B.; Zhu, Q.; Cao, T.; He, L. PARP inhibitor veliparib and HDAC inhibitor SAHA synergistically co-target the UHRF1/BRCA1 DNA damage repair complex in prostate cancer cells. *J. Exp. Clin. Cancer Res.* **2018**, *37*, 153. [CrossRef] [PubMed]
162. Cruz, O.V.; Prudnikova, T.Y.; Araiza-Olivera, D.; Perez-Plasencia, C.; Johnson, N.; Bernhardy, A.J.; Slifker, M.; Renner, C.; Chernoff, J.; Arias, L.E. Reduced PAK1 activity sensitizes FA/BRCA-proficient breast cancer cells to PARP inhibition. *Oncotarget* **2016**, *7*, 76590.
163. Li, L.; Karanika, S.; Yang, G.; Wang, J.; Park, S.; Broom, B.; Manyam, G.C.; Wu, W.; Luo, Y.; Basourakos, S. Enzalutamide-induced "BRCAness" and PARP inhibition is a synthetic lethal therapy for castration-resistant prostate cancer. *Sci. Signal.* **2017**, *10*. [CrossRef]
164. Liu, L.; Zhou, W.; Cheng, C.-T.; Ren, X.; Somlo, G.; Fong, M.Y.; Chin, A.R.; Li, H.; Yu, Y.; Xu, Y. TGFβ induces "BRCAness" and sensitivity to PARP inhibition in breast cancer by regulating DNA-repair genes. *Mol. Cancer Res.* **2014**, *12*, 1597–1609. [CrossRef] [PubMed]

cancers

MDPI

Review

The Multifactorial Role of PARP-1 in Tumor Microenvironment

Juan Manuel Martí, Mónica Fernández-Cortés, Santiago Serrano-Sáenz, Esteban Zamudio-Martinez, Daniel Delgado-Bellido, Angel Garcia-Diaz and Francisco Javier Oliver *

Instituto de Parasitología y Biomedicina López Neyra, CSIC and CIBERONC, Instituto de Salud Carlos III, 18016 Granada, Spain; jmmmc@ipb.csic.es (J.M.M.); monica.fernandez@ipb.csic.es (M.F.-C.); s.serrano@ipb.csic.es (S.S.-S.); estebanzm95@ipb.csic.es (E.Z.-M.); ddelgado@ipb.csic.es (D.D.-B.); agdiaz@ipb.csic.es (A.G.-D.)

* Correspondence: joliver@ipb.csic.es

Received: 18 February 2020; Accepted: 15 March 2020; Published: 20 March 2020

Abstract: Poly(ADP-ribose) polymerases (PARPs), represent a family of 17 proteins implicated in a variety of cell functions; some of them possess the enzymatic ability to synthesize and attach poly (ADP-ribose) (also known as PAR) to different protein substrates by a post-translational modification; PARPs are key components in the cellular response to stress with consequences for different physiological and pathological events, especially during neoplasia. In recent years, using PARP inhibitors as antitumor agents has raised new challenges in understanding their role in tumor biology. Notably, the function of PARPs and PAR in the dynamic of tumor microenvironment is only starting to be understood. In this review, we summarized the conclusions arising from recent studies on the interaction between PARPs, PAR and key features of tumor microenvironment such as hypoxia, autophagy, tumor initiating cells, angiogenesis and cancer-associated immune response.

Keywords: Tumor microenvironment; PARPs; PARylation; hypoxia; autophagy; PARP inhibitors

---

1. Introduction

1.1. PARP Family of Proteins

Poly (ADP-ribose) polymerases (PARPs), more recently named ADP-ribosyl transferases (ARTs) [1] are a family of enzymes characterized by the presence of a 50 amino acid sequence called "*PARP signature*". This allows some members of the family to synthesize and transfer ADP-ribose to a large number of substrates through a process referred as poly(ADP-ribosyl)ation or PARylation [2]. The most representative member of this family is PARP-1, firstly described by Chambon et al. in 1963 [3]. Since then, many studies have deciphered both structural and biological aspects of this family, as well as the pathological consequences of the misregulation of its members.

The best studied PARP enzyme is the original constituent of the family PARP-1. Encoded by the region 1q41-q42, this protein has a molecular weight of 114 kDa and it is constitutively expressed. In fact, PARP-1 promoter owns characteristics usually revealed in housekeeping genes. Consistently, in all tissues the mRNA of PARP-1 is present, albeit at varying levels [4]. Nevertheless, PARP-1 modulation is mainly settled at the protein level and not at the mRNA level.

The PARP superfamily contains 17 members sharing on their catalytic domain the conserved "PARP signature" (Figure 1).

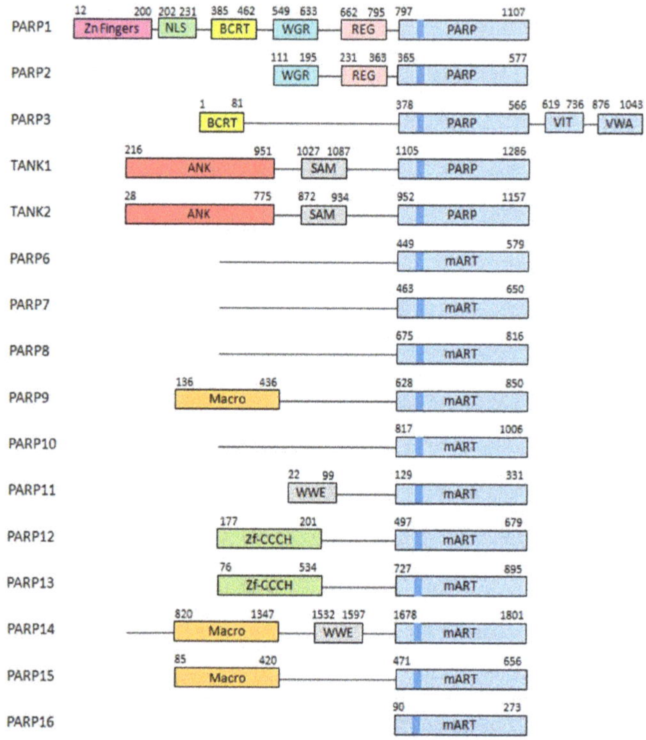

Figure 1. Poly(ADP-ribose) polymerases (PARP) family. The structure of the different members of the PARP family is described. Different domains are detailed in different colors. Brighter blue shows the PARP signature sequence, common throughout all the members of the family [5].

As this signature is not sufficient to provide a functional classification, the categorization must be established regarding their architecture and different enzymatic functions. Although firstly classified into three subgroups [6,7], this family is now grouped into five subfamilies [8,9]:

- DNA-dependent PARPs. Active during DNA damage thanks to their DNA-binding domain that consist in three zinc finger and a nuclear localization signal in the case of PARP-1 (ARTD1) [10]. Other members of this group are PARP-2 (ARTD2) and PARP-3 (ARTD3).
- Tankyrases. Containing Ankyrin-domain repeats responsible for protein-protein interactions. Very specific of this subfamily are the sterile α motifs (SAM), also related with protein-protein interactions: Tankyrase-1 (PARP-5A, ARTD5) and Tankyrase-2 (PARP-5B, ARTD6).
- CCCH PARPs. Containing CCCH motifs of the $CX_{7-11}CX_{3-9}CX_3H$ type, this domain is related with RNA-binding: TIPARP (PARP-7, ARTD7), PARP-12 (ARTD12) and PARP-13 (ARTD13).
- Macro-PARPs. Bearing macrodomain folds. They mediate the migration of the proteins to poly (and maybe also mono) ADP-ribosylation sites: BAL1 (PARP-9, ARTD9), BAL2 (PARP-14, ARTD8) and BAL3 (PARP-15, ARTD7).
- Other PARPS. Proteins that do not fit into any of the previous classifications [9]; PARP-4 (ARTD4), PARP-6 (ARTD17), PARP-8 (ARTD16), PARP-10 (ARTD10), PARP-11 (ARTD11) and PARP-16 (ARTD15).

Being the most active member of this family, the PARP-1 structure was firstly described in 1984 [11] and it is composed of three domains (Figure 2):

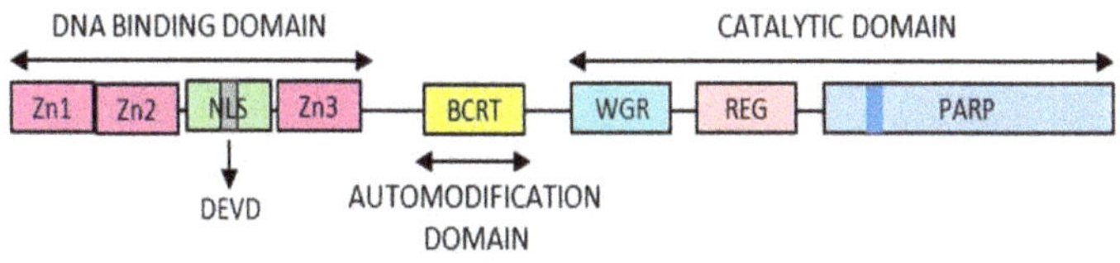

Figure 2. PARP1 structure. The different domains of PARP1 are presented here. The zinc fingers responsible for the DNA binding capacity, the automodification domain thanks to which PARP1 is modified by polymer and the catalytic domain containing the "PARP signature".

- DNA-binding domain: Involved in DNA interaction, interdomain cooperation, chromatin condensation and protein–protein binding.
- Automodification domain: Serve as acceptor during auto PARylation [12]. And mediates protein–protein interactions [13].
- NAD-binding domain: Serves as the catalytic domain, it contains the "PARP signature" sequence responsible for the PAR synthesis.

In order to perform its PARylation activity, PARP-1 and PARP-2 performs both poly(ADP-ribosyl) synthetase and transferase enzymatic activity. The PAR life cycle is described as follows [7,14] (Figure 3).

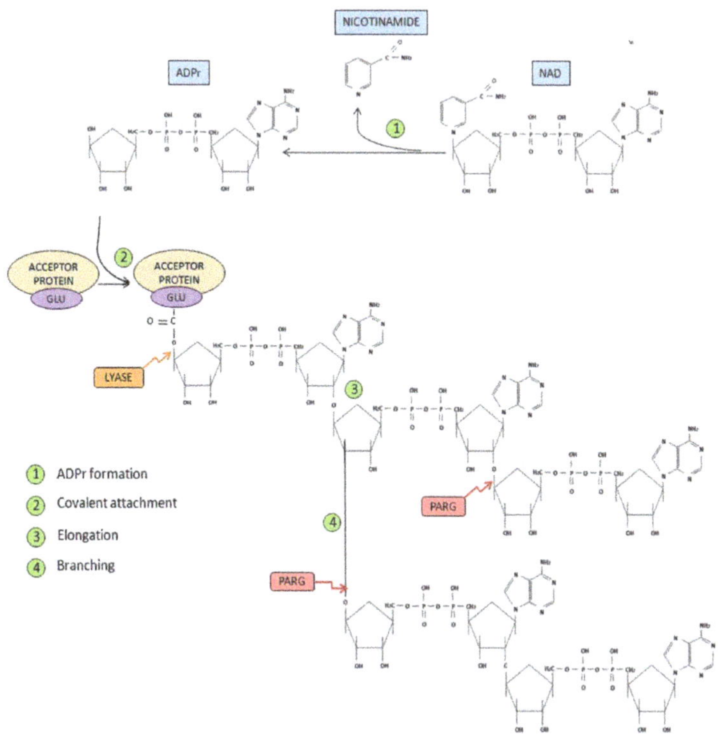

Figure 3. PARP metabolism. Different steps in polymer formation are shown in green. In contrast, the proteins involved in polymer degradation are shown in red and yellow [14,15].

- Initiation phase: Firstly, poly(ADP-ribose) synthetase activity catalyzes the breakage of the glycosidic bond between nicotinamide and ribose on the NAD+ molecule. Through this process of oxidation, ADP-ribose is formed. Subsequently, ADP-ribose is covalently attached to different acceptor proteins via formation of an ester bond between the protein (through glutamate, aspartate or lysine residues) and ADP-ribose.

- Elongation and branching reaction: In addition, PARP-1-mediated poly(ADP-ribosyl) transferase activity is able to catalyze the reactions responsible of elongation and branching, using more ADP-ribose units obtained from NAD+.
- PAR degradation: Poly(ADP-ribose) glycohydrolase (PARG) [15] mediates PAR degradation. PARG is presented in three different isoforms: PARG99 and PARG102 (of 99 and 102 kDa, respectively), both located in the cytoplasm; and PARG110 (an isoform of 111 kDa), that is located mostly in the nucleus [5].
- Ester bond breakage: Once PAR has been degraded, the firstly attached mono(ADP-ribosyl) moiety bond to the acceptor protein is removed by the ADP-ribosyl protein lyase [16].
- AMP and NAD recycling: Free poly(ADP-ribose) and ADP-ribose monomer are the final products of PAR degradation, this latter molecule can cause protein damage through glycation processes. ADP-ribose pyrophosphatase [17] converts this free ADP-ribose into AMP and ribose 5-phosphate, generating compounds much less reactive and more likely to be used in order to obtain new NAD+ [18].

As explained previously, the best described effect of PARP activation is PARylation, or covalent protein modification by PAR. This process may affect PARP itself or other proteins that become PARylated. Nevertheless, other mechanisms underlie PARP activation and PAR synthesis. They are summarized below:

- Bind of different targets to free PAR in a non-covalent way [19,20]. Consistently, since the first PAR-binding domain was described [21], two new PAR-binding motifs were discovered in recent years, i.e., a PAR-binding zinc finger motif [22] and a histone macrodomain [23].
- Free PAR can also act as a relevant signaling molecule. The discovery of a type of cell death induced by this molecule is a clear example. In this process, to as PARthanatos, the release of apoptosis-inducing factors (AIF) from the mitochondria is triggered [24–27].
- PARP activation generates an important reduction of its substrate NAD+ after high DNA breakage accumulation. This depletion has important consequences on cell survival [28].

These molecular events are responsible for the different events and function alterations that take place after PARP activation (Figure 4).

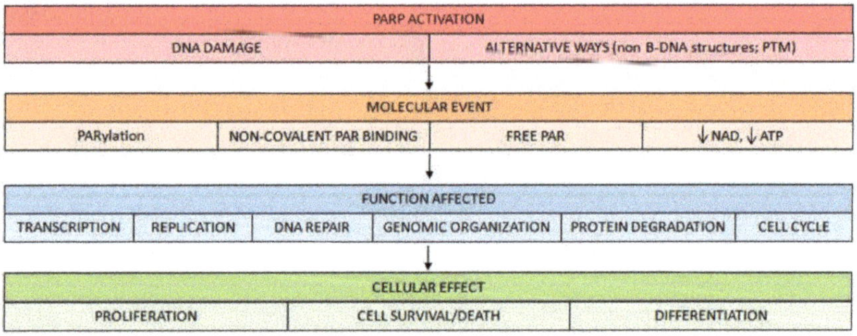

Figure 4. Molecular events following PARP activation. Once PARP is activated, downstream events of PARP signaling take place, involving either covalent PARylation of substrates, non-covalent binding of PAR polymer to proteins bearing a PAR-binding motif, release of free PAR to the cell or lowering cellular NAD+/ATP levels. Via these pathways, PARP/PARylation regulate functions such as transcription, replication, DNA repair, protein degradation and cell cycle, mediating various cellular phenomena such as proliferation, cell survival and cell death or differentiation.

The "central dogma" of PARylation asserts that during DNA damage PARP-1 is activated, on both single strand breaks (SSB) and double strand breaks (DSB). In fact, DNA strand breaks rises up to 500 times the basal activity of PARP-1 [29]. Consistently, DNA alkylating agents as well as reactive species of both oxygen and nitrogen (ROS and RNS, respectively) have been proved to act as triggers of PAR accumulation in different cell types [30]. However, different results have shown that PARP-1 is also stimulated in the absence of DNA damage. This has been observed in the presence of non-B DNA structures like bent, cruciform DNA or stably unpaired DNA [31].

In addition, PARP-1 activity can be modified in addition to DNA-related status; several post-translational modifications also alter PARP-1 activity. PARP-1 is implicated in multiple signaling pathways often involving a kinase phosphorylating different aminoacids of PARP-1, causing its activation or inhibition. Examples of this topic include PARP-1 being phosphorylated by ERK1/2 at serine 372 and threonine 373, deriving in PARP-1 activation [32], Also c-Met phosphorylates PARP-1, specifically at the Tyr907 causing its activation. [33–35]. Other phosphorylations can have the opposite effect, leading to PARP-1 inhibition, this is the case of the one performed by AMPK at Ser177 [36] and the one mediated by the protein kinase C [37]. PARP1 regulates the c-Jun N-terminal kinase (JNK) pathway, which is a driver of tumor development and treatment response. Based on that, PARP inhibition could be potentially therapeutically beneficial in ovarian cancer taken the elevated JNK activity. Furthermore, PARP1 inhibitors promote Akt [38].

Additionally, PARP-1 is regulated through other post-translational modifications different to phosphorylation but equally necessary for PARP to develop its function. PARP-1 is acetylated by p300/CBP at the lysines 498, 505, 508, 521 and 524 [39]. Also remarkable is the ubiquitination at the lysine 486 and 203 performed by the E3 ligase regulating PARP1 activity [40]. PARP-1 even undergoes ADP-ribosylation, as the described within the serine S499 [30,41].

1.2. Tumor Hypoxic Response and PARP-1

When an avascular tumor grows over a few cubic millimeters (depending on the tissue density and irrigation), the central parts of the mass experience low oxygen concentration and nutrients deprivation [42]. The response to this hypoxic environment has proven to be a key factor characterizing the tumor microenvironment and the disease outcome. It induces metabolic adaptations like glycolysis activation [43], cancer stem cell regulation [44], tumor exosome production [45] and the crucial tumor neovascularization [46]. New vessel formation supplies the hypoxic tumor with nutrients and oxygen, allowing the stressed cells to survive and divide, promoting mass formation and increasing cell plasticity, migration and aggressiveness [47–49]. The relevance of this pathway was vindicated when its discoverers were awarded the Nobel Prize of Medicine in 2019.

Furthermore, hypoxia generates resistance to different therapies. On the one hand, poorly vascularized areas make chemotherapy and immunotherapy difficult to disseminate. On the other hand, low oxygen levels make radiotherapy less effective at generating massive levels of toxic reactive oxygen species (ROS). This is the reason why hypoxic areas are known to survive treatments and are the ones to generate relapses and develop metastasis [50].

The hypoxic response is mediated by the stabilization of the hypoxia inducible factors (HIFs), a family of transcription factors composed of three alpha chains: HIF1α, HIF2α and HIF3α which are stable only during hypoxia, and one beta chain: HIF1β, which is constitutively stable. These proteins are active as heterodimers of one of the HIFα chains with the HIF1β. The heterodimers bind to the hypoxic response elements (HRE), a sequence of nucleotides located at the promoters of hypoxic inducible genes causing their overexpression. These different α/β heterodimers present similar but not identical targets and are differently expressed among tissues [51].

PARP-1 has been shown to interact with both HIF1α and HIF2α, affecting their stability and activity (Figure 5).

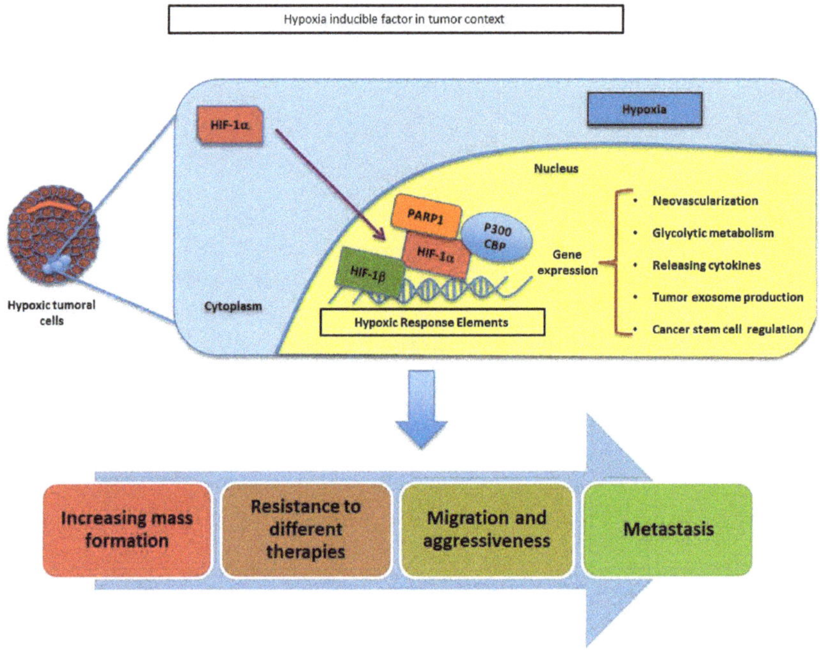

Figure 5. Schematic representation of HIF-α and PARP1 association throughout the adaptation of tumoral cells during hypoxia. The union of PARP1 to HIF-a helps to its stabilization and allows the transcription of several genes involved in tumor growth, resistance, migration and metastasis.

The crosstalk between HIF-signaling pathways and PARP-1 has been described in models of skin carcinogenesis where a reduction in HIF1α protein and mRNA is observed during PARP-1 inhibition or knockdown [52]. In the same way, in mice brains, a reduction in HIF1α accumulation during hypoxia is observed after PARP inhibition, having an impact reducing the expression of hypoxic genes like adrenomodulin or erythropoietin [53]. This dependence of HIF1α on PARP1 status can be explained by the fact that PARP-1 and HIF1α are known to form a complex during hypoxia [54]. Moreover, during hypoxia, the presence of nitrosative and oxidative stress induced via oxidative phosphorylation [55] is relevant, leading to mitochondrial inhibition and overactivation of PARP-1, promoting the stability of HIF1α [53]. It has also been described how the PARP inhibitor veliparib can sensitize both oxic and hypoxic cells in prostate and lungs to radiotherapy [56].

HIF2α is also important during tumorigenesis; tissues where HIF1α is the predominant protein become much more dependent on HIF2α during cancer progression [57]. HIF2α is related with aggressiveness and neovascularization as well [58]. Moreover, HIF2α is known to interact with PARP-1, resulting in changes in its stability and activity. HIF2α is similar to HIF1α but has different expression patterns among tissues and some of their targets do not overlap [51]. The relation between HIF2α and PARP1 has not been deeply studied but it has been proved in several cell types that PARP-1 knockdown or inhibition reduces HIF2α accumulation in the hypoxic context [59]. It has also been described that PARP-1 binds to the HIF2α promoter (but not to HIF1α promoter) controlling its transcriptional induction [60]. However, more research is needed in order to fully comprehend their interaction and its consequences.

1.3. Angiogenesis, Vasculogenic Mimicry and PARP-1

As previously described, blood vessels formation is a central aspect during tumor development. Vascularization not only distributes nutrients and oxygen, it also removes metabolic sub products and provides access for the immune system and different treatments.

The name 'angiogenesis' was used for the first time around 1900 and it was not used to refer to the tumoral context until the 1960s [61]. Since then, different varieties of angiogenesis have been depicted in cancer: Sprouting angiogenesis, micro vessel growth and microvascular proliferation [62]. Emerging studies showed the central role of this process in tumor progression [63]. Just a few tumors are able to progress without angiogenesis induction, while the vast majority of tumors present a combination of angiogenic with non-angiogenic areas [64,65]. Traditionally, treatments against angiogenesis have focused blocking new vessel growth and also trying to dismantle the existing ones, hence starving the tumors by depleting their nutrients and oxygen supply [66]. Interestingly, it has recently being reported in the PAOLA1 study (ESMO 2019) a Progression Free Survival benefit for ovarian cancer patients treated with a combination of olaparib with bevacizumab versus bevacizumab alone [67].

It is known that PARP inhibitors present antiangiogenic activity both in vitro and in vivo [52,68–70].

However, blood supply study in the cancer context became more complex when Maniotis et al. presented, in 1999, new findings describing cancer cells that coated vascular channels; these structures were composed of non-endothelial cells that contained erythrocytes and immune cells. This process was defined as vasculogenic mimicry (VM), and it is described as the de novo generation of a network created in a 3D matrix in vitro, composed by perfusable vasculogenic-like vessels, with properties similar to the matrix rich network described in aggressive tumors [71].

Through different studies performed on models of human uveal melanoma and cutaneous melanoma, the initial characterization of VM was performed. Since then, VM has been described in other malignancies like those of the kidney, lung, bladder, pancreas, prostate, gliomas, sarcoma, ovary, breast, head and neck cancer, and the list is still growing. In survival analyses, patients carrying VM in their tumors have been seen to present a reduced clinical outcome when compared with patients not expressing VM [72]. In 2000, a seminal study using arrays of gene expression in highly malignant melanoma cells (C8161, MUM 2B) vs poorly malignant melanoma cells (MUM 2C, C81-61) showed differences in gene expression among both groups [73]. Some of the many genes that showed significant differences were VE-cad. The depletion of VE-cadherin with specific siRNA and VE-Cad antibody abolished VM in a 3D in vitro model [74]. VE-cadherin is a trans-membrane protein usually located in the endothelium, necessary for the endothelial stability through its activity in cell–cell adhesion (Figure 6).

This protein presents the activities described for classic cadherins. Since the description of VM, VE-cadherin cannot be considered a marker of exclusively endothelial cells. VE-cadherin expression has been associated with aggressiveness on different tumors carrying VM. In this direction VE-cadherin can be found in highly aggressive VM tumor cells but not in poorly aggressive ones. Moreover, its down regulation in melanoma implied the loss of capacity to form VM [74].

It has been shown that activation through VEGFR-1 or through its co-receptor NRP-1 contribute to VM [75,76] while the potent inhibitor of αv integrins cilengitide displays anti-melanoma activity through the inhibition of VM [75]. A contrary role of VEGF signaling in VM has been also proposed. While is commonly accepted that VEGF and receptors like VEGFR1 or VEGFR co-receptor (NRP-1) promote VM, it has been also proposed that VM could also appear in contexts lacking this pathway: In this situation VEGF would promote angiogenesis, and its inhibition would potentiate other survival strategies including VM [77]. A study in 2013 by Rodriguez et al. gave a closer view into VM complexity, studying the interplay between PARP inhibition, epithelial-mesenchymal transition and VM. The use of the PARP inhibitors PJ-34 and KU0058948 in a murine melanoma cell line decreased the expression of VE-cadherin, as well as its phosphorylation in the residue of tyrosine 658. This resulted in the depletion of VM formation in in vitro assays [78].

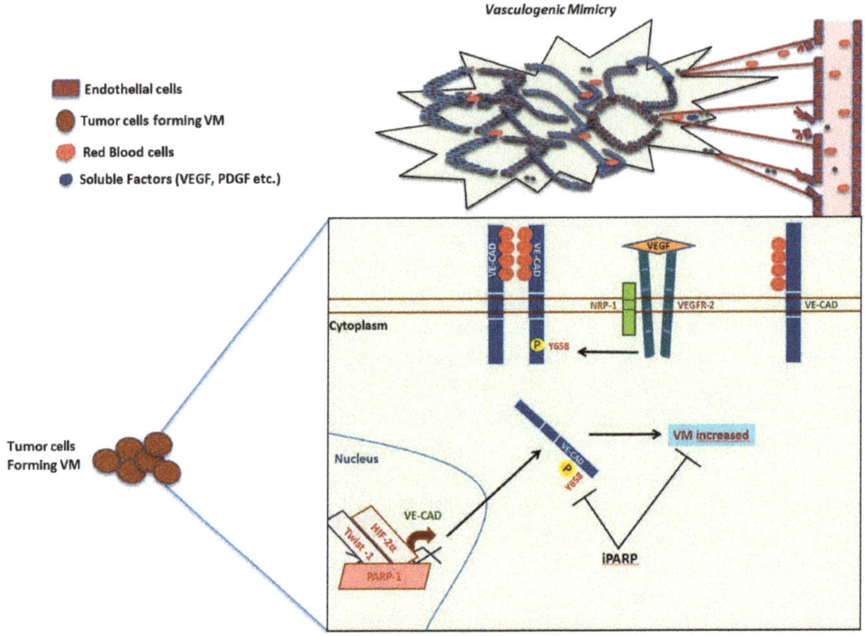

Figure 6. Vasculogenic mimicry pathway and PARP. VE-cadherin can be phosphorylated by stimuli of VEGF by VEGFR2 activity and the co-receptor (NRP-1). The soluble factor (VEGF) increases the phosphorylation of Y658 of VE-cadherin and the consequent internalization. Inhibition of PARP decreases the phosphorylation Y658 of VE-cadherin and finally the possibility to form VM in aggressive melanoma cells (B16F10 cells).

The use of angiogenesis inhibition in combination with other treatments has been approved by the FDA in some scenarios; this is the case of an anti-VEGF specific antibody (bevacizumab, Avastin, Roche) used in combination with chemotherapy or different cytokines therapies for late-stage advanced metastatic cancers (including renal cell cancer, colorectal cancer, non-squamous non-small cell lung cancer and breast cancer) [79]. On the other hand, angiogenesis inhibitors in combination with immune checkpoint in the treatment of breast cancer showed no clear benefits and are still under evaluation, so there are no current FDA-approved indications for their use in breast cancer [80]. Current studies show how the combination of angiogenesis and PARP inhibitors will be likely safe due to non-overlapping toxicities, and it might be expected that PARP inhibitors could be used in this context at full mono-therapy dosages.

2. Immuno-Response Modulation by PARP

There is increasing knowledge describing the immunological role of the PARP family that supports the combinatorial uses of PARP inhibitors (PARPi) and immunotherapies against cancer. It is known that PARPs have a role during inflammation, innate immunity and in immune cells. PARP enzymes interact with transcription and adhesion factors which are involved in the regulation of cytokines and inflammatory mediators related to different aspects of inflammation [81]. Several studies showed a PARP-1-dependent activation of NFkB, a major transcription factor during the inflammation process [82–84]. PARP-1 is required to trigger NEMO SUMOylation and monoubiquination which is necessary for NEMO and NFkB activation [84]. There are a complete set of other transcription factors involved in inflammation which are also modulated by PARPs like sp1 [85], NFAT [86] or SIRT1 [87] among others.

Otherwise, PARP-1 and PARP-2 regulate several common inflammatory factors and cytokines including Tumor Necrosis Factor-α, inducible Nitric Oxide Synthase or Interleukin1-β suggesting an overlapped mechanisms or regulation [88]. In addition, PARP enzymes are involved in regulating the expression of cytokines and chemo attractants like IL-6, IL-12 or CCL2. PARP-14 enhances the STAT6 regulation of the expression in Th2 cells through IL-4 induction, which is important for the immune function in the lung [89].

PARP-1 is involved in development and activation of different immune cell types like macrophages, microglia or dendritic cells. Moreover, PARP-1 and PARP-2 induce pro-inflammatory effects not only restricted to innate immune system cells but also important in dendritic cells and fibroblasts [90]. PARP enzymes are also involved in T-cells development. PARP-1 and PARP-2 expression is especially high in the lymphocyte-proliferative areas of the thymus. Moreover, inactivation of PARP-2 decreases the size of the thymus while reducing the number of CD4+/CD8+ (double positive) thymocytes, due to an affected survival of double positives [91]. In addition, other studies showed an impaired capacity to differentiate into Th2 cells in PARP-1 KO cells [92].

The stimulator of the interferon genes (STING) pathway was primarily described as a mechanism activated in response to microbial infections and DNA viruses, but also has relevance under certain autoimmune and inflammatory conditions. There is currently a list of new evidence suggesting a role of the STING pathway in tumor detection [93,94]. Activation of STING is produced by the accumulation of cytosolic DNA fragments which interact with the cGAMP synthase (cGAS), catalyzing the formation of the second messenger GAMP to activate STING [95]. After its induction, STING activates TBK1 which, in turn, phosphorylates STING and the interferon regulatory factor 3 (IRF3). Then IRF3 migrates to the nucleus, causing the overexpression of type I interferon genes, including interferon beta (IFNβ). STING and type I interferon beta signaling pathways are involved in T-cell priming and activation against tumor-associated antigens in the tumor microenvironment [96,97].

After the detection of cytosolic DNA and the activation of the cGAS-STING-TBK1-IRF3 axis leading to the activation of type I interferons, there is an observed induction of cytokines involved in T cell chemotaxis, as CCL5 or CXCL10. It has been reported that type I IFN production and CCL5 or CXCL10 expression correlate with the infiltration of cytotoxic lymphocytes $CD8^+$ in several cancers [98]. Jianfeng Shen et al. reported a mechanism describing PARPi modulation of immune responses against cancer, even independently of *BRCA1/2* mutational background, through the STING pathway, further enhanced by blocking different immune checkpoints [99].

It is also relevant that tumor mutational burden (TMB) serves as a good predictor of response to immunotherapy, since it correlates with the sensitivity of tumors to the immune checkpoint blockade on immunotherapies like antiPD-1/PD-L1 or CTLA4 [100]. In other words, the immunogenicity of a given tumor depends, in part, on the mutational load and subsequently on its neoantigen repertoire. Recognition of such neoantigens is considered a major factor in the efficacy of clinical immunotherapies [101]. Nonetheless, cancers presenting elevated copy numbers such as ovarian cancer and small cell lung cancer are not immunologically hot but have extraordinary levels of damaged DNA/chromosomes and there exists controversy about the neoantigen hypothesis. Some groups have shown that the vast majority of mutations within expressed genes in cancers do not lead to the formation of neoantigens that are recognised by T cells [102,103].

Mutational load range varies over several orders of magnitude between different types of tumors [104,105], but ultimately, it will be the consequence of a balance between different factors including DNA damage and DNA repair function. Conditions affecting these factors are therapeutically exploited with different approaches, using DNA chemotherapy damaging agents or radiation. This is especially important with genetic backgrounds comprising an impaired DNA repairing machinery or inhibitors of DNA repair proteins involved in DNA damage response pathways as single agents or in combination with the DNA-damaging agents [106].

As previously indicated, PARP-2, PARP3, and especially PARP-1 became catalytically active in response to DSB, recruiting proteins that are involved in chromatin remodeling and DNA repair [107]. If the tumor immune response is modulated by the mutational load of the cells and the subsequent presence of neoantigens, it seems reasonable to hypothesize that PARPi could increase the tumor mutational burden due to an impaired DNA repair pathway function, contributing to the immunotherapy efficacy. There are currently several clinical trials exploring this possibility mainly but not exclusively in ovarian cancer [106,108].

Programmed cell death-1 (PD-1) is an immune receptor mainly expressed on activated CD4+ and CD8+ T cells or peripheral B cells [109]. Interaction of PD-1 and its ligand PD-L1 is critical to control the immune response, providing its binding constitutes an immune inhibiting checkpoint which leads to immune evasion. PD-L1 can be induced in cancer cells by a variety of stimuli such as T cell interferon gamma production or ionizing radiation (IR) [110] among others. The radiation-dependent PD-L1 activation seems to be related to DSBs and DNA repair response pathways and synergistically enhances antitumor immunity if applied together with immune checkpoint inhibitors [111]. This enhanced antitumor immunity seems to be related to increased mutational burden which increases neoantigen repertoire and tumor infiltrating lymphocytes (TILs) [112].

Interestingly, several studies also reported increased intratumoral $CD8^{+}$ T cell infiltration and interferon production after PARP inhibition [99,113]. Nevertheless, this increased presence of antitumor immunity and TILs can be counterbalanced by PARPi-induced expression of PD-L1 which subsequently activates the PD-1/PD-L1 immune checkpoint pathway [114].

Taken together, these observations provide further rationale for the combinatorial uses of DNA damaging agents, PARP inhibitors, and immune checkpoint blockade, in order to increase the benefits of the enhanced antitumor immunity avoiding the immunosuppressive effects of PD-L1 overexpression (Figure 7).

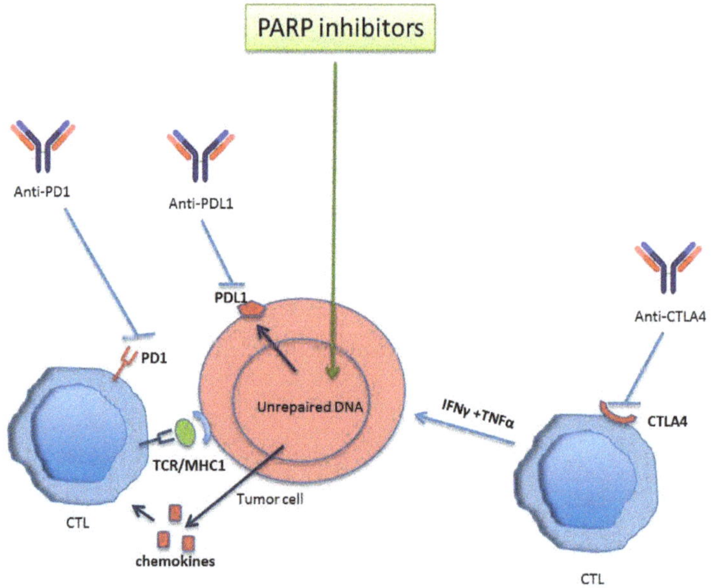

Figure 7. PARP inhibitors together with immune checkpoint inhibitors potentiate antitumor immune-mediated response. Activated T lymphocytes react against tumor antigens. The inhibition of PARP increases the number of lymphocytes infiltrating the tumor after the upregulation of chemokines, promoting an immune response mediated by CTLs. In spite PARP inhibitors modulate positively the upregulation of PDL-1 (favouring tumor scape from immune control) the anti-CTLA4 activates T cells to promote an antitumor response. Anti PD1/PDL-1 reverses CTL inhibition provoked by PARP inhibitor –induced PDL-1 expression. In this way, anti-PD1/PDL-1 can synergize with PARP inhibitors to ammeliorate antitumor immune response.

3. PARylation in Autophagy

One of the main characteristics of the metabolically hyperactive cancer cells is their high demand for nutrients and oxygen from their microenvironment. This makes them vulnerable to the deficiency of both oxygen (hypoxia) and nutrients (starvation). Both situations can induce metabolic stress leading to PARP-1 over activation; during these situations, cells must activate the autophagic pathway as a response to adapt and survive. In this context it has been shown that one of the first consequences of autophagy is the activation of poly-(ADP-ribosyl)ation. The combined effect of PARP-1 activity leading to PAR modification of AMPK (the main cell energy sensor) and the nutritional status (measured via mTORC1 activation) are fundamental during the first stages of autophagy [115].

The autophagyc pathway is present in all eukaryotic cells. It consists on a "self-eating" process necessary for the maintenance of the cell homeostasis. Through this lysosomal-dependent pathway different cellular organelles, portions of the cytosol and chaperone-associated cargoes are enfolded in double-membrane spheroids called autophagosomes, being then hydrolyzed by lysosomal enzymes [116]. Different cellular stresses can fire the autophagic response: hypoxia, pH variation, mitochondrial ROS production, DNA damage, intracellular pathogens, unfolded proteins or endoplasmic reticulum (ER) stress among others.

Expanding evidence indicates how autophagy is induced after DNA breakage. Ataxia Telangiectaxia Mutated (ATM) has been described as an important link between DNA Damage Response (DDR) and the induction of autophagy [117]. In response to DNA damage by mitochondrial ROS, external toxins or irradiation, ATM is auto phosphorylated within a Mre11, Rad50 and Nbs1 (MRN) multiprotein complex that binds DSBs. Once active ATM induces the activity of AMPk and its target tuberous sclerosis protein (TSC2), this leads to the inhibition of mTORC1, promoting the formation of autophagosomes dependent of Unc-51 like autophagy activating kinase (ULK1) [118]. This is how autophagy acts as a catabolic pathway of rapid energy recovery that the cell will use to grow and expand. However, this rapid acquisition of energy has two sides; on the one hand, it responds to a pro-survival role as long as the levels of autophagy remains biologically sustainable, this way, the degradation of excess vital components of the cell is avoided; but autophagy might also become a process of cell death due to the accumulation of autophagosomes degrading essential cellular structure;, this process is called autophagic cell death (ACD).

As we know, PARPs are important guardians of the genome integrity and they have been shown to link DNA damage response with the activation of autophagy [115,119]. Autophagy is also initiated in several settings of response to chemotherapy and mediated by PARP-1. Doxorubicin treatment leads to over-activation of PARP-1, followed by ATP and NAD+ depletion, triggering the non-toxic accumulation of autophagosomes [19]. In response to methylnitronitrosoguanidine (MNNG) double knockout Bax−/−Bak−/−MEFs activate PARP-1, reducing intracellular ATP levels and activating AMPk pathway and mTORC1 down-regulation. Suppression of AMPk pathway blocks MNNG-induced autophagy and enhances cell death [19]. The same results were found in a nasopharyngeal carcinoma model exploring over-activation of PARP-1, upon ionizing radiation there is a PARylation-dependent energy depletion and up-regulation of AMPk and ULK1 pathways [120]. All these studies have demonstrated that autophagy should be contemplated as a target in cancer during the induction of DNA damage and consequently, new strategies based on the synthetic lethality concept during PARPi must be explored.

In addition to DNA damage, nutrient deprivation can be considered as the most physiological stimulus to induce reversible autophagy. Recent studies of our group have implicated PARP-1 in autophagy induced by nutrient starvation [115,121]. The absence of PARP-1 or the use of PARP inhibitors (PJ34, DPQ and Olaparib) delay starvation-induced autophagy. Mitochondrial ROS accumulation derived from starvation promoted DNA damage and PARylation signaling from PARP-1 activation, which triggers ATP depletion, AMPk activation and mTORC1 inhibition. The lack of PARP-1 activity compensates ATP depletion compromising the cascade AMPk/mTORC1 inhibition/ autophagy. Starved PARP-1-deficient/inhibited cells showed increased apoptotic cell death [115]. In order to analyze the in vivo consequences of PARP-1 ablation on autophagy both PARP-1+/+ and PARP-1−/−pups were starved for short periods of time, concluding that PARP-1−/−neonates display a deficient liver autophagy response following acute starvation, showing that PARP-1 activity and PAR formation are key players in the decision of the cell to engage autophagy [115].

The absence of PARP-1 compromises the activation of AMPk, more precisely the isoform AMPkα, reducing phosphorylation on Thr172 [121,122]. A nuclear subpopulation of AMPkα was detected in the breast cancer cell line MCF7, where it forms a stable complex with PARP-1; moreover the activation of nuclear AMPkα requires two essential events: Firstly, starvation-induced ROS must be imported to the nucleus to generate DNA damage and PARP-1 activation. Secondly, PARP-1 modifies by PARylation the nuclear population of AMPkα. Energy depletion associated with starvation was increased during non-lethal over-activation of PARP-1 in response to ROS, resulting in a feedback loop that favored the interaction between PARP-1/AMPkα and promoted the activation of LKB1. The mechanism of modulation by PARylation of LKB1 is not clearly understood but using PARPi or during specific assays with siRNA PARP-1, LKB1 activation is compromised [121].

In non-starved cells, PARP-1 forms a complex with AMPK; after starvation PARP-1 is activated and PARylation disrupt the PARP-1/AMPK complex, this triggers the nuclear export of PARylated AMPkα to the cytoplasm. This significant PARylated AMPkα population, is recognized by the active form of LKB1 promoting the phosphorylation on regulator site Thr172 of AMPkα. Again, a positive feedback loop takes place between the cytosolic energy depletion and PARylated AMPkα to potentiate the activation of the cytosolic AMPkα population by LKB1. The presence of covalently PARylated LKB1 has not been demonstrated but the effect of PARPi using PJ34 or olaparib suggested the implication of PAR on LKB1. These data place the nucleus (a classical cellular component not associated with autophagy) and the interaction PARP-1 and AMPk as an initial sensor of the metabolic alterations derived from perturbations in the nutritional extracellular status, not necessarily related with the alterations in genomic integrity. PARylation could be considered as an alert indicator of changes in the energy balance in cancer cells (Figure 8).

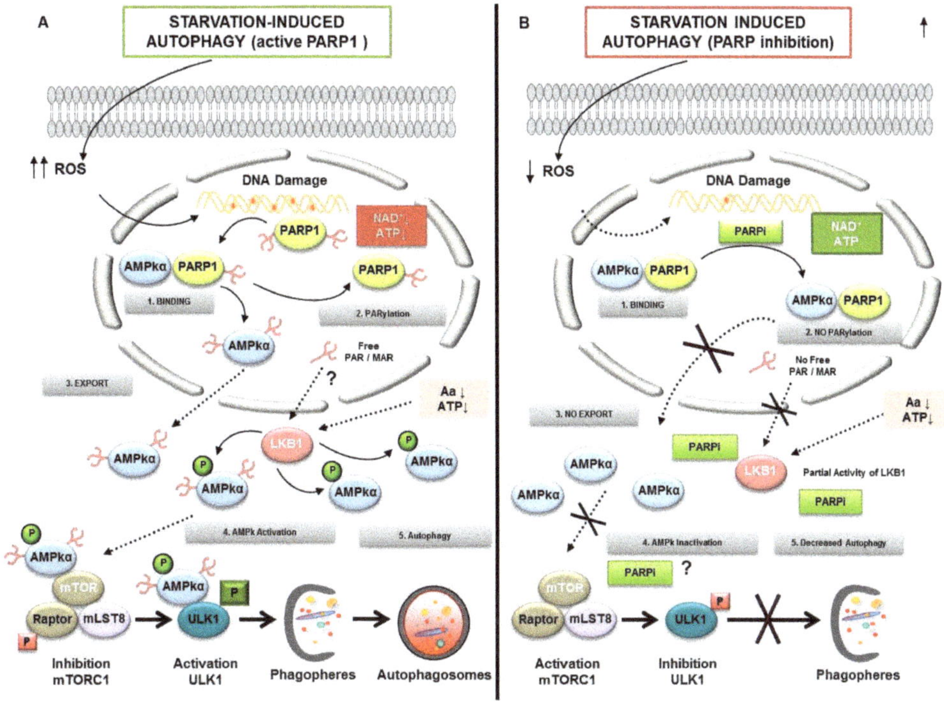

Figure 8. Regulation of autophagy by PARylation. (1) PARP1 forms a complex with AMPKα in nucleus. (**A**) PARP activation conditions: During starvation-induced autophagy, ROS release leads DNA damage and PARP1 overactivation. Self-PARylated PARP1 interact with AMPKα1 subunit (2). The complex is disrupted and PAR-AMPKα is exported to cytosol (3). The presence of PAR-AMPK and the continuous absence of amino acids and ATP depletion favor total activation of AMPKα population by LKB1, inhibition of mTORC1, interaction PAR-phospho-AMPK/ULK1, and autophagosome formation (4). LKB1 activity is presumably modified in a PARylation-dependent manner. (**B**) PARP inhibited: Starvation-induced ROS production was abrogated after PARP inhibition. Following AMPKα1/PARP1 interaction (1) the AMPKα1 subunit is not PARylated and AMPK nuclear export is inhibited (2 and 3). In spite of nutrient and energy depletion, AMPKα is inactive; mTORC1 is partially activated and interacts with ULK1, favoring its inhibition (4). Finally, the autophagosomes production will be delayed.

4. Cancer Initiating Cells

Cancer Stem Cells (CSCs) are defined as the tumor cell subpopulation that present remarkably high plasticity, considering "plasticity" as the ability of pluripotent cells to induce trans-differentiation. The first evidence of cancer initiating stem-like cells was described in 1994 by Lapidot et al. and Caceres-Cortes et al. after injecting different leukemia cell subpopulations in mice [123]. Through the years, CSCs were determined in a wide multitude of cancers like breast [124], gliomas [125], prostate [126], melanoma [127], lung [128,129], colon [130,131], pancreas [132], head and neck squamous cell [133], liver [134] and renal carcinoma [135].

CSCs are a very aggressive subpopulation of cancer cells characterized by their self-renewal capacity, tumor initiation ability, resistance to chemo- and radiotherapy and multi-lineage differentiation (stemness) [125,136,137]. The presence of this subpopulation in tumors is considered to be a marker of bad prognosis and therefore they have been proposed as target in cancer therapy.

For several years, it has been known that reprogramming processes are led by specific genetic programs. Yamanaka et al. demonstrated that the ectopic expression of four genes (Oct4, Sox2, c-Myc and KLF4) in mouse fibroblast transforms them in embryogenic stem cells. These cells, denominated by the authors as induced-pluripotent stem cells (iPSCs), were able to generate cells from any germ layer [138]. This breakthrough made Shiya Yamanaka win the Noble Price of Medicine in 2012. Today, these genes are known as Yamanaka factors and several researches link the expression of these genes with cancer [138–140]. In fact, the risk of tumor development is the most important limitation in the application of iPS cell-based therapy. Tumor cells present processes of trans-differentiation and dedifferentiation or reprogramming acquiring features of CSCs [138].

In 2012, it was demonstrated that PARP-1 is involved in the reprograming process, promoting iPSCs. An intense PARylation was detected in iPSCs. Moreover, PARP-1 knockdown reduced the capacity of iPSCs generation after Yamanaka factors overexpression. It has been found that c-Myc directly regulates PARP-1 expression and PARylation. Over-expression of PARP-1 compensates the knockdown of c-Myc in reprogramming in MEFs [141]. In 2009, Gao et al. demonstrated that PARP-1 PARylates and controls Sox2 levels, regulating its activity. They found that PARylated Sox2 increases its stability and therefore induces the expression of its target gene FGF4. PARP inhibition has been observed reverting this scenario [142,143]. To reprogram cells to iPSCs by Yamanaka factors, an epigenetic remodeling is also necessary. PARP-1 regulates modification in histones that alter the chromatin pattern driving to pluripotent cell phenotype. In this line, PARP-1 is necessary to promote the access to chromatin of Oct4 [144].

Telomerase (TERT) is a reverse transcriptase necessary for telomeres elongation in embryonic stem cells. This retrotranscriptase is highly expressed in tumors and the expression is related with stemness capacity. KLF4 directly interacts with the promoter of TERT. PARP-1 interacts with KLF4 and mediates the expression of TERT in CSCs. In fact, PARP-1 suppression dramatically reduces the recruitment of KLF4 to the promoter of TERT, reducing its expression [145]. C-Myc has also been related with TERT expression, suggesting another possible relationship between PARP-1 and telomerase [146].

Due to the high resistance to radio and chemotherapy presented by CSCs, many studies focused their interest in this subpopulation as a target for future treatment. In glioblastoma (GBM), one of the most aggressive malignancies, the presence of CSCs was proposed as the main cause of tumor relapse [138]. Furthermore, reprogramming events driving the formation of endothelial cells from tumor cells have been described [147]. The importance of CSCs in this kind of tumors led Vescovi el al. to propose the differentiation of CSCs as a target, with the aim of reducing the malignancy of this brain cancer [148]. They found that glioma stem cells treated with bone morphogenetic proteins (BMPs), cytokines belonging to TGF-β superfamily, differentiated CSCs to non-stem glioma cells. BMPs interact with membrane receptors inducing differentiation through SMADs signaling pathway. It has been demonstrated that PARP-1 negatively regulates this pathway at different levels. Ectopic expression of PARP-1 suppresses the signaling mediated by BMP. On the contrary, knockdown of PARP-1 promotes BMP-mediated differentiation. Furthermore, PARG, an enzyme responsible for PAR degradation, plays a positive role in the pathway [149,150]. On the other hand, SMAD signaling activation can be a double-edged sword in GBM because the stimulation of this pathway with TGF-β induces the expression of leukemia inhibitory factor (LIF), a cytokine that induces maintenance of "stemness" capacity [151].

5. Conclusions and Perspectives

The PARP superfamily consist of a group of proteins characterized by the presence of what is called a "PARP signature" on their sequences. Their main characteristic activity is the process referred to as PARylation. Through NAD+ and ATP consumption, ADP-Ribose is generated and then transferred as poly or mono ADP-ribose to different target factors. These modifications alter their activity or stability having relevant implications on the cellular metabolism. Moreover, excessive PARylation has an impact itself: free PAR can function as a signaling molecule and its synthesis may blunt NAD+ and ATP levels producing cell death.

PAR synthesis is activated mostly during DNA damage, altered DNA configuration or when PARP present some posttranscriptional modifications. As we summarized in this review, PARP activation has key repercussions on the cell fate affecting processes like DNA repair, transcriptional regulation, DNA remodeling, hypoxic response, epithelial mesenchymal transition, angiogenesis, autophagy, inflammation and cancer stem cell programming. All these processes lead to changes in survival, proliferation, differentiation, or even malignant transformation. Considering the relevance of the previously enumerated processes, it is easy to understand the relevance of PARP during cancer development. Cells overexpressing PARP will be more likely to repair DNA damage induced by genotoxic agents, they will adapt better to hypoxia and will be prone to produce metastasis through angiogenesis and EMT. Four PARP inhibitors have been already approved by the FDA (olaparib, rucaparib, niraparib and talazoparib) and they are used today as a result of their ability to generate "synthetic lethality" on BRCA 1/2 mutated tumors (see [152] for a review). However, ongoing advanced clinical trials will most likely expand their prescription as is the case for the combination of PARP inhibitors with classical therapies (Table 1) and with anti-angiogenic treatment (Table 2).

Currently, one of the limitations facing this therapeutic option is considering mostly on the BRCA mutated cells as HR deficient. Knowing that more than 100 genes are involved on this DNA repair pathway, it looks likely that BRCA proficient cells could be still HR defective due to other mutations. Measuring the whole genomic instability within a tumor by surveying the loss of genetic heterozygosity, telomeric allelic imbalance or the extent of somatic mutations, could be a more precise approach to define PARP inhibition sensibility, making more patients candidate of this therapeutic approach.

Another important limitation that needs to be worked out is the emergence of resistances, especially during PARP inhibition in monotherapy (Figure 9).

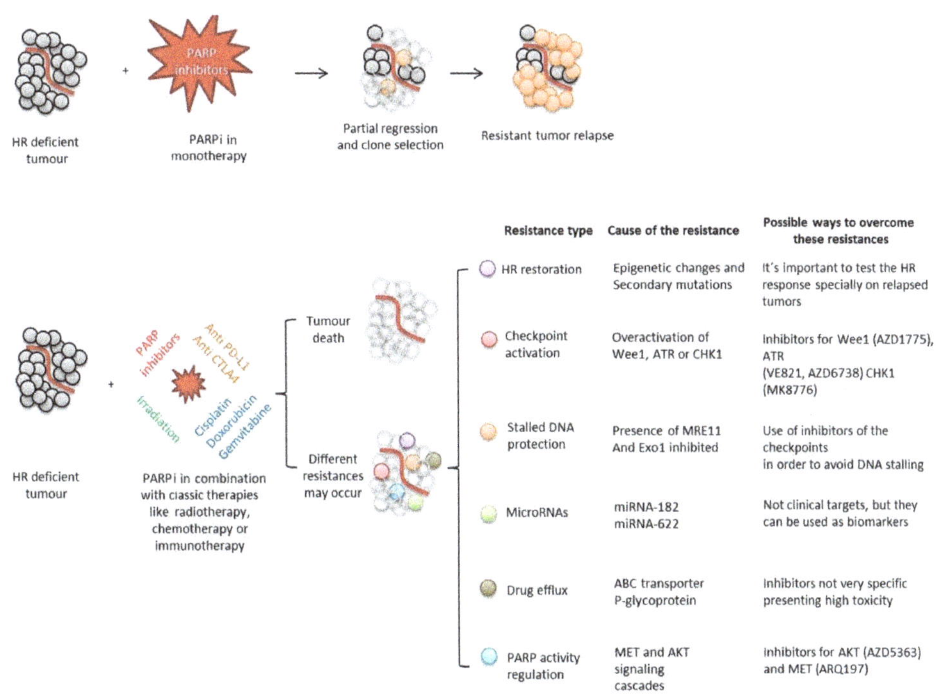

Figure 9. Summary of PARP inhibition approaches during cancer treatments, different possible outcomes and ways designed to overcome the possible appearance of resistances.

Table 1. Advanced clinical trials combining PARP inhibition with different classical therapies.

Status	Study Title	Conditions	Interventions	Phase	Number Enrolled	NCT Number
Active, not recruiting	A Phase 3 Randomized, Placebo-controlled Trial of Carboplatin and Paclitaxel With or Without Veliparib (ABT-888) in HER2-negative Metastatic or Locally Advanced Unresectable BRCA-associated Breast Cancer	Metastatic Breast Cancer	Drug: Paclitaxel Drug: Veliparib Drug: Carboplatin Other: Placebo	Phase3	513	NCT02163694
Active, not recruiting	Veliparib With Carboplatin and Paclitaxel and as Continuation Maintenance Therapy in Subjects With Newly Diagnosed Stage III or IV, High-grade Serous, Epithelial Ovarian, Fallopian Tube, or Primary Peritoneal Cancer	Ovarian Cancer Ovarian Neoplasm	Drug: Veliparib Drug: Paclitaxel Drug: Carboplatin Other: Placebo	Phase 3	1140	NCT02470585
Active, not recruiting (Has results)	Olaparib Treatment in BRCA Mutated Ovarian Cancer Patients After Complete or Partial Response to Platinum Chemotherapy	Platinum Sensitive BRCA Mutated Relapsed Ovarian Cancer Following Complete or Partial Response to Platinum Based Chemotherapy	Drug: Olaparib 300mg tablets Drug: Placebo to match olaparib 300mg	Phase 3	327	NCT01874353
Active, not recruiting	A Phase III Trial of Niraparib Versus Physician's Choice in HER2 Negative, Germline BRCA Mutation-positive Breast Cancer Patients	Carcinoma of Breast Human Epidermal Growth Factor 2 Negative Carcinoma of Breast BRCA1 Gene Mutation BRCA2 Gene Mutation	Drug: niraparib Drug: Physician's choice	Phase 3	306	NCT01905592
Active, not recruiting	A Study of Niraparib Maintenance Treatment in Patients With Advanced Ovarian Cancer Following Response on Front-Line Platinum-Based Chemotherapy	Ovarian Cancer	Drug: Niraparib Drug: Placebo	Phase 3	620	NCT02655016
Active, not recruiting (Has results)	A Study of Rucaparib as Switch Maintenance Following Platinum-Based Chemotherapy in Patients With Platinum-Sensitive, High-Grade Serous or Endometrioid Epithelial Ovarian, Primary Peritoneal or Fallopian Tube Cancer	Ovarian Cancer Fallopian Tube Cancer Peritoneal Cancer	Drug: Rucaparib Drug: Placebo	Phase 3	564	NCT01968213
Active, not recruiting (Has results)	Assessment of the Efficacy and Safety of Olaparib Monotherapy Versus Physicians Choice Chemotherapy in the Treatment of Metastatic Breast Cancer Patients With Germline BRCA1/2 Mutations.	Breast Cancer Metastatic BRCA 1 Gene Mutation BRCA 2 Gene Mutation	Drug: Olaparib Drug: Physician's choice chemotherapy	Phase 3	302	NCT02000622

Table 1. *Cont.*

Status	Study Title	Conditions	Interventions	Phase	Number Enrolled	NCT Number
Active, not recruiting	Olaparib Maintenance Monotherapy in Patients With BRCA Mutated Ovarian Cancer Following First Line Platinum Based Chemotherapy.	Newly Diagnosed Advanced Ovarian Cancer FIGO Stage III-IV(and 4 more)	Drug: Olaparib 300mg tablets	Phase 3	451	NCT01844986
Active, not recruiting	Olaparib or Cediranib Maleate and Olaparib Compared With Standard Platinum-Based Chemotherapy in Treating Patients With Recurrent Platinum-Sensitive Ovarian, Fallopian Tube, or Primary Peritoneal Cancer	BRCA Rearrangement Deleterious BRCA1 Gene Mutation Deleterious BRCA2 Gene Mutation (and 13 more)	Drug: Carboplatin Drug: Cediranib Maleate Drug: Gemcitabine Hydrochloride (and 6 more)	Phase 3	549	NCT02446600

Table 2. Different clinical trials on cancer treatment combining PARP inhibition with anti-angiogenic strategies.

Status	Study Title	Conditions	Interventions	Phase	Number Enrolled	NCT Number
Recruiting	Phase 2, A Study of Niraparib Combined With Bevacizumab Maintenance Treatment in Patients With Advanced Ovarian Cancer Following Response on Front-Line Platinum-Based Chemotherapy	Ovarian Cancer Fallopian Tube Cancer Primary Peritoneal Carcinoma	Drug: Niraparib Biological: Bevacizumab	Phase 2	90	NCT03326193
Recruiting	A Study of Cediranib and Olaparib at Disease Worsening in Ovarian Cancer	Ovarian Cancer	Drug: Cediranib Drug: Olaparib	Not Applicable	30	NCT02681237
Recruiting	A Study of Fluzoparib Given in Combination With Apatinib in Ovarian or Breast Cancer Patients	Ovarian Cancer Triple Negative Breast Cancer	Drug: Fluzoparib Drug: Apatinib	Phase 1	76	NCT03075462
Recruiting	Phase 2 Multicohort Study to Evaluate the Safety and Efficacy of Novel Treatment Combinations in Patients With Recurrent Ovarian Cancer	Ovarian Cancer	Drug: Niraparib Drug: TSR-042 Drug: Bevacizumab	Phase 2	40	NCT03574779
Recruiting	Mesothelioma Stratified Therapy (MiST): A Multi-drug Phase II Trial in Malignant Mesothelioma	Mesothelioma, Malignant	Drug: Rucaparib Drug: Abemaciclib Drug: pembrolizumab & bemcentinib Drug: Atezolizumab & Bevacizumab	Phase 2	120	NCT03654833
Recruiting	Study Evaluating the Efficacy of Maintenance Olaparib and Cediranib or Olaparib Alone in Ovarian Cancer Patients	Ovarian Cancer	Drug: Olaparib Drug: Cediranib	Phase 3	618	NCT03278717
Completed	A Study of Cediranib and Olaparib at the Time Ovarian Cancer Worsens on Olaparib	Ovarian Cancer	Drug: Olaparib Drug: Cediranib	Phase 2	4	NCT02340611

The generation of a selective force (like the caused by a drug treatment) in a highly mutagenic context can lead to the selection of resistant clones. They will then reproduce, forming a relapsed tumor resistant to the original treatment. To avoid this undesired consequence, the selective pressure can be reduced by spacing the treatments or by combining it with inhibitors for different targets, generating a constant pressure over the tumor while generating reduced selective forces.

Another way to avoid resistance is to study the genetic background and predict the more probable resistance mechanisms to arise, then combining the PARP inhibitor with secondary inhibitors for the factors related with the expected resistance. It is known that HR restoration, checkpoints activation, stalled DNA protection or drug efflux among others, can lead to PARP inhibition resistance. Combining PARP blockade with inhibitors for the possible ways of scape the treatment is being tested in clinical trials in order to make PARP inhibition more effective.

Based also in the studies summarized here, we propose that PARPi is an expansive field that may have therapeutic value beyond synthetic lethality. To this end, a precise comprehension of the implications of the different PARPs with PARylation in the complex tumor ecology is needed, including analysis of PARylation in both tumor and associated non-tumor cells, using single cell analysis and its consequences in tumor adaptation to hostile conditions. In this context, we can affirm that the use of PARP inhibitors against cancer treatment is not just a promising field but a reality and the challenge exists to widen their use by identifying new properties and deepening the role of PARP in tumor biology.

Funding: This work was supported by Ministerio de Economía y Competitividad, Spanish Ministry of Economy and Competitiveness SAF2012-40011-C02-01, SAF2015-70520-R, RTICC RD12/0036/0026, RTI2018-098968-B-I00, CIBER Cáncer ISCIII CB16/12/00421 and Fundación Domingo Martínez to FJO. JMMC, is recipient of a FPI fellowship from Spanish Ministerio de Ciencia; MFC and EZM are recipients of FPU fellowships from the Spanish Ministerio de Ciencia.

Conflicts of Interest: The authors declare "no conflict of interests".

References

1. Hottiger, M.O.; Hassa, P.O.; Luscher, B.; Schuler, H.; Koch-Nolte, F. Toward a unified nomenclature for mammalian ADP-ribosyltransferases. *Trends Biochem. Sci.* **2010**, *35*, 208–219. [CrossRef] [PubMed]
2. Peralta-Leal, A.; Rodriguez-Vargas, J.M.; Aguilar-Quesada, R.; Rodriguez, M.I.; Linares, J.L.; de Almodovar, M.R.; Oliver, F.J. PARP inhibitors: New partners in the therapy of cancer and inflammatory diseases. *Free Radic. Biol. Med.* **2009**, *47*, 13–26. [CrossRef] [PubMed]
3. Chambon, P.; Weill, J.D.; Mandel, P. Nicotinamide mononucleotide activation of new DNA-dependent polyadenylic acid synthesizing nuclear enzyme. *Biochem. Biophys. Res. Commun.* **1963**, *11*, 39–43. [CrossRef]
4. Meyer-Ficca, M.L.; Meyer, R.G.; Jacobson, E.L.; Jacobson, M.K. Poly (ADP-ribose) polymerases: Managing genome stability. *Int. J. Biochem. Cell Biol.* **2005**, *37*, 920–926. [CrossRef]
5. Meyer-Ficca, M.L.; Meyer, R.G.; Coyle, D.L.; Jacobson, E.L.; Jacobson, M.K. Human poly (ADP-ribose) glycohydrolase is expressed in alternative splice variants yielding isoforms that localize to different cell compartments. *Exp. Cell Res.* **2004**, *297*, 521–532. [CrossRef]
6. Rouleau, M.; Patel, A.; Hendzel, M.J.; Kaufmann, S.H.; Poirier, G.G. PARP inhibition: PARP1 and beyond. *Nat. Rev. Cancer* **2010**, *10*, 293–301. [CrossRef]
7. Hassa, P.O.; Hottiger, M.O. The diverse biological roles of mammalian PARPS, a small but powerful family of poly-ADP-ribose polymerases. *Front. Biosci.* **2008**, *13*, 3046–3082. [CrossRef]
8. Schreiber, V.; Dantzer, F.; Ame, J.C.; de Murcia, G. Poly (ADP-ribose): Novel functions for an old molecule. *Nat. Rev. Mol. Cell Biol.* **2006**, *7*, 517–528. [CrossRef]
9. Gibson, B.A.; Kraus, W.L. New insights into the molecular and cellular functions of poly (ADP-ribose) and PARPs. *Nat. Rev. Mol. Cell Biol.* **2012**, *13*, 411–424. [CrossRef]
10. Alemasova, E.E.; Lavrik, O.I. Poly (ADP-ribosyl) ation by PARP1: Reaction mechanism and regulatory proteins. *Nucleic Acids Res.* **2019**, *47*, 3811–3827. [CrossRef]

11. Kameshita, I.; Matsuda, Z.; Taniguchi, T.; Shizuta, Y. Poly (ADP-Ribose) synthetase. Separation and identification of three proteolytic fragments as the substrate-binding domain, the DNA-binding domain, and the automodification domain. *J. Biol. Chem.* **1984**, *259*, 4770–4776. [PubMed]
12. Desmarais, Y.; Menard, L.; Lagueux, J.; Poirier, G.G. Enzymological properties of poly (ADP-ribose) polymerase: Characterization of automodification sites and NADase activity. *Biochim. Biophys. Acta* **1991**, *1078*, 179–186. [CrossRef]
13. Bork, P.; Hofmann, K.; Bucher, P.; Neuwald, A.F.; Altschul, S.F.; Koonin, E.V. A superfamily of conserved domains in DNA damage-responsive cell cycle checkpoint proteins. *FASEB J.* **1997**, *11*, 68–76. [CrossRef] [PubMed]
14. Diefenbach, J.; Burkle, A. Introduction to poly (ADP-ribose) metabolism. *Cell. Mol. Life Sci.* **2005**, *62*, 721–730. [CrossRef]
15. Miwa, M.; Sugimura, T. Splitting of the ribose-ribose linkage of poly (adenosine diphosphate-robose) by a calf thymus extract. *J. Biol. Chem.* **1971**, *246*, 6362–6364.
16. Oka, J.; Ueda, K.; Hayaishi, O.; Komura, H.; Nakanishi, K. ADP-ribosyl protein lyase. Purification, properties, and identification of the product. *J. Biol. Chem.* **1984**, *259*, 986–995.
17. Bernet, D.; Pinto, R.M.; Costas, M.J.; Canales, J.; Cameselle, J.C. Rat liver mitochondrial ADP-ribose pyrophosphatase in the matrix space with low Km for free ADP-ribose. *Biochem. J.* **1994**, *299*, 679–682. [CrossRef]
18. Kim, M.Y.; Zhang, T.; Kraus, W.L. Poly (ADP-ribosyl) ation by PARP-1: PAR-laying NAD+ into a nuclear signal. *Genes Dev.* **2005**, *19*, 1951–1967. [CrossRef]
19. Panzeter, P.L.; Realini, C.A.; Althaus, F.R. Noncovalent interactions of poly (adenosine diphosphate ribose) with histones. *Biochemistry* **1992**, *31*, 1379–1385. [CrossRef]
20. Sauermann, G.; Wesierska-Gadek, J. Poly (ADP-ribose) effectively competes with DNA for histone H4 binding. *Biochem. Biophys. Res. Commun.* **1986**, *139*, 523–529. [CrossRef]
21. Pleschke, J.M.; Kleczkowska, H.E.; Strohm, M.; Althaus, F.R. Poly (ADP-ribose) binds to specific domains in DNA damage checkpoint proteins. *J. Biol. Chem.* **2000**, *275*, 40974–40980. [CrossRef] [PubMed]
22. Ahel, I.; Ahel, D.; Matsusaka, T.; Clark, A.J.; Pines, J.; Boulton, S.J.; West, S.C. Poly (ADP-ribose)-binding zinc finger motifs in DNA repair/checkpoint proteins. *Nature* **2008**, *451*, 81–85. [CrossRef] [PubMed]
23. Timinszky, G.; Till, S.; Hassa, P.O.; Hothorn, M.; Kustatscher, G.; Nijmeijer, B.; Colombelli, J.; Altmeyer, M.; Stelzer, E.H.; Scheffzek, K.; et al. A macrodomain-containing histone rearranges chromatin upon sensing PARP1 activation. *Nat. Struct. Mol. Biol.* **2009**, *16*, 923–929. [CrossRef] [PubMed]
24. Andrabi, S.A.; Kim, N.S.; Yu, S.W.; Wang, H.; Koh, D.W.; Sasaki, M.; Klaus, J.A.; Otsuka, T.; Zhang, Z.; Koehler, R.C.; et al. Poly (ADP-ribose) (PAR) polymer is a death signal. *Proc. Natl. Acad. Sci. USA* **2006**, *103*, 18308–18313. [CrossRef] [PubMed]
25. Wang, Y.; Kim, N.S.; Haince, J.F.; Kang, H.C.; David, K.K.; Andrabi, S.A.; Poirier, G.G.; Dawson, V.L.; Dawson, T.M. Poly (ADP-ribose) (PAR) binding to apoptosis-inducing factor is critical for PAR polymerase-1-dependent cell death (parthanatos). *Sci. Signal.* **2011**, *4*, ra20. [CrossRef]
26. Yu, S.W.; Wang, H.; Poitras, M.F.; Coombs, C.; Bowers, W.J.; Federoff, H.J.; Poirier, G.G.; Dawson, T.M.; Dawson, V.L. Mediation of poly(ADP-ribose) polymerase-1-dependent cell death by apoptosis-inducing factor. *Science* **2002**, *297*, 259–263. [CrossRef]
27. Yu, S.W.; Andrabi, S.A.; Wang, H.; Kim, N.S.; Poirier, G.G.; Dawson, T.M.; Dawson, V.L. Apoptosis-inducing factor mediates poly (ADP-ribose) (PAR) polymer-induced cell death. *Proc. Natl. Acad. Sci. USA* **2006**, *103*, 18314–18319. [CrossRef]
28. Berger, N.A.; Sims, J.L.; Catino, D.M.; Berger, S.J. Poly (ADP-ribose) polymerase mediates the suicide response to massive DNA damage: Studies in normal and DNA-repair defective cells. *Princess Takamatsu Symp.* **1983**, *13*, 219–226.
29. de Murcia, G.; Schreiber, V.; Molinete, M.; Saulier, B.; Poch, O.; Masson, M.; Niedergang, C.; Menissier de Murcia, J. Structure and function of poly(ADP-ribose) polymerase. *Mol. Cell. Biochem.* **1994**, *138*, 15–24. [CrossRef]
30. Burkle, A.; Virag, L. Poly (ADP-ribose): PARadigms and PARadoxes. *Mol. Asp. Med.* **2013**, *34*, 1046–1065. [CrossRef]

31. Lonskaya, I.; Potaman, V.N.; Shlyakhtenko, L.S.; Oussatcheva, E.A.; Lyubchenko, Y.L.; Soldatenkov, V.A. Regulation of poly(ADP-ribose) polymerase-1 by DNA structure-specific binding. *J. Biol. Chem.* **2005**, *280*, 17076–17083. [CrossRef] [PubMed]
32. Kauppinen, T.M.; Chan, W.Y.; Suh, S.W.; Wiggins, A.K.; Huang, E.J.; Swanson, R.A. Direct phosphorylation and regulation of poly(ADP-ribose) polymerase-1 by extracellular signal-regulated kinases 1/2. *Proc. Natl. Acad. Sci. USA* **2006**, *103*, 7136–7141. [CrossRef] [PubMed]
33. Du, Y.; Yamaguchi, H.; Wei, Y.; Hsu, J.L.; Wang, H.L.; Hsu, Y.H.; Lin, W.C.; Yu, W.H.; Leonard, P.G.; Lee, G.R.t.; et al. Blocking c-Met-mediated PARP1 phosphorylation enhances anti-tumor effects of PARP inhibitors. *Nat. Med.* **2016**, *22*, 194–201. [CrossRef] [PubMed]
34. Cohen-Armon, M. PARP-1 activation in the ERK signaling pathway. *Trends Pharmacol. Sci.* **2007**, *28*, 556–560. [CrossRef]
35. Cohen-Armon, M.; Visochek, L.; Rozensal, D.; Kalal, A.; Geistrikh, I.; Klein, R.; Bendetz-Nezer, S.; Yao, Z.; Seger, R. DNA-independent PARP-1 activation by phosphorylated ERK2 increases Elk1 activity: A link to histone acetylation. *Mol. Cell* **2007**, *25*, 297–308. [CrossRef]
36. Gongol, B.; Marin, T.; Peng, I.C.; Woo, B.; Martin, M.; King, S.; Sun, W.; Johnson, D.A.; Chien, S.; Shyy, J.Y. AMPKalpha2 exerts its anti-inflammatory effects through PARP-1 and Bcl-6. *Proc. Natl. Acad. Sci. USA* **2013**, *110*, 3161–3166. [CrossRef]
37. Beckert, S.; Farrahi, F.; Perveen Ghani, Q.; Aslam, R.; Scheuenstuhl, H.; Coerper, S.; Konigsrainer, A.; Hunt, T.K.; Hussain, M.Z. IGF-I-induced VEGF expression in HUVEC involves phosphorylation and inhibition of poly(ADP-ribose)polymerase. *Biochem. Biophys. Res. Commun.* **2006**, *341*, 67–72. [CrossRef]
38. Boussios, S.; Karathanasi, A.; Cooke, D.; Neille, C.; Sadauskaite, A.; Moschetta, M.; Zakynthinakis-Kyriakou, N.; Pavlidis, N. PARP Inhibitors in Ovarian Cancer: The Route to "Ithaca". *Diagnostics* **2019**, *9*, 55. [CrossRef]
39. Hassa, P.O.; Haenni, S.S.; Buerki, C.; Meier, N.I.; Lane, W.S.; Owen, H.; Gersbach, M.; Imhof, R.; Hottiger, M.O. Acetylation of poly(ADP-ribose) polymerase-1 by p300/CREB-binding protein regulates coactivation of NF-kappaB-dependent transcription. *J. Biol. Chem.* **2005**, *280*, 40450–40464. [CrossRef]
40. Martin, N.; Schwamborn, K.; Schreiber, V.; Werner, A.; Guillier, C.; Zhang, X.D.; Bischof, O.; Seeler, J.S.; Dejean, A. PARP-1 transcriptional activity is regulated by sumoylation upon heat shock. *EMBO J.* **2009**, *28*, 3534–3548. [CrossRef]
41. Loseva, O.; Jemth, A.S.; Bryant, H.E.; Schuler, H.; Lehtio, L.; Karlberg, T.; Helleday, T. PARP-3 is a mono-ADP-ribosylase that activates PARP-1 in the absence of DNA. *J. Biol. Chem.* **2010**, *285*, 8054–8060. [CrossRef] [PubMed]
42. Thomlinson, R.H.; Gray, L.H. The histological structure of some human lung cancers and the possible implications for radiotherapy. *Br. J. Cancer* **1955**, *9*, 539–549. [CrossRef] [PubMed]
43. Samanta, D.; Semenza, G.L. Metabolic adaptation of cancer and immune cells mediated by hypoxia-inducible factors. *Biochim. Biophys. Acta Rev. Cancer* **2018**, *1870*, 15–22. [CrossRef] [PubMed]
44. Tong, W.W.; Tong, G.H.; Liu, Y. Cancer stem cells and hypoxia-inducible factors (Review). *Int. J. Oncol.* **2018**, *53*, 469–476. [CrossRef] [PubMed]
45. Shao, C.; Yang, F.; Miao, S.; Liu, W.; Wang, C.; Shu, Y.; Shen, H. Role of hypoxia-induced exosomes in tumor biology. *Mol. Cancer* **2018**, *17*, 120. [CrossRef] [PubMed]
46. Yehya, A.H.S.; Asif, M.; Petersen, S.H.; Subramaniam, A.V.; Kono, K.; Majid, A.; Oon, C.E. Angiogenesis: Managing the Culprits behind Tumorigenesis and Metastasis. *Medicina* **2018**, *54*, 8. [CrossRef]
47. Sooriakumaran, P.; Kaba, R. Angiogenesis and the tumour hypoxia response in prostate cancer: A review. *Int. J. Surg.* **2005**, *3*, 61–67. [CrossRef]
48. Terry, S.; Faouzi Zaarour, R.; Hassan Venkatesh, G.; Francis, A.; El-Sayed, W.; Buart, S.; Bravo, P.; Thiery, J.; Chouaib, S. Role of Hypoxic Stress in Regulating Tumor Immunogenicity, Resistance and Plasticity. *Int. J. Mol. Sci.* **2018**, *19*, 44. [CrossRef]
49. Rofstad, E.K.; Danielsen, T. Hypoxia-induced metastasis of human melanoma cells: Involvement of vascular endothelial growth factor-mediated angiogenesis. *Br. J. Cancer* **1999**, *80*, 1697–1707. [CrossRef]
50. Graham, K.; Unger, E. Overcoming tumor hypoxia as a barrier to radiotherapy, chemotherapy and immunotherapy in cancer treatment. *Int. J. Nanomed.* **2018**, *13*, 6049–6058. [CrossRef]
51. Loboda, A.; Jozkowicz, A.; Dulak, J. HIF-1 and HIF-2 transcription factors–similar but not identical. *Mol. Cells* **2010**, *29*, 435–442. [CrossRef] [PubMed]

52. Martin-Oliva, D.; Aguilar-Quesada, R.; O'Valle, F.; Munoz-Gamez, J.A.; Martinez-Romero, R.; Garcia Del Moral, R.; Ruiz de Almodovar, J.M.; Villuendas, R.; Piris, M.A.; Oliver, F.J. Inhibition of poly(ADP-ribose) polymerase modulates tumor-related gene expression, including hypoxia-inducible factor-1 activation, during skin carcinogenesis. *Cancer Res.* **2006**, *66*, 5744–5756. [CrossRef] [PubMed]
53. Martinez-Romero, R.; Canuelo, A.; Martinez-Lara, E.; Javier Oliver, F.; Cardenas, S.; Siles, E. Poly(ADP-ribose) polymerase-1 modulation of in vivo response of brain hypoxia-inducible factor-1 to hypoxia/reoxygenation is mediated by nitric oxide and factor inhibiting HIF. *J. Neurochem.* **2009**, *111*, 150–159. [CrossRef] [PubMed]
54. Elser, M.; Borsig, L.; Hassa, P.O.; Erener, S.; Messner, S.; Valovka, T.; Keller, S.; Gassmann, M.; Hottiger, M.O. Poly(ADP-ribose) polymerase 1 promotes tumor cell survival by coactivating hypoxia-inducible factor-1-dependent gene expression. *Mol. Cancer Res.* **2008**, *6*, 282–290. [CrossRef]
55. Canuelo, A.; Martinez-Romero, R.; Martinez-Lara, E.; Sanchez-Alcazar, J.A.; Siles, E. The hypoxic preconditioning agent deferoxamine induces poly (ADP-ribose) polymerase-1-dependent inhibition of the mitochondrial respiratory chain. *Mol. Cell. Biochem.* **2012**, *363*, 101–108. [CrossRef]
56. Liu, S.K.; Coackley, C.; Krause, M.; Jalali, F.; Chan, N.; Bristow, R.G. A novel poly (ADP-ribose) polymerase inhibitor, ABT-888, radiosensitizes malignant human cell lines under hypoxia. *Radiother. Oncol.* **2008**, *88*, 258–268. [CrossRef]
57. Holmquist-Mengelbier, L.; Fredlund, E.; Lofstedt, T.; Noguera, R.; Navarro, S.; Nilsson, H.; Pietras, A.; Vallon-Christersson, J.; Borg, A.; Gradin, K.; et al. Recruitment of HIF-1alpha and HIF-2alpha to common target genes is differentially regulated in neuroblastoma: HIF-2alpha promotes an aggressive phenotype. *Cancer Cell* **2006**, *10*, 413–423. [CrossRef]
58. Raval, R.R.; Lau, K.W.; Tran, M.G.; Sowter, H.M.; Mandriota, S.J.; Li, J.L.; Pugh, C.W.; Maxwell, P.H.; Harris, A.L.; Ratcliffe, P.J. Contrasting properties of hypoxia-inducible factor 1 (HIF-1) and HIF-2 in von Hippel-Lindau-associated renal cell carcinoma. *Mol. Cell. Biol.* **2005**, *25*, 5675–5686. [CrossRef]
59. Aguilar-Quesada, R.; Munoz-Gamez, J.A.; Martin-Oliva, D.; Peralta-Leal, A.; Quiles-Perez, R.; Rodriguez-Vargas, J.M.; Ruiz de Almodovar, M.; Conde, C.; Ruiz-Extremera, A.; Oliver, F.J. Modulation of transcription by PARP-1: Consequences in carcinogenesis and inflammation. *Curr. Med. Chem.* **2007**, *14*, 1179–1187. [CrossRef]
60. Gonzalez-Flores, A.; Aguilar-Quesada, R.; Siles, E.; Pozo, S.; Rodriguez-Lara, M.I.; Lopez-Jimenez, L.; Lopez-Rodriguez, M.; Peralta-Leal, A.; Villar, D.; Martin-Oliva, D.; et al. Interaction between PARP-1 and HIF-2alpha in the hypoxic response. *Oncogene* **2014**, *33*, 891–898. [CrossRef]
61. Natale, G.; Bocci, G.; Lenzi, P. Looking for the Word "Angiogenesis" in the History of Health Sciences: From Ancient Times to the First Decades of the Twentieth Century. *World J. Surg.* **2017**, *41*, 1625–1634. [CrossRef] [PubMed]
62. Dome, B.; Hendrix, M.J.; Paku, S.; Tovari, J.; Timar, J. Alternative vascularization mechanisms in cancer: Pathology and therapeutic implications. *Am. J. Pathol.* **2007**, *170*. [CrossRef] [PubMed]
63. De Palma, M.; Biziato, D.; Petrova, T.V. Microenvironmental regulation of tumour angiogenesis. *Nat. Rev. Cancer* **2017**, *17*, 457–474. [CrossRef] [PubMed]
64. Bridgeman, V.L.; Vermeulen, P.B.; Foo, S.; Bilecz, A.; Daley, F.; Kostaras, E.; Nathan, M.R.; Wan, E.; Frentzas, S.; Schweiger, T.; et al. Vessel co-option is common in human lung metastases and mediates resistance to anti-angiogenic therapy in preclinical lung metastasis models. *J. Pathol.* **2017**, *241*, 362–374. [CrossRef]
65. Oza, A.M.; Cook, A.D.; Pfisterer, J.; Embleton, A.; Ledermann, J.A.; Pujade-Lauraine, E.; Kristensen, G.; Carey, M.S.; Beale, P.; Cervantes, A.; et al. Standard chemotherapy with or without bevacizumab for women with newly diagnosed ovarian cancer (ICON7): Overall survival results of a phase 3 randomised trial. *Lancet Oncol.* **2015**, *16*, 928–936. [CrossRef]
66. Folkman, J. Tumor angiogenesis: Therapeutic implications. *N. Engl. J. Med.* **1971**, *285*, 1182–1186. [CrossRef]
67. Boussios, S.; Karihtala, P.; Moschetta, M.; Abson, C.; Karathanasi, A.; Zakynthinakis-Kyriakou, N.; Ryan, J.E.; Sheriff, M.; Rassy, E.; Pavlidis, N. Veliparib in ovarian cancer: A new synthetically lethal therapeutic approach. *Investig. New Drugs* **2020**, *38*, 181–193. [CrossRef]
68. Tentori, L.; Lacal, P.M.; Muzi, A.; Dorio, A.S.; Leonetti, C.; Scarsella, M.; Ruffini, F.; Xu, W.; Min, W.; Stoppacciaro, A.; et al. Poly (ADP-ribose) polymerase (PARP) inhibition or PARP-1 gene deletion reduces angiogenesis. *Eur. J. Cancer* **2007**, *43*, 2124–2133. [CrossRef]
69. Pyriochou, A.; Olah, G.; Deitch, E.A.; Szabo, C.; Papapetropoulos, A. Inhibition of angiogenesis by the poly (ADP-ribose) polymerase inhibitor PJ-34. *Int. J. Mol. Med.* **2008**, *22*, 113–118. [CrossRef]

70. Lacal, P.M.; Tentori, L.; Muzi, A.; Ruffini, F.; Dorio, A.S.; Xu, W.; Arcelli, D.; Zhang, J.; Graziani, G. Pharmacological inhibition of poly(ADP-ribose) polymerase activity down-regulates the expression of syndecan-4 and Id-1 in endothelial cells. *Int. J. Oncol.* **2009**, *34*, 861–872.
71. Maniotis, A.J.; Folberg, R.; Hess, A.; Seftor, E.A.; Gardner, L.M.; Pe'er, J.; Trent, J.M.; Meltzer, P.S.; Hendrix, M.J. Vascular channel formation by human melanoma cells in vivo and in vitro: Vasculogenic mimicry. *Am. J. Pathol.* **1999**, *155*, 739–752. [CrossRef]
72. Seftor, E.A.; Meltzer, P.S.; Kirschmann, D.A.; Pe'er, J.; Maniotis, A.J.; Trent, J.M.; Folberg, R.; Hendrix, M.J. Molecular determinants of human uveal melanoma invasion and metastasis. *Clin. Exp. Metastasis* **2002**, *19*, 233–246. [CrossRef] [PubMed]
73. Bittner, M.; Meltzer, P.; Chen, Y.; Jiang, Y.; Seftor, E.; Hendrix, M.; Radmacher, M.; Simon, R.; Yakhini, Z.; Ben-Dor, A.; et al. Molecular classification of cutaneous malignant melanoma by gene expression profiling. *Nature* **2000**, *406*, 536–540. [CrossRef] [PubMed]
74. Hendrix, M.J.; Seftor, E.A.; Meltzer, P.S.; Gardner, L.M.; Hess, A.R.; Kirschmann, D.A.; Schatteman, G.C.; Seftor, R.E. Expression and functional significance of VE-cadherin in aggressive human melanoma cells: Role in vasculogenic mimicry. *Proc. Natl. Acad. Sci. USA* **2001**, *98*, 8018–8023. [CrossRef] [PubMed]
75. Ruffini, F.; Graziani, G.; Levati, L.; Tentori, L.; D'Atri, S.; Lacal, P.M. Cilengitide downmodulates invasiveness and vasculogenic mimicry of neuropilin 1 expressing melanoma cells through the inhibition of alphavbeta5 integrin. *Int. J. Cancer* **2015**, *136*, E545–E558. [CrossRef] [PubMed]
76. Pagani, E.; Ruffini, F.; Antonini Cappellini, G.C.; Scoppola, A.; Fortes, C.; Marchetti, P.; Graziani, G.; D'Atri, S.; Lacal, P.M. Placenta growth factor and neuropilin-1 collaborate in promoting melanoma aggressiveness. *Int. J. Oncol.* **2016**, *48*, 1581–1589. [CrossRef]
77. Schnegg, C.I.; Yang, M.H.; Ghosh, S.K.; Hsu, M.Y. Induction of Vasculogenic Mimicry Overrides VEGF-A Silencing and Enriches Stem-like Cancer Cells in Melanoma. *Cancer Res.* **2015**, *75*, 1682–1690. [CrossRef]
78. Rodriguez, M.I.; Peralta-Leal, A.; O'Valle, F.; Rodriguez-Vargas, J.M.; Gonzalez-Flores, A.; Majuelos-Melguizo, J.; Lopez, L.; Serrano, S.; de Herreros, A.G.; Rodriguez-Manzaneque, J.C.; et al. PARP-1 regulates metastatic melanoma through modulation of vimentin-induced malignant transformation. *PLoS Genet.* **2013**, *9*, e1003531. [CrossRef]
79. Crawford, Y.; Ferrara, N. VEGF inhibition: Insights from preclinical and clinical studies. *Cell Tissue Res.* **2009**, *335*, 261–269. [CrossRef]
80. Nakasone, E.S.; Hurvitz, S.A.; McCann, K.E. Harnessing the immune system in the battle against breast cancer. *Drugs Context* **2018**, *7*, 212520. [CrossRef]
81. Bai, P.; Virag, L. Role of poly (ADP-ribose) polymerases in the regulation of inflammatory processes. *FEBS Lett.* **2012**, *586*, 3771–3777. [CrossRef] [PubMed]
82. Stilmann, M.; Hinz, M.; Arslan, S.C.; Zimmer, A.; Schreiber, V.; Scheidereit, C. A nuclear poly (ADP-ribose)-dependent signalosome confers DNA damage-induced IkappaB kinase activation. *Mol. Cell* **2009**, *36*, 365–378. [CrossRef] [PubMed]
83. Oliver, F.J.; Menissier-de Murcia, J.; Nacci, C.; Decker, P.; Andriantsitohaina, R.; Muller, S.; de la Rubia, G.; Stoclet, J.C.; de Murcia, G. Resistance to endotoxic shock as a consequence of defective NF-kappaB activation in poly (ADP-ribose) polymerase-1 deficient mice. *EMBO J.* **1999**, *18*, 4446–4454. [CrossRef] [PubMed]
84. Hinz, M.; Stilmann, M.; Arslan, S.C.; Khanna, K.K.; Dittmar, G.; Scheidereit, C. A cytoplasmic ATM-TRAF6-cIAP1 module links nuclear DNA damage signaling to ubiquitin-mediated NF-kappaB activation. *Mol. Cell* **2010**, *40*, 63–74. [CrossRef]
85. Ha, H.C.; Hester, L.D.; Snyder, S.H. Poly (ADP-ribose) polymerase-1 dependence of stress-induced transcription factors and associated gene expression in glia. *Proc. Natl. Acad. Sci. USA* **2002**, *99*, 3270–3275. [CrossRef]
86. Olabisi, O.A.; Soto-Nieves, N.; Nieves, E.; Yang, T.T.; Yang, X.; Yu, R.Y.; Suk, H.Y.; Macian, F.; Chow, C.W. Regulation of transcription factor NFAT by ADP-ribosylation. *Mol. Cell. Biol.* **2008**, *28*, 2860–2871. [CrossRef]
87. Bai, P.; Canto, C.; Brunyanszki, A.; Huber, A.; Szanto, M.; Cen, Y.; Yamamoto, H.; Houten, S.M.; Kiss, B.; Oudart, H.; et al. PARP-2 regulates SIRT1 expression and whole-body energy expenditure. *Cell Metab.* **2011**, *13*, 450–460. [CrossRef]
88. Szanto, M.; Brunyanszki, A.; Kiss, B.; Nagy, L.; Gergely, P.; Virag, L.; Bai, P. Poly (ADP-ribose) polymerase-2: Emerging transcriptional roles of a DNA-repair protein. *Cell. Mol. Life Sci.* **2012**, *69*, 4079–4092. [CrossRef]

89. Mehrotra, P.; Riley, J.P.; Patel, R.; Li, F.; Voss, L.; Goenka, S. PARP-14 functions as a transcriptional switch for Stat6-dependent gene activation. *J. Biol. Chem.* **2011**, *286*, 1767–1776. [CrossRef]
90. Rosado, M.M.; Bennici, E.; Novelli, F.; Pioli, C. Beyond DNA repair, the immunological role of PARP-1 and its siblings. *Immunology* **2013**, *139*, 428–437. [CrossRef]
91. Yelamos, J.; Monreal, Y.; Saenz, L.; Aguado, E.; Schreiber, V.; Mota, R.; Fuente, T.; Minguela, A.; Parrilla, P.; de Murcia, G.; et al. PARP-2 deficiency affects the survival of CD4+CD8+ double-positive thymocytes. *EMBO J.* **2006**, *25*, 4350–4360. [CrossRef] [PubMed]
92. Sambucci, M.; Laudisi, F.; Novelli, F.; Bennici, E.; Rosado, M.M.; Pioli, C. Effects of PARP-1 deficiency on Th1 and Th2 cell differentiation. *Sci. World J.* **2013**, *2013*, 375024. [CrossRef] [PubMed]
93. Barber, G.N. STING: Infection, inflammation and cancer. *Nat. Rev. Immunol.* **2015**, *15*, 760–770. [CrossRef] [PubMed]
94. Mouw, K.W.; Goldberg, M.S.; Konstantinopoulos, P.A.; D'Andrea, A.D. DNA Damage and Repair Biomarkers of Immunotherapy Response. *Cancer Discov.* **2017**, *7*, 675–693. [CrossRef]
95. Ablasser, A.; Goldeck, M.; Cavlar, T.; Deimling, T.; Witte, G.; Rohl, I.; Hopfner, K.P.; Ludwig, J.; Hornung, V. cGAS produces a 2′-5′-linked cyclic dinucleotide second messenger that activates STING. *Nature* **2013**, *498*, 380–384. [CrossRef]
96. Woo, S.R.; Fuertes, M.B.; Corrales, L.; Spranger, S.; Furdyna, M.J.; Leung, M.Y.; Duggan, R.; Wang, Y.; Barber, G.N.; Fitzgerald, K.A.; et al. STING-dependent cytosolic DNA sensing mediates innate immune recognition of immunogenic tumors. *Immunity* **2014**, *41*, 830–842. [CrossRef]
97. Corrales, L.; Glickman, L.H.; McWhirter, S.M.; Kanne, D.B.; Sivick, K.E.; Katibah, G.E.; Woo, S.R.; Lemmens, E.; Banda, T.; Leong, J.J.; et al. Direct Activation of STING in the Tumor Microenvironment Leads to Potent and Systemic Tumor Regression and Immunity. *Cell Rep.* **2015**, *11*, 1018–1030. [CrossRef]
98. Muthuswamy, R.; Berk, E.; Junecko, B.F.; Zeh, H.J.; Zureikat, A.H.; Normolle, D.; Luong, T.M.; Reinhart, T.A.; Bartlett, D.L.; Kalinski, P. NF-kappaB hyperactivation in tumor tissues allows tumor-selective reprogramming of the chemokine microenvironment to enhance the recruitment of cytolytic T effector cells. *Cancer Res.* **2012**, *72*, 3735–3743. [CrossRef]
99. Shen, J.; Zhao, W.; Ju, Z.; Wang, L.; Peng, Y.; Labrie, M.; Yap, T.A.; Mills, G.B.; Peng, G. PARPi Triggers the STING-Dependent Immune Response and Enhances the Therapeutic Efficacy of Immune Checkpoint Blockade Independent of BRCAness. *Cancer Res.* **2019**, *79*, 311–319. [CrossRef]
100. Chalmers, Z.R.; Connelly, C.F.; Fabrizio, D.; Gay, L.; Ali, S.M.; Ennis, R.; Schrock, A.; Campbell, B.; Shlien, A.; Chmielecki, J.; et al. Analysis of 100,000 human cancer genomes reveals the landscape of tumor mutational burden. *Genome Med.* **2017**, *9*, 34. [CrossRef]
101. Snyder, A.; Makarov, V.; Merghoub, T.; Yuan, J.; Zaretsky, J.M.; Desrichard, A.; Walsh, L.A.; Postow, M.A.; Wong, P.; Ho, T.S.; et al. Genetic basis for clinical response to CTLA-4 blockade in melanoma. *N. Engl. J. Med.* **2014**, *371*, 2189–2199. [CrossRef] [PubMed]
102. Linnemann, C.; van Buuren, M.M.; Bies, L.; Verdegaal, E.M.; Schotte, R.; Calis, J.J.; Behjati, S.; Velds, A.; Hilkmann, H.; Atmioui, D.E.; et al. High-throughput epitope discovery reveals frequent recognition of neo-antigens by CD4+ T cells in human melanoma. *Nat. Med.* **2015**, *21*, 81–85. [CrossRef] [PubMed]
103. Lu, Y.C.; Yao, X.; Crystal, J.S.; Li, Y.F.; El-Gamil, M.; Gross, C.; Davis, L.; Dudley, M.E.; Yang, J.C.; Samuels, Y.; et al. Efficient identification of mutated cancer antigens recognized by T cells associated with durable tumor regressions. *Clin. Cancer Res.* **2014**, *20*, 3401–3410. [CrossRef] [PubMed]
104. Lawrence, M.S.; Stojanov, P.; Mermel, C.H.; Robinson, J.T.; Garraway, L.A.; Golub, T.R.; Meyerson, M.; Gabriel, S.B.; Lander, E.S.; Getz, G. Discovery and saturation analysis of cancer genes across 21 tumour types. *Nature* **2014**, *505*, 495–501. [CrossRef] [PubMed]
105. Schumacher, T.N.; Schreiber, R.D. Neoantigens in cancer immunotherapy. *Science* **2015**, *348*, 69–74. [CrossRef] [PubMed]
106. Brown, J.S.; O'Carrigan, B.; Jackson, S.P.; Yap, T.A. Targeting DNA Repair in Cancer: Beyond PARP Inhibitors. *Cancer Discov.* **2017**, *7*, 20–37. [CrossRef]
107. Li, M.; Yu, X. The role of poly (ADP-ribosyl) ation in DNA damage response and cancer chemotherapy. *Oncogene* **2015**, *34*, 3349–3356. [CrossRef]
108. Lee, J.M.; Gulley, J.L. Checkpoint and PARP inhibitors, for whom and when. *Oncotarget* **2017**, *8*, 95036–95037. [CrossRef]

109. Ishida, Y.; Agata, Y.; Shibahara, K.; Honjo, T. Induced expression of PD-1, a novel member of the immunoglobulin gene superfamily, upon programmed cell death. *EMBO J.* **1992**, *11*, 3887–3895. [CrossRef]
110. Wu, C.T.; Chen, W.C.; Chang, Y.H.; Lin, W.Y.; Chen, M.F. The role of PD-L1 in the radiation response and clinical outcome for bladder cancer. *Sci. Rep.* **2016**, *6*, 19740. [CrossRef]
111. Deng, L.; Liang, H.; Burnette, B.; Beckett, M.; Darga, T.; Weichselbaum, R.R.; Fu, Y.X. Irradiation and anti-PD-L1 treatment synergistically promote antitumor immunity in mice. *J. Clin. Investig.* **2014**, *124*, 687–695. [CrossRef] [PubMed]
112. Sato, H.; Niimi, A.; Yasuhara, T.; Permata, T.B.M.; Hagiwara, Y.; Isono, M.; Nuryadi, E.; Sekine, R.; Oike, T.; Kakoti, S.; et al. DNA double-strand break repair pathway regulates PD-L1 expression in cancer cells. *Nat. Commun.* **2017**, *8*, 1751. [CrossRef] [PubMed]
113. Huang, J.; Wang, L.; Cong, Z.; Amoozgar, Z.; Kiner, E.; Xing, D.; Orsulic, S.; Matulonis, U.; Goldberg, M.S. The PARP1 inhibitor BMN 673 exhibits immunoregulatory effects in a Brca1 (-/-) murine model of ovarian cancer. *Biochem. Biophys. Res. Commun.* **2015**, *463*, 551–556. [CrossRef]
114. Jiao, S.; Xia, W.; Yamaguchi, H.; Wei, Y.; Chen, M.K.; Hsu, J.M.; Hsu, J.L.; Yu, W.H.; Du, Y.; Lee, H.H.; et al. PARP Inhibitor Upregulates PD-L1 Expression and Enhances Cancer-Associated Immunosuppression. *Clin. Cancer Res.* **2017**, *23*, 3711–3720. [CrossRef] [PubMed]
115. Rodriguez-Vargas, J.M.; Ruiz-Magana, M.J.; Ruiz-Ruiz, C.; Majuelos-Melguizo, J.; Peralta-Leal, A.; Rodriguez, M.I.; Munoz-Gamez, J.A.; de Almodovar, M.R.; Siles, E.; Rivas, A.L.; et al. ROS-induced DNA damage and PARP-1 are required for optimal induction of starvation-induced autophagy. *Cell Res.* **2012**, *22*, 1181–1198. [CrossRef] [PubMed]
116. Dikic, I.; Elazar, Z. Mechanism and medical implications of mammalian autophagy. *Nat. Rev. Mol. Cell Biol.* **2018**, *19*, 349–364. [CrossRef]
117. Alexander, A.; Cai, S.L.; Kim, J.; Nanez, A.; Sahin, M.; MacLean, K.H.; Inoki, K.; Guan, K.L.; Shen, J.; Person, M.D.; et al. ATM signals to TSC2 in the cytoplasm to regulate mTORC1 in response to ROS. *Proc. Natl. Acad. Sci. USA* **2010**, *107*, 4153–4158. [CrossRef]
118. Tripathi, D.N.; Chowdhury, R.; Trudel, L.J.; Tee, A.R.; Slack, R.S.; Walker, C.L.; Wogan, G.N. Reactive nitrogen species regulate autophagy through ATM-AMPK-TSC2-mediated suppression of mTORC1. *Proc. Natl. Acad. Sci. USA* **2013**, *110*, E2950–E2957. [CrossRef]
119. Munoz-Gamez, J.A.; Rodriguez-Vargas, J.M.; Quiles-Perez, R.; Aguilar-Quesada, R.; Martin-Oliva, D.; de Murcia, G.; Menissier de Murcia, J.; Almendros, A.; Ruiz de Almodovar, M.; Oliver, F.J. PARP-1 is involved in autophagy induced by DNA damage. *Autophagy* **2009**, *5*, 61–74. [CrossRef]
120. Chen, Z.T.; Zhao, W.; Qu, S.; Li, L.; Lu, X.D.; Su, F.; Liang, Z.G.; Guo, S.Y.; Zhu, X.D. PARP-1 promotes autophagy via the AMPK/mTOR pathway in CNE-2 human nasopharyngeal carcinoma cells following ionizing radiation, while inhibition of autophagy contributes to the radiation sensitization of CNE-2 cells. *Mol. Med. Rep.* **2015**, *12*, 1868–1876. [CrossRef]
121. Rodriguez-Vargas, J.M.; Rodriguez, M.I.; Majuelos-Melguizo, J.; Garcia-Diaz, A.; Gonzalez-Flores, A.; Lopez-Rivas, A.; Virag, L.; Illuzzi, G.; Schreiber, V.; Dantzer, F.; et al. Autophagy requires poly(adp-ribosyl)ation-dependent AMPK nuclear export. *Cell Death Differ.* **2016**, *23*, 2007–2018. [CrossRef] [PubMed]
122. Shang, F.; Zhang, J.; Li, Z.; Zhang, J.; Yin, Y.; Wang, Y.; Marin, T.L.; Gongol, B.; Xiao, H.; Zhang, Y.Y.; et al. Cardiovascular Protective Effect of Metformin and Telmisartan: Reduction of PARP1 Activity via the AMPK-PARP1 Cascade. *PLoS ONE* **2016**, *11*, e0151845. [CrossRef] [PubMed]
123. Lapidot, T.; Sirard, C.; Vormoor, J.; Murdoch, B.; Hoang, T.; Caceres-Cortes, J.; Minden, M.; Paterson, B.; Caligiuri, M.A.; Dick, J.E. A cell initiating human acute myeloid leukaemia after transplantation into SCID mice. *Nature* **1994**, *367*, 645–648. [CrossRef] [PubMed]
124. Al-Hajj, M.; Wicha, M.S.; Benito-Hernandez, A.; Morrison, S.J.; Clarke, M.F. Prospective identification of tumorigenic breast cancer cells. *Proc. Natl. Acad. Sci. USA* **2003**, *100*, 3983–3988. [CrossRef]
125. Singh, S.K.; Hawkins, C.; Clarke, I.D.; Squire, J.A.; Bayani, J.; Hide, T.; Henkelman, R.M.; Cusimano, M.D.; Dirks, P.B. Identification of human brain tumour initiating cells. *Nature* **2004**, *432*, 396–401. [CrossRef]
126. Collins, A.T.; Berry, P.A.; Hyde, C.; Stower, M.J.; Maitland, N.J. Prospective identification of tumorigenic prostate cancer stem cells. *Cancer Res.* **2005**, *65*, 10946–10951. [CrossRef]

127. Fang, D.; Nguyen, T.K.; Leishear, K.; Finko, R.; Kulp, A.N.; Hotz, S.; Van Belle, P.A.; Xu, X.; Elder, D.E.; Herlyn, M. A tumorigenic subpopulation with stem cell properties in melanomas. *Cancer Res.* **2005**, *65*, 9328–9337. [CrossRef]
128. Kim, C.F.; Jackson, E.L.; Woolfenden, A.E.; Lawrence, S.; Babar, I.; Vogel, S.; Crowley, D.; Bronson, R.T.; Jacks, T. Identification of bronchioalveolar stem cells in normal lung and lung cancer. *Cell* **2005**, *121*, 823–835. [CrossRef]
129. Bertolini, G.; Roz, L.; Perego, P.; Tortoreto, M.; Fontanella, E.; Gatti, L.; Pratesi, G.; Fabbri, A.; Andriani, F.; Tinelli, S.; et al. Highly tumorigenic lung cancer CD133+ cells display stem-like features and are spared by cisplatin treatment. *Proc. Natl. Acad. Sci. USA* **2009**, *106*, 16281–16286. [CrossRef]
130. O'Brien, C.A.; Pollett, A.; Gallinger, S.; Dick, J.E. A human colon cancer cell capable of initiating tumour growth in immunodeficient mice. *Nature* **2007**, *445*, 106–110. [CrossRef]
131. Ricci-Vitiani, L.; Lombardi, D.G.; Pilozzi, E.; Biffoni, M.; Todaro, M.; Peschle, C.; De Maria, R. Identification and expansion of human colon-cancer-initiating cells. *Nature* **2007**, *445*, 111–115. [CrossRef] [PubMed]
132. Li, C.; Heidt, D.G.; Dalerba, P.; Burant, C.F.; Zhang, L.; Adsay, V.; Wicha, M.; Clarke, M.F.; Simeone, D.M. Identification of pancreatic cancer stem cells. *Cancer Res.* **2007**, *67*, 1030–1037. [CrossRef] [PubMed]
133. Prince, M.E.; Sivanandan, R.; Kaczorowski, A.; Wolf, G.T.; Kaplan, M.J.; Dalerba, P.; Weissman, I.L.; Clarke, M.F.; Ailles, L.E. Identification of a subpopulation of cells with cancer stem cell properties in head and neck squamous cell carcinoma. *Proc. Natl. Acad. Sci. USA* **2007**, *104*, 973–978. [CrossRef] [PubMed]
134. Ma, S.; Chan, K.W.; Hu, L.; Lee, T.K.; Wo, J.Y.; Ng, I.O.; Zheng, B.J.; Guan, X.Y. Identification and characterization of tumorigenic liver cancer stem/progenitor cells. *Gastroenterology* **2007**, *132*, 2542–2556. [CrossRef]
135. Bussolati, B.; Bruno, S.; Grange, C.; Ferrando, U.; Camussi, G. Identification of a tumor-initiating stem cell population in human renal carcinomas. *FASEB J.* **2008**, *22*, 3696–3705. [CrossRef]
136. Islam, F.; Qiao, B.; Smith, R.A.; Gopalan, V.; Lam, A.K. Cancer stem cell: Fundamental experimental pathological concepts and updates. *Exp. Mol. Pathol.* **2015**, *98*, 184–191. [CrossRef]
137. Bao, S.; Wu, Q.; McLendon, R.E.; Hao, Y.; Shi, Q.; Hjelmeland, A.B.; Dewhirst, M.W.; Bigner, D.D.; Rich, J.N. Glioma stem cells promote radioresistance by preferential activation of the DNA damage response. *Nature* **2006**, *444*, 756–760. [CrossRef]
138. Friedmann-Morvinski, D.; Verma, I.M. Dedifferentiation and reprogramming: Origins of cancer stem cells. *EMBO Rep.* **2014**, *15*, 244–253. [CrossRef]
139. Kumar, S.M.; Liu, S.; Lu, H.; Zhang, H.; Zhang, P.J.; Gimotty, P.A.; Guerra, M.; Guo, W.; Xu, X. Acquired cancer stem cell phenotypes through Oct4-mediated dedifferentiation. *Oncogene* **2012**, *31*, 4898–4911. [CrossRef]
140. Corominas-Faja, B.; Cufi, S.; Oliveras-Ferraros, C.; Cuyas, E.; Lopez-Bonet, E.; Lupu, R.; Alarcon, T.; Vellon, L.; Iglesias, J.M.; Leis, O.; et al. Nuclear reprogramming of luminal-like breast cancer cells generates Sox2-overexpressing cancer stem-like cellular states harboring transcriptional activation of the mTOR pathway. *Cell Cycle* **2013**, *12*, 3109–3124. [CrossRef]
141. Chiou, S.H.; Jiang, B.H.; Yu, Y.L.; Chou, S.J.; Tsai, P.H.; Chang, W.C.; Chen, L.K., Chen, L.H.; Chien, Y.; Chiou, G.Y. Poly (ADP-ribose) polymerase 1 regulates nuclear reprogramming and promotes iPSC generation without c-Myc. *J. Exp. Med.* **2013**, *210*, 85–98. [CrossRef]
142. Weber, F.A.; Bartolomei, G.; Hottiger, M.O.; Cinelli, P. Artd1/Parp1 regulates reprogramming by transcriptional regulation of Fgf4 via Sox2 ADP-ribosylation. *Stem Cells* **2013**, *31*, 2364–2373. [CrossRef]
143. Gao, F.; Kwon, S.W.; Zhao, Y.; Jin, Y. PARP1 poly (ADP-ribosyl) ates Sox2 to control Sox2 protein levels and FGF4 expression during embryonic stem cell differentiation. *J. Biol. Chem.* **2009**, *284*, 22263–22273. [CrossRef]
144. Doege, C.A.; Inoue, K.; Yamashita, T.; Rhee, D.B.; Travis, S.; Fujita, R.; Guarnieri, P.; Bhagat, G.; Vanti, W.B.; Shih, A.; et al. Early-stage epigenetic modification during somatic cell reprogramming by Parp1 and Tet2. *Nature* **2012**, *488*, 652–655. [CrossRef]
145. Hsieh, M.H.; Chen, Y.T.; Chen, Y.T.; Lee, Y.H.; Lu, J.; Chien, C.L.; Chen, H.F.; Ho, H.N.; Yu, C.J.; Wang, Z.Q.; et al. PARP1 controls KLF4-mediated telomerase expression in stem cells and cancer cells. *Nucleic Acids Res.* **2017**, *45*, 10492–10503. [CrossRef]
146. Marion, R.M.; Strati, K.; Li, H.; Tejera, A.; Schoeftner, S.; Ortega, S.; Serrano, M.; Blasco, M.A. Telomeres acquire embryonic stem cell characteristics in induced pluripotent stem cells. *Cell Stem Cell* **2009**, *4*, 141–154. [CrossRef]

147. Wang, R.; Chadalavada, K.; Wilshire, J.; Kowalik, U.; Hovinga, K.E.; Geber, A.; Fligelman, B.; Leversha, M.; Brennan, C.; Tabar, V. Glioblastoma stem-like cells give rise to tumour endothelium. *Nature* **2010**, *468*, 829–833. [CrossRef]
148. Piccirillo, S.G.; Reynolds, B.A.; Zanetti, N.; Lamorte, G.; Binda, E.; Broggi, G.; Brem, H.; Olivi, A.; Dimeco, F.; Vescovi, A.L. Bone morphogenetic proteins inhibit the tumorigenic potential of human brain tumour-initiating cells. *Nature* **2006**, *444*, 761–765. [CrossRef]
149. Watanabe, Y.; Papoutsoglou, P.; Maturi, V.; Tsubakihara, Y.; Hottiger, M.O.; Heldin, C.H.; Moustakas, A. Regulation of Bone Morphogenetic Protein Signaling by ADP-ribosylation. *J. Biol. Chem.* **2016**, *291*, 12706–12723. [CrossRef]
150. Lonn, P.; van der Heide, L.P.; Dahl, M.; Hellman, U.; Heldin, C.H.; Moustakas, A. PARP-1 attenuates Smad-mediated transcription. *Mol. Cell* **2010**, *40*, 521–532. [CrossRef]
151. Penuelas, S.; Anido, J.; Prieto-Sanchez, R.M.; Folch, G.; Barba, I.; Cuartas, I.; Garcia-Dorado, D.; Poca, M.A.; Sahuquillo, J.; Baselga, J.; et al. TGF-beta increases glioma-initiating cell self-renewal through the induction of LIF in human glioblastoma. *Cancer Cell* **2009**, *15*, 315–327. [CrossRef]
152. Faraoni, I.; Graziani, G. Role of BRCA Mutations in Cancer Treatment with Poly (ADP-ribose) Polymerase (PARP) Inhibitors. *Cancers* **2018**, *10*, 487. [CrossRef]

cancers

MDPI

Review

ATM-Deficient Cancers Provide New Opportunities for Precision Oncology

Nicholas R. Jette [1], Mehul Kumar [1], Suraj Radhamani [1], Greydon Arthur [1], Siddhartha Goutam [1], Steven Yip [2], Michael Kolinsky [3], Gareth J. Williams [1], Pinaki Bose [1] and Susan P. Lees-Miller [1,*]

[1] Department of Biochemistry and Molecular Biology, Robson DNA Science Centre, Charbonneau Cancer Institute, Cumming School of Medicine, University of Calgary, 3330 Hospital Drive NW, Calgary, AB T2N 1N4, Canada; nrjette@ucalgary.ca (N.R.J.); mehul.kumar@ucalgary.ca (M.K.); suraj.radhamani@ucalgary.ca (S.R.); greydon.arthur@ucalgary.ca (G.A.); goutam@ualberta.ca (S.G.); gareth.williams2@ucalgary.ca (G.J.W.); pbose@ucalgary.ca (P.B.)

[2] Tom Baker Cancer Centre, 1331 29 St NW, Calgary, AB T2N 4N2, Canada; smyip@ualberta.ca

[3] Cross Cancer Institute, 11560 University Avenue NW, Edmonton, AB T6G 1Z2, Canada; michael.kolinsky@albertahealthservices.ca

* Correspondence: leesmill@ucalgary.ca

Received: 14 February 2020; Accepted: 12 March 2020; Published: 14 March 2020

Abstract: Poly-ADP ribose polymerase (PARP) inhibitors are currently used in the treatment of several cancers carrying mutations in the breast and ovarian cancer susceptibility genes *BRCA1* and *BRCA2*, with many more potential applications under study and in clinical trials. Here, we discuss the potential for extending PARP inhibitor therapies to tumours with deficiencies in the DNA damage-activated protein kinase, Ataxia-Telangiectasia Mutated (ATM). We highlight our recent findings that PARP inhibition alone is cytostatic but not cytotoxic in ATM-deficient cancer cells and that the combination of a PARP inhibitor with an ATR (ATM, Rad3-related) inhibitor is required to induce cell death.

Keywords: ATM; olaparib; ATR; PARP; PARP inhibitor; prostate cancer; pancreatic cancer; lung cancer

1. PARP and PARP Inhibitors

Genome instability, characterized by the accumulation of mutations and chromosomal alterations in the genome, is both a hallmark and a driver of cancer [1,2]. Yet the same genomic alterations that predispose a cell to cancer may also render cells susceptible to targeted therapies. Accordingly, a goal of precision oncology is to achieve better cancer control by targeting therapy to specific genetic defects or aberrations in the tumour, while causing less damage to normal tissue and consequently, fewer side-effects. One of the most dramatic examples of success in this area has been the use of poly-ADP-ribose polymerase (PARP) inhibitors in the treatment of patients with tumours that harbour inactivating mutations in the breast and ovarian cancer susceptibility genes, *BRCA1* and *BRCA2*.

PARP was identified in the 1960s as an enzyme that metabolizes nicotinamide adenine dinucleotide, NAD^+ [3]. Early studies indicated that inhibition of PARP blocked repair of DNA strand breaks and PARP inhibitors were soon considered as potential radiation sensitizers [4]. The first PARP inhibitor, a simple analogue of nicotinamide was generated in 1971 [5] and, given the reported roles of PARP in cell death and ischemia as well as DNA repair, there was increasing interest in the clinical applications of PARP inhibition [6,7]. PARP is now established as one of a family of poly-ADP polymerases of which PARPs 1, 2 and 3 have roles in the DNA damage response [8]. Of these, PARP 1 is the most well-studied and the one to which we will refer in this review. PARP1 binds avidly to ends of DNA that occur at single and double DNA strand breaks and then, in a process called PARylation, auto-modifies to create long polymers of poly-ADP-ribose (PAR). PAR chains interact with proteins involved in DNA repair and other pathways, recruiting them to sites of DNA damage [9]. Importantly, PARylation also serves to disengage PARP from DNA ends [10]. Most PARP inhibitors in use today prevent PARylation, leading to trapping of PARP at DNA ends [11]. It was originally proposed that PARP inhibition blocked base excision repair and single strand break repair pathways thus increasing reliance on BRCA-dependent repair. However, this model has been challenged [12] and more recent studies have shown that olaparib reduces cell proliferation by inducing replication stress [13] and that olaparib sensitivity is due to engagement of homologous recombination repair (HRR) at replication forks [14].

In 2005, two seminal papers were published demonstrating that breast cancer cells with siRNA depletion of BRCA1 or BRCA2 were exquisitely sensitive to the PARP inhibitor NU1025 [15,16]. *BRCA1* and *BRCA2* are cancer predisposition genes that are inactivated in ~25% of inherited breast cancers, ~15% of all ovarian cancers and several other cancers, suggesting that PARP inhibitors might have potential in treating a wide-range of patients with BRCA-deficient tumours [17]. The PARP inhibitor AZD2881, also known as olaparib or Lynparza™, showed promise in mouse models of breast cancer [18] and quickly moved into clinical trials, showing anti-tumour activity in BRCA-mutated cancers, even in phase I studies [19]. Olaparib, the first PARP inhibitor to gain regulatory approval, is now FDA-approved in advanced ovarian [20], breast [21], pancreatic [22] and prostate cancers [23], with the PARP inhibitors rucaparib [24], niraparib [25] and talazoparib [26] also FDA-approved in varying indications [17,27].

BRCA1 and 2 play critical roles in detection, signalling and repair of DNA double strand breaks (DSBs) via the HRR pathway. HRR is active in S phase at stalled replication forks and in G2 phase of the cell cycle after DSBs have been resected to contain long ssDNA overhangs on their 3′ ends [28]. These long regions of ssDNA are bound by replication protein A (RPA) and BRCA2 plays a role in the replacement of RPA with RAD51, the protein that initiates strand invasion and the search for a homologous DNA sequence during HRR [28]. BRCA1 interacts with BRCA2 via the PALB2 protein, and is recruited to DNA damage-induced foci where it participates in activating DNA repair and cell signalling pathways [29]. Given the encouraging early results showing PARP inhibitor sensitivity in BRCA-deficient cells, screens were initiated to identify other proteins that when knocked down with siRNA might confer sensitivity to PARP inhibitors [30–32]. One of these was Ataxia Telangiectasia Mutated (ATM).

2. ATM

ATM is a member of the phosphatidylinositol-3 kinase-like (PIKK) family of serine/threonine protein kinases with critical roles in the cellular response to DNA damaging agents, such as ionizing radiation (IR), that produce DSBs [33]. Like other members of the PIKK family, ATM is a large protein of over 350 kDa that is composed of an extended N-terminal region containing multiple HEAT (Huntingtin, Elongation factor 3, A subunit of protein phosphatase 2A and mammalian Target of rapamycin) repeats and a C-terminal kinase domain that has amino acid similarity to phosphatidyl inositol-3 kinase (PI3K) and is flanked and stabilized by conserved FRAP–ATM–TRRAP (FAT) and FAT-C domains. Generation of DSBs and/or changes in chromatin structure lead to activation of ATM and its autophosphorylation on serine 1981 [34]. Activated ATM phosphorylates a multitude of downstream targets including p53, checkpoint kinase 2 (Chk2) and histone H2AX [35]. Indeed, phospho-proteomics studies have identified hundreds of PIKK-dependent, DNA damage-induced phosphorylation events in cells [36,37]. Consistent with its role in the repair of IR-induced DSBs, cell lines with loss or inactivation of ATM are radiation sensitive, have cell cycle checkpoint defects [38] and have defects in slow repair of complex DNA damage lesions and DSBs in the context of heterochromatin [39]. Recently, roles in preventing premature aging [40] and in reactive oxygen sensing [41] have also been reported.

Germline inactivation of both copies of the *ATM* gene causes Ataxia-Telangiectasia (A–T), a devastating childhood condition characterized by ataxia (wobbly gait), telangiectasia (blood vessel abnormalities) and progressive neurodegeneration, particularly in the cerebellum, that renders its victims wheelchair-bound. A–T patients also have immune defects and cancer predisposition and usually succumb to their condition in their early twenties [42]. Accordingly, cell lines derived from A–T patients and ATM knock out mice are hypersensitive to IR and other chemotherapeutic agents [43,44], raising the possibility that cancers with loss of ATM may be more sensitive to DNA damaging agents than their ATM-proficient counterparts [45].

3. Targeting ATM-Deficient Cancers

Genome sequencing has revealed that *ATM* is mutated in a variety of human cancers, including mantle cell lymphoma (MCL), colorectal, lung and prostate cancers. Analysis of *ATM* mutation frequency in The Cancer Genome Atlas (TCGA) cohort using c-Bioportal [46,47] indicates that ATM is mutated in approximately 5% of all cancers, with some, such as MCL, with a much higher mutation frequency of ~40% (Figure 1A). Similarly, ATM is mutated in ~20% of colorectal and uterine cancers and approximately 10% of prostate and lung cancers (Figure 1A). The vast majority of these mutations are missense mutations and are scattered throughout the coding region (Figure 1B and [45]). An exception is R377C/H, which occurred in 74 of the 2263 (~3%) of the cancers queried (Figure 1B), and has been identified as a cancer mutation hotspot [48,49]. The R337C/H mutation was prominent in colorectal cancer, but not in prostate, lung or pancreatic cancer (Figure 1C–F). Although the functional consequences of this and most other *ATM* mutations is not known, given that in A–T many mutations in *ATM* induce protein truncation, protein destabilization and resulting loss of function [50], combined with the fact that siRNA-mediated loss of ATM in cancer cell lines results in PARP inhibitor sensitivity [30–32], it seems likely that many cancers with *ATM* mutation that lead to loss of function could be candidates for PARP inhibitor treatment.

Given that *ATM* is mutated or lost in over 40% of MCL [51], we and others examined the effects of ATM loss on PARP inhibitor sensitivity in human lymphoma cell lines that lack ATM protein expression. These cell lines were more sensitive to olaparib than their ATM-proficient counterparts in both cell line and animal models [52–54]. Moreover, the ATM kinase inhibitor KU55933 enhanced sensitivity to olaparib in ATM-proficient cells indicating that ATM kinase activity protects from PARP inhibitor sensitivity [53]. Similar results were observed in gastric cancer cell lines [55], and in colorectal cancer cell lines with shRNA depletion of ATM [56]. Deletion of ATM in mouse models of lung cancer and pancreatic cancer also induced sensitivity to PARP inhibitors and/or DNA damaging agents, as did inhibitors of the related protein kinase ATR (ATM and Rad3-related) [57,58].

We observed that in MCL, gastric and colorectal cancer cell lines with loss or down regulation of ATM, sensitivity to olaparib was enhanced when *TP53* was also mutated or deleted [52,53,55,56]. However, in mouse models, both *TP53*-proficient and deficient cells were sensitive to olaparib [58]. Therefore, the effect of p53 status on PARP inhibitor sensitivity requires further clarification. Although co-mutation of both *ATM* and *TP53* is rare [59], co-mutation has been shown to occur in 2–3% of non-small cell lung cancer where it increases tumour mutation burden and correlates with better response to immune checkpoint therapy [60], suggesting additional opportunities for targeted therapy for ATM-deficient tumours.

To address the mechanism of olaparib-induced cell sensitivity in human cells lacking ATM, we used CRISPR/Cas9 to delete ATM from the p53-proficient lung adenocarcinoma cell line, A549. In keeping with recent findings, olaparib alone reduced cell proliferation [13], but surprisingly, did not induce cell death [61]. Rather, olaparib was found to induce reversible G2 arrest in ATM-deficient A549 cells [61]. Since the related protein kinase ATR plays a critical role in the G2 checkpoint [62], and given ATM-deficient cells are sensitive to ATR inhibitors [58,63–65], we asked whether inhibition of ATR using VE-821 [66,67], would ablate G2 arrest and induce cell death in olaparib-treated ATM-deficient cells. This was indeed found to be the case. Combined treatment with olaparib plus ATR inhibition with VE821 induced cell death only in ATM-deficient A549 cells, suggesting that patients with ATM-deficient tumours could benefit from a combination of PARP and ATR inhibitors [61].

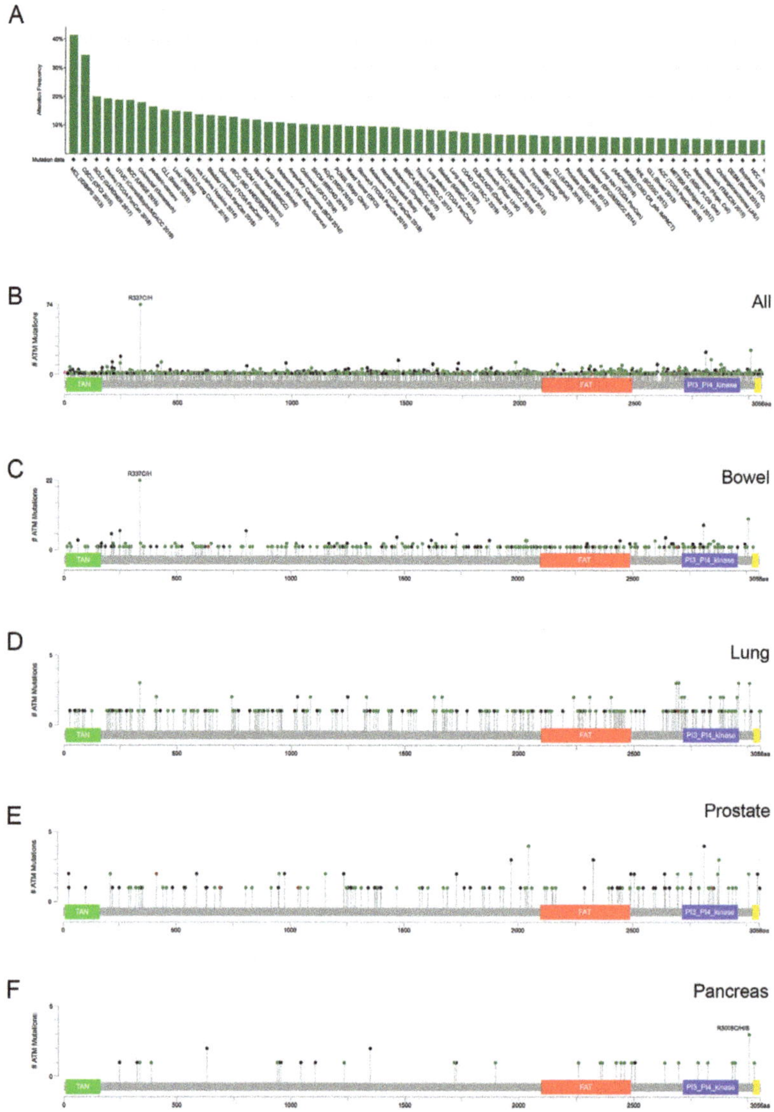

Figure 1. Frequency of Ataxia-Telangiectasia Mutated (*ATM)* mutations in human cancer. (**A**) *ATM* was queried against all entries in the curated non-redundant data set on c-Bioportal (references [46,47]) accessed January 2020. Duplicate studies were removed and copy number variations are not included. The frequency of *ATM* alteration in various cancers is shown. (**B**) ATM is a 3056 amino acid protein consisting of a N-terminal TAN (telomere length maintenance and DNA damage repair) domain (residues 7–165), and a C-terminal kinase domain (residues 2714–2961) flanked by FAT (2097–2488) and FATC (3205–3055) domains. The location of mutations in ATM from all samples in the curated non-redundant data site available on c-Bioportal (references [46,47]), accessed January 2020 (duplicate sets removed and copy number variation not include) is shown. Mutations were distributed across the entire the coding region however, one mutation R337C/H was detected in 74 out of 2263 samples, across all cancers. (**C**) The R337C/H mutation was frequent in bowel cancer (22 out of 331 samples) but less so in lung (panel **D**), prostate (panel **E**) and pancreatic cancers (Panel **F**).

ATM is frequently mutated in prostate cancer [68], as well as somatic and hereditary forms of pancreatic cancers [69,70], suggesting that patients with these cancers might also benefit from treatment with a PARP inhibitor. Nevertheless, PROFOUND, a phase III trial that examined the clinical efficacy of olaparib versus standard treatment (abiraterone acetate/enzalutamide) in patients with metastatic, castration-resistant prostate cancer (mCRPC) and HRR gene alterations, revealed less impressive rates of radiographic progression free survival benefit with olaparib in patients with *ATM* alterations, in contrast to patients with other HRR gene alterations (e.g., *BRCA2*), based upon exploratory, hypothesis-generating gene-by-gene subgroup analysis [23]. Thus, it appeared necessary to consider therapeutic approaches that may enhance the efficacy of PARP inhibition in ATM-deficient cancers.

We therefore examined whether the combination of olaparib plus an ATR inhibitor would be effective in cell line models of prostate and pancreatic cancer. We depleted ATM from the prostate cancer cell line PC-3 using CRISPR/Cas9 and found that although olaparib reduced cell proliferation, ATM-deficient cells did not undergo apoptosis unless olaparib was combined with an ATR inhibitor, either VE-821, as in [61] or AZD6738 [71], an ATR inhibitor in clinical trials [72]. Significantly, olaparib and AZD6738 had little effect on ATM-proficient cells either alone or in combination [71]. Similar results were seen with the pancreatic cancer cell line, Panc 10.05, in which ATM was depleted by shRNA [71]. Thus, the combination of PARP and ATR inhibitors could be beneficial in a number of ATM-deficient cancers, including lung, prostate and pancreatic [71]. While A549 has wild type p53 [73], PC-3 are *TP53* null [74] and Panc 10.05 contain a homozygous mutation at I225N [75,76], therefore these data suggest that the sensitivity of ATM-deficient cells to the combination of PARP inhibitor and ATR inhibitor is not dependent on p53 status.

Our results also highlight the importance of the method used to assess cell viability in determining sensitivity of a cell line to a particular agent. Although ATM-deficient cells were highly sensitive to olaparib in clonogenic survival assays and the number of viable cells was decreased compared to ATM-proficient cells measured using the trypan blue exclusion assay, analysis of sub-G1 DNA or annexin staining did not reveal evidence of cell death [61,71]. Rather, olaparib-treated, ATM-deficient cells underwent reversible G2 arrest, and did not undergo cell death until an ATR inhibitor was also added [61,71]. Moreover, similar results have been seen in a patient-derived xenograft model of *BRCA*-mutant high-grade serous ovarian cancer, suggesting that PARP inhibition also increases reliance on ATR-dependent G2 arrest in BRCA-deficient cells [77], thus the combination of PARP and ATR inhibition may have benefits in other HDR-deficient cancers.

4. ATM Mutation Versus Loss of Function: Identifying Patients Who May Benefit from PARP Inhibitor Treatment

To date, most work from our lab and others has centred on the effects of olaparib on cell lines or mice in which ATM has been deleted [52–54,57,58,61,71] or inhibited with KU55933 [53,55]. An important difference is that in cancer, ATM is mutated, but the effects of these mutations on ATM function are, for the most part, unknown. As shown in Figure 1, literally hundreds of mutations have been identified in ATM and these mutations are scattered throughout the coding region. Apart from R337H/C, a hotspot mutation prevalent in colorectal cancer, other individual mutations are seen less frequently and their effects on ATM function is not known. Recently, three papers describe cryo-electron microscopy structures of Tel1, the well-conserved ATM homolog in lower eukaryotes, showing that ATM forms an autoinhibited dimer [78–80], providing insight into the conformational changes necessary for ATM activation. Three of these structures provide atomic models of the C-terminal kinase domain, and one, from a thermophilic fungus, also provides an atomic model of the majority of the N-terminal heat repeat domain in open and closed conformations. These structures provide the molecular basis to begin to understand cancer-associated *ATM* mutations. Initial analyses predict many cancer-associated mutations in the kinase domain are likely to impact ATM activity or protein folding [78,79], and the equivalent residue to R337 appears to stabilize the packing of two helices in the N-terminal domain, with the R337C/H mutation possibly destabilizing this region of ATM [78]. The methods developed to purify wildtype ATM homologs pave the way for more comprehensive studies to test the effect of cancer-associated mutations on ATM activities and stability.

For the treatment of cancer patients based on *ATM* status, it will be critical to identify those patients who have mutations that impact ATM activity or stability and are therefore most likely to benefit from PARP/ATR inhibitor combination treatment. This may be challenging by DNA sequencing alone, given the number of mutations identified and that their effects on ATM function are, for the most part, not known. That being said, preliminary clinical trials have shown promising, if not controversial results. In an initial trial of 60 men with mCRPC, 5 were shown to have mutation of *ATM* and 4 responded to olaparib [81]. A subsequent larger trial in a similar patient population demonstrated 7 of 19 patients with ATM mutation met at least one response criteria [82]. In contrast, a multicentre retrospective review of patients with mCRPC treated with olaparib showed no patients with *ATM* mutations achieved a PSA response and had significantly worse progression-free survival (PFS) and overall survival compared to *BRCA*-mutated patients. The addition of olaparib to paclitaxel failed to improve overall survival over paclitaxel alone in Asian patients with recurrent gastric cancer, both in the overall patient population as well as those with low or absent ATM expression, as determined by immunohistochemistry (IHC) [83]. Our findings that olaparib alone is cytostatic in ATM-deficient cells and that PARP and ATR inhibitors need to be combined to kill ATM-deficient cells is in line with these findings and suggest that this combinatorial approach could improve long-term survival in patients.

Despite the large number of uncharacterized mutations in *ATM* in cancer, many mutations known to impact ATM function, such as those resulting in A–T, frequently induce protein destabilization [50]. Therefore, determining the presence of ATM protein in cancer patient biopsy samples through IHC or other approaches may prove useful. Interestingly, although data from TCGA reveals that the *ATM* gene is mutated in 12–14% of patient samples with lung adenocarcinoma, another study has reported that ~40% of lung adenocarcinoma patients have low ATM protein expression by IHC [84]. A possible explanation for the apparent difference between *ATM* gene alteration and ATM protein expression could be methylation of the *ATM* promoter leading to transcription silencing [85]. Indeed, our analysis of methylation data in TCGA datasets revealed significant negative correlations between *ATM* promoter methylation and *ATM* gene expression in prostate adenocarcinoma ($p = 1.962 \times 10^{-7}$, Spearman's rho = −0.23), lung adenocarcinoma ($p = 0.001159$, Spearman's rho = −0.15) and colon adenocarcinoma ($p = 0.0372$, Spearman's rho = −0.13) (Figure 2). These findings suggest that ATM promoter methylation may play a role in the regulation of *ATM* gene expression in these cancer types.

Another possibility for stratifying patients who will benefit from a combination of PARP and ATR inhibitors would be to use a surrogate marker of ATM functionality, either through the use of phosphospecific antibodies to ATM itself or its downstream targets or an RNA signature specific to loss of ATM functionality. Indeed, the search for identifying tumours which exhibit "BRCAness" or HRR deficiency is an area of active investigation [27,86]. Answers to these questions and more may become apparent over the next few years as PARP inhibitor therapy is tested in more patients with defects in DNA damage response genes.

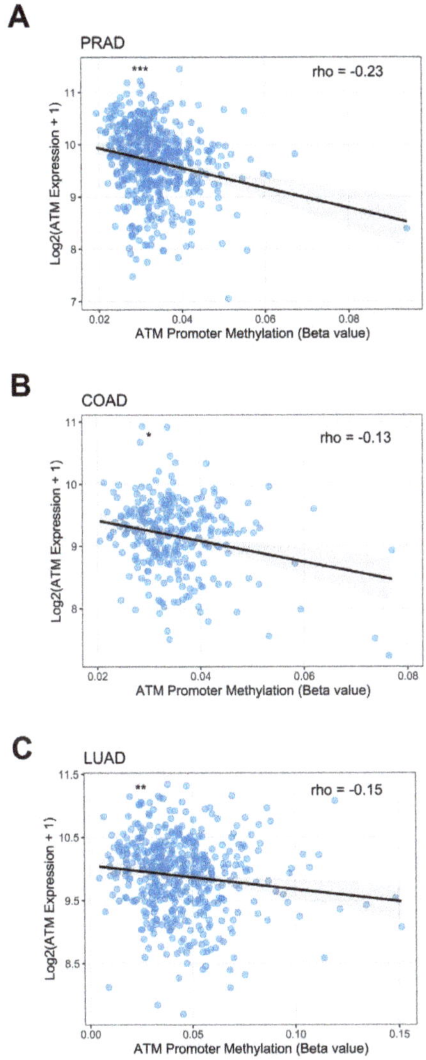

Figure 2. *ATM* promoter methylation and *ATM* gene expression in adenocarcinomas. Scatter plots showing the correlation between methylation beta values of the *ATM* promoter probe cg01756564 and *ATM* gene expression in TCGA datasets of (**A**) prostate adenocarcinoma (PRAD, $n = 496$), (**B**) colon adenocarcinoma (COAD, $n = 276$) and (**C**) lung adenocarcinoma (LUAD, $n = 454$). Spearman's rho values are indicated in the top right. Grey region of linear fit indicates 95% confidence interval. Asterisks in the top left indicate significance. * = $p < 0.05$, ** = $p < 0.01$, *** = $p < 0.001$.

5. Concluding Remarks

In conclusion, basic research into DNA damage repair biochemistry led to the identification of PARP, the inhibition of which, almost 40 years later, is showing great promise in the treatment of *BRCA*-deficient ovarian, breast, prostate and pancreatic cancer patients [87–89]. Work from our lab and others discussed here has shown that PARP inhibitors may also have potential in treating patients with ATM-deficient tumours. Our recent studies have revealed that in ATM-deficient cancer cell lines, olaparib is cytostatic not cytotoxic and that combination of olaparib with an ATR inhibitor is needed to induce cell death [61,71]. Moreover, similar results were seen in models of BRCA-mutant ovarian cancer, suggesting that inhibition of ATR potentiates the effects PARP inhibition in *BRCA*-deficient cells [77], thus the combination of PARP and ATR inhibition may have benefits in other HDR-deficient cancers. It will be exciting to see whether this finding will also apply to other PARP inhibitors and, most importantly, in the clinic for improving outcomes for cancer patients. Indeed, several clinical trials using a PARP inhibitor in combination with an ATR inhibitor are ongoing (Table 1).

Table 1. List of ongoing clinical trials combining a poly-ADP ribose polymerase (PARP) inhibitor with an ATR inhibitor. Information obtained from https://clinicaltrials.gov, accessed March 9 2020.

Clinical Trial Number	PARP Inhibitor	ATR Inhibitor	Other Therapy/Status	Cancer Type
NCT02723864	Veliparib/ABT-888	VX-970	Cisplatin	Refractory Solid Tumours
NCT034R2342	Olaparib	AZD6738	Platinum-sensitive or platinum-resistant	Recurrent ovarian cancer (CAPRI trial)
NCT03682289	Olaparib	AZD6738	None stated	Renal cell carcinoma, urothelial carcinoma, pancreatic cancers and other solid tumours
NCT03787680	Olaparib	AZD6738	DNA repair proficient/DNA repair deficient	Metastatic Castration-Resistant Prostate Cancer (TRAP trial)
NCT04065269	Olaparib	AZD6738	ARID1A loss versus no loss	Relapsed gynaecological cancers
NCT04267939	Niraparib	BAY1895344		Recurrent Advanced Solid Tumours and Ovarian Cancer

It is also possible that ongoing preclinical studies will reveal new synthetic lethal interactions within the DNA damage response. Indeed, inhibitors to other proteins in the DNA damage response are being developed [27]. Identification of alternative DNA damage response genes to target in cancer could be useful as therapies in their own right and also in cases where tumours become resistant to other therapies, such as is currently observed in *BRCA*-deficient tumours treated with PARP inhibitors [90,91]. Finally, ATM has widespread cellular roles outside the DNA damage response, including roles in cellular redox signalling [41] and regulation of autophagy [92,93], apoptosis and other cell death pathways [94–98], further increasing potential opportunities to target ATM-deficient cells with novel therapies.

Author Contributions: Conceptualization, N.R.J., M.K. (Michael Kolinsky), S.P.L.-M.; investigation, N.R.J., M.K. (Mehul Kumar), S.R., G.A., S.G., S.Y., M.K. (Michael Kolinsky), G.J.W., P.B., S.P.L.-M.; writing—original draft preparation, N.R.J., S.R., G.A., S.G., S.Y., M.K. (Michael Kolinsky), S.P.L.-M.; writing—review and editing, N.R.J., M.K. (Mehul Kumar), S.R., G.A., S.G., S.Y., M.K. (Mehul Kumar), G.J.W., P.B., S.P.L.-M.; funding acquisition, S.P.L.-M. All authors have read and agreed to the published version of the manuscript.

Funding: S.P.L.-M. thanks the Alberta Cancer Foundation, the Cancer Research Society and the Pancreatic Cancer Society of Canada for support. Work in the S.P.L.-M. and G.J.W. laboratories is also supported by NIH program project grant CA92584.

Conflicts of Interest: S.Y. has received honoraria and/or consulting fees from Janssen, Pfizer, Roche, BMS, Merck, AstraZeneca, Bayer, and Novartis. M.K. (Michael Kolinsky) has accepted honoraria and/or consulting fees from Janssen, Ipsen, Astellas, BMS, Merck, AstraZeneca, Bayer, and travel support from Novartis. N.R.J., M.K. (Mehul Kumar), S.R., G.A., S.G., G.J.W., P.B. and S.P.L.-M. declare no conflict of interest.

References

1. Hanahan, D.; Weinberg, R.A. Hallmarks of cancer: The next generation. *Cell* **2011**, *144*, 646–674. [CrossRef]
2. Hanahan, D.; Weinberg, R.A. The hallmarks of cancer. *Cell* **2000**, *100*, 57–70. [CrossRef]
3. Chambon, P.; Weill, J.D.; Mandel, P. Nicotinamide mononucleotide activation of new DNA-dependent polyadenylic acid synthesizing nuclear enzyme. *Biochem. Biophys. Res. Commun.* **1963**, *11*, 39–43. [CrossRef]

4. Ben-Hur, E.; Utsumi, H.; Elkind, M.M. Inhibitors of poly (ADP-ribose) synthesis enhance radiation response by differentially affecting repair of potentially lethal versus sublethal damage. *Br. J. Cancer Suppl.* **1984**, *6*, 39–42.
5. Clark, J.B.; Ferris, G.M.; Pinder, S. Inhibition of nuclear NAD nucleosidase and poly ADP-ribose polymerase activity from rat liver by nicotinamide and 5′-methyl nicotinamide. *Biochim. Biophys. Acta* **1971**, *238*, 82–85. [CrossRef]
6. Tentori, L.; Portarena, I.; Graziani, G. Potential clinical applications of poly(ADP-ribose) polymerase (PARP) inhibitors. *Pharm. Res.* **2002**, *45*, 73–85. [CrossRef] [PubMed]
7. Curtin, N.J.; Szabo, C. Therapeutic applications of PARP inhibitors: Anticancer therapy and beyond. *Mol. Asp. Med.* **2013**, *34*, 1217–1256. [CrossRef] [PubMed]
8. Schreiber, V.; Dantzer, F.; Ame, J.C.; de Murcia, G. Poly(ADP-ribose): Novel functions for an old molecule. *Nat. Rev. Mol. Cell Biol.* **2006**, *7*, 517–528. [CrossRef]
9. Izhar, L.; Adamson, B.; Ciccia, A.; Lewis, J.; Pontano-Vaites, L.; Leng, Y.; Liang, A.C.; Westbrook, T.F.; Harper, J.W.; Elledge, S.J. A Systematic Analysis of Factors Localized to Damaged Chromatin Reveals PARP-Dependent Recruitment of Transcription Factors. *Cell Rep.* **2015**, *11*, 1486–1500. [CrossRef]
10. D'Amours, D.; Desnoyers, S.; D'Silva, I.; Poirier, G.G. Poly(ADP-ribosyl)ation reactions in the regulation of nuclear functions. *Biochem. J.* **1999**, *342*, 249–268. [CrossRef]
11. Murai, J.; Huang, S.Y.; Das, B.B.; Renaud, A.; Zhang, Y.; Doroshow, J.H.; Ji, J.; Takeda, S.; Pommier, Y. Trapping of PARP1 and PARP2 by Clinical PARP Inhibitors. *Cancer Res.* **2012**, *72*, 5588–5599. [CrossRef]
12. Helleday, T. The underlying mechanism for the PARP and BRCA synthetic lethality: Clearing up the misunderstandings. *Mol. Oncol.* **2011**, *5*, 387–393. [CrossRef] [PubMed]
13. Maya-Mendoza, A.; Moudry, P.; Merchut-Maya, J.M.; Lee, M.; Strauss, R.; Bartek, J. High speed of fork progression induces DNA replication stress and genomic instability. *Nature* **2018**, *10*. [CrossRef] [PubMed]
14. Balmus, G.; Pilger, D.; Coates, J.; Demir, M.; Sczaniecka-Clift, M.; Barros, A.C.; Woods, M.; Fu, B.; Yang, F.; Chen, E.; et al. ATM orchestrates the DNA-damage response to counter toxic non-homologous end-joining at broken replication forks. *Nat. Commun.* **2019**, *10*, 87. [CrossRef] [PubMed]
15. Farmer, H.; McCabe, N.; Lord, C.J.; Tutt, A.N.; Johnson, D.A.; Richardson, T.B.; Santarosa, M.; Dillon, K.J.; Hickson, I.; Knights, C.; et al. Targeting the DNA repair defect in BRCA mutant cells as a therapeutic strategy. *Nature* **2005**, *434*, 917–921. [CrossRef] [PubMed]
16. Bryant, H.E.; Schultz, N.; Thomas, H.D.; Parker, K.M.; Flower, D.; Lopez, E.; Kyle, S.; Meuth, M.; Curtin, N.J.; Helleday, T. Specific killing of BRCA2-deficient tumours with inhibitors of poly(ADP-ribose) polymerase. *Nature* **2005**, *434*, 913–917. [CrossRef] [PubMed]
17. Kamel, D.; Gray, C.; Walia, J.S.; Kumar, V. PARP Inhibitor Drugs in the Treatment of Breast, Ovarian, Prostate and Pancreatic Cancers: An Update of Clinical Trials. *Curr. Drug Targets* **2018**, *19*, 21–37. [CrossRef]
18. Evers, B.; Drost, R.; Schut, E.; de Bruin, M.; van der Burg, E.; Derksen, P.W.; Holstege, H.; Liu, X.; van Drunen, E.; Beverloo, H.B.; et al. Selective inhibition of BRCA2-deficient mammary tumor cell growth by AZD2281 and cisplatin. *Clin. Cancer Res.* **2008**, *14*, 3916–3925. [CrossRef]
19. Fong, P.C.; Boss, D.S.; Yap, T.A.; Tutt, A.; Wu, P.; Mergui-Roelvink, M.; Mortimer, P.; Swaisland, H.; Lau, A.; O'Connor, M.J.; et al. Inhibition of poly(ADP-ribose) polymerase in tumors from BRCA mutation carriers. *N. Engl. J. Med.* **2009**, *361*, 123–134. [CrossRef]
20. Moore, K.; Colombo, N.; Scambia, G.; Kim, B.G.; Oaknin, A.; Friedlander, M.; Lisyanskaya, A.; Floquet, A.; Leary, A.; Sonke, G.S.; et al. Maintenance Olaparib in Patients with Newly Diagnosed Advanced Ovarian Cancer. *N. Engl. J. Med.* **2018**, *379*, 2495–2505. [CrossRef]
21. Robson, M.; Im, S.A.; Senkus, E.; Xu, B.; Domchek, S.M.; Masuda, N.; Delaloge, S.; Li, W.; Tung, N.; Armstrong, A.; et al. Olaparib for Metastatic Breast Cancer in Patients with a Germline BRCA Mutation. *N. Engl. J. Med.* **2017**, *377*, 523–533. [CrossRef] [PubMed]
22. Golan, T.; Hammel, P.; Reni, M.; Van Cutsem, E.; Macarulla, T.; Hall, M.J.; Park, J.O.; Hochhauser, D.; Arnold, D.; Oh, D.Y.; et al. Maintenance Olaparib for Germline BRCA-Mutated Metastatic Pancreatic Cancer. *N. Engl. J. Med.* **2019**, *381*, 317–327. [CrossRef] [PubMed]
23. Hussain, M.; Mateo, J.; Fizazi, K.; Saad, F.; Shore, N.D.; Sandhu, S.; Chi, K.N.; Sartor, O.; Agarwal, N.; Olmos, D.A.; et al. LBA12_PR—PROfound: Phase III study of olaparib versus enzalutamide or abiraterone for metastatic castration-resistant prostate cancer (mCRPC) with homologous recombination repair (HRR) gene alterations. In Proceedings of the ESMO Congress 2019, Barcelona, Spain, 27 September–1 October 2019.

24. Coleman, R.L.; Oza, A.M.; Lorusso, D.; Aghajanian, C.; Oaknin, A.; Dean, A.; Colombo, N.; Weberpals, J.I.; Clamp, A.; Scambia, G.; et al. Rucaparib maintenance treatment for recurrent ovarian carcinoma after response to platinum therapy (ARIEL3): A randomised, double-blind, placebo-controlled, phase 3 trial. *Lancet* **2017**, *390*, 1949–1961. [CrossRef]
25. Moore, K.N.; Secord, A.A.; Geller, M.A.; Miller, D.S.; Cloven, N.; Fleming, G.F.; Wahner Hendrickson, A.E.; Azodi, M.; DiSilvestro, P.; Oza, A.M.; et al. Niraparib monotherapy for late-line treatment of ovarian cancer (QUADRA): A multicentre, open-label, single-arm, phase 2 trial. *Lancet Oncol.* **2019**, *20*, 636–648. [CrossRef]
26. Litton, J.K.; Scoggins, M.E.; Hess, K.R.; Adrada, B.E.; Murthy, R.K.; Damodaran, S.; DeSnyder, S.M.; Brewster, A.M.; Barcenas, C.H.; Valero, V.; et al. Neoadjuvant Talazoparib for Patients With Operable Breast Cancer With a Germline BRCA Pathogenic Variant. *J. Clin. Oncol.* **2019**. [CrossRef]
27. Pilie, P.G.; Tang, C.; Mills, G.B.; Yap, T.A. State-of-the-art strategies for targeting the DNA damage response in cancer. *Nat. Rev. Clin. Oncol.* **2019**, *16*, 81–104. [CrossRef]
28. Wright, W.D.; Shah, S.S.; Heyer, W.D. Homologous recombination and the repair of DNA double-strand breaks. *J. Biol. Chem.* **2018**, *293*, 10524–10535. [CrossRef]
29. Roy, R.; Chun, J.; Powell, S.N. BRCA1 and BRCA2: Different roles in a common pathway of genome protection. *Nat. Rev. Cancer* **2012**, *12*, 68–78. [CrossRef]
30. Lord, C.J.; McDonald, S.; Swift, S.; Turner, N.C.; Ashworth, A. A high-throughput RNA interference screen for DNA repair determinants of PARP inhibitor sensitivity. *DNA Repair* **2008**, *7*, 2010–2019. [CrossRef]
31. Turner, N.C.; Lord, C.J.; Iorns, E.; Brough, R.; Swift, S.; Elliott, R.; Rayter, S.; Tutt, A.N.; Ashworth, A. A synthetic lethal siRNA screen identifying genes mediating sensitivity to a PARP inhibitor. *EMBO J.* **2008**, *27*, 1368–1377. [CrossRef]
32. McCabe, N.; Turner, N.C.; Lord, C.J.; Kluzek, K.; Bialkowska, A.; Swift, S.; Giavara, S.; O'Connor, M.J.; Tutt, A.N.; Zdzienicka, M.Z.; et al. Deficiency in the repair of DNA damage by homologous recombination and sensitivity to poly(ADP-ribose) polymerase inhibition. *Cancer Res.* **2006**, *66*, 8109–8115. [CrossRef] [PubMed]
33. Blackford, A.N.; Jackson, S.P. ATM, ATR, and DNA-PK: The Trinity at the Heart of the DNA Damage Response. *Mol. Cell* **2017**, *66*, 801–817. [CrossRef] [PubMed]
34. Bakkenist, C.J.; Kastan, M.B. DNA damage activates ATM through intermolecular autophosphorylation and dimer dissociation. *Nature* **2003**, *421*, 499–506. [CrossRef] [PubMed]
35. Ciccia, A.; Elledge, S.J. The DNA damage response: Making it safe to play with knives. *Mol. Cell* **2010**, *40*, 179–204. [CrossRef]
36. Matsuoka, S.; Ballif, B.A.; Smogorzewska, A.; McDonald, E.R., III; Hurov, K.E.; Luo, J.; Bakalarski, C.E.; Zhao, Z.; Solimini, N.; Lerenthal, Y.; et al. ATM and ATR substrate analysis reveals extensive protein networks responsive to DNA damage. *Science* **2007**, *316*, 1160–1166. [CrossRef]
37. Bennetzen, M.V.; Larsen, D.H.; Bunkenborg, J.; Bartek, J.; Lukas, J.; Andersen, J.S. Site-specific phosphorylation dynamics of the nuclear proteome during the DNA damage response. *Mol. Cell. Proteom. MCP* **2010**, *9*, 1314–1323. [CrossRef]
38. Shiloh, Y. ATM: Expanding roles as a chief guardian of genome stability. *Exp. Cell Res.* **2014**. [CrossRef]
39. Goodarzi, A.A.; Jeggo, P.A. The heterochromatic barrier to DNA double strand break repair: How to get the entry visa. *Int. J. Mol. Sci.* **2012**, *13*, 11844–11860. [CrossRef]
40. Shiloh, Y.; Lederman, H.M. Ataxia-telangiectasia (A-T): An emerging dimension of premature ageing. *Ageing Res. Rev.* **2016**. [CrossRef]
41. Ditch, S.; Paull, T.T. The ATM protein kinase and cellular redox signaling: Beyond the DNA damage response. *Trends Biochem. Sci.* **2012**, *37*, 15–22. [CrossRef]
42. Rothblum-Oviatt, C.; Wright, J.; Lefton-Greif, M.A.; McGrath-Morrow, S.A.; Crawford, T.O.; Lederman, H.M. Ataxia telangiectasia: A review. *Orphanet J. Rare Dis.* **2016**, *11*, 159. [CrossRef] [PubMed]
43. Barlow, C.; Hirotsune, S.; Paylor, R.; Liyanage, M.; Eckhaus, M.; Collins, F.; Shiloh, Y.; Crawley, J.N.; Ried, T.; Tagle, D.; et al. Atm-deficient mice: A paradigm of ataxia telangiectasia. *Cell* **1996**, *86*, 159–171. [CrossRef]
44. Shiloh, Y.; Tabor, E.; Becker, Y. Abnormal response of ataxia-telangiectasia cells to agents that break the deoxyribose moiety of DNA via a targeted free radical mechanism. *Carcinogenesis* **1983**, *4*, 1317–1322. [CrossRef] [PubMed]
45. Choi, M.; Kipps, T.; Kurzrock, R. ATM Mutations in Cancer: Therapeutic Implications. *Mol. Cancer Ther.* **2016**, *15*, 1781–1791. [CrossRef] [PubMed]

46. Gao, J.; Aksoy, B.A.; Dogrusoz, U.; Dresdner, G.; Gross, B.; Sumer, S.O.; Sun, Y.; Jacobsen, A.; Sinha, R.; Larsson, E.; et al. Integrative analysis of complex cancer genomics and clinical profiles using the cBioPortal. *Sci. Signal.* **2013**, *6*, pl1. [CrossRef] [PubMed]
47. Cerami, E.; Gao, J.; Dogrusoz, U.; Gross, B.E.; Sumer, S.O.; Aksoy, B.A.; Jacobsen, A.; Byrne, C.J.; Heuer, M.L.; Larsson, E.; et al. The cBio cancer genomics portal: An open platform for exploring multidimensional cancer genomics data. *Cancer Discov.* **2012**, *2*, 401–404. [CrossRef] [PubMed]
48. Chang, M.T.; Bhattarai, T.S.; Schram, A.M.; Bielski, C.M.; Donoghue, M.T.A.; Jonsson, P.; Chakravarty, D.; Phillips, S.; Kandoth, C.; Penson, A.; et al. Accelerating Discovery of Functional Mutant Alleles in Cancer. *Cancer Discov.* **2018**, *8*, 174–183. [CrossRef]
49. Chang, M.T.; Asthana, S.; Gao, S.P.; Lee, B.H.; Chapman, J.S.; Kandoth, C.; Gao, J.; Socci, N.D.; Solit, D.B.; Olshen, A.B.; et al. Identifying recurrent mutations in cancer reveals widespread lineage diversity and mutational specificity. *Nat. Biotechnol.* **2016**, *34*, 155–163. [CrossRef]
50. Gilad, S.; Khosravi, R.; Shkedy, D.; Uziel, T.; Ziv, Y.; Savitsky, K.; Rotman, G.; Smith, S.; Chessa, L.; Jorgensen, T.J.; et al. Predominance of null mutations in ataxia-telangiectasia. *Hum. Mol. Genet.* **1996**, *5*, 433–439. [CrossRef]
51. Greiner, T.C.; Dasgupta, C.; Ho, V.V.; Weisenburger, D.D.; Smith, L.M.; Lynch, J.C.; Vose, J.M.; Fu, K.; Armitage, J.O.; Braziel, R.M.; et al. Mutation and genomic deletion status of ataxia telangiectasia mutated (ATM) and p53 confer specific gene expression profiles in mantle cell lymphoma. *Pro. Natl. Acad. Sci. USA* **2006**, *103*, 2352–2357. [CrossRef]
52. Williamson, C.T.; Muzik, H.; Turhan, A.G.; Zamo, A.; O'Connor, M.J.; Bebb, D.G.; Lees-Miller, S.P. ATM deficiency sensitizes mantle cell lymphoma cells to poly(ADP-ribose) polymerase-1 inhibitors. *Mol. Cancer Ther.* **2010**, *9*, 347–357. [CrossRef] [PubMed]
53. Williamson, C.T.; Kubota, E.; Hamill, J.D.; Klimowicz, A.; Ye, R.; Muzik, H.; Dean, M.; Tu, L.; Gilley, D.; Magliocco, A.M.; et al. Enhanced cytotoxicity of PARP inhibition in mantle cell lymphoma harbouring mutations in both ATM and p53. *EMBO Mol. Med.* **2012**, *4*, 515–527. [CrossRef] [PubMed]
54. Weston, V.J.; Oldreive, C.E.; Skowronska, A.; Oscier, D.G.; Pratt, G.; Dyer, M.J.; Smith, G.; Powell, J.E.; Rudzki, Z.; Kearns, P.; et al. The PARP inhibitor olaparib induces significant killing of ATM-deficient lymphoid tumor cells in vitro and in vivo. *Blood* **2010**, *116*, 4578–4587. [CrossRef] [PubMed]
55. Kubota, E.; Williamson, C.T.; Ye, R.; Elegbede, A.; Peterson, L.; Lees-Miller, S.P.; Bebb, D.G. Low ATM protein expression and depletion of p53 correlates with olaparib sensitivity in gastric cancer cell lines. *Cell Cycle* **2014**, *13*, 2129–2137. [CrossRef] [PubMed]
56. Wang, C.; Jette, N.; Moussienko, D.; Bebb, D.G.; Lees-Miller, S.P. ATM-Deficient Colorectal Cancer Cells Are Sensitive to the PARP Inhibitor Olaparib. *Transl Oncol* **2017**, *10*, 190–196. [CrossRef] [PubMed]
57. Perkhofer, L.; Schmitt, A.; Romero Carrasco, M.C.; Ihle, M.; Hampp, S.; Ruess, D.A.; Hessmann, E.; Russell, R.; Lechel, A.; Azoitei, N.; et al. ATM Deficiency Generating Genomic Instability Sensitizes Pancreatic Ductal Adenocarcinoma Cells to Therapy-Induced DNA Damage. *Cancer Res.* **2017**, *77*, 5576–5590. [CrossRef] [PubMed]
58. Schmitt, A.; Knittel, G.; Welcker, D.; Yang, T.P.; George, J.; Nowak, M.; Leeser, U.; Buttner, R.; Perner, S.; Peifer, M.; et al. ATM Deficiency Is Associated with Sensitivity to PARP1- and ATR Inhibitors in Lung Adenocarcinoma. *Cancer Res.* **2017**, *77*, 3040–3056. [CrossRef]
59. Jiang, H.; Reinhardt, H.C.; Bartkova, J.; Tommiska, J.; Blomqvist, C.; Nevanlinna, H.; Bartek, J.; Yaffe, M.B.; Hemann, M.T. The combined status of ATM and p53 link tumor development with therapeutic response. *Genes Dev.* **2009**, *23*, 1895–1909. [CrossRef]
60. Chen, Y.; Chen, G.; Li, J.; Huang, Y.Y.; Li, Y.; Lin, J.; Chen, L.Z.; Lu, J.P.; Wang, Y.Q.; Wang, C.X.; et al. Association of Tumor Protein p53 and Ataxia-Telangiectasia Mutated Comutation With Response to Immune Checkpoint Inhibitors and Mortality in Patients With Non-Small Cell Lung Cancer. *JAMA Netw. Open* **2019**, *2*, e1911895. [CrossRef]
61. Jette, N.R.; Radhamani, S.; Arthur, G.; Ye, R.; Goutam, S.; Bolyos, A.; Petersen, L.F.; Bose, P.; Bebb, D.G.; Lees-Miller, S.P. Combined poly-ADP ribose polymerase and ataxia-telangiectasia mutated/Rad3-related inhibition targets ataxia-telangiectasia mutated-deficient lung cancer cells. *Br. J. Cancer* **2019**, *121*, 600–610. [CrossRef]
62. Shaltiel, I.A.; Krenning, L.; Bruinsma, W.; Medema, R.H. The same, only different—DNA damage checkpoints and their reversal throughout the cell cycle. *J. Cell Sci.* **2015**, *128*, 607–620. [CrossRef] [PubMed]

63. Lecona, E.; Fernandez-Capetillo, O. Targeting ATR in cancer. *Nat. Rev. Cancer* **2018**, *18*, 586–595. [CrossRef] [PubMed]
64. Min, A.; Im, S.A.; Jang, H.; Kim, S.; Lee, M.; Kim, D.K.; Yang, Y.; Kim, H.J.; Lee, K.H.; Kim, J.W.; et al. AZD6738, A Novel Oral Inhibitor of ATR, Induces Synthetic Lethality with ATM Deficiency in Gastric Cancer Cells. *Mol. Cancer Ther.* **2017**, *16*, 566–577. [CrossRef] [PubMed]
65. Vendetti, F.P.; Lau, A.; Schamus, S.; Conrads, T.P.; O'Connor, M.J.; Bakkenist, C.J. The orally active and bioavailable ATR kinase inhibitor AZD6738 potentiates the anti-tumor effects of cisplatin to resolve ATM-deficient non-small cell lung cancer in vivo. *Oncotarget* **2015**, *6*, 44289–44305. [CrossRef] [PubMed]
66. Charrier, J.D.; Durrant, S.J.; Golec, J.M.; Kay, D.P.; Knegtel, R.M.; MacCormick, S.; Mortimore, M.; O'Donnell, M.E.; Pinder, J.L.; Reaper, P.M.; et al. Discovery of potent and selective inhibitors of ataxia telangiectasia mutated and Rad3 related (ATR) protein kinase as potential anticancer agents. *J. Med. Chem.* **2011**, *54*, 2320–2330. [CrossRef] [PubMed]
67. Reaper, P.M.; Griffiths, M.R.; Long, J.M.; Charrier, J.D.; Maccormick, S.; Charlton, P.A.; Golec, J.M.; Pollard, J.R. Selective killing of ATM- or p53-deficient cancer cells through inhibition of ATR. *Nat. Chem. Biol.* **2011**, *7*, 428–430. [CrossRef]
68. Mateo, J.; Carreira, S.; Sandhu, S.; Miranda, S.; Mossop, H.; Perez-Lopez, R.; Nava Rodrigues, D.; Robinson, D.; Omlin, A.; Tunariu, N.; et al. DNA-Repair Defects and Olaparib in Metastatic Prostate Cancer. *N. Engl. J. Med.* **2015**, *373*, 1697–1708. [CrossRef]
69. Roberts, N.J.; Jiao, Y.; Yu, J.; Kopelovich, L.; Petersen, G.M.; Bondy, M.L.; Gallinger, S.; Schwartz, A.G.; Syngal, S.; Cote, M.L.; et al. ATM mutations in patients with hereditary pancreatic cancer. *Cancer Discov.* **2012**, *2*, 41–46. [CrossRef]
70. Kim, H.; Saka, B.; Knight, S.; Borges, M.; Childs, E.; Klein, A.; Wolfgang, C.; Herman, J.; Adsay, V.N.; Hruban, R.H.; et al. Having pancreatic cancer with tumoral loss of ATM and normal TP53 protein expression is associated with a poorer prognosis. *Clin. Cancer Res.* **2014**, *20*, 1865–1872. [CrossRef]
71. Jette, N.; Radhamani, S.; Ye, R.; Yu, Y.; Kumar, M.; Arthur, G.; Goutam, S.; Bizmar, T.A.; Bose, P.; Yip, S.; et al. ATM-deficient lung, prostate and pancreatic cancer cells are acutely sensitive to the combination of olaparib and the ATR inhibitor AZD6738. MS ID#: BIORXIV/2020/991166.
72. Mei, L.; Zhang, J.; He, K.; Zhang, J. Ataxia telangiectasia and Rad3-related inhibitors and cancer therapy: Where we stand. *J. Hematol. Oncol.* **2019**, *12*, 43. [CrossRef]
73. Yusein-Myashkova, S.; Stoykov, I.; Gospodinov, A.; Ugrinova, I.; Pasheva, E. The repair capacity of lung cancer cell lines A549 and H1299 depends on HMGB1 expression level and the p53 status. *J. Biochem.* **2016**, *160*, 37–47. [CrossRef] [PubMed]
74. Chappell, W.H.; Lehmann, B.D.; Terrian, D.M.; Abrams, S.L.; Steelman, L.S.; McCubrey, J.A. p53 expression controls prostate cancer sensitivity to chemotherapy and the MDM2 inhibitor Nutlin-3. *Cell Cycle* **2012**, *11*, 4579–4588. [CrossRef] [PubMed]
75. Forbes, S.A.; Bindal, N.; Bamford, S.; Cole, C.; Kok, C.Y.; Beare, D.; Jia, M.; Shepherd, R.; Leung, K.; Menzies, A.; et al. COSMIC: Mining complete cancer genomes in the Catalogue of Somatic Mutations in Cancer. *Nucleic Acids Res.* **2011**, *39*, D945–D950. [CrossRef] [PubMed]
76. Bamford, S.; Dawson, E.; Forbes, S.; Clements, J.; Pettett, R.; Dogan, A.; Flanagan, A.; Teague, J.; Futreal, P.A.; Stratton, M.R.; et al. The COSMIC (Catalogue of Somatic Mutations in Cancer) database and website. *Br. J. Cancer* **2004**, *91*, 355–358. [CrossRef]
77. Kim, H.; George, E.; Ragland, R.; Rafail, S.; Zhang, R.; Krepler, C.; Morgan, M.; Herlyn, M.; Brown, E.; Simpkins, F. Targeting the ATR/CHK1 Axis with PARP Inhibition Results in Tumor Regression in BRCA-Mutant Ovarian Cancer Models. *Clin. Cancer Res.* **2017**, *23*, 3097–3108. [CrossRef]
78. Jansma, M.; Linke-Winnebeck, C.; Eustermann, S.; Lammens, K.; Kostrewa, D.; Stakyte, K.; Litz, C.; Kessler, B.; Hopfner, K.P. Near-Complete Structure and Model of Tel1ATM from Chaetomium thermophilum Reveals a Robust Autoinhibited ATP State. *Structure* **2020**, *28*, 83–95. [CrossRef]
79. Yates, L.A.; Williams, R.M.; Hailemariam, S.; Ayala, R.; Burgers, P.; Zhang, X. Cryo-EM Structure of Nucleotide-Bound Tel1(ATM) Unravels the Molecular Basis of Inhibition and Structural Rationale for Disease-Associated Mutations. *Structure* **2020**, *28*, 96–104. [CrossRef]
80. Xin, J.; Xu, Z.; Wang, X.; Tian, Y.; Zhang, Z.; Cai, G. Structural basis of allosteric regulation of Tel1/ATM kinase. *Cell Res.* **2019**, *29*, 655–665. [CrossRef]

81. Mateo, J.; Boysen, G.; Barbieri, C.E.; Bryant, H.E.; Castro, E.; Nelson, P.S.; Olmos, D.; Pritchard, C.C.; Rubin, M.A.; de Bono, J.S. DNA Repair in Prostate Cancer: Biology and Clinical Implications. *Eur. Urol.* **2017**, *71*, 417–425. [CrossRef]
82. Mateo, J.; Porta, N.; Bianchini, D.; McGovern, U.; Elliott, T.; Jones, R.; Syndikus, I.; Ralph, C.; Jain, S.; Varughese, M.; et al. Olaparib in patients with metastatic castration-resistant prostate cancer with DNA repair gene aberrations (TOPARP-B): A multicentre, open-label, randomised, phase 2 trial. *Lancet Oncol.* **2019**. [CrossRef]
83. Bang, Y.J.; Im, S.A.; Lee, K.W.; Cho, J.Y.; Song, E.K.; Lee, K.H.; Kim, Y.H.; Park, J.O.; Chun, H.G.; Zang, D.Y.; et al. Randomized, Double-Blind Phase II Trial With Prospective Classification by ATM Protein Level to Evaluate the Efficacy and Tolerability of Olaparib Plus Paclitaxel in Patients With Recurrent or Metastatic Gastric Cancer. *J. Clin. Oncol.* **2015**, *33*, 3858–3865. [CrossRef] [PubMed]
84. Villaruz, L.C.; Jones, H.; Dacic, S.; Abberbock, S.; Kurland, B.F.; Stabile, L.P.; Siegfried, J.M.; Conrads, T.P.; Smith, N.R.; O'Connor, M.J.; et al. ATM protein is deficient in over 40% of lung adenocarcinomas. *Oncotarget* **2016**. [CrossRef] [PubMed]
85. Kim, W.J.; Vo, Q.N.; Shrivastav, M.; Lataxes, T.A.; Brown, K.D. Aberrant methylation of the ATM promoter correlates with increased radiosensitivity in a human colorectal tumor cell line. *Oncogene* **2002**, *21*, 3864–3871. [CrossRef] [PubMed]
86. Byrum, A.K.; Vindigni, A.; Mosammaparast, N. Defining and Modulating 'BRCAness'. *Trends Cell Biol.* **2019**, *29*, 740–751. [CrossRef] [PubMed]
87. Bradbury, A.; Hall, S.; Curtin, N.; Drew, Y. Targeting ATR as Cancer Therapy: A new era for synthetic lethality and synergistic combinations? *Pharmacol. Ther.* **2019**. [CrossRef] [PubMed]
88. Ashworth, A.; Lord, C.J. Synthetic lethal therapies for cancer: what's next after PARP inhibitors? *Nat. Rev. Clin. Oncol.* **2018**, *15*, 564–576. [CrossRef] [PubMed]
89. Mateo, J.; Lord, C.J.; Serra, V.; Tutt, A.; Balmana, J.; Castroviejo-Bermejo, M.; Cruz, C.; Oaknin, A.; Kaye, S.B.; de Bono, J.S. A decade of clinical development of PARP inhibitors in perspective. *Ann. Oncol. Off. J. Eur. Soc. Med. Oncol. ESMO* **2019**, *30*, 1437–1447. [CrossRef]
90. Noordermeer, S.M.; van Attikum, H. PARP Inhibitor Resistance: A Tug-of-War in BRCA-Mutated Cells. *Trends Cell Biol.* **2019**, *29*, 820–834. [CrossRef]
91. D'Andrea, A.D. Mechanisms of PARP inhibitor sensitivity and resistance. *DNA Repair* **2018**, *71*, 172–176. [CrossRef]
92. Tripathi, D.N.; Zhang, J.; Jing, J.; Dere, R.; Walker, C.L. A new role for ATM in selective autophagy of peroxisomes (pexophagy). *Autophagy* **2016**, *12*, 711–712. [CrossRef]
93. Zhang, J.; Tripathi, D.N.; Jing, J.; Alexander, A.; Kim, J.; Powell, R.T.; Dere, R.; Tait-Mulder, J.; Lee, J.H.; Paull, T.T.; et al. ATM functions at the peroxisome to induce pexophagy in response to ROS. *Nat. Cell Biol.* **2015**, *17*, 1259–1269. [CrossRef] [PubMed]
94. Ngoi, N.Y.L.; Choong, C.; Lee, J.; Bellot, G.; Wong, A.L.A.; Goh, B.C.; Pervaiz, S. Targeting Mitochondrial Apoptosis to Overcome Treatment Resistance in Cancer. *Cancers* **2020**, *12*, 574. [CrossRef] [PubMed]
95. Sia, J.; Szmyd, R.; Hau, E.; Gee, H.E. Molecular Mechanisms of Radiation-Induced Cancer Cell Death: A Primer. *Front. Cell Dev. Biol.* **2020**, *8*, 41. [CrossRef]
96. Wu, Q.; Allouch, A.; Martins, I.; Brenner, C.; Modjtahedi, N.; Deutsch, E.; Perfettini, J.L. Modulating Both Tumor Cell Death and Innate Immunity Is Essential for Improving Radiation Therapy Effectiveness. *Front. Immunol.* **2017**, *8*, 613. [CrossRef] [PubMed]
97. Zhang, D.; Tang, B.; Xie, X.; Xiao, Y.F.; Yang, S.M.; Zhang, J.W. The interplay between DNA repair and autophagy in cancer therapy. *Cancer Biol. Ther.* **2015**, *16*, 1005–1013. [CrossRef] [PubMed]
98. Boice, A.; Bouchier-Hayes, L. Targeting apoptotic caspases in cancer. *Biochim. Biophys. Acta Mol. Cell Res.* **2020**, *1867*, 118688. [CrossRef] [PubMed]

cancers

MDPI

Review

The Controversial Roles of ADP-Ribosyl Hydrolases MACROD1, MACROD2 and TARG1 in Carcinogenesis

Karla L.H. Feijs [1,*], Christopher D.O. Cooper [2] and Roko Žaja [1]

[1] Institute of Biochemistry and Molecular Biology, RWTH Aachen University, Pauwelsstrasse 30, 52074 Aachen, Germany; rzaja@ukaachen.de

[2] Department of Biological and Geographical Sciences, School of Applied Sciences, University of Huddersfield, Queensgate, Huddersfield West Yorkshire HD3 4AP, UK; c.d.cooper@hud.ac.uk

* Correspondence: kfeijs@ukaachen.de; Tel.: +49-(241)-8080962

Received: 15 January 2020; Accepted: 27 February 2020; Published: 5 March 2020

Abstract: Post-translational modifications (PTM) of proteins are crucial for fine-tuning a cell's response to both intracellular and extracellular cues. ADP-ribosylation is a PTM, which occurs in two flavours: modification of a target with multiple ADP-ribose moieties (poly(ADP-ribosyl)ation or PARylation) or with only one unit (MARylation), which are added by the different enzymes of the PARP family (also known as the ARTD family). PARylation has been relatively well-studied, particularly in the DNA damage response. This has resulted in the development of PARP inhibitors such as olaparib, which are increasingly employed in cancer chemotherapeutic approaches. Despite the fact that the majority of PARP enzymes catalyse MARylation, MARylation is not as well understood as PARylation. MARylation is a dynamic process: the enzymes reversing intracellular MARylation of acidic amino acids (MACROD1, MACROD2, and TARG1) were discovered in 2013. Since then, however, little information has been published about their physiological function. MACROD1, MACROD2, and TARG1 have a 'macrodomain' harbouring the catalytic site, but no other domains have been identified. Despite the lack of information regarding their cellular roles, there are a number of studies linking them to cancer. However, some of these publications oppose each other, some rely on poorly-characterised antibodies, or on aberrant localisation of overexpressed rather than native protein. In this review, we critically assess the available literature on a role for the hydrolases in cancer and find that, currently, there is limited evidence for a role for MACROD1, MACROD2, or TARG1 in tumorigenesis.

Keywords: ADP-ribosylation; cancer; macrodomain; ADP-ribosyl hydrolase; PARP; ARTD; MACROD1; MACROD2; TARG1

1. ADP-Ribosylation Reactions

Post-translational modifications of proteins can have a myriad of consequences: changing interactomes, stability, localisation, or activity are just a few examples. To date, more than 200 types of modification are known [1]. The addition of ADP-ribose moieties to a protein was first reported in the 1960s when chains of (poly)ADP-ribose (PAR) were identified on target proteins, termed poly(ADP-ribosyl)ation (PARylation) [2]. Since then, PARylation has been intensively studied, leading to the identification of crucial roles in the repair of DNA damage and to the development of specific inhibitors that are utilised in the clinic to treat diverse cancers [3]. In contrast, much less is known about its smaller sibling, mono(ADP-ribosyl)ation (MARylation), where only single ADP-ribose units (MAR) are conjugated into proteins. In 2008, it was first noted that PARP10/ARTD10 has MARylation rather than PARylation activity [4] and, lately, it has become clear that only the minority of PARP enzymes are capable of PARylation [5]. Moreover, not only protein, but also DNA has been identified as a target for ADP-ribosylation catalysed by PARP3 [6,7]. Following PARP1 and PARP2, PARP3 has also been described as a potential therapeutic target in certain cancers [8–11]. The most recent addition to the spectrum of ADP-ribosylated molecules is RNA, which can be modified in vitro by multiple PARPs [12].

One of the best-studied but still barely understood MARylating enzymes is PARP10. PARP10 was initially identified as an interactor of MYC [13] and, later, of ubiquitinated proliferating cell nuclear antigen (PCNA) [14,15]. Its overexpression leads to cell death in HeLa cells through an unknown mechanism [16]. In contrast, in non-transformed RPE-1 cells, overexpression leads to stimulation of cell growth [17]. It has been given a role in replication and DNA damage [15,17,18] as well as mitochondrial metabolism [19], but it is not clear what the relevant substrates of this enzyme are. Potential substrates were identified using protein microarrays [20], which was performed for the PARylating PARP2 [21]. However, partially due to a lack of antibodies at that time, these modifications were not verified to occur in cells. PARP10 was also reported to be a regulator of NF-kB signaling [22]. PARP14 functions as a co-activator of STAT6 and has been described as an anti-cancer and anti-inflammatory target, reviewed in more detail elsewhere [23,24]. PARP16 is located at the endoplasmic reticulum (ER) and plays a role in the unfolded protein response [25,26]. The most recent function identified for MARylation is its role in the anti-viral defence. PARP7, PARP10, and PARP12 gene expression is induced by interferon-α and ADP-ribosylation mediated by these PARPs may function to counteract viral proliferation [27,28]. In addition to enzymes of the PARP family, several sirtuins (SIRTs) appear to be able to ADP-ribosylate substrates in addition to their better-studied NAD^+-dependent deacylase activity [29,30]. The first research showing that SIRTs possess MARylation activity described it as a very weak activity [31]. Other papers, focusing on single enzymes, showed that the mitochondrial SIRT4 MARylates glutamate dehydrogenase to downregulate insulin secretion [32]. The nuclear SIRT6 modifies PARP1 [33] and BAF10 [34] to regulate their activities and, lastly, the nucleolar SIRT7 was reported to MARylate histones p53 and ELK4 [35]. To date, very little information is available about the MARylating activity of the SIRTs. The functional consequences of MARylation in normal cell physiology, thus, appear to be very diverse, reviewed extensively elsewhere [36]. Herein, we focus on the enzymes reversing the intracellular MARylation reaction: the macrodomain-containing hydrolases.

2. Macrodomains

Intimately linked to ADP-ribosylation is a protein structural module known as the macrodomain, named for the unusually large histone macroH2A from which it was identified [37]. The macrodomain is a protein fold formed around a central β-sheet, flanked by α-helices that forms a pocket where ADP-ribose can bind [38,39]. These pockets are slightly different between the macrodomain containing proteins with the consequence that the macrodomain-containing protein family can be subdivided. Some macrodomains bind to PAR and others bind to MAR [40–43].

Macrodomains not only serve as binding modules to mediate protein interactions depending on ADP-ribosylation, but a number of them have catalytic activity. Poly(ADP-ribosyl)glycohydrolase (PARG) is able to degrade PAR chains by cleaving the glycosidic bond between adjacent ADP-ribose moieties. However, it cannot remove the final ADP-ribose attached to the protein [44]. Three other proteins, MACROD1, MACROD2, and TARG1, are not active toward PAR but are capable of removing the ADP-ribose moiety connected to the protein partner, presumably by cleaving an ester bond between ADP-ribose and an acidic receptor amino acid [45–47]. They can, therefore, both reverse any MARylation as generated by PARP enzymes, and also remove the last ADP-ribose left behind by PARG. Of the 12 human macrodomain-containing proteins known today, 11 proteins were identified by sequence homology [39,43]. The macrodomain of PARG was only recognised after the structure had been solved [44], which indicates the possibility that more macrodomain-containing proteins remain to be identified. A recent bioinformatics approach has suggested that the largely uncharacterised protein

C12orf4 harbours a macrodomain, or at least a macro-like domain, but this remains to be confirmed experimentally [48].

The macrodomain is an evolutionary well-conserved module [39]. For instance, recent work has shown that certain alpha-viruses encode a protein with a macrodomain-like fold [49], which displays catalytic activity towards MARylated proteins [50–52]. The existence of such proteins enhances the evidence connecting existing links between MARylation and immunity [53]. In this review, we will focus specifically on potential oncogenic functions of the mono(ADP-ribosyl)hydrolases MACROD1, MACROD2, and TARG1. It should also be noted, however, that other macrodomain-containing proteins such as PARG and CHD1L (ALC1) have been previously linked closely to cancer [54,55].

3. Macrodomain-Containing Hydrolases: MACROD1, MACROD2, and TARG1

MACROD1, MACROD2, and TARG1 are relatively small proteins (Table 1) without any notable other domains beside the macrodomain. The first catalytic activity of MACROD1, MACROD2, and TARG1 that was recognised was their deacetylation of *O*-acetyl-ADP-ribose (*O*AADPR) to generate acetyl and ADP-ribose [56,57]. *O*AADPR arises as a by-product of SIRT-mediated deacetylation in which the acetyl group is transferred onto ADP-ribose while releasing nicotinamide, and may have important biological functions [58]. Two years later, three labs reported the activity of MACROD1, MACROD2, and TARG1 in removing ADP-ribose from protein substrates, which was proposed to be limited to removal of ADP-ribose from acidic amino acids [43,45–47]. Serine and arginine MARylation were suggested to be removed by ARH3 and ARH1, respectively [59–62], which are structurally different proteins that do not contain a macrodomain. In 2017, MACROD1, MACROD2, and TARG1 were reported as being capable of removing ADP-ribose from both modified single-stranded DNA [63] and, in 2018, also being capable of removing ADP-ribose from MARylated single-stranded RNA [12]. Lastly, in 2019, the hydrolysis of α-NAD was added to the list of their in vitro activities [64].

To date, it is not known which of these activities is relevant for their functionality in cells, either in normal physiology nor in pathological conditions. This is predominantly due to the fact that it is not well known how prevalent their in vitro substrates are in cells or what their cellular functions may be. This is partially due to the fact that antibodies or modules recognising MARylation have been developed only recently [65–67]. Before these technical developments, it was not possible to confirm with antibodies that MARylation of proteins takes place in cells, and, hence, reversal mediated by these macrodomain-containing proteins could also not be assessed. Currently, it is still not straightforward to estimate endogenous MARylation levels in cells, as no inhibitors were available for MACROD1, MACROD2, or TARG1. During experimental procedures such as cell fixation or lysis, MARylation could possibly be removed by these enzymes and, therefore, go undetected. One study was performed to identify MACROD1 inhibitors using an AlphaScreen assay, where inhibitors in the micromolar range were identified. These can serve as lead compounds for further optimisation, but are likely not specific enough yet to be used for studies in cells due to their other activities at this concentration [68]. Another study identified an allosteric inhibitor of macrodomain 2 of PARP14, which blocks its binding to ADP-ribose. However, it was not tested whether the identified compound might inhibit one of the macrodomains with catalytic activity [69].

Table 1. Summary of nomenclature, expression, and localisation data of MACROD1, MACROD2, and TARG1.

Protein Name	Alternative Names	Gene Name	MW Predicted	MW Observed	Intracellular Localisation	Expression	Regulation
MACROD1	LRP16	*MACROD1*	35.5 kDa	27 kDa* [70,71]	Mitochondrial [40,70,71]	Ubiquitous expression, enriched in skeletal muscle [70,71]	mRNA expression is induced by oestrogen [72–74]
MACROD2	C20orf133	*MACROD2*	47 kDa	50 kDa [70]	Diffuse nuclear, cytoplasmic [70,75]	So far detected only in the brain [70,76]	Phosphorylation by ATM upon DNA damage induces translocation to cytoplasm [75]
TARG1	C6orf130	*OARD1*	17 kDa	17 kDa [14,70]	Nuclear, nucleolar, stress granular [14,45,70]	Ubiquitous expression [70]	Leaves nucleoli upon DNA damage [14]

* The predicted molecular weight of MACROD1 is 35 kDa. However, due to cleavage of the N-terminus upon translocation into mitochondria, the protein detected in western blotting (WB) is smaller [70].

The roles of MACROD1, MACROD2, and TARG1 in normal cell biological processes are not well understood. They appear to be expressed and localised in different tissues and intracellular locations, which clearly gives them unique roles to play despite their biochemical similarity. This also explains why loss of TARG1 alone can lead to a neurodegenerative phenotype [45]. MACROD1 resides in mitochondria [40,70,71], but, depending on the tag used for overexpression, will reside exclusively in the nucleus instead [70]. MACROD2 displays a more diffuse cytoplasmic/nuclear distribution [70,75] and overexpressed TARG1 appears to be present in nucleoplasm, nucleoli, and cytoplasmic stress granules [14,70]. TARG1 and MACROD1 are expressed throughout the different tissues with an enrichment of MACROD1 in skeletal muscle at both the protein and RNA level [70]. MACROD2 has, thus far, only been detected in human neuroblastoma cells and in mouse cortical neurons [70,76]. Several reports have correlated polymorphisms or deletions within the *MACROD2* gene with autism-spectrum disorders [77–80]. It is not clear, however, whether the *MACROD2* gene product itself or surrounding genes are responsible for this association even though elevated protein expression in neurons supports a potential brain-specific function of MACROD2 [76]. However, a long non-coding RNA has been identified within an intron of the *MACROD2* gene, which is more highly expressed in most tissues investigated than the *MACROD2* mRNA, and, thereby, potentially confuses correlations of mutations in the *MACROD2* gene and phenotypes [81]. In another genome-wide association study, *MACROD2* was identified as a factor influencing vascular-adhesion protein-1 (VAP-1) levels, which the authors confirmed by knockdown of MACROD2. This leads to lower VAP-1 expression in adipocytes, presumably through the transcriptional regulation of VAP-1 [82]. MACROD2 was also reported to leave the nucleus upon DNA damage, dependent on phosphorylation by ATM [75], whereas TARG1 localises from nucleoli to nuclear sites of damage [14], which both have an unclear functional relevance. More data may be found on MACROD1, which is also known as leukaemia-related protein 16 or LRP16. MACROD1 has been attributed to a number of functions in the nucleus, such as co-activation of the androgen receptor [83], counteracting PARP7-mediated MARylation in the nucleus [84,85], and activation of NF-kB signalling [86–88]. MACROD1 was also reported as an enhancer of oestrogen receptor signalling [89], and is upregulated after stimulation of cells with oestrogen [72,74,90]. No regulatory factors have been identified yet for MACROD2 and TARG1. A BioID interaction screen, which identifies proteins in close proximity to the protein of interest [91], identified many proteins involved in nuclear/cytoplasmic and mitochondrial nucleic acid metabolism as interactors of TARG1 and MACROD1, respectively [70]. Whether these proteins are MARylated and serve as a substrate of TARG1/MACROD1 remains to be determined. It is possible that, despite spatio-temporal restrictions, MACROD1 and TARG1 in their respective compartments are involved in similar signalling networks, converging on the regulation of cellular nucleic acids. The physiological functions of the three enzymes remains elusive. Most puzzling perhaps is that, at the moment, despite the mitochondrial localisation of MACROD1, the majority of reports describe nuclear functions. Future work will need to address this apparent discrepancy.

4. Mono(ADP-ribosyl)ation in Cancer

The post-translational modification poly(ADP-ribosyl)ation has been intimately linked to cancer before [92,93], as have other post-translational modifications such as phosphorylation [94]. In BRCA1/BRCA2-deficient patients, PARP1 inhibitors have been shown effective specifically against the tumour cells applied in the clinic. However, this is one of the rare examples of a synthetic lethal interaction [95]. Better understanding of the processes regulated by MARylation will provide opportunities for further drug development, as is exemplified by current research into the potential of PARP14 as a drug target [24]. Little is known about the potential role of the mono(ADP-ribosyl)hydrolases in cancer. *MACROD1*, *MACROD2*, and *OARD1* (TARG1) exhibit mutations only in 0.9%, 2.6%, and 1% of cancer patient samples, respectively, from over 1000 samples in the cBioPortal curated dataset [96]. The fact that *MACROD2* mutation rates are twice the rates seen for *MACROD1* and *OARD1* likely reflects that the *MACROD2* gene is larger (Figure 1), and is located at a known fragile region [97]. A specific recurring deletion of exon 6 has been observed in esophageal squamous cell carcinoma and gastric cancer [98]. For *MACROD1*, a *RUNX-MACROD1* fusion was identified in leukaemia [99] and, for *MACROD2*, a *PDGFRA-MACROD2* fusion in a pleomorphic sarcoma [100]. Despite the low number of identified mutations in patient tumour samples, several reports have correlated *MACROD1* or *MACROD2* expression levels with the clinical outcome, as described in the next paragraphs.

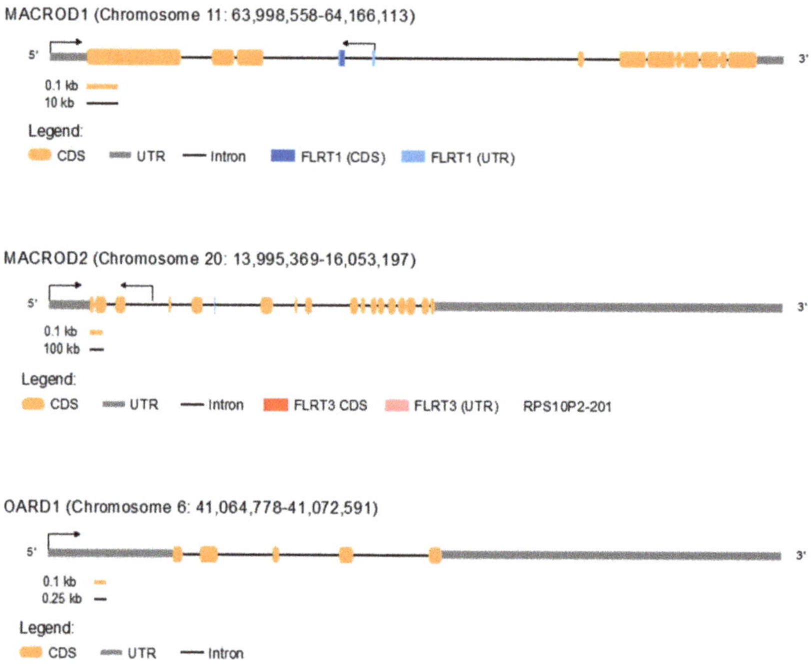

Figure 1. Overview of the *MACROD1*, *MACROD2*, and *OARD1* gene structure.

The gene structure of *MACROD1*, *MACROD2*, and *OARD1* is shown schematically. More lncRNAs are present. However, only RPS10P2-201 is displayed since it has been shown to be relevant RPS10P2-201 [81]. CDS = coding sequence. UTR = untranslated region.

5. MACROD1 in Cancer

A number of studies have addressed a potential oncogenic function of MACROD1 and started deciphering the molecular mechanism underlying observed effects. MACROD1 expression was shown to be upregulated by oestrogen [73,74], which leads to several studies of the role of MACROD1 in tumours with a differential oestrogen status. MACROD1 overexpression in the oestrogen and progesterone receptor positive Ishikawa cells, derived from an endometrial cancer, had no effect on cell proliferation. It did, however, enhance the invasiveness of these cells as measured by transwell assays [101]. Mechanistically, the authors propose a mechanism wherein MACROD1, dependent on oestrogen, blocks recruitment of ERα to the E-cadherin promoter, which lowers E-cadherin expression and, through this, enhances invasiveness. shRNA-mediated knockdown of MACROD1 achieved the opposite effect by enhancing E-cadherin expression [101]. This implies that MACROD1 can be an important factor in metastasis. These findings appear to contradict an earlier report where overexpression of MACROD1 in MCF7 cells, which are oestrogen-responsive breast cancer cells, showed an effect on cell growth. It led to enhanced proliferation [74]. A later report studying MACROD1 in 293T cells showed that knockdown of MACROD1 sensitized cells to TNFα-induced apoptosis [86]. From these studies, it is, thus, not clear what effect loss or gain of MACROD1 has on cell physiology. Recent work has demonstrated that fusing an N-terminal tag, such as GFP, to MACROD1 leads to a nuclear localisation, whereas C-terminally labelled and an endogenous protein appear to be exclusively localised in mitochondria [70]. Mass spectrometry datasets have also detected MACROD1 in the mitochondria [102]. This does not exclude the possibility that, under specific circumstances, MACROD1 may re-localise to the nucleus. In pathogenic conditions, such as the presence of a *RUNX-MACROD1* fusion protein that was identified in leukaemia [99], the protein likely also localises to the nucleus instead of the mitochondria, as the RUNX fusion will mask the mitochondrial targeting sequence, comparable to the localisation after labelling with an N-terminal GFP tag. Unfortunately, some of the studies investigating MACROD1 have either used N-terminally tagged fusion constructs [84,85,88], or have not stated clearly how the fusion proteins were generated [103]. Furthermore, the majority of applied antibodies show multiple bands in the western blot and, hence, are not suitable for immunohistochemistry (IHC) or immunofluorescence (IF) (Table 2). It will be worthwhile to repeat some of the studies of MACROD1's molecular function in carcinogenesis to clarify whether unlabeled MACROD1 overexpression leads to enhanced cell growth, or whether this effect depends on the tumour background and also to study knockdown/knockout systems more thoroughly.

Table 2. Summary of studies studying the effect of MACROD1 and MACROD2 protein levels on cell and/or tumour growth.

Cancer Tissue	Protein	Expression	Effect/Prognosis	Antibody Used	Reference
Neuroendocrine lung tumours	MACROD1	Elevated	Poorer survival	Monoclonal rabbit antibody against LRP16 Not further specified or validated	[104]
Hepatocellular carcinoma	MACROD1	Overexpressed: N- or C-tag not specified	Lower cell and tumour growth	Santa Cruz goat polyclonal This antibody is not available anymore. Whole blots are not shown. It was not validated with siRNA	[103]
Pancreatic carcinoma	MACROD1	Overexpressed: N- or C-tag not specified	Higher cell and tumour growth	Abcam rabbit polyclonal This antibody recognises multiple bands in WB and is thus not suitable for IHC/IF	[105]
Colorectal carcinoma	MACROD1	Elevated	Poorer survival	Polyclonal rabbit antibody generated by the authors' institute Not further specified or validated	[106]
Gastric carcinoma	MACROD1	Elevated	Poorer survival	Polyclonal rabbit antibody generated by the authors' institute Not further specified or validated	[107]
Breast cancer	MACROD1	MACROD1 expression quantified as either positive or negative	MACROD1 expression was higher in patients with advanced stages	LRP16 rabbit anti-human antibody, source not given Not further specified or validated	[108]
Endometrial cancer	MACROD1	Overexpressed	No effect on proliferation but enhanced invasion	Antiserum generated in rabbits against amino acids 83-324	[101]
Breast cancer	MACROD1	Overexpression	Increased proliferation	Not specified	[74]
Colorectal carcinoma	MACROD1	Non-tagged and N-terminal flag-tagged overexpression	Confers resistance to chemotherapeutics	Antibody used for IHC and WB not specified, recognises bands at ±35 and ±45 kDa	[88]
Tumours induced in mice by sublethal irradiation	MACROD2	Knockout mice	No difference between wildtype and MacroD2$^{-/-}$	Thermofisher PA5-45950 This antibody recognises bands at ±38 kDa and ±22 kDa	[109]

Despite these technical challenges, multiple studies have correlated tumour growth with MACROD1 expression in either xenograft models or in patient cohorts. Overexpression of MACROD1 in the hepatocellular carcinoma lines HepG2 and MHCC-97L leads to a decrease in cell growth and metastatic potential, as measured by transwell assays [103]. Furthermore, when delivered into nude mice, cells overexpressing MACROD1 lead to a decreased tumour volume and have a lower metastatic potential compared to cells without overexpressed MACROD1 [103]. With the information available, it cannot be distinguished whether this is a genuine effect or an artefact due to forced nuclear localisation of the overexpression construct [103]. In a similar set of experiments in the pancreatic carcinoma cell lines, Panc1, CFPAC1, Bxpc3, SW1990, AsPC1, and HPDE6-C7, opposite results were achieved [105]. Knockdown of MACROD1 leads to enhanced apoptosis and decreased cell growth. However, only one shRNA construct was used, so any effects seen could be potentially off-target. Cells overexpressing the same MACROD1 construct as in the study described before [103] grew faster and were more resistant to apoptosis. Xenograft experiments show the same trend. Cells lacking MACROD1 had a lower tumour-forming potential and higher survival rate. Cells overexpressing MACROD1 had a higher tumour volume and a lower survival rate [105]. The authors do not comment on the opposing effects in these two tumour types, but agree that larger-scale studies are required to verify these findings. A third study shows IHC of lung tumour samples with an antibody of an unknown source and specificity [104]. High MACROD1 expression, as measured by IHC, was reported to correlate with a negative outcome in colorectal carcinoma and in gastric carcinoma. However, the antibody used was generated by the authors and not validated by a western blot [106,107]. CRISPR-mediated *MACROD1* knockout rhabdomyosarcoma cells [70] and *MACROD1* knockout mice appear viable [71], which makes it unlikely that loss of MACROD1 has a drastic growth inhibitory or developmental effect.

In conclusion, most of the data available on an oncogenic function of MACROD1 rely on poorly characterised antibodies, unclear overexpression constructs or a single shRNA construct. The majority of these studies would need to be reproduced with a more thorough characterisation and description of the materials used to be able to draw deeper conclusions. Several studies agree that MACROD1 expression can be induced by oestrogen [73,74,89,90,101], but it remains unclear what the effect of this overexpression is on cells. The *RUNX-MACROD1* fusion identified in leukaemia may provide an important hint at a potential pathologic function. It is possible that the physiological mitochondrial localisation of MACROD1 can turn into a pathogenic nuclear one, where it aberrantly acts as a transcriptional activator. It will be interesting to see whether more instances can be identified where such fusions are present. Alternatively, other masking events may occur in cells, such as binding by interaction partners or PTMs, which, thereby, redirects the protein to the nucleus for a physiological function.

6. MACROD2 in Cancer

A number of analyses suggest MACROD2 may play a potential role in cancer. MACROD2 copy number is increased in three different tamoxifen-resistant MCF7 breast cancer cell lines, prompting the authors to analyse MACROD2 expression in patient samples. In oestrogen receptor-positive tissues of breast cancer patients with a recorded tamoxifen-resistance, however, three patients displayed a decreased copy number, whereas the other two patients showed an increase. Using IHC with a custom antibody, varying levels of MACROD2 were detected in primary and secondary tumour tissues of these patients, collectively showing that MACROD2 may be overexpressed in cancer tissues. MCF7 and T47D cells with exogenously overexpressed MACROD2 grow faster than control cells in media containing tamoxifen, which implies that MACROD2 confers tamoxifen-resistance to the cells and appear to be stimulated by tamoxifen, as demonstrated by faster growth in medium with tamoxifen than without [110]. Conversely, tamoxifen-resistant MCF7 clones with shRNA-mediated knockdown of MACROD2 become more sensitive even though it does not completely reverse resistance [110]. This may be mediated by the activation of oestrogen-regulated genes in response to tamoxifen in cells overexpressing MACROD2. Lastly, cells stably expressing shRNA to knockdown MACROD2 grow markedly slower in nude mice [110]. The authors argue that, as a cancer specific fragile site, the MACROD2 gene can be lost, but this fragility also allows for amplification in the specific case of ER-positive breast cancers treated with tamoxifen to incur resistance. A more recent study showed that, in approximately one-third of colorectal carcinomas investigated, heterozygous or homozygous losses were mapped within the *MACROD2* gene in which the majority are intragenic microdeletions mapping to a region in exons 4 to 5 [111]. To test whether MACROD2 deficiency can promote tumourigenesis, *MACROD2* knockout was introduced into an adenomatous polyposis coli protein (Apc) mouse model. Mutations in the APC tumour suppressor are a major driver of sporadic colorectal cancers, where it could be shown that even haploinsufficiency of *MACROD2* leads to more and larger adenoma formation [111]. Furthermore, human cells transplanted into nude mice displayed increased tumour growth when lacking *MACROD2*, but a reduced tumour growth when MACROD2 was overexpressed [111]. The underlying mechanism was suggested to be impaired PARP1 activity in the *MACROD2*$^{-/-}$ cells, which leads to increased sensitivity of DNA damage and, ultimately, causes enhanced chromosomal instability [111,112].

These findings appear paradoxical with the previous report, where loss of MACROD2 impairs cell growth from a loss of resistance to tamoxifen. *MacroD2* knockout mice do not show altered survival rates after sub-lethal irradiation compared to wildtype mice, which indicates that loss of MACROD2 alone is not sufficient to drive tumourigenesis triggered by DNA damage [109]. It is not clear whether loss or overexpression of MACROD2 contributes to tumourigenesis, or whether both can drive tumour growth dependent on conditions, such as the functionality of the DNA damage repair systems or the oestrogen receptor status. In an investigation of stage-III colon cancer, MACROD2 expression determined by immunohistochemistry was found to correlate with poor survival [113]. A human protein atlas antibody (HPA049076) was used for this work, which has been retracted by the company in the meantime and did not appear to be validated in any way, such as western blotting or siRNA-mediated knockdown. It is, thus, not clear whether this antibody recognises MACROD2 at all, or whether it recognises additional proteins, which may be upregulated in the samples analysed. Altogether, it appears that loss of MACROD2 as such is not sufficient to drive tumourigenesis, but may have an additive effect in models prone to tumour formation such as loss of APC in colorectal cells. More studies are urgently needed to clarify the exact role of MACROD2 in both the onset of cancer as well as in the response of existing tumours to therapies.

7. TARG1 in Cancer

The only phenotype associated with TARG1 expression is neurodegeneration, occurring due to a mutation that leads to a truncated protein with a disrupted macrodomain, and, thereby, loss of catalytic activity [45]. It is not clear whether this neurodegeneration is a result of a potentially toxic truncated or unfolded protein or from loss of hydrolase activity. Knockdown of TARG1 leads to a decrease in 293T cell proliferation and a slight increase in senescence in U2OS cells, which are derived from an osteosarcoma [45]. CRISPR-mediated knockdown of TARG1 does not influence HeLa or U2OS proliferation [14,70], which leaves it unclear in which setting TARG1 is required for cell growth. Overexpression does not lead to changes in cell proliferation [14,70]. We could not identify any data linking TARG1 to cancer, nor see elevated expression or mutation in databases such as COSMIC. Based on expression levels in databases and these experimental results, at this stage, it does appear unlikely that TARG1 is involved in cancer even though further experimentation is required.

8. Conclusions

Despite the presence of several publications reporting a correlation between MACROD1 or MACROD2 levels and cancer development or progression, it is not clear at the moment that they have a causative role. MACROD2 is potentially relevant in ER-positive, tamoxifen resistant breast cancers where it may confer resistance to treatment. However, larger cohorts of patient samples need to be analysed to further substantiate these initial findings. Loss of *MACROD2* in an APC null background potentially stimulates tumour formation. However, loss of *MacroD2* alone does not have any effect in a knockout mouse model. Detailed studies are required to determine how loss of MACROD2 might cooperate with loss of APC to drive cancer. MACROD1 levels are upregulated by oestrogen with unclear consequences for cells. Yet, it may down-regulate E-cadherin and, thereby, promote metastasis. If confirmed, this may be an attractive therapeutic target intending to keep the cancer dormant and prevent metastasis.

Conflicting data show that overexpression and knockdown of MACROD1 have no effect, no stimulus, and no inhibition of cell growth, which implies that, perhaps, inhibiting MACROD1 may not have detrimental effects for the whole organism but rather may be dependent on the tumour background and, thus, may represent a valid drug target. The effects observed may, however, be dependent on the constructs used as well as on the specific cell-types and have to be studied in more detail. Future work will have to dissect in which context loss or gain of MACROD1 may be driving aspects of cancer growth. The recent development and characterisation of more specific antibodies, the ongoing improvements of mass spectrometric measurements of MARylation, and also the attempts at making specific inhibitors for both transferases and hydrolases should allow a more detailed analysis of their (patho-)physiological function in the near future. The partially paradoxical findings described will undoubtedly be clarified with better validated tools to determine the extent to which MACROD1, MACROD2, and TARG1 are relevant for tumourigenesis in order to establish their potential as drug candidates.

Author Contributions: Writing, review, and editing: K.L.H.F., C.D.O.C., and R.Ž. All authors have read and agreed to the published version of the manuscript.

Funding: The START Program of the Medical Faculty, RWTH Aachen University to K.L.H.F. (10/18) and to R.Z (13/20) funded this research.

Conflicts of Interest: The authors declare no conflict of interest.

References

1. Minguez, P.; Letunic, I.; Parca, L.; Garcia-Alonso, L.; Dopazo, J.; Huerta-Cepas, J.; Bork, P. PTMcode v2: A resource for functional associations of post-translational modifications within and between proteins. *Nucleic Acids Res.* **2015**, *43*, D494–D502. [CrossRef] [PubMed]
2. Virag, L. 50 years of poly(ADP-ribosyl)ation. *Mol. Asp. Med.* **2013**, *34*, 1043–1045. [CrossRef] [PubMed]
3. Slade, D. PARP and PARG inhibitors in cancer treatment. *Genes Dev.* **2020**. [CrossRef] [PubMed]
4. Kleine, H.; Poreba, E.; Lesniewicz, K.; Hassa, P.O.; Hottiger, M.O.; Litchfield, D.W.; Shilton, B.H.; Luscher, B. Substrate-Assisted catalysis by PARP10 limits its activity to mono-ADP-ribosylation. *Mol. Cell* **2008**, *32*, 57–69. [CrossRef] [PubMed]
5. Vyas, S.; Matic, I.; Uchima, L.; Rood, J.; Zaja, R.; Hay, R.T.; Ahel, I.; Chang, P. Family-Wide analysis of poly(ADP-ribose) polymerase activity. *Nat. Commun.* **2014**, *5*, 4426. [CrossRef] [PubMed]
6. Belousova, E.A.; Ishchenko, A.A.; Lavrik, O.I. DNA is a New Target of Parp3. *Sci. Rep.* **2018**, *8*, 4176. [CrossRef]
7. Zarkovic, G.; Belousova, E.A.; Talhaoui, I.; Saint-Pierre, C.; Kutuzov, M.M.; Matkarimov, B.T.; Biard, D.; Gasparutto, D.; Lavrik, O.I.; Ishchenko, A.A. Characterization of DNA ADP-ribosyltransferase activities of PARP2 and PARP3: New insights into DNA ADP-ribosylation. *Nucleic Acids Res.* **2018**, *46*, 2417–2431. [CrossRef]
8. Rodriguez-Vargas, J.M.; Nguekeu-Zebaze, L.; Dantzer, F. PARP3 comes to light as a prime target in cancer therapy. *Cell Cycle* **2019**, *18*, 1295–1301. [CrossRef]
9. Beck, C.; Rodriguez-Vargas, J.M.; Boehler, C.; Robert, I.; Heyer, V.; Hanini, N.; Gauthier, L.R.; Tissier, A.; Schreiber, V.; Elofsson, M.; et al. PARP3, a new therapeutic target to alter Rictor/mTORC2 signaling and tumor progression in BRCA1-associated cancers. *Cell Death Differ.* **2019**, *26*, 1615–1630. [CrossRef]
10. Sharif-Askari, B.; Amrein, L.; Aloyz, R.; Panasci, L. PARP3 inhibitors ME0328 and olaparib potentiate vinorelbine sensitization in breast cancer cell lines. *Breast Cancer Res. Treat.* **2018**, *172*, 23–32. [CrossRef]

11. Oplustil O'Connor, L.; Rulten, S.L.; Cranston, A.N.; Odedra, R.; Brown, H.; Jaspers, J.E.; Jones, L.; Knights, C.; Evers, B.; Ting, A.; et al. The PARP inhibitor AZD2461 provides insights into the role of PARP3 inhibition for both synthetic lethality and tolerability with chemotherapy in preclinical models. *Cancer Res.* **2016**, *76*, 6084–6094. [CrossRef] [PubMed]
12. Munnur, D.; Bartlett, E.; Mikolcevic, P.; Kirby, I.T.; Matthias Rack, J.G.; Mikoc, A.; Cohen, M.S.; Ahel, I. Reversible ADP-ribosylation of RNA. *Nucleic Acids Res.* **2019**, *47*, 5658–5669. [CrossRef] [PubMed]
13. Yu, M.; Schreek, S.; Cerni, C.; Schamberger, C.; Lesniewicz, K.; Poreba, E.; Vervoorts, J.; Walsemann, G.; Grotzinger, J.; Kremmer, E.; et al. PARP-10, a novel Myc-interacting protein with poly(ADP-ribose) polymerase activity, inhibits transformation. *Oncogene* **2005**, *24*, 1982–1993. [CrossRef] [PubMed]
14. Butepage, M.; Preisinger, C.; von Kriegsheim, A.; Scheufen, A.; Lausberg, E.; Li, J.; Kappes, F.; Feederle, R.; Ernst, S.; Eckei, L.; et al. Nucleolar-Nucleoplasmic shuttling of TARG1 and its control by DNA damage-induced poly-ADP-ribosylation and by nucleolar transcription. *Sci. Rep.* **2018**, *8*, 6748. [CrossRef]
15. Shahrour, M.A.; Nicolae, C.M.; Edvardson, S.; Ashhab, M.; Galvan, A.M.; Constantin, D.; Abu-Libdeh, B.; Moldovan, G.L.; Elpeleg, O. PARP10 deficiency manifests by severe developmental delay and DNA repair defect. *Neurogenetics* **2016**, *17*, 227–232. [CrossRef]
16. Herzog, N.; Hartkamp, J.D.; Verheugd, P.; Treude, F.; Forst, A.H.; Feijs, K.L.; Lippok, B.E.; Kremmer, E.; Kleine, H.; Luscher, B. Caspase-Dependent cleavage of the mono-ADP-ribosyltransferase ARTD10 interferes with its pro-apoptotic function. *FEBS J.* **2013**, *280*, 1330–1343. [CrossRef]
17. Schleicher, E.M.; Galvan, A.M.; Imamura-Kawasawa, Y.; Moldovan, G.L.; Nicolae, C.M. PARP10 promotes cellular proliferation and tumorigenesis by alleviating replication stress. *Nucleic Acids Res.* **2018**, *46*, 8908–8916. [CrossRef]
18. Nicolae, C.M.; Aho, E.R.; Vlahos, A.H.; Choe, K.N.; De, S.; Karras, G.I.; Moldovan, G.L. The ADP-ribosyltransferase PARP10/ARTD10 interacts with proliferating cell nuclear antigen (PCNA) and is required for DNA damage tolerance. *J. Biol. Chem.* **2014**, *289*, 13627–13637. [CrossRef]
19. Marton, J.; Fodor, T.; Nagy, L.; Vida, A.; Kis, G.; Brunyanszki, A.; Antal, M.; Luscher, B.; Bai, P. PARP10 (ARTD10) modulates mitochondrial function. *PLoS ONE* **2018**, *13*, e0187789. [CrossRef]
20. Feijs, K.L.; Kleine, H.; Braczynski, A.; Forst, A.H.; Herzog, N.; Verheugd, P.; Linzen, U.; Kremmer, E.; Luscher, B. ARTD10 substrate identification on protein microarrays: Regulation of GSK3beta by mono-ADP-ribosylation. *Cell Commun. Signal.* **2013**, *11*, 5. [CrossRef]
21. Troiani, S.; Lupi, R.; Perego, R.; Depaolini, S.R.; Thieffine, S.; Bosotti, R.; Rusconi, L. Identification of candidate substrates for poly(ADP-ribose) polymerase-2 (PARP2) in the absence of DNA damage using high-density protein microarrays. *FEBS J.* **2011**, *278*, 3676–3687. [CrossRef] [PubMed]
22. Verheugd, P.; Forst, A.H.; Milke, L.; Herzog, N.; Feijs, K.L.; Kremmer, E.; Kleine, H.; Luscher, B. Regulation of NF-kappaB signalling by the mono-ADP-ribosyltransferase ARTD10. *Nat. Commun.* **2013**, *4*, 1683. [CrossRef] [PubMed]
23. Schweiker, S.S.; Tauber, A.L.; Sherry, M.E.; Levonis, S.M. Structure, function and inhibition of poly(ADP-ribose)polymerase, member 14 (PARP14). *Mini Rev. Med. Chem.* **2018**, *18*, 1659–1669. [CrossRef] [PubMed]
24. Qin, W.; Wu, H.J.; Cao, L.Q.; Li, H.J.; He, C.X.; Zhao, D.; Xing, L.; Li, P.Q.; Jin, X.; Cao, H.L. Research progress on PARP14 as a drug target. *Front. Pharmacol.* **2019**, *10*, 172. [CrossRef]
25. Jwa, M.; Chang, P. PARP16 is a tail-anchored endoplasmic reticulum protein required for the PERK-and IRE1α-mediated unfolded protein response. *Nat. Cell Biol.* **2012**, *14*, 1223–1230. [CrossRef]
26. Di Paola, S.; Micaroni, M.; Di Tullio, G.; Buccione, R.; Di Girolamo, M. PARP16/ARTD15 is a novel endoplasmic-reticulum-associated mono-ADP-ribosyltransferase that interacts with, and modifies karyopherin-ss1. *PLoS ONE* **2012**, *7*, e37352. [CrossRef]
27. Atasheva, S.; Frolova, E.I.; Frolov, I. Interferon-Stimulated poly(ADP-Ribose) polymerases are potent inhibitors of cellular translation and virus replication. *J. Virol.* **2014**, *88*, 2116–2130. [CrossRef]
28. Atasheva, S.; Akhrymuk, M.; Frolova, E.I.; Frolov, I. New PARP gene with an anti-alphavirus function. *J. Virol.* **2012**, *86*, 8147–8160. [CrossRef]
29. Imai, S.I.; Guarente, L. It takes two to tango: NAD+ and sirtuins in aging/longevity control. *NPJ Aging Mech. Dis.* **2016**, *2*, 16017. [CrossRef]
30. Feldman, J.L.; Dittenhafer-Reed, K.E.; Denu, J.M. Sirtuin catalysis and regulation. *J. Biol. Chem.* **2012**, *287*, 42419–42427. [CrossRef]

31. Du, J.; Jiang, H.; Lin, H. Investigating the ADP-ribosyltransferase activity of sirtuins with NAD analogues and 32P-NAD. *Biochemistry* **2009**, *48*, 2878–2890. [CrossRef] [PubMed]
32. Haigis, M.C.; Mostoslavsky, R.; Haigis, K.M.; Fahie, K.; Christodoulou, D.C.; Murphy, A.J.; Valenzuela, D.M.; Yancopoulos, G.D.; Karow, M.; Blander, G.; et al. SIRT4 inhibits glutamate dehydrogenase and opposes the effects of calorie restriction in pancreatic beta cells. *Cell* **2006**, *126*, 941–954. [CrossRef] [PubMed]
33. Mao, Z.; Hine, C.; Tian, X.; Van Meter, M.; Au, M.; Vaidya, A.; Seluanov, A.; Gorbunova, V. SIRT6 promotes DNA repair under stress by activating PARP1. *Science* **2011**, *332*, 1443–1446. [CrossRef] [PubMed]
34. Rezazadeh, S.; Yang, D.; Tombline, G.; Simon, M.; Regan, S.P.; Seluanov, A.; Gorbunova, V. SIRT6 promotes transcription of a subset of NRF2 targets by mono-ADP-ribosylating BAF170. *Nucleic Acids Res.* **2019**, *47*, 7914–7928. [CrossRef] [PubMed]
35. Mitra, N.; Dey, S. Biochemical characterization of mono ADP ribosyl transferase activity of human sirtuin SIRT7 and its regulation. *Arch. Biochem. Biophys.* **2019**, *680*, 108226. [CrossRef] [PubMed]
36. Luscher, B.; Butepage, M.; Eckei, L.; Krieg, S.; Verheugd, P.; Shilton, B.H. ADP-Ribosylation, a multifaceted posttranslational modification involved in the control of cell physiology in health and disease. *Chem. Rev.* **2018**, *118*, 1092–1136. [CrossRef] [PubMed]
37. Pehrson, J.R.; Fried, V.A. MacroH2A, a core histone containing a large nonhistone region. *Science* **1992**, *257*, 1398–1400. [CrossRef]
38. Timinszky, G.; Till, S.; Hassa, P.O.; Hothorn, M.; Kustatscher, G.; Nijmeijer, B.; Colombelli, J.; Altmeyer, M.; Stelzer, E.H.; Scheffzek, K.; et al. A macrodomain-containing histone rearranges chromatin upon sensing PARP1 activation. *Nat. Struct. Mol. Biol.* **2009**, *16*, 923–929. [CrossRef]
39. Rack, J.G.; Perina, D.; Ahel, I. Macrodomains: Structure, function, evolution, and catalytic activities. *Annu. Rev. Biochem.* **2016**, *85*, 431–454. [CrossRef]
40. Neuvonen, M.; Ahola, T. Differential activities of cellular and viral macro domain proteins in binding of ADP-ribose metabolites. *J. Mol. Biol.* **2009**, *385*, 212–225. [CrossRef]
41. Karras, G.I.; Kustatscher, G.; Buhecha, H.R.; Allen, M.D.; Pugieux, C.; Sait, F.; Bycroft, M.; Ladurner, A.G. The macro domain is an ADP-ribose binding module. *EMBO J.* **2005**, *24*, 1911–1920. [CrossRef] [PubMed]
42. Kustatscher, G.; Hothorn, M.; Pugieux, C.; Scheffzek, K.; Ladurner, A.G. Splicing regulates NAD metabolite binding to histone macroH2A. *Nat. Struct. Mol. Biol.* **2005**, *12*, 624–625. [CrossRef]
43. Feijs, K.L.; Forst, A.H.; Verheugd, P.; Luscher, B. Macrodomain-Containing proteins: Regulating new intracellular functions of mono(ADP-ribosyl)ation. *Nat. Rev. Mol. Cell Biol.* **2013**, *14*, 443–451. [CrossRef] [PubMed]
44. Slade, D.; Dunstan, M.S.; Barkauskaite, E.; Weston, R.; Lafite, P.; Dixon, N.; Ahel, M.; Leys, D.; Ahel, I. The structure and catalytic mechanism of a poly(ADP-ribose) glycohydrolase. *Nature* **2011**, *477*, 616–620. [CrossRef] [PubMed]
45. Sharifi, R.; Morra, R.; Appel, C.D.; Tallis, M.; Chioza, B.; Jankevicius, G.; Simpson, M.A.; Matic, I.; Ozkan, E.; Golia, B.; et al. Deficiency of terminal ADP-ribose protein glycohydrolase TARG1/C6orf130 in neurodegenerative disease. *EMBO J.* **2013**, *32*, 1225–1237. [CrossRef] [PubMed]
46. Jankevicius, G.; Hassler, M.; Golia, B.; Rybin, V.; Zacharias, M.; Timinszky, G.; Ladurner, A.G. A family of macrodomain proteins reverses cellular mono-ADP-ribosylation. *Nat. Struct. Mol. Biol.* **2013**, *20*, 508–514. [CrossRef] [PubMed]
47. Rosenthal, F.; Feijs, K.L.; Frugier, E.; Bonalli, M.; Forst, A.H.; Imhof, R.; Winkler, H.C.; Fischer, D.; Caflisch, A.; Hassa, P.O.; et al. Macrodomain-Containing proteins are new mono-ADP-ribosylhydrolases. *Nat. Struct. Mol. Biol.* **2013**, *20*, 502–507. [CrossRef]
48. Dudkiewicz, M.; Pawlowski, K. A novel conserved family of macro-like domains-putative new players in ADP-ribosylation signaling. *PeerJ* **2019**, *7*, e6863. [CrossRef]
49. Egloff, M.P.; Malet, H.; Putics, A.; Heinonen, M.; Dutartre, H.; Frangeul, A.; Gruez, A.; Campanacci, V.; Cambillau, C.; Ziebuhr, J.; et al. Structural and functional basis for ADP-ribose and poly(ADP-ribose) binding by viral macro domains. *J. Virol.* **2006**, *80*, 8493–8502. [CrossRef]
50. Eckei, L.; Krieg, S.; Butepage, M.; Lehmann, A.; Gross, A.; Lippok, B.; Grimm, A.R.; Kummerer, B.M.; Rossetti, G.; Luscher, B.; et al. The conserved macrodomains of the non-structural proteins of Chikungunya virus and other pathogenic positive strand RNA viruses function as mono-ADP-ribosylhydrolases. *Sci. Rep.* **2017**, *7*, 41746. [CrossRef]

51. Fehr, A.R.; Channappanavar, R.; Jankevicius, G.; Fett, C.; Zhao, J.; Athmer, J.; Meyerholz, D.K.; Ahel, I.; Perlman, S. The conserved Coronavirus macrodomain promotes virulence and suppresses the innate immune response during severe acute respiratory syndrome Coronavirus infection. *mBio* **2016**, *7*. [CrossRef] [PubMed]
52. Li, C.; Debing, Y.; Jankevicius, G.; Neyts, J.; Ahel, I.; Coutard, B.; Canard, B. Viral macro domains reverse protein ADP-ribosylation. *J. Virol.* **2016**, *90*, 8478–8486. [CrossRef] [PubMed]
53. Daugherty, M.D.; Young, J.M.; Kerns, J.A.; Malik, H.S. Rapid evolution of PARP genes suggests a broad role for ADP-ribosylation in host-virus conflicts. *PLoS Genet.* **2014**, *10*, e1004403. [CrossRef] [PubMed]
54. Tanuma, S.I.; Shibui, Y.; Oyama, T.; Uchiumi, F.; Abe, H. Targeting poly(ADP-ribose) glycohydrolase to draw apoptosis codes in cancer. *Biochem. Pharm.* **2019**, *167*, 163–172. [CrossRef] [PubMed]
55. Ma, N.F.; Hu, L.; Fung, J.M.; Xie, D.; Zheng, B.J.; Chen, L.; Tang, D.J.; Fu, L.; Wu, Z.; Chen, M.; et al. Isolation and characterization of a novel oncogene, amplified in liver cancer 1, within a commonly amplified region at 1q21 in hepatocellular carcinoma. *Hepatology* **2008**, *47*, 503–510. [CrossRef] [PubMed]
56. Peterson, F.C.; Chen, D.; Lytle, B.L.; Rossi, M.N.; Ahel, I.; Denu, J.M.; Volkman, B.F. Orphan macrodomain protein (human C6orf130) is an O-acyl-ADP-ribose deacylase: Solution structure and catalytic properties. *J. Biol. Chem.* **2011**, *286*, 35955–35965. [CrossRef]
57. Chen, D.; Vollmar, M.; Rossi, M.N.; Phillips, C.; Kraehenbuehl, R.; Slade, D.; Mehrotra, P.V.; von Delft, F.; Crosthwaite, S.K.; Gileadi, O.; et al. Identification of macrodomain proteins as novel O-acetyl-ADP-ribose deacetylases. *J. Biol. Chem.* **2011**, *286*, 13261–13271. [CrossRef]
58. Tong, L.; Denu, J.M. Function and metabolism of sirtuin metabolite O-acetyl-ADP-ribose. *Biochim. Biophys. Acta* **2010**, *1804*, 1617–1625. [CrossRef]
59. Wang, M.; Yuan, Z.; Xie, R.; Ma, Y.; Liu, X.; Yu, X. Structure-Function analyses reveal the mechanism of the ARH3-dependent hydrolysis of ADP-ribosylation. *J. Biol. Chem.* **2018**, *293*, 14470–14480. [CrossRef]
60. Abplanalp, J.; Leutert, M.; Frugier, E.; Nowak, K.; Feurer, R.; Kato, J.; Kistemaker, H.V.A.; Filippov, D.V.; Moss, J.; Caflisch, A.; et al. Proteomic analyses identify ARH3 as a serine mono-ADP-ribosylhydrolase. *Nat. Commun.* **2017**, *8*, 2055. [CrossRef]
61. Fontana, P.; Bonfiglio, J.J.; Palazzo, L.; Bartlett, E.; Matic, I.; Ahel, I. Serine ADP-ribosylation reversal by the hydrolase ARH3. *eLife* **2017**, *6*. [CrossRef] [PubMed]
62. Moss, J.; Stanley, S.J.; Nightingale, M.S.; Murtagh, J.J., Jr.; Monaco, L.; Mishima, K.; Chen, H.C.; Williamson, K.C.; Tsai, S.C. Molecular and immunological characterization of ADP-ribosylarginine hydrolases. *J. Biol. Chem.* **1992**, *267*, 10481–10488. [PubMed]
63. Munnur, D.; Ahel, I. Reversible mono-ADP-ribosylation of DNA breaks. *FEBS J.* **2017**, *284*, 4002–4016. [CrossRef] [PubMed]
64. Stevens, L.A.; Kato, J.; Kasamatsu, A.; Oda, H.; Lee, D.Y.; Moss, J. The ARH and macrodomain families of α-ADP-ribose-acceptor hydrolases catalyze α-NAD$^+$ hydrolysis. *ACS Chem. Biol.* **2019**, *14*, 2576–2584. [CrossRef]
65. Gibson, B.A.; Conrad, L.B.; Huang, D.; Kraus, W.L. Generation and characterization of recombinant antibody-like ADP-ribose binding proteins. *Biochemistry* **2017**, *56*, 6305–6316. [CrossRef]
66. Lu, A.Z.; Abo, R.; Ren, Y.; Gui, B.; Mo, J.R.; Blackwell, D.; Wigle, T.; Keilhack, H.; Niepel, M. Enabling drug discovery for the PARP protein family through the detection of mono-ADP-ribosylation. *Biochem. Pharmacol.* **2019**, *167*, 97–106. [CrossRef]
67. Butepage, M.; Krieg, S.; Eckei, L.; Li, J.; Rossetti, G.; Verheugd, P.; Luscher, B. Assessment of intracellular auto-modification levels of ARTD10 using mono-ADP-ribose-specific macrodomains 2 and 3 of murine Artd8. *Methods Mol. Biol.* **2018**, *1813*, 41–63. [CrossRef]
68. Haikarainen, T.; Maksimainen, M.M.; Obaji, E.; Lehtio, L. Development of an inhibitor screening assay for mono-ADP-ribosyl hydrolyzing macrodomains using AlphaScreen technology. *SLAS Discov.* **2018**, *23*, 255–263. [CrossRef]
69. Moustakim, M.; Riedel, K.; Schuller, M.; Gehring, A.P.; Monteiro, O.P.; Martin, S.P.; Fedorov, O.; Heer, J.; Dixon, D.J.; Elkins, J.M.; et al. Discovery of a novel allosteric inhibitor scaffold for polyadenosine-diphosphate-ribose polymerase 14 (PARP14) macrodomain 2. *Bioorg. Med. Chem.* **2018**, *26*, 2965–2972. [CrossRef]
70. Žaja, R.; Aydin, G.; Lippok, B.E.; Lüscher, B.; Feijs, K.L. A comparative analysis of MacroD1, MACROD2 and TARG1 expression, localisation and interactome. *Sci. Rep.* under review.

71. Agnew, T.; Munnur, D.; Crawford, K.; Palazzo, L.; Mikoc, A.; Ahel, I. MacroD1 is a promiscuous ADP-ribosyl hydrolase localized to mitochondria. *Front. Microbiol.* **2018**, *9*, 20. [CrossRef] [PubMed]
72. Han, W.D.; Si, Y.L.; Zhao, Y.L.; Li, Q.; Wu, Z.Q.; Hao, H.J.; Song, H.J. GC-Rich promoter elements maximally confers estrogen-induced transactivation of LRP16 gene through ERalpha/Sp1 interaction in MCF-7 cells. *J. Steroid Biochem. Mol. Biol.* **2008**, *109*, 47–56. [CrossRef] [PubMed]
73. Zhao, Y.L.; Han, W.D.; Li, Q.; Mu, Y.M.; Lu, X.C.; Yu, L.; Song, H.J.; Li, X.; Lu, J.M.; Pan, C.Y. Mechanism of transcriptional regulation of LRP16 gene expression by 17-beta estradiol in MCF-7 human breast cancer cells. *J. Mol. Endocrinol.* **2005**, *34*, 77–89. [CrossRef] [PubMed]
74. Han, W.D.; Mu, Y.M.; Lu, X.C.; Xu, Z.M.; Li, X.J.; Yu, L.; Song, H.J.; Li, M.; Lu, J.M.; Zhao, Y.L.; et al. Up-Regulation of LRP16 mRNA by 17beta-estradiol through activation of estrogen receptor alpha (ERalpha), but not ERbeta, and promotion of human breast cancer MCF-7 cell proliferation: A preliminary report. *Endocr. Relat. Cancer* **2003**, *10*, 217–224. [CrossRef] [PubMed]
75. Golia, B.; Moeller, G.K.; Jankevicius, G.; Schmidt, A.; Hegele, A.; Preisser, J.; Tran, M.L.; Imhof, A.; Timinszky, G. ATM induces MACROD2 nuclear export upon DNA damage. *Nucleic Acids Res.* **2017**, *45*, 244–254. [CrossRef]
76. Ito, H.; Morishita, R.; Mizuno, M.; Kawamura, N.; Tabata, H.; Nagata, K.I. Biochemical and morphological characterization of a neurodevelopmental disorder-related mono-ADP-ribosylhydrolase, MACRO domain containing 2. *Dev. Neurosci.* **2018**, *40*, 278–287. [CrossRef]
77. Lombardo, B.; Esposito, D.; Iossa, S.; Vitale, A.; Verdesca, F.; Perrotta, C.; Di Leo, L.; Costa, V.; Pastore, L.; Franze, A. Intragenic deletion in MACROD2: A family with complex phenotypes including microcephaly, intellectual disability, polydactyly, renal and pancreatic malformations. *Cytogenet. Genome Res.* **2019**, *158*, 25–31. [CrossRef]
78. Jones, R.M.; Cadby, G.; Blangero, J.; Abraham, L.J.; Whitehouse, A.J.O.; Moses, E.K. MACROD2 gene associated with autistic-like traits in a general population sample. *Psychiatr. Genet.* **2014**, *24*, 241–248. [CrossRef]
79. Curran, S.; Bolton, P.; Rozsnyai, K.; Chiocchetti, A.; Klauck, S.M.; Duketis, E.; Poustka, F.; Schlitt, S.; Freitag, C.M.; Lee, I.; et al. No association between a common single nucleotide polymorphism, rs4141463, in the MACROD2 gene and autism spectrum disorder. *Am. J. Med. Genet. B Neuropsychiatr. Genet.* **2011**, *156*, 633–639. [CrossRef]
80. Anney, R.; Klei, L.; Pinto, D.; Regan, R.; Conroy, J.; Magalhaes, T.R.; Correia, C.; Abrahams, B.S.; Sykes, N.; Pagnamenta, A.T.; et al. A genome-wide scan for common alleles affecting risk for autism. *Hum. Mol. Genet.* **2010**, *19*, 4072–4082. [CrossRef]
81. Bilinovich, S.M.; Lewis, K.; Grepo, N.; Campbell, D.B. The long noncoding RNA RPS10P2-AS1 is implicated in autism spectrum disorder risk and modulates gene expression in human neuronal progenitor cells. *Front. Genet.* **2019**, *10*, 970. [CrossRef] [PubMed]
82. Chang, Y.C.; Hee, S.W.; Lee, W.J.; Li, H.Y.; Chang, T.J.; Lin, M.W.; Hung, Y.J.; Lee, I.T.; Hung, K.Y.; Assimes, T.; et al. Genome-Wide scan for circulating vascular adhesion protein-1 levels: MACROD2 as a potential transcriptional regulator of adipogenesis. *J. Diabetes Investig.* **2018**, *9*, 1067–1074. [CrossRef] [PubMed]
83. Yang, J.; Zhao, Y.L.; Wu, Z.Q.; Si, Y.L.; Meng, Y.G.; Fu, X.B.; Mu, Y.M.; Han, W.D. The single-macro domain protein LRP16 is an essential cofactor of androgen receptor. *Endocr. Relat. Cancer* **2009**, *16*, 139–153. [CrossRef] [PubMed]
84. Bindesboll, C.; Tan, S.; Bott, D.; Cho, T.; Tamblyn, L.; MacPherson, L.; Gronning-Wang, L.; Nebb, H.I.; Matthews, J. TCDD-Inducible poly-ADP-ribose polymerase (TIPARP/PARP7) mono-ADP-ribosylates and co-activates liver X receptors. *Biochem. J.* **2016**, *473*, 899–910. [CrossRef]
85. Ahmed, S.; Bott, D.; Gomez, A.; Tamblyn, L.; Rasheed, A.; Cho, T.; MacPherson, L.; Sugamori, K.S.; Yang, Y.; Grant, D.M.; et al. Loss of the mono-ADP-ribosyltransferase, Tiparp, increases sensitivity to dioxin-induced steatohepatitis and lethality. *J. Biol. Chem.* **2015**, *290*, 16824–16840. [CrossRef]
86. Wu, Z.; Li, Y.; Li, X.; Ti, D.; Zhao, Y.; Si, Y.; Mei, Q.; Zhao, P.; Fu, X.; Han, W. LRP16 integrates into NF-kappaB transcriptional complex and is required for its functional activation. *PLoS ONE* **2011**, *6*, e18157. [CrossRef]
87. Wu, Z.; Wang, C.; Bai, M.; Li, X.; Mei, Q.; Li, X.; Wang, Y.; Fu, X.; Luo, G.; Han, W. An LRP16-containing preassembly complex contributes to NF-kappaB activation induced by DNA double-strand breaks. *Nucleic Acids Res.* **2015**, *43*, 3167–3179. [CrossRef]

88. Li, X.; Wu, Z.; An, X.; Mei, Q.; Bai, M.; Hanski, L.; Li, X.; Ahola, T.; Han, W. Blockade of the LRP16-PKR-NF-kappaB signaling axis sensitizes colorectal carcinoma cells to DNA-damaging cytotoxic therapy. *eLife* **2017**, *6*. [CrossRef]
89. Han, W.D.; Zhao, Y.L.; Meng, Y.G.; Zang, L.; Wu, Z.Q.; Li, Q.; Si, Y.L.; Huang, K.; Ba, J.M.; Morinaga, H.; et al. Estrogenically regulated LRP16 interacts with estrogen receptor alpha and enhances the receptor's transcriptional activity. *Endocr. Relat. Cancer* **2007**, *14*, 741–753. [CrossRef]
90. Tian, L.; Wu, Z.; Zhao, Y.; Meng, Y.; Si, Y.; Fu, X.; Mu, Y.; Han, W. Differential induction of LRP16 by liganded and unliganded estrogen receptor alpha in SKOV3 ovarian carcinoma cells. *J. Endocrinol.* **2009**, *202*, 167–177. [CrossRef]
91. Roux, K.J.; Kim, D.I.; Burke, B. BioID: A screen for protein-protein interactions. *Curr. Protoc. Protein Sci.* **2013**, *74*. [CrossRef] [PubMed]
92. Curtin, N.J. PARP inhibitors for cancer therapy. *Expert Rev. Mol. Med.* **2005**, *7*, 1–20. [CrossRef] [PubMed]
93. Davar, D.; Beumer, J.H.; Hamieh, L.; Tawbi, H. Role of PARP inhibitors in cancer biology and therapy. *Curr. Med. Chem.* **2012**, *19*, 3907–3921. [CrossRef] [PubMed]
94. Rush, J.; Moritz, A.; Lee, K.A.; Guo, A.; Goss, V.L.; Spek, E.J.; Zhang, H.; Zha, X.M.; Polakiewicz, R.D.; Comb, M.J. Immunoaffinity profiling of tyrosine phosphorylation in cancer cells. *Nat. Biotechnol.* **2005**, *23*, 94–101. [CrossRef] [PubMed]
95. Ashworth, A.; Lord, C.J. Synthetic lethal therapies for cancer: What's next after PARP inhibitors? *Nat. Rev. Clin. Oncol.* **2018**, *15*, 564–576. [CrossRef] [PubMed]
96. Gao, J.; Aksoy, B.A.; Dogrusoz, U.; Dresdner, G.; Gross, B.; Sumer, S.O.; Sun, Y.; Jacobsen, A.; Sinha, R.; Larsson, E.; et al. Integrative analysis of complex cancer genomics and clinical profiles using the cBioPortal. *Sci. Signal.* **2013**, *6*, pl1. [CrossRef]
97. Bradley, W.E.; Raelson, J.V.; Dubois, D.Y.; Godin, E.; Fournier, H.; Prive, C.; Allard, R.; Pinchuk, V.; Lapalme, M.; Paulussen, R.J.; et al. Hotspots of large rare deletions in the human genome. *PLoS ONE* **2010**, *5*, e9401. [CrossRef]
98. Hu, N.; Kadota, M.; Liu, H.; Abnet, C.C.; Su, H.; Wu, H.; Freedman, N.D.; Yang, H.H.; Wang, C.; Yan, C.; et al. Genomic landscape of somatic alterations in esophageal squamous cell carcinoma and gastric cancer. *Cancer Res.* **2016**, *76*, 1714–1723. [CrossRef]
99. Imagama, S.; Abe, A.; Suzuki, M.; Hayakawa, F.; Katsumi, A.; Emi, N.; Kiyoi, H.; Naoe, T. LRP16 is fused to RUNX1 in monocytic leukemia cell line with t(11;21)(q13;q22). *Eur. J. Haematol.* **2007**, *79*, 25–31. [CrossRef]
100. Zheng, B.; Zhang, S.; Cai, W.; Wang, J.; Wang, T.; Tang, N.; Shi, Y.; Luo, X.; Yan, W. Identification of novel fusion transcripts in undifferentiated pleomorphic sarcomas by transcriptome sequencing. *Cancer Genom. Proteom.* **2019**, *16*, 399–408. [CrossRef]
101. Meng, Y.G.; Han, W.D.; Zhao, Y.L.; Huang, K.; Si, Y.L.; Wu, Z.Q.; Mu, Y.M. Induction of the LRP16 gene by estrogen promotes the invasive growth of Ishikawa human endometrial cancer cells through the downregulation of E-cadherin. *Cell Res.* **2007**, *17*, 869–880. [CrossRef] [PubMed]
102. Smith, A.C.; Robinson, A.J. MitoMiner v4.0: An updated database of mitochondrial localization evidence, phenotypes and diseases. *Nucleic Acids Res.* **2019**, *47*, D1225–D1228. [CrossRef] [PubMed]
103. Shao, L.; Jing, W.; Wang, L.; Pan, F.; Wu, L.; Zhang, L.; Yang, P.; Hu, M.; Fan, K. LRP16 prevents hepatocellular carcinoma progression through regulation of Wnt/β-catenin signaling. *J. Mol. Med.* **2018**, *96*, 547–558. [CrossRef] [PubMed]
104. Shao, Y.; Li, X.; Lu, Y.; Liu, L.; Zhao, P. Aberrant LRP16 protein expression in primary neuroendocrine lung tumors. *Int. J. Clin. Exp. Pathol.* **2015**, *8*, 6560–6565. [PubMed]
105. Mo, Z.; Hu, M.; Yu, F.; Shao, L.; Fan, K.; Jiao, S. Leukemia-Related protein 16 (LRP16) promotes tumor growth and metastasis in pancreatic cancer. *OncoTargets Ther.* **2018**, *11*, 1215–1222. [CrossRef]
106. Xi, H.Q.; Zhao, P.; Han, W.D. Clinicopathological significance and prognostic value of LRP16 expression in colorectal carcinoma. *World J. Gastroenterol.* **2010**, *16*, 1644–1648. [CrossRef] [PubMed]
107. Li, Y.Z.; Zhao, P.; Han, W.D. Clinicopathological significance of LRP16 protein in 336 gastric carcinoma patients. *World J. Gastroenterol.* **2009**, *15*, 4833–4837. [CrossRef]
108. Yao, D.J.; Qiao, S.; Zhang, Y.; Zhao, Y.T.; Yuan, C.H. Correlation between expression of LRP16, Ki67 and EGFR and breast cancer clinical pathologic factors and prognosis. *Eur. Rev. Med. Pharmacol. Sci.* **2017**, *21*, 47–51.
109. Lo Re, O.; Mazza, T.; Vinciguerra, M. Mono-ADP-ribosylhydrolase MACROD2 is dispensable for murine responses to metabolic and genotoxic insults. *Front. Genet.* **2018**, *9*, 654. [CrossRef]

110. Mohseni, M.; Cidado, J.; Croessmann, S.; Cravero, K.; Cimino-Mathews, A.; Wong, H.Y.; Scharpf, R.; Zabransky, D.J.; Abukhdeir, A.M.; Garay, J.P.; et al. MACROD2 overexpression mediates estrogen independent growth and tamoxifen resistance in breast cancers. *Proc. Natl. Acad. Sci. USA* **2014**, *111*, 17606–17611. [CrossRef]
111. Sakthianandeswaren, A.; Parsons, M.J.; Mouradov, D.; MacKinnon, R.N.; Catimel, B.; Liu, S.; Palmieri, M.; Love, C.; Jorissen, R.N.; Li, S.; et al. MACROD2 haploinsufficiency impairs catalytic activity of PARP1 and promotes chromosome instability and growth of intestinal tumors. *Cancer Discov.* **2018**, *8*, 988–1005. [CrossRef] [PubMed]
112. Sakthianandeswaren, A.; Parsons, M.J.; Mouradov, D.; Sieber, O.M. MACROD2 deletions cause impaired PARP1 activity and chromosome instability in colorectal cancer. *Oncotarget* **2018**, *9*, 33056–33058. [CrossRef] [PubMed]
113. Van den Broek, E.; den Uil, S.H.; Coupe, V.M.H.; Delis-van Diemen, P.M.; Bolijn, A.S.; Bril, H.; Stockmann, H.; van Grieken, N.C.T.; Meijer, G.A.; Fijneman, R.J.A. MACROD2 expression predicts response to 5-FU-based chemotherapy in stage III colon cancer. *Oncotarget* **2018**, *9*, 29445–29452. [CrossRef] [PubMed]

cancers

MDPI

Review

The Development of Rucaparib/Rubraca®: A Story of the Synergy Between Science and Serendipity

Nicola J Curtin

Professor of Experimental Cancer Therapeutics, Translational and Clinical Research Institute, Newcastle University Centre for Cancer, Faculty of Medical Sciences, Medical School, Paul O'Gorman Building, Newcastle University, Newcastle upon Tyne NE2 4HH, UK; nicola.curtin@newcastle.ac.uk or nicola.curtin@ncl.ac.uk; Tel.: +44-0-191-208-4415

Received: 15 January 2020; Accepted: 26 February 2020; Published: 29 February 2020

Abstract: The poly(ADP-ribose) polymerase (PARP) inhibitor, Rubraca®, was given its first accelerated approval for BRCA-mutated ovarian cancer by the FDA at the end of 2016, and further approval by the FDA, EMA and NICE followed. Scientists at Newcastle University initiated the early stages, and several collaborations with scientists in academia and the pharmaceutical industry enabled its final development to the approval stage. Although originally considered as a chemo- or radiosensitiser, its current application is as a single agent exploiting tumour-specific defects in DNA repair. As well as involving intellectual and physical effort, there have been a series of fortuitous occurrences and coincidences of timing that ensured its success. This review describes the history of the relationship between science and serendipity that brought us to the current position.

Keywords: PARP; drug development; synthetic lethality; clinical trials

1. Introduction: Rationale for the Development of PARP Inhibitors

Poly(ADP-ribose) polymerase (PARP) inhibitors (PARPi) have been the most significant addition to the armoury for the treatment of ovarian cancer since the introduction of platinum therapy in the 1970s and represent a paradigm shift in the way cancers may be treated. This is the story of the development of the first PARPi to enter anticancer clinical trials, Rubraca®, formerly known as rucaparib or AG014699, the nomenclature used here largely reflects the name of the drug at the time of use. The development of Rubraca was the result of a substantial intellectual and physical effort by many individuals working as a team but it is hard to avoid the conclusion that luck has played a significant role, and this story describes how serendipity has contributed to the final success. This review describes the development of rucaparib in the context of other advances in the understanding of DNA repair and highlights the lucky breaks along the way.

The development of PARPi that ultimately led to rucaparib began in Newcastle University in 1990 but the story begins before that. The first observation that suggested the existence of PARP was the profound NAD^+ depletion following exposure of cells to ethyleneimine in 1956 [1]. Coming so soon after the discovery of DNA, this effect was not attributed to the effect of DNA methylation, rather an impact on cellular metabolism. The product ADP-ribose polymers, or poly(ADP-ribose) (PAR), was identified in 1963 and at first thought to be some new kind of nucleic acid [2], with elegant experiments a few years later showing the time course of the disappearance of NAD^+ and the corresponding appearance of polymers and nicotinamide [3]. The first suggestion of the involvement of PARP in DNA repair was made by Edward Miller in 1975 [4]. To test this hypothesis, Purnell and Whish [5] developed the first inhibitors, which were based on the catalytic mechanism of PARP and the observation that the by-product, nicotinamide, exerts some feedback inhibition. These early benzamide inhibitors included 3-amino benzamide (3AB), which is still used as a PARPi today. The pivotal study, conducted by

Barbara Durkacz in Sydney Shall's lab, was published in 1980, showing that 3AB prevented the repair of DNA and increased cytotoxicity following exposure to the DNA methylating agent, DMS [6].

2. Early Studies in Newcastle

Barbara Durkacz moved to Newcastle University in the mid-1980s to establish PARP-related studies, where she was joined by Mike Purnell for the synthesis of inhibitors. However, it was the appointment of Hilary Calvert as the new director of the Cancer Research Unit in Newcastle, and his collaboration with Bernard Golding, Professor of Organic Chemistry at Newcastle University, that really got inhibitor development going. They established a drug development programme in October 1990, with funding from the North of England Cancer Research Campaign. Other members of the team were Roger Griffin (chemist), Herbie Newell (pharmacologist) and me (biologist/biochemist). PARP was one of three initial targets for drug discovery, tapping into Barbara Durkacz's expertise. A succession of post-graduate chemistry students embarked on inhibitor synthesis guided by Roger Griffin and Bernard Golding. However, to establish any kind of structure–activity relationship (SAR) a robust, reproducible and quantitative activity assay was needed and herein lay the problem. The results using a published assay [7,8] were variable in the extreme, with replicates wildly different from each other, and no better than random. Months were spent trying to generate inhibition data until finally, in exasperation, I decided to try to get to the bottom of the problem. The assay involved incubating permeabilised cells with ^{32}P-NAD$^+$ and broken DNA to activate the cellular PARP then precipitating the polymer with TCA, collecting it onto filters, washing surplus NAD$^+$ away and counting the acid-precipitated radioactivity. Investigating each step of the assay to identify the source of the irreproducibility by taking replicates at different stages in the assay revealed that it was nothing to do with the cell permeabilisation or collection of the product but was the reaction itself. A eureka moment led me to discover a very simple solution: switching to plastic tubes. Because of the charged nature of NAD$^+$ and the polymer we had been siliconising glass tubes for the reaction to negate the charge on the glass and prevent binding. It turned out that the siliconising agent itself was causing highly variable interference with the activity of the enzyme. By switching to plastic this interference was removed but another problem was identified, i.e., the poor thermal conductivity of the plastic, which was overcome by a pre-warming step. Thereafter, this assay provided the robust, reproducible and quantitative data needed for SAR generation. This is the first example of the synergy of science and serendipity (#1).

The next example of serendipity (#2) came on the chemistry side. Inhibitors were designed to incorporate the nicotinamide pharmacophore, which was thought to make major interactions within the catalytic site of PARP. Studies with other NAD-dependent enzymes suggested that binding was favoured when the carboxamide was anti to the aromatic ring. In 3AB, it can freely rotate to the cis-orientation, meaning that binding is relatively weak (Ki = 10 μM, IC50 = 30 μM). Parallel SAR studies led by Judy Sebolt-Leopold at Warner Lambert and an "analogue by catalogue" approach in Kunihiro Ueda's lab identified that compounds where the carboxamide group was held anti to the aromatic ring, by constraining it through incorporation into a ring structure, had increased potency [9,10]. Our approach was to hold the carboxamide in the anti- conformation through hydrogen bonding with an oxazole or imidazole group. However, during the synthesis of the target compound, 2-methylbenzoxazole-4-carboxamide, a molecular rearrangement occurred generating 8-Hydroxy-2-methylquinazolin-4[3H]-one, NU1025, where the carboxamide was constrained in the desired orientation through incorporation into a ring [11] (Figure 1). NU1025 (Ki = 48 nM IC50 = 400 nM) had comparable activity to compounds identified by the Sebolt-Leopold and Ueda groups and was substantially more potent than the compound we had been aiming to make, which had an IC50 of 10 μM. This lucky accident therefore gave us our first hit compound and allowed us to explore the cellular effects of PARP inhibition.

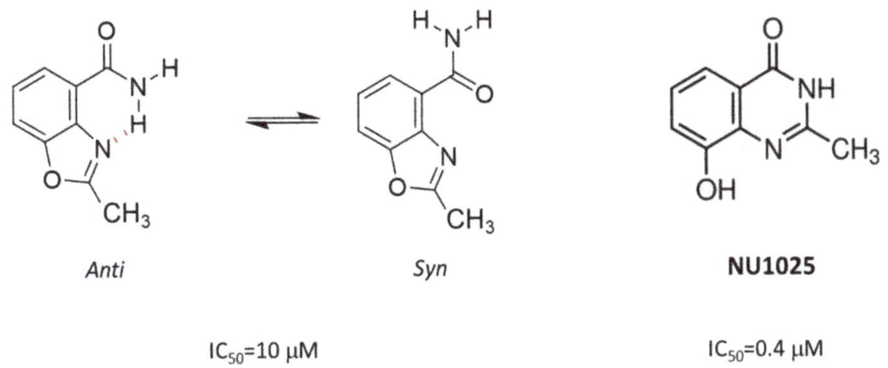

Figure 1. Accidental synthesis of first "hit" compound, NU1025. During the attempted synthesis of 2-methylbenzoxazole-4-carboxamide, which could exist in the desired anti- conformation or the syn- conformation (left), a molecular rearrangement occurred, resulting in NU1025, a much more potent compound.

My first PhD student, Karen Bowman, was tasked not only with determining the potency of the compounds against PARP activity (aided and abetted by Louise Pemberton who also synthesised some of the early inhibitors) but also with evaluating the ability of NU1025 and the benzimidazole, NU1064 (2-methylbenzimidazole-4-carboxamide, IC50 = 1.1 μM) to increase the cytotoxicity of a variety of anticancer drugs. She confirmed that PARP inhibition increases the cytotoxicity of DNA alkylating agents using temozolomide and its active derivative, MTIC. She identified that it was the DNA repair phase that was critical, rather than the DNA damage induction phase as NU1025 was as active when it was added after a 20 min pulse with MTIC as when the 2 compounds were given simultaneously. Karen also confirmed the radiosensitisation data that had been generated with the early benzamides. She showed that PARP inhibition was more effective at preventing recovery from potentially lethal damage than radiosensitising exponentially growing cells. Potentially lethal recovery is when cells are given a high dose of irradiation and their survival is estimated immediately or after a recovery period where they are allowed to repair the DNA damage. This is significant because the non-replicating fraction of a tumour is generally radioresistant. We were unable to see any sensitisation effect of NU1025 or NU1064 on antimetabolite-induced cytotoxicity [12]. Using NU1025, we were the first to demonstrate that PARP inhibition increased the DNA damage and cytotoxicity caused by topoisomerase I poisons, but not topoisomerase II poisons [13]. We continued synthesising new compounds and testing them for PARP activity with the next major step in potency being the synthesis of the benzimidazole NU1085 (2-(4-hydroxyphenyl)-benzimidazole-4-carboxamide Ki = 6 nM, IC50 = 80 nM) by Sheila Srinivasan [14]. Since our compounds were numbered sequentially, this means that our 25th compound NU1025 was 200x more potent than 3AB and our 85th compound was 1000x more potent.

The generation and characterization of PARP-knockout mice that both were viable and fertile by three groups independently in the late 1990s [15–17], initially caused us to think that PARP was non-essential, and so inhibiting it would not compromise the viability of cancer cells. However, I recalled the data of Satoh and Lindahl [18] demonstrating that nuclear extracts depleted of PARP were able to repair nicked DNA but nuclear extracts with PARP but incubated either in the absence of NAD^+ or presence of 3AB could not. These authors proposed that binding of PARP to the DNA break caused an obstruction unless it could auto poly (ADP-ribosylate) and dissociate to allow access of repair proteins to fill in and re-seal the gap. This was the first demonstration of PARP "trapping", although this phrase was not coined until at least 20 years later [19]. These observations meant inhibited PARP was likely to be more effective in inhibiting DNA repair than no PARP enzyme. The ability of cells derived from these mice to generate PAR led to the discovery of PARP2 [20] and subsequently other

PARPs by structural analogy [21], with the original most abundant PARP subsequently being called PARP1. It is important to note here that PARP1 is the most abundant PARP and most important in terms of DNA repair, with PARP2 having a similar role. Most inhibitors inhibit both PARP1 and 2, and the double knockout of PARP1 and PARP2 is lethal [22].

3. Advanced Compound Development and Pre-clinical Studies Resulting from the Newcastle–Agouron Collaboration

Our next piece of good luck (serendipity #3) was to meet Zdenek Hostomsky at the 9th NCI/EORTC annual meeting in Amsterdam in 1996. He was proposing to initiate a PARP programme at Agouron Pharmaceuticals (in La Jolla, California, USA), a company specialising in crystal structure-based drug design. He had obtained the plasmid containing the construct for the PARP1 catalytic domain from Gilbert deMurcia and the co-ordinates of its crystal structure from Georg Schultz [23]. All he needed was highly potent inhibitors. We struck up an agreement and gave him NU1085 to co-crystalise with the PARP1 catalytic domain. Initial promising results meant that their 6 month pilot funding to Newcastle was extended for a further 2 years to support two post-doctoral chemists and a technician, Alex White, Sarah Mellor and Richard Davies, and two post-doctoral biologists and a technician, Carol Delaney, Chris Calabrese and Lan Zhen Wang. At Agouron, analysis of the interactions made between the inhibitor and the PARP catalytic domain by Bob Almassy led the lead chemist on the project, Steve Webber, to propose tricyclic compounds with the carboxamide constrained within a 7 membered ring. Synthetic medicinal chemistry at Agouron was undertaken by Don Skalitzky and Stacie Canan, working with Steve Webber. Karen Maegley ran the biochemical evaluation, with cell-based target inhibition and cytotoxicity assays being done by Jianke Li. Ted Boritzki and Bob Kumpf led the conventional and computational pharmacology, respectively at Agouron in collaboration with Chris Calabrese and Huw Thomas in Newcastle. At the end of the 2 year period, several potent tricyclic indole and benzimidazole inhibitors had been made at Agouron and they extended the funding to Newcastle for a further 3 years to cover the in vitro and in vivo pre-clinical evaluation of these inhibitors, as well as the development of pharmacodynamic biomarkers until the clinical trial started in 2003.

Extensive investigation of the cellular activity of these tricyclic compounds in terms of their ability to enhance the cytotoxicity of the DNA methylating agent, temozolomide, and the topoisomerase I poison, topotecan was undertaken in Newcastle, principally in two colorectal cell lines, LoVo and SW620 by Lan Wang and Suzanne Kyle, with Jianke Li investigating the effects in A549 lung cancer cells. My group have always maintained the highest possible stringency regarding the cultivation of cells, confirming they are mycoplasma free by routine regular testing (2 monthly intervals) and handling each cell line separately (including reagents) to eliminate the possibility of cross-contamination. Although these studies preceded routine authentication, our stocks have been authenticated since and all human cell lines are either purchased new and placed in an authenticated bank at first or second passage or authenticated and then placed in the authenticated bank. In vivo efficacy and pharmacokinetic/pharmacodynamic (PK/PD) studies were also undertaken by Chris Calabrese working with Huw Thomas and Mike Batey [24]. The first potential clinical candidate was AG14361 (1-(4-((dimethylamino)methyl)phenyl)-8,9-dihydro-2,7,9a-triazabenzo[cd]azulen-6(7H)-one: Ki < 5 nM) [25–29]. AG14361 caused profound sensitisation of temozolomide and was able to overcome temozolomide resistance due to defects in DNA mismatch repair [30]. The enhancement of topotecan and ionising radiation cytotoxicity in vitro was less pronounced but again the prevention of recovery from potentially lethal irradiation was substantial. In vivo radiosensitisation studies were conducted in Manchester by Kaye Williams and Ian Stratford [29]. Investigations in NIH 3T3 cells from PARP1-knockout mice enabled my PhD student, Lisa Smith, co-supervised by a topoisomerase expert, Caroline Austin, to confirm that topotecan was more cytotoxic in PARP-null cells and sensitisation by AG14361 was indeed due to PARP1 inhibition [31].

Investigation of in vivo activity against SW620 and Lovo xenografts revealed some interesting data. AG14361 caused an approximately 3-fold increase in temozolomide-induced anticancer activity against Lovo xenografts, but the combination caused sustained complete tumour regressions of SW620 xenografts. This was curious because in cell cultures, Lovo cells were the most sensitised to temozolomide by AG14361 and SW620 cells were not sensitised at all. SW620 cells were particularly sensitive to temozolomide alone (due to functional mismatch repair and lack of methylguanine methyltransferase). This led to another example of the synergy of science of serendipity: the discovery of the vasoactivity of this class of compounds. I reasoned that the discrepancy between the in vivo and in vitro effects meant that the tumour regression had to be something to do with the tumour micro-environment. Nicotinamide was known to be a vasodilator [32], and so I hypothesised that AG14361, which, like all PARPi contains a nicotinamide pharmacophore, might improve the delivery of temozolomide to more of the tumour cells. In collaboration with Kaye Williams and Ian Stratford in Manchester, we showed that indeed AG14361 did increase the areas perfused in the tumour, a further example of the synergy of science and serendipity (#4). [29]. This effect may have also contributed to the enhancement of the anticancer activity of the topoisomerase I poison, irinotecan, by AG14361, which was greater in vivo than in vitro. Similar vasoactive effects were reported by others with olaparib, which was accompanied by increased radiosensitivity of tumour xenografts due to better oxygenation [33]. Vasoactivity may contribute to the in vivo radiosensitisation by AG14361 but we have not investigated whether this is the case. It is possible therefore that it is a class effect and that PARPi may increase the anticancer activity of other drugs by increasing delivery to the tumour. It should be noted that these experiments were done with subcutaneous xenografts and the effect on in situ tumours may be somewhat different, especially as PARPi in the clinic have not been reported to cause hypotension except rarely in the elderly population [34], as might be expected from a vasodilator. Nevertheless, there may be effects that are specific to the tumour vasculature, but this has never been investigated.

The exciting data with AG14361 led to a proposal for a clinical trial with this compound being designed at a meeting of the team early 2000 and drafted for submission to CRUK's New Agent Committee. The teams in Newcastle (CR Calabrese, AH Calvert, NJ Curtin, BW Durkacz, BT Golding, RJ Griffin and DR Newell), Manchester (J Monaghan, I Stratford, and K Williams) and Agouron (R Almassy, T Boritzki, S Canan-Koch, L DiMolfetto, G Furman, Z Hostomsky, A Johnston, R Kumpf, J Li, K Maegley, D. Skalitzky, SE Webber and K Zhang) all contributed to the proposal. Meanwhile, there had been continued evaluation of a number of potential inhibitors. Many of the studies that had originally been conducted with AG14361 were also replicated with several other compounds, in vitro by Suzanne Kyle and Lan-Zhen Wang and in vivo by Chris Calabrese and Huw Thomas. Computational studies by Bob Kumpf suggested a fluoro substituent at the 8 position would improve activity with the ultimate compound being 8-fluoro-5-(4-((methylamino)methyl)phenyl)-2,3,4,6-tetrahydro-1*H*-azepino[5,4,3-*cd*]indol-1-one, or AG014447. AG014447 and/or the phosphate salt—called AG014699—had been included in the pre-clinical in vitro and in vivo studies and was selected for clinical investigation [35]. AG014699 (rucaparib) was indeed more potent than AG14361 in in vivo studies and, whereas 10 mg/kg AG14361 caused complete tumour regressions in combination with temozolomide, 1 mg/kg AG014699 was equally effective [29,35]. It also had a similar vasoactive effect [36]. Pharmacokinetic (PK) studies with AG14699, conducted by Huw Thomas and Mike Batey, showed that although it was cleared quite quickly from the bloodstream, it accumulated in tumour xenografts and, at the efficacious dose, it suppressed PARP activity in the tumour by >50% for 24 h. The clinical trial proposal was re-written with AG014699 data, with input from the team above. The involvement of the team who had developed and studied the drug pre-clinically ensured that the first clinical trial was as rationally designed, with a full understanding of the pre-clinical data as it is possible to have and supported by translational pharmacodynamic, as well as PK measurements. The proposal was submitted to CRUK's New Agent committee in 2001, with a successful outcome. AG014699, now called rucaparib, was the first PARPi to be given to a patient with cancer in May 2003.

4. The First Clinical Trial and Supporting Science

Based on the pre-clinical data, the target dose clinically (PARP inhibitory dose: PID) was one that inhibited PARP activity by >50% in a surrogate tissue (lymphocytes) for at least 24 h. To support this pharmacodynamically guided clinical trial, we started adapting the ^{32}P NAD^+ incorporation assay we had used to measure PARP activity and inhibition in permeabilised cells as part of the early inhibitor development. In these endeavours, we were blessed with two more items of good fortune. Firstly, experiments conducted early in 2002 by Suzanne Kyle revealed that PARP inhibition was not lost as soon as the drug was removed Figure 2A), meaning that pre-treated cells could be harvested and processed without loss of inhibition. This was an instance of real serendipity (# 5), as it is quite rare to see such persistence. Secondly, at the same time, Suzanne showed that PARP activity was maintained in cells that had been cryopreserved at -80 °C for up to 14 weeks. More importantly, rucaparib's inhibition of PARP activity was incredibly stable: cells that had been pre-treated with rucaparib could be harvested and cryopreserved for up to 14 weeks without loss of inhibition, an even more remarkable example of serendipity (#6); Figure 2B.

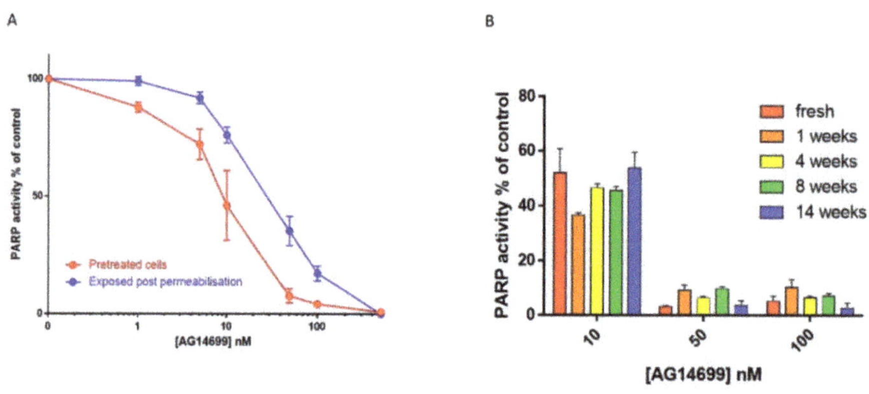

Figure 2. Inhibition of cellular poly (ADP-ribose) polymerase (PARP) activity. (**A**) PARP inhibition by increasing concentrations of AG14699 (rucaparib) following pre-treatment of L1210 cells (30 min) with drug prior to harvesting (red symbols and line) compared with drug added to the permeabilised cells in the reaction mixture (blue). Data are the mean and SD of three replicates from a representative experiment. (**B**) Stability of PARP inhibition by rucaparib with storage. Cells were exposed to rucaparib for 30 min prior to harvesting and cryopreservation. PARP activity was measured by ^{32}P NAD^+ incorporation. Previously unpublished data.

We then started harvesting lymphocytes from healthy volunteers (i.e., our colleagues in the lab). However, we soon discovered that PARP activity in human lymphocytes is orders of magnitude lower than in cultured cells and we would have needed approximately 50 ml blood pre- and post-treatment with rucaparib to be able to detect inhibition reliably. This is where yet another example of good fortune came in (serendipity #7). Alex Burkle had recently moved to Newcastle with his post-doc Ragen Pfeiffer and they had developed a much more sensitive immunoblot assay using the 10H antibody to the product, PAR [37]. They shared this technology with us and Ruth Plummer, who was my clinical fellow at that time, validated this pharmacodynamic (PD) assay to GCLP standard. It was initially intended that she would carry out the rucaparib Phase I trial. However, there was a substantial delay in the start of the trial, partially due to the take-over of Agouron by Warner Lambert (in January 1999) that was followed less than a year later by the take-over of Warner Lambert by Pfizer, who did not at that time have an oncology programme. Instead of conducting the rucaparib trial, Ruth used the PD assay to monitor the effects of temozolomide on PARP activity in patients with metastatic melanoma instead [38]. It was my next clinical fellow, Chris Jones, who conducted the trial under the guidance of Ruth Plummer and Hilary Calvert, the principal and chief investigators, respectively. Chris made further improvements to the assay and monitored PARP inhibition in the lymphocytes and tumour biopsies of patients treated with rucaparib [39]. Using this assay, another student of mine, Thomas Zaremba, set out to study PARP1 genomics, expression and activity in healthy volunteers and cancer patients with the aim of determining a) whether PARP genotype contributed to reduced PARP activity and the likelihood of developing cancer and b) whether PARP1 genotype, expression or activity contributed to unexpected toxicities in patients treated with temozolomide or radiotherapy. He did not manage to answer these questions but what he did find was that a) only 20% of the variation in PARP activity is due to the variation in PARP1 expression, b) that men have on average 40% more PARP activity than women and c) that it is androgen-driven, suggesting post-translational modification or co-activators/repressors that are androgen regulated [40]. The question this raises is: when used as a chemo- or radiosensitiser, should PARPi doses be lower in women?

The first trial of rucaparib was a dose escalation trial in cohorts of 3 patients with a variety of solid tumours receiving 1, 2, 4, 8 and 12 mg/m^2 in combination with half-dose temozolomide (100 mg/m^2). There was a dose-dependent decrease in PARP activity, with the desired ≥50% inhibition of PARP activity for 24 h being consistently achieved above 8 mg/m^2. In the absence of increased toxicity, 12 mg/m^2 was established as the safe dose to begin escalation of the temozolomide up to full dose (200 mg/m^2) in melanoma patients consenting to a pre-treatment and post-treatment biopsy. There was a dose-dependent decrease in PARP activity in the tumours too, with approx. 90% inhibition 4–8 h after administration of 12 mg/m^2. Escalating the dose to 18 mg/m^2 led to increased haematological toxicity and 12 mg/m^2 was established as the recommended Phase 2 dose to take forward in combination with temozolomide. Unfortunately, although this combination may have shown an improvement on historical data of response to temozolomide alone in patients with melanoma, there was also more haematological toxicity [41]. This may have doomed rucaparib to the same failure to progress beyond phase II studies, as with other DNA repair inhibitors such as the MGMT inhibitors O6 benzylguanine and Patrin [42], had it not been for the next and most significant example of serendipity on this project.

5. Discovery of the Synthetic Lethality of PARPi

The synergy between science and serendipity (#8) reached its peak with the discovery of synthetic lethality in BRCA mutant and other cancers defective in homologous recombination DNA repair (HRR). This came about through lucky coincidences with timing of various discoveries and meetings as described below. At the end of 2001, Herbie Newell was invited to speak at Sheffield University about the drug development programme in Newcastle University. Present at this meeting was Thomas Helleday, who had been recruited to Sheffield from Tomas Lindahl's lab. Tomas Lindahl (co-recipient of the Nobel prize for chemistry 2015 for his work on DNA repair) had a long-standing interest in PARP, its function and the effects of inhibitors. Tomas Lindahl had noted that in the presence of PARP inhibitors, there was an increase in recombination events. This led him to propose that the negative charge on the ADP-ribose polymer helped to repel negatively charged DNA from sites of DNA breakage to avoid unwanted recombination events [43]. Thomas Helleday had investigated further and found that cells lacking XRCC1 (the scaffold protein recruited to DNA breaks by ADP-ribose polymers [44], as well as those lacking PARP1, accumulated more γH2AX foci (a marker of DNA damage) and RAD51 foci (a marker of HRR). As a result of Herbie's talk in Sheffield Thomas requested some of our more potent and selective PARPi. We gave him some NU1025 and invited him to give a talk in Newcastle to present his work in January 2002. He convincingly showed that XRCC1 defects had similar effects in increasing HRR compared to PARP1 deletion or inhibition with NU1025. This suggested the polymers were not acting to inhibit recombination but instead a failure of BER led to an increase in HRR. He proposed that cell lacking BER/PARP function would be dependent on HRR for survival and as such cells lacking HRR function may be dependent on PARP activity for survival. He tested his hypothesis in cells defective in HRR (IRS1 lacking XRCC2 and irs1SF lacking XRCC3), which were killed by concentrations of NU1025 that did not kill wt or XRCC2/3 corrected cells.

By this time, I had widened my interest in DNA repair and as well as leading the pre-clinical biological studies on PARPi, I also led the DNA-PK and ATM inhibitor projects at Newcastle. As a result of this, I attended a meeting early in 2002, where Ashok Venkitaraman described the data supporting the key role of BRCA2 in HRR. BRCA 1 and BRCA 2 were discovered in the early 1990s as breast cancer susceptibility genes but at first it was not clear why. Their function in DNA repair began to emerge towards the end of that decade and it was confirmed that they played key roles in HRR around the turn of the millennium [45–48]. More reading around the topic convinced me that our PARP inhibitors might have therapeutic potential in BRCA mutant hereditary breast cancer, which I put to the drug development team in May 2002. The BRCA2 mutant Capan-1 pancreatic cells along with the BRCA wt BxPC-3 pancreatic cells, as control, were duly purchased and their sensitivity to AG14361 was determined by Suzanne Kyle. Frustratingly, these initial experiments, conducted in July 2002, did not show that the BRCA mutant cells were more sensitive. However, experiments that had been conducted 3 to 4 months previously in wt (AA8) and the HRD derivative (XRCC3 mutant: irs1SF) Chinese hamster ovary cells had indeed shown a very marked difference in the sensitivity of the wt and HRD cells (Figure 3). To this day, I am not sure why the data were so disappointing with the Capan-1 cells as we got more impressive results both in vitro and in vivo with this cell line using rucaparib (see later). Maybe it was the design of the experiment and the exposure period (24 h) was insufficient, as less than a complete cell cycle, or the moribund cells were lost during harvesting or maybe the BxPC-3 cells were also HRD.

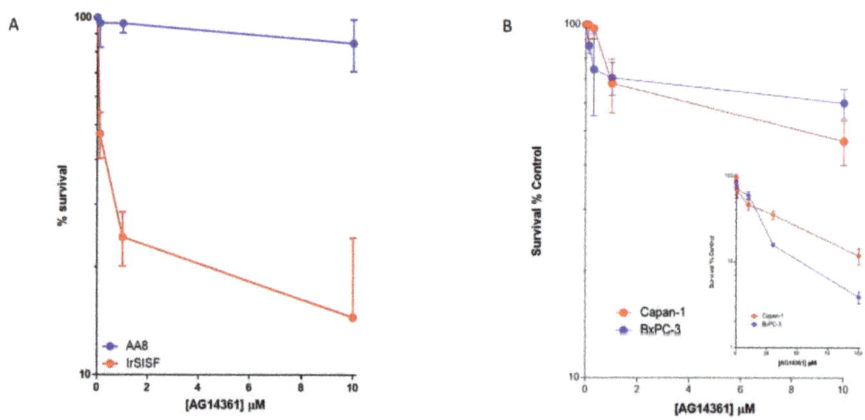

Figure 3. AG14361 is more cytotoxic in XRCC3 mutant irs1SF cells compared to wild-type AA8 Chinese hamster ovary cells (**A**) but BRCA2 mutant Capan-1 cells are not more sensitive than BRCA wild-type BxPC-3 pancreatic cancer cells (**B**). Cells were exposed to increasing concentrations of AG14361 for 24 h prior to seeding for colony formation in fresh medium. The insert shows an extended concentration range for pancreatic cell lines. Previously unpublished data pooled from three independent experiments.

We communicated our findings to Thomas Helleday, with whom we had established a collaboration, in the autumn of 2002. By Spring 2003, he had obtained the VC8 BRCA2 mutant derivatives of Chinese hamster lung fibroblast V79 cells, which he later shared with us. Suzanne got her first result with these cells in June 2003 and they were so impressive we could hardly believe it. She replicated the result several times and confirmed that 98.5% of BRCA2 mutant cells were killed by 10 μM AG14361, a concentration that had no significant effect on the viability of either parental wt cells or the BRCA2-corrected cells. Conducting in vivo experiments with these cells proved more of a challenge. The V-C8 cells would not grow subcutaneously and, conversely, the V79 cells generated very haemorrhagic tumours that rapidly made the mice unwell such that they had to be humanely killed. However, Huw Thomas eventually succeeded in growing the V-C8 cells and the BRCA2 corrected V-C8.B2 cells intramuscularly. These experiments showed that treatment of mice with AG14361 caused a complete regression of the BRCA2 mutant tumour but growth of the BRCA2 corrected tumour was unaffected. The data generated by Suzanne and Huw went into the paper that was published in Nature in 2005 [49]. These key experiments, which have led to the current use of PARPi clinically, were actually undertaken outside of the remit of the drug development programme. This was because at Herbie Newell's suggestion, the rest of the drug development management team at Newcastle had decided that PARP was no longer to be considered a drug development project to be discussed at subsequent meetings in June 2003 as the clinical trial (albeit in combination with temozolomide) was underway.

The data with the BRCA mutant and HRD models, where a tumour-specific defect in DNA repair is exploited by inhibiting a complementary DNA repair pathway were indeed ground-breaking. The findings completely changed the way we think about treating cancer, from trying to overcome some property of the tumour that gave it an advantage to exploiting a vulnerability. Indeed, the search for additional examples of synthetic lethality has become somewhat of a holy grail. However, I believe PARP inhibition in HRD is unlikely to be bettered as an example of synthetic lethality for 2 reasons: (1) there is a very high level of DNA SSB, that depend on PARP for their repair, due to the endogenous ROS produced by normal metabolism, and this is increased in cancer by virtue of the associated inflammation, and (2) HRR defects are relatively common in cancer, not only due to BRCA mutations. This is yet another example of the synergy of science and serendipity (#9)

Understanding the importance of the synthetic lethality of PARPi in HRD cancers, Thomas Helleday was keen to patent our findings before publication. Following meetings with Cancer Research Technology, we undertook work to protect rucaparib and these data were included in the patent (WO 2005/012305 A2) filed 16th April 2004.

6. The Clinical Development of Rubraca®as A Single Agent and Predictive Biomarker Development in Ovarian Cancer

During this time, the Phase I clinical trial of rucaparib with temozolomide was ongoing and there was clinical interaction with Pfizer who later conduct a single agent dose escalation study with rucaparib. This study was led by Hilary Calvert and Ruth Plummer, who recruited Yvette Drew, a clinical fellow, in 2006 to undertake the trial. In parallel, Yvette conducted pre-clinical and translational studies with me. The translational studies included taking blood samples from the patients before and at intervals after rucaparib adminiatration. These samples were used to determine the rucaparib PK and also PARP1 genotype, protein levels and inhibition by rucaparib in the patients' PBMCs (lymphocytes).She also undertook studies in a panel of BRCA wt and mutant human cancer cell lines to confirm the differential sensitivity we had seen in the matched Chinese hamster cells (AA8/irs1SF and V-C8/V-C8.B2). This panel included the UACC3199, which had epigenetically silenced BRCA1 due to promoter methylation and the Capan-1 cells that had been disappointing in the AG14361 experiments. Such studies were needed because there were reports that other BRCA mutant human cancer cells were insensitive to PARPi, including AG14361 [50,51]. Fortunately, rucaparib sensitivity was found to be substantially greater in the BRCA mutant cells (including the Capan-1 cells) than the wt and, importantly, the BRCA silenced cells were comparable with the mutant ones in this respect [52].

At the same time, I had established an interaction with a gynaecological surgeon, Richard Edmondson, who had a number of trainees keen to undertake lab-based projects for MD or PhD. This turned out to be yet another example of the synergy between science and serendipity (#10). I was aware that BRCA mutation/HRD was a determinant of sensitivity to cisplatin/carboplatin from discussions I had had with Paul Harkin at a meeting in 2003. The standard of care for ovarian cancer is surgery followed by carboplatin + paclitaxel chemotherapy with an initial response rate of approximately 60%. Knowing that only approximately 15%–25% of ovarian cancer is associated with BRCA mutations, that left 35%–45% of the responses unaccounted for. I reasoned that they must be HRD for some other reason and that we could test this hypothesis by measuring γH2AX and RAD51 foci. γH2AX foci are formed at collapsed replication forks and DNA DSBs and RAD51 coats the SS DNA to form the nucleoprotein filament needed to invade the complementary DNA on the sister chromatid during HRR resulting in RAD51 foci. Therefore, γH2AX foci indicate the lesions generated by PARP inhibition and RAD51 foci indicate when HRR occurs. We first met with Pfizer representatives (including Zdenek Hostomsky and Gerrit Los) in September 2006 to suggest that Yvette Drew, the clinical fellow undertaking the trial of single agent rucaparib, would not only work with cell lines to establish the spectrum of rucaparib sensitivity (described above) but also to develop an assay for HRR function, based on measuring γH2AX and RAD51 foci. We also proposed to evaluate the HRR status in primary cultures of ovarian cancer ascites collected during primary/debulking surgery with another clinical fellow Asima Mukhopadhyay, a trainee gynaecological oncology surgeon supervised by Richard Edmondson and me. Thanks to Zdenek Hostomsky, ever our PARP champion even though he had been diverted to other projects, I secured funding from Pfizer in 2008 to support these studies with an additional technician, Evan Mulligan, to help with planned in vivo efficacy studies..

These studies led to two key papers published in 2010/11 [52,53]. The data Yvette generated in the cell lines not only demonstrated the difference in rucaparib cytotoxicity between HRD and HRR functional cells but also that RAD51 foci were a useful discriminant/potential biomarker of HRR function. That is, the induction of RAD51 foci in the HRD cells was not significant whereas there was a significant increase in the BRCA wt cells. Looking at γH2AX foci really brought it home to us how much endogenous damage cells sustain without the addition of exogenous genotoxic agents. The level of the foci after only 24 h of inhibition of endogenous damage repair with rucaparib was the same as the standard radiotherapy fraction of 2 Gy. After some significant work to establish the optimum conditions to grow primary cultures from ascites, Asima successfully translated this HRR function assay to the ascites cultures and her initial studies showed that approximately 60% of the primary cultures were indeed HRD and that this corresponded to a greater rucaparib-induced inhibition of cell growth. This was the first demonstration that >50% of ovarian cancers are HRD and, at significantly less expense than similar findings by the TCGA, which was published the following year [54].

The in vivo experiments in mice bearing BRCA1 mutant (MDA-MB-436) or BRCA2 mutant (Capan-1) xenografts, conducted by Yvette Drew and Evan Mulligan, showed that rucaparib caused significant tumour growth delay and this was more pronounced when the treatment period was 5x weekly for 6 weeks compared with daily x10. This was not entirely unexpected from an understanding of the underlying mechanism whereby to kill the cells PARP must be inhibited as the cell progresses through S-phase. These findings were recapitulated in the clinical study that Yvette was conducting where responses were seen when the patients switched from daily treatment for 5 days every 3 weeks to continuous dosing [55]. This study, initiated in 2007, was the first single-agent trial of rucaparib, sponsored by CRUK and Pfizer. It began very cautiously, starting at low dose (escalating from 4 to 18 mg/m2 i.v) on an intermittent schedule (daily for 5 days every 3 weeks) based on the trial in combination with temozolomide. During the study, there was a switch to an oral formulation and daily dosing, starting at 92 mg (equivalent to 18 mg/m2 based on an oral bioavailability of approximately 30% and average body surface area of 1.7 m2) once a day for 7 days and escalating to 600 mg 2x/day continuously. Many more responses were seen once dosing was continuous. It is not altogether clear whether it was the increase in the dose or the intensity of the schedule that was critical. Unfortunately, this rather slow start allowed rucaparib to be overtaken by olaparib, which went into its first clinical trial as a single agent and the dose was escalated more quickly, as there was no history of combination toxicity data to suggest a cautious approach was necessary. As a result, the first publication showing that the pre-clinical data on the synthetic lethality of PARPi in HRD/BRCA mutant cancer was translatable to the clinic was with olaparib [56].

In a way, this result could have been avoided with more careful scrutiny of the pre-clinical data. Pre-clinically much lower (at least 10x) concentrations (in cell cultures) and doses (in tumour xenograft studies in mice) were needed to sensitise cells and tumours to temozolomide than were needed for single agent activity in BRCA mutated/HRD cancer cells. Furthermore, increasing the dose of rucaparib to one that was active against BRCA mutant xenografts was so toxic that the mice had to be humanely killed a few days into the drug administration [29,35]. These differences are easily understood by reference to the mode of action of the PARPi in the different scenarios. Because DNA integrity is so important, the cell has more than adequate levels of repair enzymes to cope with endogenous and most environmental DNA damage. Couple this with the fact that single base lesions and SSB are the commonest form of endogenous lesions, and these are dependent on PARP for their repair, and it is easy to see that cells will have a surplus of PARP to cope with daily levels of DNA damage. In order for the damage to overwhelm this repair capacity and cause death in cells lacking the back-up pathway of HRR, there has to be a total suppression of PARP activity and relatively high levels of drug are needed to achieve this. In addition, PARP must be inhibited long enough for all tumour cells to enter S-phase with unrepaired DNA lesions and so long/continuous exposures are necessary. In the alternative scenario where, e.g., temozolomide, induces orders of magnitude greater numbers of base lesions and SSBs, it is not necessary to inhibit their repair completely to render the remaining DNA damage sufficient to kill cells and so much lower levels and shorter duration of PARP inhibition are needed. Also, when used as a single agent, the PARPi is exploiting a tumour-specific defect, and side effects are not expected so higher doses can be used. When the PARPi is synergistic with the cytotoxic in cancer cells the combination is likely to also be synergistically toxic to normal tissues so only lower doses are tolerated.

The observation that continuous therapy was much better than intermittent led us to investigate how durable PARP inhibition by rucaparib was. This study, initiated in 2011 in collaboration with and support from Zdenek Hostomsky and Gerrit Los, was partially based on our observations in the first clinical trial: as the dose of rucaparib was increased PARP activity was suppressed in PBMCs for 24 h after the 30 min infusion, even though the drug was no longer detectable in the plasma. Indeed, at the 12 mg/m2 dose, PARP activity was still way below pre-dose levels on the Monday after the last Friday dose in many patients [39]. Pre-clinically we found that rucaparib accumulated above the extracellular concentration and was retained after drug removal in cancer cells and that PARP was inhibited for at least 72 h after a 30 min exposure. In in vivo studies, a single oral dose of rucaparib inhibited PARP activity in xenografts for at least 7 days even though it was cleared from the blood within 48 h and was undetectable in the tumour within 72 h. Moreover, tumour growth inhibition following weekly dosing with rucaparib was equivalent to daily dosing [57]. This, I believe, is another example of serendipity (#11) ensuring the continuous suppression of PARP activity necessary for antitumour activity as a single agent. It is possible that this might be exploited clinically by weekly dosing schedules.

Pfizer initiated a Phase I study in combination with carboplatin alone or with paclitaxel or pemetrexed or a combination of cyclophosphamide and epirubicin NCT01009190 in November 2009 [58]. There was little pre-clinical rationale for such a study, as PARPi do not sensitise cells to either pemetrexed (an antifolate), cyclophosphamide (DNA cross-linking agent), epirubicin (DNA intercalating topoisomerase II poison) or paclitaxel (an antitubulin agent, not a DNA damaging agent) and the data with cisplatin or carboplatin are variable and cell line dependent, most probably additive in HRD cells only [59]. The triple combination arms were dropped during the study and the MTD in combination with carboplatin (AUC 5) was 240 mg one daily (carboplatin day 1, rucaparib days 1–14 of a 21 day cycle) due largely to haematological toxicities [58]. Thus, the MTD for rucaparib in combination with carboplatin (with which is does not synergise) is approximately four times higher than with temozolomide and 2.5x lower than as a single agent.

7. Studies That Led to FDA Approval

In June 2011, Pfizer licenced rucaparib to Clovis oncology and in November, study 10, (NCT01482715) a phase I–II trial of oral rucaparib as a single agent, was initiated [60,61]. By this time, Hilary Calvert had moved from Newcastle to UCL and he supported another clinician, Rebecca

Kristeleit, who led the study. The Phase I component was a dose escalation of the oral formulation, based on the intermittent vs. continuous dosing study described above, in patients with solid tumours to establish the recommended phase II dose (RP2D) based on tolerability and efficacy. The RP2D of 600 mg 2x daily was then used to treat women with BRCA1/2 mutated high-grade serous ovarian cancer as the Phase II part of the study. The objective response rate was 60% and the toxicities were generally mild with only 5 out of 98 patients having to discontinue treatment due to an adverse event.

Ongoing studies with other PARPi clearly demonstrated that BRCA mutation was not synonymous with a response in ovarian cancer [62], with cases with BRCA mutations in the non-responder group as well as cases without BRCA mutations in the responder group. Indeed, our earlier studies (described above) clearly indicated that BRCA mutation significantly underestimated the HRD and PARPi sensitive population. Several companies were developing predictive biomarkers of HRD and we shared some of our samples that we had previously characterised by the functional assay (by this time, our collection had grown substantially) with Tom Harding at Clovis and Foundation Medicine from 2011 to 2013. However, it seemed that the normal cellular component (that died off during culture leaving us with just the cancer cells to classify) diluted the cancer component too much for the samples to be of much use. Nevertheless, based on the observation that BRCA mutations result in substantial genome-wide loss of heterozygosity (LOH), Foundation Medicine [63] developed a genomic profiling test to identify HRD independent of BRCA mutation to be used as a companion diagnostic in the Ariel 2 trial NCT01891344 [64]. Similar tests have been developed by Myriad. Although these tests represent genetic "scarring" caused by an HRR defect at some time during the cancer's development, but not necessarily at the time of treatment, they are substantially less challenging than functional assays, which by their nature require viable fresh tumour tissue. Screening for mutations in known HRR genes has also been used.

Ariel 2 was an innovative and pivotal trial that led to the rucaparib being given breakthrough status with the FDA. In this trial, 198 patients with high-grade serous ovarian cancer (HGSOC) were characterised: 40 had BRCA mutations (germline or somatic), 80 had high levels of LOH (presumed HRD) and the remaining 70 had neither BRCA mutation nor LOH. The median progression-free survival for the BRCA mutant and high LOH groups was significantly longer than the biomarker negative group. Accelerated approval was given by the FDA in December 2016. The follow-on Ariel 3 study led to the FDA approval of Rubraca (in April 2018) as maintenance therapy in platinum-sensitive ovarian, fallopian tube or primary peritoneal cancer, along with Foundation Medicine's complementary diagnostic CDx $_{\text{BRCA LOH}}$ in 2016. Ariel 3 (NCT01968213) was a randomised double-blind placebo-controlled trial in 564 patients that had responded to platinum-based chemotherapy. This trial showed a significant improvement in median progression-free survival in all patients treated with rucaparib (10.8 months) compared to placebo (5.4 months), but particularly the 130 BRCA mutated (16.6 months) and 106 BRCA wild-type, LOH-high HRD patient (9.7 months months) subgroups compared to the BRCA wt LOH low (6.7 months) [65]. More recently, Rubraca has been approved by the European Medicines Agency in January 2019 and by the National Institute for Health and Care Excellence in October 2019 for maintenance therapy in epithelial ovarian, fallopian tube, or primary peritoneal cancer who are in a complete or partial response to platinum-based chemotherapy, without the need for a companion diagnostic. Foundation Medicine's CDx $_{\text{BRCA LOH}}$ may still be used as a complementary diagnostic. This is because the Ariel 3 trial showed that platinum sensitivity was a sufficiently good indicator of response to rucaparib, which was indeed the basis of our hypothesis that led us to investigate HRD status in ovarian cancer in 2008.

Rubraca is in several advanced clinical trials as a single agent in other cancers associated with BRCA mutations: breast, pancreatic and prostate cancer. It has breakthrough designation with the FDA for castrate-resistant prostate cancer. It seems as though the best responses are observed in the ovarian setting. Whether this is because BRCA mutations, or even LOH are not entirely synonymous with HRD [66], or whether something within the tumour microenvironment influences the outcome is anybody's guess at the current time. If it is the former, then our small studies on the frequency of cancers lacking HRR function, as measured in viable tumour tissue, may be relevant [67] and suggest more complex readouts of HRR function may be necessary. Combinations with other molecularly targeted agents are also being investigated in other tumour types. For most up to date data, visit https://www.clovisoncology.com/pipeline/rucaparib/ and https://clinicaltrials.gov/.

8. Conclusions

In conclusion, the story of the development of rucaparib/Rubraca shows not only the necessary input of intellectual (creative, deductive and analytical) and physical effort needed over a protracted period of time for the successful development of a drug (summarised in the timeline, Figure 4) but also how luck played a significant role all the way through from the initial hit right through to the clinical development in ovarian cancer. The role of luck is by no means unique to the development of Rubraca®, luck has probably contributed to many of the scientific advances that we take for granted. One prime example was the production of penicillin during the Second World War, which would not have been possible without the necessary contacts and proximity of the potteries, a fact acknowledged by Norman Heatley himself in his lecture "Penicillin and Luck" [68]. There have been some setbacks in the development of Rubraca®but, on the whole, luck has been on our side—most significantly, the fact that PARPi block the repair of the most common forms of endogenously generated DNA damage and that this is synthetically lethal with a relatively common DNA repair defect in cancer, HRR. This has caused a paradigm shift in how we think about treating cancer and there are numerous investigations to identify other instances of synthetic lethality. It seems unlikely that anything as profound as the synthetic lethality of PARPi against HRD cancers will be identified. Some examples of serendipity are yet to be exploited to the full—in particular, the discovery of the durability of PARP inhibition by rucaparib warrants further exploration clinically in pharmacodynamically guided intermittent schedules. The outcome of such a study may prove to be another paradigm shift in our handling of molecularly targeted drugs, where the biologically effective dose rather than the tolerable dose is adopted as the appropriate measure to decide the recommended phase II dose.

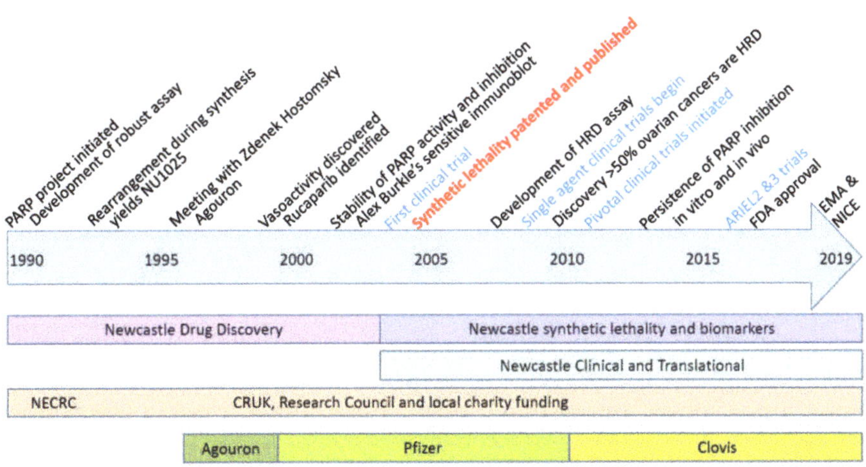

Figure 4. Timeline of the development of Rubraca®. Key milestones are indicated at the top of the timeline, the nature of the work undertaken at Newcastle, the funding sources and pharmaceutical companies involved are shown below the timeline.

9. Patents

9.1. Expired Patents

Griffin, R. J., Calvert, A. H., Curtin, N. J., Newell, D. R., and Golding, B. T. Benzamide Analogues ADPRT (PARP) Inhibitors. Patent Application Number PCT/GB95/00513 (1995)

Griffin, R. J., Calvert, A. H., Curtin, N. J., Newell, D. R. and Golding, B. T. Benzimidazole PARP Inhibitors.Patent Application Number PCT/GB96/01832.

Griffin, R. J., Calvert, A. H., Curtin, N. J., Newell, D. R. and Golding, B. T. Prodrugs of PARP Inhibitors.Patent Application Number PCT/GB97/02701.5

9.2. Active Patents

Helleday T and Curtin NJ. Therapeutic Compounds (PARP inhibitors in homologous repair/BRCA defective cancer) Patent Application Number PCT/GB2004/003183. Publication number WO 2005/012305 A2 Divisional application 16th April 2004 GB 0408524. WO2005012305A3

Boritzki TJ, Calvert AH, Curtin NJ, Dewji MR, Hostomsky Z, Jones C, Kaufman R, Klamerus KJ, Newell DR, Plummer ER , Reich SD, Steinfeldt HM, Stratford IJ, Thomas HR Williams KJ.. Therapeutic Combinations Comprising PARP inhibitor US application No. 60/612,458 Filed 22nd September 2004 WO/2006/033006) THERAPEUTIC COMBINATIONS COMPRISING POLY(ADP-RIBOSE) POLYMERASES INHIBITOR

Funding: This research has been funded by several sources over the nearly 30 years it has been going—all of which have been acknowledged in the primary research papers cited here.

Acknowledgments: People who have contributed to the body of work have been acknowledged in the text but special thanks go to my students, clinical fellows and technical staff who worked so hard to generate the data and to Zdenek Hostomsky who initiated the collaboration with Agouron and continued to champion my work until he left Pfizer.

Conflicts of Interest: I have had funding from Agouron, Pfizer and Clovis, for some of the work described in this review. By virtue of the two active patents and an agreement between Cancer Research Technology, Newcastle University, and Agouron, now Pfizer, and Clovis, I am in receipt of royalty payments. I do not take these personally, modest sums have contributed to my research accounts and a recent large sum has been used to set up a charitable fund within the local Community Foundation. https://www.communityfoundation.org.uk/group_grant/passionate-about-realising-your-potential/

References

1. Roitt, I.M. The inhibition of carbohydrate metabolism in ascites-tumour cells by ethyleneimines. *Biochem. J.* **1956**, *63*, 300–307. [CrossRef] [PubMed]
2. Chambon, P.; Weill, J.D.; Mandel, P. Nicotinamide mononucleotide activation of new DNA-dependent polyadenylic acid synthesizing nuclear enzyme. *Biochem. Biophys. Res. Commun.* **1963**, *11*, 39–43. [CrossRef]
3. Nishizuka, Y.; Ueda, K.; Nakazawa, K.; Hayashi, O. Studies on the polymer of adenosine diphosphate ribose. *JBC* **1967**, *242*, 3164–3171.
4. Miller, E.G. Stimulation of nuclear poly (adenosine diphosphate-ribose) polymerase activity from HeLa cells by endonucleases. *Biochim. Biophys. Acta* **1975**, *395*, 191–200. [CrossRef]
5. Purnell, M.R.; Whish, W.J. Novel inhibitors of poly(ADP-ribose) synthetase. *Biochem. J.* **1980**, *185*, 775–777. [CrossRef]
6. Durkacz, B.W.; Omidiji, O.; Gray, D.A.; Shall, S. (ADP-ribose)n participates in DNA excision repair. *Nature* **1980**, *283*, 593–596. [CrossRef]
7. Halldorsson, H.; Gray, D.A.; Shall, S. Poly(ADP-ribose)polymerase activity in nucleotide permeable cells. *FEBS Lett.* **1978**, *85*, 349–352. [CrossRef]
8. Grube, K.; Kupper, J.H.; Burkle, A. Direct stimulation of poly(ADP-ribose)polymerase in permeabilised cells by double-stranded DNA oligomers. *Anal. Biochem.* **1991**, *193*, 236–239. [CrossRef]
9. Suto, M.J.; Turner, W.R.; Arundel-Suto, C.M.; Werbel, L.M.; Sebolt-Leopold, J.S. Dihydroisoquinolinones: The design and synthesis of a new series of potent inhibitors of poly(ADP-ribose)polymerase. *Anti-Cancer Drug Design* **1991**, *7*, 107–117.

10. Banasik, M.; Komura, H.; Shimoyama, M.; Ueda, K. Specific inhibitors of poly(ADP-ribose)synthetase and mono(ADP-ribosyl)transferase. *J. Biol. Chem.* **1992**, *267*, 1569–1575.
11. Griffin, R.J.; Srinivasan, S.; Bowman, K.; Calvert, A.H.; Curtin, N.J.; Newell, D.R.; Pemberton, L.C.; Golding, B.T. Resistance-modifying agents. 5. Synthesis and biological properties of quinazolinone inhibitors of the DNA repair enzyme poly(ADP-ribose) polymerase (PARP). *J. Med. Chem.* **1998**, *41*, 5247–5256. [CrossRef] [PubMed]
12. Bowman, K.J.; White, A.; Golding, B.T.; Griffin, R.J.; Curtin, N.J. Potentiation of anticancer agent cytotoxicity by the potent poly(ADP-ribose) polymerase inhibitors, NU1025 and NU1064. *Br. J. Cancer* **1998**, *78*, 1269–1277. [CrossRef] [PubMed]
13. Bowman, K.J.; Newell, D.R.; Calvert, A.H.; Curtin, N.J. Differential effects of the poly(ADP-ribose) polymerase (PARP) inhibitor NU1025 on topoisomerase I and II inhibitor cytotoxicity. *Br. J Cancer* **2001**, *84*, 106–112. [CrossRef] [PubMed]
14. Griffin, R.J.; Srinivasan, S.; White, A.W.; Bowman, K.; Calvert, A.H.; Curtin, N.J.; Newell, D.R.; Golding, B.T. Novel benzimidazole and quinazolinone inhibitors of the DNA repair enzyme, poly (ADP-ribose)polymerase. *Pharm. Pharmacol. Commun.* **1996**, *2*, 43–47.
15. Wang, Z.-Q.; Auer, B.; Stingl, L.; Berghammer, H.; Haidacher, D.; Schweiger, M.; Wagner, E.R. Mice lacking ADPRT and poly(ADP-ribosyl)ation develop normally but are susceptable to skin disease. *Genes Dev.* **1995**, *9*, 509–520. [CrossRef]
16. Menissier de Murcia, J.; Masson, M.; Trucco, C.; Dantzer, F.; Oliver, J.; Flatter, E.; Niedergang, C.; Ricoul, M.; Dutrillaux, B.; Dierich, A.; et al. Requirement of poly(ADP-ribose)polymerase in recovery from DNA damage in mice and in cells. *Proc. Natl. Acad. Sci. USA* **1997**, *94*, 7303–7307. [CrossRef]
17. Masutani, M.; Suzuki, H.; Kamada, N.; Watanabe, M.; Ueda, O.; Nozaki, T.; Jishage, K.; Watanabe, T.; Sugimoto, T.; Nakagama, H.; et al. Poly(ADP-ribose) polymerase gene disruption conferred mice resistant to streptozotocin-induced diabetes. *Proc. Natl. Acad. Sci. USA* **1999**, *96*, 2301–2304. [CrossRef]
18. Satoh, M.S.; Lindahl, T. Role of poly(ADP-ribose) formation in DNA repair. *Nature* **1992**, *356*, 356–358. [CrossRef]
19. Murai, J.; Huang, S.Y.; Das, B.B.; Renaud, A.; Zhang, Y.; Doroshow, J.H.; Ji, J.; Takeda, S.; Pommier, Y. Trapping of PARP1 and PARP2 by Clinical PARP Inhibitors. *Cancer Res.* **2012**, *72*, 5588–5599. [CrossRef]
20. Amé, J.C.; Rolli, V.; Schreiber, V.; Niedergang, C.; Apiou, F.; Decker, P.; Muller, S.; Höger, T.; Ménissier-de Murcia, J.; de Murcia, G. PARP-2, A novel mammalian DNA damage-dependent poly(ADP-ribose) polymerase. *J. Biol. Chem.* **1999**, *274*, 17860–17868. [CrossRef]
21. Schreiber, V.; Dantzer, F.; Ame, J.C.; de Murcia, G. Poly(ADP-ribose): Novel functions for an old molecule. *Nat. Rev. Mol. Cell Biol.* **2006**, *7*, 517–528. [CrossRef] [PubMed]
22. Ménissier de Murcia, J.; Ricoul, M.; Tartier, L.; Niedergang, C.; Huber, A.; Dantzer, F.; Schreiber, V.; Amé, J.C.; Dierich, A.; LeMeur, M.; et al. Functional interaction between PARP-1 and PARP-2 in chromosome stability and embryonic development in mouse. *EMBO J.* **2003**, *22*, 2255–2263. [CrossRef] [PubMed]
23. Ruf, A.; Mennissier de Murcia, J.; de Murcia, G.; Schulz, G.E. Structure of the catalytic fragment of poly(AD-ribose) polymerase from chicken. *Proc. Natl. Acad. Sci. USA* **1996**, *93*, 7481–7485. [CrossRef] [PubMed]
24. Calabrese, C.R.; Batey, M.A.; Thomas, H.D.; Durkacz, B.D.; Wang, L.-Z.; Kyle, S.; Skalitzky, D.; Li, L.; Boritzki, T.; Maegley, K.; et al. Identification of potent non-toxic poly(ADP-ribose) polymerase-1 (PARP-1) inhibitors: Chemopotentiation and pharmacological studies. *Clin. Cancer Res.* **2003**, *9*, 2711–2718. [PubMed]
25. Canan Koch, S.S.; Thoresen, L.H.; Tikhe, J.G.; Maegley, K.A.; Almassy, R.J.; Li, J.; Yu, X.-H.; Zook, S.E.; Kumpf, R.A.; Zhang, C.; et al. Novel Tricyclic Poly(ADP-Ribose) Polymerase-1 Inhibitors with Potent Anticancer Chemopotentiating Activity: Design, Synthesis, and X-ray Co-Crystal Structure. *J. Med. Chem.* **2002**, *45*, 4961–4974. [CrossRef]
26. Skalitzky, D.J.; Marakovits, J.T.; Maegley, K.A.; Ekker, A.; Yu, X.-H.; Hostomsky, Z.; Webber, S.E.; Eastman, B.W.; Almassy, R.J.; Li, J.; et al. Tricyclic benzimidazoles as potent PARP-1 inhibitors. *J. Med. Chem.* **2003**, *46*, 210–213. [CrossRef]
27. White, A.W.; Curtin, N.J.; Eastman, B.W.; Golding, B.T.; Hostomsky, Z.; Kyle, S.; Li, J.; Maegley, K.; Skalitzky, D.; Webber, S.E.; et al. Potentiation of Cytotoxic Drug Activity in Human Tumour Cell Lines, by Amine-Substituted 2-Arylbenzimidazole-4-carboxamide PARP-1 Inhibitors. *Biooranic Med. Chem. Lett.* **2004**, *14*, 2433–2437. [CrossRef]

28. Tikhe, J.G.; Webber, S.E.; Hosomsky, Z.; Maegley, K.A.; Ekkers, A.E.; Li, J.; Yu, X.-H.; Almassy, R.J.; Kumpf, R.A.; Boritzki, T.J.; et al. Design, synthesis and evaluation of tricyclic inhibitors of poly(ADP-ribose) polymerase. *J. Med. Chem.* **2004**, *47*, 5467–5481. [CrossRef]
29. Calabrese, C.R.; Almassy, R.; Barton, S.; Batey, M.A.; Calvert, A.H.; Canan-Koch, S.; Durkacz, B.W.; Hostomsky, Z.; Kumpf, R.A.; Kyle, S.; et al. Preclinical evaluation of a novel poly(ADP-ribose) polymerase-1 (PARP-1) inhibitor, AG14361, with significant anticancer chemo- and radio-sensitization activity. *JNCI* **2004**, *96*, 56–67. [CrossRef]
30. Curtin, N.J.; Wang, L.-Z.; Yiakouvaki, A.; Kyle, S.; Arris, C.E.; Canan-Koch, S.; Webber, S.E.; Durkacz, B.W.; Calvert, A.H.; Newell, D.R.; et al. Novel PARP-1 inhibitor, AG14361, restores sensitivity to temozolomide in mismatch repair deficient cells. *Clin. Cancer Res.* **2004**, *10*, 881–889. [CrossRef]
31. Smith, L.M.; Willmore, E.; Austin, C.A.; Curtin, N.J. The novel PARP-1 inhibitor, AG14361, sensitises cells to topoisomerase I poisons by increasing the persistence of DNA strand breaks. *Clin. Cancer Res.* **2005**, *11*, 8449–8457. [CrossRef] [PubMed]
32. Horsman, M.R.; Chaplin, D.J.; Brown, J.M. Tumor radiosensitization by nicotinamide: A result of improved perfusion and oxygenation. *Radiat Res.* **1989**, *118*, 139–150. [CrossRef] [PubMed]
33. Senra, J.M.; Telfer, B.A.; Cherry, K.E.; McCrudden, C.M.; Hirst, D.G.; O'Connor, M.J.; Wedge, S.R.; Stratford, I.J. Inhibition of PARP-1 by olaparib (AZD2281) increases the radiosensitivity of a lung tumor xenograft. *Mol. Cancer Ther.* **2011**, *10*, 1949–1958. [CrossRef] [PubMed]
34. Available online: https://www.ema.europa.eu/en/documents/variation-report/lynparza-h-c-3726-ii-0020-epar-assessment-report-variation_en.pdf (accessed on 28 February 2020).
35. Thomas, H.D.; Calabrese, C.R.; Batey, M.A.; Canan, S.; Hostomsky, Z.; Kyle, S.; Maegley, K.A.; Newell, D.R.; Skalitzky, D.; Wang, L.-Z.; et al. Preclinical selection of a novel poly(ADP-ribose) polymerase (PARP) inhibitor for clinical trial. *Mol. Cancer Ther.* **2007**, *6*, 945–956. [CrossRef] [PubMed]
36. Ali, M.; Telfer, B.A.; McCrudden, C.; O'Rourke, M.; Thomas, H.D.; Kamjoo, M.; Kyle, S.; Robson, T.; Shaw, C.; Hirst, D.G.; et al. Vasoactivity of AG014699, a clinically active small molecule inhibitor of poly(ADP-ribose) polymerase: A contributory factor to chemopotentiation *in vivo*? *Clin. Cancer Res.* **2009**, *15*, 6106–6112. [CrossRef] [PubMed]
37. Pfeiffer, R.; Brabeck, C.; Burkle, A. Quantitiative nonisotopic immune-dot-blot method for the assessment of cellular poly(ADP-ribosyl)ation capacity. *Anal. Biochem.* **1999**, *275*, 118–122. [CrossRef] [PubMed]
38. Plummer, E.R.; Middleton, M.R.; Jones, C.; Olsen, A.; Hickson, I.; McHugh, P.; Margison, G.; McGown, G.; Thorncroft, M.; Watson, A.J.; et al. Temozolomide pharmacodynamics in patients with metastatic melanoma: DNA damage and activity of repair enzymes ATase and PARP-1. *Clin. Cancer Res.* **2005**, *11*, 3402–3409. [CrossRef]
39. Plummer, R.; Jones, C.; Middleton, M.; Wilson, R.; Evans, J.; Olsen, A.; Curtin, N.; Boddy, A.; McHugh, P.; Newell, D.; et al. Phase I Study of the Poly(ADP-Ribose) Polymerase Inhibitor, AG014699, in Combination with Temozolomide in Patients with Advanced Solid Tumors. *Clin. Cancer Res.* **2008**, *14*, 7917–7923. [CrossRef]
40. Zaremba, T.; Thomas, H.D.; Cole, M.; Coulthard, S.A.; Plummer, E.R.; Curtin, N.J. Poly(ADP-ribose) polymerase-1 (PARP-1) pharmacogenetics, activity and expression analysis in cancer patients and healthy volunteers. *Biochem. J.* **2011**, *436*, 671–679. [CrossRef]
41. Plummer, R.; Lorigan, P.; Steven, N.; Scott, L.; Middleton, M.R.; Wilson, R.H.; Mulligan, E.; Curtin, N.; Wang, D.; Dewji, R.; et al. (A phase II study of the potent PARP inhibitor, Rucaparib (PF-01367338, AG014699), with temozolomide in patients with metastatic melanoma demonstrating evidence of chemopotentiation. *Cancer Chemotherpay Pharmacol.* **2013**, *71*, 71–1199. [CrossRef]
42. Khan, O.; Middleton, M.R. The therapeutic potential of O6-alkylguanine DNA alkyltransferase inhibitors. *Expert Opin. Investig. Drugs.* **2007**, *16*, 1573–1584. [CrossRef] [PubMed]
43. Lindahl, T.; Satoh, M.S.; Poirier, G.G.; Klungland, A. Post-translational modification of poly(ADP-ribose) polymerase induced by DNA strand breaks. *Trends Biochem. Sci.* **1995**, *20*, 405–411. [CrossRef]
44. El-Khamisy, S.F.; Masutani, M.; Suzuki, H.; Caldecott, K.W. A requirement for PARP-1 for the assembly or stability of XRCC1 nuclear foci at sites of oxidative DNA damage. *Nucleic Acids Res.* **2003**, *31*, 5526–5533. [CrossRef] [PubMed]

45. Chen, J.; Silver, D.P.; Walpita, D.; Cantor, S.B.; Gazdar, A.F.; Tomlinson, G.; Couch, F.J.; Weber, B.L.; Ashley, T.; Livingston, D.M.; et al. Stable interaction between the products of the BRCA1 and BRCA2 tumor suppressor genes in mitotic and meiotic cells. *Mol. Cell.* **1998**, *2*, 317–328. [CrossRef]
46. Liu, Y.; West, S.C. Distinct functions of BRCA1 and BRCA2 in double-strand break repair. *Breast Cancer Res.* **2002**, *4*, 9–13. [CrossRef] [PubMed]
47. Venkitaraman, A.R. Functions of BRCA1 and BRCA2 in the biological response to DNA damage. *J. Cell Sci.* **2001**, *114 Pt 20*, 3591–3598.
48. Venkitaraman, A.R. Cancer susceptibility and the functions of BRCA1 and BRCA2. *Cell* **2002**, *108*, 171–182. [CrossRef]
49. Bryant, H.E.; Schultz, N.; Thomas, H.D.; Parker, K.M.; Flower, D.; Lopez, E.; Kyle, S.; Meuth, M.; Curtin, N.J.; Helleday, T. Specific killing of BRCA2-deficient tumours with inhibitors of poly(ADP-ribose)polymerase. *Nature* **2005**, *434*, 913–917. [CrossRef]
50. De Soto, J.A.; Wang, X.; Tominaga, Y.; Wang, R.H.; Cao, L.; Qiao, W.; Li, C.; Xu, X.; Skoumbourdis, A.P.; Prindiville, S.A.; et al. The inhibition and treatment of breast cancer with poly (ADP-ribose) polymerase (PARP-1) inhibitors. *Int. J. Biol. Sci.* **2006**, *2*, 179–185. [CrossRef]
51. Gallmeier, E.; Kern, S.E. Absence of specific cell killing of the BRCA2-deficient human cancer cell line CAPAN1 by poly(ADP-ribose) polymerase inhibition. *Cancer Biol. Ther.* **2005**, *4*, 703–706. [CrossRef]
52. Drew, Y.; Mulligan, E.A.; Vong, W.-T.; Thomas, H.D.; Kahn, S.; Kyle, S.; Mukhopadhyay, A.; Los, G.; Hostomsky, Z.; Plummer, E.R.; et al. Therapeutic potential of PARP inhibitor AG014699 in human cancer with mutated or methylated BRCA. *JNCI* **2011**, *103*, 334–346. [CrossRef] [PubMed]
53. Mukhopadhyay, A.; Elatter, A.; Cerbinskaite, A.; Wilkinson, S.J.; Drew, Y.; Kyle, S.; Los, G.; Hostomsky, Z.; Edmondson, R.J.; Curtin, N.J. Development of a functional assay for homologous recombination status in primary cultures of epithelial ovarian tumor and correlation with sensitivity to PARP inhibitors. *Clin. Cancer Res.* **2010**, *16*, 2344–2351. [CrossRef] [PubMed]
54. The Cancer Genome Atlas Research Network. Integrated genomic analyses of ovarian carcinoma. *Nature* **2011**, *474*, 609–615. [CrossRef] [PubMed]
55. Drew, Y.; Ledermann, J.; Hall, G.; Rea, D.; Glasspool, R.; Highley, M.; Jayson, G.; Sludden, J.; Murray, J.; Jamieson, D.; et al. Phase 2 multicentre trial investigating intermittent and continuous dosing schedules of the poly(ADP-ribose) polymerase inhibitor rucaparib in germline BRCA mutation carriers with advanced ovarian and breast cancer. *Br. J. Cancer.* **2016**, *114*, 723–730. [CrossRef] [PubMed]
56. Fong, P.C.; Boss, D.S.; Yap, T.A.; Tutt, A.; Wu, P.; Mergui-Roelvink, M.; Mortimer, P.; Swaisland, H.; Lau, A.; O'Connor, M.J.; et al. Inhibition of poly(ADP-ribose) polymerase in tumors from BRCA mutation carriers. *N. Engl. J. Med.* **2009**, *361*, 123–134. [CrossRef]
57. Murray, J.; Thomas, H.; Berry, P.; Kyle, S.; Patterson, M.; Jones, C.; Los, G.; Hostomsky, Z.; Plummer, E.R.; Boddy, A.; et al. Tumour cell retention of rucaparib, sustained PARP inhibition and efficacy of weekly as well as daily schedules. *Br. J. Cancer* **2014**, *110*, 1977–1984. [CrossRef]
58. Wilson, R.H.; Evans, T.J.; Middleton, M.R.; Molife, L.R.; Spicer, J.; Dieras, V.; Roxburgh, P.; Giordano, H.; Jaw-Tsai, S.; Goble, S.; et al. A phase I study of intravenous and oral rucaparib in combination with chemotherapy in patients with advanced solid tumours. *Br. J. Cancer.* **2017**, *116*, 884–892. [CrossRef]
59. Evers, B.; Drost, R.; Schut, E.; de Bruin, M.; van der Burg, E.; Derksen, P.W.; Holstege, H.; Liu, X.; van Drunen, E.; Beverloo, H.B.; et al. Selective inhibition of BRCA2-deficient mammary tumor cell growth by AZD2281 and cisplatin. *Clin. Cancer Res.* **2008**, *14*, 3916–3925. [CrossRef]
60. Kristeleit, R.; Shapiro, G.I.; Burris, H.A.; Oza, A.M.; LoRusso, P.; Patel, M.R.; Domchek, S.M.; Balmaña, J.; Drew, Y.; Chen, L.M.; et al. A Phase I-II Study of the Oral PARP Inhibitor Rucaparib in Patients with Germline *BRCA1/2*-Mutated Ovarian Carcinoma or Other Solid Tumors. *Clin. Cancer Res.* **2017**, *23*, 4095–4106. [CrossRef]
61. Oza, A.M.; Tinker, A.V.; Oaknin, A.; Shapira-Frommer, R.; McNeish, I.A.; Swisher, E.M.; Ray-Coquard, I.; Bell-McGuinn, K.; Coleman, R.L.; O'Malley, D.M.; et al. Antitumor activity and safety of the PARP inhibitor rucaparib in patients with high-grade ovarian carcinoma and a germline or somatic BRCA1 or BRCA2 mutation: Integrated analysis of data from Study 10 and ARIEL2. *Gynecol. Oncol.* **2017**, *147*, 267–275. [CrossRef]

62. Gelmon, K.A.; Tischkowitz, M.; Mackay, H.; Swenerton, K.; Robidoux, A.; Tonkin, K.; Hirte, H.; Huntsman, D.; Clemons, M.; Gilks, B.; et al. Olaparib in patients with recurrent high-grade serous or poorly differentiated ovarian carcinoma or triple-negative breast cancer: A phase 2, multicentre, open-label, non-randomised study. *Lancet Oncol.* **2011**, *12*, 852–861. [CrossRef]
63. Available online: https://www.accessdata.fda.gov/cdrh_docs/pdf16/P160018S001c.pdf (accessed on 28 February 2020).
64. Swisher, E.M.; Lin, K.K.; Oza, A.M.; Scott, C.L.; Giordano, H.; Sun, J.; Konecny, G.E.; Coleman, R.L.; Tinker, A.V.; O'Malley, D.M.; et al. Rucaparib in relapsed, platinum-sensitive high-grade ovarian carcinoma (ARIEL2 Part 1): An international, multicentre, open-label, phase 2 trial. *Lancet Oncol.* **2017**, *18*, 75–87. [CrossRef]
65. Coleman, R.L.; Oza, A.M.; Lorusso, D.; Aghajanian, C.; Oaknin, A.; Dean, A.; Colombo, N.; Weberpals, J.I.; Clamp, A.; Scambia, G.; et al. ARIEL3 investigators. Rucaparib maintenance treatment for recurrent ovarian carcinoma after response to platinum therapy (ARIEL3): A randomised, double-blind, placebo-controlled, phase 3 trial. *Lancet* **2017**, *390*, 1949–1961. [CrossRef]
66. Jonsson, P.; Bandlamudi, C.; Cheng, M.L.; Srinivasan, P.; Chavan, S.S.; Friedman, N.D.; Rosen, E.Y.; Richards, A.L.; Bouvier, N.; Selcuklu, S.D.; et al. Tumour lineage shapes BRCA-mediated phenotypes. *Nature* **2019**, *571*, 576–579. [CrossRef] [PubMed]
67. Gentles, L.; Goranov, B.; Matheson, E.; Herriott, A.; Kaufmann, A.; Hall, S.; Mukhopadhyay, A.; Drew, Y.; Curtin, N.J.; O'Donnell, R.L. Exploring the Frequency of Homologous Recombination DNA Repair Dysfunction in Multiple Cancer Types. *Cancers* **2019**, *11*, 354. [CrossRef] [PubMed]
68. Heatley, N. *Penicillin and Luck: Good Fortune in the Development of the Miracle Drug*; Huxley Scientific Press: Oxford, UK, 2012; ISBN 10 1909214000.

cancers

MDPI

Review

Role of Akt Activation in PARP Inhibitor Resistance in Cancer

Ferenc Gallyas Jr. [1,2,3,*], Balazs Sumegi [1,2,3,†] and Csaba Szabo [1,4]

1 Department of Biochemistry and Medical Chemistry, University of Pecs Medical School, 7624 Pecs, Hungary; balazs.sumegi@aok.pte.hu (B.S.); szabocsaba@aol.com (C.S.)
2 Szentagothai Research Centre, University of Pcs, 7624 Pecs, Hungary
3 HAS-UP Nuclear-Mitochondrial Interactions Research Group, 1245 Budapest, Hungary
4 Chair of Pharmacology, Department of Medicine, University of Fribourg, 1700 Fribourg, Switzerland
* Correspondence: ferenc.gallyas@aok.pte.hu; Tel.: +36-72-536-278
† Deceased on 1 August 2019.

Received: 2 January 2020; Accepted: 24 February 2020; Published: 25 February 2020

Abstract: Poly(ADP-ribose) polymerase (PARP) inhibitors have recently been introduced in the therapy of several types of cancers not responding to conventional treatments. However, *de novo* and acquired PARP inhibitor resistance is a significant limiting factor in the clinical therapy, and the underlying mechanisms are not fully understood. Activity of the cytoprotective phosphatidylinositol-3 kinase (PI3K)-Akt pathway is often increased in human cancer that could result from mutation, expressional change, or amplification of upstream growth-related factor signaling elements or elements of the Akt pathway itself. However, PARP-inhibitor-induced activation of the cytoprotective PI3K-Akt pathway is overlooked, although it likely contributes to the development of PARP inhibitor resistance. Here, we briefly summarize the biological role of the PI3K-Akt pathway. Next, we overview the significance of the PARP-Akt interplay in shock, inflammation, cardiac and cerebral reperfusion, and cancer. We also discuss a recently discovered molecular mechanism that explains how PARP inhibition induces Akt activation and may account for apoptosis resistance and mitochondrial protection in oxidative stress and in cancer.

Keywords: PARP-Akt interplay; PI3K; mTOR; cytoprotection; apoptosis resistance; oxidative stress; mitochondrial protection

---

1. Introduction

Inhibitors of the constitutive mammalian enzyme poly(ADP-ribose) polymerase (PARP)1 received renewed interest in recent years due to their importance in cancer therapy [1]. Mostly, their use is based on the synthetic lethality theory [2], i.e., to block the single-strand DNA break repair process in double-strand DNA break repair (BRCA1/2) deficient cancer cells that selectively eliminates the cancer cells via DNA damage accumulation while not affecting normal cells possessing intact double-strand DNA break repair system [3,4]. Accordingly, review articles focusing on the role of PARP inhibitors in DNA repair vastly dominate over those that at least mention the effect of PARP inhibition on kinase signaling, although the latter effect is an important limiting factor in the cancer therapy [5], and a pivotal mechanism in non-cancerous therapeutic applications of the PARP inhibitors [6].

The phosphatidylinositol-3 kinase (PI3K)-Akt-mammalian target of rapamycin (mTOR) pathway is very often activated in human cancer [7]. The pathway is often utilized by various cytokine and growth factor receptor signaling, and mutation, expressional change or amplification of any of its elements could cause resistance against tumor therapy [7,8]. Accordingly, all members of this pathway (including Akt) were suggested as targets in monotherapy or combination therapy for various cancers. In turn, several inhibitors of the pathway were approved by the Food and Drug Administration [9].

In previous research papers, we demonstrated that PARP inhibition leads to Akt activation resulting in cyto- and mitochondria protecting actions [10]. In turn, we proposed that Akt inhibitors may be used to prevent limitation of PARP inhibitor based anti-cancer therapy [11]. Moreover, we have recently unveiled a novel molecular mechanism for the Akt activating effect of PARP inhibition [12]. In the present review, we seek to increase awareness of the therapeutic limitations that the PARP-Akt interplay represent. In addition, we discuss the role of the PARP-Akt interplay in non-oncological disease conditions and their experimental models.

2. The PARP-Akt Interplay

2.1. Cellular Role and Regulation of Akt

Protein kinase B/Akt, a 57-kDa serine/threonine kinase, is regarded as a pivotal mediator of cell survival promoting effects of various growth factors and other factors, including cAMP, hypoxia, and cytokines (Figure 1). In mammals, three Akt genes were identified that are differentially expressed at both the mRNA and protein levels [13]. Akt1 can phosphorylate thereby inactivate various components of the apoptotic machinery, therefore, it is involved in cellular survival pathways. Akt2's function seems to be closely associated with the insulin receptor signaling pathway while the specific role of Akt3 is less clear [14]. Structure of Akts' amino terminus pleckstrin homology (PH) domain and central kinase domain is conserved across evolution, while the carboxyl terminus hydrophobic and proline-rich domain is more variable. The kinase activity of Akt is regulated by phosphorylation of Thr^{308} and Ser^{473} situated in the PH domain, although phosphorylation of Ser^{124} and Thr^{450} sites are needed for sensitizing Akt toward regulatory stimuli [13]. In the classical activation pathway of Akt (Figure 1), the upstream element PI3K phosphorylates phosphatidylinositol-4,5-bisphosphate to phosphatidylinositol-3,4,5-trisphosphate (PIP3) that binds to PH domain of the constitutively active 3-phosphoinositide-dependent protein kinase (PDK)1. Binding of PIP3 to PDK1 results in translocation of the enzyme to the cell membrane's inner surface, a crucially important step in Akt's activation process. PIP3's binding to Akt's PH domain renders Akt more accessible to the PDK1 mediated phosphorylation of Thr^{308} regulation site [13]. Negative regulation of the pathway is mediated by the phosphatase and tensin homolog PTEN that removes the 3-phosphate from PIP3 [15].

Besides the classical pathway, Akt can be activated by various signaling processes. PDK1 can interact with the downstream element protein kinase C related kinase2, rendering it capable to phosphorylate both Thr^{308} and Ser^{473} of Akt1 in a PIP3 dependent manner [16]. The mammalian target of rapamycin complex 2 (mTORC2) that phosphorylates Ser^{473} is a PDK1 independent crucial mechanism of Akt activation (Figure 1). Additionally, mTORC2 and cyclin-dependent kinase 2/cyclin can phosphorylate Ser^{477} and Thr^{479}, also resulting in Akt activation [17]. Integrin-linked kinase and calcium/calmodulin-dependent kinase-kinase that responds to the increases in intracellular calcium, activate Akt also by phosphorylating Ser^{473} [18,19]. When activated, Akt can phosphorylate various downstream targets that contain the consensus phosphorylation sequence RXRXXS/T, although Akt's substrate specificity also depends on sequence determinants outside the consensus site.

By phosphorylating its downstream targets, Akt can regulate metabolism, apoptosis, transcription factors, angiogenesis, and cell cycle progression (Figure 1). The constitutively active glycogen synthase kinase (GSK)-3 phosphorylates thereby inactivates glycogen synthase, the enzyme responsible for the synthesis of glycogen from excess glucose in the liver. When high glucose levels in the blood triggers insulin secretion, the insulin signaling activates Akt that inactivate GSK-3 by Ser^{9} phosphorylation enabling glycogen synthesis that contribute to decreasing the blood glucose level [20]. In parallel, Akt activation via Akt-targeted Rab GTPase-activating proteins and downstream Rab GTPases along with the input of Rac1 and actin filaments, molecular motors, and membrane fusion regulators leads to cell surface expression of glucose transporters GLUT4 that allow the entry of excess blood glucose into muscle and adipose tissues (Figure 1). Attenuation of these processes results in symptoms of metabolic syndrome and type 2 diabetes [21].

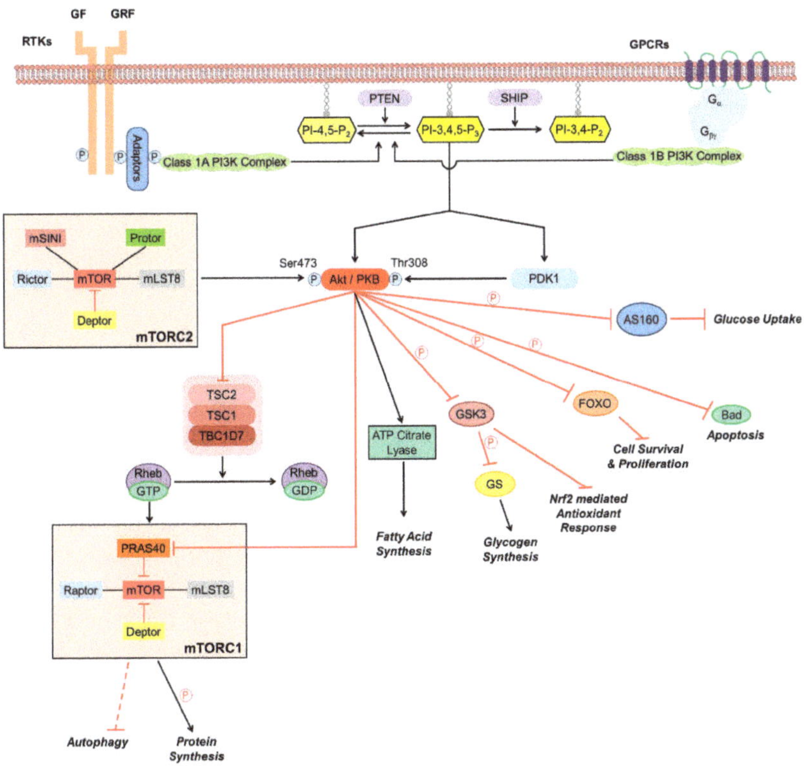

Figure 1. Schematic representation of the Akt signaling pathway. Pointed arrows denote activation while arrows with flat end represent inhibition. P indicates phosphorylation. AS160, Akt substrate of 160 kDa; Bad, Bcl-2 associated agonist of cell death; Deptor, dishevelled, Egl-10, and Pleckstrin domain-containing mTOR-interacting protein; FOXO, class O of forkhead box transcription factors; $G_{\alpha\beta\gamma}$, G protein subunits; GF, growth factor; GRF, growth-related factor; GPCR, G protein-coupled receptor; GS, glycogen synthase; GSK, GS kinase; mLST8, mammalian lethal with Sec13 protein 8; mSINI, mammalian stress-activated protein kinase interacting protein; mTOR, mammalian target of rapamycin; mTORC, mTOR complex; Nrf2, nuclear factor erythroid 2-related factor 2; PDK, 3-phosphoinositide-dependent protein kinase; PI3K, phosphatidylinositol-3 kinase; PIP_2 phosphatidylinositol-bisphosphate; PIP_3, phosphatidylinositol-3,4,5-trisphosphate; PKB, protein kinase B; PRAS40, proline-rich Akt substrate of 40 kDa; Protor, protein observed with rictor; PTEN, phosphatase and tensin homolog; Raptor, regulatory protein associated with mTOR; Rheb, Ras homolog enriched in brain; Rictor, rapamycin-insensitive companion of mTOR; RTK, Tyrosine kinase receptor; SHIP, phosphoinositide phosphatase; TBC1D7, TBC1 domain family member 7; TSC, tuberous sclerosis protein.

Activated Akt regulates various downstream signaling proteins that promote cell survival [22]. Phosphorylation thereby inactivation of the pro-apoptotic B-cell lymphoma 2 (Bcl-2) family member Bcl-2 associated agonist of cell death (Bad) by Akt prevents Bad to form heterodimers with other pro-apoptotic Bcl-2 family members, translocate to the mitochondria, and, by labilizing its outer membrane, facilitate the release of cytochrome C, apoptosis inducing factor and endonuclease G from the mitochondrial intermembrane space (Figure 1). Instead, phosphorylated Bad forms a complex with the cytoplasmic scaffolding protein 14-3-3, shifting the balance toward dominance of anti-apoptotic Bcl-2 family members, stabilization of mitochondrial integrity, and promoting cell survival [23]. Besides blocking cytochrome C mediated activation of caspase 9, Akt effectively inactivates it by directly phosphorylating its Ser^{196} [24].

Inactivation of GSK-3 by Akt is another major survival pathway since, in addition to its metabolic role, GSK-3 extensively participates in signaling pathways. Under hypoxic condition, its Tyr^{216} is phosphorylated, resulting in activation of the enzyme [25]. The activated GSK-3 contributes to hypoxic-ischemic tissue injury by multiple mechanisms. It downregulates nuclear factor erythroid 2-related factor 2 (Nrf2) expression, Nrf2 translocation from the cytoplasm to the nucleus, and Nrf2 binding to antioxidant response element (ARE) DNA sequence, thereby limiting expression of antioxidant proteins encoded in Nrf2/ARE regulated genes (Figure 1) [26]. Apart from its pro-oxidative activity, GSK-3 activation causes decreased nuclear translocation of cAMP response element-binding protein (CREB) transcription factor [27], leading to increased expression of pro-inflammatory cytokines via altering the interaction of CREB and the pro-inflammatory transcription factor nuclear factor (NF)κB with the co-activator CREB binding protein (CBP) in the nucleus [28].

After activation, Akt translocates to the nucleus where it modulates by phosphorylation the activity of various transcription factors [29]. When phosphorylated by Akt, forkhead family transcription factors are retained in the cytoplasm in form of a complex with the 14-3-3 protein and cannot promote expression of genes encoding growth factors, stress response proteins, and enzymes of carbohydrate and lipid synthesis [30]. Akt not only prevents GSK-3's negative effect on CREB activation [28], it phosphorylates CREB at its Ser^{133} that enhances CREB's binding to CBP thereby increases CREB-mediated transcription of critical survival genes [31]. Inhibitor of NFκB (IκB) retains NFκB in the cytoplasm by forming a complex with it. Phosphorylation by its kinase (IKK), causes the release of NFκB from the complex, and marks IκB for ubiquitination and proteosomal degradation. Akt regulates the process by phosphorylating IKK at Thr^{23} thereby activating it that results eventually in expression of NF-κB dependent anti-apoptotic genes [32,33]. Additionally, Akt activates mTOR, a major node of the PI3K-Akt pathway, by direct phosphorylation. Activation of mTOR trigger translation initiation factors, hypoxia and angiogenesis associated factors, and cell survival and growth associated transcription factors [34]. All in all, the PI3K-Akt pathway promotes cell proliferation and survival by various mechanisms, therefore, hyperactivation of the pathway leads to apoptosis resistance and malignant transformation [13,35].

2.2. Nuclear Effects of PARP Activation

PARP1, the original (and principal) member of a family of 18 enzymes [36], binds to various DNA structures, including single- and double-strand breaks, crossovers, cruciforms, supercoils, and some specific double-stranded sequences [37]. This binding stimulates PARP1's very low basal enzymatic activity. It cleaves its substrate, NAD^+ to nicotinamide and an ADP-ribose unit, which it covalently attaches on glutamate, aspartate, arginine, lysine, or serine residues of target proteins. PARP1 can further catalyze formation of 2′-1″-O-glycosidic or α(1‴–2″)-ADP-ribose bonds, resulting in linear or branched ADP-ribose polymer (PAR) chains, respectively [38]. PAR chains are heterogeneous either in length (up to 200 ADP-ribose units) or in the extent of branching (one branch per 20–50 ADP-ribose units), however the role of this heterogeneity has not yet been elucidated [39]. PAR chains are catabolized very shortly after their synthesis to free ADP-ribose units by exo- and endoglycosidase activities of poly-(ADP-ribose) glycohydrolases (PARGs) [39]. The mono(ADP-ribose) unit bound directly to the protein that cannot be removed by PARG is cleaved off by terminal ADP-ribose protein glycohydrolases (TARGs) [40]. The PARP1 enzyme exists in a million-copy number in the nucleus and is responsible for the majority of PARylation occurring in mammalian cells [41]. In addition to the nuclear PARP, there is also a mitochondrial form of this enzyme in many cell types, the functional role of which is currently less understood [42]. The targets of PARP1's enzymatic activity include core histones, the linker histone H1, a variety of transcription-related factors, and PARP1 itself, which is the primary target in vivo [43]. Due to its localization, activation characteristics, and enzymatic activity, PARP1 is considered to play a pivotal role in various DNA repair systems, and in the maintenance of genome integrity [44]. When PARP1 is knocked out, PARP2 takes over its role and vice versa, however the PARP1/PARP2 double knock out mutation results in embryonic lethality [41].

Recent reports indicate a scaffolding role for PARP1 in genotoxic stress induced apoptotic response [45,46]. PARP1 activated by DNA strand breaks PARylates itself and ataxia telangiectasia mutated kinase (ATM) to recruit a small ubiquitin-like modifier (SUMO)-1 ligase, PIASy, and nuclear IKKγ (NEMO). Concertedly, PIASy SUMOylates and ATM phosphorylates NEMO, leading to IKK activation and NFκB regulated resistance toward apoptosis [45].

Because of PARP1's high sensitivity toward DNA breaks and its copiousness, genotoxic stress can induce hyperactivation of the enzyme leading to deleterious consequences for the organism [47]. Experimentally induced massive DNA damage triggers excessive PARP1 activation that causes depletion of the substrate NAD^+ within minutes [47]. Therefore, the metabolic pathways that utilize NAD^+ as substrate such as glycolysis, citric acid cycle and mitochondrial respiration become compromised, resulting in reduced ATP production and cellular dysfunction. To aggravate the precarious metabolic situation, phosphoribosyl pyrophosphate synthetase and nicotinamide mononucleotide adenylyl transferase consume ATP to resynthesize NAD^+ leading to ATP depletion and predominantly necrotic cell death [48,49]. This mechanism (Figure 2) explains the detrimental effect of DNA-damaging agents, such as high dose of N-methyl-N'-nitro-N-nitrosoguanidine (MNNG) on cultured cells, and is consistent with the therapeutic efficacy of PARP-1 inhibition in animal models of oxidative stress related diseases. However, some studies indicated the necessity of including other mechanisms to explain the cytoprotective effect of PARP inhibition. Namely, MNNG was reported to cause similar NAD^+ depletion in $PARP1^{-/-}$ and $PARP1^{+/+}$ fibroblasts [50], and PARP1 inhibition failed to prevent NAD^+ and ATP depletion in carcinogen-treated hamster cells [51].

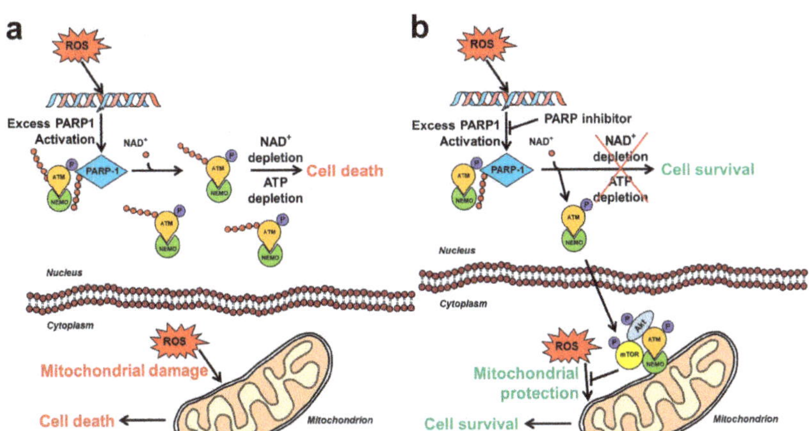

Figure 2. Schematic mechanisms of the PARP1/Akt interplay in oxidative stress. (**a**) Effects of oxidative stress in the absence of PARP inhibition. Oxidative stress causes DNA strand breaks that activate nuclear PARP. Excess PARP activity exhausts the substrate NAD^+ that impairs ATP production. Attempted restoration of NAD+ levels leads to ATP depletion and eventually cell death. ATM-NEMO complexes, even if formed, are retained in the nucleus, therefore, Akt activated cytoprotective mechanisms are inactive. The oxidative stress is free to damage the mitochondria resulting in the release of pro-apoptotic intermembrane proteins. (**b**) Effects of oxidative stress in the presence of PARP inhibition. The PARP inhibitor blocks excess PARP activity, thereby prevents NAD^+ and consequent ATP depletion. Activated ATM-NEMO complex is able to translocate to the cytoplasm and associate with the mitochondrial outer membrane. Akt and mTOR are recruited to form the ATM-NEMO-Akt-mTOR cytoprotective signalosome that prevents oxidative stress induced mitochondrial damage and activates various cell survival pathways. Pointed arrows denote activation while arrows with flat end represent inhibition. P indicates phosphorylation.

2.3. Role of the PARP-Akt Interplay in Shock and Inflammation

PARP is known to regulate a host of signal transduction pathways and PARP inhibitors have been shown to have the ability to suppress the production of various pro-inflammatory mediators (including cytokines and chemokines). These effects are thought to be important contributors to the beneficial effects of PARP inhibitors in various experimental models of local and systemic inflammation, as well as in various models of critical illness [52–59]. However, in addition to effects on pro-inflammatory signaling, PARP inhibitors' anti-inflammatory and beneficial effects in various models of inflammation and shock may also be related, at least in part, to activation of Akt. In 2003 we reported that the protective effects of the PARP inhibitor PJ34 in a model of endotoxic shock in mice (which included improved survival and protection against multiorgan dysfunction) were associated with Akt activation in various organs including the liver and the gut [60], and similar results were subsequently obtained with another PARP inhibitor, 4-hydroxyquinazoline [61]. Moreover, in a model of neuroinflammation and demyelination in mice (induced by the mitochondrial toxin cuprizone), the PARP inhibitor 4-hydroxy-quinazoline protected against many of the observed pathophysiological changes (weight loss, demyelination): these effects were associated with a modulation of Akt activation [62]. In this model, the effect of cuprizone alone was associated with an increase in Akt phosphorylation; however, Akt phosphorylation was further and markedly enhanced in the presence of the PARP inhibitor [62]. It is conceivable that Akt activation contributes to the beneficial effect of PARP inhibition in these models; however, this remains to be directly tested in future experiments.

Additional inflammatory conditions where PARP-inhibitor/Akt "connection" has been evaluated include various in vitro and/or in vivo models of arthritis and dermatitis. In an experiment by Garcia and colleagues utilizing CD95/Apo-1 (Fas)-induced apoptosis in fibroblast-like synoviocytes (FLS) from rheumatoid arthritis patients, PARP-1-deficient FLS were protected from cell death compared to wild-type cells. This was associated with the activation of the Akt-GSK survival pathway in the PARP1-deficient cells, but inhibition of PI3K/Akt pathway did not affect the difference between PARP1-competent or -deficient cells in terms of viability [63]. Thus—although in this experimental setting the same authors have demonstrated that Akt activation can, in fact confer cytoprotection [64]—in these experimental conditions, pathways other than Akt must be responsible for the beneficial effect of PARP1 deficiency. In contrast, in models of sulfur mustard induced dermatitis, the beneficial effects of the PARP inhibitor ABT-88, as well as the beneficial effects of PARP1 deficiency, appears to involve a significant modulation of the Akt pathway. In these experiments, sulfur mustard exposure decreased Akt phosphorylation in the control group, while in the PARP1/deficiency + sulfur mustard groups, a marked increase in Akt phosphorylation was observed [65].

2.4. Role of the PARP-Akt Interplay in Reperfusion Injury

In one of the earliest studies related to the subject of PARP-Akt interplay, PARP inhibitors of two different structural class (HO-3089 and 4-hydroxyquinazoline) were evaluated on energy metabolism, contractile function, and cell death of isolated rat hearts during ischemia-reperfusion. In line with earlier studies demonstrating the beneficial effects of PARP inhibitors (or genetic PARP1 deficiency) in myocardial ischemia-reperfusion [66–68], both of the tested PARP inhibitors significantly improved the recovery of high-energy phosphate intermediates in the heart [69], directly confirming the well-known PARP-activation-cellular energetic failure concept. Both of the tested PARP inhibitors also improved cardiac contractility [69]. In addition, ischemia-reperfusion induced a slight, but detectable phosphorylation of Akt at Ser437 (i.e., evidence of Akt activation). Importantly, treatment with either of the two tested PARP inhibitors induced a marked further enhancement of Akt activation [69].

Subsequent studies, utilizing similar experimental designs, confirmed and expanded these findings and directly tested the functional importance of Akt activation in the cardioprotective effect of these two PARP inhibitors (HO-3089 and 4-hydroxyquinazoline). Pharmacological inhibition of PI3-kinase by wortmannin or LY294002 reduced the PARP inhibitor-elicited increase of Akt phosphorylation during ischemia-reperfusion, and significantly diminished the PARP-inhibitor-mediated recovery of ATP and creatine phosphate in the reperfused hearts. Moreover, pharmacological inhibition of PI3-kinase also diminished the protective effect of PARP inhibitors on infarct size and the recovery of cardiac function in this model [70]. In the same study, the Akt downstream target GSK-3β was also evaluated; this effector was found to be activated by the PARP inhibitors during ischemia-reperfusion [70]. Interestingly, wortmannin or LY294002, on their own (i.e., in the absence of PARP inhibitor treatment) failed to increase infarct size or suppress cardiac function during ischemia-reperfusion, nor did they affect the basal level of Akt phosphorylation suggesting that (a) the basal and the PARP-inhibitor-induced Akt phosphorylation involves fundamentally different mechanisms; the former not being dependent on PI3-kinase, while the latter is PI3-kinase dependent and (b) the low basal level of Akt activation that occurs during ischemia-reperfusion is unable to significantly counteract cardiac reperfusion injury (as opposed to the PARP-inhibitor-induced activation of Akt, which is clearly cardioprotective) [70].

Similar findings were obtained with PARP inhibitors of various structural classes in hypoxic-reoxygenated cardiac myocytes (in vitro) and in ischemic-reperfused hearts (in vitro or in vivo) in several additional studies. For example, Palfi and colleagues evaluated the effect of the PARP inhibitor L-2286 in an ischemia-reperfusion model in Langendorff perfused rat hearts and in an isoproterenol-induced myocardial infarction model in rats. In addition to cardiac energy metabolism, oxidative damage and (in the in vivo experiment, infarct size), and the phosphorylation state of Akt and MAPK cascades were also monitored. As expected, L-2286 exerted a significant protective effect against ischemia-reperfusion-induced myocardial injury in both experimental models. In addition, in line with prior observations made with other PARP inhibitors, L-2286 enhanced the ischemia-reperfusion-induced activation of Akt. Moreover, it also induced an activation of extracellular signal-regulated kinase and p38-MAPK, while c-Jun N-terminal kinase activation was repressed [71]. Similarly, the PARP inhibitor DPQ's protective effect in another rat isolated heart model against ischemia-reperfusion was found to be associated with a marked increase in Akt phosphorylation [72]. Moreover, in H9c2 cardiomyocytes in vitro, the PARP inhibitor 5-AIQ enhanced the oxidative stress-induced activation (phosphorylation) of Akt (as well as of GSK-3β) and these actions conferred a cytoprotective effect; inhibiting the Akt/GSK-3β pathway by LY294002 significantly attenuated the cytoprotective effect of 5-AIQ [73] and similar effects were subsequently reported with the PARP inhibitor KR-33889 in the same cardiomyocyte cell line in vitro; the data presented in the paper show that presence of the PARP inhibitor extended the duration of Akt activation after hydrogen peroxide exposure [74].

The cardioprotective effect of PARP inhibitors goes beyond protection in ischemic or hypoxic cardiomyocytes or hearts, and includes storage-mediated injury, as well as various forms of chronic heart failure (including septic, diabetic, and cardiotoxic drug-induced) [75–85]. Several studies have evaluated whether this protective action in these models is also associated with (and/whether it is dependent on) Akt activation. In a model of cardiac dysfunction induced by cold storage (6 h of storage in Celsior solution), the presence of the PARP inhibitor INO-1153 in the storage fluid afforded a better subsequent functional recovery, and this effect was associated with an increased Akt activation in these hearts [86]. The cardioprotective effect of the PARP inhibitor was abrogated by wortmannin co-treatment [86]. Moreover, in a model of doxorubicin-induced heart failure, the protective effect of L-2286 was associated with an activation of Akt [87], and similar effects were noted with a PARP inhibitor of a different structural class (BPG-15) in a model of imatinib-induced cardiomyopathy [88]. In addition, in a model of "hyperglycemia" in vitro (elevated extracellular glucose exposure of H9c2 cells), siRNA-mediated PARP1 depletion was found to increase the levels of phosphorylated Akt (PARP1 depletion also provided significant cytoprotective effects) [89]. The beneficial cardiac effects of L-2286 discussed earlier were also found to extend to the post-infarction cardiac remodeling stage (8 weeks of follow-up was performed); however, at these later time points, the Akt activating effect was no longer detectable (in contrast, the protective effects were attributable to effects on various protein kinase C isoforms) [90]. However, in a model of hypertension-induced cardiac remodeling in the rat, the beneficial effects of the PARP inhibitor L-2286 against cardiac remodeling and vascular dysfunction could be attributed, at least in part, to Akt activation (and in part to effects of MAP kinase pathways) [91,92].

It should be stressed that the phenomenon of Akt-activation-mediated cardioprotection is not at all unique to PARP inhibitors. The fact that Akt activation, in general, activates cytoprotective pathways is a well-known phenomenon (see prior sections); and specifically, the cardioprotective (or, more generally, organ-protective as it also extends to various organs including the brain, kidney, and others) effect of a wide set of therapeutic interventions, including ischemic pre- and post-conditioning, calcium antagonists, beta blockers, caffeine, erythropoietin, insulin, vitamin E, and many others has been attributed, partially or wholly, to a set of cellular effects (including protection against mitochondrial dysfunction, reduction of cellular reactive oxygen species generation, induction of cytoprotective, and antiapoptotic pathways, etc.) that are, wholly or partly, the consequences of Akt activation [93–96].

The phenomenon of PARP-inhibitor mediated Akt activation (and associated cytoprotective and organ-protective effects) has also been demonstrated in various non-cardiac experimental systems exposed to ischemia or hypoxia. For instance, the first-generation PARP inhibitor 3-aminobenzamide (3-AB) was found to exert beneficial effects that are (at least in part) dependent on Akt activation in a rodent model of ischemic stroke. In Sprague-Dawley rats subjected to middle cerebral artery occlusion, 3-AB increased and prolonged Akt activation in the brain, and this effect was associated with smaller infarct volume, lower brain water content, less TUNEL staining, less cleavage of caspases, and lower induction of various adhesion molecules [97]. Similarly, in an in vitro study, nicotinamide—which is a PARP inhibitor of low potency, but it also has a host of additional pharmacological actions [98]—exerted cytoprotective effects in an anoxia model in cultured rat hippocampal neurons; its beneficial effects were associated with Akt activation, and were attenuated by inhibition of Akt phosphorylation using pretreatment with wortmannin or LY294002 [99]. Application of wortmannin (500 nM) or LY294002 (10 μM) without anoxia was not toxic to the neurons, suggesting that basal Akt activity does not regulate cell viability or survival (in contrast to the PARP-inhibitor-inducible increase in Akt activity, which does) [99].

Acute urinary retention is a condition with a significant hypoxia-associated component. In a rat model of acute urinary retention induced by bladder distension, bladder apoptosis was reduced by inhibition of PARP activation with 3-aminobenzamide treatment. The PARP inhibitor also increased the levels of ATP and NAD^+, phosphorylation of Akt, and Bcl-2/Bax ratio, and significantly reduced the activation of caspase 3 [100]. Since in this study it was not evaluated whether pharmacological inhibition of Akt activation counteracts the effects of 3-aminobenzamide, it is not possible to determine to what extent Akt activation contributed to the observed bladder-protective effects.

The cytoprotective effect of PARP inhibition (using 4-hydroxy-quinazoline) was also associated with Akt activation in an acute rat kidney rejection model [101]. Moreover, the cytoprotective effect of a clinically approved (for oncological indications) PARP inhibitor (olaparib) was also associated with Akt activation in hypoxia-reoxygenation induced acute retinal injury model [102]. However, similarly to the acute urinary retention experiments discussed above, the functional importance of Akt activation was not directly tested in these studies. In the latter model, olaparib treatment induced alterations in a host of additional pathways (MAPKs, HIF1α, Nrf2, and NFκB) [102]; thus, it is conceivable that these effects may have also contributed to its beneficial effects.

2.5. Mechanism of PARP Inhibition Induced Akt Activation

All aforementioned in vitro, ex vivo, and in vivo results indicate that inhibition of nuclear protein PARylation reaction catalyzed by PARP1 somehow triggers the activation of cytoplasmically localized Akt. Recently, we provided in vitro experimental evidences for a mechanism (Figure 2), which may account for the Akt activating effect of PARP1 inhibition under oxidative stress conditions, and is consistent with the experimental and clinical data on PARP inhibitors [12]. According to it, PARP1 activated by DNA strand breaks PARylates itself, thereby creating a scaffold that recruits ataxia telangiectasia mutated kinase (ATM) and NEMO [45]. Binding to PAR strands activates ATM that phosphorylates itself and NEMO [46]. If excess PARP activation is prevented by a pharmacological inhibitor or genetic manipulation, ATM is not PARylated [103], translocates to the cytoplasm in the form of a complex with NEMO, and at least in part associates with the mitochondrial outer membrane. There, the complex recruits Akt and mTOR, and together with them forms a signalosome that activates Akt, leading to activation of downstream pro-survival pathways (Figure 2) [12].

Several elements of this model have been described previously. Independent studies reported that ATM mediates Akt activation in cancer cell lines, in insulin treated myocytes, and under cellular stress conditions [104–106]. Defective ATM linked activation of Akt was found to be involved in diabetes and neurodegenerative diseases [106]. Furthermore, direct PARylation of ATM by PARP1 was indicated in independent reports [103,107]. PARP activity dependent formation and nuclear-to-cytosolic translocation of ATM-NEMO complex was suggested as the molecular mechanism underlying genotoxicity induced NFκB regulated apoptosis [45,108]. As for the model's remaining elements, we demonstrated an at least partial mitochondrial localization for the proposed ATM-NEMO-Akt-mTOR signalosome, and proved that PARP1 inhibitor's cytoprotective effect is abolished by silencing any components of the signalosome [12]. As a partial independent confirmation of latter results, PARP1 inhibition was shown to induce synthetic lethality in ATM deficient cells [109]. However, by showing that continuously active Akt rescued these cells [12], we demonstrated that the said synthetic lethality is predominantly resulted from insufficient Akt activation in ATM suppressed cells. These data emphasize the crucial importance of Akt activation in PARP1 inhibition's cytoprotective property that can lead to undermining PARP1 inhibition's cytostatic effect on breast cancer gene (BRCA) mutations carrying cancers [110]. Independent reports support this notion by demonstrating the increase of PARP1 inhibitors' cytotoxicity by Akt inhibitors [8,11,111].

3. PARP-Akt Interactions in Cancer Biology

Based on PARP1's role in DNA repair, early prognoses for the role of its inhibitors in cancer therapy were to enhance antitumor activity of radio- or chemotherapy [112]. Although these prognoses have proved to be correct for combination therapy, present use of PARP inhibitors is mostly based on their synthetic lethality with homologous recombination repair deficiency of DNA double strand breaks [113,114]. PARP1 binds to DNA strand breaks, and is trapped on the DNA when inactivated by the inhibitor causing replication fork collapse [115]. The resulting DNA double strand breaks are repaired by the error prone non-homologous end joining repair system when homologous recombination is defective due to mutation in its key elements, such as BRCA1 and BRCA2 [113,114]. Both mechanisms are deleterious for homologous recombination deficient cancer cells. Accordingly, in 2014, the United States Food and Drug Administration (FDA) approved the first PARP inhibitor, olaparib monotherapy for treating germline BRCA1/2 mutated, advanced stage ovarian cancer [116]. Since that time, three additional PARP inhibitors were approved and the therapy was extended to other types of BRCA deficient cancers (Table 1). Clinical trials expanded the use of PARP inhibitor monotherapy to cancers with defects other than BRCA mutation in the double-strand repair pathway and other forms of genomic instability, since these cancers seem to be more dependent on PARP1 to maintain genomic integrity [117].

Combination therapy uses PARP inhibitors together with radiotherapy, immune therapy, a cytotoxic agent, an angiogenesis blocker, a signaling pathway inhibitor, or a blocker of the DNA damage-response pathway [118]. In these therapeutic strategies, the PARP inhibitor sensitizes the cancer cells for the cytostatic effect of the agent(s) used in combination with it by limiting DNA damage repair. Alternatively, a targeted treatment, such as a signaling pathway inhibitor, could relieve the resistance of tumor cells toward the PARP inhibitor [119]. Up to now, no combination therapy with PARP inhibition was approved by FDA for the clinical practice, although a number of ongoing clinical trials utilize this strategy for various types of cancers [118].

Table 1. Food and drug administration (FDA) approved inhibitors of enzymes relevant for the PARP-Akt interplay.

Target	Name	Company	Cancer Type
mTOR	Everolimus	Novartis	Neuroendocrine tumor, breast cancer, gastrointestinal and lung neuroendocrine tumor
mTOR	Sirolimus	Pfizer	Lymphangioleiomyomatosis
mTOR	Temsirolimus	Wyeth	Renal cell carcinoma
PI3Kδ	Idelalisib	Gilead Sciences	Chronic lymphocytic leukemia, follicular B-cell non-Hodgkin lymphoma, small lymphocytic lymphoma
PI3Kδγ	Duvelisib	Intellikine	Chronic lymphocytic leukemia/small lymphocytic lymphoma
PI3Kα	Alpelisib	Novartis	HR-positive and HER2/neu-negative breast cancer
PI3Kαδ	Copanlisib	Bayer	Relapsed follicular lymphoma
PARP1/2	Oliparib	AstraZeneca	BRCA1/2 mutated ovarian, breast, and prostate cancer
PARP1/2	Rucaparib	Clovis	BRCA1/2 mutated ovarian cancer
PARP1/2	Niraparib	Tesaro	Recurrent epithelial ovarian, fallopian tube, or primary peritoneal cancer
PARP1/2	Talazoparib	Pfizer	BRCA1/2 mutated metastatic breast cancer

Solid tumors grow in hypoxic condition and undergo metabolic reprograming to provide energy and nutrients for proliferation and survival. Accordingly, increased anaerobic glycolysis, modified oxidative phosphorylation, and mitochondrial remodeling are hallmarks of cancer cells' phenotype [120]. This metabolic reprograming is often accompanied by an elevated production of reactive oxygen and nitrogen species (RONS) that can directly or indirectly activate Akt [121]. Additionally, various genetic events lead to activation of this pro-survival pathway. Activating mutations in PI3K, mTOR, and the Akt isoforms, and loss-of-function mutations and deletions in PTEN occur in about 27%, 8%, 3%, and 19% of all cancer cases, respectively [122]. PIK3CA encodes the p110α catalytic subunit of PI3K, and is the most frequently altered oncogene in human cancers, including endometrial, ovarian, colorectal, and breast cancers [123]. Mutations of mTOR and Akt are much less frequent, although, they were demonstrated in melanoma, renal carcinoma, bladder tumor, lung cancer, breast cancers, head and neck squamous cell carcinomas, and endometrial cancer [124]. PTEN loss represent the second most mutated tumor suppressor gene frequently occurring in various human cancers including colorectal cancer, breast cancer, glioblastoma, endometrial cancer, malignant melanoma, and prostate cancer [125]. Although their mutation frequency is quite low, gene amplification and overexpression of Akt isoforms occur more frequently. Akt1 gene amplification was found in breast, colon, pancreatic, gastric, esophageal, ovarian, and thyroid cancers, and glioblastoma while amplification of Akt2 occurred in pancreatic, ovarian, and breast cancers [126–128]. Akt2 overexpression was present in 40% of hepatocellular carcinoma and 57% in colorectal cancers, while Akt3 overexpression was found in prostate and breast cancers [126–128]. Furthermore, Akt was hyperactivated in many human cancers due to activating mutations of upstream elements such as PI3K and RAS, or loss of function mutations of p27, PTEN [129,130], and inositol polyphosphate-4-phosphatase2 [131]. An additional activating mechanism, N^6-methyladenosine mRNA methylation of PTEN, leads to Akt activation in many cancers [132]. Certainly, Akt activation can also be triggered by upstream signaling elements, such as growth factors or cross-talking signaling pathways [130], or via the retrograde activating mTOR-PI3K feedback loop [133]. However, PI3K/Akt pathway's oxidative regulation via glutathion, peroxiredoxins, glutaredoxins, and oxidative inactivation of PTEN remains a significant regulatory factor of Akt mediated malignant transformation [134,135]. Although inhibitors of the pathway are attractive agents in the cancer therapy, so far no Akt inhibitor has been approved by FDA. On the other hand, FDA approved four PI3K and three mTOR inhibitors (Table 1) for clinical cancer therapy, and two Akt inhibitors are in phase III clinical trials for combination therapy of breast and colorectal cancers [124].

One of the most important limitations of PARP inhibition based cytostatic therapy is the resistance of cancer cells against either PARP inhibition monotherapy or combination therapy. Various mechanisms can account for this resistance, including down-regulation of PARP1 expression, reduced cellular availability of the inhibitor, reactivation of the homologous recombination DNA repair pathway, exploiting altered cell cycle regulation, and changing the miRNA environment [136]. Additionally, activation of the PI3K-Akt pathway by PARP inhibition, a mechanism originally suggested for reducing side effects of platinum compounds and alkylating agents [10], represent a significant limitation of PARP inhibitor utilizing chemotherapy by protecting mitochondrial integrity and function, and mitigating apoptosis in the cancer cells (Figure 2). To support this notion, activation of Akt increased paclitaxel resistance in vitro [11], and negative synergism was found between the PARP inhibitor PJ-34 and cisplatin or temozolomide during a short-term combination treatment of B16F10 metastasizing melanoma cell line [137]. Additionally, in clinical trials, inhibitors of the P13K/AKT pathway potentiated cytostatic effect of PARP inhibitors in a combination therapy, indicating the clinical relevance of the concept [8,111].

4. Open Questions and Future Directions

At present, using PARP inhibitors in homologous recombination deficient malignancies is a well-established practice. Judging from the increasing number of clinical trials, extension of the strategy of synthetic lethality to more types of cancers can be expected. There are a number of ongoing clinical trials aiming at improving the efficacy of PARP inhibitors by combining them with platinum, taxane or alkylating chemotherapy, radiotherapy, and inhibitors of signaling pathways [138]. They intend to exploit synergistic effects of the agents or to overcome cytostatic resistance of the cancer cells. In the latter respect, ataxia telangiectasia and Rad3-related protein, mTOR, and NFκB pathway inhibitors were found to be effective [119,139,140].

Another line of improving therapeutic efficacy of PARP inhibition addresses removal of PAR polymers. Importance of the process is highlighted by the finding that complete PARG deficiency results in embryonic lethality [141]. Furthermore, PARG inhibition together with homologous recombination deficiency results in synthetic lethality [142], the same way as PARP1 inhibition does. Mutation of TARG1, the other enzyme that together with PARG removes PAR polymers from the target proteins, was indicated in a neurodegenerative and seizure disorder, and its deficiency sensitizes the cells to DNA alkylating agents [143]. These data suggest that removal of PAR polymers could be essential in DNA double strand break repair, and the PARG and TARG enzymes may represent targets for cancer therapy. More research and specific inhibitors are needed to elucidate the particulars concerning these enzymes.

Recent clinical trials aim to expand therapeutic application of PARP inhibitors to cancers of intact homologous recombination repair. To this end, the PARP inhibitors are proposed to be applied in combination with immune checkpoint blockade. The cancer cells are frequently resistant against immune checkpoint blockade that could be alleviated by the PARP inhibitor via promoting cross-presentation and modifying immune microenvironment. The PARP inhibitors are expected to increase T cell's tumor-killing efficacy, and to activate the cancer-immunity cycle thereby increasing the sensitivity of tumor cells toward immune checkpoint blockade [144]. However, considering the diversity of the involved mechanisms, extensive research is needed to elucidate the possible interference by various networks including the PI3K-Akt cytoprotective signaling pathway.

With the advancement of whole exome sequencing, screening of cancer-related genetic aberrations is possible at a reasonable price. The technique provides genome-wide, and high-throughput results that can be used to identify synthetic lethal pairs involving epigenetic-related synthetic lethal genes. For the development of novel rational combination therapies, enhanced genetic-interaction screens are needed that assumes integration of functional interaction data with orthogonal methods [145]. Additionally, heterogeneity of the patient population limits therapy development, and questions optimality of the applied treatment, a problem which can be amended by determining microsatellite instability and tumor mutation burden [146]. All these techniques, as they become affordable, pave the way toward the realm of personalized medicine [147].

5. Conclusions

Accumulating clinical evidence indicates that PARP inhibitors can be successfully applied in cancers not responding to conventional treatments. To date, therapeutic application of the FDA approved PARP inhibitors utilize homologous recombination repair deficiency in these tumors. At the same time, ongoing clinical trials combining PARP inhibitors with agents targeting non-homologous recombination DNA repair systems, signaling pathways, angiogenesis, or immune checkpoint mechanisms intend to extend the therapeutic potential of PARP inhibitors well beyond their present one. However, because of complexity and redundancy of the mechanisms that regulate PARP activity, we still fail to comprehend fully the processes leading to de novo and acquired resistance toward PARP inhibitor therapy. As in vitro and in vivo experimental evidences indicate, the cytoprotective PI3K-Akt pathway is activated by PARP1 inhibition, and via multiple mechanisms, its activation induce apoptosis resistance and mitochondrial protection that limit cytostatic efficacy of PARP inhibitor mono- or combination therapy. Because of the importance of the PI3K-Akt pathway in all branches of medicine associated with oxidative stress including shock, inflammation, cardiac, and cerebral reperfusion injury besides cancer, extensive research is going on to fully comprehend its mechanisms and its interplay with other signaling pathways including those regulated by PARP inhibition. The availability of FDA approved PI3K-Akt pathway inhibitors should certainly facilitate these research efforts. Additionally, recent advancements in the field of next generation sequencing and bioinformatics could contribute to improving treatment strategies and outcomes of PARP inhibitor therapy.

Author Contributions: All authors have read and agreed to the published version of the manuscript. Conceptualization, S.B.; writing—original draft preparation, F.G.J. and C.S; writing—review and editing, F.G.J.; funding acquisition, F.G.J., B.S. and C.S.

Funding: This research was funded by Hungarian grants GINOP-2.3.3-15-2016-00025, GINOP-2.3.2-15-2016-00049, EFOP-3.6.1-16-2016-00004, and Higher Education Institutional Excellence Program to F.G.J. and B.S. and by the Swiss National Foundation (SNF 310030_188525) to C.S.

Acknowledgments: We apologize to those authors whose study we could not cite due to space restriction reasons.

Conflicts of Interest: The authors declare no conflict of interest. The funders had no role in the design of the study; in the collection, analyses, or interpretation of data; in the writing of the manuscript, or in the decision to publish the results.

References

1. Yi, M.; Dong, B.; Qin, S.; Chu, Q.; Wu, K.; Luo, S. Advances and perspectives of PARP inhibitors. *Exp. Hematol. Oncol.* **2019**, *11*, 8–29. [CrossRef]
2. Wu, J.; Starr, S. Low fidelity compensatory backup alternative DNA repair pathways may unify current carcinogenesis theories. *Future Oncol.* **2014**, *10*, 1239–1253. [CrossRef]
3. Aparicio, T.; Baer, R.; Gautier, J. DNA double-strand break repair pathway choice and cancer. *DNA Repair* **2014**, *19*, 169–175. [CrossRef]
4. Kaniecki, K.; De Tullio, L.; Greene, E.C. A change of view: Homologous recombination at single-molecule resolution. *Nat. Rev. Genet.* **2018**, *19*, 191–207. [CrossRef]
5. Bai, P.; Nagy, L.; Fodor, T.; Liaudet, L.; Pacher, P. Poly (ADP-ribose) polymerases as modulators of mitochondrial activity. *Trends Endocrinol. Metab.* **2015**, *26*, 75–83. [CrossRef]
6. Henning, R.J.; Bourgeois, M.; Harbison, R.D. Poly (ADP-ribose) Polymerase (PARP) and PARP Inhibitors: Mechanisms of Action and Role in Cardiovascular Disorders. *Cardiovasc. Toxicol.* **2018**, *18*, 493–506. [CrossRef]
7. Song, M.; Bode, A.M.; Dong, Z.; Lee, M.-H. AKT as a Therapeutic Target for Cancer. *Cancer Res.* **2019**, *79*, 1019–1031. [CrossRef] [PubMed]
8. Papa, A.; Caruso, D.; Strudel, M.; Tomao, S.; Tomao, F. Update on poly-ADP-ribose polymerase inhibition for ovarian cancer treatment. *J. Transl. Med.* **2016**, *14*, 267. [CrossRef]
9. Alzahrani, A.S. PI3K/Akt/mTOR inhibitors in cancer: At the bench and bedside. *Semin. Cancer Biol.* **2019**, *59*, 125–132. [CrossRef] [PubMed]

10. Tapodi, A.; Debreceni, B.; Hanto, K.; Bognar, Z.; Wittmann, I.; Gallyas, F., Jr.; Varbiro, G.; Sumegi, B. Pivotal Role of Akt Activation in Mitochondrial Protection and Cell Survival by Poly(ADP-ribose)polymerase-1 Inhibition in Oxidative Stress. *J. Biol. Chem.* **2005**, *280*, 35767–35775. [CrossRef] [PubMed]
11. Szanto, A.; Hellebrand, E.E.; Bognar, Z.; Tucsek, Z.; Szabo, A.; Gallyas, F., Jr.; Sumegi, B.; Varbiro, G. PARP-1 inhibition-induced activation of PI-3-kinase-Akt pathway promotes resistance to taxol. *Biochem. Pharmacol.* **2009**, *77*, 1348–1357. [CrossRef] [PubMed]
12. Tapodi, A.; Bognar, Z.; Szabo, C.; Gallyas, F.; Sumegi, B. PARP inhibition induces Akt-mediated cytoprotective effects through the formation of a mitochondria-targeted phospho-ATM-NEMO-Akt-mTOR signalosome. *Biochem. Pharmacol.* **2019**, *162*, 98–108. [CrossRef] [PubMed]
13. Datta, S.R.; Brunet, A.; Greenberg, M.E. Cellular survival: A play in three akts. *Genes Dev.* **1999**, *13*, 2905–2927. [CrossRef] [PubMed]
14. Yang, Z.Z.; Tschopp, O.; Baudry, A.; Dümmler, B.; Hynx, D.; Hemmings, B.A. Physiological functions of protein kinase B/Akt. *Biochem. Soc. Trans.* **2004**, *32*, 350–354. [CrossRef] [PubMed]
15. Simpson, L.; Parsons, R. PTEN: Life as a tumor suppressor. *Exp. Cell Res.* **2001**, *264*, 29–41. [CrossRef] [PubMed]
16. Balendran, A.; Casamayor, A.; Deak, M.; Paterson, A.; Gaffney, P.; Currie, R.; Downes, C.P.; Alessi, D.R. PDK1 acquires PDK2 activity in the presence of a synthetic peptide derived from the carboxyl terminus of PRK2. *Curr. Biol.* **1999**, *9*, 393–404. [CrossRef]
17. Liu, P.; Begley, M.; Michowski, W.; Inuzuka, H.; Ginzberg, M.; Gao, D.; Tsou, P.; Gan, W.; Papa, A.; Kim, B.M.; et al. Cell-cycle-regulated activation of Akt kinase by phosphorylation at its carboxyl terminus. *Nature* **2014**, *508*, 541. [CrossRef]
18. Delcommenne, M.; Tan, C.; Gray, V.; Rue, L.; Woodgett, J.; Dedhar, S. Phosphoinositide-3-OH kinase-dependent regulation of glycogen synthase kinase 3 and protein kinase B/AKT by the integrin-linked kinase. *Proc. Natl. Acad. Sci. USA* **1998**, *95*, 11211–11216. [CrossRef]
19. Yano, S.; Tokumitsu, H.; Soderling, T.R. Calcium promotes cell survival through CaM-K kinase activation of the protein-kinase-B pathway. *Nature* **1998**, *396*, 584. [CrossRef]
20. Pap, M.; Cooper, G.M. Role of glycogen synthase kinase-3 in the phosphatidylinositol 3-kinase/Akt cell survival pathway. *J. Biol. Chem.* **1998**, *273*, 19929–19932. [CrossRef]
21. Jaldin-Fincati, J.R.; Pavarotti, M.; Frendo-Cumbo, S.; Bilan, P.J.; Klip, A. Update on GLUT4 Vesicle Traffic: A Cornerstone of Insulin Action. *Trends Endocrinol. Metab.* **2017**, *28*, 597–611. [CrossRef] [PubMed]
22. Sugden, P.H.; Fuller, S.J.; Weiss, S.; Clerk, A. Glycogen synthase kinase 3 (GSK3) in the heart: A point of integration in hypertrophic signalling and a therapeutic target? A critical analysis. *Br. J. Pharmacol.* **2008**, *153*, 137–153. [CrossRef] [PubMed]
23. Datta, S.R.; Dudek, H.; Tao, X.; Masters, S.; Fu, H.; Gotoh, Y.; Greenberg, M.E. Akt phosphorylation of BAD couples survival signals to the cell-intrinsic death machinery. *Cell* **1997**, *91*, 231–241. [CrossRef]
24. Cardone, M.H.; Roy, N.; Stennicke, H.R.; Salvesen, G.S.; Franke, T.F.; Stanbridge, E.; Frisch, S.; Reed, J.C. Regulation of cell death protease caspase-9 by phosphorylation. *Science* **1998**, *282*, 1318–1321. [CrossRef]
25. Patel, S.; Woodgett, J. Glycogen synthase kinase-3 and cancer: Good cop, bad cop? *Canc. Cell* **2008**, *14*, 351–353. [CrossRef] [PubMed]
26. Chen, X.; Liu, Y.; Zhu, J.; Lei, S.; Dong, Y.; Li, L.; Jiang, B.; Tan, L.; Wu, J.; Yu, S.; et al. GSK-3b downregulates Nrf2 in cultured cortical neurons and in a rat model of cerebral ischemia-reperfusion. *Sci. Rep.* **2016**, *6*, 20196. [CrossRef]
27. Gotschel, F.; Kern, C.; Lang, S.; Sparna, T.; Markmann, C.; Schwager, J.; McNelly, S.; von Weizsacker, F.; Laufer, S.; Hecht, A.; et al. Inhibition of GSK3 differentially modulates NF-kB, CREB, AP-1 and b-catenin signaling in hepatocytes, but fails to promote TNF-a-induced apoptosis. *Exp. Cell Res.* **2008**, *314*, 1351–1366. [CrossRef]
28. Martin, M.; Rehani, K.; Jope, R.S.; Michalek, S.M. Toll-like receptor-mediated cytokine production is differentially regulated by glycogen synthase kinase 3. *Nat. Immunol.* **2005**, *6*, 777–784. [CrossRef]
29. Meier, R.; Alessi, D.R.; Cron, P.; Andjelković, M.; Hemmings, B.A. Mitogenic activation, phosphorylation, and nuclear translocation of protein kinase Bβ. *J. Biol. Chem.* **1997**, *272*, 30491–30497. [CrossRef]
30. Brunet, A.; Bonni, A.; Zigmond, M.J.; Lin, M.Z.; Juo, P.; Hu, L.S.; Anderson, M.J.; Arden, K.C.; Blenis, J.; Greenberg, M.E. Akt promotes cell survival by phosphorylating and inhibiting a Forkhead transcription factor. *Cell* **1999**, *96*, 857–868. [CrossRef]

31. Du, K.; Montminy, M. CREB is a regulatory target for the protein kinase Akt/PKB. *J. Biol. Chem.* **1998**, *273*, 32377–32379. [CrossRef] [PubMed]
32. Ozes, O.N.; Mayo, L.D.; Gustin, J.A.; Pfeffer, S.R.; Pfeffer, L.M.; Donner, D.B. NF-κB activation by tumour necrosis factor requires the Akt serine–threonine kinase. *Nature* **1999**, *401*, 82. [CrossRef] [PubMed]
33. Zong, W.X.; Edelstein, L.C.; Chen, C.; Bash, J.; Gélinas, C. The prosurvival Bcl-2 homolog Bfl-1/A1 is a direct transcriptional target of NF-κB that blocks TNFα-induced apoptosis. *Genes Dev.* **1999**, *13*, 382–387. [CrossRef] [PubMed]
34. Kauffmann-Zeh, A.; Rodriguez-Viciana, P.; Ulrich, E.; Gilbert, C.; Coffer, P.; Downward, J.; Evan, G. Suppression of c-Myc-induced apoptosis by Ras signaling through PI(3)K and PKB. *Nature* **1997**, *385*, 544. [CrossRef]
35. Stemke-Hale, K.; Gonzalez-Angulo, A.M.; Lluch, A.; Neve, R.M.; Kuo, W.L.; Davies, M.; Carey, M.; Hu, Z.; Guan, Y.; Sahin, A.; et al. An integrative genomic and proteomic analysis of PIK3CA, PTEN, and AKT mutations in breast cancer. *Cancer Res.* **2008**, *68*, 6084–6091. [CrossRef]
36. Amé, J.C.; Spenlehauer, C.; de Murcia, G. The PARP superfamily. *Bioessays* **2004**, *26*, 882–893. [CrossRef]
37. Bouchard, V.J.; Rouleau, M.; Poirier, G.G. PARP-1, a determinant of cell survival in response to DNA damage. *Exp. Hematol.* **2003**, *31*, 446–454. [CrossRef]
38. Miwa, M.; Saikawa, N.; Yamaizumi, Z.; Nishimura, S.; Sugimura, T. Structure of poly (adenosine diphosphate ribose): Identification of 2′-[1-ribosyl-2-(or 3-) (1′-ribosyl)] adenosine-5′,5,5′-tris (phosphate) as a branch linkage. *Proc. Natl. Acad. Sci. USA* **1979**, *76*, 595–599. [CrossRef]
39. D'Amours, D.; Desnoyers, S.; D'Silva, I.; Poirier, G.G. Poly (ADP-ribosyl) ation reactions in the regulation of nuclear functions. *Biochem. J.* **1999**, *342*, 249–268. [CrossRef]
40. Steffen, J.D.; Pascal, J.M. New players to the field of ADP-ribosylation make the final cut. *EMBO J.* **2013**, *32*, 1205–1207. [CrossRef]
41. Huber, A.; Bai, P.; de Murcia, J.M.; de Murcia, G. PARP-1, PARP-2 and ATM in the DNA damage response: Functional synergy in mouse development. *DNA Repair* **2004**, *3*, 1103–1108. [CrossRef] [PubMed]
42. Brunyanszki, A.; Szczesny, B.; Virág, L.; Szabo, C. Mitochondrial poly (ADP-ribose) polymerase: The Wizard of Oz at work. *Free Radic. Biol. Med.* **2016**, *100*, 257–270. [CrossRef]
43. Kraus, W.L.; Lis, J.T. PARP goes transcription. *Cell* **2003**, *113*, 677–683. [CrossRef]
44. Bürkle, A. Poly (APD-ribosyl) ation, a DNA damage-driven protein modification and regulator of genomic instability. *Cancer Lett.* **2001**, *163*, 1–5. [CrossRef]
45. Stilmann, M.; Hinz, M.; Arslan, S.Ç.; Zimmer, A.; Schreiber, V.; Scheidereit, C. A Nuclear Poly (ADP-Ribose)-Dependent Signalosome Confers DNA Damage-Induced IkB Kinase Activation. *Mol. Cell.* **2009**, *36*, 365–378. [CrossRef] [PubMed]
46. Kozlov, S.V.; Graham, M.E.; Jakob, B.; Tobias, F.; Kijas, A.W.; Tanuji, M.; Chen, P.; Robinson, P.J.; Taucher-Scholz, G.; Suzuki, K.; et al. Autophosphorylation and ATM activation: Additional sites add to the complexity. *J. Biol. Chem.* **2011**, *286*, 9107–9119. [CrossRef] [PubMed]
47. Berger, N.A. Poly (ADP-ribose) in the cellular response to DNA damage. *Radiat. Res.* **1985**, *101*, 4–15. [CrossRef]
48. Virág, L.; Salzman, A.L.; Szabó, C. Poly (ADP-ribose) synthetase activation mediates mitochondrial injury during oxidant-induced cell death. *J. Immunol.* **1998**, *161*, 3753–3759. [PubMed]
49. Ha, H.C.; Snyder, S.H. Poly (ADPribose) polymerase is a mediator of necrotic cell death by ATP depletion. *Proc. Natl. Acad. Sci. USA* **1999**, *69*, 13978–13982. [CrossRef]
50. Shieh, W.M.; Amé, J.C.; Wilson, M.V.; Wang, Z.Q.; Koh, D.W.; Jacobson, M.K.; Jacobson, E.L. Poly (ADPribose) polymerase null mouse cells synthesize ADP-ribose polymers. *J. Biol. Chem.* **1998**, *273*, 30069–30072. [CrossRef]
51. Küpper, J.H.; Müller, M.; Wolf, I. NAD consumption in carcinogen-treated hamster cells overexpressing a dominant negative mutant of poly (ADPribose) polymerase. *Biochem. Biophys. Res. Commun.* **1999**, *265*, 525–529. [CrossRef] [PubMed]
52. Jagtap, P.; Szabo, C. Poly (ADP-ribose) polymerase and the therapeutic effects of its inhibitors. *Nat. Rev. Drug Discov.* **2005**, *4*, 421–440. [CrossRef] [PubMed]
53. Szabo, C. Poly (ADP-ribose) polymerase activation and circulatory shock. *Novartis Found. Symp.* **2007**, *280*, 92–103. [PubMed]

54. Giansanti, V.; Donà, F.; Tillhon, M.; Scovassi, A.I. PARP inhibitors: New tools to protect from inflammation. *Biochem. Pharmacol.* **2010**, *80*, 1869–1877. [CrossRef]
55. Luo, X.; Kraus, W.L. On PAR with PARP: Cellular stress signaling through poly (ADP-ribose) and PARP-1. *Genes Dev.* **2012**, *26*, 417–432. [CrossRef] [PubMed]
56. Bai, P.; Virág, L. Role of poly (ADP-ribose) polymerases in the regulation of inflammatory processes. *FEBS Lett.* **2012**, *586*, 3771–3777. [CrossRef]
57. García, S.; Conde, C. The role of poly (ADP-ribose) polymerase-1 in rheumatoid arthritis. *Mediat. Inflamm.* **2015**, *2015*, 837250. [CrossRef]
58. Sethi, G.S.; Dharwal, V.; Naura, A.S. Poly (ADP-Ribose) Polymerase-1 in Lung Inflammatory Disorders: A Review. *Front. Immunol.* **2017**, *8*, 1172. [CrossRef]
59. Berger, N.A.; Besson, V.C.; Boulares, A.H.; Bürkle, A.; Chiarugi, A.; Clark, R.S.; Curtin, N.J.; Cuzzocrea, S.; Dawson, T.M.; Dawson, V.L.; et al. Opportunities for the repurposing of PARP inhibitors for the therapy of non-oncological diseases. *Br. J. Pharmacol.* **2018**, *175*, 192–222. [CrossRef]
60. Veres, B.; Gallyas, F., Jr.; Varbiro, G.; Berente, Z.; Osz, E.; Szekeres, G.; Szabo, C.; Sumegi, B. Decrease of the inflammatory response and induction of the Akt/protein kinase B pathway by poly-(ADP-ribose) polymerase 1 inhibitor in endotoxin-induced septic shock. *Biochem. Pharmacol.* **2003**, *65*, 1373–1382. [CrossRef]
61. Veres, B.; Radnai, B.; Gallyas, F., Jr.; Varbiro, G.; Berente, Z.; Osz, E.; Sumegi, B. Regulation of kinase cascades and transcription factors by a poly (ADP-ribose) polymerase-1 inhibitor, 4-hydroxyquinazoline, in lipopolysaccharide-induced inflammation in mice. *J. Pharmacol. Exp. Ther.* **2004**, *310*, 247–255. [CrossRef] [PubMed]
62. Veto, S.; Acs, P.; Bauer, J.; Lassmann, H.; Berente, Z.; Setalo, G., Jr.; Borgulya, G.; Sumegi, B.; Komoly, S.; Gallyas, F., Jr.; et al. Inhibiting poly(ADP-ribose) polymerase: A potential therapy against oligodendrocyte death. *Brain* **2010**, *133*, 822–834. [CrossRef] [PubMed]
63. García, S.; Mera, A.; Gómez-Reino, J.J.; Conde, C. Poly (ADP-ribose) polymerase suppression protects rheumatoid synovial fibroblasts from Fas-induced apoptosis. *Rheumatology Oxford* **2009**, *48*, 483–489. [CrossRef] [PubMed]
64. García, S.; Liz, M.; Gómez-Reino, J.J.; Conde, C. Akt activity protects rheumatoid synovial fibroblasts from Fas-induced apoptosis by inhibition of Bid cleavage. *Arthritis Res. Ther.* **2010**, *12*, R33. [CrossRef]
65. Liu, F.; Jiang, N.; Xiao, Z.; Cheng, J.; Mei, Y.; Zheng, P.; Wang, L.; Zhang, X.; Zhou, X.; Zhou, W.; et al. Effects of poly (ADP-ribose) polymerase-1 (PARP-1) inhibition on sulfur mustard-induced cutaneous injuries in vitro and in vivo. *PeerJ* **2016**, *4*, e1890. [CrossRef]
66. Gilad, E.; Zingarelli, B.; Salzman, A.L.; Szabo, C. Protection by inhibition of poly (ADP-ribose) synthetase against oxidant injury in cardiac myoblasts In vitro. *J. Mol. Cell Cardiol.* **1997**, *29*, 2585–2597. [CrossRef]
67. Zingarelli, B.; Cuzzocrea, S.; Zsengellér, Z.; Salzman, A.L.; Szabo, C. Protection against myocardial ischemia and reperfusion injury by 3-aminobenzamide, an inhibitor of poly (ADP-ribose) synthetase. *Cardiovasc. Res.* **1997**, *36*, 205–215. [CrossRef]
68. Zingarelli, B.; Salzman, A.L.; Szabo, C. Genetic disruption of poly (ADP-ribose) synthetase inhibits the expression of P-selectin and intercellular adhesion molecule-1 in myocardial ischemia/reperfusion injury. *Circ. Res.* **1998**, *83*, 85–94. [CrossRef]
69. Kovacs, K.; Toth, A.; Deres, P.; Kalai, T.; Hideg, K.; Sumegi, B. Myocardial protection by selective poly(ADP-ribose) polymerase inhibitors. *Exp. Clin. Cardiol.* **2004**, *9*, 17–20.
70. Kovacs, K.; Toth, A.; Deres, P.; Kalai, T.; Hideg, K.; Gallyas, F., Jr.; Sumegi, B. Critical role of PI3-kinase/Akt activation in the PARP inhibitor induced heart function recovery during ischemia-reperfusion. *Biochem. Pharmacol.* **2006**, *71*, 441–452. [CrossRef]
71. Pálfi, A.; Tóth, A.; Kulcsár, G.; Hantó, K.; Deres, P.; Bartha, E.; Halmosi, R.; Szabados, E.; Czopf, L.; Kálai, T.; et al. The role of Akt and mitogen-activated protein kinase systems in the protective effect of poly (ADP-ribose) polymerase inhibition in Langendorff perfused and in isoproterenol-damaged rat hearts. *J. Pharmacol. Exp. Ther.* **2005**, *315*, 273–282. [CrossRef] [PubMed]
72. Song, Z.F.; Chen, D.Y.; Du, B.; Ji, X.P. Poly (ADP-ribose) polymerase inhibitor reduces heart ischaemia/reperfusion injury via inflammation and Akt signalling in rats. *Chin. Med. J.* **2013**, *126*, 1913–1917. [PubMed]

73. Park, E.S.; Kang, J.C.; Kang, D.H.; Jang, Y.C.; Yi, K.Y.; Chung, H.J.; Park, J.S.; Kim, B.; Feng, Z.P.; Shin, H.S. 5-AIQ inhibits H2O2-induced apoptosis through reactive oxygen species scavenging and Akt/GSK-3β signaling pathway in H9c2 cardiomyocytes. *Toxicol. Appl. Pharmacol.* **2013**, *268*, 90–98. [CrossRef] [PubMed]
74. Park, E.S.; Kang, D.H.; Kang, J.C.; Jang, Y.C.; Lee, M.J.; Chung, H.J.; Yi, K.Y.; Kim, D.E.; Kim, B.; Shin, H.S. Cardioprotective effect of KR-33889, a novel PARP inhibitor, against oxidative stress-induced apoptosis in H9c2 cells and isolated rat hearts. *Arch. Pharm. Res.* **2017**, *40*, 640–654. [CrossRef]
75. Garcia Soriano, F.; Virág, L.; Jagtap, P.; Szabó, E.; Mabley, J.G.; Liaudet, L.; Marton, A.; Hoyt, D.G.; Murthy, K.G.; Salzman, A.L.; et al. Diabetic endothelial dysfunction: The role of poly(ADP-ribose) polymerase activation. *Nat. Med.* **2001**, *7*, 108–113. [CrossRef]
76. Szabó, G.; Bährle, S.; Stumpf, N.; Sonnenberg, K.; Szabó, E.E.; Pacher, P.; Csont, T.; Schulz, R.; Dengler, T.J.; Liaudet, L.; et al. Poly (ADP-Ribose) polymerase inhibition reduces reperfusion injury after heart transplantation. *Circ. Res.* **2002**, *90*, 100–106. [CrossRef]
77. Pacher, P.; Liaudet, L.; Bai, P.; Virag, L.; Mabley, J.G.; Haskó, G.; Szabo, C. Activation of poly (ADP-ribose) polymerase contributes to development of doxorubicin-induced heart failure. *J. Pharmacol. Exp. Ther.* **2002**, *300*, 862–867. [CrossRef]
78. Pacher, P.; Liaudet, L.; Mabley, J.G.; Komjáti, K.; Szabo, C. Pharmacologic inhibition of poly (adenosine diphosphate-ribose) polymerase may represent a novel therapeutic approach in chronic heart failure. *J. Am. Coll. Cardiol.* **2002**, *40*, 1006–1016. [CrossRef]
79. Pacher, P.; Liaudet, L.; Soriano, F.G.; Mabley, J.G.; Szabó, E.; Szabo, C. The role of poly (ADP-ribose) polymerase activation in the development of myocardial and endothelial dysfunction in diabetes. *Diabetes* **2002**, *51*, 514–521. [CrossRef]
80. Szabo, C.; Pacher, P.; Zsengellér, Z.; Vaslin, A.; Komjáti, K.; Benkö, R.; Chen, M.; Mabley, J.G.; Kollai, M. Angiotensin II-mediated endothelial dysfunction: Role of poly (ADP-ribose) polymerase activation. *Mol. Med.* **2004**, *10*, 28–35. [CrossRef]
81. Pacher, P.; Vaslin, A.; Benko, R.; Mabley, J.G.; Liaudet, L.; Haskó, G.; Marton, A.; Bátkai, S.; Kollai, M.; Szabó, C. A new, potent poly (ADP-ribose) polymerase inhibitor improves cardiac and vascular dysfunction associated with advanced aging. *J. Pharmacol. Exp. Ther.* **2004**, *311*, 485–491. [CrossRef] [PubMed]
82. Szenczi, O.; Kemecsei, P.; Holthuijsen, M.F.; van Riel, N.A.; van der Vusse, G.J.; Pacher, P.; Szabo, C.; Kollai, M.; Ligeti, L.; Ivanics, T. Poly (ADP-ribose) polymerase regulates myocardial calcium handling in doxorubicin-induced heart failure. *Biochem. Pharmacol.* **2005**, *69*, 725–732. [CrossRef] [PubMed]
83. Xiao, C.Y.; Chen, M.; Zsengellér, Z.; Li, H.; Kiss, L.; Kollai, M.; Szabó, C. Poly (ADP-Ribose) polymerase promotes cardiac remodeling, contractile failure, and translocation of apoptosis-inducing factor in a murine experimental model of aortic banding and heart failure. *J. Pharmacol. Exp. Ther.* **2005**, *312*, 891–898. [CrossRef] [PubMed]
84. Pacher, P.; Liaudet, L.; Mabley, J.G.; Cziráki, A.; Haskó, G.; Szabo, C. Beneficial effects of a novel ultrapotent poly (ADP-ribose) polymerase inhibitor in murine models of heart failure. *Int. J. Mol. Med.* **2006**, *17*, 369–375. [CrossRef] [PubMed]
85. Korkmaz-Icöz, S.; Radovits, T.; Loganathan, S.; Li, S.; Ruppert, M.; Benke, K.; Brlecic, P.; Szabó, C.; Karck, M.; Szabó, G. Prolonging hypothermic ischaemic cardiac and vascular storage by inhibiting the activation of the nuclear enzyme poly(adenosine diphosphate-ribose) polymerase. *Eur. J. Cardiothorac. Surg.* **2017**, *51*, 829–835. [CrossRef] [PubMed]
86. Gao, L.; Kwan, J.C.; Macdonald, P.S.; Yang, L.; Preiss, T.; Hicks, M. Improved poststorage cardiac function by poly (ADP-ribose) polymerase inhibition: Role of phosphatidylinositol 3-kinase Akt pathway. *Transplantation* **2007**, *84*, 380–386. [CrossRef]
87. Bartha, E.; Solti, I.; Szabo, A.; Olah, G.; Magyar, K.; Szabados, E.; Kalai, T.; Hideg, K.; Toth, K.; Gero, D.; et al. Regulation of kinase cascade activation and heat shock protein expression by poly (ADP-ribose) polymerase inhibition in doxorubicin-induced heart failure. *J. Cardiovasc. Pharmacol.* **2011**, *58*, 380–391. [CrossRef]
88. Sarszegi, Z.; Bognar, E.; Gaszner, B.; Kónyi, A.; Gallyas, F., Jr.; Sumegi, B.; Berente, Z. BGP-15, a PARP-inhibitor, prevents imatinib-induced cardiotoxicity by activating Akt and suppressing JNK and p38 MAP kinases. *Mol. Cell Biochem.* **2012**, *365*, 129–137. [CrossRef]
89. Qin, W.D.; Liu, G.L.; Wang, J.; Wang, H.; Zhang, J.N.; Zhang, F.; Ma, Y.; Ji, X.Y.; Li, C.; Zhang, M.X. Poly(ADP-ribose) polymerase 1 inhibition protects cardiomyocytes from inflammation and apoptosis in diabetic cardiomyopathy. *Oncotarget* **2016**, *7*, 35618–35631. [CrossRef]

90. Palfi, A.; Toth, A.; Hanto, K.; Deres, P.; Szabados, E.; Szereday, Z.; Kulcsar, G.; Kalai, T.; Hideg, K.; Gallyas, F., Jr.; et al. PARP inhibition prevents postinfarction myocardial remodeling and heart failure via the protein kinase C/glycogen synthase kinase-3beta pathway. *J. Mol. Cell Cardiol.* **2006**, *41*, 149–159. [CrossRef]
91. Magyar, K.; Deres, L.; Eros, K.; Bruszt, K.; Seress, L.; Hamar, J.; Hideg, K.; Balogh, A.; Gallyas, F., Jr.; Sumegi, B.; et al. A quinazoline-derivative compound with PARP inhibitory effect suppresses hypertension-induced vascular alterations in spontaneously hypertensive rats. *Biochim. Biophys. Acta* **2014**, *1842*, 935–944. [CrossRef] [PubMed]
92. Deres, L.; Bartha, E.; Palfi, A.; Eros, K.; Riba, A.; Lantos, J.; Kalai, T.; Hideg, K.; Sumegi, B.; Gallyas, F.; et al. PARP-inhibitor treatment prevents hypertension induced cardiac remodeling by favorable modulation of heat shock proteins, Akt-1/GSK-3β and several PKC isoforms. *PLoS ONE* **2014**, *9*, e102148. [CrossRef] [PubMed]
93. Hausenloy, D.J.; Yellon, D.M. Survival kinases in ischemic preconditioning and postconditioning. *Cardiovasc. Res.* **2006**, *70*, 240–253. [CrossRef] [PubMed]
94. Zhang, Y.; Park, T.S.; Gidday, J.M. Hypoxic preconditioning protects human brain endothelium from ischemic apoptosis by Akt-dependent survivin activation. *Am. J. Physiol. Heart Circ. Physiol.* **2007**, *292*, H2573–H2581. [CrossRef]
95. Kovacs, K.; Hanto, K.; Bognar, Z.; Tapodi, A.; Bognar, E.; Kiss, G.N.; Szabo, A.; Rappai, G.; Kiss, T.; Sumegi, B.; et al. Prevalent role of Akt and ERK activation in cardioprotective effect of Ca(2+) channel- and beta-adrenergic receptor blockers. *Mol. Cell Biochem.* **2009**, *321*, 155–164. [CrossRef]
96. Covington, S.M.; Bauler, L.D.; Toledo-Pereyra, L.H. Akt: A therapeutic target in hepatic ischemia-reperfusion injury. *J. Investig. Surg.* **2017**, *30*, 47–55. [CrossRef]
97. Koh, S.H.; Park, Y.; Song, C.W.; Kim, J.G.; Kim, K.; Kim, J.; Kim, M.H.; Lee, S.R.; Kim, D.W.; Yu, H.J.; et al. The effect of PARP inhibitor on ischaemic cell death, its related inflammation and survival signals. *Eur. J. Neurosci.* **2004**, *20*, 1461–1472. [CrossRef]
98. Szabo, C. Nicotinamide: A jack of all trades (but master of none?). *Intensive Care Med.* **2003**, *29*, 863–866. [CrossRef]
99. Chong, Z.Z.; Lin, S.H.; Li, F.; Maiese, K. The sirtuin inhibitor nicotinamide enhances neuronal cell survival during acute anoxic injury through AKT, BAD, PARP, and mitochondrial associated "anti-apoptotic" pathways. *Curr. Neurovasc. Res.* **2005**, *2*, 271–285. [CrossRef]
100. Li, W.J.; Oh, S.J. PARP inhibition prevents oxidative injury of bladder induced by acute urinary retention and subsequent emptying. *Apoptosis* **2011**, *16*, 574–580. [CrossRef]
101. Kalmar-Nagy, K.; Degrell, P.; Szabo, A.; Sumegi, K.; Wittmann, I.; Gallyas, F., Jr.; Sumegi, B. PARP inhibition attenuates acute kidney allograft rejection by suppressing cell death pathways and activating PI-3K-Akt cascade. *PLoS ONE* **2013**, *8*, e81928. [CrossRef] [PubMed]
102. Kovacs, K.; Vaczy, A.; Fekete, K.; Kovari, P.; Atlasz, T.; Reglodi, D.; Gabriel, R.; Gallyas, F.; Sumegi, B. PARP inhibitor protects against chronic hypoxia/reoxygenation-induced retinal injury by regulation of MAPKs, HIF1α, Nrf2, and NFκB. *Investig. Ophthalmol. Vis. Sci.* **2019**, *60*, 1478–1490. [CrossRef] [PubMed]
103. Aguilar-Quesada, R.; Munoz-Gamez, J.A.; Martin-Oliva, D.; Peralta, A.; Valenzuela, M.T.; Matinez-Romero, R.; Quiles-Perez, R.; Menissier-de Murcia, J.; de Murcia, G.; Ruiz de Almodovar, M.; et al. Interaction between ATM and PARP-1 in response to DNA damage and sensitization of ATM deficient cells through PARP inhibition. *BMC Mol. Biol.* **2007**, *8*, 29. [CrossRef] [PubMed]
104. Halaby, M.J.; Hibma, J.C.; He, J.; Yang, D.Q. ATM protein kinase mediates full activation of Akt and regulates glucose transporter 4 translocation by insulin in muscle cells. *Cell Signal.* **2008**, *20*, 1555–1563. [CrossRef] [PubMed]
105. Stagni, V.; Manni, I.; Oropallo, V.; Mottolese, M.; Di Benedetto, A.; Piaggio, G.; Falcioni, R.; Giaccari, D.; Di Carlo, S.; Sperati, F.; et al. ATM kinase sustains HER2 tumorigenicity in breast cancer. *Nat. Commun.* **2015**, *6*, 6886. [CrossRef]
106. Yang, D.Q.; Halaby, M.J.; Li, Y.; Hibma, J.C.; Burn, P. Cytoplasmic ATM protein kinase: An emerging therapeutic target for diabetes, cancer and neuronal degeneration. *Drug Discov. Today* **2011**, *16*, 332–338. [CrossRef]
107. Lee, S.S.; Bohrson, C.; Pike, A.M.; Wheelan, S.J.; Greider, C.W. ATM kinase is required for telomere elongation in mouse and human cells. *Cell Rep.* **2015**, *13*, 1623–1632. [CrossRef]

108. Wu, Z.H.; Shi, Y.; Tibbetts, R.S.; Miyamoto, S. Molecular linkage between the kinase ATM and NF-kappaB signaling in response to genotoxic stimuli. *Science* **2006**, *311*, 1141–1146. [CrossRef]
109. Williamson, C.T.; Muzik, H.; Turhan, A.G.; Zamo, A.; O'Connor, M.J.; Bebb, D.G.; Lees-Miller, S.P. ATM deficiency sensitizes mantle cell lymphoma cells to poly (ADPribose) polymerase-1 inhibitors. *Mol. Cancer Ther.* **2010**, *9*, 347–357. [CrossRef]
110. Rajawat, J.; Shukla, N.; Mishra, D.P. Therapeutic targeting of poly (ADP-ribose) polymerase-1 (PARP1) in cancer: Current developments, therapeutic strategies, and future opportunities. *Med. Res. Rev.* **2017**, *37*, 1461–1491. [CrossRef]
111. Wang, D.; Wang, M.; Jiang, N.; Zhang, Y.; Bian, X.; Wang, X.; Roberts, T.M.; Zhao, J.J.; Liu, P.; Cheng, H. Effective use of PI3K inhibitor BKM120 and PARP inhibitor Olaparib to treat PIK3CA mutant ovarian cancer. *Oncotarget* **2016**, *7*, 13153–13166. [CrossRef] [PubMed]
112. Griffin, R.J.; Curtin, N.J.; Newell, D.R.; Golding, B.T.; Durkacz, B.W.; Calvert, A.H. The role of inhibitors of poly (ADP-ribose) polymerase as resistance-modifying agents in cancer therapy. *Biochimie* **1995**, *77*, 408–422. [CrossRef]
113. Farmer, H.; McCabe, N.; Lord, C.J.; Tutt, A.N.; Johnson, D.A.; Richardson, T.B.; Santarosa, M.; Dillon, K.J.; Hickson, I.; Knights, C.; et al. Targeting the DNA repair defect in BRCA mutant cells as a therapeutic strategy. *Nature* **2005**, *434*, 917–921. [CrossRef] [PubMed]
114. Bryant, H.E.; Schultz, N.; Thomas, H.D.; Parker, K.M.; Flower, D.; Lopez, E.; Kyle, S.; Meuth, M.; Curtin, N.J.; Helleday, T. Specific killing of BRCA2-deficient tumours with inhibitors of poly (ADP-ribose) polymerase. *Nature* **2005**, *434*, 913–917. [CrossRef]
115. Murai, J.; Huang, S.Y.; Das, B.B.; Renaud, A.; Zhang, Y.; Doroshow, J.H.; Ji, J.; Takeda, S.; Pommier, Y. Trapping of PARP1 and PARP2 by clinical PARP inhibitors. *Cancer Res.* **2012**, *72*, 5588–5599. [CrossRef]
116. Kim, G.; Ison, G.; McKee, A.E.; Zhang, H.; Tang, S.; Gwise, T.; Sridhara, R.; Lee, E.; Tzou, A.; Philip, R.; et al. FDA approval summary. Olaparib monotherapy in patients with deleterious germline BRCA mutated advanced ovarian cancer treated with three or more lines of chemotherapy. *Clin. Cancer Res.* **2015**, *21*, 4257–4261. [CrossRef]
117. Lord, C.J.; Ashworth, A. BRCAness revisited. *Nat. Rev. Cancer* **2016**, *16*, 110–120. [CrossRef]
118. Sachdev, E.; Tabatabai, R.; Roy, V.; Rimel, B.J.; Mita, M.M. PARP Inhibition in Cancer: An Update on Clinical Development. *Target. Oncol.* **2019**, *14*, 657–679. [CrossRef]
119. Shao, N.; Shi, Y.; Yu, L.; Ye, R.; Shan, Z.; Zhang, Z.; Zhang, Y.; Lin, Y. Prospect for Application of PARP Inhibitor in Patients with HER2 Negative Breast Cancer. *Int. J. Biol. Sci.* **2019**, *15*, 962–972. [CrossRef]
120. DeBerardinis, R.J.; Chandel, N.S. Fundamentals of cancer metabolism. *Sci. Adv.* **2016**, *2*, e1600200. [CrossRef]
121. Smith, K.A.; Waypa, G.B.; Schumacker, P.T. Redox signaling during hypoxia in mammalian cells. *Redox Biol.* **2017**, *13*, 228–234. [CrossRef] [PubMed]
122. National Cancer Institute GDC Data Portal. Available online: https://portal.gdc.cancer.gov/exploration?facetTab=genes&filters=%7B%22content%22%3A%5B%7B%22op%22%3A%22in%22%2C%22content%22%3A%7B%22field%22%3A%22genes.biotype%22%2C%22value%22%3A%5B%22protein_coding%22%5D%7D%7D%2C%7B%22op%22%3A%22in%22%2C%22content%22%3A%7B%22field%22%3A%22genes.gene_id%22%2C%22value%22%3A%5B%22ENSG00000082701%22%2C%22ENSG00000121879%22%2C%22ENSG00000142208%22%2C%22ENSG00000171862%22%2C%22ENSG00000198793%22%5D%7D%7D%2C%7B%22op%22%3A%22in%22%2C%22content%22%3A%7B%22field%22%3A%22genes.is_cancer_gene_census%22%2C%22value%22%3A%5B%22true%22%5D%7D%7D%5D%2C%22op%22%3A%22and%22%7D&searchTableTab=genes (accessed on 14 February 2020).
123. Lawrence, M.S.; Stojanov, P.; Mermel, C.H.; Robinson, J.T.; Garraway, L.A.; Golub, T.R.; Meyerson, M.; Gabriel, S.B.; Lander, E.S.; Getz, G. Discovery and saturation analysis of cancer genes across 21 tumour types. *Nature* **2014**, *505*, 495–501. [CrossRef] [PubMed]
124. Revathidevi, S.; Munirajan, A.K. Akt in cancer: Mediator and more. *Semin. Cancer Biol.* **2019**, *59*, 80–91. [CrossRef] [PubMed]
125. Bonneau, D.; Longy, M. Mutations of the human PTEN gene. *Hum. Mutat.* **2000**, *16*, 109–122. [CrossRef]
126. Nakatani, N.; Thompson, D.A.; Barthel, A.; Sakaue, H.; Liu, W.; Weigel, R.J.; Roth, R.A. Up-regulation of Akt3 in estrogen receptor-deficient breast cancers and androgen independent prostate cancer lines. *J. Biol. Chem.* **1999**, *274*, 21528–21532. [CrossRef]

127. Xu, X.; Sakon, M.; Nagano, H.; Hiraoka, N.; Yamamoto, H.; Hayashi, N.; Dono, K.; Nakamori, S.; Umeshita, k.; Ito, Y.; et al. Akt2 expression correlates with prognosis of human hepatocellular carcinoma. *Oncol. Rep.* **2004**, *11*, 25–32. [CrossRef]
128. Roy, H.K.; Olusola, B.F.; Clemens, D.L.; Karolski, W.J.; Ratashak, A.; Lynch, H.T.; Smyrk, T.C. AKT proto-oncogene overexpression is an early event during sporadic colon carcinogenesis. *Carcinogenesis* **2002**, *23*, 201–205. [CrossRef]
129. Downward, J. Targeting RAS signalling pathways in cancer therapy. *Nat. Rev. Cancer* **2003**, *3*, 11. [CrossRef]
130. Bellacosa, A.; Kumar, C.C.; Di Cristofano, A.; Testa, J.R. Activation of AKT kinases in cancer: Implications for therapeutic targeting. *Adv. Cancer Res.* **2005**, *94*, 29–86.
131. Gewinner, C.; Wang, Z.C.; Richardson, A.; Teruya-Feldstein, J.; Etemadmoghadam, D.; Bowtell, D.; Barretina, J.; Lin, W.M.; Rameh, L.; Salmena, L.; et al. Evidence that inositol polyphosphate 4-phosphatase type II is a tumor suppressor that inhibits PI3K signaling. *Cancer Cell.* **2009**, *16*, 115–125. [CrossRef]
132. Nakakido, M.; Deng, Z.; Suzuki, T.; Dohmae, N.; Nakamura, Y.; Hamamoto, R. Dysregulation of AKT pathway by SMYD2-mediated lysine methylation on PTEN. *Neoplasia* **2015**, *17*, 367–373. [CrossRef] [PubMed]
133. Mancini, M.; Petta, S.; Martinelli, G.; Barbieri, E.; Santucci, M.A. RAD 001 (everolimus) prevents mTOR and Akt late re-activation in response to imatinib in chronic myeloid leukemia. *J. Cell Biochem.* **2010**, *109*, 320–328. [CrossRef] [PubMed]
134. Okoh, V.O.; Felty, Q.; Parkash, J.; Poppiti, R.; Roy, D. Reactive oxygen species via redox signaling to PI3K/AKT pathway contribute to the malignant growth of 4-hydroxyestradiol-transformed mammary epithelial cells. *PLoS ONE* **2013**, *8*, e54206. [CrossRef] [PubMed]
135. Kwon, J.; Lee, S.R.; Yang, K.S.; Ahn, Y.; Kim, Y.J.; Stadtman, E.R.; Rhee, S.G. Reversible oxidation and inactivation of the tumor suppressor PTEN in cells stimulated with peptide growth factors. *Proc. Natl. Acad. Sci. USA* **2004**, *101*, 16419–16424. [CrossRef]
136. Noordermeer, S.M.; van Attikum, H. PARP Inhibitor Resistance: A Tug-of-War in BRCA-Mutated Cells. *Trends Cell Biol.* **2019**, *29*, 820–834. [CrossRef]
137. Cseh, A.M.; Fabian, Z.; Quintana-Cabrera, R.; Szabo, A.; Eros, K.; Soriano, M.E.; Gallyas, F.; Scorrano, L.; Sumegi, B. PARP Inhibitor PJ34 Protects Mitochondria and Induces DNA-Damage Mediated Apoptosis in Combination With Cisplatin or Temozolomide in B16F10 Melanoma Cells. *Front. Physiol.* **2019**, *10*, 538. [CrossRef]
138. Hou, W.H.; Chen, S.H.; Yu, X. Poly-ADP ribosylation in DNA damage response and cancer therapy. *Mutat. Res.* **2019**, *780*, 82–91. [CrossRef]
139. Yazinski, S.A.; Comaills, V.; Buisson, R.; Genois, M.M.; Nguyen, H.D.; Ho, C.K.; Todorova Kwan, T.; Morris, R.; Lauffer, S.; Nussenzweig, A.; et al. ATR inhibition disrupts rewired homologous recombination and fork protection pathways in PARP inhibitor-resistant BRCA-deficient cancer cells. *Genes. Dev.* **2017**, *31*, 318–332. [CrossRef]
140. Nakagawa, Y.; Sedukhina, A.S.; Okamoto, N.; Nagasawa, S.; Suzuki, N.; Ohta, T.; Hattori, H.; Roche-Molina, M.; Narváez, A.J.; Jeyasekharan, A.D.; et al. NF-kappaB signaling mediates acquired resistance after PARP inhibition. *Oncotarget* **2015**, *6*, 3825–3839. [CrossRef]
141. Koh, D.W.; Lawler, A.M.; Poitras, M.F.; Sasaki, M.; Wattler, S.; Nehls, M.C.; Stöger, T.; Poirier, G.G.; Dawson, V.L.; Dawson, T.M. Failure to degrade poly(ADP-ribose) causes increased sensitivity to cytotoxicity and early embryonic lethality. *Proc. Natl. Acad. Sci. USA* **2004**, *101*, 17699–17704. [CrossRef]
142. Fathers, C.; Drayton, R.M.; Solovieva, S.; Bryant, H.E. Inhibition of poly (ADP-ribose) glycohydrolase (PARG) specifically kills BRCA2-deficient tumor cells. *ABBV Cell Cycle* **2012**, *11*, 990–997. [CrossRef] [PubMed]
143. Sharifi, R.; Morra, R.; Appel, C.D.; Tallis, M.; Chioza, B.; Jankevicius, G.; Simpson, M.A.; Matic, I.; Ozkan, E.; Golia, B.; et al. Deficiency of terminal ADP-ribose protein glycohydrolase TARG1/C6orf130 in neurodegenerative disease. *EMBO J.* **2013**, *32*, 1225–1237. [CrossRef] [PubMed]
144. Chabanon, R.M.; Muirhead, G.; Krastev, D.B.; Adam, J.; Morel, D.; Garrido, M.; Lamb, A.; Hénon, C.; Dorvault, N.; Rouanne, M.; et al. PARP inhibition enhances tumor cell-intrinsic immunity in ERCC1-deficient non-small cell lung cancer. *J. Clin. Investig.* **2019**, *129*, 1211–1228. [CrossRef] [PubMed]
145. Tutuncuoglu, B.; Krogan, N.J. Mapping genetic interactions in cancer: A road to rational combination therapies. *Genome Med.* **2019**, *11*, 62. [CrossRef]

146. Bartha, Á.; Győrffy, B. Comprehensive Outline of Whole Exome Sequencing Data Analysis Tools Available in Clinical Oncology. *Cancers* **2019**, *11*, 1725. [CrossRef] [PubMed]
147. Corey, L.; Valente, A.; Wade, K. Personalized Medicine in Gynecologic Cancer: Fact or Fiction? *Surg. Oncol. Clin.* **2020**, *29*, 105–113. [CrossRef]

cancers

MDPI

Review

ARH1 in Health and Disease

Hiroko Ishiwata-Endo, Jiro Kato, Linda A. Stevens and Joel Moss *

Pulmonary Branch, National Heart, Lung, and Blood Institute, National Institutes of Health, Bethesda, MD 20892-1590, USA; endohiro@mail.nih.gov (H.I.-E.); katoj@nhlbi.nih.gov (J.K.); stevensl@nhlbi.nih.gov (L.A.S.)

* Correspondence: mossj@nhlbi.nih.gov; Tel.: +301-496-1597

Received: 27 December 2019; Accepted: 15 February 2020; Published: 19 February 2020

Abstract: Arginine-specific mono-adenosine diphosphate (ADP)-ribosylation is a nicotinamide adenine dinucleotide $(NAD)^+$-dependent, reversible post-translational modification involving the transfer of an ADP-ribose from NAD^+ by bacterial toxins and eukaryotic ADP-ribosyltransferases (ARTs) to arginine on an acceptor protein or peptide. ADP-ribosylarginine hydrolase 1 (ARH1) catalyzes the cleavage of the ADP-ribose-arginine bond, regenerating (arginine)protein. Arginine-specific mono-ADP-ribosylation catalyzed by bacterial toxins was first identified as a mechanism of disease pathogenesis. Cholera toxin ADP-ribosylates and activates the α subunit of Gαs, a guanine nucleotide-binding protein that stimulates adenylyl cyclase activity, increasing cyclic adenosine monophosphate (cAMP), and resulting in fluid and electrolyte loss. Arginine-specific mono-ADP-ribosylation in mammalian cells has potential roles in membrane repair, immunity, and cancer. In mammalian tissues, ARH1 is a cytosolic protein that is ubiquitously expressed. ARH1 deficiency increased tumorigenesis in a gender-specific manner. In the myocardium, in response to cellular injury, an arginine-specific mono-ADP-ribosylation cycle, involving ART1 and ARH1, regulated the level and cellular distribution of ADP-ribosylated tripartite motif-containing protein 72 (TRIM72). Confirmed substrates of ARH1 in vivo are Gαs and TRIM72, however, more than a thousand proteins, ADP-ribosylated on arginine, have been identified by proteomic analysis. This review summarizes the current understanding of the properties of ARH1, e.g., bacterial toxin action, myocardial membrane repair following injury, and tumorigenesis.

Keywords: arginine-specific mono-ADP-ribosylation; bacterial toxin; cholera toxin; ART1; ARH1; tumorigenesis; loss of heterozygosity; membrane repair; gender bias

1. ARH Family

1.1. Properties of ARHs

ADP-ribosylarginine hydrolase 1 (ARH1) is a member of an ADP-ribosyl-acceptor hydrolase (ARH) family, which is composed of three 39-kDa proteins, ARH1-3, based on sequence and size similarity. These hydrolases differ in enzymatic activities and biological functions [1]. Human ARH1 is a 357 amino acid (aa) protein, which shares significant sequence and size conservation with ARH2 (354 aa protein, 47% identity, 68% similarity) and ARH3 (363 aa protein, 22% identity, 41% similarity). The mouse ARH1 amino acid sequence shows 83% identity and 91% similarity to human ARH1 [2]. ARH1 requires Mg^{2+} for catalytic activity [1,3,4]. The active sites of ARH1 (human, mouse, or rat) contain aspartates residues for coordination of Mg^{2+} binding [5,6]. Critical residues involved in ARH1 enzymatic activity and binding of ADP-ribose include the conserved vicinal aspartates 55 and 56 in humans [5,6] and 60 and 61 in rats [7]. Replacement of Asp 55 and/or Asp56 in humans and Asp60 and/or Asp61 in rats with alanine, glutamine, or asparagine significantly reduced hydrolase activity (ADP-ribosylarginine hydrolase, α-NADase) [5–7].

1.2. Function and Substrates of ARHs

Mono- and poly-ADP-ribosylation are reversible post-translational modifications of proteins, DNA and RNA [8,9]. The ARTs in mammals, i.e., ART1, ART2, ART5, are confirmed mono-ADP-ribosyltransferases that modify only arginine residues [10–13]. Target residues of the poly (ADP-ribose) polymerase (PARP) family include aspartate, glutamate, and serine [14–16]. Hydrolases, such as the ARHs and macrodomain proteins, also show amino acid-specific hydrolytic reactions. Mono-ADP-ribosylation on arginine, serine, and glutamate, respectively, was hydrolyzed by ARH1, ARH3, and macrodomain proteins, i.e., MacroD1, MacroD2, and terminal ADP-ribose protein glycohydrolase (TARG1)/C6orf130 [5,17]. Poly(ADP-ribose) glycohydrolase (PARG) cleaves poly(ADP-ribose) (PAR) chains of PARP-1 but did not cleave terminal ADP-ribose linked directly to amino acids of PARP-1. ARH3 releases ADP-ribose from serine on PARP-1 [18,19], but did not hydrolyze ADP-ribose-glutamate/aspartate linkages [16,20]. ADP-ribosylated DNA and RNA were hydrolyzed by ARH3, but not ARH1 [5,8,9]. Furthermore, ARH1 and ARH3 hydrolyzed the α-*O*-glycosidic bonds of the poly-ADP-ribose polymer attached to PARP1, although ARH3 has approximately 50 times more activity than ARH1 [1,21,22]. In addition, ARH1 and ARH3 hydrolyzed *O*-acetyl-ADP-ribose (*O*AADPr), the product of sirtuin deacetylases [23]. Recent reports showed that ARH2 is primarily expressed in the heart and appears to be involved in the regulation of heart chamber outgrowth [24]; although ARH2 has significant amino acid identity and homology to ARH1, it has not been shown to be an active hydrolase [2].

Arh1-deficient mouse embryonic fibroblasts (MEFs) and mouse tissues are unable to hydrolyze the ADP-ribose-arginine bond, supporting the view that ARH1 is the only mammalian arginine-specific hydrolase [25,26]. ARH1 has a role in controlling TRIM72 ADP-ribosylation, and thereby TRIM72 activity in membrane repair: Membrane injury may cause the mixing of intracellular with extracellular factors, e.g., small molecule such as NAD^+, and proteins [27]. Another opportunity for ARH1 interaction with ART1 substrates might be through lipid raft-mediated endocytosis, which is enhanced by UV light and H_2O_2-induced stress [28]. Interestingly, glycosylphosphatidylinositol (GPI)-anchored ARTs and caveoline-3, a cofactor of TRIM72, exist within lipid rafts [10,29]. To address the question of how cytoplasmic protein ARH1 contributes to the reversion of arginine-ADP-ribosylation catalyzed by extracellular enzyme ARTs, localization of arginine ADP-ribosylated proteins needs to be determined.

Arginine-specific ADP-ribosylation has been identified in the endoplasmic reticulum (ER) protein binding immunoglobulin protein (BIP) also known as 78-kDa glucose-regulated protein (GRP78) [30], mitochondrial protein heat shock protein 75 (HSP75) also known as tumor necrosis factor receptor-associated protein 1TRAP1 [31], secreted protein tumor necrosis factor-α [32], and external plasma membrane protein integrin α7 [31], although, none of them has been identified as a substrate of ARH1. Under stress conditions, BIP/GRP78 and HSP75/TRAP1 are released to the cell surface and cytoplasm, respectively [33,34]. By this mechanism, cytoplasmic protein ARH1 might be able to interact with its target proteins. However, there is still the question as to how ecto-enzyme ARTs modify intracellular proteins. Further studies are needed to understand (1) whether extracellularly modified substrates can be internalized into cytoplasm and cleaved by ARH1, and (2) whether an intracellular transferase catalyzes arginine ADP-ribosylation.

Not all arginine ADP-ribosylated proteins have been reported to be hydrolyzed by ARH1. For example, human neutrophil peptide-1 (HNP-1) is ADP-ribosylated by ART1. HNP-1 is an antibacterial peptide contributing to host defense immunity, as well as being toxic for host epithelial cells. HNP-1 ADP-ribosylated on arginines 14 and 24 shows reduced antimicrobial and cytotoxic activities [35]. Non-enzymatic replacement of the ADP-ribosylated arginines of HNP-1 with ornithine resulted in a peptide with less cytotoxicity than unmodified HNP-1 but with retention of its antibacterial activity [36]. Thus, the ornithine-containing HNP-1, which is no longer a substrate for ARH1, may have therapeutic potential as an ARH1-resistant molecule.

Proteomic analysis has revealed that more than 1000 proteins are ADP-ribosylated on arginine residues by ARTs [31,37–39], the confirmed substrate proteins of ARH1 in vivo are only ADP-ribose-Gαs [26] and ADP-ribose-TRIM72 [27]. The amount and duration of cellular

ADP-ribosylation may be controlled by substrate-specific transferases and hydrolases, which catalyze opposing arms of ADP-ribosylation cycles. Both ADP-ribosylation and de-ADP-ribosylation alter protein activities involved in cell signaling and viability [10,27,40].

2. Structure and Enzymatic Activity of ARH1

2.1. Structure and Stereospecific Activity of ARH1

ARH1 hydrolyzes the *N*-glycosidic linkage of ADP-ribosyl-arginine (e.g., arginine-Gαs, arginine-TRIM72) [10,26,27] and also cleaves the *O*-glycosidic linkage of poly-ADP-ribose, *O*-acetyl-ADP-ribose, and α-NAD$^+$ [1,21–23]. The crystal structure of *h*ARH1 is available in a complex with ADP-ribose [5,6]. ARH1 requires Mg^{2+} for maximal catalytic activity [1,4,7]. The rat, mouse, and turkey ARH1 activities were stimulated by Mg^{2+} and dithiothreitol (DTT), whereas pig and calf ARH1 showed Mg^{2+}, but not DTT dependence. Human ARH1 was DTT independent, as well [3,25]. The difference in DTT dependence appears to reside in the number of cysteine residues in the ARH1 proteins. ARH1 has a high structural similarity with ARH3 [6]. One major difference between ARH1 and ARH3 is the binding of the adenosine ribose moiety of ADP-ribose, resulting in a difference in ARH1 binding for ADP-ribose, which is 70× less than ARH3 [6]. ARH1 has four phosphorylation sites, tyrosine (Tyr)-4, Tyr-19, Tyr-20, and Tyr-205 [41]. Phosphorylation of ARH1 leads to conformational changes of the catalytic pocket, facilitating ADP-ribose-arginine binding to ARH1, which may affect ARH1 hydrolytic activity [41]. We speculate that the different affinities for ADP-ribose between ARH1 and ARH3 might be due to phosphorylation at the ADP-ribose binding pocket. ARH1 phosphorylation, at Tyr-4 and Tyr-19, was identified in a highly metastatic hepatocellular carcinoma (HCC) cell line, MHCC97H, but not in a nonmetastatic HCC cell line, Hep3B, implying that tyrosine phosphorylation of ARH1 was associated with HCC metastasis [42].

Mammalian ART1 catalyzes the transfer of ADP-ribose from β-NAD$^+$ to proteins in a stereospecific manner, forming α-ADP-ribosylated-(arginine) proteins, which alters the protein's function in cellular biological processes [43]. Arginine-specific ADP-ribosylation is a reversible modification, however, in vitro, α-ADP-ribosyl-arginine anomerizes to the β-form, which is not hydrolyzed by ARH1. Anomerization of α- to β-ADP-ribosyl-arginine interrupts a stereospecific ADP-ribosylation cycle in vitro [17,44–46]. Mono-ADP-ribosylated proteins are de-modified by the stereospecific α-ADP-ribose acceptor hydrolases, e.g., ADP-ribosyl-acceptor hydrolases (ARHs), MacroD1, MacroD2, Af1521, TARG1/C6orf130 [5,17]. ARH1 is the only ADP-ribosylated arginine-specific hydrolase identified in mammals that cleaves the *N*-glycosidic bond at the C1 linkage of α-anomeric ADP-ribose-arginine [4,17,25]. Currently, ARH1 has four stereospecific hydrolytic activities using as substrates α-NAD, α-OAADPr, poly(ADP-ribose), and α-ADP-ribose-(arginine) protein and generating free ADP-ribose as a product.

2.2. ARH1 Protein Expression and Cellular Distribution

Among rat tissues, ADP-ribosylarginine-specific hydrolase activity was greatest in the brain, spleen, and testis [25]. ARH1 is a cytosolic protein, the product of a single gene that is ubiquitously expressed in mammalian tissues [25]. Similar to the finding with glycosylphosphatidylinositol (GPI)-linked ART1 protein expression [47], the amount of ARH1 protein increased during C2C12 myoblast differentiation into myotubes, indicating that it may play a role in myoblast differentiation [27]. ARH1 protein levels in lung adenocarcinoma and lymphoma in *Arh1*$^{+/-}$ mice were lower than detectable levels by Western blotting. However, ARH1 existed in surrounding *Arh1*$^{+/-}$ nontumorous lung tissue [40], suggesting that the loss of ARH1 activity enhanced tumor formation. These data are consistent with a role for inactivation or loss of the functioning *Arh1* gene or protein in the mouse tumorigenesis model. According to the human cancer database Oncomine (www.oncomine.org) [48], *ARH1* mRNA expression in human lung adenocarcinoma was significantly lower than in that of normal lung tissue [40,49], consistent with a tumor-suppressor function of ARH1.

2.3. Arginine-Specific Mono-ADP-Ribosylation Cycle

Arginine-specific ADP-ribosyltransferases and ADP-ribosylarginine hydrolase 1 (ARH1) are opposing arms of a mono-ADP-ribosylation cycle [50]. ADP-ribosylation of arginine was first discovered as a mechanism of action of bacterial toxins, e.g., *vibrio cholerae* cholera toxin, *Escherichia coli* (*E. coli*) heat-labile enterotoxin, *pseudomonas aeruginosa* exoenzyme S (ExoS), which catalyze the NAD^+-dependent disruption of the signal transduction pathway by ADP-ribosylation of critical proteins, e.g., G protein alpha subunit that is stimulating for adenylyl cyclase (Gαs), rat sarcoma viral oncogene homolog (Ras), and Ras-related protein in brain (Rab) [51–54]. Cholera toxin produces arginine ADP-ribosylated Gαs, resulting in intoxication, which is terminated by ARH1 cleaving ADP-ribose from ADP-ribosylated Gαs. Thus, arginine-specific mono-ADP-ribosylation is a reversible reaction.

Other possible substrates of ARH1 might be Ras and Rab. It has been reported that ExoS catalyzes arginine ADP-ribosylation of Ras and Rab inhibiting nerve growth factor-stimulated neurite formation of PC-12 cells and disrupting normal vesicle trafficking, respectively, however, it has not known whether ARH1 cleaves ADP-ribose from ADP-ribosylated Ras and Rab [53,54].

In mammalian cells, endogenous ADP-ribosyltransferases (ARTs), extracellular GPI-anchored ART1 and ART2, and secreted ART5, catalyze arginine-specific ADP-ribosylation similar to those of the bacterial toxins, e.g., cholera toxin, *E. coli* toxin [10–13]. Ecto-ART proteins show tissue-specific expression such as heart and skeletal muscle for ART1, lymphocytes for mouse ART2, and testis for ART5. In humans and chimpanzees, however, ART2 is a pseudogene [10]. Furthermore, ART5 is primarily an NAD^+ glycohydrolase; NADase activity of ART5 is 10x higher than its ADP-ribosyltransferase activity [10,55]. Therefore, GPI-linked ART1 seems to be a primary contributor to arginine-specific ADP-ribosylation in humans [35,56]. Under normal conditions, there is no difference in NAD^+ levels of lung, heart, and brain between wild-type and *Arh1*-deficient mice, suggesting that ARH1 does not consume NAD^+ [49]. However, ARH1 decreased the levels of ADP-ribose-(arginine) content [27]. Arginine-specific ADP-ribosyltransferase activity in mouse heart did not differ between wild-type and *Arh1*-deficient mice, suggesting that the accumulation of ADP-ribosylarginine content is due to ARH1 deficiency in $Arh1^{-/-}$ mice [27]. These data imply that cardiomyocytes are undergoing an arginine-specific ADP-ribosylation cycle in vivo. It is not surprising that this hypothesis raises questions regarding how an extracellular protein, GPI-linked ART1, catalyzes ADP-ribosylation in the extracellular space where the NAD^+ concentration is 0.1 μM [11,13,57] and how cytoplasmic protein ARH1 hydrolyzes ADP-ribosyated proteins synthesized by an extracellular enzyme ART1. There is evidence that cellular NAD^+ may be released into the extracellular matrix during inflammation and under pathological circumstances where cells may be killed by ischemic stress, thus providing substrate for the ADP-ribosyltransferases [58–60]. Additional evidence suggests that serum TRIM72 is released from injured or dead cells and is detectable following muscle injury induced by cardiac ischemia-reperfusion and treadmill exercise [61,62]. Furthermore, ART1, ARH1, caveolin-3, and cytoplasmic membrane repair protein TRIM72 were detected in macromolecule complexes [27]. Altogether, cytoplasmic ARH1 may leak with TRIM72 and NAD^+ into the extracellular space where ART1 resides. In the last step of the cycle, the release of ADP-ribose from ADP-ribosylated TRIM72 by ARH1 promotes oligomerization of TRIM72 and recruitment of TRIM72 to the site of injury [27]. Thus, ART1-TRIM72-ARH1 appears to constitute an ADP-ribosylation cycle.

3. Functions of ARH1 in Disease

3.1. Defense Mechanism against the Action of Cholera Toxin

According to the World Health Organization (WHO), each year, there are still about 1.3 million to 4 million cases of cholera and 21,000 to 143,000 deaths worldwide [63]. Cholera toxin produced by *Vibrio cholerae* consists of a catalytic A-subunit, which dissociates from its B-subunits in the ER and catalyzes ADP-ribosylation of the α subunit of the intestinal Gs protein (Gαs) [64,65]. ADP-ribosylated Gαs at arginine 187 is incapable of hydrolyzing GTP and remains in an active state, resulting in stimulation of adenylyl cyclase (AC) and increased cyclic AMP (cAMP) formation. In this model, increased cAMP activates the cystic fibrosis transmembrane conductance regulator (CFTR) chloride channel, which leads to a loss of Cl^-, Na^+, and water in the intestinal lumen, causing the devastating diarrhea characteristic of cholera [26,64,66,67].

Cholera toxin ADP-ribosylates arginine moieties in a number of proteins, e.g., unidentified 18-, 98-, and 200-kDa proteins, however, Gαs including Gαs-S and Gαs-L appears to be the predominant protein that is ADP-ribosylated on arginine by cholera toxin [68]. The amount of ADP-ribosyl Gαs in the presence of cholera toxin was greater in *Arh1*-deficient mice, which suggests that, in wild-type mice, ARH1 cleaves ADP-ribose from Gαs, thereby generating unmodified Gαs and reducing fluid accumulation caused by cholera toxin [26]. Indeed, *Arh1*-deficient mice were more sensitive to cholera toxin-stimulated fluid accumulation in intestinal loops than wild-type mice [69]. In addition, the ARH1-based host defense mechanism occurs in a gender-specific manner. Female *Arh1*-deficient mice were more sensitive to the cholera toxin than were male mice [69]. Male and female wild-type mice, however, did not show a difference in cholera toxin sensitivity. The knockout mice but not the wild-type mice data supported the finding that women had a higher prevalence of cholera than men [70]. In addition to differences in ARH1 in humans and mice, these gender effects in humans may result from other factors such as societal norms (e.g., domestic responsibility for caring of the sick, time spent at home, and accessibility to health care), rather than biological differences in reaction to cholera and/or cholera toxin.

3.2. Tumor-Suppressor Function of ARH1

3.2.1. Increased Tumor Formation in *Arh1*-Deficient and *Arh1*-Heterozygous Mice

Increased tumorigenesis was seen in *Arh1*-deficient and *Arh1*-heterozygous mice [2,40]. During a 24-months observation period, 20.5% (32 out of 156 mice) of *Arh1*-deficient and 11% (19 out of 169 mice) of *Arh1*-heterozygous mice showed increased frequency and extent of tumors in multiple organs, e.g., adenocarcinoma in lung, uterus, and mammary gland; hepatocellular carcinoma; hepatic and gastrointestinal lymphoma; hemangiosarcoma [40]. Tumors between $Arh1^{-/-}$ and $Arh1^{+/-}$ mice differed in the age of appearance with $Arh1^{-/-}$ and $Arh1^{+/-}$ mice, showing tumors at 3 months and 6 months, respectively [40]. Consistent with increased in vivo tumorigenesis, mouse embryonic fibroblasts (MEFs) generated from *Arh1* knockout and heterozygous mice showed increased cell proliferation and tumor formation in nude mice compared to wild-type MEFs [40]. Furthermore, *Arh1*-knockout MEFs transformed with an inactive double-mutant (D60, 61A) *Arh1* gene [7,40] did not rescue the *Arh1* knockout MEFs and showed increased cell proliferation as well as tumor formation in nude mice [2]. In agreement, overexpression of active ARH1 protein in *Arh1*-deficient MEFs partially reversed the tendency to develop tumors [2]. Other $Arh1^{+/-}$ MEFs that developed tumors in nude mice showed loss of heterozygosity of the remaining *Arh1* gene. Tumorigenic MEFs with *Arh1* gene heterozygosity showed a mutation in the remaining allele and expressed a low level of ARH1 activity [2]. These data are consistent with a tumor-suppressor function of ARH1.

The transmembrane ecto-enzyme CD38 functions as a NAD glycohydrolase and an ADP-ribosyl cyclase and is an NAD^+-dependent oncogene [71]. Consistent with this hypothesis, CD38 was overexpressed in 41% (11 out of 27 human lung tumor samples) of tumor cells [49]. The anti-CD38 monoclonal antibody, daratumumab (DARA), is an approved treatment for patients with multiple myeloma [72]. CD38 activities were inhibited by ADP-ribosylation on arginine [71]; it is not known whether ADP-ribosylated CD38 is a substrate of ARH1. Deletion of the *Cd38* gene reduced tumor formation in both *Arh1*-deficient and wild-type mice [49], with significant reductions in the incidence of lymphomas, adenocarcinoma, and hemangio/histolytic sarcomas [49]. Knockout of CD38 in A549 human adenocarcinoma cells inhibited anchorage-independent cell growth, cell invasion, and xenograft growth in nude mice [49]. In contrast, *Arh1*-deficiency in MEFs affected cell cycle progression, resulting in increased cell proliferation [40]. These data suggest that ARH1 affects the cell cycle, preventing tumor formation, rather than control cell migration, as is the case with CD38-mediated metastasis [22,49]. In addition, estrogen promoted the survival rate of *Arh1*-deficient MEFs in the murine circulation and increased tumor metastasis to the lung [73]. As described above, increased tumor formation was dependent on the loss of ARH1 activity. Thus, ARH1 plays a role in cell proliferation in response to modifiers of tumorigenesis, e.g., CD38, estrogen.

3.2.2. ARH1 Heterozygosity and Tumorigenesis

As noted earlier, a 24-month observation of ARH1 littermates from birth revealed that *Arh1*-deficient mice showed a 1.8× higher incidence of tumor formation than *Arh1*-heterozygous mice. However, between the ages of 24 and 33 months, the frequency of tumors seen in *Arh1*-deficient and heterozygous mice was similar, 31% and 28%, respectively. This age-dependent increased occurrence of malignancy in *Arh1*-heterozygous mice resulted in a loss of heterozygosity (LOH) and an absence or mutation of the *Arh1* gene. Mutation of the good allele in the ARH1 heterozygous mice resulted in an ARH1 protein whose activity was between 4% to 55% of the wild-type ARH1 [2,40]. *Arh1* gene mutation in MEFs and $Arh1^{+/-}$ heterozygous mice tended to be in exons 2 and 3 that is comparable to the human ARH1 catalytic site in exons 3 and 4 [2,40]. In the human cancer database, LOH of the *ARH1* gene was identified in the lung (15%) and kidney (18%) [2]. According to the human somatic tumor mutation database, human *ARH1* gene mutations observed in cancer were also located in the human ARH1 catalytic region that corresponds to the mutation sites in mouse tumors [2]. Based on these findings, ARH1 in the murine model appears to be applicable to human cancer studies. Together, these data support the hypothesis that ARH1 is a tumor-suppressor gene that participates in the pathogenesis of both human and mouse cancers.

3.3. *Membrane Repair Function of ARH1*

Arh1-deficient 8-month-old mice developed cardiomyopathy with myocardial fibrosis. Cardiac fibrosis occurred in a gender-specific manner with fibrosis in *Arh1*-knockout male mice being 10× greater than in female mice [27]. Cardiac fibrosis is characterized by increased collagen type I deposition due to aging or as a result of injury, e.g., myocardial infarction, hypertensive heart disease, idiopathic dilated cardiomyopathy, and diabetic hypertrophic cardiomyopathy [74]. In contrast, regardless of sex, during dobutamine-induced stress, *Arh1*-knockout mice showed significantly lower ejection fraction and fractional shortening than wild-type mice, consistent with systolic dysfunction [27]. The membrane repair protein TRIM72 was identified as a substrate for ARH1 and ART1 [27]. TRIM72 had been described as an essential molecule of the membrane repair process, recruiting intracellular vesicles to sites of membrane disruption [75,76]. Cytoplasmic protein TRIM72 leaks from injured cardiac tissue into serum [61], thus serum TRIM72 may be a potential biomarker of acute cardiac injury. TRIM72 was ADP-ribosylated on arginines 207 and 260 [37,39]. The endogenous ADP-ribosylated TRIM72 level was elevated in *Arh1*-deficient mice following cardiac ischemia-reperfusion injury.

In rat cardiac myocytes, greater than 80% of cellular NAD^+ is in mitochondria [77]. Indeed, NAD^+ release from mitochondria in cytosol protects myocytes from post-ischemic reperfusion injury [77]. The importance of ARH1, ART1, and TRIM72 ADP-ribosylation cycle in plasma membrane repair and wound healing was demonstrated using a laser injury model and scratch wound-healing assay in C2C12 myotubes after stable transformation with TRIM72, ARH1, ART1, and ARH1 plus ART1 shRNA, and transient transformation with wild-type TRIM72-GFP and double mutant TRIM72 (R207K, R260K)-GFP that is not ADP-ribosylated [27]. In addition, heterogeneous complexes containing TRIM72 with components of a reversible ADP-ribosylation cycle included ART1, ARH1, caveolin-3 [27,75,78,79]. Notably, the mono-ADP-ribosyltransferase inhibitors vitamin K_1 and novobiocin, as well as the loss of ARH1 activity, inhibited the oligomerization of TRIM72, the essential mechanism by which TRIM72 is recruited to the site of membrane injury [27]. Taken together, the arginine mono-ADP-ribosylation cycle controlled by ART1 and ARH1 is fundamental to the oligomerization of TRIM72 during the membrane repair process in cardiomyocytes (Figure 1).

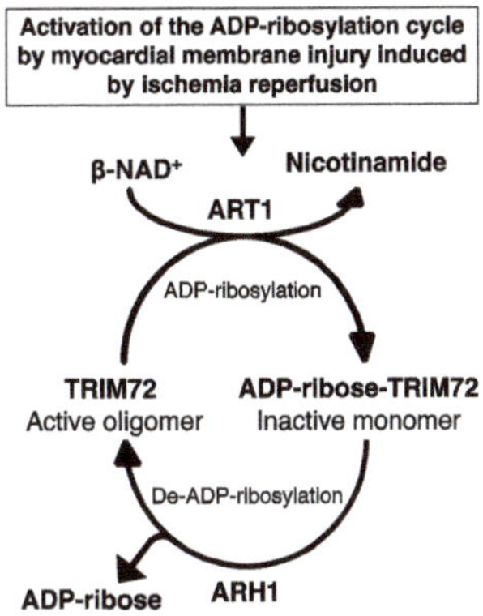

Figure 1. Tripartite motif-containing protein 72 (TRIM72) adenosine diphosphate (ADP)-ribosylation cycle in membrane repair. Ischemia-reperfusion-induced membrane disruption increased the ADP-ribosylation of TRIM72 by ADP-ribosyltransferase (ART) 1 at the sites of membrane damage, facilitating binding of TRIM72 and caveolin-3 to the membrane. Oligomerization of TRIM72 is essential for acute membrane repair and involves the recruitment of TRIM72 and intracellular vesicles at the injury sites [78]. ADP-ribosylarginine hydrolase (ARH) 1 catalyzes de-ADP-ribosylation of modified TRIM72, cleaving the ADP-ribose from ADP-ribosylated (arginine)TRIM72, promoting oligomerization of TRIM72 at the sites of injury.

4. Proteomics

Reanalysis of phosphoproteomic data identified arginine mono-ADP-ribosylation sites on 79 proteins [37]. Using a peptide-based enrichment strategy, 830 proteins containing ADP-ribosylated arginine were identified in mouse liver after H_2O_2-induced oxidative stress [31]. Arginine was the major (86%) ADP-ribosylated amino acid in mouse liver lysed in RIPA buffer with 40 μM PJ-34 and 1 μM adenosine diphosphate (hydroxymethyl) pyrrolidinediol (ADP-HPD), a specific PARP inhibitor (IC_{50} 120 nM) [6,31]. Following overexpression of murine arginine-specific ADP-ribosyltransferase 2 (ART2) in microglia, 33 arginine ADP-ribosylated proteins were identified [38]. Most recently, 354 proteins ADP-ribosylated on arginine were identified from wild-type mouse heart, mouse skeletal muscle, and C2C12 myotubes (Figure 2) [39]. Six ADP-ribosylated proteins overlapping with mouse heart, mouse skeletal muscle, and C2C12 myotubes were identified (Figure 2), e.g., Golgi apparatus protein 1 (Glg1), basement membrane-specific heparan sulfate proteoglycan core protein (Hspg2), integrin α-7, nidogen-1, protein disulfide-isomerase A3, and TRIM72 (Table 1). HNP-1, integrin α-7, and TRIM72 represent the only previously reported ART1 target proteins [5,27,36,80,81]. Among these 354 proteins, only 6 proteins were identified in *Art1*-deficient mouse skeletal muscle (Table 2) [39]. These data support the hypothesis that ART1 is the main contributor to skeletal muscle and heart arginine ADP-ribosylation.

Together, recent advances in proteomic analysis uncovered more than 1000 proteins ADP-ribosylated on arginine residues [31,32,37–39]. ADP-ribosylome data, analyzed by the gene ontology, suggested that ADP-ribosylated proteins of the wild-type mouse heart, but not *Art1*-deficient mouse heart (i.e., ART1-independent ADP-ribosylated proteins), are involved in the regulation of muscle contraction and apoptotic processes [39]. Further analysis by STRING-based protein-protein interaction analysis identified arginine ADP-ribosylated protein interaction networks that are involved in stress response, wounding response, and regulation of the heart rate [39]. Thus, arginine-specific ADP-ribosylation cycle controlled by ART1 and ARH1 is important in muscle physiology and pathophysiology.

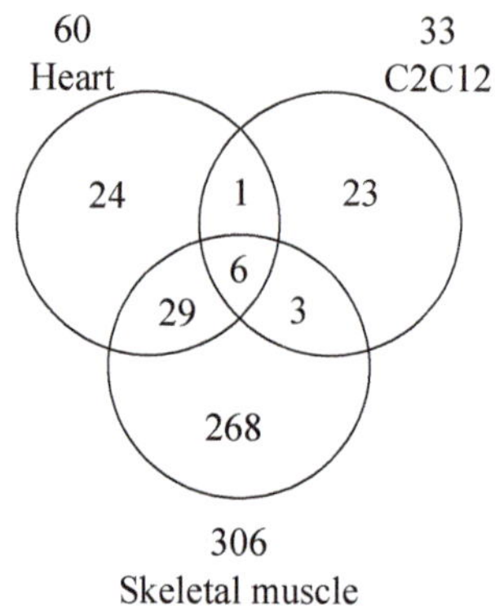

Figure 2. Venn diagram of proteins ADP-ribosylated on arginine in wild-type mouse heart, skeletal muscle, and C2C12 myotubes. Data are modified from Leutert et al. [39].

Table 1. Proteins adenosine diphosphate (ADP)-ribosylated on arginine residues in C2C12 myotubes, mouse skeletal muscle, and mouse heart. Data derived from Leutert et al. [39].

Gene Name	Sample	ADP-Ribosylation Sites	Protein Accesion	Protein Description
Glg1	C2C12 myotubes	R94	tr|F8WHM5|F8WHM5_MOUSE	Golgi apparatus protein 1 (Fragment)
	Skeletal muscle	R94, R313, R909, R1071	tr|F8WHM5|F8WHM5_MOUSE	Golgi apparatus protein 1 (Fragment)
	Heart	R94	tr|F8WHM5|F8WHM5_MOUSE	Golgi apparatus protein 1 (Fragment)
Hspg2	C2C12 myotubes	R588	tr|B1B0C7|B1B0C7_MOUSE	Basement membrane-specific heparan sulfate proteoglycan core protein
	Skeletal muscle	R588, R1956, R2957, R4018, R4148	tr|B1B0C7|B1B0C7_MOUSE	Basement membrane-specific heparan sulfate proteoglycan core protein
	Heart	R588	tr|B1B0C7|B1B0C7_MOUSE	Basement membrane-specific heparan sulfate proteoglycan core protein
Itga7	C2C12 myotubes	R149, R898	sp|Q61738-2|ITA7_MOUSE	Isoform Alpha-7X1A of Integrin alpha-7
	Skeletal muscle	R548	tr|G3X9Q1|G3X9Q1_MOUSE	Integrin alpha 7
	Heart	R608, R896	sp|Q61738-2|ITA7_MOUSE	Isoform Alpha-7X1A of Integrin alpha-7
Nid1	C2C12 myotubes	R318	sp|P10493|NID1_MOUSE	Nidogen-1
	Skeletal muscle	R318, R349, R799	sp|P10493|NID1_MOUSE	Nidogen-1
	Heart	R318	sp|P10493|NID1_MOUSE	Nidogen-1
Pdia3	C2C12 myotubes	R39, R62	tr|F6Q404|F6Q404_MOUSE	Protein disulfide-isomerase A3 (Fragment)
	Skeletal muscle	R39	tr|F6Q404|F6Q404_MOUSE	Protein disulfide-isomerase A3 (Fragment)
	Heart	R62	sp|P27773|PDIA3_MOUSE	Protein disulfide-isomerase A3
Trim72	C2C12 myotubes	R118	sp|Q1XH17|TRI72_MOUSE	Tripartite motif-containing protein 72
	Skeletal muscle	R115, R118, R207, R371	sp|Q1XH17|TRI72_MOUSE	Tripartite motif-containing protein 72
	Heart	R118, R260, R207	sp|Q1XH17|TRI72_MOUSE	Tripartite motif-containing protein 72

Table 2. Characterization of ADP-ribosylation by ADP-ribosyltransferase (ART)1 of mouse heart and skeletal muscle proteins. *Art1*-deficient mouse heart and skeletal muscle contained very few proteins ADP-ribosylated on arginine residues compared to the wild-type, which suggested that ART1 is a major contributor to ADP-ribosylation in skeletal muscle and heart. Data derived from Figure 3B from Leutert et al. [39].

Number of ADP-Ribosylated Protein		**ADP-Ribosylated Amino Acid**					
		Arg	**Glu**	**Lys**	**Ser**	**Asp**	**Met**
C2C12 cells	Wild-type	33	0	1	1	1	0
Skeletal muscle	Wild-type	303	0	2	1	1	0
	Art1-KO	6	2	2	4	1	0
Heart	Wild-type	60	5	5	7	2	1
	Art1-KO	0	4	4	7	2	0

5. Conclusions

Increasing evidence from knockout mouse models, where the ADP-ribosylation cycle is disrupted, shows the importance of arginine-specific ADP-ribosyltion cycles in disease, e.g., cancer [2,40], bacterial toxin-mediated infection [26,69], cardiomyopathy with myocardial fibrosis [27], and muscle weakness [39]. Data are consistent with ART1 and ARH1 serving as opposing arms of an arginine-specific ADP-ribosylation cycle. The COSMIC database analysis of human somatic mutations in cancer revealed 32 *ARH1* mutations in human lung, breast, and colon cancers, overlapping with the mutations found in *Arh1*-heterozygous mice after 6 months of age [2], demonstrating that ARH1 is an age-related cancer risk factor. In addition, *Arh1* deficiency resulted in gender-biased phenotypes. *Arh1*-deficient mice showed a female-biased increase in tumorigenicity and susceptibility to cholera [26,49,82]. In contrast, cardiomyopathy with myocardial fibrosis was seen in male more than female *Arh1*-deficient mice [27]. Recently, advances in mass spectrometry-based proteomics identified numerous arginine ADP-ribosylated proteins as well as the location of their modification sites in vitro. The role of the modification on function has been demonstrated for a limited number of proteins such as Gαs, TRIM72, and HNP-1, probably due to difficulties in reproducing mono-ADP-ribosylation, in vivo. Finding substrates of ARH1 can lead to studies of physiologically and pathologically relevant conditions for deciphering ARH1 functions in health and disease.

Author Contributions: Writing—original draft preparation, H.I.-E.; Writing—review and editing, H.I.-E., L.A.S., J.K. and J.M.; Supervision—J.M. All authors have read and agreed to the published version of the manuscript.

Funding: This research was supported by the Intramural Research Program of the National Institutes of Health, National Heart, Lung, and Blood Institute.

Conflicts of Interest: All authors declare no conflict of interest.

References

1. Oka, S.; Kato, J.; Moss, J. Identification and characterization of a mammalian 39-kDa poly(ADP-ribose) glycohydrolase. *J. Biol. Chem.* **2006**, *281*, 705–713. [CrossRef]
2. Kato, J.; Vekhter, D.; Heath, J.; Zhu, J.; Barbieri, J.T.; Moss, J. Mutations of the functional ARH1 allele in tumors from ARH1 heterozygous mice and cells affect ARH1 catalytic activity, cell proliferation and tumorigenesis. *Oncogenesis* **2015**, *4*, e151. [CrossRef]
3. Takada, T.; Iida, K.; Moss, J. Cloning and site-directed mutagenesis of human ADP-ribosylarginine hydrolase. *J. Biol. Chem.* **1993**, *268*, 17837–17843.
4. Moss, J.; Tsai, S.C.; Adamik, R.; Chen, H.C.; Stanley, S.J. Purification and characterization of ADP-ribosylarginine hydrolase from turkey erythrocytes. *Biochemistry* **1988**, *27*, 5819–5823. [CrossRef]
5. Stevens, L.A.; Kato, J.; Kasamatsu, A.; Oda, H.; Lee, D.Y.; Moss, J. The ARH and Macrodomain Families of alpha-ADP-ribose-acceptor Hydrolases Catalyze alpha-NAD(+) Hydrolysis. *ACS Chem. Biol.* **2019**, *4*, 2576–2584. [CrossRef]

6. Rack, J.G.M.; Ariza, A.; Drown, B.S.; Henfrey, C.; Bartlett, E.; Shirai, T.; Hergenrother, P.J.; Ahel, I. (ADP-ribosyl)hydrolases: Structural Basis for Differential Substrate Recognition and Inhibition. *Cell Chem. Biol.* **2018**, *25*, 1533–1546. [CrossRef]
7. Konczalik, P.; Moss, J. Identification of critical, conserved vicinal aspartate residues in mammalian and bacterial ADP-ribosylarginine hydrolases. *J. Biol. Chem.* **1999**, *274*, 16736–16740. [CrossRef]
8. Munnur, D.; Ahel, I. Reversible mono-ADP-ribosylation of DNA breaks. *FEBS J.* **2017**, *284*, 4002–4016. [CrossRef]
9. Munnur, D.; Bartlett, E.; Mikolcevic, P.; Kirby, I.T.; Matthias Rack, J.G.; Mikoc, A.; Cohen, M.S.; Ahel, I. Reversible ADP-ribosylation of RNA. *Nucleic Acids Res.* **2019**, *47*, 5658–5669. [CrossRef]
10. Okazaki, I.J.; Moss, J. Glycosylphosphatidylinositol-anchored and secretory isoforms of mono-ADP-ribosyltransferases. *J. Biol. Chem.* **1998**, *273*, 23617–23620. [CrossRef]
11. Okazaki, I.J.; Kim, H.J.; Moss, J. Cloning and characterization of a novel membrane-associated lymphocyte NAD:arginine ADP-ribosyltransferase. *J. Biol. Chem.* **1996**, *271*, 22052–22057. [CrossRef]
12. Okazaki, I.J.; Moss, J. Mono-ADP-ribosylation: A reversible posttranslational modification of proteins. *Adv. Pharmacol.* **1996**, *35*, 247–280.
13. Corda, D.; Di Girolamo, M. Mono-ADP-ribosylation: A tool for modulating immune response and cell signaling. *Sci. STKE* **2002**, *2002*, pe53. [CrossRef]
14. Vyas, S.; Matic, I.; Uchima, L.; Rood, J.; Zaja, R.; Hay, R.T.; Ahel, I.; Chang, P. Family-wide analysis of poly(ADP-ribose) polymerase activity. *Nat. Commun.* **2014**, *5*, 4426. [CrossRef]
15. Abplanalp, J.; Hottiger, M.O. Cell fate regulation by chromatin ADP-ribosylation. *Semin. Cell Dev. Biol.* **2017**, *63*, 114–122. [CrossRef]
16. Rosenthal, F.; Feijs, K.L.; Frugier, E.; Bonalli, M.; Forst, A.H.; Imhof, R.; Winkler, H.C.; Fischer, D.; Caflisch, A.; Hassa, P.O.; et al. Macrodomain-containing proteins are new mono-ADP-ribosylhydrolases. *Nat. Struct. Mol. Biol.* **2013**, *20*, 502–507. [CrossRef]
17. Moss, J.; Oppenheimer, N.J.; West, R.E., Jr.; Stanley, S.J. Amino acid specific ADP-ribosylation: Substrate specificity of an ADP-ribosylarginine hydrolase from turkey erythrocytes. *Biochemistry* **1986**, *25*, 5408–5414. [CrossRef]
18. Fontana, P.; Bonfiglio, J.J.; Palazzo, L.; Bartlett, E.; Matic, I.; Ahel, I. Serine ADP-ribosylation reversal by the hydrolase ARH3. *Elife* **2017**, *6*, e28533. [CrossRef]
19. Abplanalp, J.; Leutert, M.; Frugier, E.; Nowak, K.; Feurer, R.; Kato, J.; Kistemaker, H.V.A.; Filippov, D.V.; Moss, J.; Caflisch, A.; et al. Proteomic analyses identify ARH3 as a serine mono-ADP-ribosylhydrolase. *Nat. Commun.* **2017**, *8*, 2055. [CrossRef]
20. Feijs, K.L.; Forst, A.H.; Verheugd, P.; Luscher, B. Macrodomain-containing proteins: Regulating new intracellular functions of mono(ADP-ribosyl)ation. *Nat. Rev. Mol. Cell Biol.* **2013**, *14*, 443–451. [CrossRef]
21. Kasamatsu, A.; Nakao, M.; Smith, B.C.; Comstock, L.R.; Ono, T.; Kato, J.; Denu, J.M.; Moss, J. Hydrolysis of O-acetyl-ADP-ribose isomers by ADP-ribosylhydrolase 3. *J. Biol. Chem.* **2011**, *286*, 21110–21117. [CrossRef]
22. Bu, X.; Kato, J.; Moss, J. Emerging roles of ADP-ribosyl-acceptor hydrolases (ARHs) in tumorigenesis and cell death pathways. *Biochem. Pharmacol.* **2019**, *167*, 44–49. [CrossRef]
23. Ono, T.; Kasamatsu, A.; Oka, S.; Moss, J. The 39-kDa poly(ADP-ribose) glycohydrolase ARH3 hydrolyzes O-acetyl-ADP-ribose, a product of the Sir2 family of acetyl-histone deacetylases. *Proc. Natl. Acad. Sci. USA* **2006**, *103*, 16687–16691. [CrossRef]
24. Smith, S.J.; Towers, N.; Saldanha, J.W.; Shang, C.A.; Mahmood, S.R.; Taylor, W.R.; Mohun, T.J. The cardiac-restricted protein ADP-ribosylhydrolase-like 1 is essential for heart chamber outgrowth and acts on muscle actin filament assembly. *Dev. Biol.* **2016**, *416*, 373–388. [CrossRef]
25. Moss, J.; Stanley, S.J.; Nightingale, M.S.; Murtagh, J.J., Jr.; Monaco, L.; Mishima, K.; Chen, H.C.; Williamson, K.C.; Tsai, S.C. Molecular and immunological characterization of ADP-ribosylarginine hydrolases. *J. Biol. Chem.* **1992**, *267*, 10481–10488.
26. Kato, J.; Zhu, J.; Liu, C.; Moss, J. Enhanced sensitivity to cholera toxin in ADP-ribosylarginine hydrolase-deficient mice. *Mol. Cell. Biol.* **2007**, *27*, 5534–5543. [CrossRef]
27. Ishiwata-Endo, H.; Kato, J.; Tonouchi, A.; Chung, Y.W.; Sun, J.; Stevens, L.A.; Zhu, J.; Aponte, A.M.; Springer, D.A.; San, H.; et al. Role of a TRIM72 ADP-ribosylation cycle in myocardial injury and membrane repair. *JCI Insight* **2018**, *3*, e97898. [CrossRef]

28. Massaeli, H.; Viswanathan, D.; Pillai, D.G.; Mesaeli, N. Endoplasmic reticulum stress enhances endocytosis in calreticulin deficient cells. *Biochim. Biophys. Acta Mol. Cell. Res.* **2019**, *1866*, 727–736. [CrossRef]
29. Clarke, M.S.; Caldwell, R.W.; Chiao, H.; Miyake, K.; McNeil, P.L. Contraction-induced cell wounding and release of fibroblast growth factor in heart. *Circ. Res.* **1995**, *76*, 927–934. [CrossRef]
30. Fabrizio, G.; Di Paola, S.; Stilla, A.; Giannotta, M.; Ruggiero, C.; Menzel, S.; Koch-Nolte, F.; Sallese, M.; Di Girolamo, M. ARTC1-mediated ADP-ribosylation of GRP78/BiP: A new player in endoplasmic-reticulum stress responses. *Cell. Mol. Life Sci.* **2015**, *72*, 1209–1225. [CrossRef]
31. Martello, R.; Leutert, M.; Jungmichel, S.; Bilan, V.; Larsen, S.C.; Young, C.; Hottiger, M.O.; Nielsen, M.L. Proteome-wide identification of the endogenous ADP-ribosylome of mammalian cells and tissue. *Nat. Commun.* **2016**, *7*, 12917. [CrossRef]
32. Laing, S.; Koch-Nolte, F.; Haag, F.; Buck, F. Strategies for the identification of arginine ADP-ribosylation sites. *J. Proteomics* **2011**, *75*, 169–176. [CrossRef]
33. Tsai, Y.L.; Zhang, Y.; Tseng, C.C.; Stanciauskas, R.; Pinaud, F.; Lee, A.S. Characterization and mechanism of stress-induced translocation of 78-kilodalton glucose-regulated protein (GRP78) to the cell surface. *J. Biol. Chem.* **2015**, *290*, 8049–8064. [CrossRef]
34. Amoroso, M.R.; Matassa, D.S.; Laudiero, G.; Egorova, A.V.; Polishchuk, R.S.; Maddalena, F.; Piscazzi, A.; Paladino, S.; Sarnataro, D.; Garbi, C.; et al. TRAP1 and the proteasome regulatory particle TBP7/Rpt3 interact in the endoplasmic reticulum and control cellular ubiquitination of specific mitochondrial proteins. *Cell Death Differ.* **2012**, *19*, 592–604. [CrossRef]
35. Paone, G.; Stevens, L.A.; Levine, R.L.; Bourgeois, C.; Steagall, W.K.; Gochuico, B.R.; Moss, J. ADP-ribosyltransferase-specific modification of human neutrophil peptide-1. *J. Biol. Chem.* **2006**, *281*, 17054–17060. [CrossRef]
36. Stevens, L.A.; Barbieri, J.T.; Piszczek, G.; Otuonye, A.N.; Levine, R.L.; Zheng, G.; Moss, J. Nonenzymatic conversion of ADP-ribosylated arginines to ornithine alters the biological activities of human neutrophil peptide-1. *J. Immunol.* **2014**, *193*, 6144–6151. [CrossRef]
37. Matic, I.; Ahel, I.; Hay, R.T. Reanalysis of phosphoproteomics data uncovers ADP-ribosylation sites. *Nat. Methods* **2012**, *9*, 771–772. [CrossRef]
38. Rissiek, B.; Menzel, S.; Leutert, M.; Cordes, M.; Behr, S.; Jank, L.; Ludewig, P.; Gelderblom, M.; Rissiek, A.; Adriouch, S.; et al. Ecto-ADP-ribosyltransferase ARTC2.1 functionally modulates FcgammaR1 and FcgammaR2B on murine microglia. *Sci. Rep.* **2017**, *7*, 16477. [CrossRef]
39. Leutert, M.; Menzel, S.; Braren, R.; Rissiek, B.; Hopp, A.K.; Nowak, K.; Bisceglie, L.; Gehrig, P.; Li, H.; Zolkiewska, A.; et al. Proteomic Characterization of the Heart and Skeletal Muscle Reveals Widespread Arginine ADP-Ribosylation by the ARTC1 Ectoenzyme. *Cell Rep.* **2018**, *24*, 1916–1929. [CrossRef]
40. Kato, J.; Zhu, J.; Liu, C.; Stylianou, M.; Hoffmann, V.; Lizak, M.J.; Glasgow, C.G.; Moss, J. ADP-ribosylarginine hydrolase regulates cell proliferation and tumorigenesis. *Cancer Res.* **2011**, *71*, 5327–5335. [CrossRef]
41. Zhu, J.; Lv, Y.; Han, X.; Xu, D.; Han, W. Understanding the differences of the ligand binding/unbinding pathways between phosphorylated and non-phosphorylated ARH1 using molecular dynamics simulations. *Sci. Rep.* **2017**, *7*, 12439. [CrossRef] [PubMed]
42. Li, H.; Ren, Z.; Kang, X.; Zhang, L.; Li, X.; Wang, Y.; Xue, T.; Shen, Y.; Liu, Y. Identification of tyrosine-phosphorylated proteins associated with metastasis and functional analysis of FER in human hepatocellular carcinoma cells. *BMC Cancer* **2009**, *9*, 366. [CrossRef] [PubMed]
43. Butepage, M.; Eckei, L.; Verheugd, P.; Luscher, B. Intracellular Mono-ADP-Ribosylation in Signaling and Disease. *Cells* **2015**, *4*, 569–595. [CrossRef] [PubMed]
44. Moss, J.; Jacobson, M.K.; Stanley, S.J. Reversibility of arginine-specific mono(ADP-ribosyl)ation: Identification in erythrocytes of an ADP-ribose-L-arginine cleavage enzyme. *Proc. Natl. Acad. Sci. USA* **1985**, *82*, 5603–5607. [CrossRef] [PubMed]
45. Tsuchiya, M.; Tanigawa, Y.; Mishima, K.; Shimoyama, M. Determination of ADP-ribosyl arginine anomers by reverse-phase high-performance liquid chromatography. *Anal. Biochem.* **1986**, *157*, 381–384. [CrossRef]
46. Moss, J.; Zolkiewska, A.; Okazaki, I. ADP-ribosylarginine hydrolases and ADP-ribosyltransferases. Partners in ADP-ribosylation cycles. *Adv. Exp. Med. Biol.* **1997**, *419*, 25–33. [CrossRef]
47. Friedrich, M.; Bohlig, L.; Kirschner, R.D.; Engeland, K.; Hauschildt, S. Identification of two regulatory binding sites which confer myotube specific expression of the mono-ADP-ribosyltransferase ART1 gene. *BMC Mol. Biol.* **2008**, *9*, 91. [CrossRef]

48. Rhodes, D.R.; Yu, J.; Shanker, K.; Deshpande, N.; Varambally, R.; Ghosh, D.; Barrette, T.; Pandey, A.; Chinnaiyan, A.M. ONCOMINE: A cancer microarray database and integrated data-mining platform. *Neoplasia* **2004**, *6*, 1–6. [CrossRef]
49. Bu, X.; Kato, J.; Hong, J.A.; Merino, M.J.; Schrump, D.S.; Lund, F.E.; Moss, J. CD38 knockout suppresses tumorigenesis in mice and clonogenic growth of human lung cancer cells. *Carcinogenesis* **2018**, *39*, 242–251. [CrossRef]
50. Okazaki, I.J.; Zolkiewska, A.; Takada, T.; Moss, J. Characterization of mammalian ADP-ribosylation cycles. *Biochimie* **1995**, *77*, 319–325. [CrossRef]
51. Moss, J.; Vaughan, M. Mechanism of action of choleragen. Evidence for ADP-ribosyltransferase activity with arginine as an acceptor. *J. Biol. Chem.* **1977**, *252*, 2455–2457. [PubMed]
52. Vaughan, M.; Moss, J. Mechanism of action of choleragen. *J. Supramol. Struct.* **1978**, *8*, 473–488. [CrossRef] [PubMed]
53. Coburn, J.; Gill, D.M. ADP-ribosylation of p21ras and related proteins by Pseudomonas aeruginosa exoenzyme S. *Infect. Immun.* **1991**, *59*, 4259–4262. [CrossRef] [PubMed]
54. Ganesan, A.K.; Frank, D.W.; Misra, R.P.; Schmidt, G.; Barbieri, J.T. Pseudomonas aeruginosa exoenzyme S ADP-ribosylates Ras at multiple sites. *J. Biol. Chem.* **1998**, *273*, 7332–7337. [CrossRef]
55. Weng, B.; Thompson, W.C.; Kim, H.J.; Levine, R.L.; Moss, J. Modification of the ADP-ribosyltransferase and NAD glycohydrolase activities of a mammalian transferase (ADP-ribosyltransferase 5) by auto-ADP-ribosylation. *J. Biol. Chem.* **1999**, *274*, 31797–31803. [CrossRef]
56. Glowacki, G.; Braren, R.; Firner, K.; Nissen, M.; Kuhl, M.; Reche, P.; Bazan, F.; Cetkovic-Cvrlje, M.; Leiter, E.; Haag, F.; et al. The family of toxin-related ecto-ADP-ribosyltransferases in humans and the mouse. *Protein Sci.* **2002**, *11*, 1657–1670. [CrossRef]
57. Murry, C.E.; Wiseman, R.W.; Schwartz, S.M.; Hauschka, S.D. Skeletal myoblast transplantation for repair of myocardial necrosis. *J. Clin. Investig.* **1996**, *98*, 2512–2523. [CrossRef]
58. Verderio, C.; Bruzzone, S.; Zocchi, E.; Fedele, E.; Schenk, U.; De Flora, A.; Matteoli, M. Evidence of a role for cyclic ADP-ribose in calcium signalling and neurotransmitter release in cultured astrocytes. *J. Neurochem.* **2001**, *78*, 646–657. [CrossRef]
59. Adriouch, S.; Hubert, S.; Pechberty, S.; Koch-Nolte, F.; Haag, F.; Seman, M. NAD+ released during inflammation participates in T cell homeostasis by inducing ART2-mediated death of naive T cells in vivo. *J. Immunol.* **2007**, *179*, 186–194. [CrossRef]
60. Hubert, S.; Rissiek, B.; Klages, K.; Huehn, J.; Sparwasser, T.; Haag, F.; Koch-Nolte, F.; Boyer, O.; Seman, M.; Adriouch, S. Extracellular NAD+ shapes the Foxp3+ regulatory T cell compartment through the ART2-P2X7 pathway. *J. Exp. Med.* **2010**, *207*, 2561–2568. [CrossRef]
61. Lemckert, F.A.; Bournazos, A.; Eckert, D.M.; Kenzler, M.; Hawkes, J.M.; Butler, T.L.; Ceely, B.; North, K.N.; Winlaw, D.S.; Egan, J.R.; et al. Lack of MG53 in human heart precludes utility as a biomarker of myocardial injury or endogenous cardioprotective factor. *Cardiovasc. Res.* **2016**, *110*, 178–187. [CrossRef] [PubMed]
62. Weisleder, N.; Takizawa, N.; Lin, P.; Wang, X.; Cao, C.; Zhang, Y.; Tan, T.; Ferrante, C.; Zhu, H.; Chen, P.J.; et al. Recombinant MG53 protein modulates therapeutic cell membrane repair in treatment of muscular dystrophy. *Sci. Transl. Med.* **2012**, *4*, 139ra185. [CrossRef] [PubMed]
63. Meeting of the Strategic Advisory Group of Experts on immunization, April 2017—Conclusions and recommendations. *Wkly. Epidemiol. Rec.* **2017**, *92*, 301–320.
64. Cassel, D.; Pfeuffer, T. Mechanism of cholera toxin action: Covalent modification of the guanyl nucleotide-binding protein of the adenylate cyclase system. *Proc. Natl. Acad. Sci. USA* **1978**, *75*, 2669–2673. [CrossRef]
65. Vanden Broeck, D.; Horvath, C.; De Wolf, M.J. Vibrio cholerae: Cholera toxin. *Int. J. Biochem. Cell Biol.* **2007**, *39*, 1771–1775. [CrossRef]
66. Moss, J.; Vaughan, M. Activation of adenylate cyclase by choleragen. *Annu. Rev. Biochem.* **1979**, *48*, 581–600. [CrossRef]
67. Holmgren, J. Actions of cholera toxin and the prevention and treatment of cholera. *Nature* **1981**, *292*, 413–417. [CrossRef]
68. Gill, D.M.; Meren, R. ADP-ribosylation of membrane proteins catalyzed by cholera toxin: Basis of the activation of adenylate cyclase. *Proc. Natl. Acad. Sci. USA* **1978**, *75*, 3050–3054. [CrossRef]

69. Watanabe, K.; Kato, J.; Zhu, J.; Oda, H.; Ishiwata-Endo, H.; Moss, J. Enhanced sensitivity to cholera toxin in female ADP-ribosylarginine hydrolase (ARH1)-deficient mice. *PLoS ONE* **2018**, *13*, e0207693. [CrossRef]
70. Sevilimedu, V.; Pressley, K.D.; Snook, K.R.; Hogges, J.V.; Politis, M.D.; Sexton, J.K.; Duke, C.H.; Smith, B.A.; Swander, L.C.; Baker, K.K.; et al. Gender-based differences in water, sanitation and hygiene-related diarrheal disease and helminthic infections: A systematic review and meta-analysis. *Trans. R. Soc. Trop. Med. Hyg.* **2016**, 637–648. [CrossRef]
71. Han, M.K.; Cho, Y.S.; Kim, Y.S.; Yim, C.Y.; Kim, U.H. Interaction of two classes of ADP-ribose transfer reactions in immune signaling. *J. Biol. Chem.* **2000**, *275*, 20799–20805. [CrossRef] [PubMed]
72. Laubach, J.P.; Tai, Y.T.; Richardson, P.G.; Anderson, K.C. Daratumumab granted breakthrough drug status. *Expert Opin. Investig. Drugs* **2014**, *23*, 445–452. [CrossRef] [PubMed]
73. Kato, J.R.; Bu, X.N.; Moss, J. Estrogen promotes tumorigenesis by ADP-ribosyl-acceptor hydrolase 1 (ARH1)-deficient cells and mice. *Cancer Res.* **2014**, *74*, 2445. [CrossRef]
74. Hinderer, S.; Schenke-Layland, K. Cardiac fibrosis—A short review of causes and therapeutic strategies. *Adv. Drug Deliv. Rev.* **2019**, *146*, 77–82. [CrossRef]
75. Cai, C.; Masumiya, H.; Weisleder, N.; Pan, Z.; Nishi, M.; Komazaki, S.; Takeshima, H.; Ma, J. MG53 regulates membrane budding and exocytosis in muscle cells. *J. Biol. Chem.* **2009**, *284*, 3314–3322. [CrossRef]
76. Jia, Y.; Chen, K.; Lin, P.; Lieber, G.; Nishi, M.; Yan, R.; Wang, Z.; Yao, Y.; Li, Y.; Whitson, B.A.; et al. Treatment of acute lung injury by targeting MG53-mediated cell membrane repair. *Nat. Commun.* **2014**, *5*, 4387. [CrossRef]
77. Di Lisa, F.; Menabo, R.; Canton, M.; Barile, M.; Bernardi, P. Opening of the mitochondrial permeability transition pore causes depletion of mitochondrial and cytosolic NAD+ and is a causative event in the death of myocytes in postischemic reperfusion of the heart. *J. Biol. Chem.* **2001**, *276*, 2571–2575. [CrossRef]
78. Hwang, M.; Ko, J.K.; Weisleder, N.; Takeshima, H.; Ma, J. Redox-dependent oligomerization through a leucine zipper motif is essential for MG53-mediated cell membrane repair. *Am. J. Physiol. Cell Physiol.* **2011**, *301*, C106–C114. [CrossRef]
79. Ma, L.L.; Zhang, F.J.; Qian, L.B.; Kong, F.J.; Sun, J.F.; Zhou, C.; Peng, Y.N.; Xu, H.J.; Wang, W.N.; Wen, C.Y.; et al. Hypercholesterolemia blocked sevoflurane-induced cardioprotection against ischemia-reperfusion injury by alteration of the MG53/RISK/GSK3beta signaling. *Int. J. Cardiol.* **2013**, *168*, 3671–3678. [CrossRef]
80. Zolkiewska, A.; Moss, J. Integrin alpha 7 as substrate for a glycosylphosphatidylinositol-anchored ADP-ribosyltransferase on the surface of skeletal muscle cells. *J. Biol. Chem.* **1993**, *268*, 25273–25276.
81. Stevens, L.A.; Levine, R.L.; Gochuico, B.R.; Moss, J. ADP-ribosylation of human defensin HNP-1 results in the replacement of the modified arginine with the noncoded amino acid ornithine. *Proc. Natl. Acad. Sci. USA* **2009**, *106*, 19796–19800. [CrossRef] [PubMed]
82. Shim, B.; Pacheco-Rodriguez, G.; Kato, J.; Darling, T.N.; Vaughan, M.; Moss, J. Sex-specific lung diseases: Effect of oestrogen on cultured cells and in animal models. *Eur. Respir. Rev.* **2013**, *22*, 302–311. [CrossRef] [PubMed]

cancers

MDPI

Review

PARP Inhibitors as Therapeutics: Beyond Modulation of PARylation

Ahrum Min [1,2] and Seock-Ah Im [1,2,3,4,*]

1 Cancer Research Institute, Seoul National University College of Medicine, Seoul 03080, Korea; mar6716@hanmail.net
2 Biomedical Research Institute, Seoul National University Hospital, Seoul 03080, Korea
3 Department of Internal Medicine, Seoul National University Hospital, Seoul 03080, Korea
4 Translational Medicine, Seoul National University College of Medicine, Seoul 03080, Korea
* Correspondence: moisa@snu.ac.kr; Tel.: +82-2-2072-0850; Fax: +82-2-765-7081

Received: 30 December 2019; Accepted: 5 February 2020; Published: 8 February 2020

Abstract: Poly (ADP-ribose) polymerase (PARP) 1 is an essential molecule in DNA damage response by sensing DNA damage and docking DNA repair proteins on the damaged DNA site through a type of posttranslational modification, poly (ADP-Ribosyl)ation (PARylation). PARP inhibitors, which inhibit PARylation through competitively binding to NAD+ binding site of PARP1 and PARP2, have improved clinical benefits for BRCA mutated tumors, leading to their accelerated clinical application. However, the antitumor activities of PARP inhibitors in clinical development are different, due to PARP trapping activity beyond blocking PARylation reactions. In this review, we comprehensively address the current state of knowledge regarding the mechanisms of action of PARP inhibitors. We will also discuss the different effects of PARP inhibitors in combination with cytotoxic chemotherapeutic agents regarding the mechanism of regulating PARylation.

Keywords: PARP; PARP inhibitors; PARylation; trapping; cancer therapeutic strategy

1. Introduction

All cells have more than tens of thousands of events that damage DNA in multiple ways, ranging from single base mismatches, bulky adducts in DNA bases, intra- and inter-strand DNA crosslinks, to single- and double-strand breaks (SSBs and DSBs) [1,2]. This DNA damage threatens genomic stability. There are DNA damage responses (DDRs), the sophisticated mechanisms of genome protection in cells, that function to activate the cell cycle checkpoint pathway to maintain genome stability by stopping or delaying the cell cycle during DNA damage or unstable DNA replication to allow the repair of damaged DNA lesions. DDRs activate transcription of a repair molecule or pro-apoptotic molecule to cause overexpression of the related molecule. DDRs activate mechanisms to remove uncontrolled damaged cells and to repair DNA damage from apoptosis caused by chromatid instability [3]. Regulation of the DDR pathway is induced by post-translational modification (PTM); poly ADP-Ribosylation (PARylation) is the pivotal PTM that occurs rapidly at the damage site during DDR [4,5].

PARylation is the reaction of transferring ADP-ribose residues to target substrates by ADP-ribosyl transferase using NAD+. It rapidly recognizes multiple types of DNA damage, including SSBs, and is recruited to the damaged site to induce the recruitment of DDR molecules so that the poly (ADP-Ribose) polymerase (PARP), specifically PARP1, PARP2, PARP3, PARP5a, and PARP5b, which are known as the major molecules of DDR, performs poly (ADP-ribose) (PAR) synthesis in humans [6]. PARP1 was proposed as a new treatment for cancer, as the synthetic lethality concept suggested that its depletion in breast-cancer patients with germline mutations in the *BRCA1* or *BRCA2* genes, key molecules in the homologous recombination (HR) pathway, could cause cancer cell death [7,8]. Since it was proven to be true, PARP inhibitors that inhibit DDR resulted in improved clinical benefits and became standard therapy [9–11]. To date, four PARP inhibitors have been approved by the FDA and are being applied clinically. However, while all PARP inhibitors inhibit PARP catalytic activities, they have different cytotoxicities. Therefore, the anti-tumor effects of the PARP inhibitors have been suggested to be due to PARP trapping, as well as the inhibition of the enzymatic activities [12,13].

The catalytic inhibition and trapping effects of PARP are tightly regulated, and the cytotoxicity of each mechanism can cause different reactivities. Therefore, in this review, based on mechanisms of PARP, we intend to examine the difference of anti-tumor effect of the PARP inhibitors and the current aspect of the roles in combination treatment.

2. PARPs and PARylation

Poly (ADP-ribose) polymerase (PARP) is a family of 17 proteins in mammals, encoded by different genes, but with a conserved catalytic domain. Other than the catalytic domain, PARP family members contain one or more other motifs or domains, including zinc fingers, a breast cancer-susceptibility protein (BRCA) C-terminus-like (BRCT) motifs, ankyrin repeats, macro domains, and WWE domains [14] (Figure 1A). PARP1 was the first family member identified and has a critical role in SSB repair through the metabolism of recruiting and dissociating repair proteins by PARylation. In addition to DNA damage repair, PARP1 has important roles in a various range of cellular processes from cell proliferation to cell death, due to having diverse substrates like nuclear proteins involved in transcriptional regulation, apoptotic cell death, chromatin decondensation, inflammation, and cell cycle regulation [15,16]. PARP1 has a total molecular weight of 113 kDa and contains seven independent domains (Figure 1B) [5,17]. The N-terminus is the DNA binding domain (residues 1-353), which contains three zinc-finger DNA-binding domains, ZnFI, ZnFII, and ZnFIII, which are responsible for recognizing sites of damaged DNA and binding through allosteric activation. In the N-terminus there is a nuclear localization sequence (NLS) that places PARP1 in the nucleus with the KRK-X(11)-KKKSKK sequence. Between residues 211 and 214, there is a DEVD site that is cleaved by caspase into fragments of 23 and 89 KDa during apoptosis [18]. Residues 373 to 662 are the auto-modification domain consists of BRCA C-terminus-like (BRCT) domain serving sites of auto-ADP ribosylation and functioning in protein-protein interaction, and a WGR domain which roles in activating DNA damage repair by interaction with ZnFI, ZnFII, and catalytic domain. The auto-modification domain is rich in glutamate and lysine residues and is the site of self-PARylation. Finally, the C-terminus (residues 662–1014) is the catalytic domain, and the (ADP-Ribosyl) transferase (ART) domain is a NAD+ acceptor site where the His-Try-Glu residues called ART signatures are preserved well [19–21]. The helical subdomain (HD), an auto-inhibitory domain in the C-terminus, inhibits the binding of PARP1 and β-nicotinamide adenine dinucleotide without binding to DNA. When PARP1 binds to the DNA damage site, the auto-inhibitory function of HD is removed. The activation of the catalytic activity of ART and the generation of PAR chains in the target protein lead to the recruitment of DNA repair molecules. Thereafter, PARP1 is dissociated from DNA by auto-PARylation of PARP1, resulting in DNA repair [22].

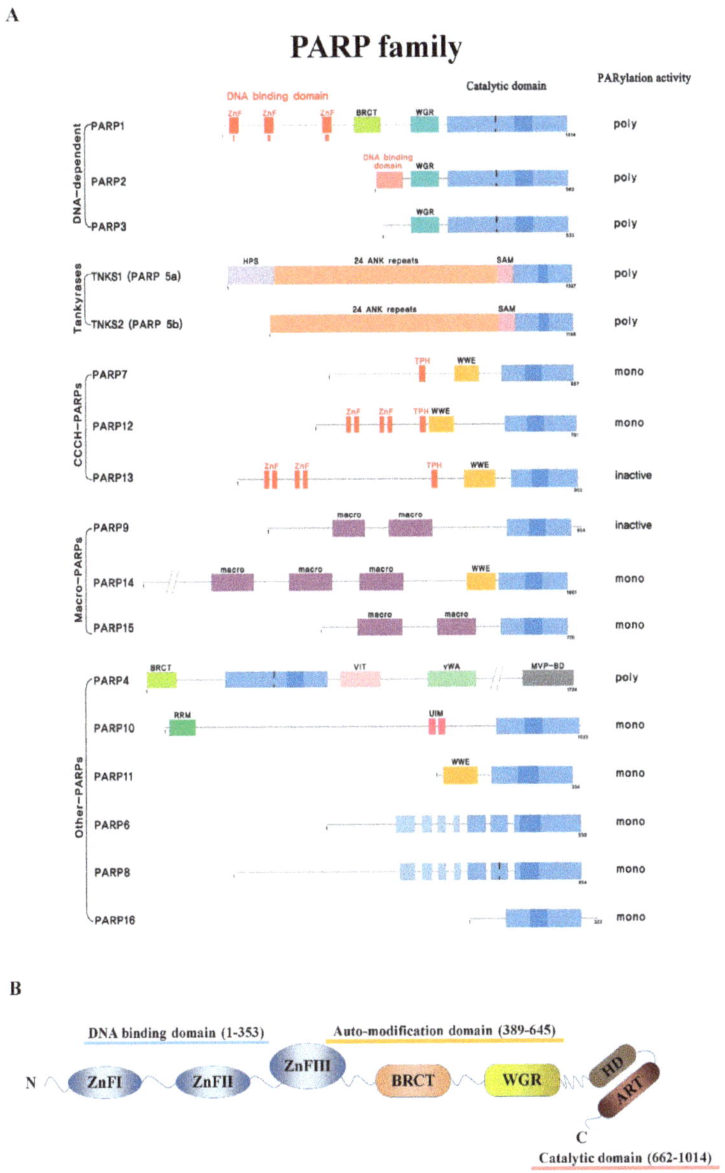

Figure 1. PARPs structure (**A**) The PARP family consists of 17 members, divided into five subgroups according to domain structure and function: DNA damage-dependent PARPs (PARP1, PARP2, and PARP3), tankyrases (tankyrase1/PARP5 and tankyrase2/PARP5b), CCCH-type PARPs (PARP7, PARP12, and PARP13), macro-PARPs [B-aggressive lymphoma 1 (BAL1)/PARP9, BAL2/PARP14, and BAL3/PARP15], and other PARPs (PARP4, PARP6, PARP8, PARP10, PARP11, and PARP16). The catalytic domain at the C-terminus is conserved in all members and contains additional zinc fingers, BRCA C-terminus-like (BRCT) motifs, ankyrin repeats, macro domains, and WWE domains. (**B**) The seven major domains of PARP1 include three zinc-finger domains in the DNA binding domain, the BRCT domain in the auto-modification domain, and the pADPr accepting WGR domain (W), located centrally. The C-terminus has two catalytic domains: ART and a helical domain (HD).

This series of reactions is caused by PARylation. While the catalytic domain is conserved in the PARP family, only PARP1/2/3/4/5a/5b activates PARylation by possessing the His-Tyr-Glu motif called the "ART signature" [5,15,23–26]. The role of PARP3 as an (ADP-ribosyl) transferase is controversial. PARP4 is the largest protein in the PARP family, and PARP5a and PARP5b, classified as tankyrase1/2, have a SAM (Sterile Alpha motif) domain that interacts between proteins with the ability to homo- and hetero-oligomerize PARP1 and 2. PARP1 transfers ADP-ribose residues from NAD+ to acidic amino acid residues such as glutamates (E), lysine (K), arginine (R), serine (S), and aspartate (D), forming the negative poly (ADP-ribose) (PAR) chain [5,24,26]. PARP1 is believed to perform more than 90% of total PARylation in response to DNA damage. As soon as DNA damage occurs, ADP-ribosylation is covalently bound to the carbonyl group of the acidic residues of the target protein via ester bonds. PARP then forms a PAR chain by cleaving the glycosidic bond between nicotinamide and ribose of NAD+ by catalytic activity and binding ADP-ribosylation to the target protein via a 2′,1″-O-glycosidic bond [22,27]. PARylation in the DNA damage repair pathway plays a role throughout DNA strand breaks repair through rapid DNA repair molecules recruitment to the DNA damage site, DNA damage signal transduction, causing apoptosis and protein degradation. Typically, the BRCT domain of X-ray repair cross-complementing protein 1 (XRCC1) binds directly to the PAR chain to be recruited to the DNA damage site. Upon XRCC1 binding to DNA damage sites, the PAR formation is increased by sequestering poly (ADP-ribose) glycohydrolase (PARG) from the interaction with PARP1 and PARG, resulting in causing dissociation of PARP1 from DNA damage site and increasing repair signal transduction [26–28]. In addition, PARP1 has been reported to promote DNA repair by interacting with DNA glycosylase 8-oxoguanine glycosylase 1 (OGG1), XRCC1, DNA polymerase (DNAP) β, DNA ligase III, proliferating cell nuclear antigen (PCNA), aprataxin, and condensin I involved in BER and SSBR through PARylation of PARP1 [29,30].

3. Clinical Development of PARP Inhibitors

PARP inhibitors are nicotinamid analogs that inhibit PARylation through competitively binding to the NAD+ binding sites of PARP1 and PARP2. As a cancer treatment drug, olaparib is first defined as the HR deficient tumor treatment, and it is fully approved by FDA for serous ovarian cancer and breast cancer treatment with germline *BRCA1* or *BRCA2* mutations [9,31,32]. To date, four PARP inhibitors (olaparib, niraparib, rucaparib, and talazoparib) have been FDA approved, and veliparib is waiting for FDA approval with promising results of phase III trial showing significantly extended Progression-free survival (PFS) in combination with carboplatin plus paclitaxel in serous ovarian cancer, and the drugs are compared in Table 1.

Table 1. Comparison of available PARP inhibitors in the clinics.

Agents	Company	Target	Application in Clinics		Dosage	Mean Half-Life (Hours)	Catalytic Inhibition (IC_{50} in Wild-Type DT40 Cells; nM) [13]	PARP Trapping Potency (Relative to Olaparib) [33]	Cytotoxicity (EC_{50} in BRCA2 Mutated Capan-1 Cells; nM) [34]
			Indication	Clinical Trials Based on FDA Approval					
Olaparib	AstraZeneca	PARP1 PARP2 PARP3	Maintenance treatment of germline BRCA-mutated advanced ovarian cancer.	**Study 42** (NCT01078662) [35]	300mg BID	14.9 ± 8.2	6	1	259
			Maintenance treatment of recurrent serous ovarian cancer regardless of BRCA mutations	**SOLO-2** (NCT01874353) [11] **Study 19** (NCT00753545) [36]					
			Treatment of germline BRCA-mutated HER2-negative locally advanced or metastatic breast cancers	**OlympiAD** (NCT02000622) [9]					
			First-line maintenance treatment of germline BRCA-mutated metastatic pancreatic cancer	**POLO trial** (NCT02184195) [37]					
Rucaparib	Clovis Oncology	PARP1 PARP2 PARP3	Treatment of germline and/or somatic BRCA-mutated advanced ovarian cancer	**ARIEL2** (NCT01891344) [38] **Study 10** (NCT01482715) [39]	600mg BID	18 ± 1	21	1	609
			Maintenance treatment in a platinum-sensitive recurrent epithelial ovarian, fallopian tube, or primary peritoneal cancer	**ARIEL3** (NCT01968213) [40]					
Niraparib (MK4827)	Tesaro	PARP1 PARP2	Maintenance treatment of platinum-sensitive, recurrent ovarian cancer	**ENGOT-OV16/NOVA** (NCT01847274) [41]	300mg QD	36	60	2	650
			Treatment of homologous recombination deficiency (HRD) positive advanced ovarian, fallopian tube, or primary peritoneal cancer	**QUADRA** (NCT02354586) [42]					

Table 1. *Cont.*

Agents	Company	Target	Application in Clinics			Mean Half-Life (Hours)	Catalytic Inhibition (IC_{50} in Wild-Type DT40 Cells; nM) [13]	PARP Trapping Potency (Relative to Olaparib) [33]	Cytotoxicity (EC_{50} in BRCA2 Mutated Capan-1 Cells; nM) [34]
			Indication	Clinical Trials Based on FDA Approval	Dosage				
Talazoparib (BMN-673)	Pfizer	PARP1 PARP2	Treatment of germline BRCA-mutated HER2-negative locally advanced or metastatic breast cancers	**EMBRACA** (NCT01945775) [43]	1mg QD	90	4	100	5
Veliparib (ABT-888)	Abbott Laboratories	PARP1 PARP2	Not yet approved for any indication, but in 2014. FDA awards orphan drug designation to Veliparib for advanced non-small cell lung cancer (NSCLC). In phase III trial, veliparib significantly improved progression-free survival in BRCA-mutated or HRD cohort compared to carboplatin plus paclitaxel (NCT02470585) [44]. In phase III trial of veliparib with carboplatin and paclitaxel in advanced HER2-negative breast cancer with germline BRCA mutation, median PFS in patients treated with veliparib plus carboplatin and paclitaxel was 14.5 months compared to 12.6 months in placebo plus carboplatin and paclitaxel (BROCADE3; NCT02163694) [45].			5.2	30	<0.2	>10,000

It has been reported that the PARP inhibitor olaparib causes cell death by synthetic lethality in BRCA-deficient breast, ovarian, and prostate cancer. In a phase 2 trial, olaparib maintenance treatment in platinum-sensitive ovarian cancer patients improved progression-free survival (PFS) by 7 months [46]. In the SOLO-2 study, median PFS increased from 5.5 to 19.1 months over placebo in patients with BRCA1/2 mutations [11]. The OlympiAD trial, which was the global phase 3 trial for metastatic HER2-negative breast cancer with germline *BRCA1* or *BRCA2* mutations, confirmed that olaparib increased PFS by 2.8 months to 7.0 months (hazard ratio 0.58; P<0.001) and received FDA approval [9]. Not only for BRCA-deficient tumors, in the phase 2 STUDY-19 showed olaparib maintenance prolonged PFS in BRCA wild-type relapsed, platinum-sensitive serous ovarian cancer patients, extending the FDA approval to olaparib maintenance treatment in platinum-sensitive patients regardless of the BRCA status [47,48]. Moreover, PARP inhibitors were applied in prostate and pancreatic cancer, as well beyond the breast and ovarian cancers. The developed genomics technology verified that about 20 percent of prostate cancers have defects in DNA repair genes, resulting in a good candidate for PARP inhibitors [49]. In the phase II TOPARP-A trial, olaparib showed an 88 percent response rate in metastatic castrate-resistant prostate cancer (mCRPC) with BRCA1/2, ATM, or PALB2 mutation [50]. The PFS of patients with DNA repair gene defects increased from 2.7 to 9.8 months, and overall survival was also extended from 7.5 to 13.8 months. Based on these results, the FDA granted olaparib in breakthrough status in prostate cancer treatment, and several clinical trials are conducted in prostate cancer using PARP inhibitors. In addition, the median PFS was increased from 3.8 to 7.4 months with olaparib maintenance in the phase 3 POLO trial for metastatic pancreatic cancer patients with germline BRCA mutations who were sensitive to the first-line platinum-based therapy [37]. Recently, the FDA approved olaparib plus bevacizumab for the maintenance treatment of advanced ovarian cancer patients, who showed a response to the first-line platinum-based chemotherapy regardless of BRCA mutation. It is based on phase III PAOLA-1 trial results, in which the addition of olaparib to bevacizumab improved PFS significantly, to 37.2 months compared with 17.7 months in the placebo group among Homologous recombination deficiency (HRD)-positive ovarian cancer patients (HR, 0.33; 95% CI, 0.25–0.45) [51].

Rucaparib received FDA approval by demonstrating efficacy in a phase 2 trial of relapsed platinum-sensitive ovarian cancer patients who had previously received at least two platinum-based chemotherapies [38]. In addition, based on the phase 3 ARIEL3 trial, it received FDA approval for maintenance treatment of recurrent epithelial ovarian, fallopian tube, or primary peritoneal cancer with a partial or complete response to platinum-based chemotherapy [40].

In the case of niraparib, a randomized and double-blind phase 3 trial in 553 platinum-sensitive, recurrent ovarian cancer patients showed an increase in PFS from 5.5 to 21.0 months relative to placebo in the presence of germline BRCA mutations [41]. The FDA approval was granted based on the phase 3 trial on patients with recurrent epithelial ovarian, fallopian tube, or primary peritoneal cancer who were in complete or partial response to platinum-based chemotherapy [41].

Talazoparib has been FDA-approved for its efficacy against cancers with germline *BRCA1* or *BRCA2* mutations and HER2-negative metastatic breast cancer [52]. In the phase 3 EMBRACA trial, upon which FDA approval was based, talazoparib significantly increased median PFS from 5.6 to 8.6 months compared to physician's choice standard-of-care chemotherapy. The objective response rate was more than doubled over that of the control arm (62.6% for talazoparib vs. 27.2% for chemotherapy [OR: 4.99 (95% CI: 2.9–8.8), $p < 0.0001$]) [53].

Veliparib is another potent inhibitor of PARP1 and PARP2 in the developmental stage. In the phase III VELIA trial, which involved adding veliparib to first-line induction chemotherapy with carboplatin and paclitaxel followed by maintenance monotherapy in serous ovarian cancer increased median PFS from 17.3 to 23.5 months regardless of BRCA or HRD status [54]. Besides, veliparib 120mg bid plus carboplatin AUC6 and paclitaxel 200 mg/m^2 treatment showed improved PFS from 4.2 to 5.8 months and OS from 8.4 to 10.3 months in phase II trial of Non-small-cell lung cancer (NSCLC) [55]. These results contributed that FDA grants orphan drug designation to veliparib for advanced NSCLC.

Veliparib, in addition to carboplatin and paclitaxel, also significantly improved PFS to patients with HER2-negative advanced or metastatic breast cancer with germline BRCA-mutation in a phase III trial. The rate of 3-year PFS was 26 percent on veliparib addition group when the placebo group showed 11 percent [45]. Although veliparib is not yet approved to FDA on any indications, these data could accelerate the application of veliparib in the clinics.

PARP inhibitors in the clinic showed improved clinical benefits, not only for tumors with *BRCA* mutations, but also for platinum-sensitive tumors caused by HRD, leading to their accelerated clinical application. However, the effects of these PARP inhibitors are difficult to understand as an inhibition of catalytic activity that simply inhibits PARylation. Although all four PARP inhibitors can inhibit the catalysis by PARP1 and PARP2, as shown in Table 1, each PARP inhibitor shows different cytotoxicity. PARP inhibition by PARP inhibitors induces a cytotoxicity far superior to the cytotoxicity induced by the knockout of PARP genes, suggesting that their antitumor effects are due to mechanisms other than the catalytic inhibition of PARP [12,13,34,56].

This difference can be conceptualized as PARP trapping: the ability of PARP inhibitors to trap PARP-DNA complexes while increasing the stability of the binding between PARP and DNA. As shown in Table 1, each PARP inhibitor has a different cytotoxicity that correlates with its PARP trapping activity. In other words, talazoparib, with the strongest PARP trapping effect, also is the most cytotoxic. Therefore, PARP trapping should be considered as a mechanism for the application of PARP inhibitors in clinical trials [34,57,58]. These differences in PARP trapping capacity may have different effects on combination therapy, as well as on monotherapy. Indeed, reactivities differ with different combination partners for each drug.

4. Combination Effect of Conventional Chemotherapy according to the Mechanism of Action of PARP Inhibitors

PARP inhibitors inhibit the catalytic activity of NAD+ depletion through competitive binding with NAD+, thereby inhibiting PARP itself as well as PARylation of target proteins in the nucleus. Not only do they cause cytotoxicity via irreparable damage caused by inhibiting repair protein recruitment to DNA damage sites, but also, they block cellular replication by inducing stalling or collapsing of the replication fork, as the PARP protein is continuously trapped in SSBs due to suppressed dissociation of PARP from DNA. This results in more deleterious DSBs, which in turn leads to cell death (Figure 2) [59–61].

PARP inhibitors were originally developed to sensitize tumors to the effects of DNA damaging agents, including ionizing radiation, temozolomide, and topotecan. In fact, PARP inhibitors have successfully sensitized tumors to radiation and to camptothecin, a topoisomerase I inhibitor [62,63]. However, the combination of gemcitabine, doxorubicin, and taxan showed no significant synergies [64,65]. It was observed that the efficacy of combination therapy with the same chemoagent was dependent on PARP inhibitors [66]. PARP inhibitors are thought to have different synergies in combinations with different chemoagents, depending on the mechanism and activity of cytotoxicity, as well as the regulation of PARylation. The combined effect of PARP inhibitors with different classes of chemoagents in the clinic is summarized in Table 2, in accordance with the mechanism of action of PARP inhibitors.

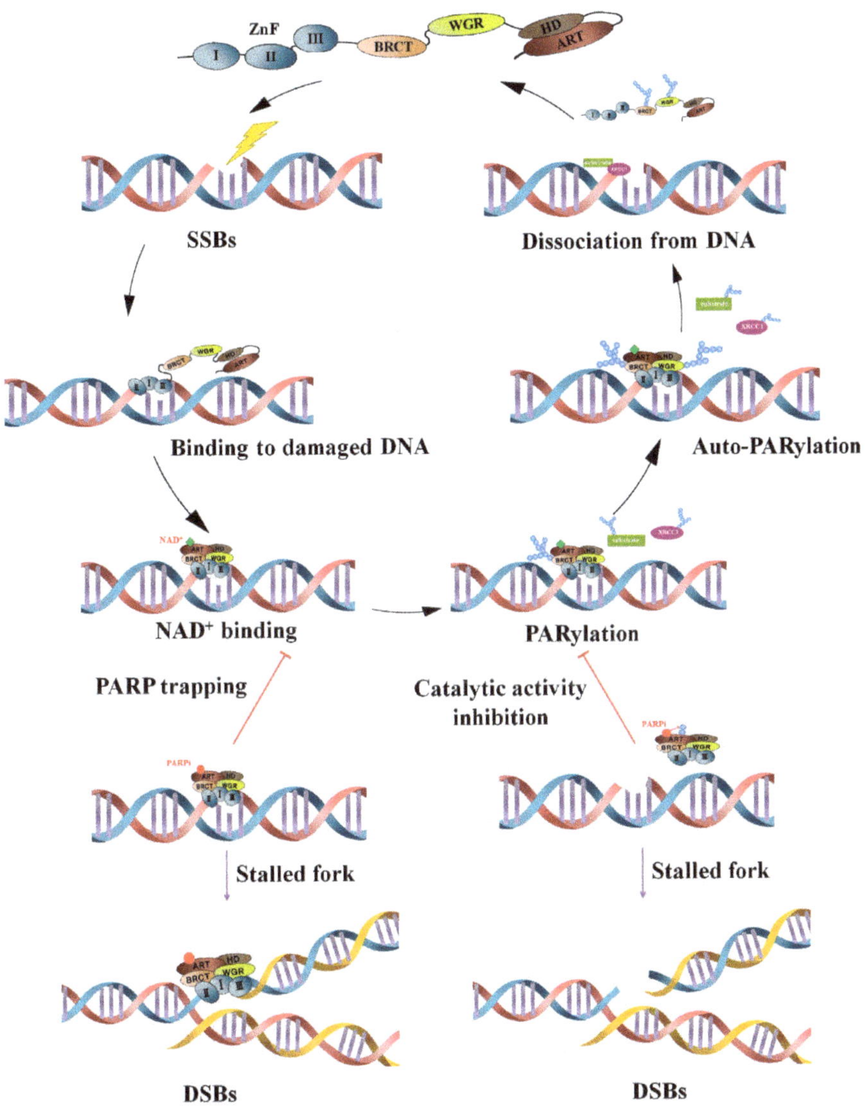

Figure 2. Mechanisms of PARylation and PARP inhibition in the DNA damage response. When a DNA single-strand break occurs, PARP quickly binds to the damage site using a zinc finger domain. It causes recruitment of DNA repair proteins to DNA damage sites by catalyzing PARylation between PARP and its target proteins XRCC1, DNA ligase III, etc., by the ART catalytic domain using NAD+ as a substrate. PARP auto-PARylation then decreases the affinity for DNA, resulting in dissociation from DNA so that the repair protein can bind. At this time, the PARP inhibitor binds to the pocket instead of NAD+, causing PARP to be trapped in the DNA. It encounters a replication fork, causing it to stall, and is converted into double-strand breaks (DSBs) leading to cell death; alternatively, it blocks the recruitment of repair proteins by blocking enzymatic activity where PARP PARylation occurs.

Table 2. Combination effects of PARP inhibitors and chemotherapeutics according to the action mechanism of PARP inhibitors on the combination strategies.

Action Mechanism of PARP Inhibitors on the Combination Effects	Combined Chemotherapeutics: Class of Agents	Combined Chemotherapeutics: Chemotherapy Agents	PARP Inhibitor	Tumor Type	Trial	Phase	Outcome
Inhibition of PARP catalytic activity	Topoisomerase I inhibitors	Topotecan	Olaparib	Advanced solid tumors	NCT00516438 [67]	I	The Maximum Tolerated Dose (MTD) was determined as topotecan 1.0 mg/m^2/day × 3 days plus olaparib 100 mg bid.
			Veliparib	Recurrent cervix cancer	NCT01266447 [68]	II	Topotecan 0.6 mg/m^2/day × 5 days plus veliparib 10 mg bid treatment resulted in 7% partial response (PR).
	Platinum-based inhibitors	Carboplatin	Olaparib	Refractory or recurrent breast and ovarian cancer	NCT01237067 [69]	I	The MTD as olaparib 200 mg bid plus carboplatin AUC4; The responses including CRs and PRs was higher in BRCA mutation carriers compared with nonmutation carriers (68% vs 19%)
			Veliparib	HER2-negative metastatic breast cancer	NCT01251874 [70]	I	The MTD was established as veliparib 250 mg bid plus carboplatin AUC5.
			Rucaparib	Advanced solid tumors	NCT01009190 [71]	I	The MTD for combination was established as 240 mg/day oral rucaparib and carboplatin AUC5.
PARP trapping	Alkylating agents	Temozolomide	Olaparib	Relapsed glioblastoma	NCT01390571 [72]	I	The temozolomide 75 mg/m^2 daily plus olaparib 150 mg/day × 21 days treatment was well tolerated and encouraged 6 months progression-free survival rates.
			Veliparib	Metastatic breast cancer and BRCA1/2 mutated breast cancer	NCT01506609 [73]	II	The responses in the BRCA mutation carriers showed a total response rate (RR) 25% (7/28) and clinical benefit rate (CBR) 50%.
		Mitomycin C	Veliparib	Metastatic, unresectable or recurrent solid tumors	NCT01017640 [74]	I	Veliparib 200 mg bid after treated mitomycin C 10 mg/m^2/day × 21 days was recommended for combination treatment.
Inhibition of PARP transcription cofactor function	Antimetabolite	Gemcitabine	Olaparib	Pancreatic cancer	NCT00515866 [75]	I	Olaparib 100 mg bid plus gemcitabine 600 mg/m^2/week was tolerated and recommended for the phase II trial.
Unknown mechanism	taxanes	paclitaxel	Olaparib	Metastatic triple-negative breast cancer (TNBC)	NCT00707707 [64]	I	Olaparib 200 mg bid daily in combination with paclitaxel 90 mg/m^2/week × 3 of 4 weeks was tolerated; The overall response rate (ORR) was 33.3%.
				Advanced gastric cancer	NCT01063517 [76]	II	Olaparib 100 mg bid plus paclitaxel 80 mg/m^2 treatment showed overall survival and progression-free survival benefit in ATM enriched phase II study.
					NCT01924533 [32]	III	In phase III study, median overall survival was 8.8 months in the weekly paclitaxel 80 mg/m^2 + olaparib 100 mg bid vs 6.9 months in the paclitaxel + placebo group; HR 0.79, $p = 0.026$, without statistical significance.

The table is based on the details from http://www.clinicaltrials.gov/.

Their synergistic effect on chemoagents by inhibition of catalytic activity is well defined by topoisomerase I inhibitors, topotecan and camptothecin. The synergistic effect of combination therapy with alkylating agents such as MMS is well defined by the PARP-DNA trapping activity of PARP inhibitors (Figure 3) [77–79]. First, alkylating agents generate basic sites that consist of a 1-nucleotide gap with 3′-OH and 5′-deoxyribose phosphate (5′-dRP) groups at the ends of the breaks created by APEX1 endonuclease. PARP1 binds directly to 5′-dRP and recruits BER proteins to induce repair. However, PARP inhibitors bind PARP to the 5-dRP end, trapping and maintaining the PARP-DNA complex so that the accumulation of SSBs leads to DSBs, and ultimately, cell death [60,61,66,78,80]. This synergistic effect based on PARP trapping is the strongest for talazoparib. On the other hand, a topoisomerase I inhibitor causes SSBs by the endonuclease activity of TOP1 and causes DNA damage by trapping TOP1cc covalently bonded with TOP1 at the DNA 3′ end. At this time, the PARP inhibitor sustains the trapping of TOP1cc by inhibiting PARP1 recruitment of TDP1 through PARylation, so that TDP1 can remove the covalent attachment of TOP1 by phosphodiesterase activity. At this point, the PARP inhibitor lets the trapping of TOP1cc continue by preventing TDP1 from removing the covalent attachment of TOP1, as PARP1 recruits TDP1 through PARylation. This is due to the regulation of PARylation activity by PARP inhibitors, which has the same effect on all PARP inhibitors developed in the clinic [63,77,79]. Those applying combination therapies of PARP inhibitors and chemotherapy within the clinic should carefully consider the mechanism(s) of action prior to selecting the drug.

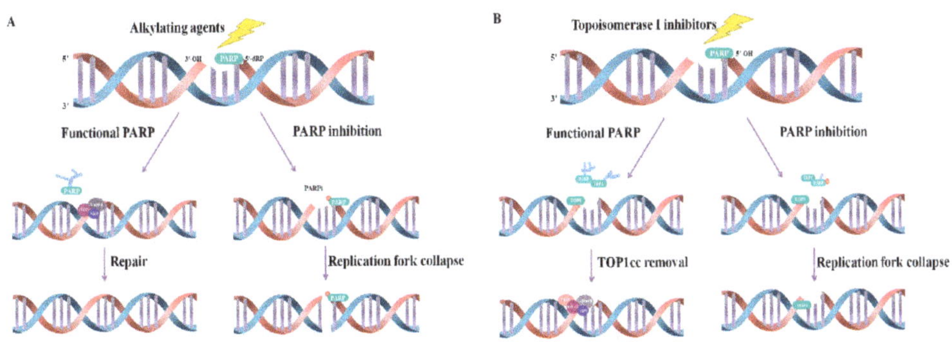

Figure 3. Principle of combination therapy with chemoagents, based on trapping effects and inhibition of catalytic activity. (**A**) An alkylating agent forms a single-nucleotide gap with 5′-deoxyribose phosphate (5-dRP). PARP1/2 senses and binds it, inducing recruitment of BER molecules. A PARP inhibitor prevents repair by inhibiting dissociation of PARP via trapping the PARP-DNA complex, which is PARP bound to 5-dRP. (**B**) TOP1cc, a covalent binding state of TOP1 induced by Top1 and a DNA 3′-end, is repaired by TDP1 recruited to TOP1cc by PARylation and PAR transferase of PARP1 to induce TOP1-DNA complex excision. Topoisomerase I inhibitors continuously induce TOP1cc. PARP inhibitors inhibit catalytic activity, suppressing the recruitment of TDP1 and TOP1-DNA covalent complex repair, resulting in a synergistic effect.

5. Conclusions

PARP performs a variety of functions from the transcriptional level, to activation and localization through post-translational modification. In the DNA damage response, PARP contributes to the activation of itself or its target protein through the regulation of PARylation. PARP inhibitors block this catalytic activity of PARP, preventing the activation of normal repair pathways. These PARP inhibitors have demonstrated dramatic anti-tumor effects for tumors with HRD, such as those with BRCA mutations. To date, four PARP inhibitors have been approved by the FDA and applied in clinical practice. However, these four have different effects on the trapping of the PARP-DNA complex, despite inhibiting the common catalytic activity of PARP. This PARP trapping leads to improved cytotoxicity via replication fork collapse, leading to conversion to DSBs. When PARP inhibitors are used in combination with alkylating agents, synergistic effects are achieved. In contrast, inhibition of the catalytic activity of PARP has a synergistic effect when combined with topoisomerase I inhibitors. In other words, the synergies of combination therapies with PARP inhibitors can be induced differently depending on the mechanism of action of individual PARP inhibitors. Understanding the characteristics of each PARP inhibitor to strategically select synergistic partners is an important matter that must be considered to produce maximum antitumor effects.

Author Contributions: A.M. and S.-A.I. wrote the manuscript. All authors have read and agreed to the published version of the manuscript.

Funding: This research received no external funding.

Conflicts of Interest: Seock-Ah Im is a recipient of research funds from AstraZeneca Inc., Pfizer and Roche, has consultant and advisory roles for AstraZeneca, Amgen, Eisai, Hanmi Corp, Lilly, MediPacto, Novartis, Pfizer, and Roche. Ahrum Min declared that she has no conflict of interest.

References

1. Broustas, C.G.; Lieberman, H.B. DNA Damage Response Genes and the Development of Cancer Metastasis. *Radiat. Res.* **2014**, *181*, 111–130. [CrossRef] [PubMed]
2. Ciccia, A.; Elledge, S.J. The DNA Damage Response: Making It Safe to Play with Knives. *Mol. Cell* **2010**, *40*, 179–204. [CrossRef] [PubMed]
3. Hoeijmakers, J.H.J. Genome maintenance mechanisms for preventing cancer. *Nature* **2001**, *411*, 366–374. [CrossRef] [PubMed]
4. Perina, D.; Mikoc, A.; Ahel, J.; Cetkovic, H.; Žaja, R.; Ahel, I. Distribution of protein poly(ADP-ribosyl)ation systems across all domains of life. *DNA Repair* **2014**, *23*, 4–16. [CrossRef]
5. Gibson, B.A.; Kraus, W.L. New insights into the molecular and cellular functions of poly(ADP-ribose) and PARPs. *Nat. Rev. Mol. Cell Boil.* **2012**, *13*, 411–424. [CrossRef]
6. Chambon, P.; Weill, J.; Mandel, P. Nicotinamide mononucleotide activation of a new DNA-dependent polyadenylic acid synthesizing nuclear enzyme. *Biochem. Biophys. Res. Commun.* **1963**, *11*, 39–43. [CrossRef]
7. O'Neil, N.J.; Bailey, M.L.; Hieter, P. Synthetic lethality and cancer. *Nat. Rev. Genet.* **2017**, *18*, 613–623. [CrossRef]
8. Beijersbergen, R.L.; Wessels, L.F.; Bernards, R. Synthetic Lethality in Cancer Therapeutics. *Annu. Rev. Cancer Boil.* **2017**, *1*, 141–161. [CrossRef]
9. Robson, M.; Im, S.-A.; Senkus, E.; Xu, B.; Domchek, S.M.; Masuda, N.; Delaloge, S.; Li, W.; Tung, N.; Armstrong, A.; et al. Olaparib for Metastatic Breast Cancer in Patients with a Germline BRCA Mutation. *N. Engl. J. Med.* **2017**, *377*, 523–533. [CrossRef]
10. Tappenden, P.; Harnan, S.; Ren, S.; Thokala, P.; Wong, R.; Mukuria, C.; Green, C.; Pledge, S.; Tidy, J. Olaparib for Maintenance Treatment of BRCA 1 or 2 Mutated, Relapsed, Platinum-Sensitive Ovarian, Fallopian Tube and Peritoneal Cancer in People Whose Relapsed Disease has Responded to Platinum-Based Chemotherapy: An Evidence Review Group Perspective of a NICE Single Technology Appraisal. *Pharmacoeconomics* **2017**, *35*, 97–109.
11. Pujade-Lauraine, É.; Ledermann, J.A.; Gebski, V.; Penson, R.T.; Oza, A.M.; Poveda, A.; Fujiwara, K.; Liu, J.; Lowe, E.S.; Bloomfield, R.; et al. Olaparib tablets as maintenance therapy in patients with platinum-sensitive, relapsed ovarian cancer and a BRCA1/2 mutation (SOLO2/ENGOT-Ov21): A double-blind, randomised, placebo-controlled, phase 3 trial. *Lancet Oncol.* **2017**, *18*, 1274–1284. [CrossRef]
12. Murai, J.; Huang, S.Y.; Das, B.B.; Renaud, A.; Zhang, Y.; Doroshow, J.H.; Ji, J.; Takeda, S.; Pommier, Y. Trapping of PARP1 and PARP2 by Clinical PARP Inhibitors. *Cancer Res.* **2012**, *72*, 5588–5599. [CrossRef]
13. Murai, J.; Huang, S.Y.N.; Renaud, A.; Zhang, Y.; Ji, J.; Takeda, S.; Morris, J.; Teicher, B.; Doroshow, J.H.; Pommier, Y. Stereospecific PARP trapping by BMN 673 and comparison with olaparib and rucaparib. *Mol. Cancer Ther.* **2014**, *13*, 433–443. [CrossRef]
14. Amé, J.C.; Spenlehauer, C.; de Murcia, G. The PARP superfamily. *Bioessays* **2014**, *26*, 882–893. [CrossRef] [PubMed]
15. Luo, X.; Kraus, W.L. On PAR with PARP: Cellular stress signaling through poly(ADP-ribose) and PARP-1. *Genes Dev.* **2012**, *26*, 417–432. [CrossRef] [PubMed]

16. Bai, P. Biology of Poly(ADP-Ribose) Polymerases: The Factotums of Cell Maintenance. *Mol. Cell* **2015**, *58*, 947–958. [CrossRef]
17. Hottiger, M.O.; Hassa, P.O.; Lüscher, B.; Schüler, H.; Koch-Nolte, F. Toward a unified nomenclature for mammalian ADP-ribosyltransferases. *Trends Biochem. Sci.* **2010**, *35*, 208–219. [CrossRef]
18. Loeffler, P.A.; Cuneo, M.J.; Mueller, G.A.; Derose, E.F.; Gabel, S.A.; London, R.E. Structural studies of the PARP-1 BRCT domain. *BMC Struct. Boil.* **2011**, *11*, 37. [CrossRef]
19. Eustermann, S.; Wu, W.-F.; Langelier, M.-F.; Yang, J.-C.; Easton, L.E.; Riccio, A.A.; Pascal, J.M.; Neuhaus, D. Structural Basis of Detection and Signaling of DNA Single-Strand Breaks by Human PARP-1. *Mol. Cell* **2015**, *60*, 742–754. [CrossRef]
20. Langelier, M.F.; Servent, K.M.; Rogers, E.E.; Pascal, J.M. A third zinc-binding domain of human poly(ADP-ribose) polymerase-1 coordinates DNA-dependent enzyme activation. *J. Biol. Chem.* **2008**, *283*, 4105–4114. [CrossRef]
21. Langelier, M.-F.; Planck, J.L.; Roy, S.; Pascal, J.M. Structural basis for DNA damage-dependent poly(ADP-ribosyl)ation by human PARP-1. *Science* **2012**, *336*, 728–732. [CrossRef] [PubMed]
22. Leung, A.K. Poly(ADP-ribose): An organizer of cellular architecture. *J. Cell Boil.* **2014**, *205*, 613–619. [CrossRef] [PubMed]
23. Schuhwerk, H.; Atteya, R.; Siniuk, K.; Wang, Z.-Q. PARPing for balance in the homeostasis of poly(ADP-ribosyl)ation. *Semin. Cell Dev. Boil.* **2017**, *63*, 81–91. [CrossRef] [PubMed]
24. Eisemann, T.; Pascal, J.M. Poly(ADP-ribose) polymerase enzymes and the maintenance of genome integrity. *Cell. Mol. Life Sci.* **2019**, *77*, 19–33. [CrossRef] [PubMed]
25. Kleine, H.; Poreba, E.; Lesniewicz, K.; Hassa, P.O.; Hottiger, M.O.; Litchfield, D.W.; Shilton, B.H.; Lüscher, B. Substrate-Assisted Catalysis by PARP10 Limits Its Activity to Mono-ADP-Ribosylation. *Mol. Cell* **2008**, *32*, 57–69. [CrossRef] [PubMed]
26. Kamaletdinova, T.; Fanaei-Kahrani, Z.; Wang, Z.-Q. The Enigmatic Function of PARP1: From PARylation Activity to PAR Readers. *Cells* **2019**, *8*, 1625. [CrossRef]
27. Wei, H.; Yu, X. Functions of PARylation in DNA Damage Repair Pathways. *Genom. Proteom. Bioinform.* **2016**, *14*, 131–139. [CrossRef]
28. Kim, I.K.; Stegeman, R.A.; Brosey, C.A.; Ellenberger, T. A quantitative assay reveals ligand specificity of the DNA scaffold repair protein XRCC1 and efficient disassembly of complexes of XRCC1 and the poly(ADP-ribose) polymerase 1 by poly(ADP-ribose) glycohydrolase. *J. Biol. Chem.* **2015**, *290*, 3775–3783. [CrossRef]
29. Noren Hooten, N.; Lohani, A.; Evans, M.K.; Evans, M.K. Poly(ADP-ribose) polymerase 1 (PARP-1) binds to 8-oxoguanine-DNA glycosylase (OGG1). *J. Biol. Chem.* **2011**, *286*, 44679–44690. [CrossRef]
30. Dantzer, F.; De La Rubia, G.; Murcia, J.M.-D.; Hostomsky, Z.; De Murcia, G.; Schreiber, V. Base excision repair is impaired in mammalian cells lacking Poly(ADP-ribose) polymerase-1. *Biochemistry* **2000**, *39*, 7559–7569. [CrossRef]
31. Gyawali, B. The OlympiAD trial: Who won the gold? *Ecancermedicalscience* **2017**, *11*, ed75. [CrossRef] [PubMed]
32. Bang, Y.-J.; Xu, R.-H.; Chin, K.; Lee, K.-W.; Park, S.H.; Rha, S.Y.; Shen, L.; Qin, S.; Xu, N.; Im, S.-A.; et al. Olaparib in combination with paclitaxel in patients with advanced gastric cancer who have progressed following first-line therapy (GOLD): A double-blind, randomised, placebo-controlled, phase 3 trial. *Lancet Oncol.* **2017**, *18*, 1637–1651. [CrossRef]
33. Hopkins, T.A.; Shi, Y.; Rodriguez, L.E.; Solomon, L.R.; Donawho, C.K.; DiGiammarino, E.L.; Panchal, S.C.; Wilsbacher, J.L.; Gao, W.; Olson, A.M.; et al. Mechanistic Dissection of PARP1 Trapping and the Impact on In Vivo Tolerability and Efficacy of PARP Inhibitors. *Mol. Cancer Res.* **2015**, *13*, 1465–1477. [CrossRef] [PubMed]
34. Shen, Y.; Aoyagi-Scharber, M.; Wang, B. Trapping Poly(ADP-Ribose) Polymerase. *J. Pharmacol. Exp. Ther.* **2015**, *353*, 446–457. [CrossRef]
35. Kaufman, B.; Shapira-Frommer, R.; Schmutzler, R.K.; Audeh, M.W.; Friedlander, M.; Balmaña, J.; Mitchell, G.; Fried, G.; Stemmer, S.M.; Hubert, A.; et al. Olaparib Monotherapy in Patients With Advanced Cancer and a Germline BRCA1/2 Mutation. *J. Clin. Oncol.* **2015**, 33, 244–250. [CrossRef]

36. Ledermann, J.; Harter, P.; Gourley, C.; Friedlander, M.; Vergote, I.; Rustin, G.; Scott, C.L.; Meier, W.; Shapira-Frommer, R.; Safra, T.; et al. Olaparib maintenance therapy in patients with platinum-sensitive relapsed serous ovarian cancer: A preplanned retrospective analysis of outcomes by BRCA status in a randomised phase 2 trial. *Lancet Oncol.* **2014**, *15*, 852–861. [CrossRef]
37. Golan, T.; Hammel, P.; Reni, M.; Van Cutsem, E.; Macarulla, T.; Hall, M.J.; Park, J.-O.; Hochhauser, D.; Arnold, D.; Oh, Y.; et al. Maintenance Olaparib for Germline BRCA-Mutated Metastatic Pancreatic Cancer. *N. Engl. J. Med.* **2019**, *381*, 317–327. [CrossRef]
38. Swisher, E.M.; Lin, K.K.; Oza, A.M.; Scott, C.L.; Giordano, H.; Sun, J.; Konecny, G.E.; Coleman, R.L.; Tinker, A.V.; O'Malley, D.M.; et al. Rucaparib in relapsed, platinum-sensitive high-grade ovarian carcinoma (ARIEL2 Part 1): An international, multicentre, open-label, phase 2 trial. *Lancet Oncol.* **2017**, *18*, 75–87. [CrossRef]
39. Kristeleit, R.; Shapiro, G.I.; Burris, H.A.; Oza, A.M.; Lorusso, P.; Patel, M.R.; Domchek, S.M.; Balmaña, J.; Drew, Y.; Chen, L.-M.; et al. A Phase I–II Study of the Oral PARP Inhibitor Rucaparib in Patients with Germline BRCA1/2-Mutated Ovarian Carcinoma or Other Solid Tumors. *Clin. Cancer Res.* **2017**, *23*, 4095–4106. [CrossRef]
40. Coleman, R.L.; Oza, A.M.; Lorusso, M.; Aghajanian, C.; Oaknin, A.; Dean, A.; Colombo, N.; Weberpals, J.I.; Clamp, A.; Scambia, G.; et al. Rucaparib maintenance treatment for recurrent ovarian carcinoma after response to platinum therapy (ARIEL3): A randomised, double-blind, placebo-controlled, phase 3 trial. *Lancet* **2017**, *390*, 1949–1961. [CrossRef]
41. Mirza, M.R.; Monk, B.J.; Herrstedt, J.; Oza, A.M.; Mahner, S.; Redondo, A.; Fabbro, M.; Ledermann, J.A.; Lorusso, D.; Vergote, I.; et al. Niraparib Maintenance Therapy in Platinum-Sensitive, Recurrent Ovarian Cancer. *N. Engl. J. Med.* **2016**, *375*, 2154–2164. [CrossRef] [PubMed]
42. Moore, K.N.; Secord, A.A.; Geller, M.A.; Miller, D.S.; Cloven, N.; Fleming, G.F.; Hendrickson, A.E.W.; Azodi, M.; DiSilvestro, P.; Oza, A.M.; et al. Niraparib monotherapy for late-line treatment of ovarian cancer (QUADRA): A multicentre, open-label, single-arm, phase 2 trial. *Lancet Oncol.* **2019**, *20*, 636–648. [CrossRef]
43. Litton, J.K.; Rugo, H.S.; Ettl, J.; Hurvitz, S.A.; Gonçalves, A.; Lee, K.-H.; Fehrenbacher, L.; Yerushalmi, R.; Mina, L.A.; Martin, M.; et al. Talazoparib in Patients with Advanced Breast Cancer and a Germline BRCA Mutation. *N. Engl. J. Med.* **2018**, *379*, 753–763. [CrossRef] [PubMed]
44. Nishikawa, T.; Matsumoto, K.; Tamura, K.; Yoshida, H.; Imai, Y.; Miyasaka, A.; Onoe, T.; Yamaguchi, S.; Shimizu, C.; Yonemori, K.; et al. Phase 1 dose-escalation study of single-agent veliparib in Japanese patients with advanced solid tumors. *Cancer Sci.* **2017**, *108*, 1834–1842. [CrossRef] [PubMed]
45. Diéras, V.C.; Han, H.S.; Kaufman, B.; Wildiers, H.; Friedlander, M.; Ayoub, J.P.; Puhalla, S.L.; Bondarenko, I.; Campone, M.; Jakobsen, E.H.; et al. LBA9Phase III study of veliparib with carboplatin and paclitaxel in HER2-negative advanced/metastatic gBRCA-associated breast cancer. *Ann. Oncol.* **2019**, *30*, mdz394-008. [CrossRef]
46. Oza, A.M.; Cibula, D.; Benzaquen, A.O.; Poole, C.; Mathijssen, R.H.J.; Sonke, G.S.; Colombo, N.; Spacek, J.; Vuylsteke, P.; Hirte, H.; et al. Olaparib combined with chemotherapy for recurrent platinum-sensitive ovarian cancer: A randomised phase 2 trial. *Lancet Oncol.* **2015**, *16*, 87–97. [CrossRef]
47. Ledermann, J.A.; Pujade-Lauraine, E. Olaparib as maintenance treatment for patients with platinum-sensitive relapsed ovarian cancer. *Ther. Adv. Med Oncol.* **2019**, *11*, 1758835919849753. [CrossRef]
48. Matsumoto, K.; Nishimura, M.; Onoe, T.; Sakai, H.; Urakawa, Y.; Onda, T.; Yaegashi, N. PARP inhibitors for BRCA wild type ovarian cancer; gene alterations, homologous recombination deficiency and combination therapy. *Jpn. J. Clin. Oncol.* **2019**, *49*, 703–707. [CrossRef]
49. Robinson, D.; Van Allen, E.M.; Wu, Y.M.; Schultz, N.; Lonigro, R.J.; Mosquera, J.M.; Montgomery, B.; Taplin, M.E.; Pritchard, C.C.; Attard, G.; et al. Integrative Clinical Genomics of Advanced Prostate Cancer. *Cell* **2015**, *162*, 454. [CrossRef]
50. Mateo, J.; Carreira, S.; Sandhu, S.; Miranda, S.; Mossop, H.; Perez-Lopez, R.; Rodrigues, D.N.; Robinson, D.; Omlin, A.; Tunariu, N.; et al. DNA-Repair Defects and Olaparib in Metastatic Prostate Cancer. *N. Engl. J. Med.* **2015**, *373*, 1697–1708. [CrossRef]
51. Ray-Coquard, I.; Pautier, P.; Pignata, S.; Pérol, D.; González-Martín, A.; Berger, R.; Fujiwara, K.; Vergote, I., Colombo, N., Mäenpää, J.; et al. Olaparib plus Bevacizumab as First-Line Maintenance in Ovarian Cancer. *N. Engl. J. Med.* **2019**, *381*, 2416–2428. [CrossRef] [PubMed]

52. McCann, K.E. Advances in the use of PARP inhibitors for BRCA1/2-associated breast cancer: Talazoparib. *Futur. Oncol.* **2019**, *15*, 1707–1715. [CrossRef]
53. Litton, J.; Rugo, H.; Ettl, J.; Hurvitz, S.; Gonçalves, A.; Lee, K.-H.; Fehrenbacher, L.; Yerushalmi, R.; Mina, L.; Martin, M.; et al. Abstract GS6-07: EMBRACA: A phase 3 trial comparing talazoparib, an oral PARP inhibitor, to physician's choice of therapy in patients with advanced breast cancer and a germline BRCA mutation. *Gen. Sess. Abstr.* **2018**, *78*, GS6-07.
54. Coleman, R.L.; Fleming, G.F.; Brady, M.F.; Swisher, E.M.; Steffensen, K.D.; Friedlander, M.; Okamoto, A.; Moore, K.N.; Ben-Baruch, N.E.; Werner, T.L.; et al. Veliparib with First-Line Chemotherapy and as Maintenance Therapy in Ovarian Cancer. *N. Engl. J. Med.* **2019**, *381*, 2403–2415. [CrossRef]
55. Ramalingam, S.; Blais, N.; Mazieres, J.; Reck, M.; Jones, C.; Juhász, E.; Urban, L.; Orlov, S.; Barlesi, F.; Kio, E.; et al. A Randomized, Double-Blind, Phase 2 Trial of Veliparib (ABT-888) With Carboplatin and Paclitaxel in Previously Untreated Metastatic or Advanced Non-Small Cell Lung Cancer. *Int. J. Radiat. Oncol.* **2014**, *90*, S4–S5. [CrossRef]
56. Strom, C.E.; Uhlen, M.; Szigyarto, C.A.-K.; Erixon, K.; Helleday, T.; Helleday, T. Poly (ADP-ribose) polymerase (PARP) is not involved in base excision repair but PARP inhibition traps a single-strand intermediate. *Nucleic Acids Res.* **2011**, *39*, 3166–3175. [CrossRef]
57. Lord, C.J.; Ashworth, A. PARP inhibitors: Synthetic lethality in the clinic. *Science* **2017**, *355*, 1152–1158. [CrossRef]
58. Yi, M.; Dong, B.; Qin, S.; Chu, Q.; Wu, K.; Luo, S. Advances and perspectives of PARP inhibitors. *Exp. Hematol. Oncol.* **2019**, *8*, 1–12. [CrossRef]
59. Rouleau, M.; Patel, A.; Hendzel, M.J.; Kaufmann, S.H.; Poirier, G.G. PARP inhibition: PARP1 and beyond. *Nat. Rev. Cancer* **2010**, *10*, 293–301. [CrossRef]
60. Murai, J.; Pommier, Y. PARP Trapping Beyond Homologous Recombination and Platinum Sensitivity in Cancers. *Annu. Rev. Cancer Boil.* **2019**, *3*, 131–150. [CrossRef]
61. Lu, Y.; Liu, Y.; Pang, Y.; Pacak, K.; Yang, C. Double-barreled gun: Combination of PARP inhibitor with conventional chemotherapy. *Pharmacol. Ther.* **2018**, *188*, 168–175. [CrossRef] [PubMed]
62. Curtin, N.J.; Szabó, C. Therapeutic applications of PARP inhibitors: Anticancer therapy and beyond. *Mol. Asp. Med.* **2013**, *34*, 1217–1256. [CrossRef] [PubMed]
63. Matsuno, Y.; Hyodo, M.; Fujimori, H.; Shimizu, A.; Yoshioka, K.-I. Sensitization of Cancer Cells to Radiation and Topoisomerase I Inhibitor Camptothecin Using Inhibitors of PARP and Other Signaling Molecules. *Cancers* **2018**, *10*, 364. [CrossRef] [PubMed]
64. Dent, R.A.; Lindeman, G.J.; Clemons, M.; Wildiers, H.; Chan, A.; McCarthy, N.J.; Singer, C.F.; Lowe, E.S.; Watkins, C.L.; Carmichael, J. Phase I trial of the oral PARP inhibitor olaparib in combination with paclitaxel for first- or second-line treatment of patients with metastatic triple-negative breast cancer. *Breast Cancer Res.* **2013**, *15*, R88. [CrossRef]
65. Ihnen, M.; Zu Eulenburg, C.; Kolarova, T.; Qi, J.W.; Manivong, K.; Chalukya, M.; Dering, J.; Anderson, L.; Ginther, C.; Meuter, A.; et al. Therapeutic potential of the poly(ADP-ribose) polymerase inhibitor rucaparib for the treatment of sporadic human ovarian cancer. *Mol. Cancer Ther.* **2013**, *12*, 1002–1015. [CrossRef]
66. Matulonis, U.A.; Monk, B.J. PARP inhibitor and chemotherapy combination trials for the treatment of advanced malignancies: Does a development pathway forward exist? *Ann. Oncol.* **2017**, *28*, 443–447. [CrossRef]
67. Samol, J.; Ranson, M.; Scott, E.; Macpherson, E.; Carmichael, J.; Thomas, A.; Cassidy, J. Safety and tolerability of the poly(ADP-ribose) polymerase (PARP) inhibitor, olaparib (AZD2281) in combination with topotecan for the treatment of patients with advanced solid tumors: A phase I study. *Investig. New Drugs* **2012**, *30*, 1493–1500. [CrossRef]
68. Kunos, C.; Deng, W.; Dawson, D.; Lea, J.S.; Zanotti, K.M.; Gray, H.J.; Bender, D.P.; Guaglianone, P.P.; Carter, J.S.; Moore, K.N. A phase I-II evaluation of veliparib (NSC #737664), topotecan, and filgrastim or pegfilgrastim in the treatment of persistent or recurrent carcinoma of the uterine cervix: An NRG Oncology/Gynecologic Oncology Group study. *Int. J. Gynecol. Cancer* **2015**, *25*, 484–492.
69. Lee, J.-M.; Peer, C.J.; Yu, M.; Amable, L.; Gordon, N.; Annunziata, C.M.; Houston, N.; Goey, A.K.L.; Sissung, T.M.; Parker, B.; et al. Sequence-Specific Pharmacokinetic and Pharmacodynamic Phase I/Ib Study of Olaparib Tablets and Carboplatin in Women's Cancer. *Clin. Cancer Res.* **2017**, 23, 1397. [CrossRef]

70. Wesolowski, R.; Zhao, M.; Geyer, S.M.; Lustberg, M.B.; Mrozek, E.; Layman, R.M.; Macrae, E.M.; Zhang, J.; Hall, N.; Schregel, K.; et al. Phase I trial of the PARP inhibitor veliparib (V) in combination with carboplatin (C) in metastatic breast cancer (MBC). *J. Clin. Oncol.* **2014**, *32*, 1074. [CrossRef]
71. Wilson, R.H.; Evans, T.J.; Middleton, M.R.; Molife, L.R.; Spicer, J.; Diéras, V.; Roxburgh, P.; Giordano, H.; Jaw-Tsai, S.; Goble, S.; et al. A phase I study of intravenous and oral rucaparib in combination with chemotherapy in patients with advanced solid tumours. *Br. J. Cancer* **2017**, *116*, 884–892. [CrossRef] [PubMed]
72. Halford, S.E.R.; Cruickshank, G.; Dunn, L.; Erridge, S.; Godfrey, L.; Herbert, C.; Jefferies, S.; Lopez, J.S.; McBain, C.; Pittman, M.; et al. Results of the OPARATIC trial: A phase I dose escalation study of olaparib in combination with temozolomide (TMZ) in patients with relapsed glioblastoma (GBM). *J. Clin. Oncol.* **2017**, *35*, 2022. [CrossRef]
73. Isakoff, S.J.; Puhalla, S.; Domchek, S.M.; Friedlander, M.; Kaufman, B.; Robson, M.; Telli, M.L.; Diéras, V.; Han, H.S.; Garber, J.E.; et al. A randomized Phase II study of veliparib with temozolomide or carboplatin/paclitaxel versus placebo with carboplatin/paclitaxel in BRCA1/2 metastatic breast cancer: Design and rationale. *Futur. Oncol.* **2017**, *13*, 307–320. [CrossRef] [PubMed]
74. Villalona-Calero, M.A.; Duan, W.; Zhao, W.; Shilo, K.; Schaaf, L.J.; Thurmond, J.; Westman, J.A.; Marshall, J.; Xiaobai, L.; Ji, J.; et al. Veliparib Alone or in Combination with Mitomycin C in Patients with Solid Tumors With Functional Deficiency in Homologous Recombination Repair. *J. Natl. Cancer Inst.* **2016**, *108*, djv437. [CrossRef]
75. Bendell, J.; O'Reilly, E.M.; Middleton, M.R.; Chau, I.; Hochster, H.; Fielding, A.; Burke, W.; Burris, H. Phase I study of olaparib plus gemcitabine in patients with advanced solid tumours and comparison with gemcitabine alone in patients with locally advanced/metastatic pancreatic cancer. *Ann. Oncol.* **2015**, *26*, 804–811. [CrossRef]
76. Bang, Y.-J.; Im, S.-A.; Lee, K.-W.; Cho, J.Y.; Song, E.-K.; Lee, K.H.; Kim, Y.H.; Park, J.O.; Chun, H.G.; Zang, D.Y.; et al. Randomized, Double-Blind Phase II Trial With Prospective Classification by ATM Protein Level to Evaluate the Efficacy and Tolerability of Olaparib Plus Paclitaxel in Patients With Recurrent or Metastatic Gastric Cancer. *J. Clin. Oncol.* **2015**, *33*, 3858–3865. [CrossRef]
77. Pommier, Y.; O'Connor, M.J.; De Bono, J. Laying a trap to kill cancer cells: PARP inhibitors and their mechanisms of action. *Sci. Transl. Med.* **2016**, *8*, 362. [CrossRef]
78. Murai, J.; Zhang, Y.; Morris, J.; Ji, J.; Takeda, S.; Doroshow, J.H.; Pommier, Y. Rationale for poly(ADP-ribose) polymerase (PARP) inhibitors in combination therapy with camptothecins or temozolomide based on PARP trapping versus catalytic inhibition. *J. Pharmacol. Exp. Ther.* **2014**, *349*, 408–416. [CrossRef]
79. Patel, A.G.; Schneider, P.A.; Dai, N.T.; McDonald, J.S.; Poirier, G.G.; Kaufmann, S.H. Enhanced killing of cancer cells by poly(ADP-ribose) polymerase inhibitors and topoisomerase I inhibitors reflects poisoning of both enzymes. *J. Biol. Chem.* **2012**, *287*, 4198–4210. [CrossRef]
80. Murai, J.; Pommier, Y. Classification of PARP Inhibitors Based on PARP Trapping and Catalytic Inhibition, and Rationale for Combinations with Topoisomerase I Inhibitors and Alkylating Agents. In *PARP Inhibitors for Cancer Therapy*; Curtin, N.J., Sharma, R.A., Eds.; Springer International Publishing: Cham, Switzerland, 2015; Volume 83, pp. 261–274.

cancers

MDPI

Review

Immunomodulatory Roles of PARP-1 and PARP-2: Impact on PARP-Centered Cancer Therapies

José Yélamos [1,2,*], Lucia Moreno-Lama [1], Jaime Jimeno [3] and Syed O. Ali [4]

1 Cancer Research Program, Hospital del Mar Medical Research Institute (IMIM), 08003 Barcelona, Spain; morenolamalucia@gmail.com
2 Immunology Unit, Department of Pathology, Hospital del Mar, 08003 Barcelona, Spain
3 Department of General Surgery, Breast Unit, Hospital Universitario Marqués de Valdecilla, 39008 Santander, Spain; jaime.jimeno@scsalud.es
4 Oxford University Hospitals, NHS, Oxford OX3 9DU, UK; omar.ali1795@gmail.com
* Correspondence: jyelamos@imim.es; Tel.: +34-93-3160411; Fax: +34-93-3160410

Received: 30 December 2019; Accepted: 6 February 2020; Published: 8 February 2020

Abstract: Poly(ADP-ribose) polymerase-1 (PARP-1) and PARP-2 are enzymes which post-translationally modify proteins through poly(ADP-ribosyl)ation (PARylation)—the transfer of ADP-ribose chains onto amino acid residues—with a resultant modulation of protein function. Many targets of PARP-1/2-dependent PARylation are involved in the DNA damage response and hence, the loss of these proteins disrupts a wide range of biological processes, from DNA repair and epigenetics to telomere and centromere regulation. The central role of these PARPs in DNA metabolism in cancer cells has led to the development of PARP inhibitors as new cancer therapeutics, both as adjuvant treatment potentiating chemo-, radio-, and immuno-therapies and as monotherapy exploiting cancer-specific defects in DNA repair. However, a cancer is not just made up of cancer cells and the tumor microenvironment also includes multiple other cell types, particularly stromal and immune cells. Interactions between these cells—cancerous and non-cancerous—are known to either favor or limit tumorigenesis. In recent years, an important role of PARP-1 and PARP-2 has been demonstrated in different aspects of the immune response, modulating both the innate and adaptive immune system. It is now emerging that PARP-1 and PARP-2 may not only impact cancer cell biology, but also modulate the anti-tumor immune response. Understanding the immunomodulatory roles of PARP-1 and PARP-2 may provide invaluable clues to the rational development of more selective PARP-centered therapies which target both the cancer and its microenvironment.

Keywords: PARP; immunomodulation; tumor microenvironment

1. Introduction

Poly(ADP-ribose) polymerase-1 (PARP-1) and PARP-2 are two enzymes of the PARP family of proteins that, in response to DNA damage, catalytically cleave β-NAD$^+$ and transfer ADP-ribose moieties onto specific amino residues of acceptor proteins. This process, termed poly(ADP-ribosyl)ation (PARylation), forms poly(ADP-ribose) (PAR) polymers varying in size and branching, which have diverse functional and structural effects on target proteins [1–3]. The deletion of either PARP-1 or PARP-2 in mice is associated with disturbances of DNA integrity and repair, supporting key shared functions of these proteins that are pivotal to DNA repair [4]. Indeed, combined PARP-1 and PARP-2 deficiency leads to embryonic lethality [5], which is likely due to their central role in the DNA damage response (DDR) [2,4].

Studies based on the role of these PARPs in the DDR in cancer cells have led to the development of PARP inhibitors as new therapeutic tools in cancer, both as adjuvant treatment potentiating chemotherapy, radiotherapy, and immunotherapy and as monotherapy exploiting cancer cell-specific defects in DNA repair, such as BRCA mutations [6–9]. However, the tumor microenvironment is formed from more than just tumor cells, and also includes stromal cells and infiltrating cells of the innate and adaptive immune system, which are likely to also be affected by PARP inhibition. These cells communicate with each other through direct contact and/or indirect signals that can alter the functionality of immune cells so that they either favor or limit tumor growth [10,11]. Emerging evidence supporting the immunomodulatory roles of PARP-1 and PARP-2 has raised the prospect of harnessing PARP inhibition to not only target the cancer itself, but also therapeutically modify its microenvironment.

In this review, we highlight the functions of PARP-1 and PARP-2 in the immune system and how their immunomodulatory roles might impact the response to tumors. We will examine recent data suggesting specific and redundant roles of PARP-1 and PARP-2 in the innate and adaptive immune responses and the immunological potential of PARP inhibitors. Understanding the immunomodulatory roles of PARP-1 and PARP-2 may provide invaluable clues for the rational development and exploitation of more selective anti-cancer PARP inhibitor drugs, both as new monotherapeutic approaches and in combinations with immunotherapy.

2. Impact of PARP-1 and PARP-2 on T Cell Development and Function

T cell development is a highly regulated process beginning in the thymus from bone marrow-derived lymphoid precursors, and giving rise to mature T cells through well-characterized sequential maturation steps involving a complex transcriptional network orchestrating cell proliferation, survival, and differentiation [12]. The earliest thymic progenitors are named double-negative (DN) cells, comprising four fractions (DN1 to DN4), which are characterized by a lack of CD4 and CD8 surface markers. DN2 and DN3 thymocytes express recombination-activating genes (Rag) and undergo extensive T cell receptor (TCR) β, γ, and δ gene rearrangement to express functional TCR chains. A successful recombination of TCRγ and TCRδ promotes the generation of $\gamma\delta$ T cells. In contrast, the generation of $\alpha\beta$ T cells requires additional differentiation steps. A successfully rearranged TCRβ chain associates with CD3 chains to form a pre-TCR. The expression of a pre-TCR drives DN4 differentiation into double-positive (DP) thymocytes—the most abundant population in the thymus—expressing both CD4 and CD8 surface markers. During this stage of development, the thymocytes re-express the Rag genes, which allows multiple rounds of TCRα gene rearrangements to increase the probability of forming a functional $\alpha\beta$ TCR. DP thymocytes undergo a very strict selection process, such that those that express a TCR which is not able to interact with self-major histocompatibility complex (MHC)/self-peptide complexes die due to neglect. In the same way, the DP thymocytes that bind self-MHC/self-peptide molecules with a high affinity are eliminated by negative selection. Meanwhile, those DP thymocytes expressing TCRs that bind self-MHC/self-peptide ligands with a low affinity are positively selected and differentiated into either CD4$^+$ or CD8$^+$ single-positive (SP) thymocytes [12]. At this stage of development, some CD4$^+$ thymocytes express the transcription factor Forkhead box protein 3 (FoxP3), which confers the cells an immunosuppressive function (Treg) [13]. All kinds of T cells generated in the thymus will seed the peripheral lymphoid tissues (Figure 1).

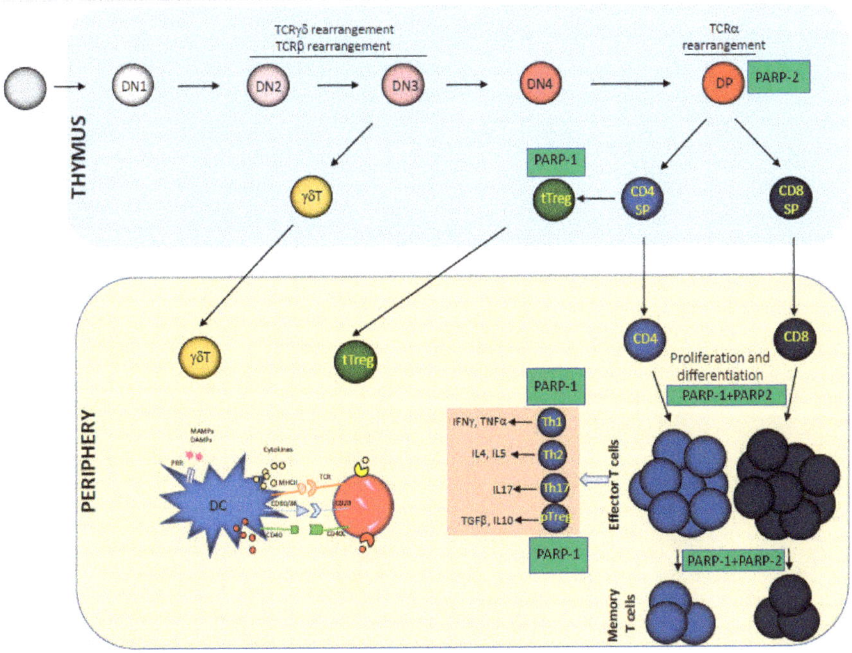

Figure 1. Schematic representation of T cell development depicting the specific stages in which PARP-1 and/or PARP-2 are playing a role. TCR, T cell receptor; DN, double negative; DP, double positive; SP, single positive; DC, dendritic cells; Treg, regulatory T cells.

Although both PARP-1 and PARP-2 proteins are expressed in thymocytes [14], only PARP-2 plays a significant role in thymocyte development. Therefore, PARP-2-deficient, but not PARP-1-deficient, mice show a significant reduction in the number of DP thymocytes. This phenotype is associated with the role of PARP-2 in preventing the accumulation of DNA double-strand breaks (DSBs) and the resulting activation of a DNA damage-induced apoptotic response during TCRα rearrangements [14,15]. In fact, p53 deficiency restores thymocyte populations in PARP-2-deficient mice [15]. In contrast, PARP-1 regulates tTreg development [16], while PARP-2 does not seem to play any role in tTreg development [17] (Figure 1).

Once the T cells in the thymus have matured, they migrate to the peripheral lymphoid tissues forming the naïve T cell pool, where they continue their differentiation to become fully immunocompetent to mount appropriate immune T cell responses to antigen challenge [18]. Naïve T cells proliferate both in situations of lymphopenia (homeostatic proliferation) driven by TCR/self-peptide–MHC interactions, and in response to antigen challenges driven by TCR/foreign-peptide–MHC interactions and co-stimulation, accompanied by differentiation into effector T cells (Th1, Th2, Th17, Treg, and cytotoxic T cells) and the final generation of a memory T cell population [19] (Figure 1). The control of T cell homeostasis is not only mediated by MHC–TCR interactions and cytokine-mediated signals, but also processes which regulate essential T cell functions to maintain genomic stability, such as cell-cycle checkpoints, DNA repair, and apoptosis [20,21].

Although PARP-1 deficiency or PARP-2 deficiency alone does not affect the number of T cells in peripheral lymphoid tissues [14,17], double deficiency in the T cell compartment results in a significant decrease in both CD4$^+$ and CD8$^+$ peripheral T cells [17]. The T cell lymphopenia present in mice with double PARP-1 and PARP-2 deficiency indicates that these proteins act in a coordinated manner to prevent the accumulation of unrepaired DNA breaks upon homeostatic proliferation or in response to antigen challenge, but not under basal conditions, avoiding T cell death [17]. PARP-1 and PARP-2 likely act through the principle of synthetic lethality [22], whereby they regulate two independent, but functionally linked, processes. T cell lymphopenia in double-deficient mice for PARP-1 and PARP-2 blunts the anti-viral immune response and the response to other T cell-dependent antigens [17]. Furthermore, double deficiency of PARP-1 and PARP-2 in the T cell compartment in mice affects the T cell response to tumors [23]. Although PARP inhibitors do not achieve the persistence of inhibition as obtained in T lymphocytes with double genetic deficiency of PARP-1 and PARP-2, these pharmacological inhibitors can still impact the T cell compartment and thus the T cell immune response. Indeed, in a mouse breast tumor model induced by the AT-3 cell line, which is sensitive to the PARP inhibitor olaparib, the anti-tumor effect of olaparib is blunted by an intact immune system [23]. As such, it would be interesting to study how PARP inhibitors used in the clinic affect the immune compartment in patients.

Transcriptional activation via different signaling pathways is fundamental to the differentiation of T cells. Among the key transcription factors in T cell development and function are the nuclear factor of activated T cells (NFAT) family of transcription factors (NFAT1 to NFAT5) [24]. After antigenic recognition by the TCR, a signaling cascade is initiated in the T cell, leading to the activation and nuclear translocation of NFAT1, NFAT2, and NFAT4, where, in combination with other transcription factors such as AP1, they regulate the expression of cytokines and lineage-specific transcription factors to control pathways of T cell differentiation into Th1 or Th2 types [24]. Of note, our group has demonstrated that PARP-1 is activated during T cell activation, where it modulates the activity of NFAT through PARylation, as evidenced by PARP inhibitors causing an increase in NFAT-dependent transactivation [25]. Moreover, PARP-1 plays a critical role in the gene expression reprogramming that takes place in T cells upon activation [26]. Indeed, PARP-1-deficiency seems to bias the T cell response to a Th1 phenotype [26] and has been shown to reduce differentiation into Th2 cells in different experimental models [26,27]. The pharmacological inhibition of PARP has led to more controversial results, where, in one case, the PARP inhibitor led to an increase in Th1 cytokine production and a reduction in Th2 cytokines [28], while in another case, the inhibition led to a decrease in Th1 cells [29]. These discrepancies may be associated with the type of inhibitor used or the experimental model.

PARP-1 also plays a role in the generation of Treg cells in the periphery (pTreg) from $CD4^+$ T cells that express FoxP3, and PARP-1-deficient mice have been found to display an increased number of Treg cells [16]. Moreover, PARP-1 negatively regulates the suppressive function of Treg cells at the posttranslational level through FoxP3 PARylation [30]. PARP-1 can also regulate the generation of Treg cells through its role in regulating the expression of transforming growth factor β receptors (TGFβR) in $CD4^+$ T cells, and therefore affects TGFβ signaling in T cells [30]. Interestingly, the inhibition of TGFβRI expression by PARP-1 is dependent on PARP enzymatic activity, while the inhibition of TGFβRII expression depends on the interaction of PARP-1 with the promoter of the TGFβRII gene [31]. In contrast, the function of PARP-2 in transcriptional regulation in T cells remains unclear.

3. Impact of PARP-1 and PARP-2 on B cell Development and Function

As with the development of T lymphocytes, the development of B cells, which takes place in the bone marrow, is also a precisely regulated process that starts from pluripotent hematopoietic stem cells. In the first step, the hematopoietic stem cells differentiate into pro-B cells that transiently express the Rag genes, which mediate immunoglobulin (Ig) heavy-chain gene rearrangements to assemble a V_HDJ_H-Cμ protein. The association of this protein with Vpre-B and λ5 surrogate light chain proteins leads to the formation of the pre-B cell receptor (BCR) complex in the large pre-B cell population. Pre-BCR signaling results in differentiation into small pre-B cells, which re-express the Rag genes, allowing Ig light chain V_LJ_L gene rearrangement. The succeeding association of V_LJ_L with V_HDJ_H-Cμ generates fully functional membrane-bound IgM receptors in immature B cells, which further differentiate into transitional B cells which co-express IgM and IgD receptors on their surface [32] (Figure 2). Despite the role of PARP-1 and PARP-2 in DNA repair, their role in Ig V(D)J gene recombination has remained unclear or unknown [33–35]. Recent data obtained by our group show that mice with dual, but not individual, PARP-1 and PARP-2 deficiency exhibit a reduced number of B cells in the bone marrow [36] (Figure 2). A possible explanation for this bone marrow hypocellularity is that the V(D)J recombination process is defective in these cells. However, a detailed analysis showed that neither single nor double PARP-1/PARP-2 deficiency affected Ig V(D)J gene recombination [36]. As in the T cell compartment, B cell lymphopenia in dually PARP-1- and PARP-2-deficient mice is associated with an accumulation of unrepaired DNA damage in proliferating B cells leading to cell death, suggesting a potential model whereby coordinated signals from PARP-1 and PARP-2 are required to maintain genomic integrity during lymphoid proliferation. This is consistent with recent data showing that dual PARP-1 and PARP-2 deficiency results in the accumulation of replication-associated DNA damage due to the impaired stabilization of Rad51 at damaged DNA replication forks and uncontrolled DNA resection thereafter [37].

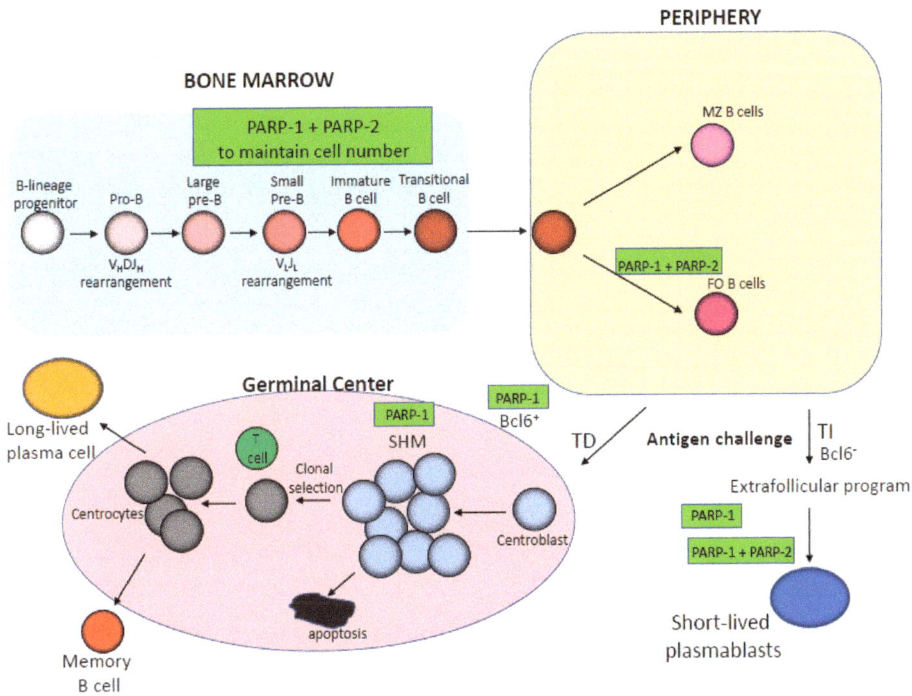

Figure 2. Schematic representation of B cell development depicting the specific stages in which PARP-1 and/or PARP-2 play a role. MZ, marginal zone; FO, follicular B cells; TD, T cell-dependent antigen; TI, T cell-independent antigen; SHM, somatic hypermutation.

The differentiation of transitional B cells leaving the bone marrow continues in the peripheral lymphoid organs, which gives rise to mature marginal zone (MZ) or mature follicular (FO) B cells [32]. After their interaction with antigens, mature B cells will either proliferate and differentiate extra follicularly (Bcl-6^-) or form germinal centres (Bcl-6^+), ultimately generating short-lived antibody secreting plasmablasts or long-lived memory B cells and antibody-secreting plasma cells, respectively [38–40] (Figure 2). Antibody diversity is in part achieved during these responses through DNA editing via both Ig class-switching recombination (CSR) and somatic hypermutation (SHM), which are mediated by activation-induced cytidine deaminase (AID) [41].

While single PARP-1 or PARP-2 deficiency does not affect the B cell compartment in peripheral lymphoid tissues [14,17], combined PARP-1 and PARP-2 deficiency impairs peripheral B cell homeostasis [36]. This lymphopenia does not affect all B cell populations equally, such that only the number of FO B cells is dramatically reduced in mice with double PARP-1 and PARP-2 deficiency, while the number of MZ B cells is not affected. The reason for this is unclear and requires further exploration [36] (Figure 2). Interestingly, double PARP-1 and PARP-2 deficiency impairs antibody responses to T cell-independent, but not T cell-dependent, antigens [36]. In addition, T cell-independent antigens elicit IgG1- and IgG2b-predominant antibody responses in single PARP-1-deficient mice [36,42] (Figure 2). It is important to note that despite their role in DNA repair, neither PARP-1 nor PARP-2 are required for CSR [36,43] and instead, their role in B cell homeostasis underpins their importance for Ig responses to specific antigens. Another role of the PARPs in B cell development is the role of PARP-1 activation in switching off Bcl6 [44]—a transcription factor essential for the formation of germinal centers [45,46]. However, germinal center formation upon immunization is normal in mice with single or dual deficiencies of PARP-1 and PARP-2 [36]. Meanwhile, the role of PARP-1 in SHM is controversial, with some data showing a dispensable role [47], while other data indicate a role of PARP-1 in SHM [48]. Meanwhile, the role of PARP-2 in SHM is unknown.

4. Role of PARP-1 and PARP-2 in the Cellular Components of the Innate Immune System

In addition to their role during the development and function of cellular components of the adaptive immune system, PARP-1 and PARP-2 have also been involved in different functional aspects of cells involved in the innate immune response, including neutrophils, macrophages, dendritic cells, and natural killer (NK) cells. These innate immune cells serve as the front line of host protection to infection and non-infectious tissue damage. In addition, cells of innate immunity are critical for stimulating subsequent adaptive immune responses [49].

Neutrophils are key players in acute and chronic inflammatory responses through their role in phagocytosis, the recruitment of other immune cells, and the secretion of antibacterial proteins [50]. In cancer, tumor-associated neutrophils are thought to contribute to inflammation in the tumor [51]. Of note, PARP-1 is important in the recruitment and function of neutrophils in different processes related to inflammation [52–55]. Meanwhile, the role of PARP-2 in neutrophil biology remains elusive (Figure 3).

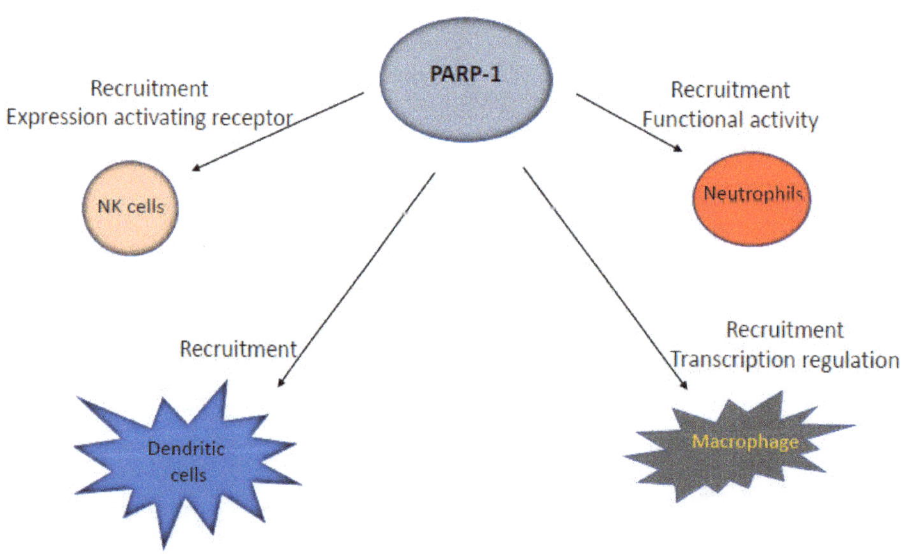

Figure 3. Schematic representation of the role played by PARP-1 in cells of the innate immune system.

Macrophages are differentiated from circulating monocytes after extravasation into tissues. Upon differentiation, macrophages are prepared to sense and respond to infection and tissue injuries through the phagocytosis of dead cells, debris, and foreign materials [56]. Besides phagocytosis, macrophages are also important as antigen-presenting cells (APC) to T cells [57]. Macrophages show considerable plasticity, which permits them to adapt their phenotype in response to different microenvironments. There are two major forms of activated macrophages, termed pro-inflammatory M1, which is characterized by the production of pro-inflammatory cytokines, and anti-inflammatory M2, which is characterized by the secretion of anti-inflammatory cytokines [58]. Of note, PARP inhibitors inhibit the expression of LPS-induced proinflammatory cytokines like tumor necrosis factor α (TNFα), interleukin-1 (IL-1), and IL-6 by macrophages [59]. Meanwhile, recent work has shown that PARP-1, but not its enzymatic activity, enhances the transcriptional activity of LPS-induced proinflammatory genes in macrophages [60]. This effect would be mediated by the modulatory role of PARP-1 on the transcription factor NF-κB [61]. In addition, functional interplay between PARP-1 and lysine-specific histone demethylase 1A (LSD1) protects pro-inflammatory M1 macrophages from death under oxidative conditions [62]. Moreover, macrophage recruitment in an airway inflammatory model was severely blocked in PARP-1-deficient mice [63]. Meanwhile, the role of PARP-2 in macrophages remains unknown (Figure 3).

Dendritic cells (DC) are specialized APC which process antigen and present it in the context of self-MHC molecules to T cells. In addition, they also upregulate cell surface receptors, including CD80, CD86, and CD40, which interact with co-receptors on the T cells surface (CD40L and CD28), in order to induce proper T cell activation [64,65]. While the role of PARP-1 in the recruitment of DC to tissues in different pathological situations seems to be well-established, its role in the differentiation and function of these cells is less clear [66–69]. On the other hand, the function of PARP-2 in DC remains unexplored (Figure 3).

NK cells have a wide array of inhibitory and stimulatory receptors on their cell surface that are used for immune surveillance. Upon activation, NK cells show potent cytolytic activity in response to infected or transformed cells by releasing cytotoxic perforin and granzyme and activating apoptotic pathways in target cells through the production of TNFα or via direct cell–cell contact through activation of the tumor necrosis factor-related apoptosis-inducing ligand (TRAIL) and Fas ligand (FASL) pathways [70,71]. Recent work has demonstrated important roles of PARP-1 in NK cell biology. For instance, PARP-1 controls NK cell recruitment to the site of viral infection [72,73]. In addition, PARP-1 is involved in the downregulation of NK cell-activating receptor ligands for immune evasion in acute myeloid leukemia [74] (Figure 3).

5. How Could the Immunomodulatory Roles of PARP-1 and PARP-2 Impact the Immune Response to Tumors?

Tumors contain not only cancer cells, but other cell types, including tissue-resident and peripherally-recruited immune cells, fibroblasts, and endothelial cells, which form the tumor microenvironment. Interactions between these cells—cancerous and non-cancerous—are known to either favor or limit tumorigenesis. Indeed, cancer progression is a dynamic process that, based on those interactions, has been divided into three stages: elimination, equilibrium, and escape [75].

During the cancer elimination phase, a competent immune response takes place in which innate and adaptive immune cells are recruited to the tumor microenvironment, where they exert a strong anti-tumor response [51] (Figure 4). The aforementioned immunomodulatory functions of PARP-1 and PARP-2 would thus be expected to have an impact on the immune response against the tumor. Indeed, we have observed a reduction in tumor growth in PARP-1-deficient host-mice and in PARP-2-deficient host-mice, compared to wild-type specimens, in both a C57 syngeneic tumor model induced by the AT-3 breast tumor cell line [23] and in a Balb/c syngeneic tumor model induced by the LP07 lung adenocarcinoma cell line [76], in which both cancer cells lines are proficient for PARP-1 and PARP-2 proteins. This effect may be associated with their immunomodulatory roles.

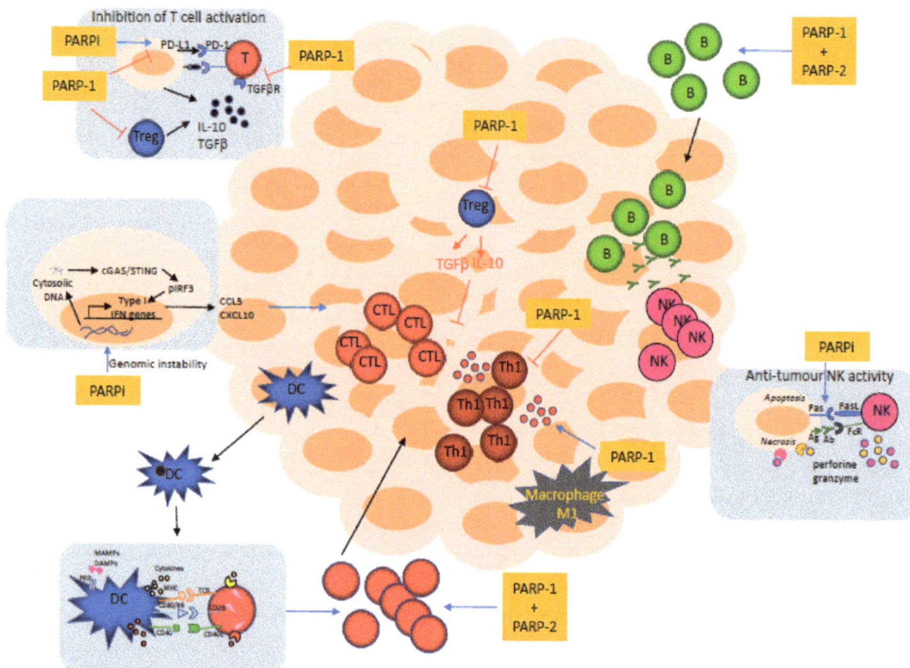

Figure 4. Schematic representation of the tumor microenvironment indicating the stages in which PARP-1 or its combination with PARP-2 or PARP inhibitors might play a role. In the boxes, we have included further details on the involvement of PARP-1 and/or PARP-2 or PARPi in certain contexts of the immune response. CTL, cytotoxic T cells; DC, dendritic cells; B, B cells; PARPi, PARP inhibitors.

T cells, in particular, $CD8^+$ cytotoxic T cells (CTL) and $CD4^+$ Th1 cells, are major contributors to the adaptive host-defense against tumors [10]. Tumor-derived antigens are processed by APCs (mainly dendritic cells), carried to draining lymph nodes and presented to naïve T cells, in order to prime them. Antigen presentation, together with the induction of co-stimulatory signals mediated by the binding of CD28 on the T cell to CD80/CD86 on the APC, leads to the differentiation of naïve $CD8^+$ T cells into tumor-specific CTLs. These, in turn, migrate to the tumor microenvironment to kill cancer cells through the secretion of perforin and granzyme [77]. The anti-tumor effect of $CD4^+$ Th1 cells is mediated through the secretion of IL-2, TNFα, and IFNγ, enhancing $CD8^+$ T cell responses and activating macrophages and NK cells [78–80] (Figure 4). Previous data from our group has indicated that coordinated signals from PARP-1 and PARP-2 are required to maintain T cell homeostasis and for the differentiation from naïve to effector T cells affecting both $CD4^+$ and $CD8^+$ lineages [17]. Accordingly, T cell lymphopenia in dual PARP-1/PARP-2-deficient mice can affect the recruitment of lymphocytes to the tumor microenvironment [23]. Moreover, a defect in the ability of dually PARP-1/PARP-2-deficient T cells to differentiate into effector cells could have consequences for the anti-tumor response (Figure 4). On the other hand, PARP-1-deficiency seems to bias T cell responses to a Th1 phenotype [26] that may also impact tumor progression. Similarly, B cell lymphopenia in dual PARP-1/PARP-2-deficient mice can affect the recruitment of B cells to the tumor microenvironment (Figure 4).

The aforementioned biological roles of PARP-1 in macrophage biology may impact the response of these cells to tumors. Classically-activated M1 macrophages can kill many tumor cells by mechanisms including the recognition of damage-associated molecular patterns (DAMPs) from dying tumor cells and the production of nitric oxide. In addition, the co-operation of T cells and macrophages through direct contact or through the secretion of cytokines is important in the anti-tumor response [77]. Recent work from Hottiger´s group shows that mice with a conditional loss of PARP-1 in myeloid lineages fail to control tumor growth in an MC-38-induced tumor model of colon cancer, which could be attributed to reduced Th1 and CD8$^+$ T cell responses [60], suggesting that PARP-1 in macrophages controls Th1 responses to tumors (Figure 4). However, this intrinsic role of PARP-1 in myeloid cells is independent of its enzymatic activity, so would be of limited utility from the point of view of a pharmacological blockade [60].

Intratumor NK cells have been shown to play a very important role in the control of tumor growth [81]. The balance between stimulatory and inhibitory receptor signals determines the activation of NK cells against tumor cells. NK cells may also be activated to kill tumor cells coated with anti-tumor antibodies by antibody-dependent cell-mediated cytotoxicity (ADCC) (Figure 4). Moreover, the tumoricidal capacity of NK cells is increased by cytokines (IL-2, IL-15, IL-12). Although the role of PARP-1 and PARP-2 in modulating NK cell activity against tumors is largely unknown, the aforementioned function of PARP-1 in controlling NK cell recruitment to the site of viral infection [72,73] and its role in the downregulation of NK cell-activating receptor ligands to evade immune surveillance in acute myeloid leukemia [74], may impact tumor progression.

Tumor cells that have escaped the immune response undergo different strategies in order to acquire immune tolerance, including (i) tumor cell-intrinsic modifications, like the loss of human leukocyte antigens (HLA) class I molecules, loss of tumor-associated antigens, and increased resistance to cell killing by immune cells, and (ii) the generation of an immuno-suppressive microenvironment through the recruitment of cells with immunosuppressive activities (Treg, macrophages M2, and myeloid-derived suppressive cells); the expression of inhibitory checkpoint molecules such as PD-L1, PD-L2, and cytotoxic T lymphocyte-associated antigen 4 (CTLA4) [82–84]; the deprivation of nutrients and oxygen; and the secretion of immunosuppressive cytokines (TGFβ, IL-10, and VEGF) [85] (Figure 4).

The previously mentioned roles of PARP-1 in Treg development and function [16,30] may impact the response to tumors. In addition, PARP-1 inhibition leads to the up-regulation of TGFβ receptor expression in CD4$^+$ T cells that subsequently affects TGFβ signal transduction [31], which may impact the response to tumors (Figure 4).

However, these tumor microenvironment escape mechanisms can be modified by different strategies in order to reactivate the immune response against tumor cells [86]. Indeed, re-activating the normal function of immune cells in the tumor microenvironment is one of the biggest challenges in oncology research. Accordingly, emerging immunotherapeutic strategies aim to reverse immune tolerance either by modulating T cell co-receptor signals or boosting the recognition of tumor-associated antigens by using monoclonal antibodies [10]. In addition to those strategies based on biological approaches, modifying the immune response through a small-molecule approach targeting intracellular signaling pathways, such as with PARP inhibitors, may represent a breakthrough that is complementary to, and potentially synergistic with, immunotherapy [87].

6. PARP Inhibitors as Immunomodulatory Agents

PARP proteins exert their function through their physical association with or by the PARylation of partner proteins [3]. Although most of the immunomodulatory roles of PARP proteins have been based on studies of mice with the genetic deletion of these proteins (Table 1), PARP inhibitors might induce similar immune cell alterations that will modify their interaction with tumor cells. Indeed, recent work has shown how PARP inhibitors might impact the mechanisms used by tumors to evade immunity, although many of these studies are focused on tumor cell-intrinsic mechanisms. These studies can provide information to rationalize the combined use of PARP inhibitors with other strategies aimed at reactivating the immune system against the tumor.

Table 1. Immunological phenotypes in PARP-1- and/or PARP-2-deficient mouse models.

Immunological Process	Parp-1$^{-/-}$	Parp-2$^{-/-}$	Parp-1$^{-/-}$CD4-Parp-2$^{f/f}$	Parp-1$^{-/-}$CD19-Parp-2$^{f/f}$	Artd1^{DMyel}	References
Thymocyte development		Decreased DP survival	Decreased DP survival			[14,15].
Peripheral T cell homeostasis			T cell lymphopenia			[17]
T cell differentiation	Bias to a Th1 phenotype Increased Treg Increased TGFβR		Compromises antibody production T cell-dependent (TD) antigens			[17] [26] [16,30,31] [31]
Central and peripheral B cell homeostasis				B cell lymphopenia		[36]
B cell function	Increased antibody production to TI antigens Enhanced somatic hypermutation(SHM)			Depletion of follicular (FO) B cells Compromises antibody production to T cell-independent (TI) antigens		[36] [36] [48]
Neutrophil function	Impaired recruitment					[52–55]
Macrophage function	Impaired recruitment				Inhibition of LPS-induced pro-inflammatory genes	[60] [63]
Dendritic cell function	Impaired recruitment					[66–69]
Natural killer (NK) cell function	Impaired recruitment					[72]

One of the most successful strategies for reinstating an existing anti-cancer T cell immune response is the use of blocking antibodies against cell surface inhibitory co-receptors like cytotoxic T lymphocyte-associated protein 4 (CTLA4) and programmed cell death 1 (PD-1), which block the engagement of PD-1 or CTLA4 with their ligand (PD-L1 and PD-L2 for PD1; CD80/CD86 for CTLA4), thus avoiding the initiation of signaling pathways leading to the suppression of T cell activation. Of note, PARP inhibitors upregulate the expression of PD-L1 in cancer cells and enhance cancer-associated immunosuppression (Figure 4). This immunosuppression is reversible by blocking the PD-1/PD-L1 interaction [88]. This study established the rationale of combining PARP inhibitors with checkpoint blockade agents [7,89–91] or agents that alter PD-L1 expression [92], which has led to numerous clinical trials (Table 2). Although the result of an early phase II clinical trial combining Durvalumab with Olaparib in patients with relapsed small cell lung cancer did not meet the preset bar for efficacy [93], we are awaiting the results of ongoing clinical trials to better judge their effectiveness. Moreover, the PARP inhibitor Niraparib has been shown to enhance type I interferon signaling and T cell infiltration in the tumor and improve the therapeutic effect of anti-PD-1 [94].

Table 2. Clinical trials with PARP inhibitors in combination with check-point blockade agents (www.clinicaltrials.gov).

PARPi	IMMUNE CHECK-POINT INHIBITOR	CLINICAL PHASE	CONDITIONS	IDENTIFIER
Talazoparib	Avelumab (anti-PD-L1)	II	Advanced or metastatic solid tumors	NCT03330405
Pamiparib	Tislelizumab (anti-PD-1)	I/Ib	Solid tumors	NCT02660034
Rucaparib	Nivolumab (anti-PD-1)	Ib/IIa	Prostate Cancer, Endometrial Cancer	NCT03572478
Olaparib	Tremelimumab (anti-CTLA-4)	I/II	Ovarian Cancer, Fallopian Tube Cancer, Peritoneal Neoplasms	NCT02571725
Talazoparib	Avelumab (anti-PD-L1)	III	Ovarian Cancer	NCT03642132
Rucaparib	Nivolumab (anti-PD-1)	II	Biliary Tract Cancer	NCT03639935
Niraparib	PD-1 Inhibitor	II	Lung Neoplasms	NCT03308942
Talazoparib	Pembrolizumab (anti-PD1)	I/II	Solid Tumor, Epithelial Ovarian Cancer, Fallopian Tube Cancer, Peritoneal Cancer, Triple Negative Breast Cancer, Small Cell Lung Cancer, Metastatic Breast Cancer, Malignant Melanoma, Non-Small Cell Lung Cancer, Urothelial Carcinoma	NCT04158336
Olaparib	Atezolizumab (anti-PD-L1)	II	Locally Advanced Unresectable Breast Carcinoma, Metastatic Breast Carcinoma, Stage III Breast Cancer AJCC v7, Stage IIIA Breast Cancer AJCC v7, Stage IIIB Breast Cancer AJCC v7, Stage IIIC Breast Cancer AJCC v7, Stage IV Breast Cancer AJCC v6 and v7	NCT02849496
Olaparib	Durvalumab (Anti-PD-L1)	II	Endometrial Neoplasms, Uterine Neoplasms, Endometrium Cancer	NCT03951415
Talazoparib	Avelumab (anti-PD-L1)	I/II	Breast Cancer	NCT03964532
Olaparib	Durvalumab (Anti-PD-L1)	II	Mismatch Repair Proficient Colorectal Cancer, Pancreatic Adenocarcinoma, Leiomyosarcoma	NCT03851614
Olaparib	Durvalumab (Anti-PD-L1)	II	Triple Negative Breast Cancer	NCT03167619
Olaparib	Durvalumab (Anti-PD-L1)	I	Anatomic Stage IV Breast Cancer AJCC v8, Estrogen Receptor Negative, HER2/Neu Negative, Progesterone Receptor Negative, Prognostic Stage IV Breast Cancer AJCC v8, Triple-Negative Breast Carcinoma	NCT03544125
Niraparib	Dostarlimab (Anti-PD-1)	II/III	Ovarian Carcinosarcoma, Endometrial Carcinosarcoma	NCT03651206

Table 2. *Cont.*

PARPi	IMMUNE CHECK-POINT INHIBITOR	CLINICAL PHASE	CONDITIONS	IDENTIFIER
Veliparib	Nivolumab (anti-PD-1)	I	Advanced Solid Neoplasm, Aggressive Non-Hodgkin Lymphoma, Recurrent Solid Neoplasm, Refractory Mantle Cell Lymphoma, T-Cell Non-Hodgkin Lymphoma, Unresectable Solid Neoplasm	NCT03061188
Rucaparib	Nivolumab (anti-PD-1)	II	Epithelial Ovarian Cancer, Fallopian Tube Cancer, Primary Peritoneal Carcinoma, High Grade Serous Carcinoma, Endometrioid Adenocarcinoma	NCT03824704
Talazoparib	Avelumab (anti-PD-L1)	II	Squamous Cell Carcinoma of the Head and Neck (SCCHN), Metastatic Castration Resistant Prostate Cancer (mCRPC)	NCT04052204
Niraparib	Pembrolizumab (anti-PD1)	I/II	Triple Negative Breast Cancer, Ovarian Cancer, Breast Cancer, Metastatic Breast Cancer, Advanced Breast Cancer, Stage IV Breast Cancer, Fallopian Tube Cancer, Peritoneal Cancer	NCT02657889
Olaparib	Durvalumab (anti-PD-L1)	II	Metastatic Triple Negative Breast Cancer, Breast Cancer, ER-Negative PR-Negative HER2-Negative Breast Cancer, ER-Negative PR-Negative HER2-Negative Breast Neoplasms, Triple-Negative Breast Cancer, Triple-Negative Breast Neoplasm	NCT03801369
Olaparib	Durvalumab (anti-PD-L1)	II	Prostate Cancer	NCT03810105
Rucaparib	Nivolumab (anti-PD1)	II	Small Cell Lung Cancer	NCT03958045
veliparib	Nivolumab (anti-PD1)	I	Non-Small Cell Lung Cancer	NCT02944396
Olaparib	Pembrolizumab (anti-PD1)	III	Prostatic Neoplasms	NCT03834519
Olaparib	Durvalumab (anti-PD-L1) and Tremelimumab (anti-CTLA-4)	II	BRCA1 Gene Mutation, BRCA2 Gene Mutation, Ovarian Serous Adenocarcinoma, Recurrent Fallopian Tube Carcinoma, Recurrent Ovarian Carcinoma, Recurrent Primary Peritoneal Carcinoma	NCT02953457
Olaparib	Durvalumab (anti-PD-L1)	II	Squamous Cell Carcinoma of the Head and Neck	NCT02882308
Olaparib	Durvalumab (anti-PD-L1)	II	Glioma, Cholangiocarcinoma, Solid Tumor, IDH Mutation	NCT03991832
Niraparib	Dostarlimab (anti-PD1)	III	Ovarian Cancer	NCT03602859
Olaparib	Durvalumab (anti-PD-L1)	I	Advanced Malignant Solid Neoplasm, Metastatic Malignant Solid Neoplasm, Unresectable Malignant Solid Neoplasm	NCT03842228
Niraparib	Nivolumab (anti-PD1) or Ipilimumab (anti-CTLA4)	I II	Pancreatic Adenocarcinoma	NCT03404960
Niraparib	Atezolizumab (anti-PD-L1)	II	Solid Tumor	NCT04185831

NK cells can kill cancer cells by inducing death receptor-mediated apoptosis through the expression of FasL or TRAIL [71]. PARP inhibitors have been shown to sensitize cancer cells to death receptor-mediated apoptosis by upregulating death receptor surface expression [95,96] (Figure 4). In addition, the inhibition of PARP-1 upheld the capacity of NK cells to kill myeloid leukemic cells, and restored the proliferation and cytokine production of NK cells and cytotoxic T cells [97].

Recent work has revealed the intriguing link between genomic instability and the accumulation of DNA in the cytoplasm, which triggers the activation of innate immune responses through the cyclic GMP-AMP synthase (cGAS)/stimulator of interferon genes (STING) pathway that evolved to signal the presence of exogenous DNA [98]. Accordingly, it has been shown that PARP inhibitors promote the accumulation of cytosolic DNA, which activates the DNA-sensing cGAS–STING pathway and stimulates type I interferon (IFNs) gene expression to induce anti-tumor immunity independent of the BRCA status, providing a rationale for using PARP inhibitors as immunomodulatory agents [99–101]. Moreover, treatment with PARP inhibitors stimulates the type I IFN response in cells and tumors lacking BRCA2 [102]. Furthermore, PARP inhibition seems to augment cytotoxic T cell tumor infiltration through activation of the cGAS/STING innate immune pathway, leading to increased levels of chemokines, such as CXCL10 and CCL5, that induce the activation and function of cytotoxic $CD8^+$ T cells [103,104]. The effect of PARP inhibition-induced T cell recruitment to tumors is more noticeable in homologous recombination-deficient compared with homologous recombination-proficient triple negative breast cancer (TNBC) cells [104].

7. Conclusions and Future Prospects

The promise of PARP inhibitors in cancer therapy was initially based on proposed effects on genomic integrity in the cancer cell itself. Since then, it has been uncovered that PARPs play additional roles in other important aspects of cellular biology which could be of significance for both tumor physiology and its microenvironment. Here, we can see that the immunomodulatory roles of PARP-1 and PARP-2 are complex, with specific and overlapping roles which vary by cellular compartment and context. Future work will be needed to consider how this effect of PARP inhibition on the tumor microenvironment differs by tumor type, grade, and stage. PARP inhibition may serve as an important adjuvant to immunotherapeutic strategies or indeed benefit from the checkpoint blockade itself, but will require further elucidation of the precise mechanism by which it interacts with immune pathways.

Author Contributions: J.Y., L.M.-L., J.J., and S.O.A. wrote the manuscript. All authors have read and agreed to the published version of the manuscript.

Funding: This research was funded by The Fundación Científica de la Asociación Española Contra el Cáncer (AECC), grant number PROYEI6018YÉLA, and the Spanish Ministerio de Economía, Industria y Competitividad, grant number SAF2017-83565-R.

Acknowledgments: The authors thank C. Ampurdanés, A. Ali., and N. Lutfi. for their helpful discussions.

Conflicts of Interest: The authors declare no conflicts of interest.

References

1. Daniels, C.M.; Ong, S.-E.; Leung, A.K.L. The Promise of Proteomics for the Study of ADP-Ribosylation. *Mol. Cell* **2015**, *58*, 911–924. [CrossRef]
2. Yelamos, J.; Farres, J.; Llacuna, L.; Ampurdanes, C.; Martin-Caballero, J. PARP-1 and PARP-2: New players in tumour development. *Am. J. Cancer Res.* **2011**, *1*, 328–346. [PubMed]
3. Alemasova, E.E.; Lavrik, O.I. Poly(ADP-ribosyl)ation by PARP1: Reaction mechanism and regulatory proteins. *Nucleic Acids Res.* **2019**, *47*, 3811–3827. [CrossRef] [PubMed]
4. Bai, P. Biology of Poly(ADP-Ribose) Polymerases: The Factotums of Cell Maintenance. *Mol. Cell* **2015**, *58*, 947–958. [CrossRef] [PubMed]
5. Ménissier de Murcia, J.; Ricoul, M.; Tartier, L.; Niedergang, C.; Huber, A.; Dantzer, F.; Schreiber, V.; Amé, J.-C.; Dierich, A.; LeMeur, M.; et al. Functional interaction between PARP-1 and PARP-2 in chromosome stability and embryonic development in mouse. *EMBO J.* **2003**, *22*, 2255–2263. [CrossRef]
6. Mateo, J.; Lord, C.J.; Serra, V.; Tutt, A.; Balmaña, J.; Castroviejo-Bermejo, M.; Cruz, C.; Oaknin, A.; Kaye, S.B.; de Bono, J.S. A decade of clinical development of PARP inhibitors in perspective. *Ann. Oncol.* **2019**, *30*, 1437–1447. [CrossRef]
7. Stewart, R.A.; Pilie, P.G.; Yap, T.A. Development of PARP and immune-checkpoint inhibitor combinations. *Cancer Res.* **2018**, *78*, 6717–6725. [CrossRef]
8. Yélamos, J.; Galindo, M.; Navarro, J.; Albanell, J.; Rovira, A.; Rojo, F.; Oliver, J. Enhancing tumor-targeting monoclonal antibodies therapy by PARP inhibitors. *Oncoimmunology* **2016**, *5*, e1065370. [CrossRef]

9. Ashworth, A.; Lord, C.J. Synthetic lethal therapies for cancer: What´s next after PARP inhibitors? *Nat. Rev. Clin. Oncol.* **2018**, *15*, 564–576. [CrossRef]
10. Baumeister, S.H.; Freeman, G.J.; Dranoff, G.; Sharpe, A.H. Coinhibitory Pathways in Immunotherapy for Cancer. *Annu. Rev. Immunol.* **2016**, *34*, 539–573. [CrossRef]
11. Hinshaw, D.C.; Shevde, L.A. The Tumor Microenvironment Innately Modulates Cancer Progression. *Cancer Res.* **2019**, *79*, 4557–4566. [CrossRef] [PubMed]
12. Koch, U.; Radtke, F. Mechanisms of T Cell Development and Transformation. *Annu. Rev. Cell Dev. Biol.* **2011**, *27*, 539–562. [CrossRef] [PubMed]
13. Koizumi, S.; Ishikawa, H. Transcriptional Regulation of Differentiation and Functions of Effector T Regulatory Cells. *Cells* **2019**, *8*, 939. [CrossRef] [PubMed]
14. Yélamos, J.; Monreal, Y.; Saenz, L.; Aguado, E.; Schreiber, V.; Mota, R.; Fuente, T.; Minguela, A.; Parrilla, P.; De Murcia, G.; et al. PARP-2 deficiency affects the survival of CD4 + CD8+ double-positive thymocytes. *EMBO J.* **2006**, *25*, 4350–4360. [CrossRef]
15. Nicolás, L.; Martínez, C.; Baró, C.; Rodríguez, M.; Baroja-Mazo, A.; Sole, F.; Flores, J.M.; Ampurdanés, C.; Dantzer, F.; Martin-Caballero, J.; et al. Loss of poly(ADP-ribose) polymerase-2 leads to rapid development of spontaneous T-cell lymphomas in p53-deficient mice. *Oncogene* **2010**, *29*, 2877–2883. [CrossRef]
16. Nasta, F.; Laudisi, F.; Sambucci, M.; Rosado, M.M.; Pioli, C. Increased Foxp3 + Regulatory T Cells in Poly(ADP-Ribose) Polymerase-1 Deficiency. *J. Immunol.* **2010**, *184*, 3470–3477. [CrossRef]
17. Navarro, J.; Gozalbo-López, B.; Méndez, A.C.; Dantzer, F.; Schreiber, V.; Martínez, C.; Arana, D.M.; Farrés, J.; Revilla-Nuin, B.; Bueno, M.F.; et al. PARP-1/PARP-2 double deficiency in mouse T cells results in faulty immune responses and T lymphomas. *Sci. Rep.* **2017**, *7*, 41962. [CrossRef]
18. Grossman, Z.; Paul, W.E. Dynamic Tuning of Lymphocytes: Physiological Basis, Mechanisms, and Function. *Annu. Rev. Immunol.* **2015**, *33*, 677–713. [CrossRef]
19. Jameson, S.C. Maintaining the norm: T-cell homeostasis. *Nat. Rev. Immunol.* **2002**, *2*, 547–556. [CrossRef]
20. Baek, K.-H.; Shin, H.-J.; Yoo, J.-K.; Cho, J.-H.; Choi, Y.-H.; Sung, Y.-C.; McKeon, F.; Lee, C.-W. p53 deficiency and defective mitotic checkpoint in proliferating T lymphocytes increase chromosomal instability through aberrant exit from mitotic arrest. *J. Leukoc. Biol.* **2003**, *73*, 850–861. [CrossRef]
21. Prochazkova, J.; Sakaguchi, S.; Owusu, M.; Mazouzi, A.; Wiedner, M.; Velimezi, G.; Moder, M.; Turchinovich, G.; Hladik, A.; Gurnhofer, E.; et al. DNA Repair Cofactors ATMIN and NBS1 Are Required to Suppress T Cell Activation. *PloS Genet.* **2015**, *11*, e1005645. [CrossRef] [PubMed]
22. Hartwell, L.H.; Szankasi, P.; Roberts, C.J.; Murray, A.W.; Friend, S.H. Integrating genetic approaches into the discovery of anticancer drugs. *Science* **1997**, *278*, 1064–1068. [CrossRef]
23. Moreno-Lama, L.; Galindo-Campos, M.A.; Martínez, C.; Comerma, L.; Vazquez, I.; Vernet-Tomas, M.; Ampurdanés, C.; Lutfi, N.; Martin-Caballero, J.; Dantzer, F.; et al. Coordinated signals from PARP-1 and PARP-2 are required to establish a proper T cell immune response to breast tumors in mice. *Oncogene* **2020**. [CrossRef] [PubMed]
24. Macian, F. NFAT proteins: Key regulators of T-cell development and function. *Nat. Rev. Immunol.* **2005**, *5*, 472–484. [CrossRef] [PubMed]
25. Valdor, R.; Schreiber, V.; Saenz, L.; Martínez, T.; Muñoz-Suano, A.; Dominguez-Villar, M.; Ramírez, P.; Parrilla, P.; Aguado, E.; García-Cózar, F.; et al. Regulation of NFAT by poly(ADP-ribose) polymerase activity in T cells. *Mol. Immunol.* **2008**, *45*, 1863–1871. [CrossRef]
26. Saenz, L.; Lozano, J.J.; Valdor, R.; Baroja-Mazo, A.; Ramirez, P.; Parrilla, P.; Aparicio, P.; Sumoy, L.; Yélamos, J. Transcriptional regulation by poly(ADP-ribose) polymerase-1 during T cell activation. *BMC Genom.* **2008**, *9*, 171. [CrossRef]
27. Sambucci, M.; Laudisi, F.; Novelli, F.; Bennici, E.; Rosado, M.M.; Pioli, C. Effects of PARP-1 deficiency on Th1 and Th2 cell differentiation. *Sci. World J.* **2013**, *2013*, 375024. [CrossRef]
28. Ghonim, M.A.; Pyakurel, K.; Ibba, S.V.; Al-Khami, A.A.; Wang, J.; Rodriguez, P.; Rady, H.F.; El-Bahrawy, A.H.; Lammi, M.R.; Mansy, M.S.; et al. PARP inhibition by olaparib or gene knockout blocks asthma-like manifestation in mice by modulating CD4+ T cell function. *J. Transl. Med.* **2015**, *13*, 225. [CrossRef]
29. Gonzalez-Rey, E.; Martínez-Romero, R.; O'Valle, F.; Aguilar-Quesada, R.; Conde, C.; Delgado, M.; Oliver, F.J. Therapeutic effect of a poly(ADP-ribose) polymerase-1 inhibitor on experimental arthritis by downregulating inflammation and Th1 response. *PLoS ONE* **2007**, *2*, e1071. [CrossRef]

30. Luo, X.; Nie, J.; Wang, S.; Chen, Z.; Chen, W.; Li, D.; Hu, H.; Li, B. Poly(ADP-ribosyl)ation of FOXP3 Protein Mediated by PARP-1 Protein regulates the function of regulatory t cells. *J. Biol. Chem.* **2015**, *290*, 28675–28682. [CrossRef]
31. Zhang, P.; Nakatsukasa, H.; Tu, E.; Kasagi, S.; Cui, K.; Ishikawa, M.; Konkel, J.E.; Maruyama, T.; Wei, G.; Abbatiello, B.; et al. PARP-1 regulates expression of TGF-β receptors in T cells. *Blood* **2013**, *122*, 2224–2232. [CrossRef] [PubMed]
32. Hardy, R.R.; Hayakawa, K. B C ELL D EVELOPMENT P ATHWAYS. *Annu. Rev. Immunol.* **2001**, *19*, 595–621. [CrossRef] [PubMed]
33. Wang, Z.Q.; Stingl, L.; Morrison, C.; Jantsch, M.; Los, M.; Schulze-Osthoff, K.; Wagner, E.F. PARP is important for genomic stability but dispensable in apoptosis. *Genes Dev.* **1997**, *11*, 2347–2358. [CrossRef] [PubMed]
34. Morrison, C.; Smith, G.C.M.; Stingl, L.; Jackson, S.P.; Wagner, E.F.; Wang, Z.Q. Genetic interaction between PARP and DNA-PK in V(D)J recombination and tumorigenesis. *Nat. Genet.* **1997**, *17*, 479–482. [CrossRef] [PubMed]
35. Brown, M.L.; Franco, D.; Burkle, A.; Chang, Y. Role of poly(ADP-ribosyl)ation in DNA-PKcs- independent V(D)J recombination. *Proc. Natl. Acad. Sci. USA* **2002**, *99*, 4532–4537. [CrossRef] [PubMed]
36. Galindo-Campos, M.A.; Bedora-Faure, M.; Farrés, J.; Lescale, C.; Moreno-Lama, L.; Martínez, C.; Martín-Caballero, J.; Ampurdanés, C.; Aparicio, P.; Dantzer, F.; et al. Coordinated signals from the DNA repair enzymes PARP-1 and PARP-2 promotes B-cell development and function. *Cell Death Differ.* **2019**, *26*, 2667–2681. [CrossRef]
37. Ronson, G.E.; Piberger, A.L.; Higgs, M.R.; Olsen, A.L.; Stewart, G.S.; McHugh, P.J.; Petermann, E.; Lakin, N.D. PARP1 and PARP2 stabilise replication forks at base excision repair intermediates through Fbh1-dependent Rad51 regulation. *Nat. Commun.* **2018**, *9*, 746. [CrossRef]
38. García de Vinuesa, C.; O'Leary, P.; Sze, D.M.-Y.; Toellner, K.-M.; MacLennan, I.C.M. T-independent type 2 antigens induce B cell proliferation in multiple splenic sites, but exponential growth is confined to extrafollicular foci. *Eur. J. Immunol.* **1999**, *29*, 1314–1323. [CrossRef]
39. Inoue, T.; Moran, I.; Shinnakasu, R.; Phan, T.G.; Kurosaki, T. Generation of memory B cells and their reactivation. *Immunol. Rev.* **2018**, *283*, 138–149. [CrossRef]
40. Tas, J.M.J.; Mesin, L.; Pasqual, G.; Targ, S.; Jacobsen, J.T.; Mano, Y.M.; Chen, C.S.; Weill, J.-C.; Reynaud, C.-A.; Browne, E.P.; et al. Visualizing antibody affinity maturation in germinal centers. *Science* **2016**, *351*, 1048–1054. [CrossRef]
41. Muramatsu, M.; Kinoshita, K.; Fagarasan, S.; Yamada, S.; Shinkai, Y.; Honjo, T. Class switch recombination and hypermutation require activation-induced cytidine deaminase (AID), a potential RNA editing enzyme. *Cell* **2000**, *102*, 553–563. [CrossRef]
42. Ambrose, H.E.; Willimott, S.; Beswick, R.W.; Dantzer, F.; de Murcia, J.M.; Yelamos, J.; Wagner, S.D. Poly(ADP-ribose) polymerase-1 (Parp-1)-deficient mice demonstrate abnormal antibody responses. *Immunology* **2009**, *127*, 178–186. [CrossRef] [PubMed]
43. Robert, I.; Dantzer, F.; Reina-San-Martin, B. Parp1 facilitates alternative NHEJ, whereas Parp2 suppresses IgH/c-myc translocations during immunoglobulin class switch recombination. *J. Exp. Med.* **2009**, *206*, 1047–1056. [CrossRef] [PubMed]
44. Ambrose, H.E.; Papadopoulou, V.; Beswick, R.W.; Wagner, S.D. Poly-(ADP-ribose) polymerase-1 (Parp-1) binds in a sequence-specific manner at the Bcl-6 locus and contributes to the regulation of Bcl-6 transcription. *Oncogene* **2007**, *26*, 6244–6252. [CrossRef]
45. Dent, A.L.; Shaffer, A.L.; Yu, X.; Allman, D.; Staudt, L.M. Control of inflammation, cytokine expression, and germinal center formation by BCL-6. *Science* **1997**, *276*, 589–592. [CrossRef]
46. Ye, B.H.; Cattoretti, G.; Shen, Q.; Zhang, J.; Hawe, N.; De Waard, R.; Leung, C.; Nouri-Shirazi, M.; Orazi, A.; Chaganti, R.S.K.; et al. The BCL-6 proto-oncogene controls germinal-centre formation and Th2- type inflammation. *Nat. Genet.* **1997**, *16*, 161–170. [CrossRef] [PubMed]
47. Jacobs, H.; Fukita, Y.; Van Der Horst, G.T.J.; De Boer, J.; Weeda, G.; Essers, J.; De Wind, N.; Engelward, B.P.; Samson, L.; Verbeek, S.; et al. Hypermutation of immunoglobulin genes in memory B cells of DNA repair-deficient mice. *J. Exp. Med.* **1998**, *187*, 1735–1742. [CrossRef] [PubMed]
48. Tepper, S.; Mortusewicz, O.; Członka, E.; Bello, A.; Schmidt, A.; Jeschke, J.; Fischbach, A.; Pfeil, I.; Petersen-Mahrt, S.K.; Mangerich, A.; et al. Restriction of AID activity and somatic hypermutation by PARP-1. *Nucleic Acids Res.* **2019**, *47*, 7418–7429. [CrossRef]

49. Thompson, E.C. Innate immune cells in motion. *Trends Immunol.* **2011**, *32*, 451. [CrossRef]
50. Kolaczkowska, E.; Kubes, P. Neutrophil recruitment and function in health and inflammation. *Nat. Rev. Immunol.* **2013**, *13*, 159–175. [CrossRef]
51. Gonzalez, H.; Hagerling, C.; Werb, Z. Roles of the immune system in cancer: From tumor initiation to metastatic progression. *Genes Dev.* **2018**, *32*, 1267–1284. [CrossRef] [PubMed]
52. Szabó, C.; Lim, L.H.; Cuzzocrea, S.; Getting, S.J.; Zingarelli, B.; Flower, R.J.; Salzman, A.L.; Perretti, M. Inhibition of poly (ADP-ribose) synthetase attenuates neutrophil recruitment and exerts antiinflammatory effects. *J. Exp. Med.* **1997**, *186*, 1041–1049. [CrossRef] [PubMed]
53. Dharwal, V.; Naura, A.S. PARP-1 inhibition ameliorates elastase induced lung inflammation and emphysema in mice. *Biochem. Pharm.* **2018**, *150*, 24–34. [CrossRef]
54. Wang, S.; Yang, F.-J.; Wang, X.; Zhou, Y.; Dai, B.; Han, B.; Ma, H.-C.; Ding, Y.-T.; Shi, X.-L. PARP-1 promotes tumor recurrence after warm ischemic liver graft transplantation via neutrophil recruitment and polarization. *Oncotarget* **2017**, *8*, 88918–88933. [CrossRef]
55. Mota, R.A.; Sánchez-Bueno, F.; Saenz, L.; Hernández-Espinosa, D.; Jimeno, J.; Tornel, P.L.; Martínez-Torrano, A.; Ramírez, P.; Parrilla, P.; Yélamos, J. Inhibition of poly(ADP-ribose) polymerase attenuates the severity of acute pancreatitis and associated lung injury. *Lab. Investig.* **2005**, *85*, 1250–1262. [CrossRef] [PubMed]
56. Lavin, Y.; Mortha, A.; Rahman, A.; Merad, M. Regulation of macrophage development and function in peripheral tissues. *Nat. Rev. Immunol.* **2015**, *15*, 731–744. [CrossRef]
57. Mosser, D.M.; Edwards, J.P. Exploring the full spectrum of macrophage activation. *Nat. Rev. Immunol.* **2008**, *8*, 958–969. [CrossRef]
58. Mantovani, A.; Sozzani, S.; Locati, M.; Allavena, P.; Sica, A. Macrophage polarization: Tumor-associated macrophages as a paradigm for polarized M2 mononuclear phagocytes. *Trends Immunol.* **2002**, *23*, 549–555. [CrossRef]
59. Hauschildt, S.; Scheipers, P.; Bessler, W.; Schwarz, K.; Ullmer, A.; Flad, H.D.; Heine, H. Role of ADP-ribosylation in activated monocytes/macrophages. *Adv. Exp. Med. Biol.* **1997**, *419*, 249–252.
60. Kunze, F.A.; Bauer, M.; Komuczki, J.; Lanzinger, M.; Gunasekera, K.; Hopp, A.-K.; Lehmann, M.; Becher, B.; Müller, A.; Hottiger, M.O. ARTD1 in Myeloid Cells Controls the IL-12/18-IFN-γ Axis in a Model of Sterile Sepsis, Chronic Bacterial Infection, and Cancer. *J. Immunol.* **2019**, *202*, 1406–1416. [CrossRef]
61. Aguilar-Quesada, R.; Muñoz-Gámez, J.A.; Martín-Oliva, D.; Peralta-Leal, A.; Quiles-Pérez, R.; Rodríguez-Vargas, J.M.; Ruiz de Almodóvar, M.; Conde, C.; Ruiz-Extremera, A.; Oliver, F.J. Modulation of transcription by PARP-1: Consequences in carcinogenesis and inflammation. *Curr. Med. Chem.* **2007**, *14*, 1179–1187. [CrossRef] [PubMed]
62. Tokarz, P.; Płoszaj, T.; Regdon, Z.; Virág, L.; Robaszkiewicz, A. PARP1-LSD1 functional interplay controls transcription of SOD2 that protects human pro-inflammatory macrophages from death under an oxidative condition. *Free Radic. Biol. Med.* **2019**, *131*, 218–224. [CrossRef]
63. Zerfaoui, M.; Naura, A.S.; Errami, Y.; Hans, C.P.; Rezk, B.M.; Park, J.; Elsegeiny, W.; Kim, H.; Lord, K.; Kim, J.G.; et al. Effects of PARP-1 deficiency on airway inflammatory cell recruitment in response to LPS or TNF: Differential effects on CXCR2 ligands and Duffy Antigen Receptor for Chemokines. *J. Leukoc. Biol.* **2009**, *86*, 1385–1392. [CrossRef]
64. Murphy, T.L.; Grajales-Reyes, G.E.; Wu, X.; Tussiwand, R.; Briseño, C.G.; Iwata, A.; Kretzer, N.M.; Durai, V.; Murphy, K.M.; Edu, T. Transcriptional Control of Dendritic Cell Development. *Annu. Rev. Immunol.* **2016**, *34*, 93–119. [CrossRef]
65. Wculek, S.K.; Cueto, F.J.; Mujal, A.M.; Melero, I.; Krummel, M.F.; Sancho, D. Dendritic cells in cancer immunology and immunotherapy. *Nat. Rev. Immunol.* **2020**, *20*, 7–24. [CrossRef]
66. Echeverri Tirado, L.C.; Ghonim, M.A.; Wang, J.; Al-Khami, A.A.; Wyczechowska, D.; Luu, H.H.; Kim, H.; Sanchez-Pino, M.D.; Yélamos, J.; Yassin, L.M.; et al. PARP-1 is critical for recruitment of dendritic cells to the lung in a mouse model of asthma but dispensable for their differentiation and function. *Mediat. Inflamm.* **2019**, *2019*, 1656484. [CrossRef]
67. Cavone, L.; Aldinucci, A.; Ballerini, C.; Biagioli, T.; Moroni, F.; Chiarugi, A. PARP-1 inhibition prevents CNS migration of dendritic cells during EAE, suppressing the encephalitogenic response and relapse severity. *Mult. Scler.* **2011**, *17*, 794–807. [CrossRef]

68. Aldinucci, A.; Gerlini, G.; Fossati, S.; Cipriani, G.; Ballerini, C.; Biagioli, T.; Pimpinelli, N.; Borgognoni, L.; Massacesi, L.; Moroni, F.; et al. A key role for poly(ADP-ribose) polymerase-1 activity during human dendritic cell maturation. *J. Immunol.* **2007**, *179*, 305–312. [CrossRef]
69. Wang, J.-Q.; Tang, Y.; Li, Q.-S.; Xiao, M.; Li, M.; Sheng, Y.-T.; Yang, Y.; Wang, Y.-L. PARG regulates the proliferation and differentiation of DCs and T cells via PARP/NF-κB in tumour metastases of colon carcinoma. *Oncol. Rep.* **2019**, *41*, 2657–2666. [CrossRef]
70. Cerwenka, A.; Lanier, L.L. Natural killer cell memory in infection, inflammation and cancer. *Nat. Rev. Immunol.* **2016**, *16*, 112–123. [CrossRef]
71. Prager, I.; Watzl, C. Mechanisms of natural killer cell-mediated cellular cytotoxicity. *J. Leukoc. Biol.* **2019**, *105*, 1319–1329. [CrossRef] [PubMed]
72. Shou, Q.; Fu, H.; Huang, X.; Yang, Y. PARP-1 controls NK cell recruitment to the site of viral infection. *Jci Insight* **2019**, *4*. [CrossRef] [PubMed]
73. Mocchegiani, E.; Muzzioli, M.; Giacconi, R.; Cipriano, C.; Gasparini, N.; Franceschi, C.; Gaetti, R.; Cavalieri, E.; Suzuki, H. Metallothioneins/PARP-1/IL-6 interplay on natural killer cell activity in elderly: Parallelism with nonagenarians and old infected humans. Effect of zinc supply. *Mech. Ageing Dev.* **2003**, *124*, 459–468. [CrossRef]
74. Heyman, B.; Jamieson, C. To PARP or not to PARP?-Toward sensitizing acute myeloid leukemia stem cells to immunotherapy. *EMBO J.* **2019**, *38*, e103479. [CrossRef]
75. Dunn, G.P.; Old, L.J.; Schreiber, R.D. The three Es of cancer immunoediting. *Annu. Rev. Immunol.* **2004**, *22*, 329–360. [CrossRef]
76. Chacon-Cabrera, A.; Fermoselle, C.; Salmela, I.; Yelamos, J.; Barreiro, E. MicroRNA expression and protein acetylation pattern in respiratory and limb muscles of Parp-1-/- and Parp-2-/- mice with lung cancer cachexia. *Biochim. Biophys. Acta Gen. Subj.* **2015**, *1850*, 2530–2543. [CrossRef]
77. Matsushita, H.; Vesely, M.D.; Koboldt, D.C.; Rickert, C.G.; Uppaluri, R.; Magrini, V.J.; Arthur, C.D.; White, J.M.; Chen, Y.-S.; Shea, L.K.; et al. Cancer exome analysis reveals a T-cell-dependent mechanism of cancer immunoediting. *Nature* **2012**, *482*, 400–404. [CrossRef]
78. Kalams, S.A.; Walker, B.D. The critical need for CD4 help in maintaining effective cytotoxic T lymphocyte responses. *J. Exp. Med.* **1998**, *188*, 2199–2204. [CrossRef]
79. Pardoll, D.M.; Topalian, S.L. The role of CD4+ T cell responses in antitumor immunity. *Curr. Opin. Immunol.* **1998**, *10*, 588–594. [CrossRef]
80. Shankaran, V.; Ikeda, H.; Bruce, A.T.; White, J.M.; Swanson, P.E.; Old, L.J.; Schreiber, R.D. IFNγ, and lymphocytes prevent primary tumour development and shape tumour immunogenicity. *Nature* **2001**, *410*, 1107–1111. [CrossRef]
81. Muntasell, A.; Rojo, F.; Servitja, S.; Rubio-Perez, C.; Cabo, M.; Tamborero, D.; Costa-García, M.; Martínez-Garcia, M.; Menéndez, S.; Vazquez, I.; et al. NK Cell Infiltrates and HLA Class I Expression in Primary HER2+ Breast Cancer Predict and Uncouple Pathological Response and Disease-free Survival. *Clin. Cancer Res.* **2019**, *25*, 1535–1545. [CrossRef] [PubMed]
82. Topalian, S.L.; Drake, C.G.; Pardoll, D.M. Targeting the PD-1/B7-H1(PD-L1) pathway to activate anti-tumor immunity. *Curr. Opin. Immunol.* **2012**, *24*, 207–212. [CrossRef]
83. Böger, C.; Behrens, H.-M.; Krüger, S.; Röcken, C. The novel negative checkpoint regulator VISTA is expressed in gastric carcinoma and associated with PD-L1/PD-1: A future perspective for a combined gastric cancer therapy? *Oncoimmunology* **2017**, *6*, e1293215. [CrossRef] [PubMed]
84. Snyder, A.; Makarov, V.; Merghoub, T.; Yuan, J.; Zaretsky, J.M.; Desrichard, A.; Walsh, L.A.; Postow, M.A.; Wong, P.; Ho, T.S.; et al. Genetic basis for clinical response to CTLA-4 blockade in melanoma. *N. Engl. J. Med.* **2014**, *371*, 2189–2199. [CrossRef] [PubMed]
85. Van Der Burg, S.H.; Arens, R.; Ossendorp, F.; Van Hall, T.; Melief, C.J.M. Vaccines for established cancer: Overcoming the challenges posed by immune evasion. *Nat. Rev. Cancer* **2016**, *16*, 219–233. [CrossRef]
86. Gubin, M.M.; Zhang, X.; Schuster, H.; Caron, E.; Ward, J.P.; Noguchi, T.; Ivanova, Y.; Hundal, J.; Arthur, C.D.; Krebber, W.J.; et al. Checkpoint blockade cancer immunotherapy targets tumour-specific mutant antigens. *Nature* **2014**, *515*, 577–581. [CrossRef]
87. Adams, J.L.; Smothers, J.; Srinivasan, R.; Hoos, A. Big opportunities for small molecules in immuno-oncology. *Nat. Rev. Drug Discov.* **2015**, *14*, 603–622. [CrossRef]

88. Jiao, S.; Xia, W.; Yamaguchi, H.; Wei, Y.; Chen, M.-K.; Hsu, J.-M.; Hsu, J.L.; Yu, W.-H.; Du, Y.; Lee, H.-H.; et al. PARP Inhibitor Upregulates PD-L1 Expression and Enhances Cancer-Associated Immunosuppression. *Clin. Cancer Res.* **2017**, *23*, 3711–3720. [CrossRef]
89. Lassen, U. Combining PARP inhibition with PD-1 inhibitors. *Lancet. Oncol.* **2019**, *20*, 1196–1198. [CrossRef]
90. Li, A.; Yi, M.; Qin, S.; Chu, Q.; Luo, S.; Wu, K. Prospects for combining immune checkpoint blockade with PARP inhibition. *J. Hematol. Oncol.* **2019**, *12*, 98. [CrossRef]
91. Higuchi, T.; Flies, D.B.; Marjon, N.A.; Mantia-Smaldone, G.; Ronner, L.; Gimotty, P.A.; Adams, S.F. CTLA-4 Blockade Synergizes Therapeutically with PARP Inhibition in BRCA1-Deficient Ovarian Cancer. *Cancer Immunol. Res.* **2015**, *3*, 1257–1268. [CrossRef] [PubMed]
92. Shao, B.; Li, C.-W.; Lim, S.-O.; Sun, L.; Lai, Y.-J.; Hou, J.; Liu, C.; Chang, C.-W.; Qiu, Y.; Hsu, J.-M.; et al. Deglycosylation of PD-L1 by 2-deoxyglucose reverses PARP inhibitor-induced immunosuppression in triple-negative breast cancer. *Am. J. Cancer Res.* **2018**, *8*, 1837–1846. [PubMed]
93. Thomas, A.; Vilimas, R.; Trindade, C.; Erwin-Cohen, R.; Roper, N.; Xi, L.; Krishnasamy, V.; Levy, E.; Mammen, A.; Nichols, S.; et al. Durvalumab in Combination with Olaparib in Patients with Relapsed SCLC: Results from a Phase II Study. *J. Thorac. Oncol.* **2019**, *14*, 1447–1457. [CrossRef] [PubMed]
94. Wang, Z.; Sun, K.; Xiao, Y.; Feng, B.; Mikule, K.; Ma, X.; Feng, N.; Vellano, C.P.; Federico, L.; Marszalek, J.R.; et al. Niraparib activates interferon signaling and potentiates anti-PD-1 antibody efficacy in tumor models. *Sci. Rep.* **2019**, *9*, 1853. [CrossRef] [PubMed]
95. Meng, X.W.; Koh, B.D.; Zhang, J.-S.; Flatten, K.S.; Schneider, P.A.; Billadeau, D.D.; Hess, A.D.; Smith, B.D.; Karp, J.E.; Kaufmann, S.H. Poly(ADP-ribose) polymerase inhibitors sensitize cancer cells to death receptor-mediated apoptosis by enhancing death receptor expression. *J. Biol. Chem.* **2014**, *289*, 20543–20558. [CrossRef] [PubMed]
96. Fenerty, K.E.; Padget, M.; Wolfson, B.; Gameiro, S.R.; Su, Z.; Lee, J.H.; Rabizadeh, S.; Soon-Shiong, P.; Hodge, J.W. Immunotherapy utilizing the combination of natural killer- and antibody dependent cellular cytotoxicity (ADCC)-mediating agents with poly (ADP-ribose) polymerase (PARP) inhibition. *J. Immunother. Cancer* **2018**, *6*, 133. [CrossRef] [PubMed]
97. Aurelius, J.; Martner, A.; Riise, R.E.; Romero, A.I.; Palmqvist, L.; Brune, M.; Hellstrand, K.; Thorén, F.B. Chronic myeloid leukemic cells trigger poly(ADP-ribose) polymerase-dependent inactivation and cell death in lymphocytes. *J. Leukoc. Biol.* **2013**, *93*, 155–160. [CrossRef]
98. Bakhoum, S.F.; Cantley, L.C. The Multifaceted Role of Chromosomal Instability in Cancer and Its Microenvironment. *Cell* **2018**, *174*, 1347–1360. [CrossRef]
99. Shen, J.; Zhao, W.; Ju, Z.; Wang, L.; Peng, Y.; Labrie, M.; Yap, T.A.; Mills, G.B.; Peng, G. PARPI triggers the STING-dependent immune response and enhances the therapeutic efficacy of immune checkpoint blockade independent of BRCANEss. *Cancer Res.* **2019**, *79*, 311–319. [CrossRef]
100. Ding, L.; Kim, H.J.; Wang, Q.; Kearns, M.; Jiang, T.; Ohlson, C.E.; Li, B.B.; Xie, S.; Liu, J.F.; Stover, E.H.; et al. PARP Inhibition Elicits STING-Dependent Antitumor Immunity in Brca1-Deficient Ovarian Cancer. *Cell Rep.* **2018**, *25*, 2972–2980. [CrossRef]
101. Chabanon, R.M.; Muirhead, G.; Krastev, D.B.; Adam, J.; Morel, D.; Garrido, M.; Lamb, A.; Hénon, C.; Dorvault, N.; Rouanne, M.; et al. PARP inhibition enhances tumor cell–intrinsic immunity in ERCC1-deficient non–small cell lung cancer. *J. Clin. Invest.* **2019**, *129*, 1211–1228. [CrossRef] [PubMed]
102. Reisländer, T.; Lombardi, E.P.; Groelly, F.J.; Miar, A.; Porru, M.; Di Vito, S.; Wright, B.; Lockstone, H.; Biroccio, A.; Harris, A.; et al. BRCA2 abrogation triggers innate immune responses potentiated by treatment with PARP inhibitors. *Nat. Commun.* **2019**, *10*, 3143. [CrossRef] [PubMed]
103. Sen, T.; Rodriguez, B.L.; Chen, L.; Della Corte, C.M.; Morikawa, N.; Fujimoto, J.; Cristea, S.; Nguyen, T.; Diao, L.; Li, L.; et al. Targeting DNA damage response promotes antitumor immunity through STING-mediated T-cell activation in small cell lung cancer. *Cancer Discov.* **2019**, *9*, 646–661. [CrossRef] [PubMed]
104. Pantelidou, C.; Sonzogni, O.; Taveira, M.D.O.; Mehta, A.K.; Kothari, A.; Wang, D.; Visal, T.; Li, M.K.; Pinto, J.; Castrillon, J.A.; et al. Parp inhibitor efficacy depends on CD8+ T-cell recruitment via intratumoral sting pathway activation in brca-deficient models of triple-negative breast cancer. *Cancer Discov.* **2019**, *9*, 722–737. [CrossRef]

cancers

MDPI

Review

PARylation During Transcription: Insights into the Fine-Tuning Mechanism and Regulation

Zoltán G. Páhi, Barbara N. Borsos, Vasiliki Pantazi, Zsuzsanna Ujfaludi and Tibor Pankotai *

Department of Oral Biology and Experimental Dental Research, Faculty of Dentistry, University of Szeged, Tisza Lajos krt 83, H-6722 Szeged, Hungary; pahizol@gmail.com (Z.G.P.); borsos.barbara.nikolett@gmail.com (B.N.B.); vasopantazi@outlook.com (V.P.); odianya@bio.u-szeged.hu (Z.U.)

* Correspondence: pankotai@bio.u-szeged.hu or pankotai.tibor@stoma.szote.u-szeged.hu

Received: 5 December 2019; Accepted: 9 January 2020; Published: 11 January 2020

Abstract: Transcription is a multistep, tightly regulated process. During transcription initiation, promoter recognition and pre-initiation complex (PIC) formation take place, in which dynamic recruitment or exchange of transcription activators occur. The precise coordination of the recruitment and removal of transcription factors, as well as chromatin structural changes, are mediated by post-translational modifications (PTMs). Poly(ADP-ribose) polymerases (PARPs) are key players in this process, since they can modulate DNA-binding activities of specific transcription factors through poly-ADP-ribosylation (PARylation). PARylation can regulate the transcription at three different levels: (1) by directly affecting the recruitment of specific transcription factors, (2) by triggering chromatin structural changes during initiation and as a response to cellular stresses, or (3) by post-transcriptionally modulating the stability and degradation of specific mRNAs. In this review, we principally focus on these steps and summarise the recent findings, demonstrating the mechanisms through which PARylation plays a potential regulatory role during transcription and DNA repair.

Keywords: transcription; PARylation; PARP; DNA damage; transcription silencing

---

1. Introduction

The Mechanism of Poly-ADP-Ribosylation and the PARP Superfamily

PARylation is a reversible post-translational modification (PTM), in which writers, such as poly(ADP-ribose) polymerases (PARPs) as well as erasers, including poly(ADP-ribose) glycohydrolases (PARGs) and ADP-ribosyl hydrolase 3 (ARH3) are involved [1–7]. PARPs are NAD^+-dependent enzymes and thus require a source of NAD^+ which is provided by nicotinamide mononucleotide adenylyl transferases (NMNATs) [8]. ADP-ribosylation is a multistep process, involving initiation, elongation, branching, and the release of PAR units. First, PARP binds to nicotinamide adenine dinucleotide (NAD^+) and cleaves the nicotine amid unit, catalysing the transfer of the ADP-ribose moieties to the acceptor protein [9]. During initiation, the first ADP-ribose monomer can be covalently linked to Lys, Arg, Glu, Asp, Cys, Ser, or Thr amino acid residues of the acceptor protein [9]. During the branch formation step, 2′-1′ ribose-ribose bonds are generated between ADP-ribose units.

In human cells, PARPs are classified based on their enzymatic activity: PARP1, PARP2, PARP5a, and PARP5b catalyse PAR chain formation, while PARPs 3, 4, 6–8, 10–12, and 14–16 have been described as mono-ADP-ribosyl transferases (MARTs) [9]. PARP5a and PARP5b share a high level of similarity and are also called tankyrase 1 and tankyrase 2, respectively, due to their ankyrin repeat region and the sterile alpha motif [10]. Since PARylation is a reversible process, the covalently-attached PAR can be removed by PARGs and ARH3 as well, keeping the PAR levels in the cell under control. While PARG can efficiently cleave the PAR O-glycosidic bond, ARH3 is mainly responsible for the hydrolysis of protein-free PAR [11].

2. PARylation in Transcription Regulation

2.1. The Major Regulatory Steps of Transcription Activation

RNA synthesis requires a well-coordinated regulation of transcription at different levels. As a first step of initiation, transcription factor II (TFII)D-TFIIA-TFIIB binds to the promoter region; then TFIIF, along with RNA polymerase II (RNAPII), is recruited, resulting in the stabilisation of the pre-initiation complex (PIC) [12]. Next, TFIIE and TFIIH join to the core PIC, contributing to its association with the mediator complex [13]. Xeroderma pigmentosum type B (XPB), one of the subunits of TFIIH, induces DNA unwinding around the transcription start site (TSS) and initiates the formation of the transcription bubble [14]. Subsequently, the cyclin-dependent kinase 7 (CDK7) subunit of TFIIH phosphorylates the C-terminal domain (CTD) of RNAPII at Ser5, which is indispensable for transcription initiation. Following the synthesis of approximately 20–60 base pairs of RNA, RNAPII is stopped, which is the so-called promoter-proximal pausing [15]. During this step, negative elongation factors, including the dephosphorylated form of DRB sensitivity inducing factor (DSIF) and negative elongation factor (NELF), bind to the RNAPII, thereby hindering the elongation [16,17]. The cyclin-dependent kinase 9 (CDK9) subunit of positive transcriptional elongation factor b (P-TEFb) promotes the elongation process by phosphorylating DSIF, NELF, and RNAPII CTD at Ser2 [18,19]. Consequently, the phosphorylated NELF complex dissociates, while DSIF becomes a positive elongation factor responsible for recruiting other factors, such as cyclin-dependent kinase 12 (CDK12), which can also catalyse the phosphorylation of RNAPII CTD at Ser2 and by this, supporting the elongation step [20,21].

2.2. PARP1 Plays a Key Role in the Fine-Tune Regulation of Transcription Initiation

During transcription initiation, PARP1 can PARylate sequence-specific transcription factors, such as nuclear factor kappa-light-chain-enhancer of activated B cells (NF-κB), Myb-related protein B (B-MYB), organic cation transporter 1 (OCT1), sex determining region Y-box 2 (SOX2), and Krueppel-like factor 8 (KLF8), as well as oestrogen- and retinoic acid receptors, which can either inhibit or enhance the activity of these factors (Figure 1A) [22–29]. Along with PARP1, PARP7 has been identified as an important player in the transcription regulation of pluripotency genes and in their protection from epigenetic repression [30]. Originally, PARP1 was identified as TFIIC, which is capable of facilitating the initiation steps of mRNA synthesis through direct interaction with the basal transcription machinery [31]. Recently, it has been shown that PARP1 can also serve as a scaffold protein by stimulating the recruitment of coregulator complexes, such as p300, NF-κB, and p50 as well as the mediator complex to promoter regions (Figure 1A) [32]. It has been proven that the DNA-binding, rather than the catalytic activity of PARP1, enhances transcription by promoting the early steps of PIC formation [33]. However, only a limited amount of data is available, which suggests that any member of the basal transcription machinery is PARylated [34]. Moreover, PARP1 facilitates not only the recruitment but also the release of specific transcription co-regulators, leading to dynamic exchange between transcription factors, such as TLE family member 1 (TLE1) transcriptional corepressor complex to histone-acetyltransferase (HAT)-containing complex in neurons (Figure 1A) [35]. Additionally, PARP1 is also necessary for maintaining the relaxed chromatin structure of actively transcribed genes, supporting the active transcription of these genes [36]. In contrast, biochemical studies have revealed that PARP1 is not indispensable for the initiation of transcription on intact DNA templates, but single-stranded DNA breaks (SSBs) promote PARP1 binding, resulting in repression of nick-dependent transcription [31].

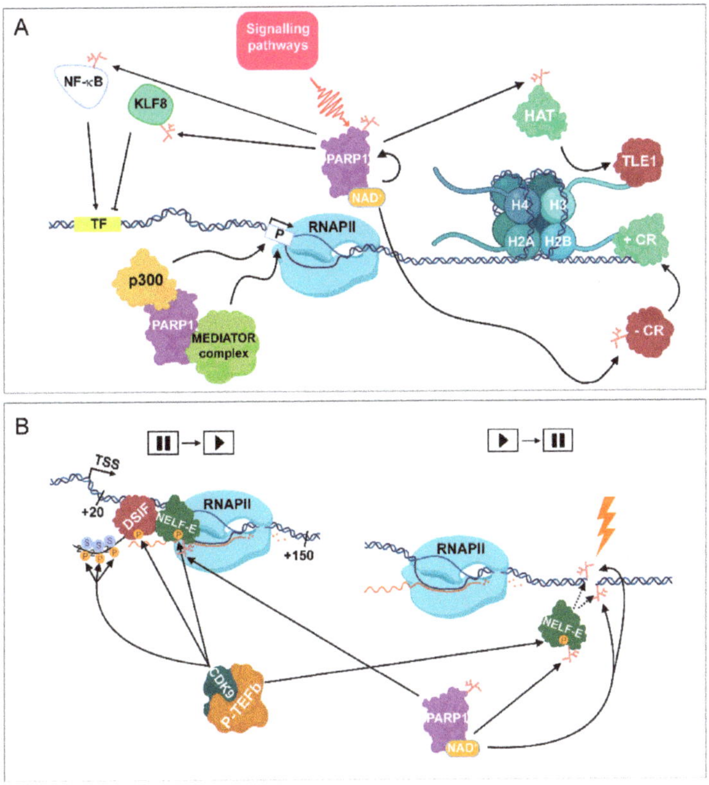

Figure 1. PARylation in transcription initiation. (**A**) As a consequence of the activation of certain signalling pathways, PARP1 catalyses the transfer of ADP-ribose through its binding to NAD$^+$ cofactor. PARP1 can PARylate numerous sequence-specific transcription factors, such as NF-κB and KLF8, which can facilitate and attenuate transcription initiation, respectively. PARP1 can also act as a scaffold protein by promoting the recruitment of various co-regulator complexes, such as p300 and the mediator complex to the promoter region (P), leading to transcription initiation catalysed by RNAPII. Moreover, PARP1 is implicated in the release of subsequent corepressor complex, such as transducin-like enhancer protein 1 (TLE1), resulting in its exchange to the PARylated HAT complex. PARP1 also participates in the exchange between negative coregulators, (–CRs) to positive ones (+ CRs). TF = transcription factor binding site. (**B**) (Left part) Following transcription initiation, RNAPII is stopped at around +20–+150 bp from the TSS. The promoter-proximal pausing process is induced by negative elongation factors, such as NELF-E and DSIF. During the release of the initial pausing, CDK9/ PTEFb phosphorylates NELF-E, DSIF, and the CTD of RPB1 (the largest subunit of RNAPII) at Ser2. Phosphorylated NELF-E (P-NELF-E) is released from RNAPII, while phosphorylated DSIF (P-DSIF) acts as a positive elongation factor, and the S2P-RNAPII becomes capable of proceeding the elongation step. Right part: Following DNA damage, PARP1 targets P-NELF-E for PARylation, thereby hindering its DNA-binding ability, leading to transcription silencing.

2.3. PARP1 Mediates Promoter-Proximal Pausing and Transcription Elongation

Following transcription initiation, RNAPII promoter-proximal pausing can be observed between the +20 and +150 region around the TSS [37]. In the early elongation step, RNAPII activity is temporarily paused by the contribution of NELF-E and DSIF [21]. During the promoter-proximal pause release, P-TEFb phosphorylates DSIF, NELF-E, and the CTD of RPB1 (the largest subunit of RNAPII) at Ser2, making it competent for transcription elongation and preventing its inhibition by DSIF and NELF-E

(Figure 1B) [37,38]. It has been also demonstrated that PARP1 directs RNAPII for PARylation shortly after DNA damage, facilitating the recruitment of NELF-E to RNAPII and by this, NELF-E plays a potential role in transcription silencing at the DNA break sites (Figure 1B) [39].

The *Drosophila* orthologue of PARP1 plays an important regulatory role during the elongation phase of heat shock-induced transcription [23,40]. Two members of the NELF complex, NELF-A and NELF-E, have been shown to be targeted by PARP1 [41]. Moreover, the phosphorylation of NELF-E catalysed by CDK9/P-TEFb, is indispensable for its subsequent PARylation, resulting in the attenuation of its DNA-binding ability (Figure 1B) [41]. Additionally, inhibition of either PARP1 or CDK9/P-TEFb results in reduced phosphorylation of RNAPII CTD at Ser2, thus leading to promoter-proximal pausing. The genome-wide distribution of PARylation is mostly enriched at actively transcribed regions, where high levels of NELF-B, RNAPII and H3K4me^3 can be observed [41]. Based on these data, PARylation plays a potential regulatory role throughout the entire transcription process. Nevertheless, further investigations are needed to highlight the proper PARP-mediated regulatory mechanisms.

2.4. PARylation Regulates Transcription Responses during DNA Damage

Genome integrity is being constantly challenged by various genotoxic stresses, which can result in different types of DNA lesions, including SSBs and double-stranded breaks (DSBs). DNA damage repair requires tight regulation, since inappropriate repair can lead to genome instability and tumourigenesis. In this regard, eukaryotes have evolved various mechanisms addressing DSBs, including homologous recombination (HR) and non-homologous end-joining (NHEJ) [42]. Although different factors are involved in these DNA repair pathways, crosstalk may occur between them. When DSBs arise within an actively transcribed unit, NELF-E and -A are rapidly accumulated at the break sites through a PARP1-dependent manner, leading to the silencing of this transcription unit [39]. Additionally, in vitro PAR-binding assays have also revealed that NELF-E has elevated binding capacity to PAR moieties, and the interaction between PARP1 and NELF-E is weakened following exposure to ionising radiation [41]. Additionally, the broken DNA region is PARylated by PARP1, facilitating the recruitment of NELF-E and silencing of transcription at the break site [39]. Furthermore, at sites of laser-induced DNA damage, PARP1 can also indirectly regulate transcription through its interaction with the TIMELESS protein. It has been also established that the PARP1–TIMELESS complex plays an essential role in HR [43].

The dual inhibition of PARP1 and PARP2 can lead to reduced binding of nucleosome remodelling deacetylase (NuRD) complex to sites of DNA damage [44]. NuRD can facilitate the recruitment of protein kinase C-binding protein 1 (ZMYND8) to the PARylated DNA damage sites [45]. Lysine-specific demethylase 5A (KDM5A) plays a key role in demethylation of H3K4 to regulate the binding of ZMYND8-NuRD complexes to the DSB (Figure 2A,B) [46]. Consequently, PARP1 can regulate transcription silencing not only by recruiting chromatin remodellers but also demethylases, which can remove the methyl groups from H3K4me^3, responsible for transcription activation. During DNA damage-induced transcription silencing, PARP1 facilitates the recruitment of the polycomb repressive complex 1 and 2 (PRC1 and PRC2), ensuring the proper chromatin structure and transcription arrest at the damaged sites [44,47]. Moreover, the chromodomain Y like (CDYL) protein interacts with PRC2 in a PARP-dependent manner [48,49]. During DSB-induced transcription silencing, PARP1 promotes the recruitment of KRAB-associated protein-1 (KAP-1), heterochromatin protein 1 (HP1), and suppressor of variegation 3–9 homolog 1 (SUV39H1) (Figure 3A) [50]. SUV39H1 methylates H3K9 around the DSB, leading to the recruitment of additional KAP-1/HP1/SUV39H1 complexes thereby contributing to the spreading of H3K9me^3 signal (Figure 3A,B) [50]. It results in the activation of histone acetyltransferase 5 (KAT5), which acetylates ataxia-telangiectasia mutated (ATM) (Figure 3C) [51,52]. ATM phosphorylates KAP-1, leading to the dissociation of the KAP-1/HP1/SUV39H1 complex from the chromatin, which allows the activation of the ATM-dependent HR repair pathway (Figure 3C,D) [50].

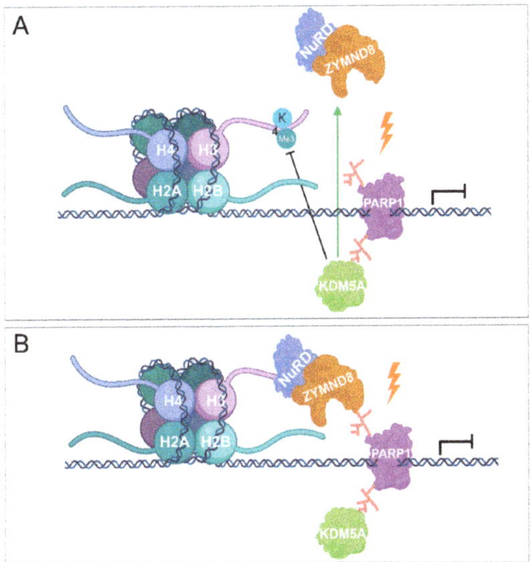

Figure 2. PARylation in transcription repression. (**A**,**B**) PARP1 is also involved in transcription silencing by recruiting demethylases, such as KDM5A, catalysing the removal of methyl groups and being responsible for transcription activation. KDM5A promotes the recruitment of NuRD and ZMYND8 to the lesion site by demethylating H3K4me^3, thereby contributing to transcription silencing.

By contrast, during NHEJ, the Ku70/Ku80 heterodimers bind to the broken DNA ends and recruit the DNA-dependent protein kinase catalytic subunit (DNA-PKcs), a key player in transcription silencing processes. Additionally, PARP1 has been shown to interact with both DNA-PKcs and Ku70/80 [53]. Furthermore, even a single DSB can lead to DNA-PKcs-dependent transcription silencing, suggesting that PARP1 could also be involved in this pathway. At the end of this process, the HECT E3 ubiquitin ligase, WWP2 directs the stalled RNAPII complex to proteasomal degradation [54].

A recent systematic analysis has demonstrated that the TFIID complex member, TBP-associated factor 15 (TAF15), is bound to laser-induced DNA damage sites in a PAR-dependent manner [55]. Moreover, a proteome-wide mass-spectrometry analysis has revealed that the RNA-binding protein, RNA-binding motif protein X-linked (RBMX), is PARylated following exposure to genotoxic stress [56,57]. The DNA damage-induced appearance of PARylated transcription factors suggests an uncommon and transient transcription bursting, which generates the so-called DDR RNAs (DDRNAs) or DSB-induced RNAs (diRNAs) [58]. Furthermore, PARylation can indirectly facilitate transcription silencing through diRNA-mediated chromatin compaction [58,59].

The diRNAs presumably contribute to the recruitment of DDR factors and chromatin modifiers or participate in transcription silencing at DNA break sites [60]. The presence of DDRNAs at the damage sites is critical for the activation of DDR, since in the case of hindering the formation of double-stranded RNAs by silencing either Drosha or Dicer, tumour suppressor P53-binding protein 1 (53BP1) foci formation is highly reduced. Additionally, it has been also shown that breast cancer type 1 (BRCA1), implicating in HR and competing with 53BP1 and DNA repair protein RAD51 homolog 1 (RAD51) foci formation, is dramatically reduced upon silencing Drosha regardless of the cell cycle phase [61].

This finding might be explained by the fact that in the absence of sister chromatids, the HR pathway can only be activated in the presence of an RNA strand, which is ensured by DNA damage induced *de novo* transcription [62]. On the contrary, the recruitment of DDRNAs is not influenced by 53BP1 but rather it is highly dependent on the presence of RNAPII. This suggests a potential role of RNAPII in DDR activation by synthesising damage-induced long non-coding RNAs (dilncRNAs) at the site of DNA damage. It seems that the broken DNA ends can serve as promoters for the transcription of dilncRNAs, while simultaneously with this process, the transcription of coding regions is prevented. The transcription of dilncRNA is bidirectional, since it can be initiated towards both directions from the break site. However, DNA:RNA hybrids should be resolved by helicases or RNase H enzymes, since these can prevent further recruitment of the repair factors to the site of DNA damage [63]. Recently, it has been demonstrated that besides the interaction with RNAPII, BRCA1 also contributes to the resolution of DNA:RNA hybrids and preserves genome integrity through the recruitment of DNA/RNA helicase senataxin (SETX) to the terminal regions of genes [64,65]. Since HR is the most accurate DSB repair pathway, the appropriate and controlled recruitment of BRCA1/2 to the damage site is indispensable for the efficient repair. At this point, the inhibition of PARPs can counteract with the ongoing repair processes, resulting in genome instability [66]. Additionally, in BRCA1/2-deficient tumour cells, higher sensitivity to PARP inhibition can be observed [67,68]. These results as well as the preclinical trials highlight the substantial role of PARP inhibitors in cancer therapy [67,69–71]. Additional factors, including ATP-dependent RNA helicase A (DHX9), PARP1, scaffold attachment factor B2 (SAFB2), multiple myeloma SET domain (MMSET) and DNA-PKcs have been recently identified as proteins that interact with RNA/DNA hybrids [72]. Furthermore, DHX9 along with PARP1 plays a remarkable role in preventing R-loop accumulation and facilitating transcription termination [72]. PARP1 contributes to the enhancement of DHX9 helicase activity [73]. DHX9 interacts with a large number of proteins related to transcription, including RNAPII, suggesting that DHX9 travels with the elongating RNAPII and contributes to the resolution of R-loops in a PARP1 dependent manner [72].

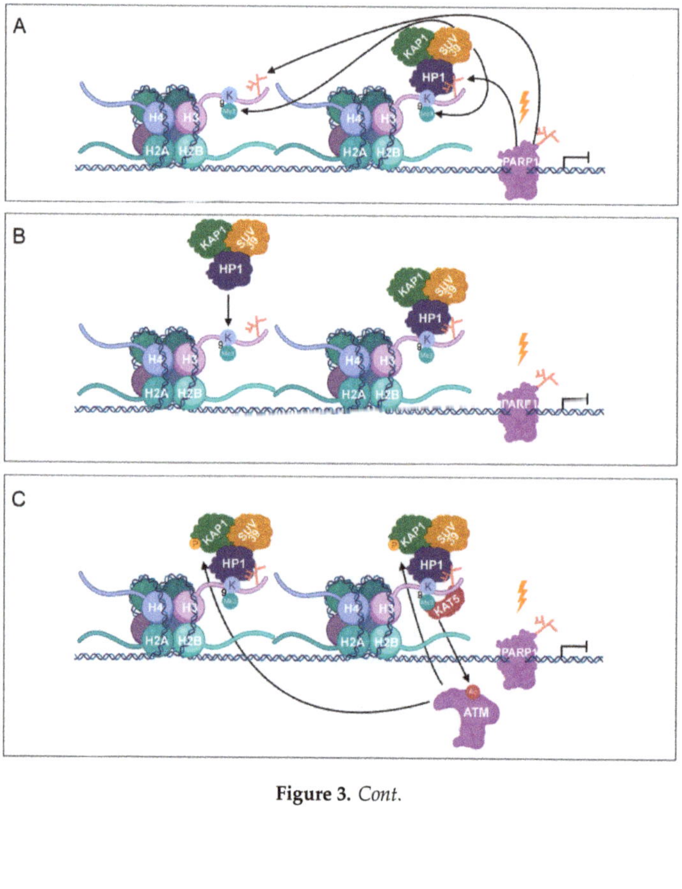

Figure 3. *Cont.*

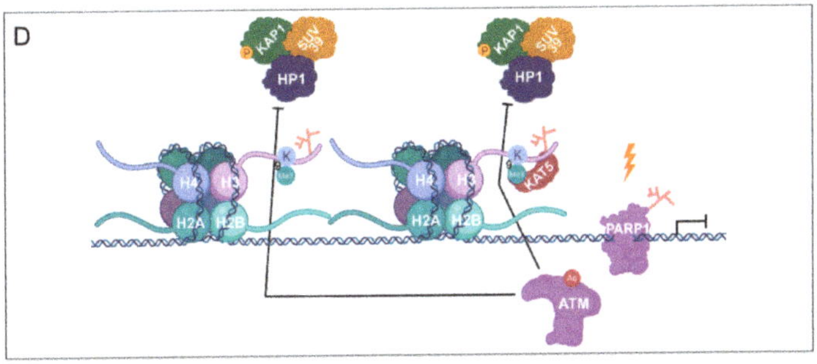

Figure 3. PARylation in transcription silencing during DNA damage. (**A**) As a response to DNA damage-induced transcription silencing, PARP1 facilitates the recruitment of SUV39H1, KAP1, and HP1. (**B**) Subsequent SUV39H1–HP1–KAP1 containing complexes are recruited, resulting in the spreading of H3K9me^3 signal. (**C**) KAT5 is activated, which acetylates ATM, being responsible for the phosphorylation of KAP-1. (**D**) Phosphorylation of KAP1 contributes to the dissociation of SUV39H1–HP1–KAP1 from the chromatin.

Moreover, it has been shown that topoisomerase I and II play an indispensable role in the RNAPII pause release [74]. Elongating RNAPII induces torsional and topological stresses in the super-helical DNA, which should be resolved by topoisomerase I and II [75–78]. Topoisomerases take part in the activation of DDR during transcription activation and elongation processes [74]. Topoisomerase I is involved in the prevention of R-loop formation, which is one of the major sources of transcription-coupled genome instability, by removing negative supercoiling structures behind the RNAPII [78,79]. Upon transcription blockage, spliceosome displacement could result in R-loop formation, leading to ATM activation in a DSB-independent manner [80,81]. As a consequence of RNA/DNA hybridisation, ssDNA strands are formed, which are more susceptible to different kinds of DNA damage. Additionally, the formation of R-loops can lead to genome instability by interfering with DNA replication [82,83]. In addition to topoisomerases, R-loop formation can also be inhibited by RNase H, RNA/DNA helicases, and suppressors of proteins promoting R-loop formation [84,85]. Although Topoisomerase I is responsible for alleviating the torsional stress in the DNA, it gets trapped and is accumulated in close proximity to the DNA lesions. This leads to failures in DNA repair, genome instability, and tumourigenesis [86]. Such phenomenon occurs during camptothecin (CPT)-induced topoisomerase I inhibition, resulting in the accumulation of antisense RNAPII transcripts and then R-loop formation at actively transcribed regions [87–89]. Following CPT treatment, PARP1 interacts with topoisomerase I in both nucleolar compartments, playing a role in eliminating covalent topoisomerase I–DNA complexes through PARylation and recruiting repair factors to these sites [90,91]. In addition to PARP1, through the PARylation of topoisomerase I, PARP2 also plays a pivotal role in the removal of the stalled enzyme. By this mechanism, both PARP1 and PARP2 have a positive impact on preserving genome stability. Furthermore, PARP enzymes interfere with the actions of CPT, resulting in drug resistance. Therefore, combining PARP inhibitors with CPT may enhance therapeutic efficacy [92].

3. PARylation in the Regulation of DNA Damage-Induced Chromatin Structural Changes

Several studies suggest that both reduced structural constraints and altered nucleosome occupancy influence the accessibility of chromatin in response to DNA damage. Additionally, as a consequence of persistent DNA damage, the unfolding and spatial expansion of certain chromatin regions can be observed. Following DNA damage, robust chromatin decondensation occurs in a PARP-dependent manner [93,94]. The addition of the highly negatively charged PAR chains to histones and to other chromatin-associated proteins results in an electrostatic repulsion with the negatively charged DNA, leading to chromatin relaxation. Following DNA damage, PARP1 can also initiate chromatin conformational changes through interaction with other chromatin-modifying factors. Therefore, the activity of PARP1 seems to be indispensable for chromatin decondensation, contributing to initiation of DDR signalling and the recruitment of repair factors [55,95]. Following laser-micro irradiation, PARP1 is recruited to the site of DNA damage within seconds, while PARP2 binds only 30 s later. This finding supports that PARP1 is mainly responsible for the transient reorganisation of the chromatin structure [96,97]. Furthermore, kinetic analyses have shown that the binding of PARP1 is necessary for the recruitment of MRE11–RAD50–NBS1 (MRN) to the DSB sites (Figure 4A) [96]. Following DSB recognition, ATM, recruited by the MRN complex, phosphorylates H2A.X at S139 (referred to as γH2A.X), which triggers the recruitment of ring finger protein 8 (RNF8) and ring finger protein 168 (RNF168) ubiquitin ligases, participating in the K63-linked polyubiquitylation of H1 histones and in K13 and K15 ubiquitylation of H2A histones, respectively [98,99]. PARP1 facilitates the recruitment of the SWItch/sucrose non-fermentable (SWI/SNF)-related matrix-associated actin-dependent regulator of chromatin subfamily A member 5 (SMARCA5/SNF2H) to the sites of DNA damage and promotes the interaction between SMARCA5 and ADP-ribosylated RNF168 (Figure 4B) [100]. These results have confirmed that PARylation is a crucial step both for chromatin reorganisation and RNF168-mediated ubiquitylation of H2A, being responsible for the recruitment of additional DDR factors [100]. Hence, a functional link can be recognised between PARylation and ubiquitylation during DNA repair.

Subsequently to the auto-activation of PARP1, amplified in liver cancer 1 (ALC1), which interacts with PARP1 and histones, is recruited via a similar kinetic to PAR and catalyses nucleosome sliding in an ATP-dependent manner [101–103]. This recruitment requires its C-terminal PAR-binding macrodomain, which recognises PARylated PARP1. Recently, it has been shown that the interaction between the ATPase catalytic domain and the C-terminal macrodomain of ALC1 is necessary to keep ALC1 in an inactive state under physiological conditions. However, the activation of PARP1 disrupts the interaction of the aforementioned domains and allows the stimulation of the remodelling and PAR-dependent binding activity of ALC1 [104,105]. Nevertheless, the interaction of ALC1 with histones is dependent on the presence of the histone chaperone aprataxin and PNK-like factor (APLF), which are localised at the site of DNA lesion in a PAR-dependent manner and are PARylated by PARP1 [106,107]. It has been still unclear how the ALC1 contributes to the relaxation of the chromatin structure around the break site; therefore, further investigations are needed to reveal novel chromatin-associated interaction partners of it.

Additionally, a strong interaction has been demonstrated between APLF and macroH2A.1.1 following hydrogen peroxide-induced DNA damage, thus indicating a potential role of APLF in chromatin rearrangement [107]. Furthermore, this histone chaperone recognises branched PAR chains, catalysed by PARP2, and mediates histone H3 removal during DNA repair [97]. With regards to histone H3, the incorporation of the histone variant H3.3 has been linked to the PARP1-mediated accumulation of the chromodomain helicase DNA binding protein 2 (CHD2), triggering chromatin expansion (Figure 4B) [108]. Although CHD2 cannot recognise PAR moieties, earlier PAR-dependent events, including the rapid localisation of ALC1 and relaxation of the chromatin structure can trigger the accumulation of CHD2 near DNA breaks.

Several studies have highlighted that transient chromatin relaxation precedes chromatin compaction for protecting regions around the DNA break, resulting in the reveal of the DSB site that needs to be restored by the contribution of DDR factors [109]. Another hypothesis concerns the inhibition of replication and transcription at DSB-flanking regions in order to prevent interference with the repair machinery and mediate faithful repair. MacroH2A acts as a tumour suppressor and is a significant player in the maintenance of the heterochromatic structure as well as in the inactivation of the X chromosome, during which macroH2A inhibits the enzymatic activity of PARP1 [110–113]. The PAR-binding ability of the macrodomain modules has been recently demonstrated, underlining the PAR-capping of macroH2A.1.1, its capability of sensing PARP1 activation, and the subsequent reorganisation of chromatin structure by establishing a compacted chromatin environment [114]. Furthermore, only macroH2A.1.1 suppresses PARP1 activity, preventing the formation of an open chromatin structure. By contrast, all three histone variants of macroH2A (macroH2A.1.1, macroH2A.1.2 and macroH2A.2) retain the ability to stabilise condensed chromatin structure via their common linker region [115]. Subsequently, macroH2A.1.1 participates in recognizing and binding PAR chains to inactivate PARP1 (Figure 4C), whereas the linker region, being present in all three isoforms, may play an additional role in chromatin compaction. Together with PARP1 inhibition, the linker region contributes to the stabilisation of this architecture.

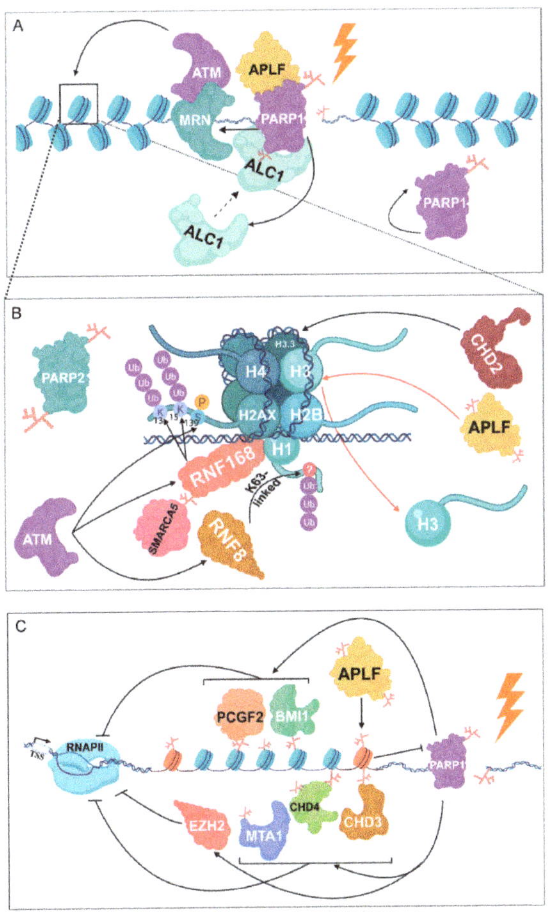

Figure 4. PARylation is required for chromatin structural changes following DNA damage. (**A**) PARP1 facilitates the binding of the MRN complex, which is known as the first DNA damage sensor, being responsible for the recruitment of ATM. Following PARP1 auto-activation, amplified in liver cancer 1 (ALC1) is implicated in nucleosome sliding by interacting with histones in a PARP- and aprataxin and PNK-like factor (APLF)-dependent manner. (**B**) ATM phosphorylates H2A.X at S139, resulting in the recruitment of RNF8 and RNF168. RNF8 catalyses the K63-linked poly-ubiquitylation of H1, while RNF168 is responsible for the ubiquitylation of H2A at K13 and K15. PARP1 promotes the recruitment of SMARCA5 to DSB sites and its subsequent interaction with RNF168. PARylated APLF, recognizing branched PAR chains catalysed by PARP2, participates in H3 removal during DNA repair. Furthermore, PARP1-mediated accumulation of chromodomain helicase DNA binding protein 2 (CHD2) leads to dynamic exchange of H3 to H3.3. (**C**) Members of the NuRD and polycomb complex, involving metastasis associated protein 1 (MTA1), CHD3, CHD4, polycomb group ring finger 2 (PCGF2), polycomb group ring finger 4 (BMI1), and enhancer of zeste homolog 2 (EZH2) take part in transcription inhibition through a PARP-mediated pathway.

Factors involved in the NuRD complex, along with members of the Polycomb complex, play an indispensable role in the PARP1-associated DDR [44,55,116]. Particularly, metastasis associated protein 1 (MTA1), chromodomain helicase 3 and 4 (CHD3, CHD4), and all members of the NuRD complex are rapidly PARylated and exhibit enhanced binding to the site of the DSB. At laser-micro-irradiated sites, PARP1-mediated complete loss of transcription can be observed [44]. While Polo and colleagues revealed similar recruitment kinetics for CHD4 and PAR, other groups demonstrated that CHD3 and CHD4 require the initial PAR-dependent chromatin relaxation, potentially mediated by ALC1. Consequently, their accumulation appears at a later time point [116]. Recruitment of polycomb group ring finger 2 (PCGF2), polycomb group ring finger 4 (BMI1), and components of the polycomb repressive complex 1 (PRC1) are largely abrogated upon inhibition of PARP1. Interestingly, enhancer of zeste homolog 2 (EZH2), belonging to the polycomb repressive complex 2 (PRC2) and associated with transcription repression (H3K27me3 formation), is recruited depending on the enzymatic activity of PARP1 (Figure 4C) [44,47]. Recent findings have indicated that PARP1 inhibits the histone methyltransferase activity of EZH2, thus resulting in a more relaxed chromatin structure at the damaged regions [117,118]. Nevertheless, these opposing observations may arise from different kinds of DNA damage source, inducing different repair pathways or even from diverse phases of the repair process.

4. Role of PARP1 in RNA Metabolism

In addition to transcription, the abundance and decay of mRNAs can be post-transcriptionally regulated by controlling splicing, polyadenylation and nuclear export. A recent study has demonstrated that PARP1 is recruited to specific nucleosomes localised at the exon/intron boundaries, corresponding to specific splice sites. Additionally, PARP1 inhibition results in changes in the alternative splicing. Emerging data have indicated that PARP1 stimulates the recruitment of U2 snRNPs (small nuclear ribonucleoproteins), therefore positively influencing the exon recognition and the further splicing procedure [119]. Moreover, heterogeneous nuclear riboproteins (hnRNPs) can tightly bind to PAR chains, promoting the dissociation of hnRNPs from RNA and subsequent intron splicing [120]. So far, 11 human hnRNP proteins have been demonstrated to be capable of recognising and binding to PARylated targets. Affinity-purification mass spectrometry assays (AP-MS) combined with Gene Ontology classifications (GO) have shown that not only PARP1, but also PARP2 plays a role in intron splicing and interacts with hnRNPs [121]. It has been also shown that PARP10 is able to PARylate specific RNA substrates, leading to the protection of RNA ends which can act as a platform for recruiting other proteins [122]. Following stress responses, auto-PARylated PARP1 can bind to several nuclear proteins and initiate their transport to Cajal bodies, contributing to the regulation of either the assembly or disassembly of transcription- and splicing-related complexes [123]. Additionally, genome-wide data suggest that PARP1 can oppositely influence the transcription elongation by altering the elongation speed [124]. PARP1 also affects the assembly of human pre-mRNA 3′-processing complex. During transcription termination, PARylation is involved in hindering polyadenylation by catalysing the ADP-ribosylation of polyadenylate-polymerase (PAP), resulting in its reduced binding to mRNA transcripts [125]. Furthermore, PARP14 seems to be involved in the posttranscriptional regulation of mRNA stability since it can promote the degradation of specific transcripts by interacting selectively with tristetrapolin (TTP) [126]. Finally, PARylation could affect transcription by regulating the transport of specific mRNAs, since the nuclear export of mature mRNAs is also regulated by PARP1. In lipopolysaccharide-treated cells, PARP1-dependent PARylation of embryonic lethal abnormal vision-like 1 (ELAV-like protein 1) triggers the RNA nuclear-cytoplasmic shuttling, leading to the enhanced stability of mRNA [127]. These data suggest that following stress responses, PARP enzymes could affect the mRNA maturation at multiple levels. PARylation may be involved in most of the RNA metabolism-related processes, such as splicing, polyadenylation and mRNA maturation. Nonetheless, further investigations are needed to the more precise understanding of these regulatory pathways.

5. Discussion

In addition to DNA damage response, PARylation regulates various processes, such as chromatin remodelling, transcription activation and repression, ubiquitylation, RNA metabolism as well as cellular stress responses. PARylation can have the following effects on these processes: (1) it can ensure a surface for protein interactions, and (2) certain proteins, such as Imitation SWI (ISWI) and ALC1, possessing a PAR-binding domain (consisting of a PAR-binding zinc finger (PBZ), a PAR-binding motif (PBM), and a WWE), can be recruited to DNA through this process. Additionally, PARP1 can act as a scaffold protein by regulating the recruitment of transcription co-regulator complexes, such as p300 or the mediator complex. These results also support that PARylation is implicated in various mechanisms by promoting complex assembly [32]. On the other hand, during DNA repair, PARPs have an indispensable function in identifying DNA breaks and participating in DNA repair pathway choice. In the absence of PARylation, the insufficient activation of proteins involved in DNA repair can result in the malfunction of the repair mechanism. Furthermore, inappropriate activation of any DNA repair pathway can contribute to genome instability, leading to tumourigenesis. For instance, in HR-deficient tumour cells, in the absence of BRCA1/BRCA2, PARP1 inhibition has been shown to have cytotoxic side-effects. The putative mechanism of this hypersensitivity can be explained by the fact that PARP inhibitors can disturb the recruitment of BRCA1 to the damaged sites, resulting in inadequate HR activation. Although PARP1 is a well-characterised protein, the precise function of other PARPs in regulating other cellular processes has remained unclear. PARP inhibitors may alleviate the speed of DNA repair, leading to the collapse of the replication fork and high therapeutic efficacy during tumour therapy [71]. Moreover, PARylation can interfere with the early recruitment of both BRCA1 and BRCA2, contributing to HR deficiency [128–131]. Interestingly, during HR, PARPs not only recruit MRE11 and NBS1 to the damage sites, but also play a regulatory role during transcription. In HR-related de novo transcription, DHX9 interacts with PARP1, thereby regulating the transcription [72,73]. Moreover, BRCA1-RNAPII interaction contributes to the resolution of DNA:RNA hybrids [64,65]. These results highlight that PARPs can act as transcription regulators in various processes, which can reveal new possibilities in applying PARP inhibitors in clinical trials [132,133]. Although several clinical reports have already demonstrated that PARP inhibitors could be beneficial during tumour therapy, we have to mention that the exact biochemical mechanisms regulated by these PARP inhibitors still remained unexplored. Therefore, further investigations are required to uncover these PARylation-mediated mechanisms to reduce the off-target effects of PARP inhibitors.

6. Conclusions

In this review, we address the role of PARylation to understand their function during the transcription-coupled cellular responses. We summarise the possible canonical mechanisms by which PARylation exerts its regulatory roles during the transcription responses. However, despite the increasing knowledge on the related topic, the extent of the contribution of PARylation needs to be elucidated. Depending on the cellular context of this PTM, it can exert opposite effects on the same cellular processes. During transcription, it was shown that PARylation can determine the transcription state either by activating or inhibiting the transcription of different sets of genes, leading to distinct biological outcomes. According to this, it is worthwhile performing studies able to address the effect of PARP inhibition on transcription responses. Further efforts should be initiated for the better understanding of the underlying mechanism of actions to achieve more effective therapeutic benefits with minimal side-effects.

Author Contributions: Wrote the paper: Z.G.P., B.N.B., V.P., Z.U., T.P. All authors have read and agreed to the published version of the manuscript

Funding: This work was supported by the National Research, Development and Innovation Office grants GINOP-2.3.2-15-2016-00020, GINOP-2.3.2-15-2016-00036, GINOP-2.2.1-15-2017-00052, and NKFI-FK 132080, and the Tempus Foundation.

Conflicts of Interest: The authors declare no conflict of interest.

References

1. Chambon, P.; Weill, J.D.; Mandel, P. Nicotinamide mononucleotide activation of a new DNA-dependent polyadenylic acid synthesizing nuclear enzyme. *Biochem. Biophys. Res. Commun.* **1963**, *11*, 39–43. [CrossRef]
2. de Murcia, G.; de Murcia, J.M. Poly(ADP-ribose) polymerase: A molecular nick-sensor. *Trends Biochem. Sci.* **1994**, *19*, 172–176. [CrossRef]
3. Kameshita, I.; Matsuda, Z.; Taniguchi, T.; Shizuta, Y. Poly(ADP-ribose) synthetase. Separation and identification of three proteolytic fragments as the substrate-binding domain, the DNA-binding domain, and the automodification domain. *J. Biol. Chem.* **1984**, *259*, 4770–4776. [PubMed]
4. Lin, W.; Amé, J.C.; Aboul-Ela, N.; Jacobson, E.L.; Jacobson, M.K. Isolation and characterization of the cDNA encoding bovine poly(ADP- ribose) glycohydrolase. *J. Biol. Chem.* **1997**, *272*, 11895–11901. [CrossRef] [PubMed]
5. Slade, D.; Dunstan, M.S.; Barkauskaite, E.; Weston, R.; Lafite, P.; Dixon, N.; Ahel, M.; Leys, D.; Ahel, I. The structure and catalytic mechanism of a poly(ADP-ribose) glycohydrolase. *Nature* **2011**, *477*, 616–622. [CrossRef] [PubMed]
6. Barkauskaite, E.; Brassington, A.; Tan, E.S.; Warwicker, J.; Dunstan, M.S.; Banos, B.; Lafite, P.; Ahel, M.; Mitchison, T.J.; Ahel, I.; et al. Visualization of poly(ADP-ribose) bound to PARG reveals inherent balance between exo- and endo-glycohydrolase activities. *Nat. Commun.* **2013**, *4*, 2164. [CrossRef] [PubMed]
7. Fontana, P.; Bonfiglio, J.J.; Palazzo, L.; Bartlett, E.; Matic, I.; Ahel, I. Serine ADP-ribosylation reversal by the hydrolase ARH3. *eLife* **2017**, *6*, e28533. [CrossRef]
8. Emanuelli, M.; Carnevali, F.; Saccucci, F.; Pierella, F.; Amici, A.; Raffaelli, N.; Magni, G. Molecular cloning, chromosomal localization, tissue mRNA levels, bacterial expression, and enzymatic properties of human NMN adenylyltransferase. *J. Biol. Chem.* **2001**, *276*, 406–412. [CrossRef]
9. Vyas, S.; Matic, I.; Uchima, L.; Rood, J.; Zaja, R.; Hay, R.T.; Ahel, I.; Chang, P. Family-wide analysis of poly(ADP-ribose) polymerase activity. *Nat. Commun.* **2014**, *5*, 4426. [CrossRef]
10. Cook, B.D.; Dynek, J.N.; Chang, W.; Shostak, G.; Smith, S. Role for the Related Poly(ADP-Ribose) Polymerases Tankyrase 1 and 2 at Human Telomeres. *Mol. Cell. Biol.* **2002**, *22*, 332–342. [CrossRef]
11. Mashimo, M.; Kato, J.; Moss, J. ADP-ribosyl-acceptor hydrolase 3 regulates poly (ADP-ribose) degradation and cell death during oxidative stress. *Proc. Natl. Acad. Sci. USA* **2013**, *110*, 18964–18969. [CrossRef] [PubMed]
12. He, Y.; Fang, J.; Taatjes, D.J.; Nogales, E. Structural visualization of key steps in human transcription initiation. *Nature* **2013**, *495*, 481–486. [CrossRef] [PubMed]
13. Bernecky, C.; Grob, P.; Ebmeier, C.C.; Nogales, E.; Taatjes, D.J. Molecular architecture of the human Mediator-RNA polymerase II-TFIIF assembly. *PLoS Biol.* **2011**, *9*, e1000603. [CrossRef] [PubMed]
14. Guzmán, E.; Lis, J.T. Transcription Factor TFIIH Is Required for Promoter Melting In Vivo. *Mol. Cell. Biol.* **1999**, *19*, 5652–5658. [CrossRef]
15. Lis, J. Promoter-associated pausing in promoter architecture and postinitiation transcriptional regulation. *Cold Spring Harb. Symp. on Quant. Biol.* **1998**, *63*, 347–356. [CrossRef]
16. Wada, T.; Takagi, T.; Yamaguchi, Y.; Ferdous, A.; Imai, T.; Hirose, S.; Sugimoto, S.; Yano, K.; Hartzog, G.A.; Winston, F.; et al. DSIF, a novel transcription elongation factor that regulates RNA polymerase II processivity, is composed of human Spt4 and Spt5 homologs. *Genes Dev.* **1998**, *12*, 343–356. [CrossRef]
17. Yamaguchi, Y.; Takagi, T.; Wada, T.; Yano, K.; Furuya, A.; Sugimoto, S.; Hasegawa, J.; Handa, H. NELF, a multisubunit complex containing RD, cooperates with DSIF to repress RNA polymerase II elongation. *Cell* **1999**, *97*, 41–51. [CrossRef]
18. Marshall, N.F.; Price, D.H. Purification of P-TEFb, a transcription factor required for the transition into productive elongation. *J. Biol. Chem.* **1995**, *270*, 12335–12338. [CrossRef]
19. Wada, T.; Takagi, T.; Yamaguchi, Y.; Watanabe, D.; Handa, H. Evidence that P-TEFb alleviates the negative effect of DSIF on RNA polymerase II-dependent transcription in vitro. *EMBO J.* **1998**, *17*, 7395–7403. [CrossRef]
20. Fujinaga, K.; Irwin, D.; Huang, Y.; Taube, R.; Kurosu, T.; Peterlin, B.M. Dynamics of Human Immunodeficiency Virus Transcription: P-TEFb Phosphorylates RD and Dissociates Negative Effectors from the Transactivation Response Element. *Mol. Cell. Biol.* **2004**, *24*, 787–795. [CrossRef]

21. Ping, Y.H.; Rana, T.M. DSIF and NELF Interact with RNA Polymerase II Elongation Complex and HIV-1 Tat Stimulates P-TEFb-mediated Phosphorylation of RNA Polymerase II and DSIF during Transcription Elongation. *J. Biol. Chem.* **2001**, *276*, 12951–12958. [CrossRef]
22. Hassa, P.O.; Hottiger, M.O. The functional role of poly(ADP-ribose)polymerase 1 as novel coactivator of NF-κB in inflammatory disorders. *Cell. Mol. Life Sci.* **2002**, *59*, 1534–1553. [CrossRef]
23. Tulin, A.; Spradling, A. Chromatin loosening by poly(ADP)-ribose polymerase (PARP) at Drosophila puff loci. *Science* **2003**, *299*, 560–562. [CrossRef]
24. Cervellera, M.N.; Sala, A. Poly(ADP)-ribose) polymerase is a B-MYB coactivator. *J. Biol. Chem.* **2000**, *275*, 10692–10696. [CrossRef]
25. Nie, J.; Sakamoto, S.; Song, D.; Qu, Z.; Ota, K.; Taniguchi, T. Interaction of Oct-1 and automodification domain of poly(ADP-ribose) synthetase. *FEBS Lett.* **1998**, *424*, 27–32. [CrossRef]
26. Lai, Y.S.; Chang, C.W.; Pawlik, K.M.; Zhou, D.; Renfrow, M.B.; Townes, T.M. SRY (sex determining region Y)-box2 (Sox2)/poly ADP-ribose polymerase 1 (Parp1) complexes regulate pluripotency. *Proc. Natl. Acad. Sci. USA* **2012**, *109*, 3772–3777. [CrossRef]
27. Lu, H.; Wang, X.; Li, T.; Urvalek, A.M.; Yu, L.; Li, J.; Zhu, J.; Lin, Q.; Peng, X.; Zhao, J. Identification of poly (ADP-ribose) polymerase-1 (PARP-1) as a novel krüppel-like factor 8-interactingand-regulating protein. *J. Biol. Chem.* **2011**, *286*, 20335–20344. [CrossRef]
28. Zhang, F.; Wang, Y.; Wang, L.; Luo, X.; Huang, K.; Wang, C.; Du, M.; Liu, F.; Luo, T.; Huang, D.; et al. Poly(ADP-ribose) polymerase 1 is a key regulator of estrogen receptor α-dependent gene transcription. *J. Biol. Chem.* **2013**, *288*, 11348–11357. [CrossRef]
29. Le May, N.; Iltis, I.; Amé, J.C.; Zhovmer, A.; Biard, D.; Egly, J.M.; Schreiber, V.; Coin, F. Poly (ADP-Ribose) Glycohydrolase Regulates Retinoic Acid Receptor-Mediated Gene Expression. *Mol. Cell* **2012**, *48*, 785–798. [CrossRef]
30. Roper, S.J.; Chrysanthou, S.; Senner, C.E.; Sienerth, A.; Gnan, S.; Murray, A.; Masutani, M.; Latos, P.; Hemberger, M. ADP-ribosyltransferases Parp1 and Parp7 safeguard pluripotency of ES cells. *Nucleic Acids Res.* **2014**, *42*, 8914–8927. [CrossRef]
31. Slattery, E.; Dignam, J.D.; Matsui, T.; Roeder, R.G. Purification and analysis of a factor which suppresses nick-induced transcription by RNA polymerase II and its identity with poly(ADP-ribose) polymerase. *J. Biol. Chem.* **1983**, *258*, 5955–5959.
32. Hassa, P.O.; Hottiger, M.O. The diverse biological roles of mammalian PARPs, a small but powerful family of poly-ADP-ribose polymerases. *Front. Biosci.* **2008**, *13*, 3046–3082. [CrossRef]
33. Meisterernst, M.; Stelzer, G.; Roeder, R.G. Poly(ADP-ribose) polymerase enhances activator-dependent transcription in vitro. *Proc. Natl. Acad. Sci. USA* **1997**, *94*, 2261–2265. [CrossRef]
34. Rawling, J.M.; Alvarez-Gonzalez, R. TFIIF, a basal eukaryotic transcription factor, is a substrate for poly(ADP-ribosyl)ation. *Biochem. J.* **1997**, *324*, 249–253. [CrossRef]
35. Ju, B.G.; Solum, D.; Song, E.J.; Lee, K.J.; Rose, D.W.; Glass, C.K.; Rosenfeld, M.G. Activating the PARP-1 sensor component of the groucho/TLE1 corepressor complex mediates a CaMKinase IIδ-dependent neurogenic gene activation pathway. *Cell* **2004**, *119*, 815–829. [CrossRef]
36. Krishnakumar, R.; Kraus, W.L. PARP-1 Regulates Chromatin Structure and Transcription through a KDM5B-Dependent Pathway. *Mol. Cell* **2010**, *39*, 736–749. [CrossRef]
37. Muse, G.W.; Gilchrist, D.A.; Nechaev, S.; Shah, R.; Parker, J.S.; Grissom, S.F.; Zeitlinger, J.; Adelman, K. RNA polymerase is poised for activation across the genome. *Nat. Genet.* **2007**, *39*, 1507–1511. [CrossRef]
38. Fujita, T.; Piuz, I.; Schlegel, W. The transcription elongation factors NELF, DSIF and P-TEFb control constitutive transcription in a gene-specific manner. *FEBS Lett.* **2009**, *583*, 2893–2898. [CrossRef]
39. Awwad, S.W.; Abu-Zhayia, E.R.; Guttmann-Raviv, N.; Ayoub, N. NELF -E is recruited to DNA double-strand break sites to promote transcriptional repression and repair. *EMBO Rep.* **2017**, *18*, 745–764. [CrossRef]
40. Petesch, S.J.; Lis, J.T. Rapid, Transcription-Independent Loss of Nucleosomes over a Large Chromatin Domain at Hsp70 Loci. *Cell* **2008**, *134*, 74–84. [CrossRef]
41. Gibson, B.A.; Zhang, Y.; Jiang, H.; Hussey, K.M.; Shrimp, J.H.; Lin, H.; Schwede, F.; Yu, Y.; Kraus, W.L. Chemical genetic discovery of PARP targets reveals a role for PARP-1 in transcription elongation. *Science* **2016**, *353*, 45–50. [CrossRef]
42. Chapman, J.R.; Taylor, M.R.G.; Boulton, S.J. Playing the End Game: DNA Double-Strand Break Repair Pathway Choice. *Mol. Cell* **2012**, *47*, 497–510. [CrossRef]

43. Xie, S.; Mortusewicz, O.; Ma, H.T.; Herr, P.; Poon, R.R.Y.; Helleday, T.; Qian, C. Timeless Interacts with PARP-1 to Promote Homologous Recombination Repair. *Mol. Cell* **2015**, *60*, 163–176. [CrossRef]
44. Chou, D.M.; Adamson, B.; Dephoure, N.E.; Tan, X.; Nottke, A.C.; Hurov, K.E.; Gygi, S.P.; Colaiácovo, M.P.; Elledge, S.J. A chromatin localization screen reveals poly (ADP ribose)-regulated recruitment of the repressive polycomb and NuRD complexes to sites of DNA damage. *Proc. Natl. Acad. Sci. USA* **2010**, *107*, 18475–18480. [CrossRef]
45. Spruijt, C.G.; Luijsterburg, M.S.; Menafra, R.; Lindeboom, R.G.H.; Jansen, P.W.T.C.; Edupuganti, R.R.; Baltissen, M.P.; Wiegant, W.W.; Voelker-Albert, M.C.; Matarese, F.; et al. ZMYND8 Co-localizes with NuRD on Target Genes and Regulates Poly(ADP-Ribose)-Dependent Recruitment of GATAD2A/NuRD to Sites of DNA Damage. *Cell Rep.* **2016**, *17*, 783–798. [CrossRef]
46. Gong, F.; Clouaire, T.; Aguirrebengoa, M.; Legube, G.; Miller, K.M. Histone demethylase KDM5A regulates the ZMY ND8-NuRD chromatin remodeler to promote DNA repair. *J. Cell Biol.* **2017**, *216*, 1959–1974. [CrossRef]
47. Campbell, S.; Ismail, I.H.; Young, L.C.; Poirier, G.G.; Hendzel, M.J. Polycomb repressive complex 2 contributes to DNA double-strand break repair. *Cell Cycle* **2013**, *12*, 2675–2683. [CrossRef]
48. Zhang, Y.; Yang, X.; Gui, B.; Xie, G.; Zhang, D.; Shang, Y.; Liang, J. Corepressor protein CDYL functions as a molecular bridge between polycomb repressor complex 2 and repressive chromatin mark trimethylated histone lysine 27. *J. Biol. Chem.* **2011**, *286*, 42414–42425. [CrossRef]
49. Abu-Zhayia, E.R.; Awwad, S.W.; Ben-Oz, B.M.; Khoury-Haddad, H.; Ayoub, N. CDYL1 fosters double-strand break-induced transcription silencing and promotes homology-directed repair. *J. Mol. Cell Biol.* **2018**, *10*, 341–357. [CrossRef]
50. Ayrapetov, M.K.; Gursoy-Yuzugullu, O.; Xu, C.; Xu, Y.; Price, B.D. DNA double-strand breaks promote methylation of histone H3 on lysine 9 and transient formation of repressive chromatin. *Proc. Natl. Acad. Sci. USA* **2014**, *111*, 9169–9174. [CrossRef]
51. Sun, Y.; Jiang, X.; Xu, Y.; Ayrapetov, M.K.; Moreau, L.A.; Whetstine, J.R.; Price, B.D. Histone H3 methylation links DNA damage detection to activation of the tumour suppressor Tip60. *Nat. Cell Biol.* **2009**, *11*, 1376–1382. [CrossRef]
52. Sun, Y.; Jiang, X.; Chen, S.; Fernandes, N.; Price, B.D. A role for the Tip60 histone acetyltransferase in the acetylation and activation of ATM. *Proc. Natl. Acad. Sci. USA* **2005**, *102*, 13182–13187. [CrossRef]
53. Spagnolo, L.; Barbeau, J.; Curtin, N.J.; Morris, E.P.; Pearl, L.H. Visualization of a DNA-PK/PARP1 complex. *Nucleic Acids Res.* **2012**, *40*, 4168–4177. [CrossRef]
54. Caron, P.; Pankotai, T.; Wiegant, W.W.; Tollenaere, M.A.X.; Furst, A.; Bonhomme, C.; Helfricht, A.; de Groot, A.; Pastink, A.; Vertegaal, A.C.O.; et al. WWP2 ubiquitylates RNA polymerase II for DNA-PK-dependent transcription arrest and repair at DNA breaks. *Genes Dev.* **2019**, *33*, 684–704. [CrossRef]
55. Izhar, L.; Adamson, B.; Ciccia, A.; Lewis, J.; Pontano-Vaites, L.; Leng, Y.; Liang, A.C.; Westbrook, T.F.; Harper, J.W.; Elledge, S.J. A Systematic Analysis of Factors Localized to Damaged Chromatin Reveals PARP-Dependent Recruitment of Transcription Factors. *Cell Rep.* **2015**, *11*, 1486–1500. [CrossRef]
56. Adamson, B.; Smogorzewska, A.; Sigoillot, F.D.; King, R.W.; Elledge, S.J. A genome-wide homologous recombination screen identifies the RNA-binding protein RBMX as a component of the DNA-damage response. *Nat. Cell Biol.* **2012**, *14*, 318–328. [CrossRef]
57. Jungmichel, S.; Rosenthal, F.; Altmeyer, M.; Lukas, J.; Hottiger, M.O.; Nielsen, M.L. Proteome-wide identification of poly(ADP-Ribosyl)ation targets in different genotoxic stress responses. *Mol. Cell* **2013**, *52*, 272–285. [CrossRef]
58. Francia, S.; Michelini, F.; Saxena, A.; Tang, D.; De Hoon, M.; Anelli, V.; Mione, M.; Carninci, P.; D'adda Di Fagagna, F. Site-specific DICER and DROSHA RNA products control the DNA-damage response. *Nature* **2012**, *488*, 231–235. [CrossRef]
59. Michelini, F.; Pitchiaya, S.; Vitelli, V.; Sharma, S.; Gioia, U.; Pessina, F.; Cabrini, M.; Wang, Y.; Capozzo, I.; Iannelli, F.; et al. Damage-induced lncRNAs control the DNA damage response through interaction with DDRNAs at individual double-strand breaks. *Nat. Cell Biol.* **2017**, *19*, 1400–1411. [CrossRef]
60. Wang, Q.; Goldstein, M. Small RNAs Recruit Chromatin-Modifying Enzymes MMSET and Tip60 to Reconfigure Damaged DNA upon Double-Strand Break and Facilitate Repair. *Cancer Res.* **2016**, *76*, 1904–1915. [CrossRef]

61. Lu, W.T.; Hawley, B.R.; Skalka, G.L.; Baldock, R.A.; Smith, E.M.; Bader, A.S.; Malewicz, M.; Watts, F.Z.; Wilczynska, A.; Bushell, M. Drosha drives the formation of DNA:RNA hybrids around DNA break sites to facilitate DNA repair. *Nat. Commun.* **2018**, *9*, 532. [CrossRef]
62. Lobrich, M.; Jeggo, P. A Process of Resection-Dependent Nonhomologous End Joining Involving the Goddess Artemis. *Trends Biochem. Sci.* **2017**, *42*, 690–701. [CrossRef]
63. El Hage, A.; French, S.L.; Beyer, A.L.; Tollervey, D. Loss of Topoisomerase I leads to R-loop-mediated transcriptional blocks during ribosomal RNA synthesis. *Genes Dev.* **2010**, *24*, 1546–1558. [CrossRef]
64. Scully, R.; Anderson, S.F.; Chao, D.M.; Wei, W.J.; Ye, L.Y.; Young, R.A.; Livingston, D.M.; Parvin, J.D. BRCA1 is a component of the RNA polymerase II holoenzyme. *Proc. Natl. Acad. Sci. USA* **1997**, *94*, 5605–5610. [CrossRef]
65. Hatchi, E.; Skourti-Stathaki, K.; Ventz, S.; Pinello, L.; Yen, A.; Kamieniarz-Gdula, K.; Dimitrov, S.; Pathania, S.; McKinney, K.M.; Eaton, M.L.; et al. BRCA1 Recruitment to Transcriptional Pause Sites Is Required for R-Loop-Driven DNA Damage Repair. *Mol. Cell* **2015**, *57*, 636–647. [CrossRef]
66. Caron, M.C.; Sharma, A.K.; O'Sullivan, J.; Myler, L.R.; Ferreira, M.T.; Rodrigue, A.; Coulombe, Y.; Ethier, C.; Gagné, J.P.; Langelier, M.F.; et al. Poly(ADP-ribose) polymerase-1 antagonizes DNA resection at double-strand breaks. *Nat. Commun.* **2019**, *10*, 2954. [CrossRef]
67. Farmer, H.; McCabe, H.; Lord, C.J.; Tutt, A.H.J.; Johnson, D.A.; Richardson, T.B.; Santarosa, M.; Dillon, K.J.; Hickson, I.; Knights, C.; et al. Targeting the DNA repair defect in BRCA mutant cells as a therapeutic strategy. *Nature* **2005**, *434*, 917–921. [CrossRef]
68. AlHilli, M.M.; Becker, M.A.; Weroha, S.J.; Flatten, K.S.; Hurley, R.M.; Harrell, M.I.; Oberg, A.L.; Maurer, M.J.; Hawthorne, K.M.; Hou, X.; et al. In vivo anti-tumor activity of the PARP inhibitor niraparib in homologous recombination deficient and proficient ovarian carcinoma. *Gynecol. Oncol.* **2016**, *143*, 379–388. [CrossRef]
69. Fong, P.C.; Boss, D.S.; Yap, T.A.; Tutt, A.; Wu, P.; Mergui-Roelvink, M.; Mortimer, P.; Swaisland, H.; Lau, A.; O'Connor, M.J.; et al. Inhibition of poly(ADP-ribose) polymerase in tumors from BRCA mutation carriers. *N. Engl. J. Med.* **2009**, *361*, 123–134. [CrossRef]
70. Fong, P.C.; Yap, T.A.; Boss, D.S.; Carden, C.P.; Mergui-Roelvink, M.; Gourley, C.; De Greve, J.; Lubinski, J.; Shanley, S.; Messiou, C.; et al. Poly(ADP)-ribose polymerase inhibition: Frequent durable responses in BRCA carrier ovarian cancer correlating with platinum-free interval. *J. Clin. Oncol.* **2010**, *28*, 2512–2519. [CrossRef]
71. Bryant, H.E.; Schultz, N.; Thomas, H.D.; Parker, K.M.; Flower, D.; Lopez, E.; Kyle, S.; Meuth, M.; Curtin, N.J.; Helleday, T. Specific killing of BRCA2-deficient tumours with inhibitors of poly(ADP-ribose) polymerase. *Nature* **2005**, *434*, 913–917. [CrossRef]
72. Cristini, A.; Groh, M.; Kristiansen, M.S.; Gromak, N. RNA/DNA Hybrid Interactome Identifies DXH9 as a Molecular Player in Transcriptional Termination and R-Loop-Associated DNA Damage. *Cell Rep.* **2018**, *23*, 1891–1905. [CrossRef]
73. von Kobbe, C.; Harrigan, J.A.; Schreiber, V.; Stiegler, P.; Piotrowski, J.; Dawut, L.; Bohr, V.A. Poly(ADP-ribose) polymerase 1 regulates both the exonuclease and helicase activities of the Werner syndrome protein. *Nucleic Acids Res.* **2004**, *32*, 4003–4014. [CrossRef]
74. Bunch, H.; Lawney, B.P.; Lin, Y.F.; Asaithamby, A.; Murshid, A.; Wang, Y.Y.E.; Chen, B.P.C.; Calderwood, S.K. Transcriptional elongation requires DNA break-induced signalling. *Nat. Commun.* **2015**, *6*, 10191. [CrossRef]
75. Giaever, G.N.; Wang, J.C. Supercoiling of intracellular DNA can occur in eukaryotic cells. *Cell* **1988**, *55*, 849–856. [CrossRef]
76. Liu, L.F.; Miller, K.G. Eukaryotic DNA topoisomerases: Two forms of type I DNA topoisomerases from HeLa cell nuclei. *Proc. Natl. Acad. Sci. USA* **1981**, *78*, 3487–3491. [CrossRef]
77. Teves, S.S.; Henikoff, S. Transcription-generated torsional stress destabilizes nucleosomes. *Nat. Struct. Mol. Biol.* **2014**, *21*, 88–94. [CrossRef]
78. Wu, H.Y.; Shyy, S.; Wang, J.C.; Liu, L.F. Transcription generates positively and negatively supercoiled domains in the template. *Cell* **1988**, *53*, 433–440. [CrossRef]
79. Massé, E.; Drolet, M. Escherichia coli DNA topoisomerase I inhibits R-loop formation by relaxing transcription-induced negative supercoiling. *J. Biol. Chem.* **1999**, *274*, 16659–16664. [CrossRef]
80. Tresini, M.; Warmerdam, D.O.; Kolovos, P.; Snijder, L.; Vrouwe, M.G.; Demmers, J.A.; van, I.W.F.; Grosveld, F.G.; Medema, R.H.; Hoeijmakers, J.H.; et al. The core spliceosome as target and effector of non-canonical ATM signalling. *Nature* **2015**, *523*, 53–58. [CrossRef]

81. Aguilera, A.; Garcia-Muse, T. R loops: From transcription byproducts to threats to genome stability. *Mol. Cell* **2012**, *46*, 115–124. [CrossRef]
82. Gan, W.; Guan, Z.; Liu, J.; Gui, T.; Shen, K.; Manley, J.L.; Li, X. R-loop-mediated genomic instability is caused by impairment of replication fork progression. *Genes Dev.* **2011**, *25*, 2041–2056. [CrossRef]
83. Houlard, M.; Artus, J.; Leguillier, T.; Vandormael-Pournin, S.; Cohen-Tannoudji, M. DNA-RNA hybrids contribute to the replication dependent genomic instability induced by Omcg1 deficiency. *Cell Cycle* **2011**, *10*, 108–117. [CrossRef]
84. Yang, Y.; McBride, K.M.; Hensley, S.; Lu, Y.; Chedin, F.; Bedford, M.T. Arginine methylation facilitates the recruitment of TOP3B to chromatin to prevent R loop accumulation. *Mol. Cell* **2014**, *53*, 484–497. [CrossRef]
85. Wahba, L.; Amon, J.D.; Koshland, D.; Vuica-Ross, M. RNase H and multiple RNA biogenesis factors cooperate to prevent RNA:DNA hybrids from generating genome instability. *Mol. Cell* **2011**, *44*, 978–988. [CrossRef]
86. Redinbo, M.R.; Stewart, L.; Kuhn, P.; Champoux, J.J.; Hol, W.G.J. Crystal structures of human topoisomerase I in covalent and noncovalent complexes with DNA. *Science* **1998**, *279*, 1504–1513. [CrossRef]
87. Marinello, J.; Chillemi, G.; Bueno, S.; Manzo, S.G.; Capranico, G. Antisense transcripts enhanced by camptothecin at divergent CpG-island promoters associated with bursts of topoisomerase I-DNA cleavage complex and R-loop formation. *Nucleic Acids Res.* **2013**, *41*, 10110–10123. [CrossRef]
88. Marinello, J.; Bertoncini, S.; Aloisi, I.; Cristini, A.; Malagoli Tagliazucchi, G.; Forcato, M.; Sordet, O.; Capranico, G. Dynamic Effects of Topoisomerase I Inhibition on R-Loops and Short Transcripts at Active Promoters. *PLoS ONE* **2016**, *11*, e0147053. [CrossRef]
89. Cristini, A.; Park, J.H.; Capranico, G.; Legube, G.; Favre, G.; Sordet, O. DNA-PK triggers histone ubiquitination and signaling in response to DNA double-strand breaks produced during the repair of transcription-blocking topoisomerase I lesions. *Nucleic Acids Res.* **2016**, *44*, 1161–1178. [CrossRef]
90. Yung, T.M.; Sato, S.; Satoh, M.S. Poly(ADP-ribosyl)ation as a DNA damage-induced post-translational modification regulating poly(ADP-ribose) polymerase-1-topoisomerase I interaction. *J. Biol. Chem.* **2004**, *279*, 39686–39696. [CrossRef]
91. Bauer, P.I.; Chen, H.J.; Kenesi, E.; Kenessey, I.; Buki, K.G.; Kirsten, E.; Hakam, A.; Hwang, J.I.; Kun, E. Molecular interactions between poly(ADP-ribose) polymerase (PARP I) and topoisomerase I (Topo I): Identification of topology of binding. *FEBS Lett.* **2001**, *506*, 239–242. [CrossRef]
92. Malanga, M.; Althaus, F.R. Poly(ADP-ribose) reactivates stalled DNA topoisomerase I and induces DNA strand break resealing. *J. Biol. Chem.* **2004**, *279*, 5244–5248. [CrossRef] [PubMed]
93. Messner, S.; Altmeyer, M.; Zhao, H.; Pozivil, A.; Roschitzki, B.; Gehrig, P.; Rutishauser, D.; Huang, D.; Caflisch, A.; Hottiger, M.O. PARP1 ADP-ribosylates lysine residues of the core histone tails. *Nucleic Acids Res.* **2010**, *38*, 6350–6362. [CrossRef] [PubMed]
94. Ray Chaudhuri, A.; Nussenzweig, A. The multifaceted roles of PARP1 in DNA repair and chromatin remodelling. *Nat. Rev. Mol. Cell Biol.* **2017**, *18*, 610–621. [CrossRef]
95. Okano, S.; Lan, L.; Yasui, A.; Mori, T.; Caldecott, K.W. Spatial and temporal cellular responses to single-strand breaks in human cells. *Mol. Cell. Biol.* **2003**, *23*, 3974–3981. [CrossRef]
96. Haince, J.F.; McDonald, D.; Rodrigue, A.; Déry, U.; Masson, J.Y.; Hendzel, M.J.; Poirier, G.G. PARP1-dependent kinetics of recruitment of MRE11 and NBS1 proteins to multiple DNA damage sites. *J. Biol. Chem.* **2008**, *283*, 1197–1208. [CrossRef] [PubMed]
97. Chen, Q.; Kassab, M.A.; Dantzer, F.; Yu, X. PARP2 mediates branched poly ADP-ribosylation in response to DNA damage. *Nat. Commun.* **2018**, *9*, 3233. [CrossRef] [PubMed]
98. Kolas, N.K.; Chapman, J.R.; Nakada, S.; Ylanko, J.; Chahwan, R.; Sweeney, F.D.; Panier, S.; Mendez, M.; Wildenhain, J.; Thomson, T.M.; et al. Orchestration of the DNA-damage response by the RNF8 ubiquitin ligase. *Science* **2007**, *318*, 1637–1640. [CrossRef] [PubMed]
99. Mattiroli, F.; Vissers, J.H.A.; Van Dijk, W.J.; Ikpa, P.; Citterio, E.; Vermeulen, W.; Marteijn, J.A.; Sixma, T.K. RNF168 ubiquitinates K13-15 on H2A/H2AX to drive DNA damage signaling. *Cell* **2012**, *150*, 1182–1195. [CrossRef]
100. Smeenk, G.; Wiegant, W.W.; Marteijn, J.A.; Luijsterburg, M.S.; Sroczynski, N.; Costelloe, T.; Romeijn, R.J.; Pastink, A.; Mailand, N.; Vermeulen, W.; et al. Poly(ADP-ribosyl)ation links the chromatin remodeler SMARCA5/SNF2H to RNF168-dependent DNA damage signaling. *J. Cell Sci.* **2013**, *126*, 889–903. [CrossRef]

101. Ahel, D.; Hořejší, Z.; Wiechens, N.; Polo, S.E.; Garcia-Wilson, E.; Ahel, I.; Flynn, H.; Skehel, M.; West, S.C.; Jackson, S.P.; et al. Poly(ADP-ribose)-dependent regulation of DNA repair by the chromatin remodeling enzyme ALC1. *Science* **2009**, *325*, 1240–1243. [CrossRef] [PubMed]
102. Gottschalk, A.J.; Timinszky, G.; Kong, S.E.; Jin, J.; Cai, Y.; Swanson, S.K.; Washburn, M.P.; Florens, L.; Ladurner, A.G.; Conaway, J.W.; et al. Poly(ADP-ribosyl)ation directs recruitment and activation of an ATP-dependent chromatin remodeler. *Proc. Natl. Acad. Sci. USA* **2009**, *106*, 13770–13774. [CrossRef] [PubMed]
103. Sellou, H.; Lebeaupin, T.; Chapuis, C.; Smith, R.; Hegele, A.; Singh, H.R.; Kozlowski, M.; Bultmann, S.; Ladurner, A.G.; Timinszky, G.; et al. The poly(ADP-ribose)-dependent chromatin remodeler Alc1 induces local chromatin relaxation upon DNA damage. *Mol. Biol. Cell* **2016**, *27*, 3791–3799. [CrossRef] [PubMed]
104. Singh, H.R.; Nardozza, A.P.; Möller, I.R.; Knobloch, G.; Kistemaker, H.A.V.; Hassler, M.; Harrer, N.; Blessing, C.; Eustermann, S.; Kotthoff, C.; et al. A Poly-ADP-Ribose Trigger Releases the Auto-Inhibition of a Chromatin Remodeling Oncogene. *Mol. Cell* **2017**, *68*, 860–871. [CrossRef] [PubMed]
105. Gottschalk, A.J.; Trivedi, R.D.; Conaway, J.W.; Conaway, R.C. Activation of the SNF2 family ATPase ALC1 by poly(ADP-ribose) in a stable ALC1·PARP1·nucleosome intermediate. *J. Biol. Chem.* **2012**, *287*, 43527–43532. [CrossRef] [PubMed]
106. Ahel, I.; Ahel, D.; Matsusaka, T.; Clark, A.J.; Pines, J.; Boulton, S.J.; West, S.C. Poly(ADP-ribose)-binding zinc finger motifs in DNA repair/checkpoint proteins. *Nature* **2008**, *451*, 81–85. [CrossRef]
107. Mehrotra, P.V.; Ahel, D.; Ryan, D.P.; Weston, R.; Wiechens, N.; Kraehenbuehl, R.; Owen-Hughes, T.; Ahel, I. DNA repair factor APLF is a histone chaperone. *Mol. Cell* **2011**, *41*, 46–55. [CrossRef]
108. Luijsterburg, M.S.; de Krijger, I.; Wiegant, W.W.; Shah, R.G.; Smeenk, G.; de Groot, A.J.L.; Pines, A.; Vertegaal, A.C.O.; Jacobs, J.J.L.; Shah, G.M.; et al. PARP1 Links CHD2-Mediated Chromatin Expansion and H3.3 Deposition to DNA Repair by Non-homologous End-Joining. *Mol. Cell* **2016**, *61*, 547–562. [CrossRef]
109. Burgess, R.C.; Burman, B.; Kruhlak, M.J.; Misteli, T. Activation of DNA Damage Response Signaling by Condensed Chromatin. *Cell Rep.* **2014**, *9*, 1703–1717. [CrossRef]
110. Fu, Y.; Lv, P.; Yan, G.; Fan, H.; Cheng, L.; Zhang, F.; Dang, Y.; Wu, H.; Wen, B. MacroH2A1 associates with nuclear lamina and maintains chromatin architecture in mouse liver cells. *Sci. Rep.* **2015**, *5*, 17186. [CrossRef]
111. Douet, J.; Corujo, D.; Malinverni, R.; Renauld, J.; Sansoni, V.; Marjanović, M.P.; Cantariño, N.; Valero, V.; Mongelard, F.; Bouvet, P.; et al. MacroH2A histone variants maintain nuclear organization and heterochromatin architecture. *J. Cell Sci.* **2017**, *130*, 1570–1582. [CrossRef] [PubMed]
112. Sporn, J.C.; Jung, B. Differential regulation and predictive potential of macroH2A1 isoforms in colon cancer. *Am. J. Pathol.* **2012**, *180*, 2516–2526. [CrossRef] [PubMed]
113. Nusinow, D.A.; Hernández-Muñoz, I.; Fazzio, T.G.; Shah, G.M.; Kraus, W.L.; Panning, B. Poly(ADP-ribose) polymerase 1 is inhibited by a histone H2A variant, macroH2A, and contributes to silencing of the inactive X chromosome. *J. Biol. Chem.* **2007**, *282*, 12851–12859. [CrossRef] [PubMed]
114. Timinszky, G.; Till, S.; Hassa, P.O.; Hothorn, M.; Kustatscher, G.; Nijmeijer, B.; Colombelli, J.; Altmeyer, M.; Stelzer, E.H.K.; Scheffzek, K.; et al. A macrodomain-containing histone rearranges chromatin upon sensing PARP1 activation. *Nat. Struct. Mol. Biol.* **2009**, *16*, 923–929. [CrossRef]
115. Kozlowski, M.; Corujo, D.; Hothorn, M.; Guberovic, I.; Mandemaker, I.K.; Blessing, C.; Sporn, J.; Gutierrez-triana, A.; Smith, R.; Portmann, T.; et al. MacroH 2 A histone variants limit chromatin plasticity through two distinct mechanisms. *EMBO Rep.* **2018**, *19*, 1–13. [CrossRef]
116. Polo, S.E.; Kaidi, A.; Baskcomb, L.; Galanty, Y.; Jackson, S.P. Regulation of DNA-damage responses and cell-cycle progression by the chromatin remodelling factor CHD4. *EMBO J.* **2010**, *29*, 3130–3139. [CrossRef]
117. Caruso, L.B.; Martin, K.A.; Lauretti, E.; Hulse, M.; Siciliano, M.; Lupey-Green, L.N.; Abraham, A.; Skorski, T.; Tempera, I. Poly(ADP-ribose) Polymerase 1, PARP1, modifies EZH2 and inhibits EZH2 histone methyltransferase activity after DNA damage. *Oncotarget* **2018**, *9*, 10585–10605. [CrossRef]
118. Yamaguchi, H.; Du, Y.; Nakai, K.; Ding, M.; Chang, S.-S.; Hsu, J.L.; Yao, J.; Wei, Y.; Nie, L.; Jiao, S.; et al. EZH2 contributes to the response to PARP inhibitors through its PARP-mediated poly-ADP ribosylation in breast cancer. *Oncogene* **2018**, *37*, 208–217. [CrossRef]
119. Matveeva, E.; Maiorano, J.; Zhang, Q.; Eteleeb, A.M.; Convertini, P.; Chen, J.; Infantino, V.; Stamm, S.; Wang, J.; Rouchka, E.C.; et al. Involvement of PARP1 in the regulation of alternative splicing. *Cell Discov.* **2016**, *2*, 15046. [CrossRef]
120. Ji, Y.; Tulin, A.V. Poly(ADP-ribosyl)ation of heterogeneous nuclear ribonucleoproteins modulates splicing. *Nucleic Acids Res.* **2009**, *37*, 3501–3513. [CrossRef]

121. Isabelle, M.; Moreel, X.; Gagné, J.P.; Rouleau, M.; Ethier, C.; Gagné, P.; Hendzel, M.J.; Poirier, G.G. Investigation of PARP-1, PARP-2, and PARG interactomes by affinity-purification mass spectrometry. *Proteome Sci.* **2010**, *8*, 22. [CrossRef] [PubMed]
122. Munnur, D.; Bartlett, E.; Mikolčević, P.; Kirby, I.T.; Matthias Rack, J.G.; Mikoč, A.; Cohen, M.S.; Ahel, I. Reversible ADP-ribosylation of RNA. *Nucleic Acids Res.* **2019**, *47*, 5658–5669. [CrossRef] [PubMed]
123. Kotova, E.; Jarnik, M.; Tulin, A.V. Poly (ADP-ribose) polymerase 1 is required for protein localization to Cajal body. *PLoS Genet.* **2009**, *5*, e1000387. [CrossRef] [PubMed]
124. Matveeva, E.A.; Al-Tinawi, Q.M.H.; Rouchka, E.C.; Fondufe-Mittendorf, Y.N. Coupling of PARP1-mediated chromatin structural changes to transcriptional RNA polymerase II elongation and cotranscriptional splicing. *Epigenetics Chromatin* **2019**, *12*, 15. [CrossRef] [PubMed]
125. Di Giammartino, D.C.; Shi, Y.; Manley, J.L. PARP1 Represses PAP and Inhibits Polyadenylation during Heat Shock. *Mol. Cell* **2013**, *49*, 7–17. [CrossRef] [PubMed]
126. Bilal Iqbal, M.; Johns, M.; Cao, J.; Liu, Y.; Yu, S.C.; Hyde, G.D.; Laffan, M.A.; Marchese, F.P.; Cho, S.H.; Clark, A.R.; et al. PARP-14 combines with tristetraprolin in the selective posttranscriptional control of macrophage tissue factor expression. *Blood* **2014**, *124*, 3646–3655. [CrossRef] [PubMed]
127. Ke, Y.; Han, Y.; Guo, X.; Wen, J.; Wang, K.; Jiang, X.; Tian, X.; Ba, X.; Boldogh, I.; Zeng, X. PARP1 promotes gene expression at the post-transcriptiona level by modulating the RNA-binding protein HuR. *Nat. Commun.* **2017**, *8*, 14632. [CrossRef]
128. Moynahan, M.E.; Cui, T.Y.; Jasin, M. Homology-directed DNA repair, mitomycin-C resistance, and chromosome stability is restored with correction of a Brca1 mutation. *Cancer Res.* **2001**, *61*, 4842–4850.
129. Moynahan, M.E.; Pierce, A.J.; Jasin, M. BRCA2 is required for homology-directed repair of chromosomal breaks. *Mol. Cell* **2001**, *7*, 263–272. [CrossRef]
130. Moynahan, M.E.; Chiu, J.W.; Koller, B.H.; Jasint, M. Brca1 controls homology-directed DNA repair. *Mol. Cell* **1999**, *4*, 511–518. [CrossRef]
131. Tutt, A.; Bertwistle, D.; Valentine, J.; Gabriel, A.; Swift, S.; Ross, G.; Griffin, C.; Thacker, J.; Ashworth, A. Mutation in Brca2 stimulates error-prone homology-directed repair of DNA double-strand breaks occurring between repeated sequences. *EMBO J.* **2001**, *20*, 4704–4716. [CrossRef] [PubMed]
132. Hegan, D.C.; Lu, Y.; Stacheleka, G.C.; Crosbya, M.E.; Bindraa, R.S.; Glazer, P.M. Inhibition of poly(ADP-ribose) polymerase down-regulates BRCA1 and RAD51 in a pathway mediated by E2F4 and p130. *Proc. Natl. Acad. Sci. USA* **2010**, *107*, 2201–2206. [CrossRef] [PubMed]
133. Wright, R.H.G.; Lioutas, A.; Dily, F.L.; Soronellas, D.; Pohl, A.; Bonet, J.; Nacht, A.S.; Samino, S.; Font-Mateu, J.; Vicent, G.P.; et al. ADP-ribose-derived nuclear ATP synthesis by NUDIX5 is required for chromatin remodeling. *Science* **2016**, *352*, 1221–1225. [CrossRef] [PubMed]

MDPI
St. Alban-Anlage 66
4052 Basel
Switzerland
Tel. +41 61 683 77 34
Fax +41 61 302 89 18
www.mdpi.com

Cancers Editorial Office
E-mail: cancers@mdpi.com
www.mdpi.com/journal/cancers

www.ingramcontent.com/pod-product-compliance
Lightning Source LLC
Chambersburg PA
CBHW061508220926
51850CB00004B/201
* 9 7 8 3 0 3 6 5 8 1 5 6 9 *